Speleothem Science

Speleothem Science

From Process to Past Environments

Ian J. Fairchild
School of Geography, Earth and Environmental Sciences
University of Birmingham
Birmingham, UK

and

Andy Baker
Connected Waters Initiative Research Centre & National Centre
for Groundwater Research and Training (NCGRT)
University of New South Wales
Manly Vale, NSW, Australia

with contributions from

Asfawossen Asrat, University of Addis Abbaba, Ethiopia
David Domínguez-Villar, University of Birmingham and CENIEH, Burgos, Spain
John Gunn, University of Birmingham
Adam Hartland, University of New South Wales
David Lowe, British Geological Survey

Series Editor: Ray Bradley, University Distinguished Professor in the Department of Geosciences and Director of the Climate System Research Center, University of Massachusetts Amherst. http://www.paleoclimate.org

A John Wiley & Sons, Ltd., Publication

This edition first published 2012 © 2012 by Ian J. Fairchild, Andy Baker.

Wiley-Blackwell is an imprint of John Wiley & Sons, formed by the merger of Wiley's global Scientific, Technical and Medical business with Blackwell Publishing.

Blackwell Publishing was acquired by John Wiley & Sons in February 2007. Blackwell's publishing program has been merged with Wiley's global Scientific, Technical and Medical business to form Wiley-Blackwell.

Registered office: John Wiley & Sons, Ltd, The Atrium, Southern Gate, Chichester, West Sussex, PO19 8SQ, UK

Editorial offices: 9600 Garsington Road, Oxford, OX4 2DQ, UK
The Atrium, Southern Gate, Chichester, West Sussex, PO19 8SQ, UK
111 River Street, Hoboken, NJ 07030-5774, USA

For details of our global editorial offices, for customer services and for information about how to apply for permission to reuse the copyright material in this book please see our website at www.wiley.com/wiley-blackwell.

Library of Congress Cataloging-in-Publication Data

Fairchild, Ian J. (Ian John)
Speleothem science : from process to past environments / Ian J. Fairchild and Andy Baker.
p. cm.
Includes bibliographical references and index.
ISBN 978-1-4051-9620-8 (cloth)
1. Speleothems. 2. Paleoclimatology. I. Baker, Andy, 1968– II. Title.
GB601.F35 2012
551.44'7–dc23

2011046018

A catalogue record for this book is available from the British Library.

Wiley also publishes its books in a variety of electronic formats. Some content that appears in print may not be available in electronic books.

Set in 9/12 pt Meridien by Toppan Best-set Premedia Limited
Printed and bound in Malaysia by Vivar Printing Sdn Bhd

1 2012

Contents

Companion Website

This book has a companion website:
www.wiley.com/go/fairchild/speleothem_with Figures and Tables from the book for downloading

Preface

This book is a response to the explosion of interest in speleothem archives of environmental change and our experience of being scientifically stretched to understand the processes that form them. Hence, in this volume we have constructed a broad syllabus, including much new material from our own research and scholarship, as an attempt to match the requirements of a core text for anyone starting research in speleothem science. We also trust that the book will be useful in relation to subjects such as Quaternary science, geochemistry and carbonate geology: for teachers, researchers and students. Quaternary science is multidisciplinary and likewise the science of speleothems draws on many subjects. Chapter 3 summarizes many facets of the global climate system and Chapter 9 focuses on issues in dating, several of which are common to other archives. Chapter 10 provides material on the calibration of proxies which should also be of general interest, while Chapters 11 and 12 draw out the special and often unique contributions made to Quaternary science by speleothem studies. Speleothem formation also illustrates and integrates many fundamental principles of geochemistry. Chapters 2, 6, 7 and 8 provide a complement to the classic marine emphasis of carbonate geology texts and provide exemplars for many core concepts in mineralogy and geochemistry. Students of water chemistry and hydrogeology will find an updated summary of many issues related to carbonate aquifers in Chapters 2, 4 and 5. There is also plenty of material here on general scientific principles, e.g. the emphasis on systems and material transfer, for example in Chapters 1 and 4, that could be effectively used as case examples in undergraduate degrees in physical geography and environmental science. Readers can find the illustrations from this book on-line at the publisher's website www.wiley.com/go/fairchild/speleothem including additional use of colour, and we have also made available on-line on our own website www.speleothemscience.info the spreadsheets that were used to produce many of the new graphics. We would be grateful for readers pointing out errors that can be corrected on-line or in a future reprinting. Finally, we hope that this book will inspire some undergraduates towards the research frontier wherever they may find it in the environmental (geo)sciences.

This volume is a tangible outcome of the lively 6 years that we were co-located at the University of Birmingham. Here, we developed a common understanding of speleothem science while also pursuing our other research agendas. We planned the book in 2008 and both benefitted from periods of study leave in 2009: AB at University College, Durham, while holding an Institute of Advanced Studies Fellowship, and IJF at the University of Newcastle, Australia, financially supported by the Leverhulme Trust. Andy moved on to the University of New South Wales (UNSW), Australia, in January 2010, and the continuation of the Leverhulme Study Abroad Fellowship allowed IJF to catch up with AB in Sydney's Northern Beaches in September 2010. Over the past 15 years at Birmingham, Exeter, Newcastle, Keele and UNSW, we have had the pleasure of supervising the work of many fine research students and fellows on speleothem-related work, several of whom are now forging their own research careers. The Natural Environment Research Council, the European Community, the UK's Royal Society, the Australian Research Council and the Leverhulme Trust have supported our work through several projects. We are grateful to several colleagues with connections to Birmingham who have provided input or specific sections to this book; their contributions are listed in the contents. We salute the generations of speleologists who made so many of the discoveries that provided a foundation for our science, several of whom went on to be professional scientists. AB especially thanks Pete Smart, Larry Edwards, Dominique Genty, John Gunn, Tim Atkinson and Paul Williams for their advice, encouragement and debate. IJF made his underground scientific debut in September 1994 in the company of Silvia Frisia and Andrea Borsato, who have remained good friends and collaborators ever since, as subsequently have been Frank McDermott, Christoph Spötl, Dave Mattey and Pauline Treble, as well as those previously mentioned, and others we should have done. We have found the global speleothem community to be highly supportive and forward-looking, and are grateful for all the insight and scientific inspiration we have found there, including through the Climate Change—the Karst Record meetings, the sixth of which was held in Birmingham in June 2011. We are indebted to Ian Francis for his encouragement and to his colleagues at Wiley-Blackwell for their helpfulness and efficiency in the production process. We also thank the publisher's three reviewers, Denis Scholz, Maurice Tucker and Ming Tan, and Gregoire Mariethoz and Bryce Kelly for their comments on different parts of our text. We are indebted to Anne Ankcorn and Kevin Burkhill of the School of Geography, Earth and Environmental Sciences at the University of Birmingham, for their excellent work on drafting many of the figures in this book. Both of us depend on the enthusiasm and support of our wives Sue and Jo. While Sue and Ian simply have a non-aggression pact regarding horses and caves, we have Jo to thank particularly for proof-editing the first drafts of our text. We trust the final version will be well received, although we are only too aware of the shortcomings and biases that books written during short and busy lives bring.

Ian J. Fairchild and Andy Baker

Acknowledgements

The authors are grateful to the copyright holders for permission to reproduce the following material:

Acta Carsologica (Figures 1.2 and 2.32)

American Association for the Advancement of Science (Plate 3.7 (part), 12.1, Figures 11.7, 12.4c)

American Association of Petroleum Geologists (Figure 2.35)

American Chemical Society (Plate 5.1, Figures 5.8, 5.9, 5.16)

American Geophysical Union (Plates 3.6 and 12.2; Figures 3.1, 3.4, 3.7, 3.9, 3.1, 3.12, 3.13, 4.23, 4.35, 11.16, 11.20)

Andrea Dutton (Figure 7.29)

Annual Reviews (Figure 4.31)

Australian Government (Plate 3.2)

British Cave Research Association (Figure 5.3a)

Chaoyong Hu (Plate 3.5, right)

CRC Press (Figures 3.8 and 5.18)

Eligio Vacca (Plate 7.9)

Elsevier-Pergamon (Plates 2.2, 2.3, 2.4, 7.6c, 11.2 and 11; Figures 1.1, 1.4 (left), 1.B3, 2.4, 2.5a, 2.6, 2.7, 2.11, 2.14, 2.17, 2.24, 2.28, 2.29a, 4.1b, 4.7, 4.15, 4.16a, 4.16c, 4.16d, 4.19, 4.22, 4.26, 4.29, 5.5, 5.6, 5.10, 5.14, 5.22, 5.23, 6.3, 6.4, 6.B1, 6.B4, 6.B5, 7.1, 7.5b-f, 7.20, 7.21, 7.26, 7.27, 7.28, 7.30, 8.2a, 8.3, 8.6, 8.7a, 8.8, 8.9, 8.12, 8.13, 8.17, 8.18, 8.19, 8.20, 8.23, 8.24, 8.25, 8.26, 9.2, 9.3, 9.5, 9.6, 9.7, 10.10, 11.2, 11.8, 11.9a, 11.10, 11.19, 12.2, 12.4a, b, 12.6, 12.7

European Mineralogical Union and the Mineralogical Society (Figure 5.17)

Fabrizio Antonioli (Plate 7.7)

Geological Association of Canada (Figures 2.19 and 2.20)

Geological Society of America (Figure 2.20, 2.22, 2.25, 3.16a, b, 6.5, 11.9b, c, 12.5)

Geological Society of London (Plate 7.3, Figures 1.5b, 2.1, 5.24)

International Glaciological Society (Figure 5.12)

International Journal of Speleology (Plate 7.5, 8.22, 11.13)

IPCC Report Climate Change 2007 The Physical Science Basis their Figure 7.5 (Figure 1.B1)

Jacqueline Shinker (Plate 3.1)

Journal of Geology (Figure 3.16d)

Jud Partin and PAGES (Figure 11.11)

Karst Waters Institute (Figure 1.10)

Mineralogical Society of America (Figures 5.7 and 8.5)

Ming Tan (Plate 7.6a; Figure 1.2a)

National Speleological Society (Figures 2.26 and 7.2)

Nature publications (Figure 11.15)

Otto de Voogd (Plate 7.1 (right))

Oxford University Press (Figure 2.3)

Paul Williams (Figure 2.2)

Phil Hopley (Figure 12.1)

Royal Society for Chemistry (Figure 6.B2)

Sage Publishers (Plate 7.4)

SEPM (Society for Sedimentary Geology) (Figures 2.16, 2.21, 7.12, 7.13, 7.17, 7.19, 7.22, 7.23)

Springer (Plate 3.7 (part), 2.15, 2.23, 3.B1, 4.32, 7.4b)

Stan Robinson (Plate 7.1 (left))

University of Chicago Press (Figure 3.3)

Verband Österreichischer Höhlenforscher (Figure 7.3)

Wedgwood Museum, Barlaston, Staffordshire (Plate 1.1 insets)

Wiley-Blackwell (Figure 2.5b, c, 2.12, 2.18, 2.27a, 2.33, 2.34, 3.14, 3.15, 3.16c, 3.17, 3.21, 4.13, 4.16b, 4.25, 4.33, 5.11, 7.11b, c, 7.18, 10.6, 11.4, 11.6)

1 Scientific and geological context

CHAPTER 1
Introduction to speleothems and systems

For paleoclimate, the past two decades have been the age of the ice core. The next two may be the age of the speleothem.

Gideon Henderson (*Science*, 2006)

1.1 What is all the fuss about?

Moore (1952) recognized a need to specify an unambiguous term for mineral deposits that grew within caves and proposed 'speleothem' (Greek: *spelaion*, cave; *thema*, deposit). In recent years speleothems have been established as one of the most valuable resources for understanding Earth surface conditions in the past, from times when glaciers waxed and waned and our human ancestors emerged, to the present day. By 'conditions' we mean not only the local context (soil, vegetation, landscape instability and climate), but also the regional to global patterns of change that characterize former environments and climates (Fig. 1.1).

Because no two speleothems are identical (Plate 1.1), they were formerly thought to be too complex to generate reliable archives of the past. It has taken a considerable effort over the past 40 years both to show that reliable records can be obtained which, in some cases, display global phenomena, and to demonstrate that speleothems form by a set of processes that can be rationalized and understood. Even in the late 1990s, textbooks on palaeoclimates and palaeoenvironments treated speleothems only briefly. Now it can be argued that certain long and well-dated records provide the most definitive archives of the global environmental system. Speleothems also provide an enormous resource for future research on past changes, with an ultimate dynamic range of eight orders of magnitude from days to a million or more years.

The rest of section 1.1 provides a summary of the essentials of speleothem science for the benefit of all those, from undergraduates to specialists in related fields, who want to get to first base. In section 1.2, we explain how the rest of the book is organized.

1.1.1 What types of speleothem are useful for generating climate archives?

Many different forms of speleothem can be distinguished (Hill & Forti, 1997). Here we consider primarily two types of deposit (*dripstones*) that grow from dripping water: *stalactites* (Greek: *stalaktós*, dripping) growing down from the cave roof and *stalagmites* (Greek: *stalagmós*, dropping) building up from the cave floor. Also pertinent are more continuous deposits (*flowstones*) that accrete beneath thin sheets of water on cave walls and floors (Fig. 1.2a). Finally, we consider some long records obtained from crystalline deposits that form underwater. Flowstones tend to have fairly continuous layers and so it is possible to duplicate records by sampling (e.g. by coring) in different places.

Speleothem Science: From Process to Past Environments, First Edition. Ian J. Fairchild, Andy Baker.

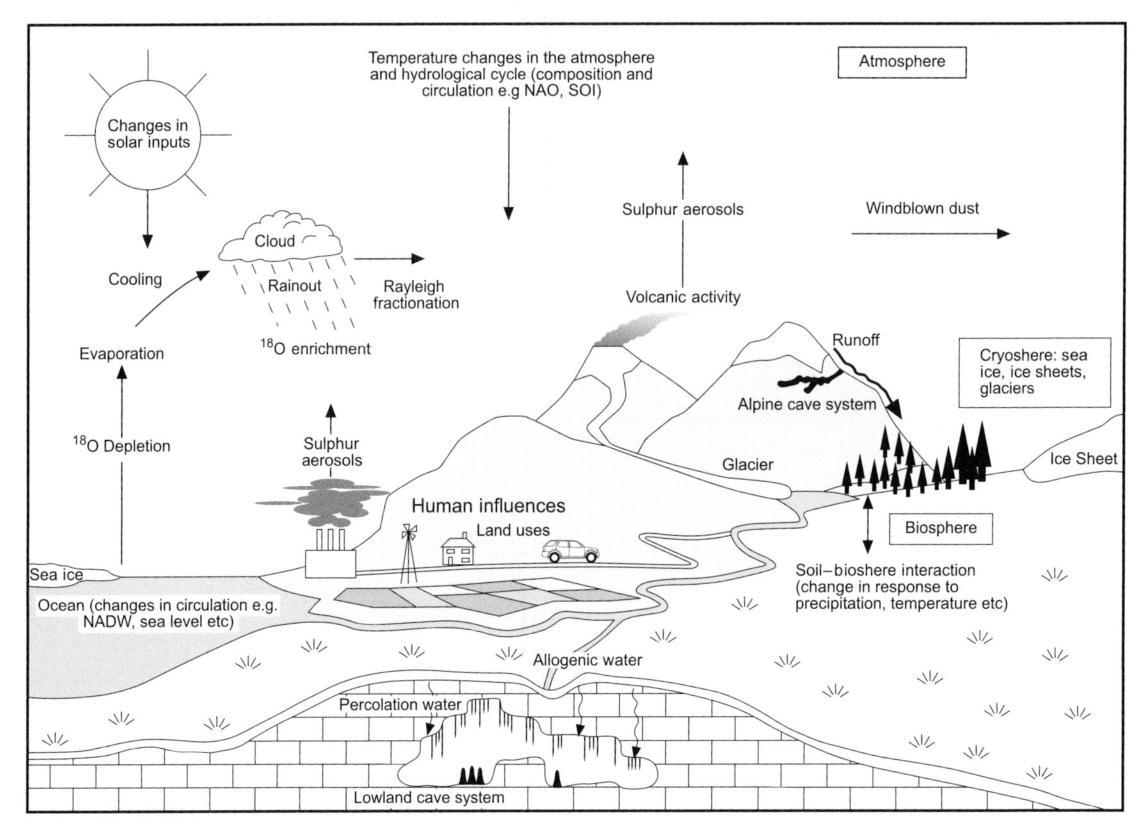

Fig. 1.1 Speleothems as underground recorders of signals related to parameters of the external Earth system (Fairchild et al., 2006a).

However, the layers can show a small-scale topography reflecting the ponding of surface water (Fig. 1.2a). Stalagmites are more commonly used to generate archives than stalactites because their internal structure is simpler. Fig. 1.2b illustrates a stalagmite cross-section illustrating that there is internal layering which tends to be flat on the top of the sample, allowing a set of observations representing different time periods in the past (*time series*) to be generated along a sub-vertical line. However, the sample illustrated in Fig. 1.2b displays lateral shifts in its growth axis, related to changes in the landing position of water drops. It is also commonly found that growth may pause for an extended period and then resume, and that these characteristics reflect either changes in climate or local processes within the aquifer in different cases.

1.1.2 Where do speleothems occur?

The speleothems used to find out about the past are almost all *calcareous* (made largely of calcium carbonate, $CaCO_3$) and composed of the minerals calcite and/or aragonite. Such speleothems occur within *carbonate rocks*, typically *limestones* ($CaCO_3$) and/or *dolomite* ($CaMg(CO_3)_2$. These rocks store, transmit and yield water readily and hence are *aquifers*. The *permeability* of aquifers refers to the ease with which they transmit water. Carbonate aquifers occur in *karstic* regions (Gunn, 2004; Ford & Williams, 2007) where there is little surface water because it drains readily into the bedrock (Fig. 1.3). The cavities in the bedrock typically range from tiny pores, through enlarged fissures to *conduits* which used to host, or still contain, underground streams. Over time, the drainage system

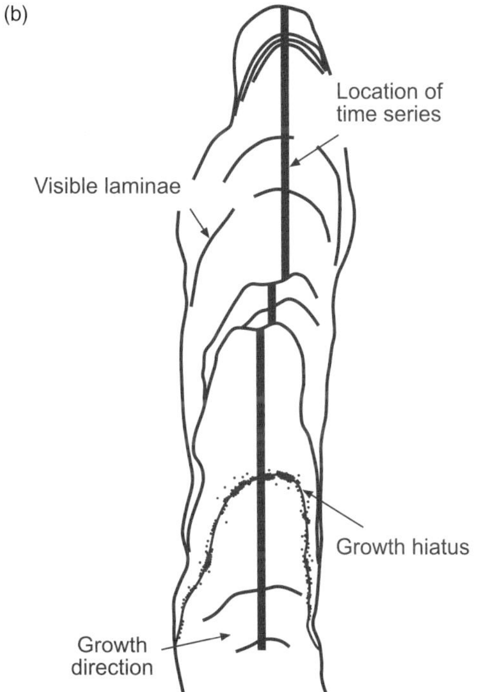

Fig. 1.2 (a) Speleothems in Grand Roc Cave, Dordogne, France. The cave floor is covered with a flowstone on which have developed walls (terracettes) and intervening gour pools. A columnar stalagmite a few centimetres wide occurs to lower right and stalactites in the upper left. (b) Cross-section through a stalagmite illustrating visible growth layers (laminae) and the shifting position of the growth axis which represents the ideal position to generate the most continuous time series.

develops and tends to lower the depth of the *water table*, the regional surface below which cavities are water-filled. The water table divides the aquifer into the underlying *phreatic* and overlying *vadose* zones, although in karst terrains the division is not always simple. Pragmatically, in this book we refer to the vadose zone, from which dripwater originates, as part of the aquifer, although conventionally the term is often restricted to the phreatic portion of the rock. Caves in vadose zones tend to cut vertically downwards, whereas in the phreatic zone they develop as passages elongated in the direction of water movement. Speleothems are normally formed in abandoned passages (Fig. 1.3) in the vadose zone and are fed by water passing through the soil into the uppermost karst, which is typically a zone of significant water storage (the *epikarst*). Speleothems form part of a long and complex history of cave spaces. Hence they can be interlayered with particulate sediments and are ultimately may be later broken by earthquakes or human disturbance, buried, flooded, eroded or dissolved.

Modern calcareous speleothems occur in nearly all continental regions and are principally limited by the availability of karstic host rocks and liquid water. However, they tend to be only weakly developed in frigid regions where there is less chemical drive for them to form (Fig. 1.4b; section 1.1.3). They are found in many modern desert regions, having formed during more humid conditions in past millennia. Speleothems that originally formed not far above the water table can be found on mountain tops in tectonically uplifting regions. Conversely, precipitates that formed during previous ice ages when sea level was much lower are observed beneath the sea and can be recovered for study by divers.

1.1.3 How do they form?

The essential chemical processes that lead to speleothem formation are illustrated in Fig. 1.4a and described in its legend. Essentially there is a carbonate dissolution region in the soil and upper epikarst and an underlying vadose region of $CaCO_3$ precipitation. These correspond to $CaCO_3$-*undersaturated* and -*oversaturated* solutions respectively. The amount

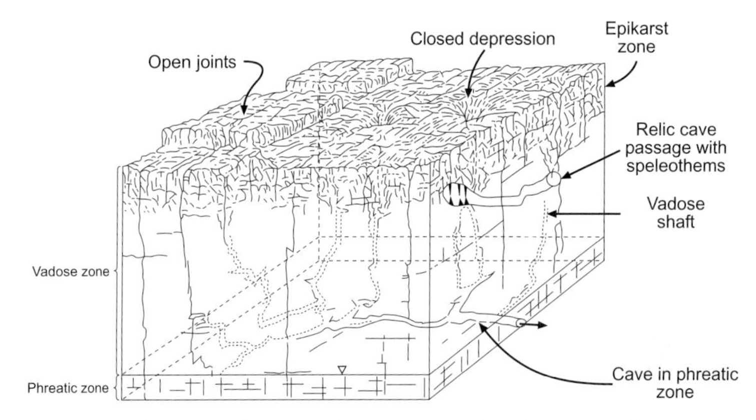

Fig. 1.3 The complex structure of an active karst aquifer in which speleothems typically occur in caves that are no longer being developed and are well above the water table (modified from Smart and Whitaker, 1991).

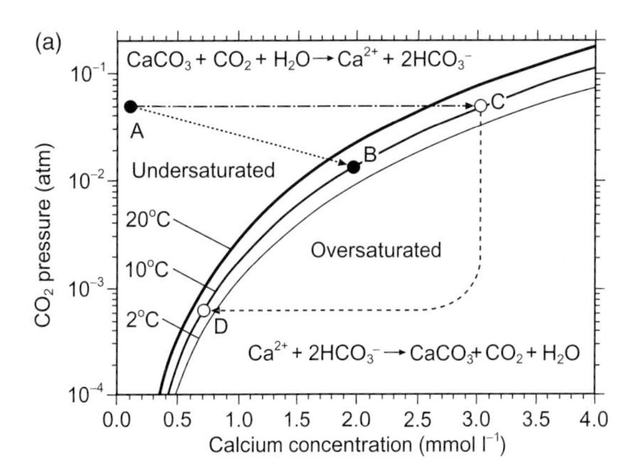

Fig. 1.4 (a) The normal chemical pathway leading to speleothem formation. High carbon dioxide partial pressures (P_{CO_2}) arise in the soil (e.g. point A) owing to respiration and decomposition of organic matter. When percolating water with a high P_{CO_2} reaches carbonate minerals, they will be dissolved, increasing the calcium concentration in solution. If there is no renewal of CO_2, the water follows a 'closed system' path to saturation point B, whereas if CO_2 is replenished to maintain a constant P_{CO_2} then saturation will be reached at point C (both B and C are specific to 10 °C as shown). As the water descends the karst system, at some point it may encounter an air space with a lower P_{CO_2} than the original soil. The water degasses CO_2 and enters the oversaturated field and tends to precipitate $CaCO_3$ (e.g. dashed line C–D). Modified from Kaufmann (2003). (b) Holocene speleothems in a cave above the Arctic circle, Norway: slow growth limited by low P_{CO_2} in the soil zone.

of calcite and dolomite that can be dissolved by water depends on its acidity, normally expressed as the partial pressure of carbon dioxide (P_{CO_2}) with which the water is stable. Although strong acid (e.g. from oxidation of sulphide minerals such as *pyrite*, FeS_2) can be important locally, normally dissolution is enhanced mainly by high P_{CO_2} values of up to several per cent (0.01–0.1 atm) generated by respiration and organic decomposition in soils. As the water descends through the karst, it ultimately encounters a gas phase with a lower P_{CO_2} compared with that which it has previously encountered. This causes degassing of CO_2 from the solution and precipitation of $CaCO_3$. Figure 1.4 shows that the difference in P_{CO_2} between soil and cave is a key control on the quantity of calcium removal from the water, and hence the rate of calcium carbonate precipitation. Growth rate, i.e. upward extension rate in the case of stalagmites, can be of the order of a millimetre per year in humid, warm regions, but is more typically less than 100 μm per year in cool temperate regions. Precipitation can happen above the site where the observer is based and such *prior calcite precipitation* leads to reduced oversaturation and can be identified by a characteristic evolution of water composition. Long-term continuous growth as slow as 1 mm per thousand years has been documented from weakly oversaturated waters.

Speleothem growth requires quite specific conditions of availability of water to supply the ingredients of growth and circulation of air to take away the waste product carbon dioxide. We use the metaphor of the *speleothem incubator* in this book to describe this life-support system that maintains speleothem growth. Although conditions in cave interiors are much more constant than above ground, they are not completely static. Temperature varies near entrances, the humidity and carbon dioxide content of the air change laterally and through the year, and the quantity of infiltrating water is a function of the passing seasons. These changing conditions normally impart an annual visible or chemical lamination within a speleothem if it grows quickly enough to be resolved. This lamination contains information about the seasonality of the climate during deposition. The commonest type is an *annual couplet* (Fig. 1.5a), reflecting warmer–cooler or more usually wetter–drier alternations, but a discrete thin *impulse lamina* (Fig. 1.5b), characteristic of seasonal influx of soil-derived material, is common in cool temperate climates.

1.1.4 How do we date them?

A time series of observations is of limited use unless we can assign real ages to it. In the case of modern, actively accumulating speleothems, their rate of growth can be determined by direct observation, e.g. on human artefacts (Fig. 1.5a). Another successful technique on these materials is to demonstrate a distinct signal within the stalagmite of enhanced levels of radiocarbon (^{14}C) resulting from atmospheric nuclear tests of the 1960s: this has been done for the sample illustrated in Fig. 1.5b (Smith et al., 2009). In both cases, these observations show that the growth lamination displayed is annual and so the duration of growth of older speleothems formed in the same setting can be derived by counting laminae. However, it would be unwise to rely solely on such a method because hiatuses in growth may not have been detected, and in any case growth may not have continued up to the present day. Hence normally an *absolute* dating method is used which allows assignment of a speleothem layer to a particular calendar age. The most common methods are *radiometric*, that is they rely on decay of a radioactive species from a defined starting point. By far the most commonly used method for samples between a few hundred and a few hundred thousand years in age is the uranium-thorium disequilibrium method, but a variety of radiometric and other methods can be used as a check or to extend the dating range to millions of years.

The development of techniques for precise and accurate U–Th dating on small amounts of sample is the single most important factor that has allowed speleothem science to become so prominent in recent years (Edwards et al., 1987; Hoffmann et al., 2007). Work that used to be laboriously undertaken on tens of grams of sample in the 1980s can now be performed on milligrams following the successive development of thermal ionization mass spectrometry (TIMS) and multi-collector inductively coupled plasma mass-spectrometry (MC-ICPMS). The core principle is that as the speleothem grows, it incorporates some uranium from aqueous solution, but fails to incorporate the insoluble element thorium. The nuclide ^{230}Th accumulates over time, by alpha-decay from ^{234}U, after that particular speleothem growth layer has been deposited. The *half-life* (time taken for half the radioactive nuclide to decay) is around 245,000 years and the method can be used to date samples up to around 500,000 years in age. Cheng et al. (2009b) achieved astonishingly good analytical precision (2σ) of 100 years or less on samples over 120,000 years old, which represents the state-of-the-art.

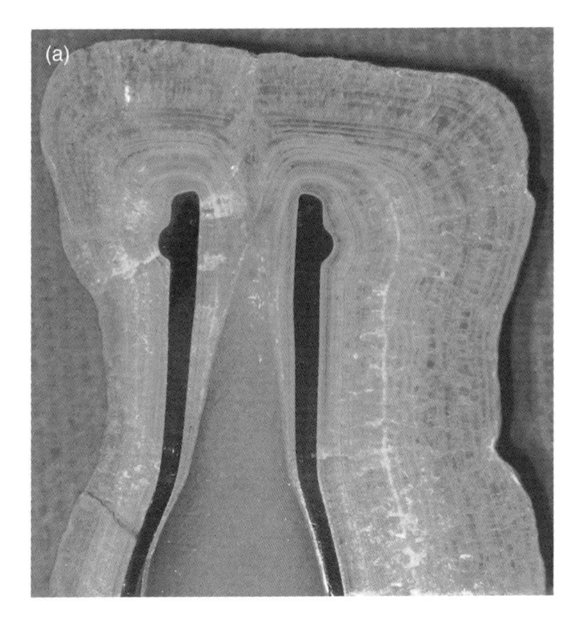

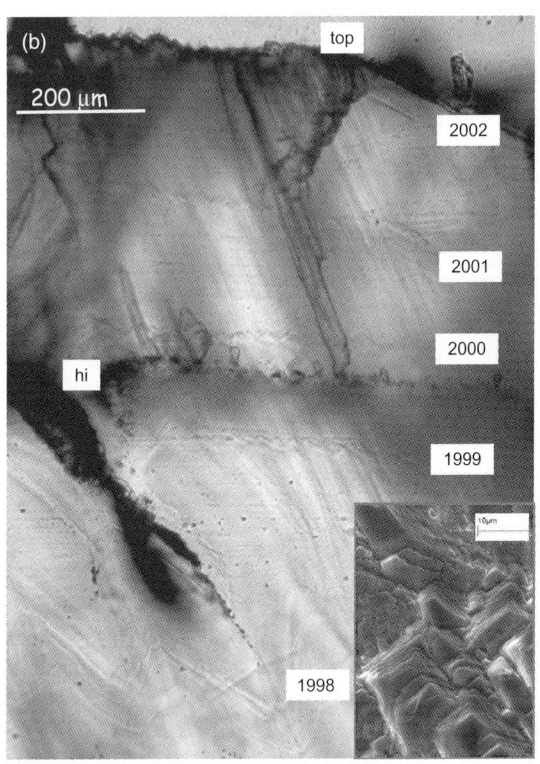

Fig. 1.5 (a) Sectioned speleothem that has grown around a bottle (black) in Proumeyssac Cave, Dordogne, France. Annual growth couplets in calcite represent alternations of clear calcite (darker) and calcite containing fluid inclusions (lighter). (b) Plate 7.2, thin section of the top of stalagmite Obi84, Obir Cave, Austria, showing growth from late 1998 to the end of 2002 by bright annual impulse laminae (plane polarized light). Black area lower left is air-filled inclusion and growth zones show that this occupies a depression on the growth surface. Hiatus surface (hi) represents a brief in-year pause in growth. Inset shows a scanning electron microscope view of the crystal growth surfaces (after Fairchild et al., 2010).

Because each radiometric data refers to one particular growth layer, what is then needed is an *age-model*: that is a continuous function of age versus distance on the sample. Fig. 1.6 illustrates data from several Chinese stalagmites. Stalagmite DA (Wang et al., 2005) displays nearly linear growth for the past 9000 years, whereas sb10 shows a slightly less regular growth with a hiatus of several thousand years. In both cases, the dates are close together and have sufficiently small errors that there is no ambiguity about the age-model, but in other cases there would a choice of how exactly to draw the lines between the dates (Scholz & Hoffmann, 2011). Sample D4 displays several step-changes in the rate of growth, but growth tends to be linear between these steps. All of these records are perfect in the sense that there are no *age-reversals* (stratigraphically younger samples with older ages). In practice, many samples are less ideal for dating because they can contain a lot of original detrital Th, as shown by the presence of ^{232}Th. Although this can be corrected, it is not always known which ^{230}Th/^{232}Th ratio to use for correction and so errors can be much larger and the data can display age-reversals. Other samples may be difficult to date simply because of low U content. Where annual laminae are present, the optimal strategy is to combine lamina-counting with the U-series chronology.

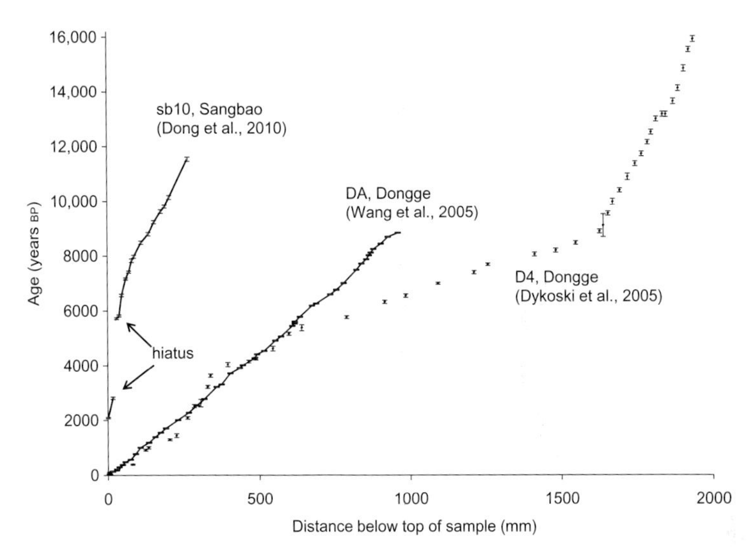

Fig. 1.6 Age models for some Chinese stalagmites included in the compilations of Plate 1.2 (Dykoski et al., 2005; Wang et al., 2005; Dong et al., 2010). DA shows continuous quasi-linear growth, sb10 illustrates a hiatus, and D4 illustrates data points defining a series of growth periods with differing linear growth rates.

1.1.5 What are the proxies for past environments and climates?

Proxies are parameters that can be measured in an *archive* (e.g. speleothems, ice cores, etc.) and which stand in, or substitute for, an environmental variable (e.g. mean annual temperature, seasonal monsoon intensity, or vegetation type). Examples of speleothem proxy parameters are growth rate, Mg concentration, $\delta^{18}O$ or $\delta^{13}C$ signature (see Box 5.3 for isotope definitions). In palaeoenvironmental analysis, the behaviour of proxy variables over time is used to interpret the changing environments or climates. In some cases this can be done quantitatively by means of an equation or process known as a *transfer function* (Fig. 1.7). Fairchild et al. (2006a) attempted to show systematically how the environmental signal becomes encoded in order to improve our understanding of how the proxy signals work. They drew attention to five realms in which signals were generated or modified:

1 Atmosphere (input of energy and matter; e.g. amount or $\delta^{18}O$ composition of rain, temperature variability).

2 Soil and upper epikarst (processes of organic decomposition, mineral dissolution, water flow and mixing; e.g characteristic $\delta^{13}C$ of vegetation, typical trace element to Ca ratio from bedrock dissolution).

3 Lower epikarst and cave (degassing and prior calcite precipitation, evaporation; e.g. changes in trace element to Ca ratios or $\delta^{13}C$ signature; influence of solution saturation state on growth rate).

4 $CaCO_3$ precipitation (partitioning or fractionation of elements and isotopes between water and $CaCO_3$ possibly dependent on growth rate or other effects; changes require use of transfer functions to derive water compositions).

5 Secondary change (e.g. change of aragonite to calcite; exchange of ^{18}O between calcite and water in inclusions).

In practice, certain parameters in particular contexts have proved especially valuable. The most commonly used proxy is $\delta^{18}O$ (McDermott, 2004; Lachniet, 2009), which has been found to be most powerful in cases where the changing atmospheric $\delta^{18}O$ composition of the original dominates the variation in $\delta^{18}O$ in $CaCO_3$; this is thought to be the case, for example, in the Chinese monsoon records in Plate 1.2. In other cases, the effects of temperature *per se* on rainfall composition and on the

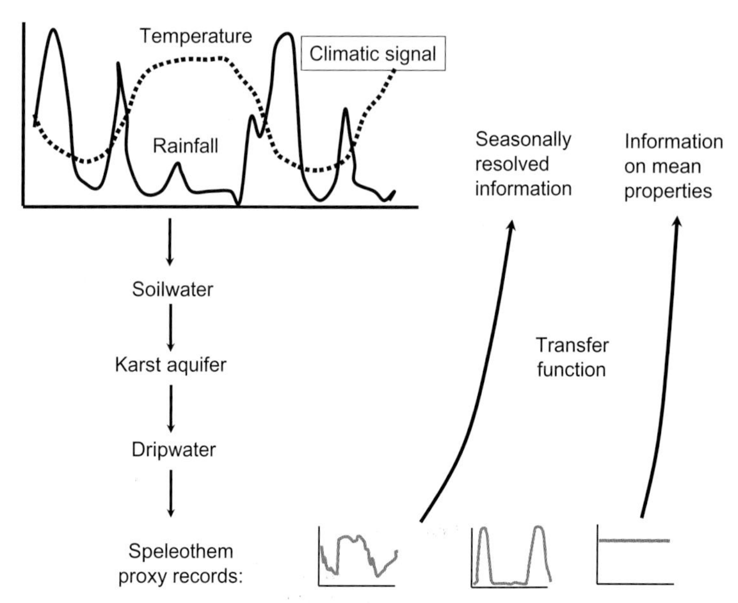

Fig. 1.7 Preservation of environmental (e.g. climatic) signals in speleothems and the use of transfer functions to inverse model the original environmental condition. An annual cycle of temperature and rainfall variability is used as an example. This is encoded in a speleothem by means of its proxy parameters. At one extreme, there may be a reasonably faithful representation, or a rectified version, of the original signal, whereas at the other, a characteristic constant value of a parameter is produced in the speleothem. Information on the original conditions can be recovered by reversing the coding process; where this is done quantitatively, an equation called a transfer function is used.

transfer function between water and speleothem composition can be opposed, and complicated by seasonal changes in rainfall composition: here difficulties in interpretation arise because of the local context. Likewise for $\delta^{13}C$, there are certain cases where the strong difference in isotopic composition between the aridity-tolerant C4 grasses and C3 vegetation is directly reflected in changing $\delta^{13}C$ signatures over time in speleothems. On the other hand, there are several other controls on $\delta^{13}C$, which has meant that such records are often left uninterpreted. Aridity can also be reflected in covarying $\delta^{13}C$, Mg and Sr signatures of speleothems, because of varying importance of prior calcite precipitation (Chapters 5 and 8). However, more generally many different patterns of variation of trace elements and isotopes with each other occur, only some of which are currently interpretable. Some specific signals such as high contents of colloid-transported elements make sense in terms of reflecting high rainfall and infiltration, but they are not expected to yield quantitative rainfall records. It is also worthwhile to mention that growth rate variations, determined from the thickness of annual layers, have been shown to reflect climatic variables in several cases and have been applied particularly prominently to modern climate calibrations and understanding the climatology of the last millennium (see Chapters 10 and 11).

1.1.6 How do speleothems compare with other archives?

The properties of Quaternary archives in general are well summarized in several texts (e.g. Bradley, 1999; Battarbee & Binney, 2008). The strengths of the speleothem archives lie in the following characteristics:

1 The common occurrence of continuous episodes of growth, thousands of years in duration, and the preservation of information representing timescales from days up to a million years. This is matched by ice cores, but other archives tend to have strengths either in their high resolution (e.g. tree rings) or long duration (e.g. deep marine sediments).

2 The excellent chronologies that can be obtained by U-series dating. This is far superior to the other proxies in which long records develop, although some other materials (e.g. coral terraces) can also be dated using the same techniques.

3 Speleothems contain several proxy parameters that can be used singly or in combination. Multiple proxies are being developed in numerous other archives too, including peat cores, lacustrine and marine sediments, ice cores, and indeed in individual components, such as foraminifera. It is the ability to record these parameters at very high-resolution which is a particular strength of speleothems and compares with marine coral records and tree rings for example.

4 They are widespread in inhabited continental areas and so their records have a direct relevance to regional climates and environments, and may also contribute to archaeological investigations. A limiting factor here may be the uniqueness of individual stalagmite samples, which makes conservation a concern (see Appendix 1), whereas other continental archives such as ice cores and lake records are typically more aerially extensive, as are flowstones.

5 Speleothems are physically and chemically robust and are relatively protected from erosion. This contrasts with other continental records, which are accordingly often not available before the Holocene. Certain lake records provide the strongest competition in this category and often have complementary information.

The weaknesses of speleothem science, in its current state of development, can be summarized as follows:

1 Insufficient inter-comparisons with other archives in well-constrained (e.g. co-located) contexts in order to improve the robustness of interpretation of both speleothem and other archives.

2 Insufficient understanding of the meaning of proxy variables. Many studies rely on one proxy and do not present data for others. Sometimes interpretations are unduly speculative or too simplistic in expecting a parameter to directly reflect a climate parameter.

1.1.7 What next for speleothem science?

In recent years there has been a rapid advance in analytical techniques to improve sensistivity and accuracy which has had the effect of reducing the destructiveness of sampling. Accordingly, even without further technical developments, there is an enormous amount to do in terms of applying state-of-the-art techniques to existing samples. In addition, there are many research frontiers, including the following:

1. The development of U–Pb methods to extend records throughout the Quaternary and into the Pliocene.

2. A more sophisticated understanding of how speleothem records contribute to understanding climatic drivers and teleconnections of climate.

3. Continued development of new types of proxy and the more effective use of multiproxy approaches.

We come back to these issues at the end of the book.

1.2 How is this book organized?

The first three chapters provide an introduction and context. Chapter 1 sets out our overall approach. The dynamic nature of speleothem-forming environments both on the human timescale and in deep (geological) time makes it attractive to apply the interdisciplinary concepts of system science as an aid to model development, as we discuss in the next section. We set out an explanatory framework for speleothems, firstly by introducing two new terms as metaphors: the *speleothem incubator*, the place in which speleothem growth is nurtured, and which lies within the *speleothem factory*, which also contains the delivery system for the raw materials for growth. We lay out the types of change occurring over different timescales and the parameters by which these changes are delivered; our task in this book is to provide a coherent connection between them. Chapter 2 focuses on the development and infilling of caves within human to geological timescales, emphasizing the various issues in carbonate geology. We outline the typical properties of aquifer carbonate rocks and their pores, the diversity of the cave systems, and the place of speleothems in the evolution and ultimate infilling or collapse of the cave. In Chapter 3, we show how the cascading and interlinked systems of climate, atmospheric moisture, soils and vegetation control many of the variables that determine the properties of speleothems. The different features of the climate system that might be recognizable in the past from speleothem studies are illustrated, as are the extent of variability of these climatic modes. We review how the biotic response to climate, modulated by the geological substrates, is reflected in the distinctive physicochemical properties of soils and their associated surface and subsurface ecosystems.

The second part of the book (Chapters 4–6) deal with the transfer processes in karst that characterize its dynamics and composition. Chapter 4 develops in detail the case for thinking of cave and karst processes as a speleothem life-support system: the speleothem incubator. We deal separately with the issues of water, air and heat movement in subsurface karst, illustrating the extent to which these can be understood quantitatively. We summarize by reviewing the integrative property of cave climatology and its influence on speleothem properties. Chapter 5 reviews the inorganic chemical components of the karst environment and its aqueous solutions in relation to speleothem composition. We deal with the fundamentally important carbonate system before considering each chemical variable, most of which have already been applied as palaeoenvironmental proxies when preserved in speleothems. Chapter 6 introduces the relatively neglected issues of the biogeochemistry of karst environments, illustrating the range of organic components that can be mobilized and incorporated into speleothems, and describing what we currently understand to be the important molecular transformations imparted on organic matter and the effects of organic assimilation, breakdown and transport on inorganic species.

The third part of the book focuses on the speleothems themselves, but we restrict attention to *calcareous* examples because these are overwhelmingly the types that are used in palaeostudies (Calaforra et al. (2008) discussed gypsum speleothems), using other analogues where appropriate. Chapter 7, on the architecture of speleothems, first discusses the fundamental controls on their shape and internal structure leading to a geometrical classification. A synthesis of their internal mineralogical, crystallographic and geometric structures sets an agenda for what needs to be explained in a given sample. Chapter 8 gives an overview of the geochemistry of speleothems, starting with the macroscopic setting, moving to the details of the water-speleothem interface and systematically describing the fractionations that occur between water and calcium carbonate during speleothem formation. This leads to two types of conceptual model: those that reflect patterns of seasonal change and those that reveal change in space within the cave environment. Chapter 9 on dating is particularly critical because a significant competitive advantage of speleothems in relation to other palaeoclimate proxies arises from one's ability to determine the absolute age of formation of a given speleothem layer. We deal systematically with the radiometric methods (those that depend on decay of specific radioactive isotopes), palaeomagnetism and annual-layer counting, before concluding with an analysis of age-depth models and uncertainties.

The most pressing argument for developing speleothem studies is their capability for contributing to palaeoscience (Roberts, 1998; Bradley, 1999; Batterbee & Binney, 2008), i.e. for recording information on past environments and climates, with an unparalleled combination of age-resolution and range of environmental proxies. Further, they are located in continental regions whose future climatic evolution is the subject of intense scrutiny. Study of the past boosts our confidence in forecasting (Skinner, 2008), particularly providing data for testing the *general circulation models* (GCMs) and other models used for prediction of future climate. We explore these aspects in the final section of this book, comprising Chapters 10–12. Chapter 10 tackles issues of calibration and validation of speleothem climatic proxies using young speleothems that have formed within the era of instrumental meteorological measurements. This is a rapidly developing field in which a range of approaches from related proxy materials are being used and there is a currently a major international effort to identify the key locations which can answer important climatological questions. In Chapter 11, we focus on the post-glacial Holocene Epoch, summarizing the range of relatively subtle climate drivers, key target time periods, quasi-periodic signals and model comparisons that have been investigated, with a particularly detailed focus on the last millennium. The Holocene is a period of immense archaeological interest as well as potentially one where disturbance by human activity is a key influence. Speleothems have already provided some intriguing pieces in the jigsaw, as well as more complete and meaningful records in some regions. Finally, in Chapter 12, we look at the longer Pleistocene Period (and beyond) in which the major Cenozoic glacial–interglacial cycles have occurred as has been documented primarily through study of ocean cores, ice cores, and some long lake and loess sections. Here speleothem records have been demonstrated in particular regions to be excellent integrators of aspects of hemispheric or global climate and have been used to refine timescales used by workers on other Quaternary proxies.

1.3 Concepts and approaches of system science

Systems science covers a diverse range of approaches to understanding the workings of natural and anthropogenic environments (Kump et al., 2009). Understanding the formation of speleothems requires a range of approaches, and systems thinking has two roles to play. Firstly, a qualitative systems analysis is useful to clarify our thinking about relationships of factors that may influence speleothem formation. Secondly, a quantitative approach allows us to test our understanding by building models that simulate aspects of the processes that lead to speleothem growth and initial steps in this direction are discussed in Chapter 10.

The science of systems has diverse origins, but threads derived from the mathematical modelling of regulatory processes (*cybernetics*, Wiener, 1948) and from organismal biology (von Bertalanffy, 1950, 1968) were highly influential in the post-war period. The detailed concepts of system theory were subsequently applied to geomorphic environments by the influential geographers Chorley and Kennedy (1971), contributing to geography's 'quantitative revolution', and were further developed by Bennett and Chorley (1978), although with no reference to karst.

Ford and Williams (1989, 2007) identified the development of structured networks of caves and drainage networks in unconfined karst as a case of a *cascading system*, which Huggett (1985) defined as '*interconnected pathways of transport of energy or matter, or both, together with such storages of energy and matter as may be required*'. Cascading systems are amenable to box-modelling (Box 1.1). In addition, we observe the development of morphological forms of defined geometry, which also show *self-organization* (the spontaneous development of ordered structures, such as calcareous deposits developing cascades of pools with rims). We might further aim to build a process–response model (Chorley & Kennedy, 1971) that brings together the quantitative mass or energy flow information with morphological parameters to give a more sophisticated understanding of the system.

A central issue in system science is the interactive nature of system components and in particular the occurrence of *feedback* behaviour, whether *positive* (reinforcing the applied change) or *negative* (resisting the change). Box 1.1 takes the well-known example of the carbon cycle to illustrate feedback concepts, showing how the issues can be downscaled to caves.

A variety of system concepts that refer to the nature of change over time (Chorley and Kennedy, 1971; White et al., 1992; Phillips, 2009) are illustrated in Box 1.1 and Fig. 1.8, and will be applied to karst in the next section. A system parameter may show an evolution over time, in which case it is described as *non-stationary* (Fig. 1.8a). Alternatively it may be *stationary*, that is, having a consistent mean value averaged over the period of interest.

Box 1.1 Box models and feedback

This box contains some basic material on flows of matter within simplified systems. Kump et al. (2009) give an accessible introduction, whereas Rodhe (1992) and Chameides and Perdue (1997) provide a more rigorous mathematical treatment.

A *reservoir* (also known as a box) contains matter (mass M) and can be considered homogeneous.

A *flux* (F) is the amount of material transferred from one reservoir to another per unit time.

Many systems are approximately *linear*, that is F_i (the flux or fluxes from reservoir i) is directly proportional to M_i, the mass in reservoir i (volume V_i can also be used instead of M_i for fluids at constant temperature and pressure).

If all relevant fluxes and masses remain constant over time, the system must be in kinetic or dynamic balance, and is said to be in *steady-state*. Under this condition, the underlying kinetic constants can be readily calculated.

A rate constant can be defined for each flux as follows:

$$F_{ij} = M_i\, k_{ij} \qquad \text{(I)}$$

where F_{ij} refers to the flux from reservoir i to reservoir j and k_{ij} is the corresponding rate constant.

$$1/k_{ij} = \tau_{ij} \qquad \text{(II)}$$

where τ_{ij} is the *residence time*, corresponding to the mean time that a given molecule resides in reservoir i before it is removed to reservoir j.

A relevant example is provided in Fig. 1.B1 by the comparison of carbon reservoirs and fluxes from the pre-industrial period (black), when the system was presumed to be in steady-state (black) with the changing situation in the 1990s (grey). Here the residence time (lifetime) of fossil fuels equals the reservoir size (3700 Gt C) divided by the current usage rate (6.4 Gt yr^{-1}), i.e. 580 years.

However, interpreting residence times when there are multiple fluxes out of a reservoir (i.e. multiple *sinks*) needs more careful discussion. For example, radioactive isotopes such as ^{14}C produced during atmospheric nuclear testing in the 1950s and early 1960s have provided important insights into element cycling. Half of the excess radiocarbon (^{14}C) disappeared in 10 years (Fig. 3.24). This observation has been used by climate-change deniers as evidence for the short lifetime of carbon in the atmosphere. Actually the *turnover time* of a reservoir (Rodhe, 1992), the time it would take to empty the atmosphere of ^{14}C if the sinks remained constant and the sources were zero, is even

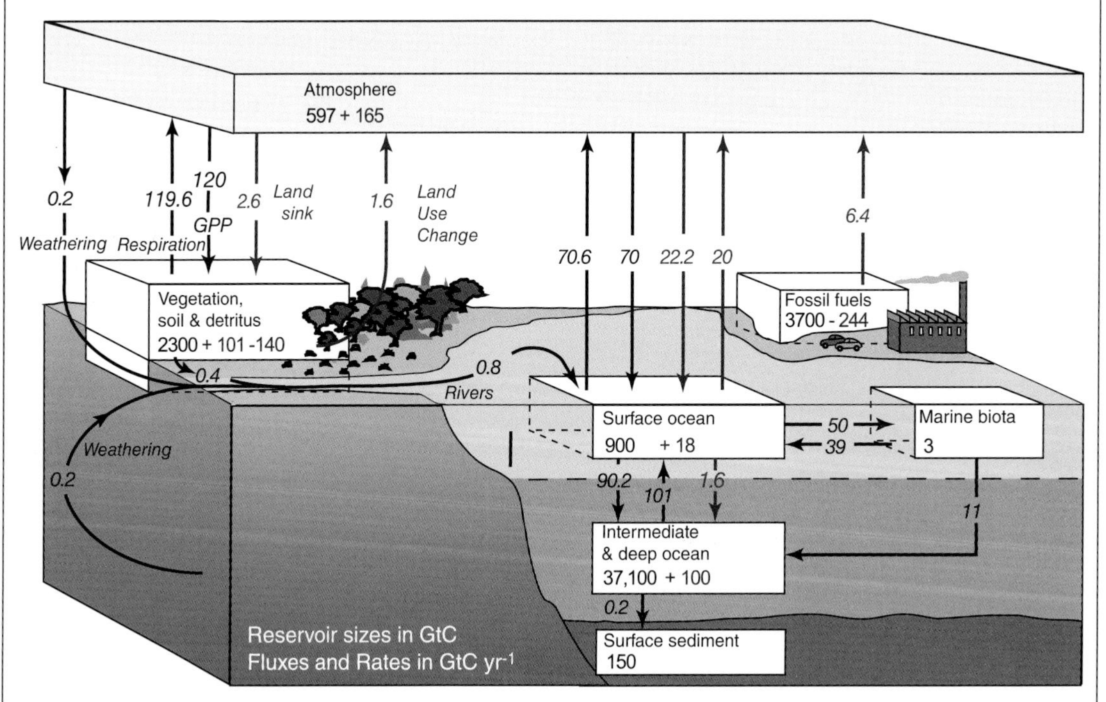

Fig. 1.B1 The carbon cycle (IPCC, 2007) showing reservoir (boxes) and fluxes (arrows). Figures in black are from the pre-industrial era whereas those in grey (red) are the changes resulting from human activity in the Anthropocene.

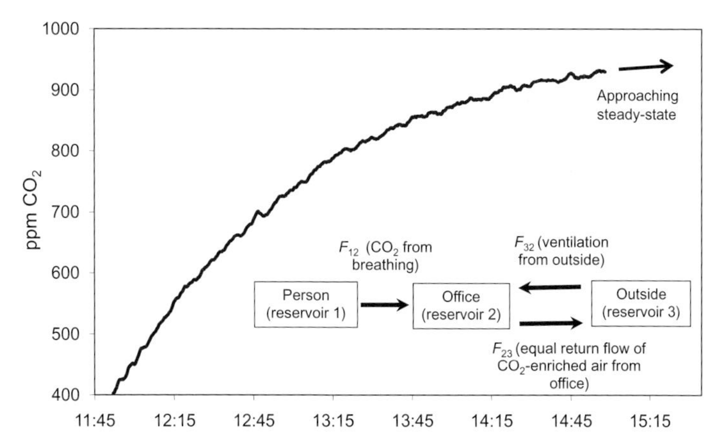

Fig. 1.B2 Results of an experiment to illustrate the concept of a system evolving to a new steady-state after a change in boundary conditions (see text for explanation).

shorter. In Fig. 1.B1, using the figures (in black) from the steady-state pre-industrial period:

Atmospheric carbon turnover time = mass/total outwards fluxes = 597/(0.2 + 120 + 70) = 3.1 yr.

The key point here is that ^{14}C is temporarily stored in vegetation and the surface ocean and is rapidly returned to the atmosphere, so that a more meaningful figure for the residence time of carbon is given by the ratio of the atmospheric mass (597 Gt C) to the 'permanent' sink in sediments (0.2 Gt C yr^{-1}), i.e. around 2500 years. Such a long period is the reason that our current carbon emissions are a cause for concern.

If the system is then perturbed by a change in the sources or sinks of matter, the rate constants can be used to estimate how the system will change. Using the pre-industrial figures and applying eqn. (I), the rate constant for transfer of carbon from the atmosphere to the surface ocean is 120/597 = 0.201 yr^{-1}. Given the enhanced mass of atmospheric C typical of the 1990s (597 + 165 = 762 Gt), application of this rate constant suggests that the flux of C to the surface ocean should then be 762 × 0.201 = 153 Gt C yr^{-1}. In fact the figure is just 92.2 (70 + 22.2), implying that more complex types of behaviour need to be considered, e.g. by more detailed consideration of the processes involved (Rodhe, 1992; see also the carbonate system in Chapter 5). Nevertheless, the approach clearly indicates the sense of change that is expected.

The above case example is from (large-scale) Earth system science, but there are analogous examples in terms of temporary sediment storage in geomorphic systems and in terms of carbon storage in soils for speleothem studies (see Chapter 8). Changing carbon dioxide concentrations in air can also be used to provide a good example of how to apply linear systems theory in a more local context. An experiment was conducted in which an investigator entered a previously well-ventilated empty office, closed the door and window, and monitored the changing carbon dioxide levels. The pattern of change is shown in Fig. 1.B2, increasing from an initial value close to that of the external atmosphere at around 400 ppm, but clearly flattening off after 3 hours towards a new steady-state value. This can be rationalized in terms of a box model (inset in Fig. 1.B2) with the person respiring CO_2 as reservoir 1, and the office (reservoir 2) exchanging air with the infinite external reservoir 3.

The mass of CO_2 in the office rises at the following rate:

$$dM_2/dt = F_{12} + F_{32} - F_{23} \tag{III}$$

It can be seen that the negative feedback that results in a stabilization of the carbon dioxide level is an inherent part of the structure of the system: as the concentration of CO_2 in the office rises, it is more efficiently removed in each parcel of air that exchanges with the exterior. This is a perfect linear system where the mass of carbon dioxide removed to the exterior is directly proportional to the mass in the room, because the rate of air exchange is constant. Ultimately a new steady state is reached where dM_2/dt is zero:

$$F_{12} = F_{23} - F_{32} \tag{IV}$$

F_{12} can be estimated from general adult physiology based on the typical CO_2 content (3.6%) and volume (8 litres) of exhaled breath per minute. With V_2 = 32,000 litres, this would lead to an initial increase in P_{CO_2} of 9 ppm min^{-1} whereas the actual observed figure was 7 ppm min^{-1} (the subject was a slow breather!). The rate of air exchange through the window frame was not known at the start of the experiment, but can be calculated as 420

(Continued)

litres min^{-1} by simulating the experiment stepwise on a spreadsheet (see www.speleothemscience.info) and adjusting parameters to fit the observed curve. Hence the approach allows a complete quantification.

Mass balance approaches have many applications in geomorphic and geochemical studies. One key example in cave environments is the CO_2 level in cave air because this directly control rates of speleothem growth, but normally varies seasonally. Figure 1.B3 illustrates the main fluxes of CO_2 in Ernesto, a small Italian cave (Frisia et al., 2011). The rate of decay of transient high CO_2 levels from visitors was used to estimate cave air residence times and exchange rates with the external atmosphere during steady-state summer conditions. The exchange times were used to demonstrate a large flux of carbon dioxide from air-filled fissures in the surrounding epikarst, much larger than that which could be released from dripwater. When considering the transition from summer to winter conditions, additional information is needed because all the fluxes are liable to change with changing temperatures and aquifer conditions. Although the system does not remain linear,

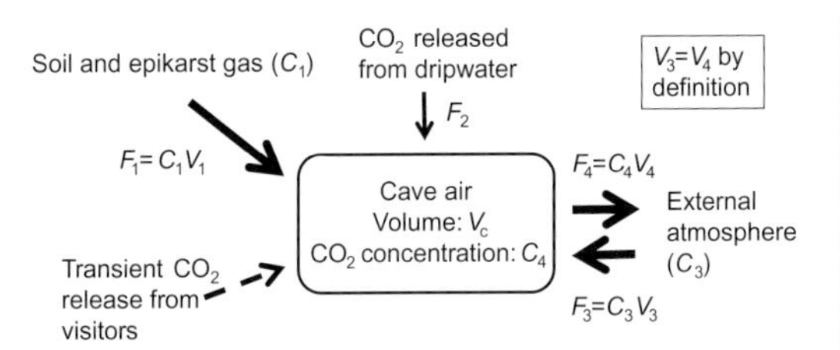

Fig. 1.B3 Structure of box model for carbon dioxide fluxes derived from a study of Ernesto Cave, Italy (Frisia et al., 2011). F, flux; V, volume; C, concentration.

the system structure remains unchanged, and enables quantitative solutions to be derived (Faimon et al., 2006; Kowalczk & Froelich, 2010; Frisia et al., 2011). A key feedback turns out to be a reduction in epikarst CO_2 flux because of an increase in water saturation of the aquifer in winter. These issues are explored further in Chapters 4, 5 and 8.

Within this period, variation can be periodic (cyclic) as shown in Fig. 1.8b, or may be subject to non-periodic fluctuations. Cyclicity can be studied by a variety of statistical techniques that are introduced in Chapter 10. In Fig. 1.8b, the sequence is *deterministic* as the cycles are all exactly the same, whereas in nature there is a random component and hence the systems are non-deterministic or *stochastic*.

Where there is memory of previous system states, stochastic processes can give rise to apparent structured trends (Fig. 1.8c) without a corresponding systematic driving force. This is a really important concept for palaeoscientists because there is a strong temptation to over-interpret plots of variables against time (e.g. Fig. 1.8d), e.g. to consider all 'wiggles' on plots of past variables as meaningful (Fairchild et al., 2006a). The property is referred to as *autocorrelation* and statistical models can be created to describe it. Such models have an *order* which represents the number of previous states influencing the current state (Huggett, 1985). For example, in Fig. 1.8c, the upper line indicates a sequence of random numbers Y_1, Y_2, Y_3 . . . (within the range 0–1) whereas the lower (spiky) line is a random walk representing an example of an autocorrelative function (Z) of order 1 that is related to the random series Y at time t as follows:

$$Z_t = Z_{t-1} + Y_t - 0.5 \text{ (the term } -0.5 \text{ ensures that the series is stationary rather than rising).} \quad (1.1)$$

Figure 1.8c shows that apparent systematic variations are observed with the same degree of noise as the random number series. In real systems, it may be a major research task to distinguish purely random from systematic behaviour.

Stationary systems show a degree of resilience to a time-limited disturbance, as illustrated in Fig. 1.8d where the original system property is eventually recovered. An external forcing leads to system change after a *reaction time* and the system behaviour responds over a characteristic period known as the *relaxation time*. This contrasts with that shown in Fig. 1.8a where the system's response to a series of transient forcing events is to move closer to a *threshold* beyond which it rapidly moves to a different state.

In the geomorphic literature, the term *sensitivity* has been related to the probability that a persistent change will result from a disturbance (Brunsden and Thornes, 1979; Phillips, 2009); that is, a sensi-

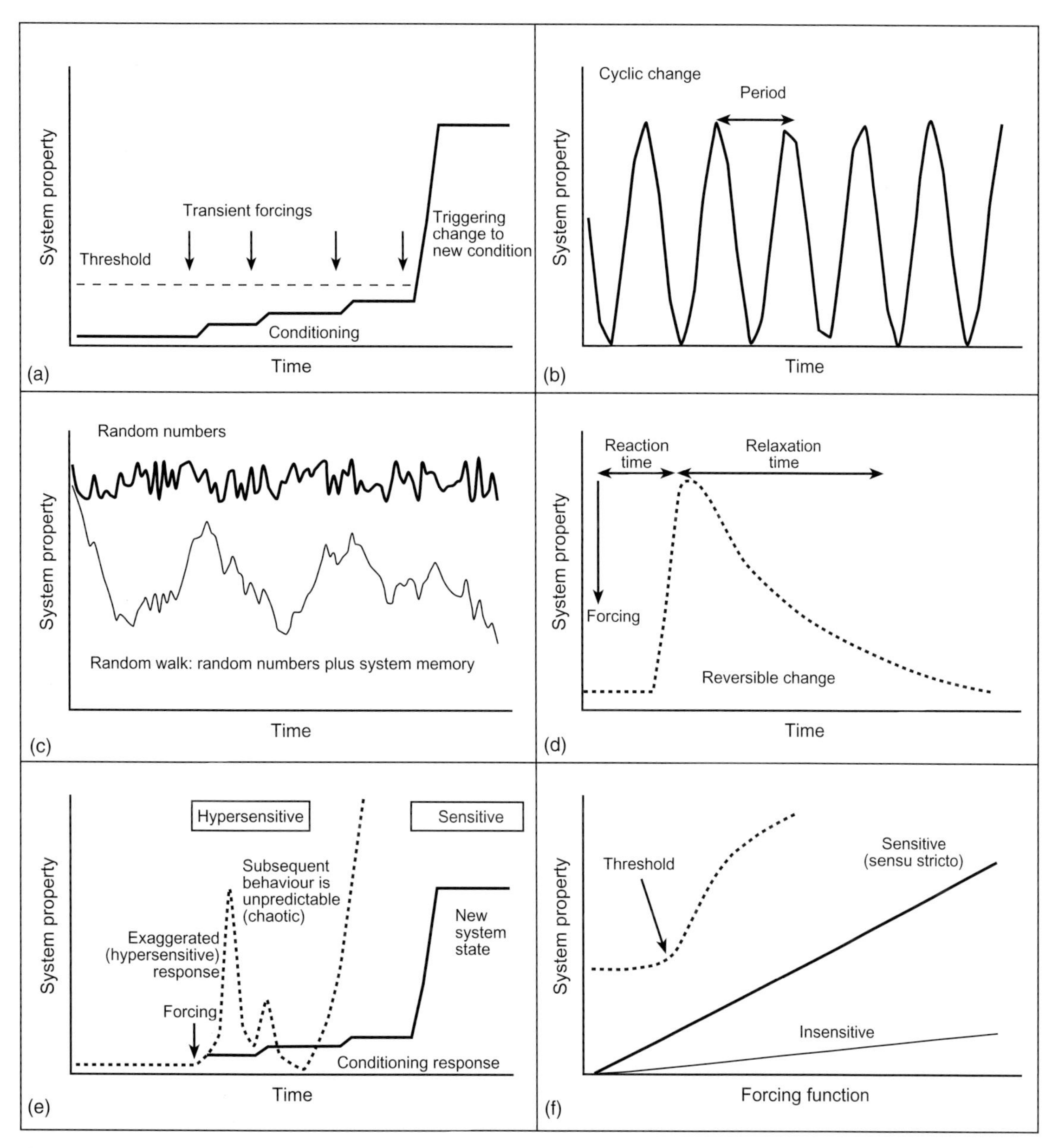

Fig. 1.8 Types of system behaviour. Diagrams (a)–(d) and (f) illustrate changes in a system property over time. In (a) there are a series of transient forcings; in (b) there is an implied cyclic forcing (although internal system dynamics can also give rise to this type of behaviour); in (c) the behaviour illustrates the addition of random forcings in a system with some memory effects; in (d) and (f) there is a single transient forcing with very different results. Diagram (e) illustrates the concepts of threshold and sensitivity in relation to a persistent forcing. See text for further explanation.

tive system (Fig. 1.8e) is opposite to the resilient system shown in Fig. 1.8d. In physical science literature more generally, sensitivity usually refers to a gradient between a system property and a long-lasting forcing function (Fig. 1.8f).

The behaviour of a system over time could show simple linearly scaled responses to disturbances (e.g. double rainfall produces double discharge), but real systems are more likely to be complex where exact physical solutions are unavailable: all we can then do is to ascribe probabilities for behaviour. A special case of complex systems are those described as *chaotic*, where there is hypersensitivity to the initial conditions such that the final result varies wildly and cannot be predicted (Fig. 1.8e). Phenomena driven by fluid turbulence, such as weather systems are of this type. Chaos theory and its relative catastrophe theory deal with nonlinear dynamics, that is, mathematically they refer to systems that cannot be described by systems of linear equations. Whereas a non-chaotic system can have a specific defined state (a defined solution to a mathematical equation) that can be described as an attractor, in chaos theory the (strange) attractors are fuzzy. For example, climate is the (strange) attractor for weather.

Catastrophe theory deals with system discontinuities, or *bifurcations*, where a small additional forcing can lead to a major change in system state that can be difficult to reverse such as the modelled behaviour of oceanic circulation in the Atlantic (Rahmstorf, 1995). The growth history of many speleothems reveals irreversible changes.

Currently, the dominant reference to systems in environmental science is Earth system science (ESS), which seeks to understand and predict the future changes in the interactions between the different major components of the Earth (geosphere, biosphere, atmosphere, hydrosphere and cryosphere), together with human interactions (Ehlers & Krafft, 2001). ESS makes most sense in the context of large-scale phenomena that ideally can be sensed remotely and which are sufficiently well understood as to be amenable to quantitative simulation: global biogeochemical cycling is an example (Chameides & Perdue, 1997). In this field, it is common to use industrial metaphors to describe the workings of the Earth, for example the text of Raiswell et al. (1980): *Environmental Chemistry: The Earth–Air–Water Factory.* The development of improved GCMs for climate underpins much large-scale research in ESS and is also highly relevant as a context for the interpretation of speleothem palaeoenvironmental research, as discussed in Chapters 10–12. To the extent that speleothems may contain encoded information on temperature, rainfall and vegetation, and are available over much of the land area of the Earth, they can also contribute to the testing of Earth system models, in which biogeochemical modules supplement the earlier purely physicochemical content of earlier models.

However, system concepts also apply at the local scale: what Richards and Clifford (2008) refer to playfully as LESS (local environmental systems science). Identifying the behaviour of individual geomorphic systems, their process-interactions, feedbacks and emergent properties provides a holistic basis for site-specific and generic studies of particular environments. As we will see, this turns out to be an enormously important insight for speleothem science.

1.4 The speleothem factory within the karst system

In this book, we propose a nested three-fold conceptual scheme (Fig. 1.9) in terms of the following:

1 The *karst system* as a whole where the emphasis is on its role as an aquifer and hence is on a much larger spatial scale than that relevant for speleothems.

2 A broader *speleothem factory* which operates as a cascading system, supplying the raw ingredients from the epikarst, soil-ecosystem, and adjoining atmosphere and transferring them to the cave environment. We develop this analogy further in this section.

3 A speleothem life-support system which we term the *speleothem incubator.* This consists of those parts of the karst and cave environment that maintain the growth of speleothems during a 'lifetime' when they are active. This maintenance includes the reg-

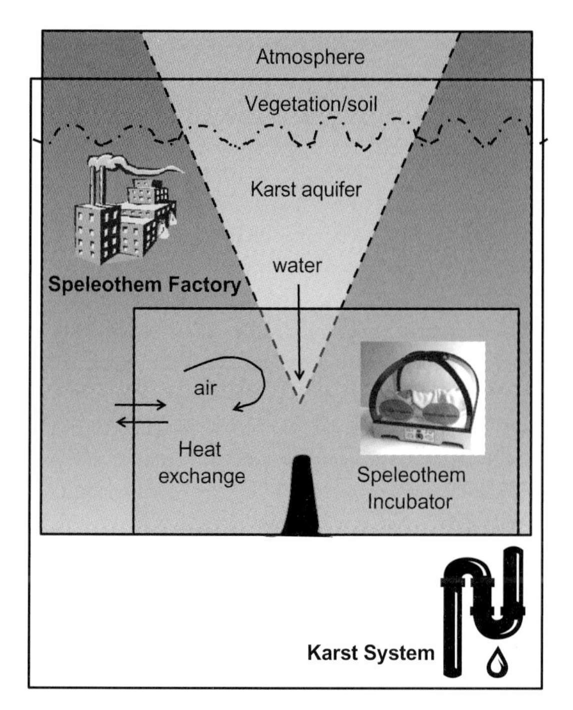

Fig. 1.9 The nested domains of (1) the karst system whose primary drive is to maximize the efficient transport of water; (2) the speleothem factory which supplies raw materials and generates speleothems; and (3) the speleothem incubator which is the cave and adjacent karst environment which maintain the conditions for speleothem growth.

ulation of air and water flows and is described in detail in Chapter 4.

Klimchouk & Ford (2000a) define the karst system as follows:

The karst system is an integrated mass-transfer system in soluble rocks with a permeability structure dominated by conduits dissolved from the rock and organized to facilitate the circulation of fluid.

White (1999) has lucidly summarized the key features of the system that need to be addressed in order to model water flow through karst aquifers. His outline structure of the system is illustrated in Fig. 1.10. Water is recharged at through autogenic sources (concentrated in closed depressions or as diffuse infiltration) or, from allogenic sources, at the surface (sinking streams) or sub-surface (perched aquifers). Karst is often rationalized as a triple-porosity aquifer (*conduits, fractures* and *matrix*) and, as shown by tracer tests, is typically organized into several groundwater basins in which discharge occurs in a single spring, or a small number of related springs. The underflow spring of Fig. 1.10 is the exit conduit at low flow, but overflow springs operate at high flow. Important nonlinearities are introduced by variable exchange between matrix and faster flow routes at different discharges, and by activation of overflow routes throughout the

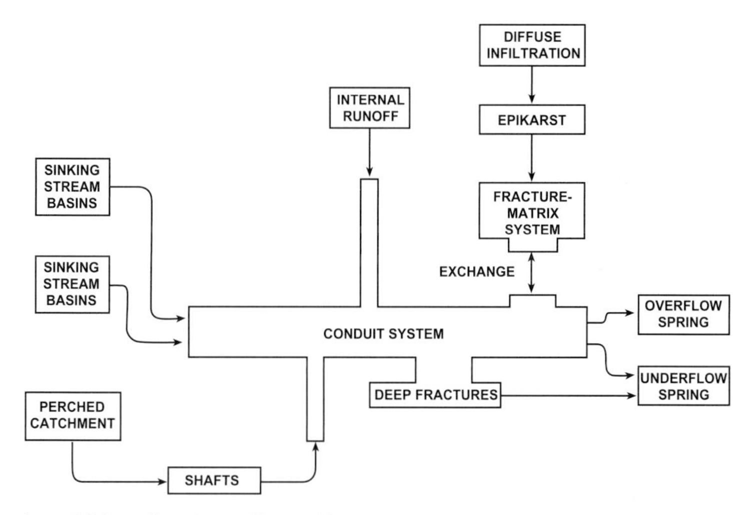

Fig. 1.10 Conceptual model for a karstic aquifer (White, 1999).

system. The key limitation is lack of knowledge about the three-dimensional layout of the drainage components, although constraints can be made by a range of observations including deconvolving discharge and chemical information at springs coupled with use of multiple tracers (see, for example, Perrin & Luetscher, 2008).

Hence it is clear that there exists a conceptual understanding of karst as an aquifer which marries with the key role of these aquifers in water supply. However, we do not currently have such an analysis for speleothem-forming environments. Here, we are often concerned with the life history of individual water droplets, while the major aggregated riverine flows do not leave a permanent chemical record inside the cave (although tufas may form externally; Andrews, 2006). Karst systems in the above sense are conceptually important for this book and so their development over time is discussed later (section 2.4). However, here our main focus concerns the conditions that determine the existence and properties of speleothems: this turns out to be something rather different.

The geomorphic literature stresses the role of physical forces in enacting change. Clearly landscapes are largely modified by the action of gravity in driving flows down slope and the dynamics of moving fluids in causing erosion and sediment transport. Aggradation occurs in areas where these forces are ameliorated. If we return to the industrial analogy, the construction of landscape involves a multitude of energetic grooving, planing and coating actions. This stands in distinct contrast to speleothem growth.

The growth of many speleothems, i.e. dripstones, involves the passage of pregnant waters drop-by-drop over surfaces accreting at rates that are so slow (micrometres to millimetres per year) that they barely overlap with those of industrial processes. Each speleothem body is unique, and in this sense they are analogous to archaeological artefacts. Indeed, they often co-exist with human artefacts with which they share great intrinsic value. This suggests that the manufacturing analogy would be that of a *workshop* where articles, tailored to individual customer requirements, are carefully constructed. Alternatively, the term *factory* is less ambiguous in terms of what work is conducted there, although there is a risk of implying mass production of uniform objects. In this book, we balance the unique aspects of individual speleothems against their generic characteristics, but in the issue of nomenclature, the generic has won out. Hence we refer to the speleothem-forming system, including the sub-system supplying necessary raw materials, as the *speleothem factory.* The great ceramic manufacturers of the Industrial Revolution were involved in a similar enterprise: the moulding from air, water and geological materials of arrays of objects, each readily identifiable by class, but with variety rendered by the detail of design or surface finish (Plate 1.1). Within the speleothem factory, a major sub-system is represented by the action of the karst aquifer and cave environment in regulating speleothem formation. This is the *speleothem incubator* of Fig. 1.9.

There are some conceptual similarities between the *speleothem factory* and the *weathering engine* of Brantley and White (2009). They are dealing with the weathering of silicate rocks in regolith and soil profiles in what has become known as the *critical zone.* Their *weathering engine* system encompasses the forcings, processes and consequences of this weathering. Carbonate rocks differ, however, in their more rapid growth and dissolution kinetics (Chapter 5) and the permeable nature of the epikarst compared with most regoliths.

In the following sections, we present an overview of the development of karst and speleothem-forming environments on different timescales as a summary account of the phenomena that are discussed in detail later in the book and hence it is expected that the reader may wish to return to this chapter and re-read it more critically once they have familiarized themselves with the evidence.

1.4.1 Long-term change

The speleothem factory represents both a distinct spatial zone within the karst system and a distinct stage or stages in the long-term evolution of cave environments (Figs. 1.11 and 1.12e). Speleothem formation is inhibited during the active stages of cave dissolution and occupancy by streams (Fig. 1.12a, b). A relative fall in base level arises natu-

Fig. 1.11 Karst phenomena and high-relief topography. (a) Clon Cave, Guangxi province (Guilin karst region), China, a region with significant tectonic uplift enhancing topography. Tributary river entering a karstic massif at the end of a blind valley. The river continues to flow along a low gradient to reach the Lijiang river on the far side of the hill. (b) Val di Tovel, Trentino province, Italy. Old relic cave close to the skyline, developed along a sub-vertical fault zone, and near the end of its lifetime.

rally by the progressive gravitationally controlled downcutting of subterranean streams in order to provide a more efficient drainage (Figs. 1.11 and 1.12c). Base-level fall could also be accentuated given tectonic uplift or climatically controlled lowering of the water table. This sets the scene for speleothem growth. Once water is free to seep into air spaces of cave chambers and passages and degas excess carbon dioxide, and once active streamflow ceases, speleothem formation can begin. Ultimately the cave passage will be filled by a combination of speleothems and clastic sediment, unless first eroded. Complexity is introduced by the relationships of cave passages to specific geological features or the land surface (Fig. 1.12). This is a long-term cycle, and Ford (1980) noted that 'caves are the longest-lived (least time-limited) elements in [karstic] landscapes' (Fig. 1.11b). Individual caves may continue to accrete speleothems over time periods of the order of 10^4–10^7 years. Some infilled caves have survived over a billion years in association with palaeokarstic surfaces within sedimentary successions (Kerans & Donaldson, 1988).

In systems terms, the long-term evolution of karstic environments is one of continued evolution through successive thresholds on spatial scales of 10^{-2}–10^3 m with details that are contingent upon the exact bedrock configuration in a given location and the succession of climatic and regional geological histories. The main feedback process relates to the evolution of cavity systems. Initially they are systematically enlarged, but may become partly or wholly infilled by sediment, hence feeding back on the pathways of water flow further up the pressure gradient. There is also a chemical aspect to this feedback through the formation of speleothems. During the progressive enlargement of cave chambers and passages, there is an increasing probability that there will be significant air space within them. As air spaces become connected to the surface, exchange of gas with the surface increases. Hence the excess carbon dioxide degassed from solution can be removed which allows speleothems to form. Carbonate precipitation in seasonally air-filled cavities can block flow routes and hence feeds back on the hydrological system. The key point about this speleothem feedback is that it is a natural consequence of the creation of caves and hence inherent to the development of karst systems.

1.4.2 Annual-scale behaviour

Now let us focus on the key parameters of the speleothem factory and the timescales over which they vary. We start by considering its functioning over a year. How would an observer living in a cave at constant temperature be able to keep track of the seasons? Based on results from increased cave monitoring activities over the past 15 years, it is now clear that air composition, water quantity and

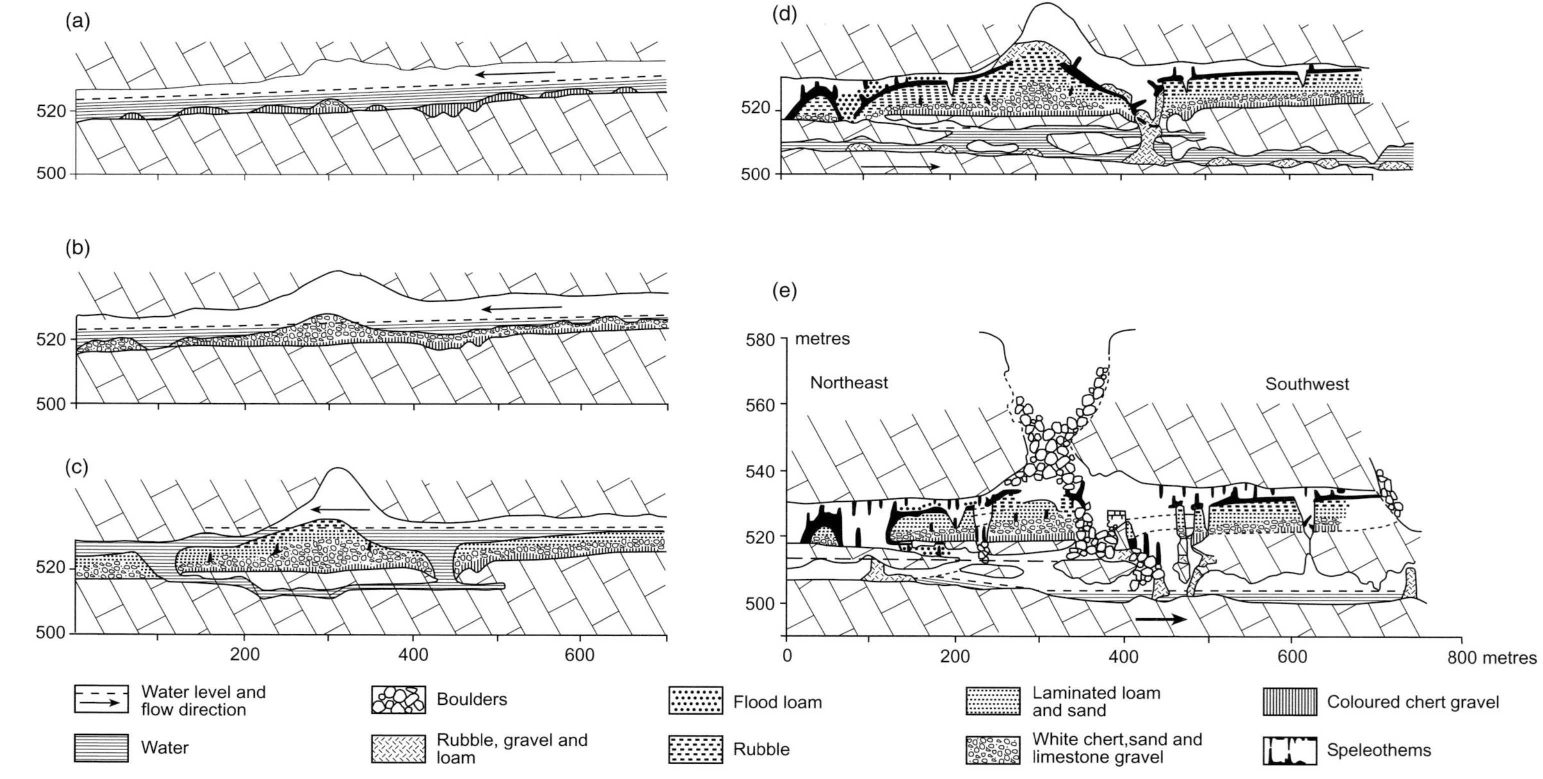

Fig. 1.12 Example of the reconstruction of long-term evolution of cave systems and their filling. (a)–(d) Development phases of the Otoska Jama cave, Postojna system, Slovenia, inferred from (e) which is the present-day configuration (after Gospodarič, 1976). A middle Quaternary sediment cone, resulting from roof breakdown, ponded water of the Pivka River and facilitated downcutting of the passage upstream of the cone. Downcutting was accelerated by continued sediment accumulation close to the ceiling of the upper passage. Phases of speleothem genesis are marked by flowstone layers and dripstone deposits with collapse of layers by erosive undercutting being evident. Reworking of sediment occurs, for example in association with debris movement from roof collapse and subsequent infill beneath the open aven (the Stora Apnenica doline).

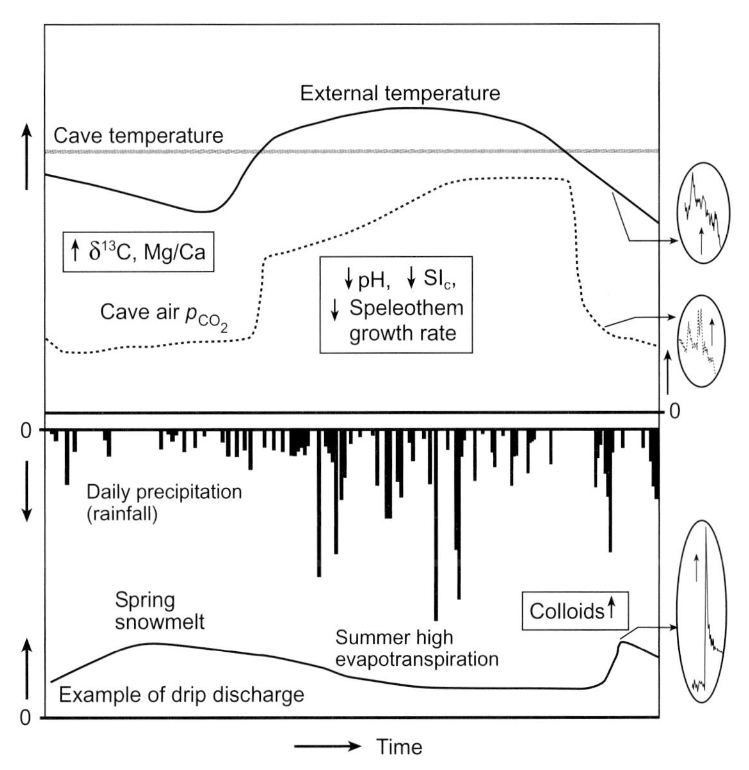

Fig. 1.13 Examples of some possible annual changes in parameters affecting the growth of speleothems, in this case based on study of Alpine environments including Miorandi et al. (2010). Insets (right) show detail of types of short-term (day to week-scale) behaviour. See text for discussion.

composition are all susceptible to annual changes: these are important to the operation of the speleothem incubator as is described in Chapter 4. Here we draw out some generalities (Fig. 1.13). The chemical terms given below are defined in Chapter 5.

It is increasingly recognized that the annual variation in temperature can have an over-riding control on the rate of air exchange between a cave and the exterior, although synoptic meteorology can also be important. Typically air-exchange is enhanced when the external temperature drops below the cave temperature and, unless there is an internal source of CO_2-rich air, the result is a lower P_{CO_2} of cave air in winter. Degassing of water in contact with the cave air is limited by its P_{CO_2}, and slow drips will attain a P_{CO_2} and pH in equilibrium with cave air. In summer, compared with winter, there is a lower supersaturation of the dripwater for calcium carbonate precipitation and hence slower growth rates. Greater degassing and $CaCO_3$ growth in winter is associated with an increase (both in dripwater and calcium carbonate), in $\delta^{13}C$, and also Mg/Ca further along a water flowline where $CaCO_3$ precipitation is occurring. Many speleothems display annual couplets in crystal texture that can be linked to bimodal cave conditions of this type.

Significant variations can also occur during the hydrological year. Each rainfall event is associated with a lag before the event water reaches a given position in the karstic aquifer and there will typically be an exponential decline in discharge over the relaxation time as in Fig. 1.8d. The cumulative impact of rainfall events (plus snowmelt during warmer spells in winter and spring) is smoothed by aquifer storage. A seasonal bias in infiltration should occur in all climate zones, although the gross year-on-year behaviour can vary to differing extents. The mean discharge of a drip is heavily dependent on the details of the water flowpaths, and is likely to relate more to the total filling of the aquifer than the time period of maximum infiltration. Even where this is the case, short-term increases in discharge can coincide with infiltration events. Where event water forms a significant proportion of dripwater, its seasonal variation can be reflected in that of the dripwater, e.g. in Alpine environments, winter rain and snowfall would be expected to be associated with light $\delta^{18}O$ signatures, and individual events could also be associated with reductions in saturation index and speleothem growth rate. Where saturation index is constant, increased water flux will result in increased speleothem growth rate. Perennially wet caves are characterized by a distinct flush of soil-derived colloids and trace elements in the autumn season. In contrast, caves associated with strong summer drought can be linked with enhanced degassing and calcite precipitation at an early stage along the water flowline during this season.

Dripwater hydrology can be approximated as a cascading system with linear characteristics, although nonlinear behaviour provides a more accurate description related, for example, to the complexities associated with air entrapment along flowpaths. The reaction times and sensitivity of individual drips are highly variable and thresholds for change are seasonally variable, being influenced by the overall state of water filling of the aquifer. Likewise, progress is being made in terms of quantifying the mass balance of air circulation as discussed in Chapter 4.

In summary, the strong drive of annual variations in temperature and quantity of infiltrating water, imprint themselves on the cave environment and dripwater characteristics, and hence on the associated speleothems. Different speleothems in the same chamber are likely to vary in terms of properties dependent on drip hydrology, but to display similarity in properties that are controlled by cave ventilation.

Inter-annual meteorological variation, both chaotic and associated with atmospheric modes of variability such as the El Niño/Southern Oscillation or the North Atlantic Oscillation, is inevitable. However, over a short timescale, the bounding system properties, as discussed in the next section, are stationary, or at least can recover their initial values; in other words, the changes are reversible. This may also apply to subtle changes over longer periods of time, perhaps up to hundreds of years, but in general the longer time periods usher in fundamental changes in system action as described in the next section.

1.4.3 Decadal- to multi-millennial-scale changes

Many of the parameters of the speleothem factory that can vary over these timescales are depicted diagrammatically in Plate 1.3 and listed in Table 1.1, together with their controlling variables and some relevant references. We have set the decadal scale as the lower limit for this section, because this is the minimum period over which we could recognize the onset of a long-term shift in climatic regime, such as that which occurred, for example, at the beginning of the Holocene 11,700 years ago (Rasmussen et al., 2006). However, many of the changes we discuss would require rather longer periods to develop. Individual speleothems are known to grow steadily for up to 10^4 years and more intermittently or with more strongly varying rates up to 10^5 years. At the 10^5–10^6 year timescale it is highly likely that structural changes to the karst system will occur, and in seismically active areas, flow rates and hence dripwater supply can change over periods more like 10^3 years. Long-term controlling variables of such time periods are listed in Table 1.1 for comparison with those that show changes over shorter periods.

The climatic controls are illustrated both directly and indirectly. A change in mean annual temperature will be reflected in a change in cave temperature (T) and a reduction in seasonal temperature contrast will tend to lessen the importance of cave air circulation (AC). A change in total atmospheric precipitation (P) will be moderated by any change in evapotranspiration (E), but $P - E$ changes will result in modifications to mean annual water infiltration to, and discharge from, the karst (Q). Increased rainfall intensity will result in relatively

Table 1.1 Parameters of the speleothem factory and their control over long and intermediate timescales. Sense of influence shown as (+) positive covariations, (o) complex, and (–) inverse variation.

	Parameter	Long-term controlling variables	Controlling variables for decadal–multi-millennial variation	Discussed in this book
AC	Strength of air circulation (with implications for contents of trace gases, aerosols and particulates)	Size of cave network (+), closeness to ground surface (+), surface relief (+)	Seasonal temperature variations (+); synoptic weather systems (+)	Sections 4.4 and 4.6
C	Soil CO_2 production	History of weathering and soil accumulation (+); position on catena (o)	Temperature (+), biomass (+)	Section 3.3
d	Thickness of water film on speleothem surface	—	*r* (–), *Q* (+), *x* (o)	Sections 7.2 and 7.3
dv	Carbon isotope composition of vegetation	—	Climate (C_4 vegetation in semi-arid grasslands) (o)	Sections 3.4
f	Drip fall height onto stalagmite or flowstone	Cave geometry	*h* (–)	Section 7.2
h	Height of speleothem	—	Duration and rate of speleothem growth (+)	Section 7.2
m	Vegetation biomass	—	Potential evapotranspiration (+)	Section 3.4
PCP	Prior calcite precipitation (calcite precipitation upflow of cave observation site)	Geometry of cavity systems in aquifer	Change in size of cavity systems (+); P – E (–)	Sections 5.5 and 8.5
P – E	Precipitation minus evaporation	—	Global climate (o); atmospheric modes (o)	Section 3.1
Q_1	Discharge via conduit flow	Existence of conduits	High-intensity rainfall events (+)	Sections 2.4 and 4.3
Q_2	Discharge via fracture flow (two alternative routes shown as 2a and 2b)	Aquifer properties	Relatively high-intensity rainfall events (+); seismic changes to aquifer properties (o)	Section 4.3
Q_3	Discharge via seepage flow	Aquifer properties	P – E (+), proportion of snowmelt (+)	Section 4.3
r	Radius of curvature of speleothem top	—	*Q* (–), *f* (–), *w* (–)	Section 7.2
R	Ratio in precipitation of stalactite versus stalagmite	Degree of degassing above cave	*PCP* (+)	Sections 7.2 and 7.3
RH	Relative humidity of cave air	Cave network geometry; location with respect to external atmosphere	*AC* (–), *P – E* (+)	Sections 4.2, 4.5 and 4.6
s	Soil depth and mineral properties	Long—term climate, bedrock properties, position on catena and aeolian input	Aeolian input (+), slope movement (o)	Section 3.3
S	Aeolian input	—	Aridity (+), glacial period (+), atmospheric circulation (o)	Section 5.3
T	Cave temperature	—	Mean annual external temperature (+)	Section 4.5
w	Width of speleothem	—	*Q* (+), *f* (+)	Sections 7.2 and 7.3
x	Crystal structure of speleothem	—	Supersaturation and impurity content of fluids (o)	Section 7.4

higher fluxes routed through conduit (Q_1) or fracture (Q_2) porosity, while increased snowmelt is likely to be reflected mainly in higher seepage flow (Q_3). The intensity and timing of rainfall may affect the entrainment of organic and other colloidal and colloidally bound components and their supply to speleothems. Meteorological/climatic changes tend to modify the mean and variation in the composition of $\delta^{18}O$ of atmospheric precipitation and hence of speleothems, but the atmospheric processes are complex and so will not be summarized here (see Chapter 3).

Many facets of the biota in the ecosystems overlying the karst are liable to change as the result of shifting climate. We have picked out the variables of total biomass (*m*) and $\delta^{13}C$ composition of biomass (*dv*) because these are parameters which have been of particular interest to palaeoenvironmental analysis. Distinctive biomarkers and relative abundances of different groups of organic entities and compounds will also be dependent on climate. The soil depth and composition (s), can vary, particularly with extraneous input from aeolian or upslope sources. The mean soil P_{CO_2} level (*C*) is an important parameter because it controls the maximum growth rate for speleothems; it is maximal at intermediate water saturations and seasonally higher temperatures.

Prior calcite precipitation (*PCP*) is a consequence of degassing of water into air spaces before the drip arrives in the cave under observation. Its sensitivity to dry conditions has been shown in several studies, but in detail depends on aquifer properties. The ratio of growth on stalactites and stalagmites (*R*) relates to *PCP*. A water that has already degassed will be ready to grow a stalactite (high *R*), whereas one that mostly degasses as it lands on a speleothem, will predominantly lead to stalagmite growth (low *R*). Factor *PCP* is associated with covarying increases in $\delta^{13}C$ and trace elements such as Mg in the stalagmite.

Various geometrical properties within the cave are interlinked. The width of the speleothem covaries with Q, but also tends to be higher because of splashing effects, when fall height (*f*) is greater. Increasing speleothem height (*h*) may significantly reduce *f*. The ratio h/w affects surface curvature (*r*) which, together with the surface roughness of the crystalline surface (*x*), affects the depth of the water film (*d*) on the stalagmite or flowstone surface, which in turn is a factor in speleothem growth rates.

The vigour of the air circulation affects the removal of internally generated gases such as carbon dioxide and radon (and its particulate daughters), but also the introduction of aerosols. More stagnant air can allow relative humidities to approach 100%, given sufficient inflow of water to the cave which leads to relatively slower growth, but closer to chemical and isotopic equilibrium.

In Plate 1.3, these parameters are shown to vary between the early stages of growth of a broad domal stalagmite in a humid climate (Plate 1.3a) and the later continued growth of a narrow form in a semi-arid climate (Plate 1.3b). In this case certain factors work together to reinforce a change, e.g. the reduction in water discharge (*Q*) and fall height (*f*) both promote the reduction in stalagmite width (*w*). Growth rate of the stalagmite is slower because of higher prior precipitation (*P* and *R*), lower discharge (*Q*) and more sluggish circulation (*AC*) as the result of higher mean temperatures (*T*) due in this hypothetical case to warmer winters.

So far in this section, we have discussed this qualitative systems analysis in terms of system parameters that in part are controlled by the setting of the cave in response to long-term geomorphological evolution and in part are linked to changing climates, as well as being inter-related. When studying speleothem samples however, the mental process needs to be inverted to examine the system from the point of view of the finished object in the speleothem factory. To what extent can we determine the processes and conditions of manufacture? Table 1.2 takes four properties commonly measured in speleothems (growth rate, $\delta^{18}O$ and $\delta^{13}C$ signatures, and content of the alkaline earth elements) and illustrates that for each of them, their baseline value is related to either geographic position or bedrock properties. However, each property is capable of being interpreted in terms of more than one controlling variable on both medium-term and annual timescales. Hence there is inherent ambiguity: this is like the situation where

Table 1.2 Identification of controlling variables for some parameters commonly measured in speleothems. For abbreviations, see Table 1.1. Sense of influence shown as (+) positive covariation, (o) complex, and (−) inverse variation.

Speleothem parameter	Long-term controlling variables	Controlling variables for decadal-multi-millennial variation	Annual-scale controlling variables	Discussed in this book
Growth rate	Karst aquifer properties	T, $P - E$	AC, Q	Chapter 7.2
$\delta^{18}O$	Location, altitude	$\delta^{18}O$ of atmospheric moisture (+); atmospheric dynamics (o), T (o)	Seasonal variation in weather systems (o)	Chapters 3.2, 3.3, 4.3, 8.4, 10.3, 11 and12.
$\delta^{13}C$	Location, altitude	m (−), dv (+)	AC (+), P (+)	Chapters 3.3, 5.4, 8.4, 11 and 12.
Mg, Sr	Bedrock composition	Aeolian input (o); P (+)	P (+), crystallographic factors (o)	Chapters 5.3, 5.5, 8, 11 and 12.

several manufacturers use different patented processes to produce a finished article with specified characteristics. Fortunately, this uncertainty can be resolved in many individual cases because of the dominance of one controlling factor, and in many cases there is a clear climatic implication.

In fact, over the past 15 years, remarkable evidence has been gained that speleothem factories on the decadal to multi-millennial timescale can behave in a coherent and reversible fashion. The strongest type of evidence is the presence of long-term cyclicity in measured parameters than can be matched with Milankovitch orbital variations, but sub-Milankovitch events are also often recognized, as is discussed in Chapter 12. On the other hand, many caves display speleothems which show extreme growth rate variations over much shorter periods of time, indicative of disruptions to aquifer properties, and curtailing their ability to record long-term processes. In between the very long-period Milankovitch variations and the expressions of the annual climatic cycle, are a range of stochastic behaviours for which we struggle, sometimes successfully, to separate random or karst-specific noise from an Earth system signal. In later chapters, we justify how these interpretations are made.

For future research, it is clear that work on monitoring cave environments, and associated soils and ecosystems, is vital to improve our understanding of speleothem-forming environments. There is also an important role for experimentation (e.g. Huang & Fairchild, 2001). In relation to the porcelain factory analogy, the experimentalist takes the goal beyond that of the antique expert studying the output of a past porcelain factory, towards that of setting up their own manufactory. Although timescales will limit what we can achieve with such control systems, they effectively complement the work of numerical models in the understanding of the workings of modern environments and the interpretation of proxy palaeoenvironmental records.

CHAPTER 2
Carbonate and karst cave geology

Our focus in this book is on *calcareous* speleothems, which are primarily located in karstified carbonate rocks. Useful systematic overviews of karst rocks focused on the properties that lead them to develop and function as aquifers are provided by Ford and Williams (2007, Chapter 2) and Klimchouk and Ford (2000a). In this section we aim both to provide key pointers to carbonate geology for readers lacking a strong geological background, and to provide a convenient summary of the relevant issues for geologists, including the nature of karst rocks and the spaces within them. We leave the detailed discussion of carbonate mineralogy and geochemistry to Chapters 7 and 8 in the context of the speleothems themselves.

As an introduction, we explain some fundamental issues about the evolution of the Earth system which help to account for the distribution and chemistry of carbonate rocks (section 2.1). We then discuss the *lithologies* (rock types) of the karstic host rocks (2.2), including their composition, origin, and the types of cavities that they contain. In section 2.3, we present an overview of carbonate *diagenesis* (the post-depositional changes to carbonates), including the mechanisms of *speleogenesis* (cave formation) in near-surface settings in young carbonate sediments. Section 2.4 overviews the most widespread type of speleogenesis: that of already lithified rocks. Finally, in section 2.5, we discuss the extraordinarily complex ways in which caves are filled and the fragile place of speleothem deposition compared with the powerful agents of sub-surface rivers and gravitational collapse.

2.1 Carbonates in the Earth system over geological time

Geological time is divided into the Archaean, Proterozoic and Phanerozoic Eons, the first two being informally grouped together as the Precambrian. Carbonate rocks form only a small proportion of preserved Archaean sediments, but shallow marine carbonates become abundant in the Proterozoic and, much later (250 Ma), became widespread in deep marine environments too. The ocean composition was dominated by interaction with the mantle via extensive volcanism in the Archaean, but by the Proterozoic significant continental crust had formed and this is reflected in a change in the composition of the oceans, as captured by the chemical tracer $^{87}Sr/^{86}Sr$ within marine carbonates (Fig. 2.1a). Sr isotope ratios diverged upwards from the mantle composition reflecting the weathering of continental crustal rocks in which ^{87}Sr had formed by radioactive decay of ^{87}Rb which is concentrated in continental crustal rocks such as granites. The subsequent evolution of $^{87}Sr/^{86}Sr$ (Fig. 2.1a) reflects the relative importance of sea-floor spreading (tending to lower the ratio) and mountain building (tending to raise it) and can be modelled quantitatively, allowing for buffering effects such as the destruction of the carbonate rocks themselves by weathering, releasing their contained

Speleothem Science: From Process to Past Environments, First Edition. Ian J. Fairchild, Andy Baker.

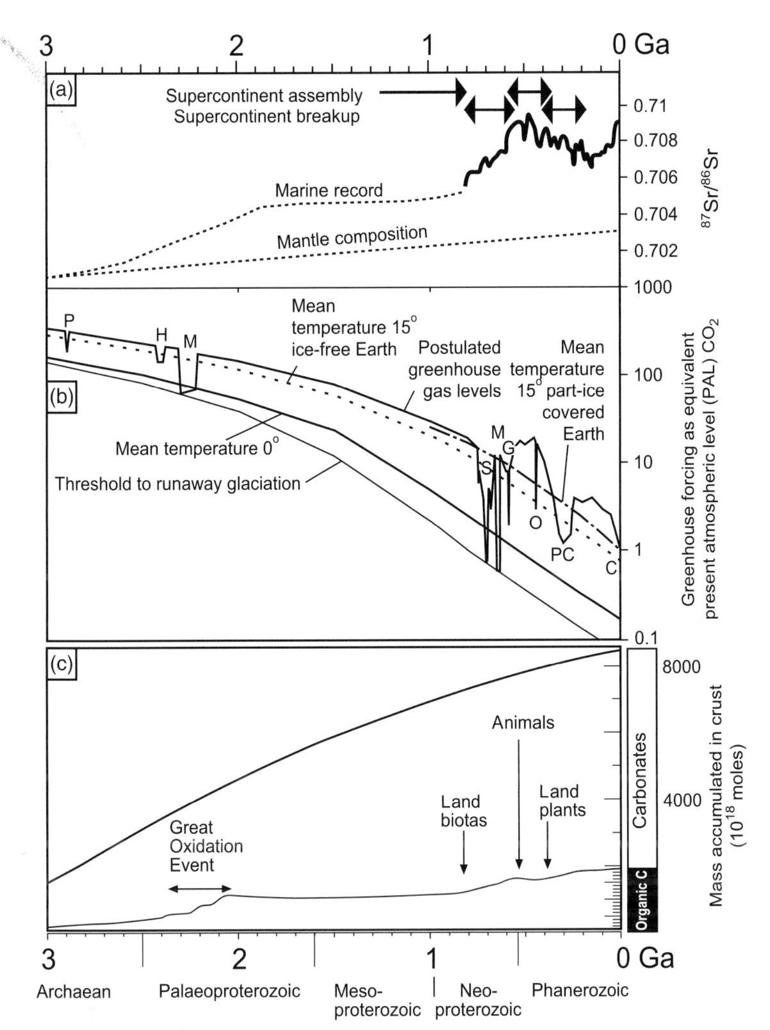

Fig. 2.1 Long-term evolution of the Earth system adapted from Fairchild and Kennedy (2007). Ga, billions of years ago. The diagram shows the last two-thirds of Earth history within which nearly all preserved carbonate rocks formed. (a) The Sr isotope ratio $^{87}Sr/^{86}Sr$ in oceanic precipitates such as carbonate rocks reflects the tectonic history of the Earth. (b) Change in greenhouse gas forcing in Earth history – the Earth has oscillated between ice-free and glacial periods and the long-term decline in forcing reflects the existence of Gaian feedback mechanisms compensating for the increase in luminosity of the Sun. (c) Increase in carbon in the crust resulting from the progressive fixation of volcanic CO_2 released from the mantle as carbonate and the reduction of around 20% of it to organic carbon in sedimentary rocks. As organic C forms, it releases oxidation capacity leading to increasing sulphate (at the expense of sulphide), oxidized iron, and stepwise increases in oxygen in the atmosphere. See text for further explanation.

Sr. The typical lifetime of a carbonate rock today from formation to destruction by weathering or subduction is around 200 Myr. In karstic areas, the Sr isotope composition of the carbonate host rock, characteristic of its time of formation (McArthur et al., 2001), but possibly modified by diagenesis, is one of the chemical sources that can be examined in speleothems.

Although radioactive decay continues as an important heat engine of the Earth, surface environments are more strongly influenced by solar radiation, which has increased by over 30% during Earth history. The Gaian view draws attention to the evidence for liquid water, and only occasional glaciation, through the sedimentary record of the past 3.8 billion years, and proposes that temperature was regulated by negative feedback mechanisms (Kump et al., 2009). As a consequence there has been a progressive decline in the abundance of atmospheric greenhouse gases (Fig. 2.1b), which probably switched from largely methane to carbon dioxide during the Neoproterozoic (but concentrations were punctuated by low values triggering ice ages). Hence, early oceans were more acidic than today, but high supersaturations for carbonate were present until organisms developed the ability to precipitate at lower saturation states at the end of the Precambrian.

One view of geological history, emphasizing the chemical changes, is that of a great titration experiment between rocks and acidic volcanic gases (Garrels & McKenzie, 1971). There has been an exponential decline in release of 'primordial' gas from the mantle over time (most volcanic gases today have been recycled through the crust). The most important gas is carbon dioxide because of its abundance and its effectiveness in weathering rocks. It turns out that the key rocks are silicates containing Ca and/or Mg because much of these elements can be converted into carbonate rocks in the geological cycle (Berner, 2004). Weathering of such silicates can be represented as

$$2CO_2 + 3H_2O + (Ca, Mg)SiO_3 \rightarrow (Ca, Mg)^{2+} + Si(OH)_4 + 2HCO_3^- \quad (2.1)$$

where the products are all solutes (a more realistic reaction would also show the presence of secondary clay minerals). The fate of nearly all the Ca^{2+} and much of the Mg^{2+} is to be precipitated as carbonate minerals:

$$2HCO_3^- + (Ca, Mg)^{2+} \rightarrow (Ca, Mg)CO_3 + H_2O + CO_2 \quad (2.2)$$

Once the precipitation of aqueous silica is also allowed for, the overall reaction can be represented as:

$$CO_2 + (Ca, Mg)SiO_3 \rightarrow (Ca, Mg)CO_3 + SiO_2 \quad (2.3)$$

The consequence has been a steady accumulation of carbonate rocks through Earth history as depicted in Fig. 2.1c. A result is a global distribution of carbonate rocks (Fig. 2.2) which means that karst and speleothems are similarly widely distributed, although rates of formation are low in very cold or arid climates.

The increasing proportion of carbonates with time has a subtle effect on the map. Carbonates are abundant in the young sediments and metasediments of Mesozoic–Cenozoic basins, both those in orogenic belts and on stable platforms (e.g. southern Europe and the Alpine mountain chain). Precambrian rocks of the continental shields do indeed have relatively little carbonate, although bias is introduced because these ancient mountain chains are for the most part eroded to their deep roots. However, Australia's interior is a good example of a shield with numerous Archaean and Proterozoic sedimentary basins, but low proportions of carbonate.

Throughout the sedimentary record, life has utilized carbon, and reduced organic carbon has also accumulated in sediments, at about 20% of the rate of carbonates, with some fluctuations including bursts of carbon accumulation associated with decreases in the oxidation state of surface environments. These and other events are recorded in the carbon isotope composition of carbonate rocks in different time periods (Fairchild & Kennedy, 2007; Prokoph et al., 2009). The carbon isotope composition of karst rocks is, like Sr isotopes, also a marker that has some influence on the speleothem composition.

During the Phanerozoic, primordial outgassing has been limited and the system approximates a steady state as depicted in Fig. 2.3. As part of the plate tectonic cycles associated with continents splitting, sliding past each other, or colliding, new mountain belts continually form, consisting of rocks uplifted from depth. Hence, carbonate rocks form part of the continental crust undergoing weathering. Carbon dioxide is also regenerated by 'reverse weathering' reactions that take place at depth as partial melting of subducted rocks occurs or sediments are metamorphosed: hence new minerals form by reactions that are the reverse of eqn.

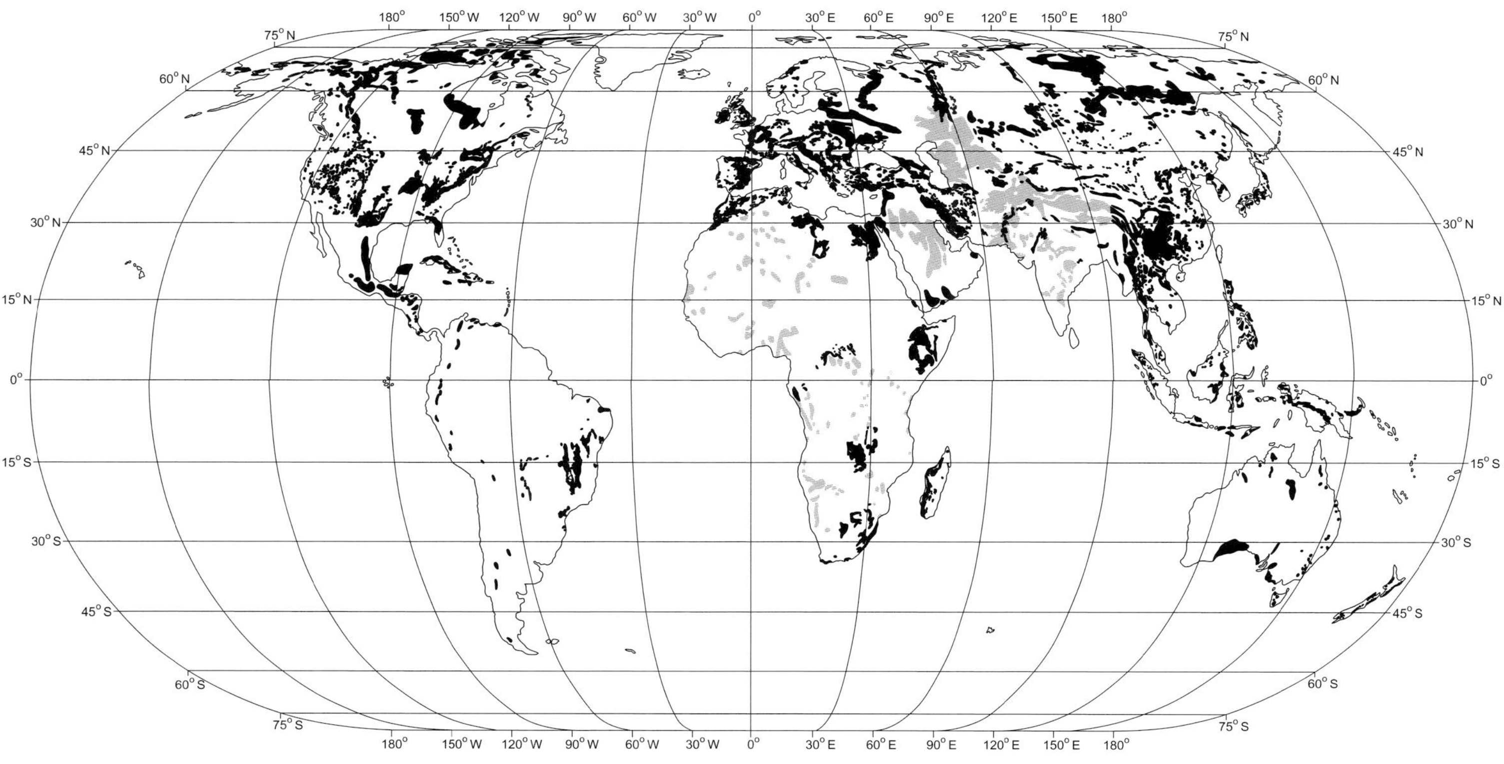

Fig. 2.2 Global distribution of carbonate rocks (Eckert IV equal area projection) redrafted from a higher resolution version compiled by Paul Williams and Yin Ting Fong from various sources (Williams, 2009). Pure carbonate rocks crop out on 12.5% of the continents (excluding Antarctica, Greenland and Iceland). This version differentiates those areas where carbonate rocks are relatively pure and continuous (black) from those where they are abundant but discontinuous or impure (grey).

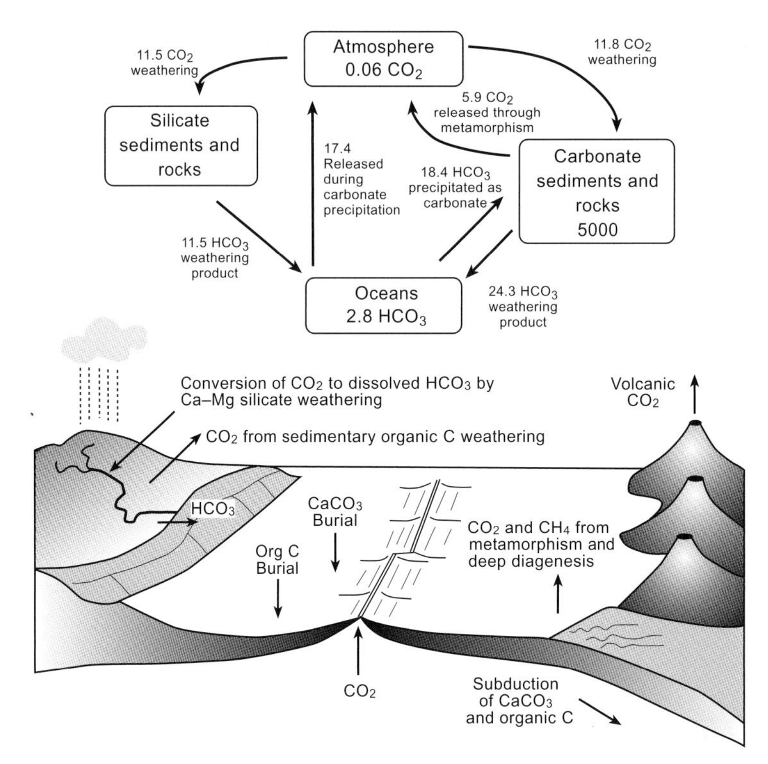

Fig. 2.3 The long-term cycle of CO_2 during the Phanerozoic eon with reservoirs of carbon shown in 10^{18} moles and fluxes of carbon shown as $10^{12} mol\,yr^{-1}$. The cartoon (after Berner, 2004) illustrates some key processes, a sub-set of which, representing the essential carbon dioxide feedback of the carbonate-silicate system, are captured in the box model (compiled from data in Berner et al., 1983). Note that CO_2 generated by metamorphism can emerge either via springs or in volcanoes. Magnesium from seawater also exchanges with Ca in newly formed ocean crust. Arvidson et al. (2006) provided a recent sophisticated example of mass balance modelling.

(2.1). Figure 2.3 illustrates the quantities involved and it is apparent that carbonate rocks in the crust contain orders of magnitude more carbon than the atmosphere and oceans. Nevertheless, on a short (human) timescale, the removal fluxes of carbon related to biological activity are sufficient to turn over the atmospheric composition in a few years (box 1.1), whereas the equivalent time for the mass balance model shown in Fig. 2.3 [$0.06 \times 10^6/(11.5 + 11.8)$] is around 2600 years. Thus, the effects of the geological cycle represent a major feedback, but one that is far too slow to influence the current human-induced greenhouse gas crisis. Nevertheless, it is notable that the dissolution of carbonate rocks, which mainly takes place in karstic areas, is quantitatively by far the most important weathering reaction occurring on the Earth's surface (Lerman, 1994; Liu et al., 2010). Also, from an ecological viewpoint, it is important in maintaining soil base status and preventing acidification.

Within Phanerozoic time (Table 2.1), a pattern of cyclic alteration of carbonate rock properties becomes apparent (Fig. 2.4 and Stanley, 2006). A combination of mass balance modelling and analysis of fluid inclusions inside marine salt deposits has enabled the reconstruction of the Ca and Mg content of the oceans. Variations in these parameters also appear to coincide with changes in the distribution of the $CaCO_3$ minerals aragonite and

Table 2.1 Phanerozoic timescale based on Gradstein et al. (2004) updated by Walker and Geissman (2009). Ma, millions of years ago. Period (qualified by early, middle, late) refers to time period, whereas System (lower, middle, upper) represents the rocks formed during the corresponding Period. Note that the base of the Quaternary was moved by the International Commission on Stratigraphy from 1.8 to 2.6 Ma in 2009.

Era	Period/System	Starting age (Ma)	Symbol
Cenozoic	Quaternary	2.6	Q
	Neogene (subdivided into Miocene 23.0, Pliocene 5.3)	23.0	Ng
	Paleogene (subdivided into Paleocene 65.5, Eocene 55.8, Oligocene 33.9)	65.5	Pg
Mesozoic	Cretaceous	146	K
	Jurassic	202	J
	Triassic	251	Tr
Palaeozoic	Permian	299	Pm
	Carboniferous (subdivided into Mississippian 359, Pennsylvanian 318)	359	C (M, P)
	Devonian	416	D
	Silurian	444	S
	Ordovician	489	O
	Cambrian	542	C

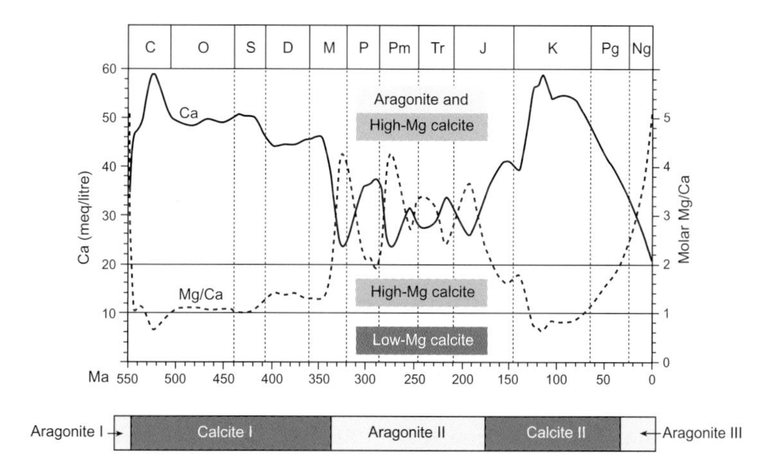

Fig. 2.4 (after Stanley, 2006). Phanerozoic evolution of sea water in terms of Ca concentration and Mg/Ca ratio compared with periods of dominance of either aragonite or calcite as inorganic precipitates and the primary mineralogy of reef-forming organisms. See text for discussion and Table 2.1 for abbreviations.

calcite as primary phases in marine carbonates (aragonite is replaced by calcite later in diagenesis). Using constraints from modelling as well as data, Mackenzie et al. (2008) generalized that there are two modes of behaviour, which approximately coincide with the 'calcite' and 'aragonite' seas depicted in Fig. 2.4. 'Aragonite seas' are marked by high Mg/Ca ratios in seawater, which inhibit calcite formation because it can only precipitate as high-Mg calcite, which is a relatively soluble form. 'Calcite seas' have low marine Mg/Ca and saturation states are also higher. The 'aragonite seas' show evidence that the primary inorganic precipitates tend to be aragonite or high-Mg calcite, and reef-building organisms are also aragonitic which may in part be an adaptation to the water chemistry (Stanley, 2006). However, when all calcareous organisms are considered, the abundance of aragonitic types (e.g. some molluscs) shows a completely different trend, increasing towards the present (Kiessling et al., 2008). Formerly aragonite-rich limestones may have distinctive chemical characteristics, such as high Sr or U content, relevant to speleothem chemistry, but the correlation with geological age is weak. This is because the abundance of primary aragonite is variable in any given geological time period and the marker elements can also be sourced in non-carbonate components in the aquifer.

It might be thought that dolomite $(CaMgCO_3)_2$ should be more abundant in times in Earth history when seawater Mg/Ca was high. This is almost certainly true in the Proterozoic when most carbonate facies were dolomitized, but the patterns in the Phanerozoic are more difficult to discern. Holland (2005) compared different studies and found large discrepancies in estimates of the abundance of dolomite versus time. The modelling approach of Mackenzie et al. (2008) implies that dolomites are relatively more abundant in the low Mg/Ca calcite seas, where factors other than Mg/Ca (such as low sulphate) also come into play to facilitate its formation. As will be discussed later, the formation of dolomite is complex. In the karstic context, study of the origin of dolomite in the karst aquifer can help understand the spatial variation in dripwater chemistry and hence speleothem properties.

One other relevant issue is the geological time span over which speleothem formation occurred. Before the development in the Devonian of vascular plants with deep roots, soil systems were more rudimentary and so CO_2-degassing to generate speleothems would have been less conspicuous, although it could have occurred, where P_{CO_2} was enhanced by strong acid released by oxidation of sulphides, as long ago as the Archaean (Brasier, 2011). Nevertheless, records of Palaeozoic and older speleothems are currently very sparse.

2.2 Lithologies of carbonate host rocks

2.2.1 Carbonate facies

The term facies refers to a sediment type associated with a particular depositional environment. Most sediments are *siliciclastic*, in that they are composed dominantly of silicate mineral grains such as quartz and feldspar in the sand fraction ($\frac{1}{16}$–2 mm) together with various clay minerals in mud ($<\frac{1}{16}$ mm). Carbonate sediments can only dominate where the input of siliciclastic sediment is low. Figure 2.5a (Bosence, 2005) summarizes the range of large-scale settings in which thick sections of carbonate rocks may accumulate as shallow-water *platforms* with widths of 10^1–10^3 km, and away from major terrigenous (land-derived) sources. Modern carbonate facies are primarily in the marine tropics, because rates of accumulation are higher here (typically 10^{-3}–10^{-1} m per year). Significant accumulation of carbonates as organic skeletons also occurs on temperate shelves, but the sediments are typically diluted by siliciclastic sediment. Analogous precipitates also form in continental settings, particularly in lakes, where virtually all dissolved carbonate entering closed lakes (those without an outlet) and some of that which enters open lakes is exported as calcareous matter in sediment. The margins of the carbonate platforms can be distinguished as either gently sloping *ramps*, or *rimmed shelves*, where the rim may consist of reefs or carbonate sandbanks. A special situation arises for unattached platforms (the Great Bahama Bank being a classic example (Fig. 2.5b)), which are physically isolated from sediment sources of major landmasses. Ramps and rimmed shelves can grade into one another over time (Fig. 2.5c), as discussed in the next section, and may coexist on different margins of the same platform.

The initial carbonate sediment that is the precursor to a karstified rock may consist of individual sand-sized particles, known as grains or *allochems* in the carbonate literature, or already be *lithified* (hardened as a rock) on the sea floor, as in the case of coral reefs. Some literature uses the simple terms calcilutite, calcarenite and calcirudite to distinguish predominantly mud-, sand- and gravel-sized parti-

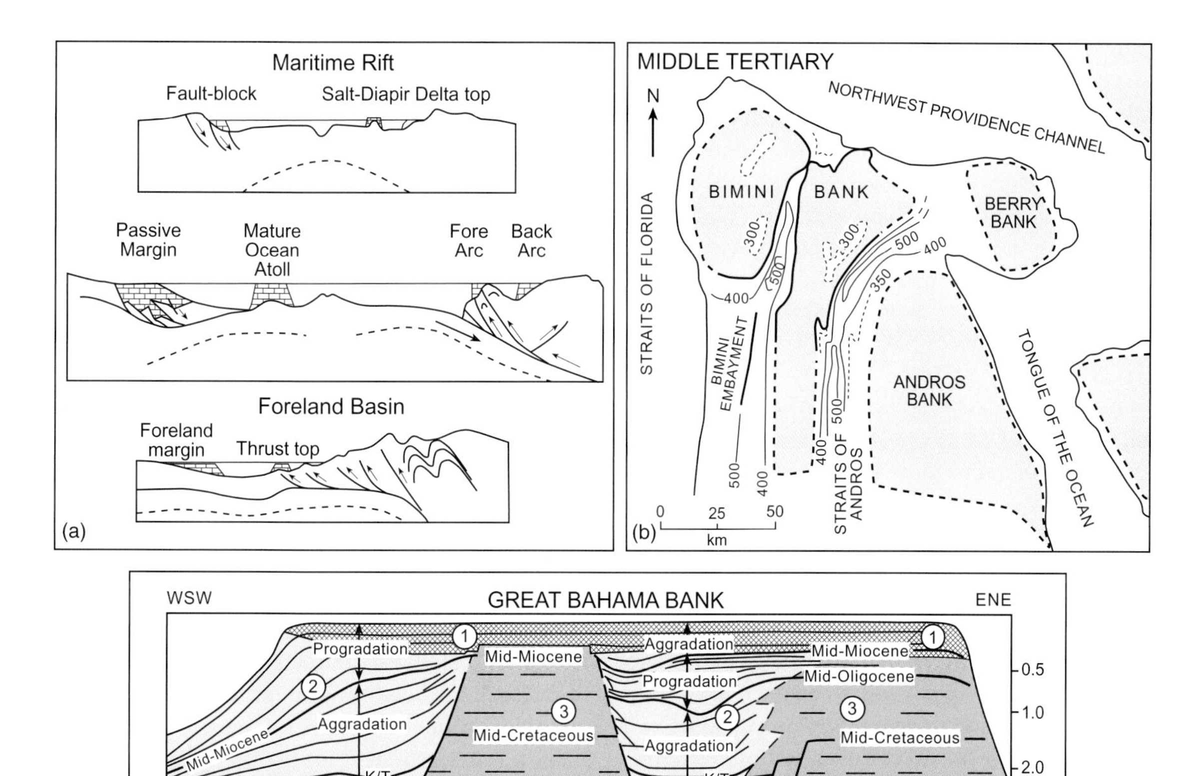

Fig. 2.5 Carbonate platforms. (a) Tectonic situations of carbonate platforms from Bosence (2005); (b) palaeogeographic map and (c) schematic cross-section based on seismic data, northwestern Great Bahama Bank (Melim & Masaferro, 1997). Note the change over time of the western margin from a ramp (2 refers to the sloping seismic reflectors) to a rimmed shelf (horizontal reflectors labelled 1). The oldest part, numbered 3, has chaotic reflectors. The rate of subsidence is currently less than the long-term mean and is no more than 1–2 m per 100 kyr over the past 300 kyr (Carew & Mylroie, 1997).

cles, but usually more specific descriptions are given. The individual particles are distinguished (Folk, 1962) as follows.

1 *Skeletal grains*: pieces of organic skeletons (secreted by a calcareous organism).

2 *Ooids*: spheroidal structures that accrete concentric growth layers and are transported and rotated during growth from agitated supersaturated waters.

3 *Intraclasts*: fragments derived by reworking of a local, already lithified, carbonate sediment. The prefix 'intra-' refers to their local derivation *within* the basin of deposition.

4 *Peloids*: rounded grains, lacking concentric structure that originate either as faecal pellets (from organisms that eat carbonate mud), or abraded intraclasts, or degraded ooids.

The carbonate sediment or rock may also contain fine carbonate *mud*, or internally precipitated *cement* crystals which imparts strength, with residual *pores* which are fluid-filled voids.

Two well-known classifications of carbonate rocks are by Folk (1962) and Dunham (1962). Folk's classification makes up composite words with a prefix for the main type of allochem as listed above (bio-, oo-, intra- and pel-) and a suffix (-micrite, -sparite) to distinguish an original muddy matrix (micrite) from a cement (sparite). *Micrite*, a contraction of microcrystalline calcite, was originally thought to represent simply fine carbonate particles (known to be the product of breakdown of many calcareous algae for example) and expected to accumulate in quiet-water environments. However, it is now known that micrite can also originate as cement in high-energy environments.

Dunham's classification is most widely used and is illustrated on the left-hand side of Figure 2.6. In general terms, the loss of carbonate mud in the sequence *mudstone* → *wackestone* → *packstone* → *grainstone* can be interpreted as reflecting increasingly energetic (turbulent) depositional environments, and this is reflected in the distribution of such carbonate types across a sloping carbonate ramp where wave and storm action diminishes with increasing water depth (Fig. 2.7). The name can be augmented by an adjective to indicate the type of allochem, e.g. skeletal wackestone or oolitic grainstone. This approach was supplemented by Embry and Klovan (1971) by several terms (Fig. 2.6, right) that have proved useful in describing the variation of carbonate lithologies associated with carbonate *buildups* (localized carbonate accumulations), including *reefs* (wave-resistant structures). *Bafflestones, bindstones* and *framestones* can be made by microbial mats as well as by higher organisms. Microbial structures are known as *stromatolites* and are the only type of build-up that existed until calcification of skeletal organisms evolved near the end of the Proterozoic. Their distinctive domed or columnar growth forms can sometimes resemble speleothem morphology.

How much does the precise type of limestone matter, when it comes to karst development? For porous limestones, the nature of pore systems often does reflect primary characteristics. In contrast, in the case of fully cemented (non-porous) limestones, there are some differences in susceptibilities to dissolution, dependent on crystal size and texture (Sweeting, 1972), but such factors are quite subtle features in relation to cave formation and poroperm properties. More important is the carbonate mineralogy as discussed below, and the range of discontinuities present in the rock mass, which focus water flow (section 1.3.2).

When modern marine carbonates are compared with their ancient equivalents, a striking difference in their mineralogy is observed: the primary sediments are mostly aragonite and magnesian calcite

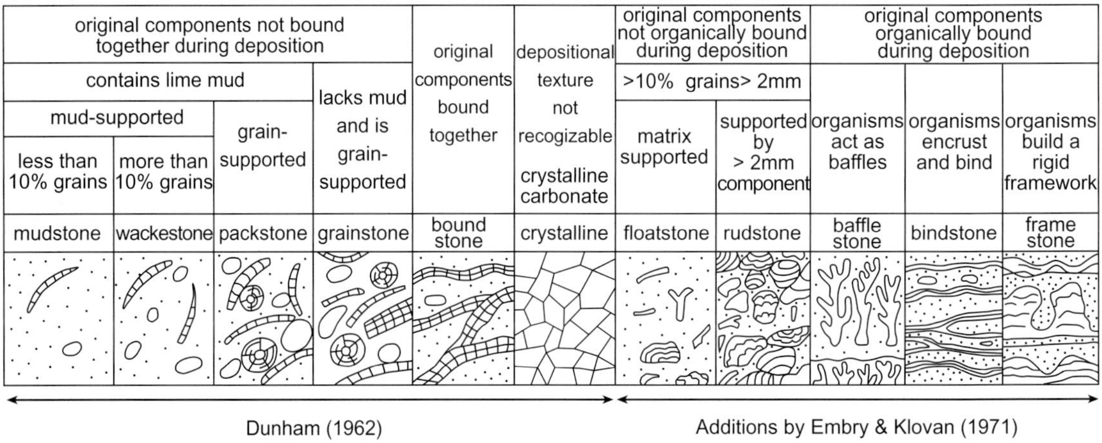

original components not bound together during deposition				original components bound together	depositional texture not recogizable	original components not organically bound during deposition		original components organically bound during deposition		
contains lime mud			lacks mud and is grain-supported			>10% grains> 2mm				
mud-supported		grain-supported				matrix supported	supported by > 2mm component	organisms act as baffles	organisms encrust and bind	organisms build a rigid framework
less than 10% grains	more than 10% grains				crystalline carbonate					
mudstone	wackestone	packstone	grainstone	bound stone	crystalline	floatstone	rudstone	baffle stone	bindstone	frame stone

Dunham (1962) — Additions by Embry & Klovan (1971)

Fig. 2.6 Classification of different types of carbonate (from Tucker, 1991 after Dunham, 1962; and Embry & Klovan, 1971).

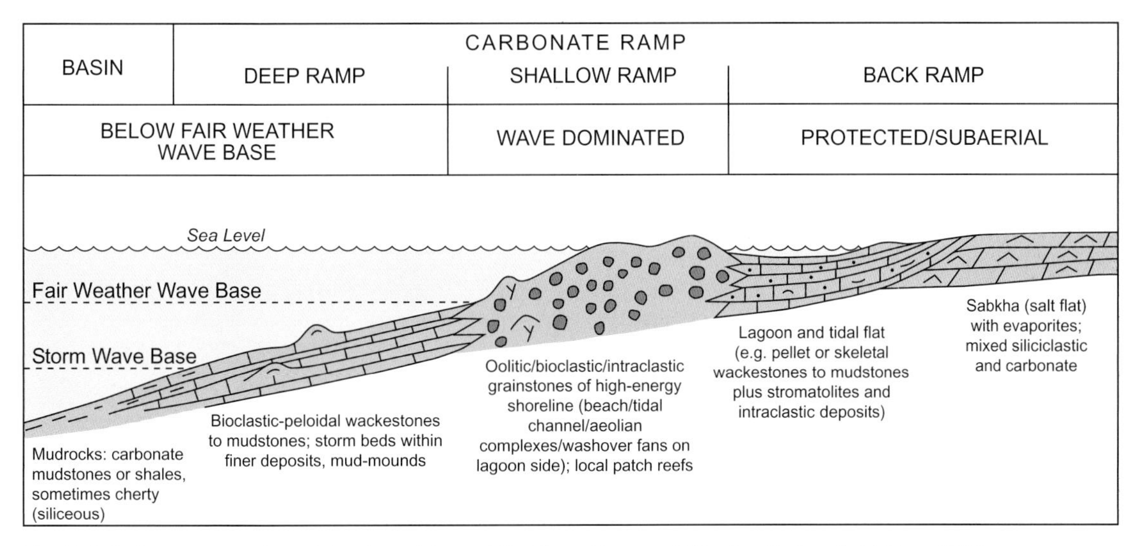

Fig. 2.7 Carbonate ramp depositional model (modified from Moore, 2001 after Tucker & Wright, 1990).

(>4 mole% $MgCO_3$ (Dickson, 1990)), whereas ancient carbonates are composed of calcite and dolomite. Seawater has a distinctive chemical composition with high Mg/Ca (Fig. 2.4) and the two elements compete to be incorporated in calcite. However, little Mg is included in aragonite because of the larger size of the cation site in that mineral, Mg^{2+} being much smaller than Ca^{2+}. Rapid precipitation of carbonates in tropical areas results in the formation of different proportions of metastable carbonates (aragonite and magnesian (high-Mg) calcite with 11–18 mole% $MgCO_3$). Also organisms exert a metabolic control on the mineralogy, chemistry and crystal fabric of their calcareous skeletons. Most shallow-water carbonate-producing organisms secrete aragonite (modern corals, many molluscs, some calcareous algae) or high-Mg calcite (most calcareous algae), whereas only few organisms secrete low-Mg calcite (e.g. brachiopods, trilobites, some molluscs, Palaeozoic corals). In deep-marine environments, on the other hand, the dominant forms are foraminifera and coccoliths, which are both low-Mg calcite. In deep-marine and sediment burial conditions, and in most meteoric waters, low-Mg calcite is more stable than magnesian calcite, and there is a tendency for the metastable carbonates to dissolve and new carbonates (low-Mg calcite and/or dolomite) to form, usually much more slowly.

Aragonite tends to persist longer than high-Mg calcite (and can survive in carbonates as old as Palaeozoic if protected by organic films), but its survival is difficult to predict precisely. Apart from the general issue of carbonate rock formation, there are two ways in which this issue is relevant to speleothems. Firstly, preferential dissolution of aragonite over calcite in low-permeability portions of karstic host rocks may occur (Fairchild et al., 2006b) and may be recognized by associated trace elements (e.g. aragonite is relatively rich in Sr). Secondly, aragonitic speleothems exist, which are susceptible to conversion to calcite, but at widely varying rates.

Dolomite formation is a relatively slow process, apart from certain lakes where it can precipitated during seasonal drying, but more usually the process of dolomitization is constrained by both the rate of dissolution of pre-existing carbonates and the supply of Mg ions. Examples of dolomitizing settings are illustrated in Figure 2.8, and Figure 2.9 illustrates the cross-cutting and porous nature of dolomite formation during deep burial. Dolomite as a host rock has several implications for speleothems: it is less prone to cave formation, it tends

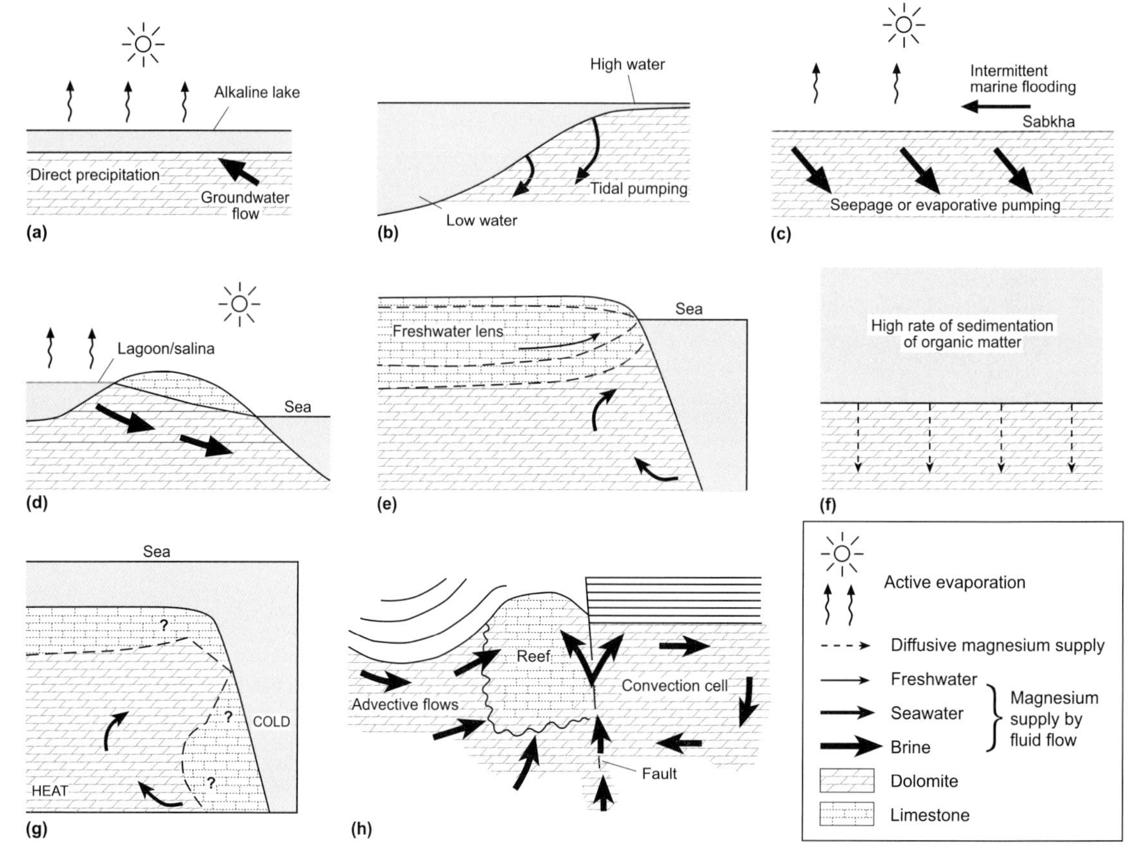

Fig. 2.8 Dolomitization models. Diagrams (a)–(c) and (f) illustrate syn-depositional modes of dolomitization where finely crystalline dolomites forms complete beds, whereas in the other situations the boundary of dolomitization can cut bedding and the dolomite crystal size will be larger, especially at higher temperatures. (a) Direct lacustrine precipitation based on lakes in the Coorong region of Australia; (b) peritidal model based on Florida Keys; (c) sabkha (salt flat); (d) seepage reflux; (e) mixing-zone—such zones are typically characterized by minor dolomitization and more major dissolution, but underlying marine fluids may cause more extensive dolomitization (Little Bahama Banks, Vahrenkampet al. (1991)); (f) dolomite in deep-sea/lagoonal organic-rich sediments; (g) dolomitization induced by geothermal convective flow on atoll/isolate carbonate platform; (h) generalized subsurface model with gravitational flow, deep basinal fluids ascending faults and convection cell.

to breakdown to yield sandy residues of dolomite crystals (Fig. 2.10) and, because its dissolution leads to relatively high Mg/Ca in cave waters, it can also be associated with the formation of aragonite speleothems.

Some varieties of dolomite, as can be seen in Figure 2.8, are associated with strong evaporation, and hence it is common to find more soluble evaporite rocks and mineral phases associated with such carbonates. The range of evaporite minerals known to occur is very large (Warren, 1989), even for a relatively constant starting solution such as is represented by sea water. However, in practice, the overwhelming association is with calcium sulphate, which is largely represented in either the hydrated form gypsum ($CaSO_4.2H_2O$) or as anhydrite ($CaSO_4$). Gypsum or anhydrite precipitates when seawater is reduced to approximately 20% of its original

Fig. 2.9 Soplao, Cantabria, Spain. Left, surface outcrop of cross-cutting diagenetic boundary between dolomite (left) and limestone (right). Right, porous dolomite exposed in the underlying cave.

Fig. 2.10 Contrasting shapes of two hills made of Devonian dolomite (left) and limestone (right) at Tanjawan, Guangxi province, China. Inset images show sand-textured weathered dolomite surface (left, 2-cm-high image) whereas limestone surface is solid with dissolution pits (karren, right 30-cm-high image).

volume, followed by halite (NaCl) at 10% and more soluble high-Mg and K-salts at 2%. In continental environments, a range of minerals can be present in different catchments, trona ($NaHCO_3.Na_2CO_3.2H_2O$) being the most common phase not found in marine-derived brines. Anhydrite is favoured over gypsum in hyperarid and hot conditions such as found in the sabkhas and lagoons of the Arabian Gulf. It forms displacive nodules in supratidal settings and replacive crystal aggregates in subtidal settings. Gypsum forms coarsely crystalline deposits in salinas and is commonly also found as dispersed crystals in siliciclastic or carbonate matrices. On burial, gypsum converts to anhydrite with increased temperature (the conversion is complete above 50 °C) and this is reversed during uplift, provided fluids can penetrate. As a result of diagenetic processes, nodules and crystals of these minerals are often found as leached voids or replaced by secondary minerals such as quartz (SiO_2), barite ($BaSO_4$) or pyrite (FeS_2). Beds of gypsum or anhydrite tend to dissolve completely in the near-surface of humid environments and are associated with extensive collapse features. Calcium-rich solutions, such as those resulting from calcium sulphate dissolution, tend to drive a process termed *dedolomitization*, the leaching and partial replacement of dolomite by calcite, which can result in unusually porous carbonate host rocks (Ayora et al., 1998). Such aquifers can provide a responsive media for generation of chemical signals in speleothems (McMillan et al., 2005).

2.2.2 The architecture of carbonate host rocks: sequence stratigraphy

A carbonate aquifer is typically not completely uniform, but displays vertical and lateral variations in lithology. Within carbonate platform successions, the key driver for change in sedimentation patterns are the relative rates of carbonate production, export of carbonate sediments to deeper-water environments, subsidence of the underlying sedimentary basin and worldwide (eustatic) sea-level change, the last particularly in response to changes in the volume and distribution of ice sheets. The rationalization of the resulting patterns of sedimentation and erosion is the subject of sequence stratigraphy, which originated with studies of sedimentation patterns observable in the sub-surface via large-scale geophysical (seismic reflection) surveys (Vail et al., 1977), and has developed since the late 1980s into the core framework for analysis of marine sedimentary successions (Emery & Myers, 1996; Moore, 2001; Schlager, 2005). The nomenclature of sequence stratigraphy is complex and in part contested (Catuneau et al., 2009), but here we just focus on a few relevant issues.

Fig. 2.11a illustrates a wide range of carbonate lithologies occurring laterally within a short vertical transect. These are quite likely to differ in impurity content and porosity characteristics, factors that could control the inception of cave passages and properties of speleothems should they subsequently develop within the rock mass. Figure 2.11b illustrates how the sequence stratigraphic approach simplifies the section into units reflecting episodes of sea-level rise (transgression) and fall (regression) with intervening highstands and lowstands of sea level. The understanding gained from this kind of analysis is now starting to be applied to carbonate aquifers and it has the useful property of allowing predictions to be made about the geometry of sedimentary units when projected beyond their region of outcrop. Additionally, there are implications for the post-depositional evolution of the rocks, in particular changes in porosity. Significant exposure would be expected to be associated with karstification and/or fracturing, together with a range of diagenetic changes as described in subsequent sections. Such phenomena may lie at specific horizons within a thick carbonate unit (e.g. Baceta et al., 2007; see also section 2.4) and the sequence stratigraphic analysis is a useful formal tool to promote the identification, and model the distribution of these structures. Paterson et al. (2008) show how computer simulation of the impact of significant sea level changes on carbonate platforms, for given subsidence and climatic conditions, allows the distribution of freshwater and mixed water diagenetic environments, and hence patterns of early diagenetic change, to be predicted.

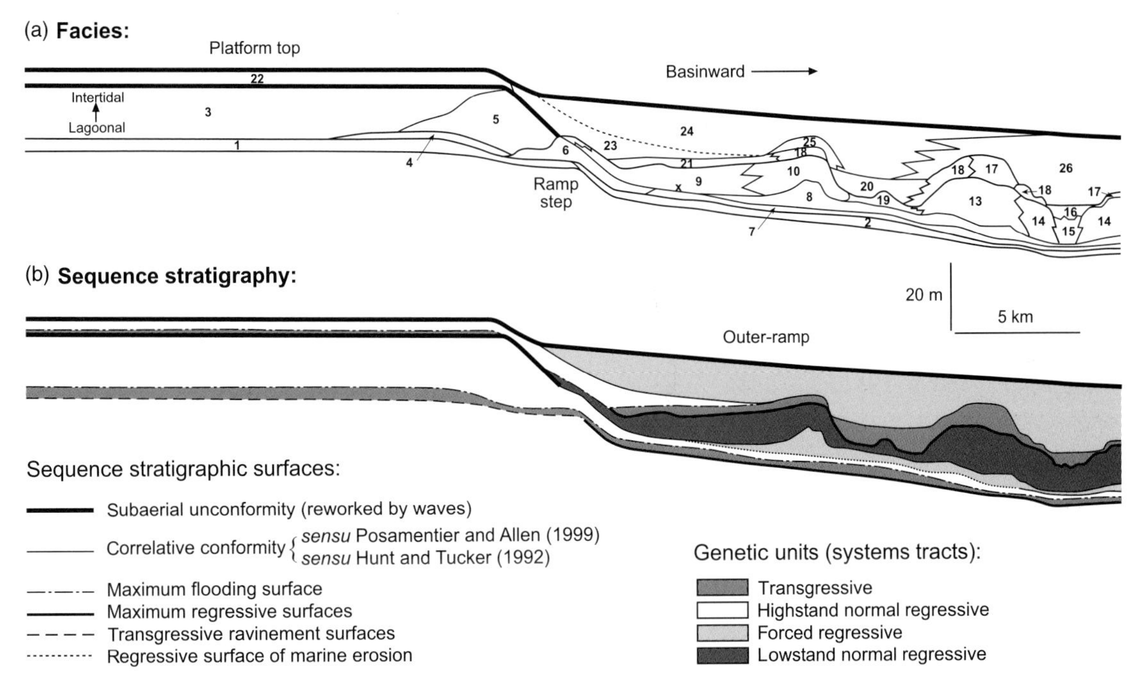

Fig. 2.11 Comparison of lithostratigraphic and sequence stratigraphy analysis of a carbonate ramp (Devonian western Canada). Simplified from Catuneau et al. (2009), based on original data of MacNeil and Jones (2006). (a) Each of the different carbonate facies occurs within a discrete region and, where thick enough, might be mappable as a lithostratigraphic entity (formation, member or bed). (b) Sequence stratigraphic analysis simplifies the detail and emphasizes that there are two phases of rising/high sea level (highstand overlying transgressive systems tracts; units 1–7 and 16–23 in (a)) associated with build-up of a rimmed shelf with interior lagoon and tidal flats, and which are truncated by erosion (subaerial unconformity). These are separated by a phase of low sea level (forced regressive and lowstand normal regressive; units 8–15 in (a)) during which the topography of the lower part of the ramp is enhanced by growth of localized (patch) reefs, persisting during the subsequent marine transgression. Finally, a strong episode of sea level fall leads to exposure of the whole region following deposition of units 24–26 in A during the regression. The subaerial unconformities would be expected to be associated with karstification.

It is not uncommon to find stratigraphic successions in which carbonate formations are bounded vertically by non-carbonates, but there are fewer cases where repeated alternations occur. An example is provided by the Yoredale cycles which are well-developed in the Lower Carboniferous of northern England and southern Scotland and which can be interpreted in terms of sea level change forced by 10^5-year (orbital eccentricity) cycles (Tucker et al., 2009). More generally, the sequence stratigraphic literature recognizes the occurrence of a hierarchy of phenomena at different spatial scales, with smaller-scale phenomena representing higher-frequency events (Catuneau et al., 2009). A wide variety of phenomena can be responsible for lithological changes on the 0.1–1 m (bed) or 1–10 m (parasequence) scale, ranging from individual depositional events, through lateral migration of bedforms or landforms, to externally forced changes in sediment supply and/or biological productivity. For example, shallowing-upwards parasequences of shallow subtidal through intertidal to supratidal sediments could have their origin in either entirely local or global phenomena (Pratt et al., 1992). Sedimentological expertise can be used to construct facies models on this smaller scale and hence predict the lateral persistence of distinctive sedimentary beds within an aquifer.

2.2.3 Impure and geologically complex host rocks

All transitions between pure limestones, dolomites and siliciclastic sedimentary rocks and their metamorphic equivalents occur in geological terrains. Although generalizations are difficult, Klimchouk and Ford (2000b) considered that karst does not develop fully (i.e. with a wholly subsurface drainage) unless carbonate composes more than 50% of the rock (with sand impurities), or 70–80% (with clay), because such impurities tend to clog developing cavity systems (White, 1988). A specific agent promoting enhanced dissolution is the sulphuric acid generated by oxidation of metal sulphides. Lead and zinc sulphides are commonly hosted in carbonates, but the commonest sulphide is pyrite (FeS_2), traces of which occur in many marine deposits, with a strong association with marine organic-rich shale, as well as in metamorphic rocks.

An extensive quantitative analysis relating cavern development to lithology in folded Ordovician sedimentary rocks in the Appalachians was undertaken by Rauch and White (1970), who found optimum development in the limestones with lowest content of silicate impurities, with the lowest proportion of sparite, and with 2–3% MgO. A weakness of this approach is that the key factor promoting karstification may be the presence of specific layers (inception horizons, section 2.4) in which dissolution is initially focused, and which may not resemble the bulk composition of the rock unit. Rauch and White (1977) performed controlled laboratory dissolution studies and confirmed the observations about Mg and clay content, although the ultimate control may have been the presence of silty laminae in the rock. The dolomite content was the key variable inhibiting dissolution.

The variety of different dolomitization mechanisms (Fig. 2.8) is associated with a range of geometries of the resulting dolomitized rocks. Where dolomitization is linked to a particular surface environment and occurs early in the diagenetic history, individual beds tend to be completely dolomitized, that is, the boundary of dolomitization lies parallel to bedding. Where related to different fluids from the depositional waters, dolomitization may be incomplete or display a boundary that cuts bedding (Fig. 2.9a), for example it can be parallel to a fault. The lower solubility of dolomite compared with calcite, and the porous nature of many dolomites (Fig. 2.9b), results in a tendency for cave systems in dolomite to be less developed than in limestone (White, 1988).

Karstic phenomena are commonly developed within metamorphic rocks in *orogenic belts* (geologically complex terrains associated with mountain-building) and in some regions it is associated with the buoyant rise of intrusions of rock salt (*salt diapirs*). Folded and faulted carbonates in such belts may still be correctly described as limestones or dolomites, but if crystal growth under heat and pressure is significant, they become *marbles* (although quarrymen use the term more loosely for a carbonate rock that can be polished). Marbles adjacent to igneous intrusions can develop crystal sizes of up to several mm and are associated with the growth of several new minerals, mainly silicates. Cave formation in such terrains is promoted by geological features. Firstly, the carbonate units may be relatively restricted in thickness or lateral distribution so that they are susceptible to attack by aggressive meteoric waters that have traversed adjoining silicate rock masses or which contain silicate layers which act as inception horizons for karst development (Faulkner, 2006). Secondly, the stresses associated with orogenic uplift inevitably result in significant fracturing which facilitate cave development. For example, a regional study in the Sierras Blanca and Mijas of southern Spain demonstrated the importance of fracture sets (both faults and joints) in specific orientations in facilitating karstification, although lithology (calcite versus dolomite) was even more important as a control (Andreo et al., 1997). Fracturing related to isostatic recovery from ice loading at the end of the last ice age can rapidly lead to cave formation as demonstrated by Faulkner (2007) for marbles in the Scandinavian Caledonides.

Active tectonics and seismic events are of course not restricted to belts of metamorphic rocks, but also cut young sediments in rift basins, where there is plenty of evidence for their role in catastrophic events at archaeological sites, including caves (Nur, 2008); less frequently they occur far from plate boundaries. Evidence for current fault movement is found even in seismically quiet regions such as Belgium (Vandycke & Quinif, 2001). Seismic activity has important implications for the continuity of speleothem growth (Box 7.1).

2.2.4 Carbonate porosity

Porosity is expressed as a volume fraction or percentage and can be determined on rock slabs, or in thin sections impregnated with dye, either qualitatively, or quantitatively using image analysis software. In the oil industry, nuclear magnetic resonance (NMR) imaging is widely used to measure porosity of the wall-rock of boreholes, and NMR instruments exist to determine the porosity of water-saturated rock cores (Arnold et al., 2006). The data are more complex to interpret in carbonates than in siliciclastic sediments (Westphal et al., 2005) for reasons given below, but total porosity and information on pore size can be recovered (Kappes et al., 2007). The open or interconnected porosity in core samples can be determined by a triple weighing approach by successively determining the weight of the water-saturated sample, the sample dried at 110 °C and the sample suspended in water within a vacuum chamber (Pulido-Bosch et al., 2004).

It is useful to distinguish *primary porosity*, which represents the cavity systems originally present in sediment at the time of deposition, from *secondary porosity*, which covers all the subsequently generated spaces. In siliciclastic sediments and rocks, the primary sediment consists of regularly shaped grains with intervening primary pores on a scale comparable to that of the grains. These tend to be progressively eliminated by a combination of physical compaction and cementation as the sediment is buried, although secondary porosity can form by leaching (selective dissolution) of particular grains and cements both in the deep subsurface and in meteoric water environments (Scholle & Schluger, 1979).

Carbonate porosity is more complex and has been of great interest to the petroleum industry because of its importance in determining the properties of

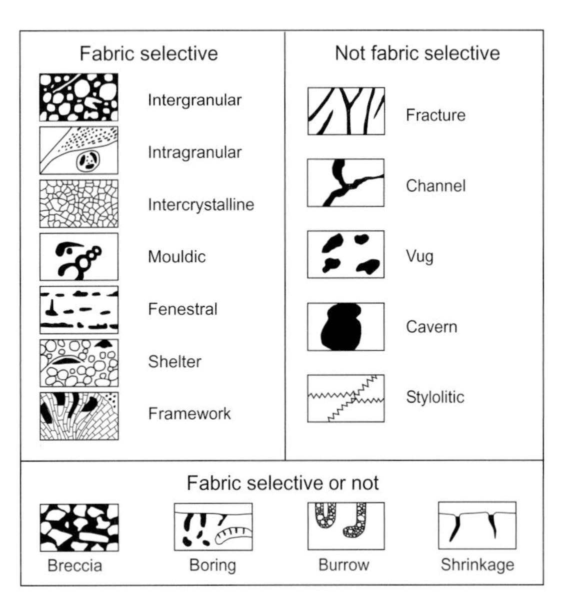

Fig. 2.12 Choquette and Pray's (1970) carbonate porosity types, modified after Tucker and Wright (1990).

reservoirs from which hydrocarbons are extracted and this research has produced many studies which are of relevance to karstified carbonates (Moore, 2001). A widely used classification by Choquette and Pray (1970) is illustrated in Figure 2.12. Pore size is referred to by the terms megapore (>4 mm), mesopore ($\frac{1}{16}$–4 mm) and micropore (<$\frac{1}{16}$ mm). The major distinction of 'fabric-selective' and 'not fabric-selective' porosity is different from that of primary and secondary porosity. Most of the fabric-selective types might be primary, but on the other hand secondary porosity can be manifested as all of the porosity types.

The most obvious primary porosity is the intergranular category, but skeletal organisms also contain pores, distinguished as intragranular or framework depending on whether they are within allochems or organic growth frameworks. Shelter cavities arise by lack of matrix beneath relatively large protecting allochems whereas fenestral cavities are open spaces produced by a range of processes such as early decay of organic matter, gas generation and desiccation. Where the sediment is a chemical precipitate, intercrystalline porosity can also be primary. Intergranular porosity of coarser carbonates tends to be transformed either by cementation or dissolution, whereas carbonate mud loses porosity more steadily by compaction. The most systematic and best understood behaviour is shown by deposits of pelagic carbonates, such as Cretaceous–Cenozoic marine oozes/chalks/limestones which consist predominantly of low-Mg calcite and retain high primary intergranular porosity. They show systematic trends of porosity loss from primary values of 60–80%, diminishing to 40% by a combination of mechanical compaction and recrystallization, but at a critical depth of 300–1000 m burial, chemical compaction (pressure dissolution) takes over with a reduction of porosity to 20–30% associated with local reprecipitation of calcite cement (Fabricius, 2003; Ando et al., 2006). Wherever metastable carbonates are present, there will be a drive towards mineralogical stabilization. This, combined with compaction and cementation, leads to low (<10%) primary porosity in Palaeozoic and many Mesozoic–Cenozoic carbonates.

Secondary porosity arises from enlargement of existing cavities and removal of more susceptible components by dissolution, or it may be the remaining porosity between secondary crystals (e.g. of dolomite) which have replaced the primary phases (Fig. 2.9). Where cavities are sufficiently large and interconnected, physical erosion, transport and deposition of sediment particles can also occur. Fabric-selective dissolution can re-create primary pore types (including burrows and borings) or be associated with preferential dissolution of fossils (e.g. aragonitic) to yield mouldic porosity. *Vugs* are pores that cut across the original fabric of the rock, but may have development by dissolution-enlargement of other pore types, such as moulds of fossils (Choquette & Pray, 1970) or aggregates of soluble mineral phases such as anhydrite. *Fractures* (Fig. 2.13) are largely two-dimensional discontinuities resulting from tensional stress (joints) or lateral displacement (faults). They are typically enlarged by dissolution, whereas *channels* (conduits of the karst literature) are more one-dimensional and necessitate significant dissolution. *Cavernous porosity* may also be associated with collapse features and hence the development of *breccia porosity*.

Fig. 2.13 Classic karstic limestone exposures at El Torcal, Almeria province, Spain, displaying abundant bedding-parallel sub-horizontal joints and less common sub-vertical joints. There are also prominent steep dissolution phenomena (rillenkarren).

Stylolites are irregular surfaces created by pressure dissolution of carbonate, leaving an insoluble residue typically of silicate and either organic or oxide phases. During later tectonic deformation these, and more planar (clay-rich) pressure-solution surfaces, can be the site of opening fissures (Graham Wall, 2006); hence Tucker and Wright (1990) added this as a porosity type. Klimchouk and Ford (2000a) use the term *fissures* as an all-encompassing term to include all planar cavities. The uplift and weathering of a landscape results in the generation of *joints* by pressure release. These can be sub-horizontal close to the land surface or, where adjacent to plateau margins, sub-vertical. Where flat-bedded strata occur next to an escarpment, they may display *cambering* phenomena where there is tilting on underlying shale units leading to the formation of steep, downwards-tapering fissures (*'gulls'* of Parks, 1991), which may become caves (Simms, 1994; Murphy & Lundberg, 2009).

Hydrogeologically, porosity types have been grouped into the three components which comprise the *triple-porosity* (or -permeability) model of karst aquifers (White, 2006). Water flow is predominantly through conduits and secondarily through fractures, whereas the matrix, containing small pores, offers only poor yields. The fabric-

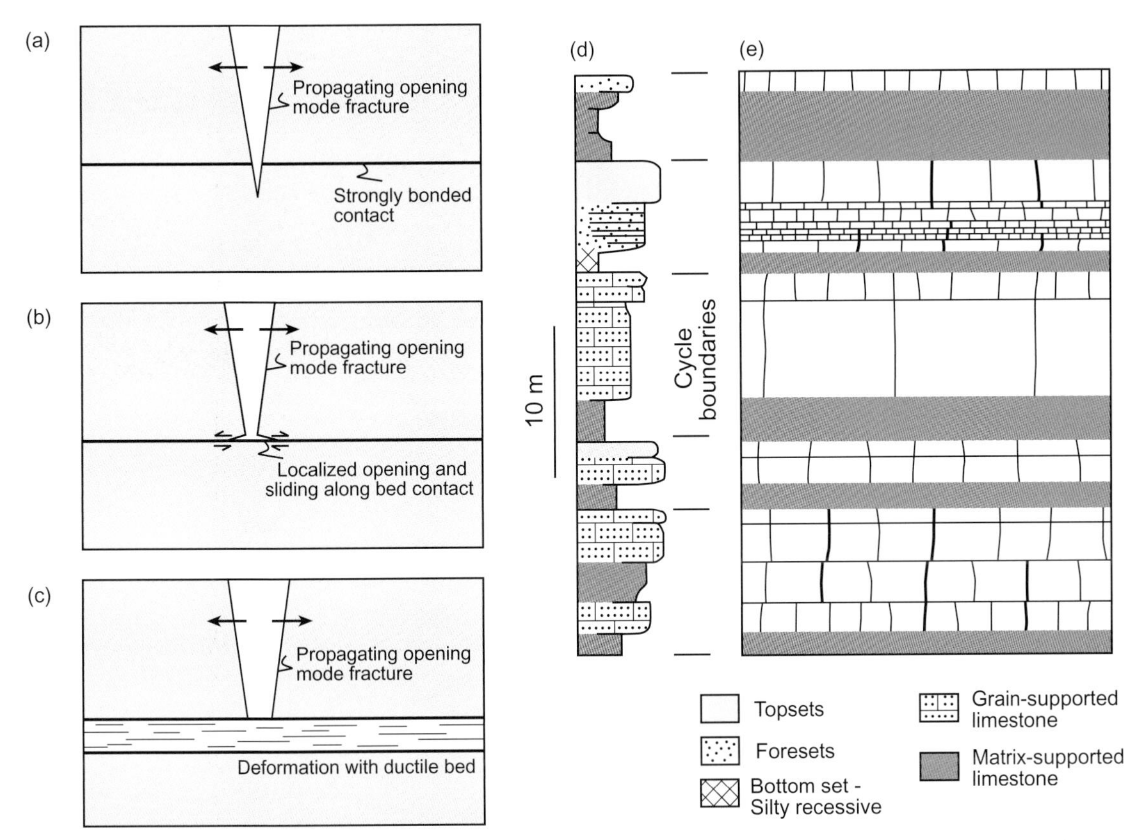

Fig. 2.14 Modelled fracture development in sedimentary successions (redrawn from Cooke et al., 2006). Part (a) shows fractures propagating across a strong interface, whereas (b) and (c) show fracture termination related to bed-surface sliding and ductile deformation respectively. (d) A stratigraphic log of Cretaceous shallowing-upwards from Spain, and (e) the corresponding modelled fracture systems which are focused in the more competent shallow-water facies.

selective elements of porosity in the matrix tend to be poorly connected and do not by themselves promote speleogenesis (Klimchouk & Ford, 2000b), except in mixing zones as described in section 2.3.

As will be discussed in Chapter 4, speleothems are typically fed by water which can be characterized as some mixture of flow from minor fractures (*quickflow*, responsive to infiltration events) and *delayed flow* from the matrix (usually thought of as a continuously permeable medium, but which in lithified carbonates be a network of minor fractures). Even in a geologically simple case, the nature of the fracture pathway can be complex. For example, observations and modelling of shallowly dipping beds demonstrates that the physical stratigraphy and the degree of weakness of the associated lithological boundaries are important control on the location and vertical continuity of fractures, thicker weak shale horizons facilitating the fracture termination (Fig. 2.14 and Cooke et al., 2006). As a result, water flow may be limited by pathways obliquely across relatively impermeable layers (often contain clay minerals) from which solutes can also be leached. Imaging of matrix porosity (Fig. 2.15 and Motyka, 1998) shows a wide range of porosity types as described above including both fabric-selective pores and microfissures. Complementing visual approaches, NMR evidence in an

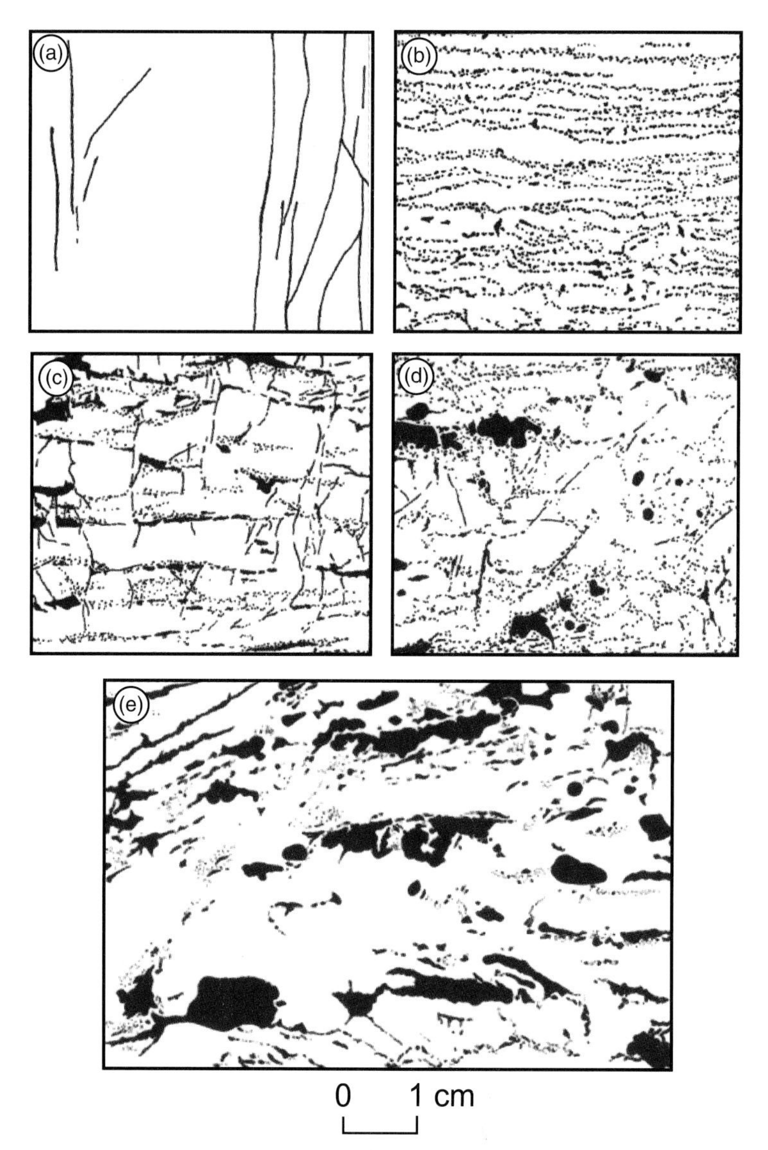

Fig. 2.15 Illustrations of matrix porosity in Polish karstic rocks (redrawn from Motyka, 1998). (a) Microfissures in limestone; (b) bedding parallel pore systems in dolomite; (c) incipient breccias porosity in dolomite containing metal ore; (d) dolomite with vuggy porosity; (e) shelter and fenestral cavities, limestone. Parts (a)–(d) are Triassic and (e) is Devonian in age.

Upper Jurassic aquifer demonstrated that even where total porosity is less than 5%, micropores and fractures were of sub-equal importance (Kappes et al., 2007). Wide ranges of matrix porosity are found in carbonates in karst of continental regions with a tendency for reduction with age and deep geological burial, but typical values of less than 1 to a few per cent are found in most Mesozoic and older carbonates (Atkinson, 1977b; Worthington et al., 2000; Pulido-Bosch et al., 2004; Rzonca, 2008). Motyka (1998) emphasizes the importance of this passive reservoir of water for solute storage. From the speleothem point of view, sampling of dripwater at low flow gives an indication of the lithologies that this water has encountered in the rock matrix, and indeed they may impart a characteristic signature to the water chemistry (Baker et al., 2000) and speleothems derived from it.

2.3 Carbonate diagenesis and eogenetic karst

In this section we discuss the alteration of carbonate sediments by processes that have some relationship to their original depositional setting. 'Original setting' allows for subsequent changes in sea level, but pre-dates significant burial under younger sediments. This is the realm of *eodiagenesis* as envisaged by Choquette and Pray (1970), and includes important examples of cave formation. Carbonate diagenesis in these settings is well-reviewed by, among others, James and Choquette (1990a, b), Tucker and Wright (1990), Tucker (1993) and Moore (2001) and is particularly well illustrated by Scholle (1978) and Scholle et al. (1983) and Schneidermann and Harris (1985).

2.3.1 Early diagenesis in marine waters and brines

Both aragonite and high-Mg calcite occur widely as cements in tropical carbonate environments in oolites, reefs and even in deep-water buildups. The Mg content of calcites increases with temperature; Mg-rich calcite is relatively more soluble, hence aragonite is relatively more important in the warmest, more evaporated water. Cementation phenomena on beaches are analogous to speleothems in that they are often caused by CO_2-degassing.

These primary marine cements typically grow as aggregates of narrow or micritic crystals, typically in an *isopachous* (equal-thickness, i.e. concentric) fringe. The more compact varieties of magnesian calcite and *botryoidal aragonite* (also *crystal-ray aragonite*), which are sometimes found in reef carbonates, have internal textures that compare with some speleothems (section 7.4). When found in ancient limestones, these cement forms are replaced by *neomorphic* calcite, which is calcite that has grown as the primary minerals dissolved, without ever leaving a significant cavity. Calcite-after-aragonite is characterized typically by equant crystals, sometimes with aragonite relics protected by organic films (Sandberg, 1985). Magnesian calcite shows very little alteration in texture, although sometimes small rhombs of dolomite (*microdolomite)* are found dispersed within the neomorphic calcite. Three types of calcite crystal orientation pattern result can be distinguished in ancient limestones: radial fibrous calcite, fascicular-optic and radiaxial calcite (Fig. 2.16 and Kendall, 1985).

Evaporative marine waters can influence marine carbonate settings both in the exposed environments of the sabkhas, or in subaqueous salinas or hypersaline lagoons, or semi-arid to arid settings (Moore, 2001). The degree of evaporation obtainable ultimately is related to the climatic aridity, which can be expressed as the mean vapour pressure or activity of water in the atmosphere (Kinsman, 1976). This concept also has application to evaporation in caves. Salt flats (*sabkhas*) are characterized by infiltration of floodwater (marine or continental) that is then evaporated, leading to extensive nucleation of gypsum or anhydrite crystals or concretions, and variable dolomitization of the carbonate. Seepage reflux of brines beneath partly restricted barred basins may occur, also leading to dolomitization. In more isolated lagoons and other settings where fine, organic-rich sediment accumulates, the more likely processes are precipitation of calcite or dolomite cements forced by reactions, such as sulphate-reduction, accompanying the bacterial decay of organic matter. A by-product is pyrite, which could later lead to such beds becoming karstic inception horizons (section 2.4).

2.3.2 Vadose diagenetic processes

Marine sediment can subsequently be affected by *meteoric water* (atmospheric precipitation that has infiltrated the ground) because of sediment accumulation to above mean sea level and/or eustatic (worldwide) sea-level fall or tectonic uplift. The meteoric environment is effective in transforming modern carbonate sediments, and the various zones within it are shown in Figure 2.17. In some cases, such as the subsurface of Anewatak atoll, a

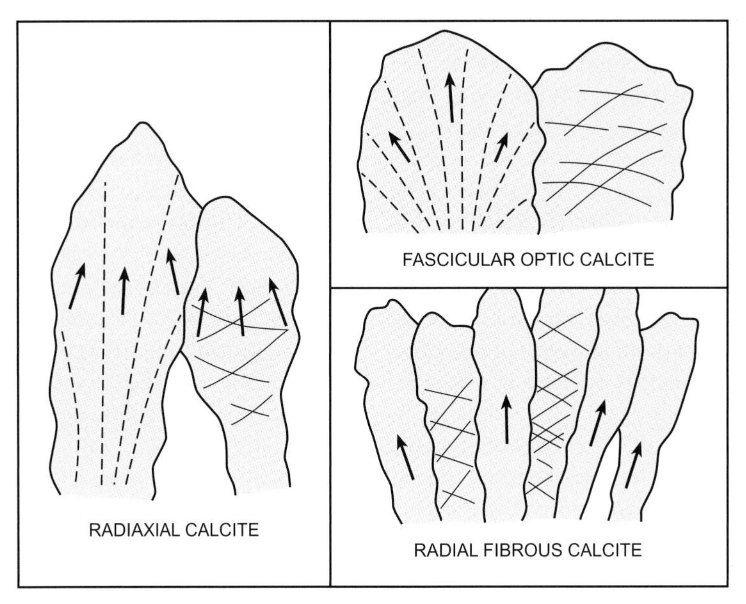

Fig. 2.16 Distinction of different types of texture of calcite (magnesian when grown in warm temperatures) cement crusts in marine limestones. Arrows indicate the orientation of the z crystallographic axis of calcite; dashed lines are sub-crystal boundaries and continuous crossed lines are twin planes (formed by secondary deformation). From Kendall (1985).

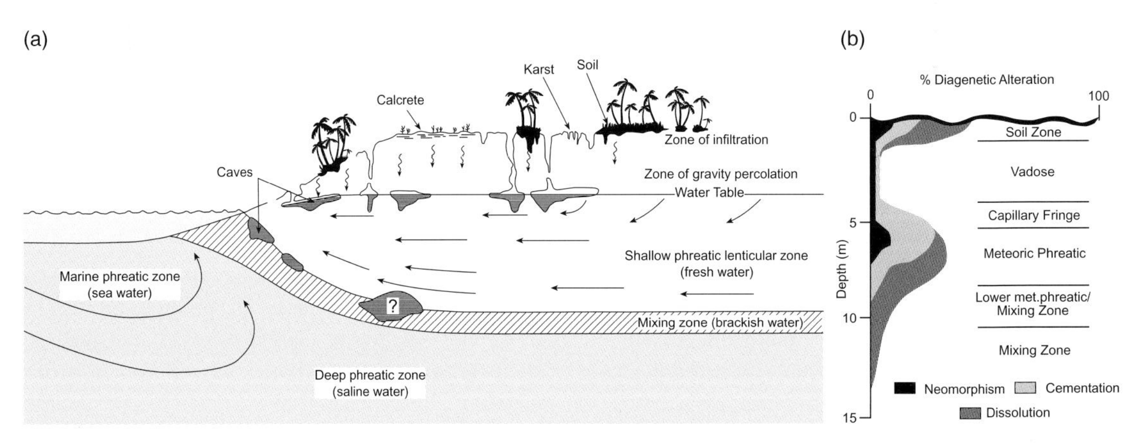

Fig. 2.17 The main zones of the meteoric diagenetic environment (James & Choquette, 1990b) compared with idealized fossil meteoric water zones based on a generalization of examples from boreholes through Anewatak atoll (Quinn & Saller, 1997).

series of discrete zones of meteoric alteration up to 10 m thick can be found, representing discrete sea level lowstands (Quinn & Saller, 1997). The vadose zone is one of water movement vertically downwards in response to infiltration, or upwards as a capillary phenomenon during evapotranspiration. As discussed quantitatively in Chapter 5, meteoric waters will readily dissolve and reach saturation with calcite (or, at a higher level, with aragonite, if sufficient is present). In climates that are not persistently arid or humid throughout the year, the relative importance of infiltration and capillary rise can reverse in different seasons, leading to alternating dissolution and precipitation of $CaCO_3$.

Where the carbonate sediment is unconsolidated (typically Holocene), it serves as a substrate for soil development. Carbon dioxide levels will become elevated by plant respiration, which promotes carbonate dissolution, and globally P_{CO_2} can be predicted by its relationship with potential evapotranspiration (Brook et al., 1983) as discussed in Chapter 3 (Fig. 3.18). However, pre-Holocene carbonate will have been lithified and its overlying soil is thin, and carbon dioxide is readily lost from it by diffusion. This is the case, for example, in the Bahamas where soils on Plio-Pleistocene limestones have low P_{CO_2} except where they locally thicken into metre-scale dissolution pockets. Here, vadose dissolution is estimated as equivalent to 1.6–3.2% of porosity generation per thousand years (Fig. 2.18 and Whitaker & Smart, 2007b), the higher figure applying when downward flow takes the form of seepage through the matrix rather than more concentrated flow through fissures.

Rates of dissolution in the meteoric zone, which may include a significant or dominant role for the vadose zone, have been quantified using information on water chemistry and flow rates, and a comparison of results from six islands has been made by Whitaker and Smart (2007b). Annual rates of dissolution equivalent to 0.04–0.064 mm yr^{-1} of surface lowering are found in islands with low to moderate infiltration (200–370 mm yr^{-1}) despite large variations in the proportions of the more soluble aragonite and magnesian calcite. Drier environments have retained more metastable carbonates, but because the water that does infiltrate is more efficient at dissolution, the two effects cancel out. High infiltration leads to much higher rates of dissolution; e.g. island Laura in Fiji has 1780 mm yr^{-1} of infiltration and calculated 0.35 mm yr^{-1} dissolution.

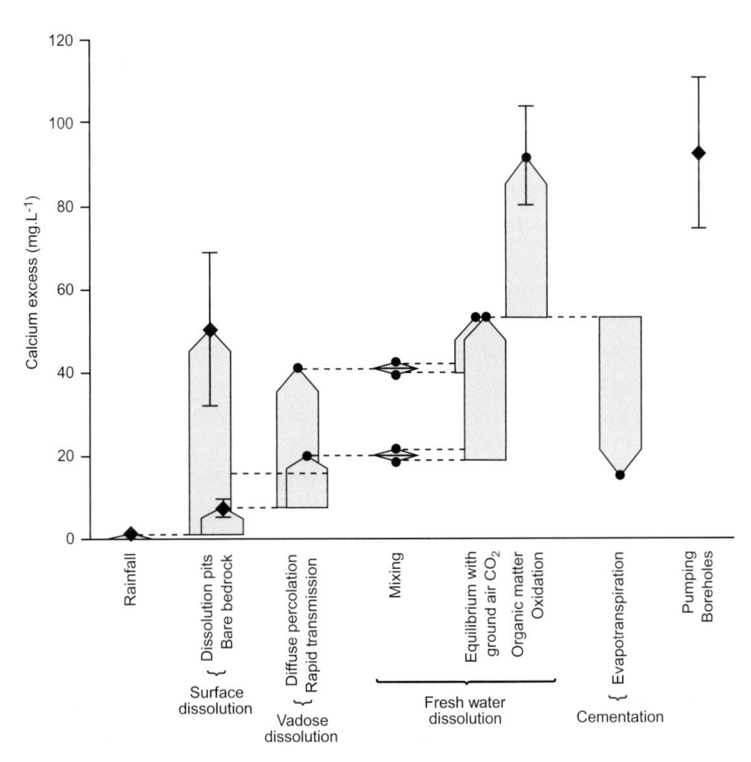

Fig. 2.18 Calcium budget for Plio-Pleistocene limestones of North Andros Island (after Whitaker & Smart, 2007b), corrected for marine sea-salt.

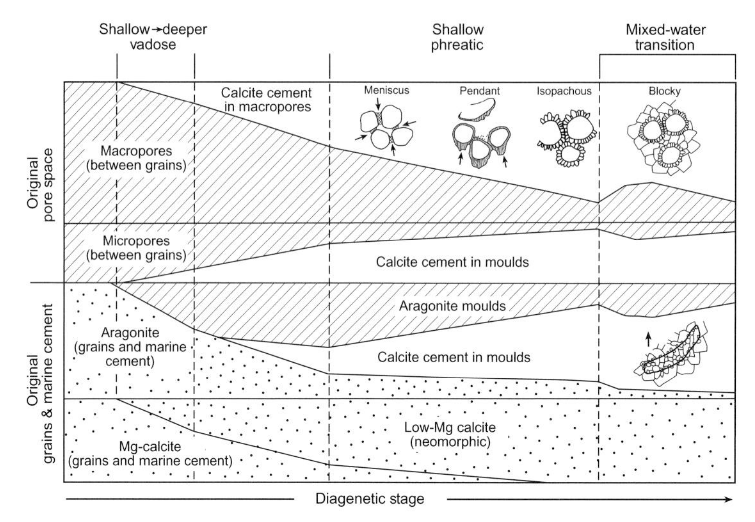

Fig. 2.19 Conceptual model of the evolution in a mixed aragonite–magnesian calcite sediment passing through diagenetic stages through vadose to phreatic to mixing zone conditions together with cartoons of the cement types (after James & Choquette, 1990b).

The most important processes stimulating calcite precipitation in the vadose zone are evapotranspiration or degassing. Evaporation can occur to a depth of a metre or so, whereas transpiration by tree roots can have a much greater reach. In Australia, *Eucalyptus marginata* can extend to a water table at 15–40m depth (Florence, 1996). Where the carbonate sediment below the soil zone and above the water table is affected by these processes, cement distribution tends to be irregular even in grainstone lithologies, and cement may define *meniscus* (at grain contacts) or *pendant* (subgrain) distributions, following the location of water films (Fig. 2.19). Although not shown on this diagram, aragonite debris can be replaced by calcite with relatively good fabric preservation associated with a micrometre-scale alteration zone (Pingitore, 1976), characteristic of the limited amount of water available.

Calcrete (also known as caliche, kankar or cornstone in different literatures) refers to a suite of macroscopic structures developing within and on top of soil horizons as a result of carbonate precipitation. It is characteristic of semi-arid or seasonally dry and warm environments (see Chapter 3, Fig. 3.21). A common form is the development of *rhizoliths* (root concretions) in dune sands, but 1- to 10-cm-scale nodular or pervasive chalky or laminar precipitates can also occur, transitional to the underlying carbonate. In thicker calcrete developments, there can be discontinuous laminated plates and crusts beneath a more continuous solid *hardpan* which be layered or contain concentrically laminated centimetre-scale structures (*pisoids* or *vadoids*), or be fragmented by root action. A range of microstructures are preserved, many of which are diagnostic of specific physicochemical or biological processes (Esteban & Klappa, 1983).

2.3.3 Meteoric phreatic diagenesis

Meteoric water on carbonate islands will tend towards the lens shape of Dupuit–Ghyben–Herzberg theory, overlying denser, marine-derived groundwater (Vacher, 1988). The shape of the lens reflects that of isostatic balance whereby the total column of freshwater-saturated sediment heaped beneath the centre of the island to the base of the lens exerts the same pressure as the adjacent column of sediment saturated with seawater. The height of the water table is determined by precipitation and groundwater flow. In the simplest case, the thickness of freshwater should be 40 times the maximum elevation of the groundwater table above sea level (because seawater is around 1/40 more dense than freshwater), but this will be reduced if porosity increases with depth (see below). The boundary with underlying salt water is not sharp because of the effects of diffusion and dispersion, the latter related to short-term tidal oscillations as well as seasonal changes (Whitaker & Smart, 1997). More complex meteoric conditions are found in association with reefs such as Anewatak atoll, where porosity increases with depth from cemented Holocene deposits into porous Pleistocene limestones (Buddemeier & Oberdorfer, 1997). However, an increase in porosity in a porous phreatic zone, giving rise to different hydrological properties, is quite a common phenomenon in carbonate islands (Whitaker & Smart, 1997; Vacher & Mylroie, 2002).

In the northern Bahamas area, Whitaker & Smart (2007a, b) showed that the higher Ca concentrations at depth (Fig. 2.18) are driven by P_{CO_2} values exceeding 10^{-2} atm, which can be attributed to oxidation of surface-derived colloidal organic matter within the sediments (Atkinson, 1977a; McClain et al., 1992), consistent with the suboxic condition of underlying phreatic waters. The dissolution in the combined vadose and phreatic zones is countered by localized evaporation-driven calcite precipitation near the water table (Fig. 2.18). During sea level lowstands, the vadose zone would be much thicker and this organic input would not be effective, in keeping with the more static behaviour of the lens (Melim, 1996), except in aquifers with significant secondary permeability and recharge from allogenic (non-carbonate) sources.

Figure 2.19 summarizes what James & Choquette (1990b) termed mineral-controlled meteoric diagenesis whereby the progressive stabilization of the mineralogy is associated with cementation processes leading to significant porosity reduction. Examples of rates of transformation of mixed biogenic sediments of Holocene ages in hydrologically active zones of a Bahamian site were given by McClain et al. (1992), which can be expressed as 4–14% for high-Mg calcite, and 0–4% for aragonite, per thousand years. At an intermediate stage in the diagenetic evolution, the effect can be that of an inversion of porosity with leaching of bioclasts and partial filling of pore space with cement, but this mouldic porosity is later filled. Once the meteoric-marine mixing zone is reached, however, porosity enhancement is expected because mixed water with two different P_{CO_2} values results in waters that are undersaturated (Wigley & Plummer, 1976). High-Mg calcite alteration results in immediate and preferential fill of intragranular porosity and the secondary calcite is often neomorphic, retaining primary textures on a fine scale. Hence, magnesian calcite alteration is likely to be very efficient in terms of carbonate reprecipitation. Also, Budd (1988) estimates an overall efficiency approaching 90% of reprecipitation of carbonate resulting from aragonite dissolution beneath oolite sand islands and indicates that this process can continue in the mixing zone given the continued presence of aragonite. Vacher et al. (1990) characterized the process of lithification as a first-order reaction with a half-life of the order of 6000–7000 years. This estimate, and others given above, indicates that partial lithification of initially unconsolidated sediment is readily accomplished within an interglacial period. Ultimately, the transformations may often be limited by the rate of redox reactions associated with organic matter oxidation, rather than simply metastable mineral stabilization (McClain et al., 1992).

Where meteoric water forms a confined aquifer beneath an impermeable lithology, the pressure head from its source region on land can drive flow well offshore and indeed offshore freshwater springs are known, for example tens of kilometres away from the Atlantic coast of Florida.

2.3.4 Eogenetic karst development

Eogenetic karst in the modern world is strongly influenced by the more than 100 m fluctuations in late Quaternary sea level as can be demonstrated, for example, by the altitudes of cave systems closely matching palaeo-shoreline indicators (Florea et al., 2007). This has been formalized as a Carbonate Island Karst Model, although the key factor is the presence of young carbonate rather than the presence specifically on an island. Mylroie and Carew (2000) summarized much research on such carbonates in terms of three conceptual models, to

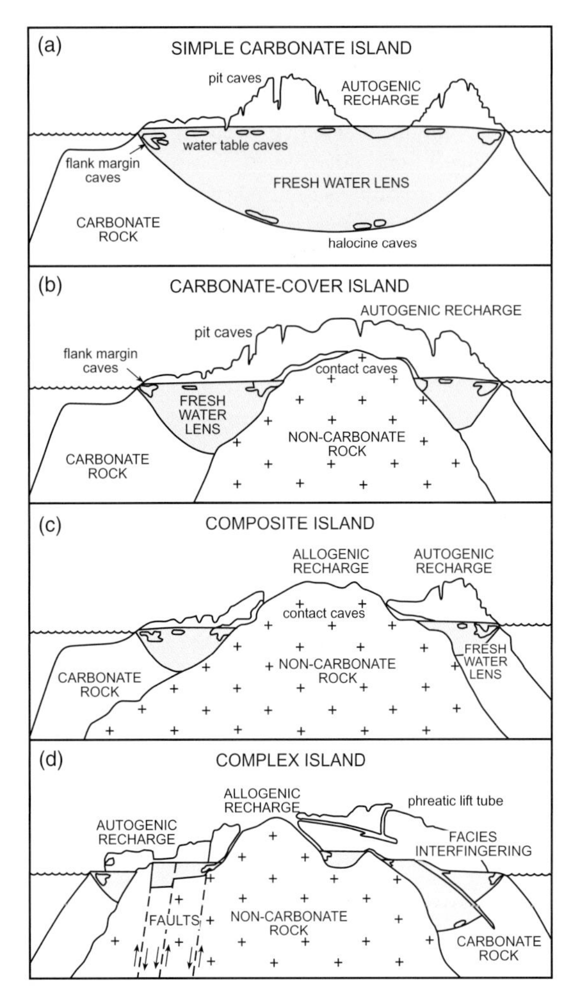

Fig. 2.20 Conceptual models of eogenetic karst evolution that make up the Carbonate Island Karst Model (Jenson et al., 2006).

which was added a fourth by Jenson et al. (2006; Fig. 2.20): simple carbonate islands (e.g. Bahamas, with pure carbonates down to a depth exceeding the −125 m of glacial sea level), carbonate-cover islands (e.g. Bermuda, in which carbonate sediment rest on a shallow non-carbonate bedrock), carbonate-rimmed islands (e.g. Barbados), and complex islands (e.g. Saipan in the Marianas Islands) where volcaniclastic sediments and locally confined aquifers exist in a context of tectonic activity. Carbonate-rimmed islands (a less extreme case of a carbonate-rimmed continental mass) are distinctive in having *allogenic recharge* (recharge from non-carbonate lithologies). Because such waters tend to be strongly corrosive, the result can be a terrain with typical karstic landforms (such as dry valleys) inland passing to a half-lens at the coast. A classic area of this type is the Yucatan peninsula of Mexico, a region where important work demonstrating dissolution processes in mixing zones was performed (Back & Hanshaw, 1970). An important common factor is that karstification is ultimately terminated by basin subsidence and replacement of meteoric by marine pore fluids, although in Quaternary times, this followed several major changes in sea level and hence shifts in diagenetic zones.

In simple carbonate islands, vertical caves (vadose shafts) can develop from local concentration of flow, for example beneath shallow surface depressions. The limited dissolution capacity of *autogenic recharge* (i.e. rainwater passing discharge directly into the carbonate aquifer) mean that shafts may not reach the water table, although can become enlarged, or joined with adjacent caves. Caves commonly develop at the water table in response to mixing of waters of different composition or oxidation of organic matter. Mylroie and Carew (2000) give typical sizes of 1–3 m deep and 3–10 m wide on the Bahamas, where the closeness of the water table to the land surface results in extensive collapse to give sinkholes. At the mixing zone, the same pair of mechanisms can operate leading to cave formation (e.g. Wigley & Plummer, 1976; Bottrell et al., 1991), with organic matter oxidation aided by deposition of colloidal organic matter at the salinity interface. Flow is concentrated at the island flanks where vadose mixing occurs and so this is the zone for preferential development of mixing zone caves. On the Bahamas, caves found at elevations 1 to 7 m above present day sea-level clearly represent the last interglacial around 125 ka when sea level was around 6 m above present (the Bahamas are tectonically stable so no correction for vertical tectonic movements is required). Such caves can be entered only when breached by erosion and lack indications of turbulent water flow. In Bermuda's moist climate, flank-margin caves formed during the last interglacial period are mostly eroded. On both carbonate-cover islands such as Bermuda, and in continental settings where calcareous dunes rest on non-carbonate bedrock, caves commonly develop as vadose collapse caves at the interface with the underlying bedrock, or at higher levels. The development of such systems in Australian aeolianites was first described by Jennings (1968) who used the term *syngenetic* karst, and described the development of a calcrete surface layer and underlying solution pipes as important facets in its evolution.

Blue holes are tidally influenced subsurface voids, below sea level for most of their depth, and developed in carbonate banks and islands. They can be filled with sea water (ocean holes) or occur inland. They can form via a variety of mechanisms, e.g. flooding of caves during sea level rise, collapse extending the vertical cave dimensions, or by failure of the bank-margin failure to produced deep flooded fractures (Mylroie & Carew, 2000). Speleothems collected from blue holes provide important data on terrestrial conditions during sea-level lowstands (Richards et al., 1994).

The cave types described above occur in specific geological or hydrogeological settings and contrast strongly with the catchment-draining caves of telogenetic karst described in section 2.4, although rare examples are known too in eogenetic karst, e.g. within the interior of Great Bahama Bank (Vacher & Mylroie, 2002). They develop within permeable aquifers which commonly possess a distinctive, mainly double porosity with touching vugs defining flow channels with intervening intergranular pores (Vacher & Mylroie, 2002). This results from the reorganization of porosity during meteoric diagenesis leading to strikingly permeable limestones without net increase in porosity. A typical result is that 10–20% of the rock consists of large, touching-vug pores (occupying about 30% of the rock porosity) and that these pore channels transmit 99.99% of the water flow, with horizontal permeability many orders of magnitude larger than vertical permeability.

A successful modelling approach to the development of flank-margin caves has been demonstrated by Labourdette et al. (2007). Closed chambers develop from undersaturated meteoric-marine

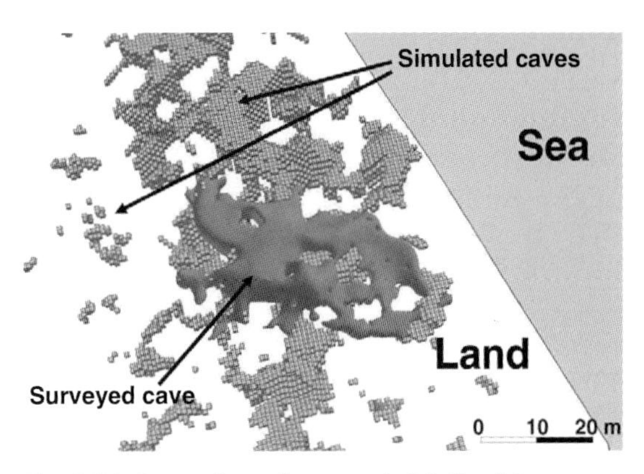

Fig. 2.21 Comparison of surveyed, Salt Pond Cave, Long Island, Bahamas, with probabilistically modelled caves (Labourdette et al., 2007).

mixed fluids entering the chamber by diffuse flow and can only develop substantially at low- and highstands of sea level, when of the order of 10^4 years is available for cave development under stable sea level conditions; enlargement of caves in the Bahamas is estimated to occur at rates of around $80\,m^3\,km^{-2}\,yr^{-1}$. Cave reconstruction used a probabilistic model to simulate the growth and coalescence of spherical voids. Several phases of growth were used to correspond with the observed multimodal distribution of cave areas. The model successfully simulated the large central chambers flanked by maze-like passages (Fig. 2.21) in detail depicting globular chamber shapes, within-cave pillars, and passages connecting chambers resembling the natural caves.

A different style of mixing-zone cave development thought to be more characteristic of higher discharge regimes in larger carbonate islands, or carbonate terrains adjacent to non-carbonate geology, and transitional to the forms of telogenetic karst, was described from the eastern Yucatan peninsula (Quintana Roo) by Smart et al. (2006). Most of the caves are flooded at the present day and form a sub-parallel anastomosing network up to 12 km long and perpendicular to the coast, with locally irregular spongework margins; their form is influenced by local joints and by roof collapse. Cross-sections through the tubes are elliptical or, near the coast, with a high aspect ratio related to

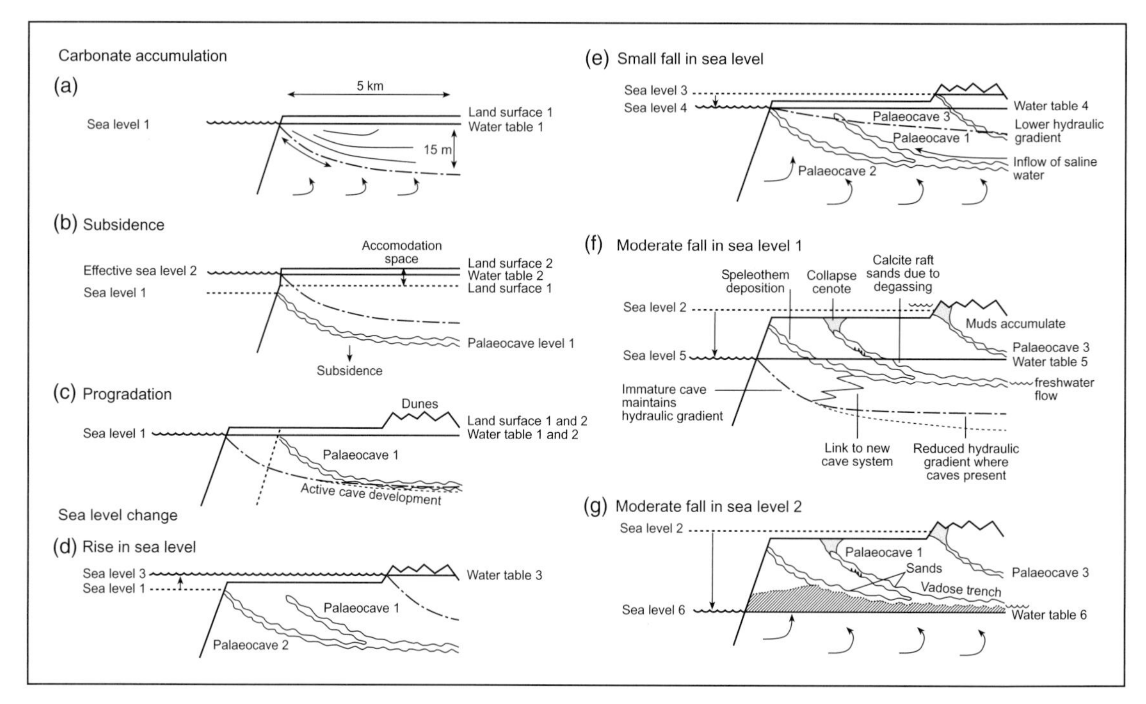

Fig. 2.22 Model of mixing zone cave development based on Quintana Roo, Yucatan peninsula (Smart et al., 2006) illustrating complex polyphase cave development in response to sea level change. Vertical exaggeration is around 120×.

fissures. The multi-stage nature of different phases of speleothem and cave sediment formation show that the caves have multiple phases of development. Figure 2.22 illustrates a model of cave development in relation to carbonate deposition and changes in sea level.

2.3.5 Burial diagenesis

Because young carbonates affected by meteoric diagenesis typically remain porous, it follows that the low porosities attained by ancient carbonates are largely achieved during burial diagenesis (Scholle & Halley, 1985). Data summarized by Moore (2001) indicate that by a burial depth of 300 m, the physical processes of mechanical particle re-orientation, dewatering and grain deformation result in porosities of 60% for pelagic oozes, 40% for shelf muds and 30% for grainstones. At greater depths, a key process is chemical compaction which operates by pressure dissolution of grains where they come into contact and therefore bear the full lithostatic pressure, which is higher by a factor of around 2–2.5 (the ratio of specific gravities of sediment particles to water) than the hydrostatic pressure on grain surfaces adjacent to water. In addition to welding of contacts between grains, impure carbonate horizons are the focus of dissolution and give rise to pressure solution seams, whereas irregular stylolitic surfaces develop within purer limestones. Thermal alteration of organic matter plays an important role in the temperature range of 60–110 °C (~1–3 km depth) where creation and migration of petroleum and methane successively occurs. Release of acidic organic radicals encourages the generation of secondary porosity and the redistribution of mineral phases as cements in other parts of the sedimentary basin. Petroleum migration may interrupt the process and solidified bitumen products are sometimes found as the final cavity filling. Associated thermal alteration of clays leads to release of Mg^{2+} which promotes dolomitization, which is in any case thermodynamically favoured

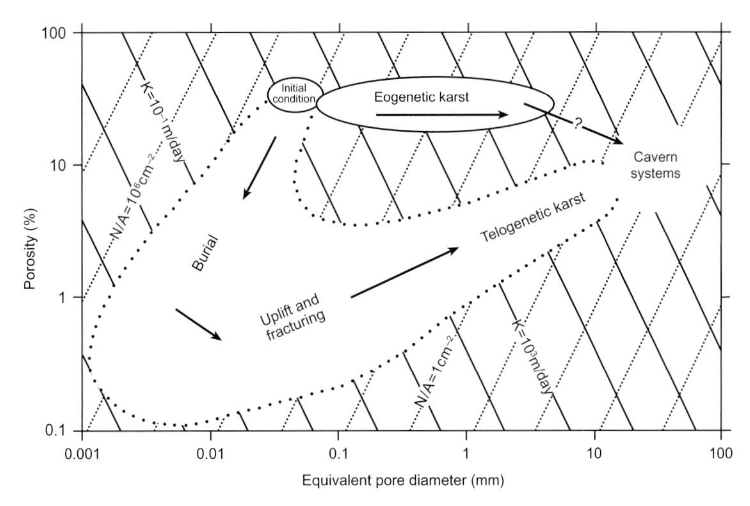

Fig. 2.23 Evolution of limestone permeability within a plot of porosity versus equivalent pore diameter (Vacher & Mylroie, 2002). *K* is hydraulic conductivity, defined in Darcy's Law with units of m day^{-1} for the equivalent uniformly porous medium. For a limestone with touching vug porosity, *N*/*A* is the number *N* of touching-vug tubes per unit area *A*.

by higher temperatures. Burial history and sometimes dolomitization are sometimes associated with mineralization events that produce metallic phases, dissolution of which imparts distinctive trace element compositions in karstic waters that can be transferred to speleothems (Fairchild et al., 2010).

Figure 2.23, which has been developed by Vacher and Mylroie (2002) from an original concept by Smith et al. (1976), contrasts the distinctively different pathway of eogenetic karst evolution from that of burial diagenesis, followed by uplift and telogenetic karst development, as described in the next section.

2.4 Speleogenesis in mesogenetic and telogenetic karst (with contributions from John Gunn and David J Lowe)

This section concerns speleogenesis in diagenetically mature limestones: those that have had their primary porosity largely eliminated through the burial processes discussed in the previous section. How are such essentially solid rock masses converted into highly permeable karst? The most common case is dissolution related to the penetration of meteoric water into the carbonate rocks following their uplift towards the Earth's surface: this is *telogenetic* karst (Choquette & Pray, 1970). However, subsurface-derived (*hypogenetic*) thermal fluids can cause void formation at depth: this is *mesogenetic* karstification (Choquette & Pray, 1970). The single most comprehensive source of data and ideas on speleogenesis is the book edited by Klimchouk et al. (2000), but there are also extensive overviews in the texts by Ford and Williams (2007) and Palmer (2007) and the encyclopedia edited by Gunn (2004). Chemical terms used in this section are defined in Box 5.1.

Caves are best defined as natural underground voids able to be entered by humans and it is such relatively large spaces that are the major arteries of water flow through karst. However, there is also a continuum, from m to sub-mm scale, of interconnected voids which route the percolating source-water for speleothems to caves (Chapter 3). Usually caves develop from such smaller cavities and there is increasing interest in understanding how and when they start to form because this may occur long before the main phase of cave growth and under different burial conditions (Lowe, 2000). A time gap arises because the early stages of cavity development can be extremely slow; there is a kinetic threshold to rapid growth once cavities reach a critical size of a few millimetres such that they can support turbulent flow (White, 1977). It is also recognized that it is common for karstification to redevelop cave systems that had initially formed under different conditions (Ford & Williams, 2007; Palmer, 2007). Such caves could have been either mesogenetic or telogenetic, and commonly resist collapse at depth, although might have been lined or filled with sediments or precipitates (see section 2.5).

A key reason for speleothem workers to understand how cave passages develop is to appreciate the context of their samples. The development of caves along particular geological horizons may also provide clues to the pathways followed by dripwater and their composition at low flow (Baker et al., 2000).

2.4.1 Chronologies of cave development

In Chapter 1, we explained the consequences of progressive karstification in terms of lowering of water levels over time, leading to deeper cave development. Such a progression would also be expected where base level is falling, most commonly because of tectonic uplift of land, or isostatic rebound after glaciation. Evidence for relative ages of cave passages can be gained by morphological study and there has also been extensive effort in dating speleothems and determining palaeomagnetic properties of sediments to test these ideas. However, in recent years, cosmogenic isotope dating of coarse-grained fluvial sediment has become the most robust method for dating phreatic conditions in cave passages. The method relies on the production of cosmogenic isotopes in minerals such as quartz by cosmic ray bombardment close to the land surface. On the 10^4–10^5 year timescale, it is the accumulation of the isotopes that is used for dating (Cockburn & Summerfield, 2004). In contrast, on the million-year timescale, the faster radioactive decay of ^{26}Al compared with ^{10}Be provides a reliable age (Stock et al., 2005). Dating of such fluvial sediments should yield results closer to the age of origin of the passage than subsequent vadose speleothem deposition. Stock et al. (2005) provided an example to illustrate this point (Fig. 2.24) demonstrating a clear time sequence from oldest to youngest from highest to lowest cave passages in the tectonically active Sierra Nevada from cosmogenic dating, whereas most dated spele-

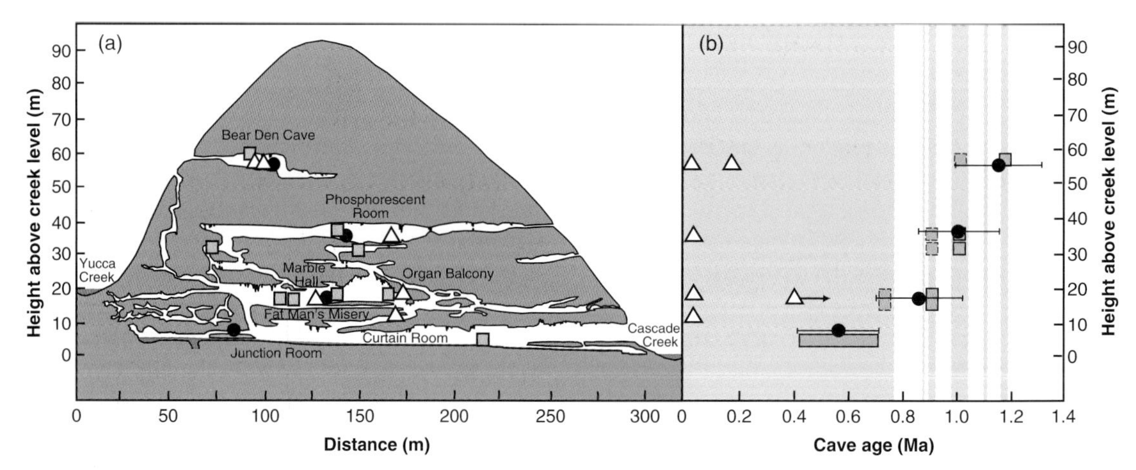

Fig.2.24 (a) Cross-section and (b) age determinations from Crystal Cave, Sierra Nevada, California. The progressively younger age of the caves at lower altitudes is shown by cosmogenic (^{26}Al–^{10}Be) dates on coarse-grained fluvial sediment (black circles with error bars) whereas speleothem U–Th ages (white triangles) show no relationship with height. Grey bars correspond to normal palaeomagnetic polarity and all (mud) samples studied (grey squares) displayed normal polarity—they may be much younger than the coarse fluvial sediment that they overlie. From Stock et al. (2005).

othems grew long after cave formation. This approach applied to cave sediments is yielding valuable information on uplift and denudation rates in diverse regions (Anthony & Granger, 2006; Haeuselmann et al., 2007; Refsnider, 2010). However, speleothems can still form an important part of a multi-technique study (e.g. Westaway et al., 2010) and for the future there is particular potential at sites where U–Pb dating is possible because this extends the age-range significantly (see Chapter 9). Polyak et al. (2008) showed that dating subaqueous speleothems by U–Pb methods provided useful constraints on timescales of landscape evolution in the Grand Canyon region.

2.4.2 Geometry of cave passages and systems

Palmer (1991, 2007) has provided a particularly lucid overview of cave morphology based on a major review of cave systems. Four main types are seen in plan view (Fig. 2.25). The *branchwork* type, resembling surface streams in plan, is the most distinct and common of these (60% according to Palmer, 2007). *Maze* caves display a complex interconnected system and of these, *network* systems are composed of linear segments guided by joints or faults, whereas *anastomosic* systems display less direct routes between junctions. *Ramiform* passages are not associated with a strong linear extension and commonly have small-scale irregularities on their walls, termed spongework. More generally spongework is associated with porous host rocks. A key insight by Palmer (1991) was to relate these patterns to the source of water for dissolution. Quantitative data (Fig. 2.25e) reveal that branchwork caves relate to point sources of recharge, whereas anastomosic systems have multiple dispersed inputs and have a link to floodwater. Both system types are exclusively telogenetic. Networks and spongework have a wide range of associations, including *hypogenic* water (warm water derived

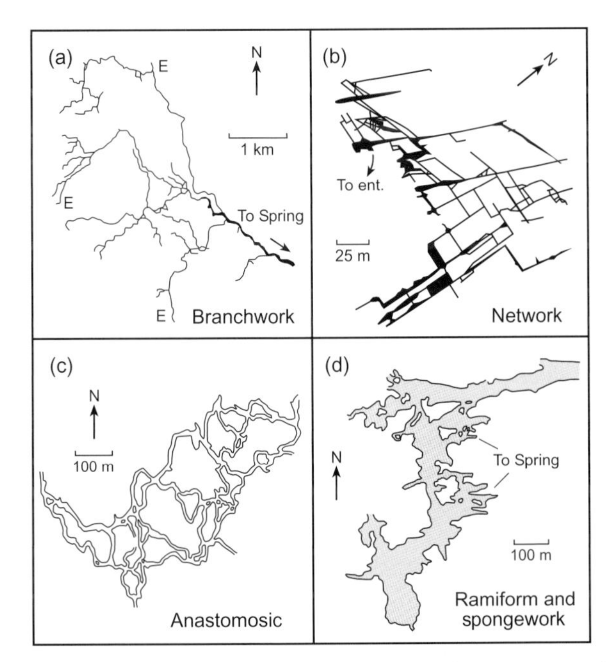

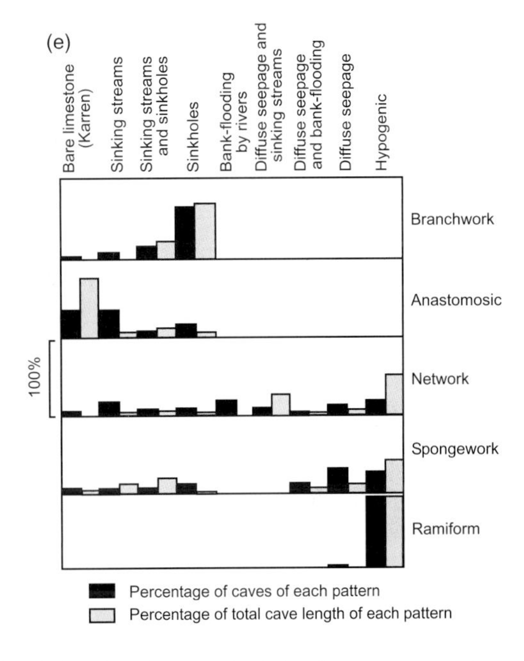

Fig. 2.25 Characteristic plan morphologies of cave systems, their relative abundance and relationship to sources of recharge. (a) Branchwork: Crevice Cave, Missouri. (b) Network: part of Crossroads Cave, Virginia. (c) Anastomotic: part of Hölloch, Switzerland. (d) Ramiform and spongework: Carlsbad Cavern New Mexico. (e) Relationship between cave pattern and type of recharge for 427 caves (total length 2315 km). After Palmer (1991).

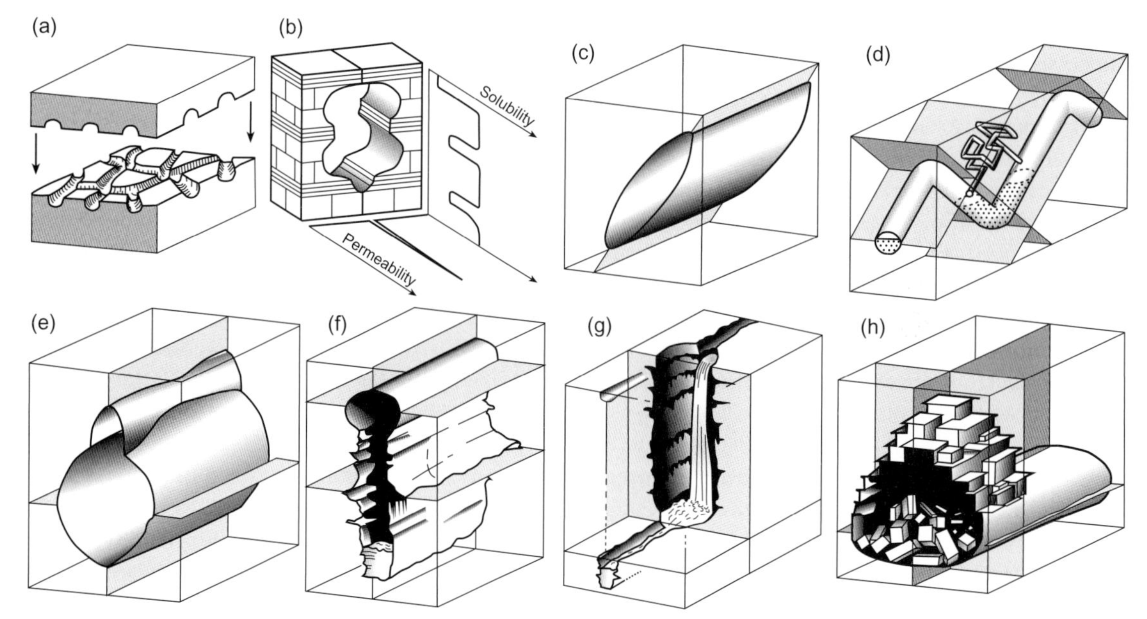

Fig. 2.26 Genetic varieties of cave morphology. (a) Anastomoses signifying slow dissolution along an inception horizon. (b) Phreatic passage cross-section with variations in width corresponding to solubility of wall rock. (c) Phreatic passage elongated along fault or joint. (d) The main zig-zag passage displays phreatic loops guided by joint planes. One region has been filled with sediment and has been bypassed by a tube with corkscrew morphology. (e) Phreatic tube with inner tube or arch on the roof generated either by mixing corrosion or dissolution along a fracture. (f) Keyhole passage generated by vadose downcutting from an initial phreatic tube. The vadose canyon varies in width depending on the solubility of the wall rock. (g) Vadose-enlarged waterfall shaft with fluted walls leading to phreatic passage. (h) Collapse dome caused by breakdown at the intersection of two joints. After Lauritzen & Lundberg (2000).

from below), whereas the ramiform morphology is exclusively associated with hypogene conditions.

Historically there has been much emphasis on the relationship of the cave system to the concept of a water table (Ford & Williams, 1989), even though it is different in nature from the water table characteristic of continuous porous media. This approach is made more difficult by the presence of both vadose and phreatic elements of cave systems, and the fact that they have often developed over an extended period of time during which the base level has changed. Waters near the land surface are likely to be particularly aggressive (undersaturated) to limestone and drainage of fractured limestone develops a self-organizing property whereby larger channels enlarge more quickly, creating a low point on the land surface (a doline) towards which drainage of perched water occurs (Palmer, 1991; Williams, 2008). Dissolution is particularly rapid where there are sinking streams and at high discharge when calcite saturation also tends to be lower. Major vadose system development (e.g. Figure 2.26g) occurs in mature karst where karstification has already led to significant lowering of the water table.

Many actively growing cave passages lie at or just below the water table as conventionally defined; this is a zone where fluid flow is rapid and waters may still be slightly undersaturated. Such passages develop along incipient routes of greatest hydraulic efficiency (Palmer, 1991), reflecting the optimization of the karst system (section 1.4). Ford (1965) drew attention to the important role played by the degree of fracturing of the bedrock: highly fractured bedrock allowed conduits to develop along the water table, whereas in little-fractured rock passages can descend to deep beneath the water table before encountering a suitably orientated fracture to loop back upwards (Fig. 2.26d). Variations within this spectrum were described as the four-state model by Ford and Ewers (1978). It is recognized that the process of dissolution peaks when high discharges of relatively undersaturated water occur and all passages close to the water table are flooded. Hence the ideal form of a phreatic tube is circular in cross-section, but variations are expected depending on the relative solubility of the walls (Fig. 2.26b, c). On the passage ceiling, a distinct narrow 'inner' tube with corrosional depressions can develop locally (Figs. 2.26e and 2.27c) which could be caused either by mixing corrosion (Bögli, 1964) from a fluid of different P_{CO_2} leaking into the cave from a fracture or by other mechanisms such as abrasion or condensation. Forms termed *cupolas*, although similar in appearance, are normally associated with rather deeper well-rounded cuspate dissolution pockets (Osborne, 2004). These structures do not have a preferred orientation and are typical of ramiform caves, generally linked to minimal current activity.

Sudden changes in passage size can result from ceiling collapse (Fig. 2.26h) whereas accumulation of impermeable sediment within a passage can increase complexity. For example, a loop may be bypassed by highly pressurized flows (Fig. 2.26d), or a passage largely filled with sediment may grow upwards by dissolution: a process referred to as *paragenesis* (Fig. 2.27a). Geometrically this is upside-down compared with *keyhole structure* (Lauritzen & Lundberg, 2000), which forms by vadose downcutting from the floor of a phreatic tube (Figs. 2.26f and 2.27b).

Where there is turbulent flowing water, asymmetric flute-like scallop forms develop whose size is inversely proportional to flow velocity (Curl, 1974) and which have been used to inform reconstructions of high flow conditions during cave formation. Recognition of the important role of flood conditions helps to explain the development of maze systems generally where several alternative routes are simultaneously exploited (Palmer, 1991). It is also apparent that flood water height is represented by the locus of the high points in a looping phreatic passage system (Ford & Williams, 1989; Haeuselmann et al., 2007) as shown for example in Fig. 2.28b. In many cave systems developed over a long period of time the system has developed at several levels, which may correspond to stratigraphic (inception) horizons as in Figure 2.28a. Each level corresponds to an inception horizon leading to karst water emergence at springs. Distinct speleogenetic phases of cave development close to the water table with intervening periods of downcutting can be recognized, as has been shown by independent dating (e.g. Fig. 2.24).

Sometimes cave passages penetrate below the water table and then rise significantly to springs. Worthington (2001) explained this phenomenon in terms of lowered fluid viscosity, but this is countered by the reduction in fracturing at depth; nevertheless it does imply that a deeper passage need not be a younger one (Palmer, 2007). However, in some cases this may simply reflect a previous lower base level: a particularly good example is the old systems draining to the desiccated Mediterranean Basin during the Messinian stage of the Miocene around 7–5 Ma (Audra et al., 2006).

2.4.3 Localization of caves: the inception horizon hypothesis

Increasing attention is now being paid to the factors that guide the location of caves. Although it has long been known that caves commonly follow specific bedding horizons, the inception horizon hypothesis of Lowe (1992, 2000) develops this much further by arguing that even for classic telogenetic cave systems, development of proto-cavities is likely to have occurred in the sub-surface during burial. Filipponi et al. (2009) reviewed literature on 18 major cave systems and demonstrated clearly that major cave passages are developed at only a small number of horizons. Work on systems in central Europe provided new data (Plate 2.2) and inspection of the field evidence shows a very specific link to particular levels. Several different characteristics can lead to a tendency for permeability development (Lowe & Gunn, 1997), for example (1) well-developed bedding horizons along which slip can occur and narrow fractures may open, (2) shale or other clastic horizons which retain water

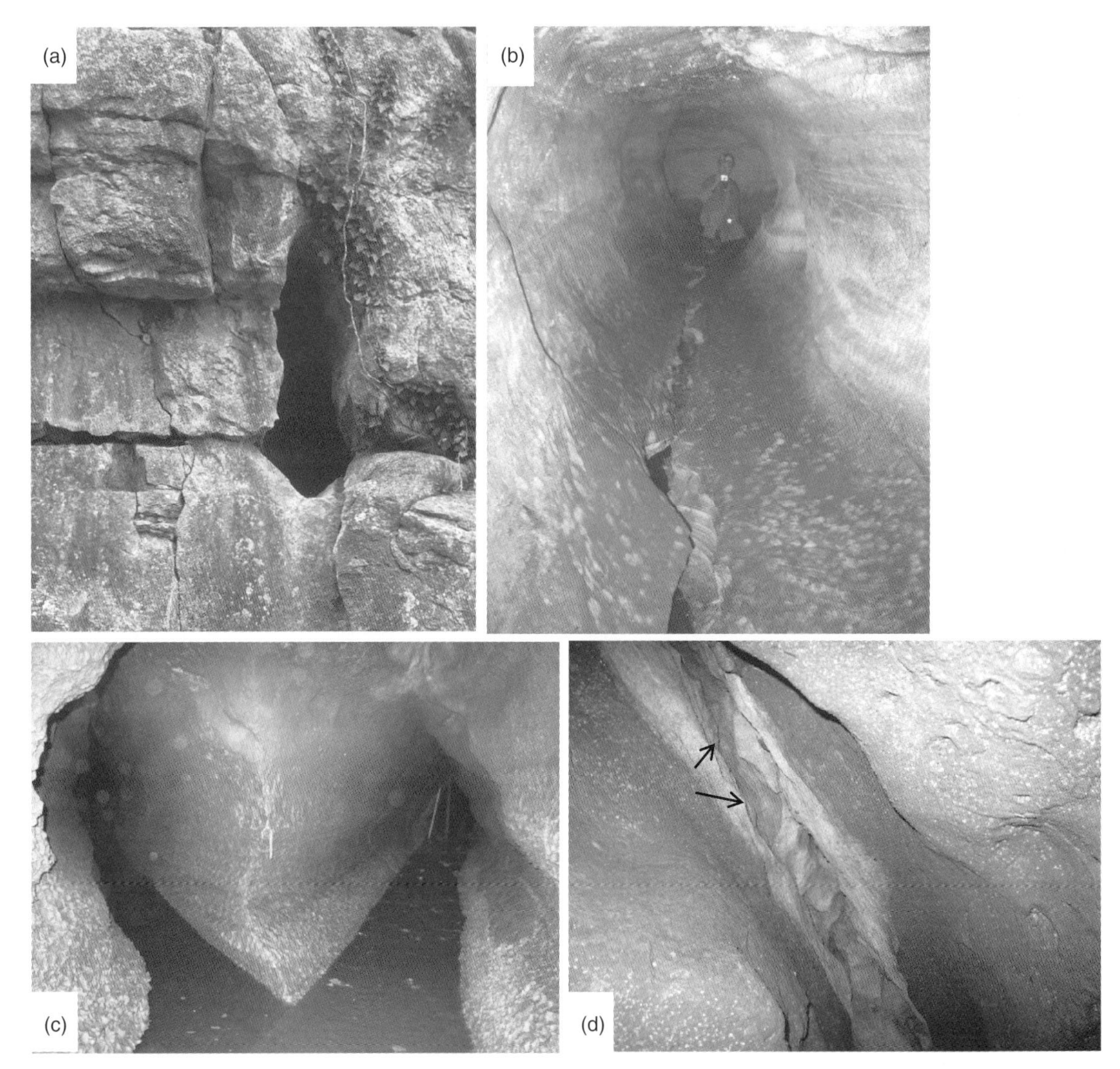

Fig. 2.27 (a) Paragenetic passage, i.e. developed upwards from a phreatic tube, Oughtdarra, County Clare, Ireland (Simms 2004). (b) Cave passage in marble layer within carbonate-poor schist, Grønligrotta, Mo i Rana, Norway. The passage has a keyhole shape of phreatic tube with a vadose notch cut in the floor. (c) Branchwork cave looking upstream. Near-horizontal joint-controlled stream passages in horizontally bedded impure (burrow-mottled) Ordovician limestones, Cherry Grove, Minnesota; tributary junction (width of view 5 m). (d) Same location as (c) looking directly upwards to cave ceiling showing inner tube around 50 cm wide with dissolution bowl morphologies in clean region above mud-streaked walls from a recent flood. Flood water penetrated 11.5 m up a nearby vertical shaft and reached 3 m in the cave, being overlain by pressurized air.

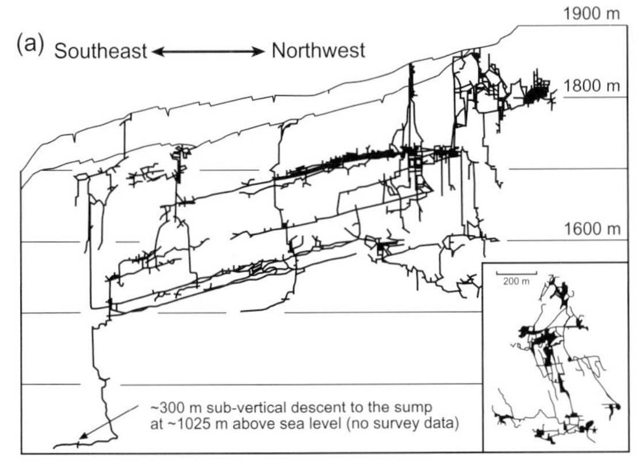

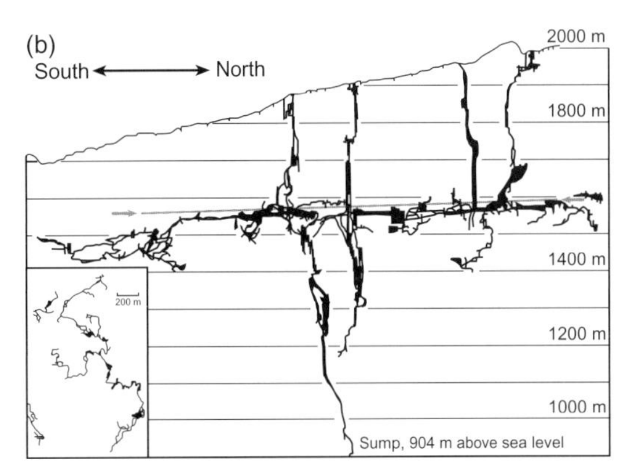

Fig. 2.28 Cave development in the Northern Calcareous Alps, Austria (Plan et al., 2009). (a) Passage development in the Burgunderschacht Cave System as seen in a vertical section looking along the bedding strike. Cave development consists dominantly of passages that follow the bedding (dipping 14° southeast) at a small number of inception horizons. Vertical elements are vadose shafts and shorter phreatic passages. Inset illustrates map of passages illustrating guidance of passage orientation by faulting. (b) Vertical section of the DÖF–Sonnenleiter Cave System. Arrows and grey line indicate the water table of a speleogenetic phase inclined with 1.6° towards the south. Inset illustrates plan view of passage.

and within which diagenetic reactions can generate acid and (3) evaporite horizons. Many of these features occur at repeated levels within cyclic successions and so can be related to the sequence stratigraphic concept. Analysis of permeability from micro-cores across 18 inception horizons by Filipponi et al. (2010) indicates that inception horizons may exhibit relatively high permeability, or sharply bound a more permeable horizon, or be the location of bedding-plane slip. A cartoon model from Filipponi et al. (2009) is illustrated in Plate 2.3 and depicts a phreatic passage developing at an inception horizon and orientated along the intersection of the bedding with joints.

One sign of proto-cave development at specific horizons is the phenomenon of anastomoses (Fig. 2.26a and Plate 2.4) which are geometrical equivalents of spongework (Lauritzen & Lundberg, 2000) and which originate in the absence of strong hydraulic heads. Such macropores are not well studied, principally because most work on burial diagenesis stems from the oil industry (e.g. Ehrenberg & Nadeau, 2005) and such phenomena would not be recovered during deep drilling. However, we do know of several burial diagenetic phenomena that might enhance carbonate dissolution. Anhydrite dissolution leads to Ca-rich waters that create porosity by dedolomitization of impure dolomite rocks (Pezdič et al., 1998). Thermochemical sulphate reduction leads to H_2S generation and petroleum geologists are also well aware of secondary porosity generated in the oil window and thought to be associated with thermal breakdown of organic matter to carbon dioxide and organic acids (Mazzullo & Harris, 1992).

The onset of a telogenetic regime is associated with rock fracturing in relation to tectonic uplift or more subtle variations in the amounts of isostatic rebound because of lateral variations in the extent of erosion. This imparts a driving hydraulic head, even in the absence of compressional folding, allowing fluid movement down stratal dip. The renewal of oxic conditions creates new possible sources of acid for dissolution, notably oxidation of pyrite, particularly focused in shaly horizons.

2.4.4 Mesogenetic caves

Mesogenetic hypogene cave systems represent a different phenomenon in that large caves of distinctive morphology can form and which are commonly associated with distinct mineral phases, such as Pb–Zn sulphides (Klimchouk, 2009).

Carbonate dissolution and metalliferous deposition may be associated with cooling of hot fluids, with the presence of acid in the form of hydrogen sulphide and/or with mixing of hot fluids with sources of water derived from higher in the sediment pile (Corbella et al., 2004). This is the setting for the extraordinary sulphurous ecosystems and diverse speleothems of the famous Lechugilla Cave, New Mexico (Hill & Forti, 1997). It is not normally thought that the mineral fillings associated with such caves have climatic significance, but it has recently been argued by Garofalo et al. (2010) that the giant gypsum crystals of Naica Mine, Mexico, apparently grew in distinct climatic phases during the Pleistocene. It is also commonly found that caves of hypogene origin later undergo modification and speleothem formation under telogenetic conditions.

2.4.5 Modelling the development of conduits and networks

Important early work was performed by forcing unsaturated water through soluble experimental analogues such as salt and gypsum (plaster of Paris) and observing the competition between enlarging conduits (Ford & Ewers, 1978). The inevitable result was the convergence of passages, leading to one particular conduit capturing the main flow and developing much further than the rest. In detail, branchwork systems can be related to different spatial patterns of the initial fissures that become enlarged, as reviewed in detail by Ford and Williams (1989, 2007).

Early work on investigating penetration of calcite-undersaturated waters (Weyl, 1958) led to the result that saturation would be achieved after a short distance and in this view elongated cylindrical cave passages were inexplicable. However, experimental work on calcite dissolution kinetics (e.g. Plummer et al., 1978) revealed that the rate of dissolution diminished greatly but did not reach zero as solutions approached saturation. This is discussed in terms of chemical activities in Chapter 5, but in the karstic modelling literature a rate expression is used of the following form:

$$F \propto (m - [\mathrm{Ca}]/[\mathrm{Ca}_{\mathrm{eq}}])^n \qquad (2.4)$$

where F is the flux of Ca to solution per unit area per unit time, [Ca] is the calcium concentration, $[\mathrm{Ca}]_{\mathrm{eq}}$ is the Ca concentration at equilibrium and m and n are constants within a certain range of undersaturation. Figure 2.29a is a graphical plot of the relationship in eqn. (2.4) illustrating how the dissolution flux varies as Ca increases in a solution with laminar flow (laminar and turbulent flow are defined in Chapter 4). Diffusive transport limits the rate of dissolution according to the thickness of the fluid film (=width of cavity) as shown. Hydration of CO_2 is also a limiting factor for films up to 1 mm thickness (Dreybrodt et al., 1996). The existence of continued slow dissolution close to saturation allows gradual widening of conduits which in turn allows faster flow, an example of positive feedback. Figure 2.29b (Kaufman, 2009) illustrates that eventually widening reaches a critical threshold throughout the length of the conduit allowing faster flow—this is *breakthrough* (Dreybrodt, 1996). Typically, the faster flow is turbulent and contains eddies that resist and slow the flow and cause it to mix thoroughly, which accelerates dissolution. In this way, the victor conduits capture the flow in the karstic system.

Such one-dimensional numerical modelling of conduits was extended to networks by Dreybrodt (1988) and Palmer (1991) and successive two- and three-dimensional models have now successfully reproduced many features of the natural systems and their experimental analogues. Plate 2.5 provides an example using the KARST model of Kaufmann et al. (2010) showing how a network of fine initial fractures, with statistical variability in their properties evolves to a major conduit whose linearity depends on the density of the fracturing. In detail, there are many ways to formulate a model which yield diverse estimates of breakthrough times, although Kaufman et al. (2010) found that several independent two-dimensional models yielded similar results when primed with comparable starting conditions. Generalized statements of model behaviour allow the model sensitivity to be explored. For example, Dreybrodt (1996) found that

$$T = 9 \times 10^{-14} (l/i)^{4/3} a_o^{-3} k_{n_2}^{1/3} c_{\mathrm{eq}}^{-4/3} \qquad (2.5)$$

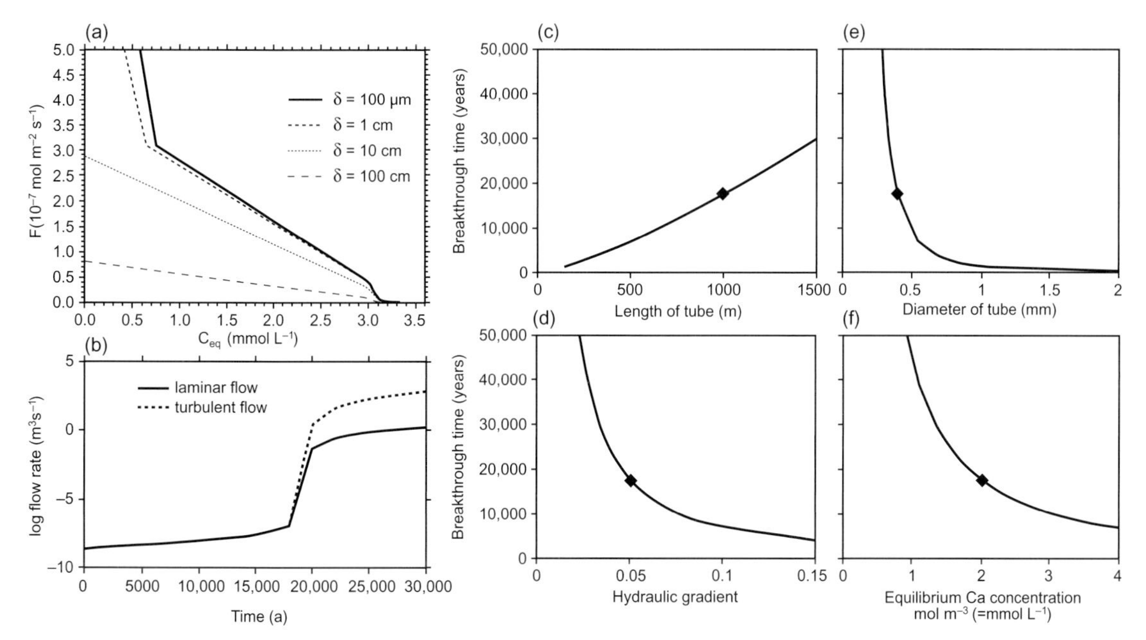

Fig. 2.29 Modelling dissolution kinetics. (a) Nonlinear relationship between dissolutional flux and Ca concentration in relation to thickness of moving fluid layer (δ). The equilibrium concentration in this case corresponds to p_{CO_2} of 0.1 atmosphere at 10 °C. (b) Modelled breakthrough of flow illustrating also strongly increased flow rates where turbulent flow occurs. From Kaufmann (2009). (c)–(f) Breakthrough times for a conduit of specified length (default 1000 m) and original diameter (default 0.4 mm) subject to a specified hydraulic gradient (default 0.05) and equilibrium Ca concentration (default 2 mmol L^{-1}) using generalized results from the model of Dreybrodt (1996). The diamond refers to the default conditions; otherwise the parameter is varied on the *x*-axis. Breakthrough times increase with the length of the tube (c) but diminish rapidly beyond a threshold initial diameter (d), and similarly at high hydraulic gradient (e) and equilibrium Ca concentration (f). Exact numerical values are highly model-dependent.

where *T* is the breakthrough time in years (all other units are in centimetres, seconds and moles), the initial constant is for a circular cross-section, *l* is the length of the system, *i* is the hydraulic gradient, a_o is the initial aperture width and k_{n_2} is around 1.6×10^{-9}. Figure 2.29c–f depicts an exploration of model sensitivity using this relationship. The default starting values are a 0.4 mm diameter tube, a hydraulic gradient of 0.05, with a head of 50 m and length of system of 1000 m and an equilibrium Ca concentration of 2 mol L^{-1}. The most sensitive parameter is the initial tube diameter, which needs more justification in the literature in relation to inception processes, whereas the least sensitive is the system length. It is notable that model results place cave breakthrough in the time frame of 500–20,000 years and subsequent enlargement of passage radius occurs at rates of 0.1–1 mm per year (Palmer, 1991). In detail the rate depends on all the parameters mentioned above, as well as the frequency of flood conditions. Such rates would yield a passage 10 m in diameter in a period of 5000–50,000 years, but over time, floods will find increasing difficulty in filling passages, which will limit the rate of enlargement. Modelling has also verified other concepts such as the role of fracture density in guiding the occurrence of phreatic loops (Kaufmann & Romanov, 2008) and the weakly competitive nature of hypogene karstification (Rehrl et al., 2008). Although modelling has given many valuable insights, the difficulty in prescribing initial and boundary conditions sufficiently precisely will always pose problems in predicting future karst development or retrodicting how quickly a current configuration has been derived. Nevertheless, there are important applications to the potential risk of catastrophe due to rapid conduit development in response to steep hydraulic heads adjacent to dams (Dreybrodt, 1996). A natural example approximating to such conditions is tubes formed in fractured, rapidly isostatically rebounding carbonates in mountainous terrains (Faulkner, 2009).

In summary, it is clear that the commonest type of cave systems, branchwork caves, have the potential to develop fully, from pre-existing inception voids, of millimetre to sub-millimetre diameter, on the 10^4-year timescale during which relatively stable base level conditions may have existed during the Pleistocene. However, repeated base level changes may be expected, potentially exposing lower inception horizons, and there continues to be much scope in using the context of dated speleothems within caves to understand karstic evolution. The relationship of speleothems to clastic sediment accumulation within caves is an important part of this context and we turn to this topic next.

2.5 Cave infilling

In this section we show how speleothems represent just one of several materials that accumulate in caves (Table 2.2). Many types of clastic sediment occur, derived from outside the cave, via transport within the cave network, and by collapse of the roof. Many of these products are emplaced by powerful forces that may also be associated with erosion of previously deposited materials. Determining the dates of start or end of growth of speleothems, or the minimum age at which a speleothem was eroded, can contribute greatly to understanding the geological history of the cave filling in relation to external forcing factors.

2.5.1 Mechanisms of cave infill and their relative power

The sedimentary fill of caves presents a bewildering complexity in terms of the variety of deposits, their rapid lateral changes, and the multi-stage history of deposition and erosion that they display (Fig. 2.30). Plate 2.6 summarizes information on the rates of deposition of different types of cave sedi-

Table 2.2 Cave sediment types (compiled from several sources including Gillieson (1996)). Chemical deposits are discussed in later chapters. An alternative grouping, from an archaeological perspective, is geogenic, biogenic and anthropogenic (Farrand, 2001).

Type	Nature	Origin
Clastic	• Angular gravel • Diamicton (poorly sorted sediment containing mud, sand and gravel) • Layered, sub-angular to sub-rounded gravel and/or sorted sands and silts • Mud (silt and clay)	• Roof collapse or scree deposit • Debris flow • Fluvial (cave stream deposits) or beach for shoreline caves • Suspension deposits (fluvial slackwater) or roof percolation deposits (insoluble residue or soil-derived)
Organic	• Crudely stratified organophosphatic deposits • Discrete organic-rich laminae or bone layers • Anthropogenic	• Bat and bird guano • Transported organic debris • Charcoal from fire; variety of imported materials
Chemical	• Speleothems • Interstitial minerals • Concentric deposits	• *In situ* accretionary precipitates • Mineral cements • Mobile mineral growths (pisoids)
Ice	• Speleothems, crusts, rime	• *In situ* precipitates from dripwater; freezing of standing water; condensation deposits

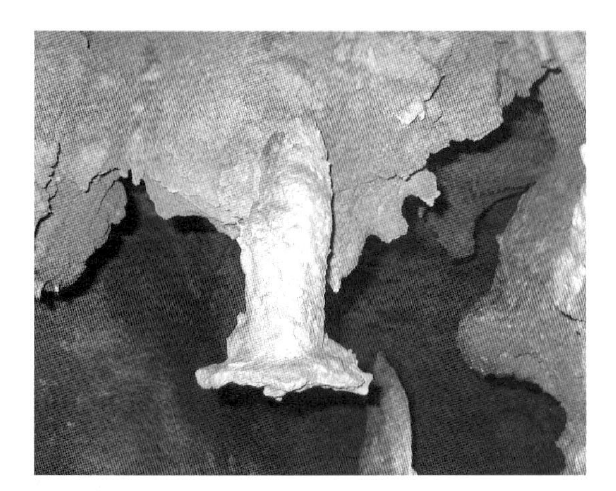

Fig. 2.30 Stalagmite that originally accumulated on a sediment floor which has since largely been eroded, leaving it suspended above the current floor secured by overlying sediment (Grotte de Villars, France).

ment. Processes that lead to quasi-instantaneous deposition of significant thicknesses of sediment take precedence over slower processes, and can be associated with erosion or deformation of pre-existing sediment. In Fig. 2.31 the slower processes, including speleothem formation, are differentiated according to cave hydration, from waterlogged through to dry. A common mode of preservation of vertebrate debris (Simms, 1994) is sub-vertical transport through cavities ranging from small crevices, to large sinkholes, either stimulated by percolating dripwater, or by gravitational instability of dry sediment. Flowstone deposition can alternate with deposition of fine clastic sediments; unsurprisingly flowstones, and sometimes stalagmites, can contain disseminated or thin-layered clastic sediment (Plate 2.7 and Fig. 6.2; Dasgupta et al., 2010; Zhornyak et al., 2011).

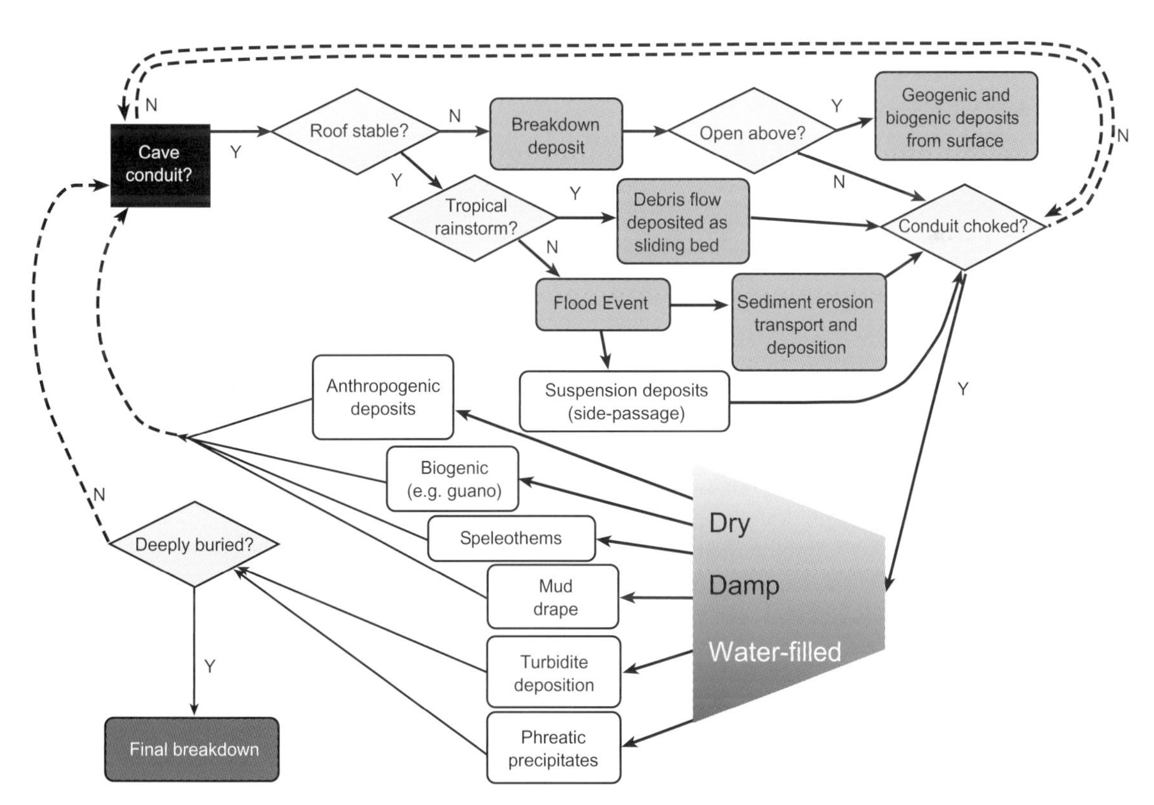

Fig. 2.31 Flow diagram to illustrate the precedence of powerful agents of sedimentation such as roof collapse, debris flow and river floods over slower depositional processes which are differentiated according to the hydrological state of the cave. See text for separate discussions of the nature of each of the types of sedimentation.

2.5.2 Dating the infills

Many caves show a history of events spanning more than 10^5–10^6 years. The longer timescales have usually been demonstrated from study of clastic sediments, but recent work on U–Pb dating of speleothems is now also proving to be important (Chapter 12). Distinction of normal and reversed magnetization of sediments, for example, has proved an important tool in demonstrating their great antiquity; for example, extensive work in modern caves in Slovenia, Slovakia and the Czech Republic has established that many sediments pre-date the traditional base of the Quaternary at 1.8 Ma (Bosák et al., 2003; Hajna et al., 2008a, b). In areas of active downcutting, where younger caves are found at lower elevations, such approaches can be used (alongside others, Farrant et al., 1995) to deduce the rate of base level lowering. For example, Sasowsky et al. (1995), working in the Cumberland plateau of Tennessee, deduced that around 50 m of fluvial incision had occurred since the end of the Jaramillo event (the penultimate period of normal polarity) at 0.9 Ma. However, more generally the age of deposition could be ambiguous because the composite sediment record may not record all the palaeomagnetic chrons (time periods defined by reversals). More commonly nowadays, cosmogenic isotopes are used to provide ages of cave formation, as discussed in section 2.4. An excellent example of the combined use of cosmogenic dates on sediments, U-Th dates on speleothems, and descriptions of stratigraphic data is the study by Lundberg and McFarlane (2007) of Kent's Cavern, Devon, UK, a site of great archaeological importance. Recurrent flowstone growth in interglacials is very well documented. It is interrupted by characteristic reworking and cold-stage sedimentation and, at Kents Cavern, evidence of periglacial disturbance. Lundberg et al. (2010) document another style of cold-stage deposition in-between warm-stage flowstones at another classic UK site (Victoria Cave, North Yorkshire): that of varve-like laminites formed under ice-damming conditions. Here the cave record as a whole provides by far the most complete Quaternary record in the region.

2.5.3 Physical sedimentology

The work of Gospodarič (1976) (Figs. 1.7 and 2.32) is a good exemplar of a classical approach to cave sediments and there has continued to be a focus on mineralogical and size distribution studies of provenance and bulk sediment properties (Sasowksy & Mylroie, 2004). A complementary aspect is the use of hydrodynamic theory to quantify flow properties (Gillieson, 1996; Bosch & White, 2004), which has proved invaluable in the study of surface sediments (Allen, 1985). However, a problem identified is the effect of rapid lateral changes in bed slope and sediment type on estimated bed roughness, which is a crucial parameter for determining sediment erosion and bedform development. Sediment textures are highly variable owing to the role of specific flood events and changes along passages, which combined with the effects of reworking, makes the correlation of individual events and estimates of flow velocity and discharge difficult. Somewhat disconcertingly, despite some careful studies by Gillieson (1986, 1996) and Valen et al. (1997) among others, the application of conventional sedimentary facies analysis is very under-developed in cave science and contrasts markedly with the range of sophisticated analyses of alluvial deposits above ground (Miall, 1996; Bridge, 2003). Although there are valid excuses (logistical difficulties in working underground, the obscuring effects of pervasive mud deposition, only sporadic presence of cleanly eroded sections, rapid lateral changes), there is much to be done. An excellent example of what can be achieved in terms of distinguishing sedimentary processes and facies is illustrated by the study of Ghinassi et al. (2009) in Romito Cave, southern Italy.

Clastic deposits show peculiarities related to grain size issues, which are accentuated compared with surface environments because of the complex geometry of caves. For example, there is a tendency for sand to be removed by reworking, whereas gravel requires a significantly higher shear stress to be transported. Nevertheless, van Gundy and White (2009) described the example of a 1995 flood through Mystic Cave, West Virginia, where, despite the local trapping of suspended cobbles by

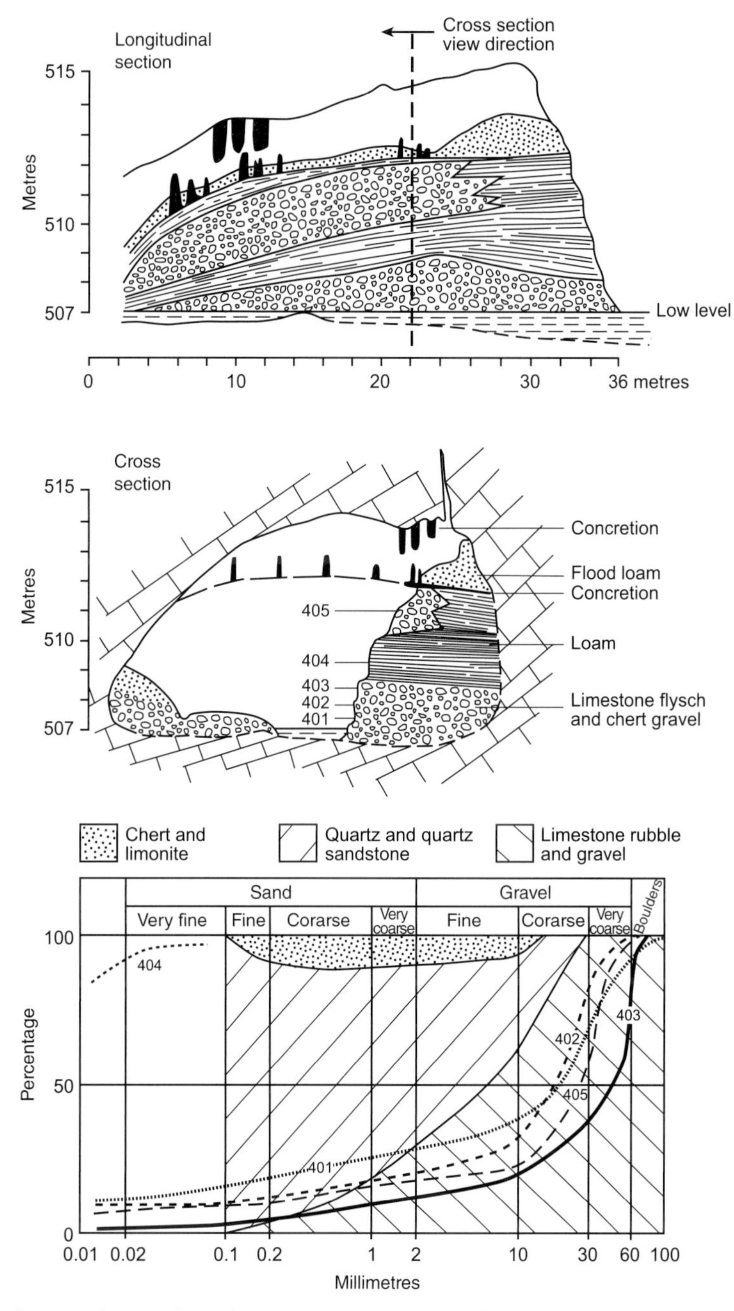

Fig. 2.32 Sections of cave sediments from the Postojna cave system, Slovenia (used to construct the model history of Fig. 1.7) together with examples of grain size distribution and lithological composition of sand and gravel fractions. Note that the grain size divisions are not identical to those conventionally used in sedimentological literature. After Gospodarič (1976) and Gillieson (1996).

active speleothems two metres above the bed, very little sediment was deposited during the event. This attests to the power of caves with an appropriate geometry to act as by-pass systems and for speleothem growth to occur in places where significant flood events occur. Whereas mud only slowly drops from suspension, it is difficult to re-erode because it presents a smooth boundary layer and becomes more cohesive on drying. Mud is also deposited in backwater settings, such as short side-passages with the result that a more complete record of sedimentary events can be retained. Renault (1967–1968) also linked mud sedimentation to paragenetic passages in which the passage grows by upward dissolution of the roof and the passage has a broad, low profile. Mud drape successions in largely filled cave passages can have some distinctive properties as described by Bull (1981). He demonstrated at a site in South Wales some 200 m below ground surface, that parallel-laminated muds accrete to the roof of cave passages and on slopes up to 90°. The drapes contain widely correlatable millimetre-scale laminae interpreted as infiltration events from multiple fissures and more minor cracks in the karstic bedrock. Continuous fine sediment sequences provide excellent material for the study of stratigraphic variations in magnetic susceptibility. The magnetic carrier is magnetite, which is synthesized by soil bacteria at rates that are climate-dependent. Ellwood et al. (1997, 2001) argued that the susceptibility displays systematic correlatable variations, owing to the inwash of soils containing amounts of magnetite which vary over time, being more abundant during warmer, moister conditions. An additional source of sediment is that of breakdown of the cave walls. Hajna (2003) described Slovenian examples and demonstrated the production of a bleached and leached (but mineralogically unaltered) surface layer on limestone walls, especially adjacent to fluvial sediment or in areas subject to condensation corrosion. The surface layer is mechanically weak and is subject to fluvial erosion, from which fine carbonate particles are derived.

Gillieson (1986) demonstrated that diamictons up to several metres thick in caves of the New Guinea Highlands arise from discrete events linked to generation of surface mudflows and were likely emplaced as sliding-beds extruded through the cave passages. These units are separated by laminated muds with interbedded flowstone horizons. Another facies likely characteristic of tropical environments with occasional intense rainfall events are turbidites (Osborne 2008), which reflect sediment surges into poorly drained cave sites that are difficult to access. Accordingly, they are best known in palaeokarsts. Graded marine sediment fills are known, for example the dolomitized semi-precious deposits known as caymanites (Jones, 1992), and can originate either during marine transgression or through storm or tsunami activity.

Cave breakdown is a crucial process generating piles of coarse angular sediment as a result of failure of the roof. When water-filled, cave systems benefit from the effects of hydrostatic pressure which distribute stress uniformly. However, an air-filled cavity represents a significant defect in the stress field of the karstic host rock (Davies, 1930; Jennings, 1985) and the cavity is overlain by a domal zone bounded by the surface of maximum shear stress (White, 1988) in which tensional stresses may result in failure. The largest cave chambers have a beehive-shaped ceiling representing the most stable configuration (White, 1988). The seminal treatment by Davies (1951) used beam mechanics to analyse the conditions for failure and, depending on the bending strength of the limestone, stability decreases with wider ceilings, thicker beds, or beds with free edges which behave as cantilevers rather than fixed beams (White, 1988). In practice, the geological structure is crucial in the style of breakdown, e.g. alternating hard and soft beds promote its occurrence (Sweeting, 1972). A critical liability to failure can arise through progressive dissolutional enlargement of fractures, undercutting, seasonal flooding, drying accompanied by salt crystallization or frost action (Sweeting, 1972; Jennings, 1985; White, 1988). Alternatively, the model of Tharp (1995) which considers that failure occurs following critical progress of creep induced by microcrack propagation, implies that no specific trigger is required (White & White, 2000). Although there is archaeological evidence in seismically active areas for human burial as a consequence of sudden roof collapse (Nur, 2008), earthquakes do not necessarily result in collapses.

2.5.4 Archaeological issues

Distinctive deposits form in limestone rock-shelters and cave-mouth environments (Figs. 2.33 and 2.34) and form a major archaeological resource (Lewin & Woodward, 2009). Woodward and Goldberg (2001) emphasized that palaeoclimatic interpretations are more feasible with rock shelters of active karstic environments where there is water and sediment supply from the karstic host rock, vegetation develops, and humid conditions promote frost-breakdown. The entrance zone is commonly the site of a sloping depositional surface (into or out of the cave) contain a variety of slope-derived sediments, including sediment produced by frost shattering, and aeolian deposits, as well as reflecting modification by human and animal activity, given the importance of caves as shelters (Plate 2.8). In contrast to optimization for palaeoclimate, passive (dry) environments are better for preservation of materials that can be accurately dated (speleothems excepted). Wind-blown detritus, where it is the dominant constituent of the cave deposits, can be dated by luminescence dating techniques, whereas tooth enamel can be dated by electron spin resonance and burned debris by radiocarbon and/or thermoluminescence (Schwarcz & Rink, 2001). These techniques show that accumulation of sediment is complex and can be episodic (Farrand, 2001); hence it is important to compare with continuous records in the region (Woodward & Goldberg, 2001).

Hiatuses in archaeological sites at cave entrances and within caves can be recognized by phenomena such as weathered zones (e.g. discoloured, leached of $CaCO_3$ or other soluble species, highly weathered mineral grains) or bioturbation (Farrand, 2001). The study of micromorphology, using thin sections of resin-impregnated sediment, including electron microscopy, and sometimes supplemented by field-adapted Fourier transform infra-red spectroscopy, clarifies the origin of sediments, helps delineate anthropogenic from natural processes, and identifies key processes such as freeze-thaw action and phosphatization of calcite to be recognized (Karkanas et al., 2000, 2008; Goldberg & Sherwood, 2006). Shahack-Gross et al. (2004) showed that such destruction of calcareous remains was promoted by acid solutions leaching from overlying guano, which itself breaks down in the process. Preservation of guano and associated salt deposits is possible under slightly drier conditions. Bird et al. (2007) gave the first description of the recovery of multiple palaeoclimate proxies from a guano accumulation (Philippines; last glacial and Holocene). Accumulation rates were of the order of 0.2 $mm\,yr^{-1}$ and the proxies included both stable carbon and nitrogen isotopes.

The diversity of archaeological sites in caves is instructive. For example, key work on early activities of *Homo sapiens* comes from study of former sea caves developed in quartzite in South Africa. The deposits, which date back to over 160 ka, were dated by a combination of U-series on flowstone, and optically stimulated fluorescence on quartz, in aeolian dunes interstratified with the chemical deposits. In this case, the very existence of calcareous speleothems owes to leaching from calcareous dunes which overlie the caves and sealed them for much of their history. The human activity is restricted in timing to intervals of relatively high sea level, but migration occurred towards the moving coast at times of lower sea level (Marean et al., 2007).

2.5.5 The long-term prognosis

The ultimate fate of caves is either (1) to be subaerially eroded (Fig. 1.12b) or (2) to persist at shallow burial depths incompletely filled or (3) to collapse and/or be cemented during deeper burial. An example of (1) is the tendency for rock shelters to evolve and degrade over time, as shown in Fig. 2.34. Another example is illustrated by the Qesem cave system of Israel, which is starting to erode away after a long history of filling during which significant sagging of sediments into underlying voids occurred (Frumkin et al., 2009).

Fates (2) and (3) relate to the phenomenon of palaeokarst (James & Choquette, 1988), which refers to the occurrence of ancient karstic phenomena within a limestone unit that is subsequently buried further in the sedimentary basin, although

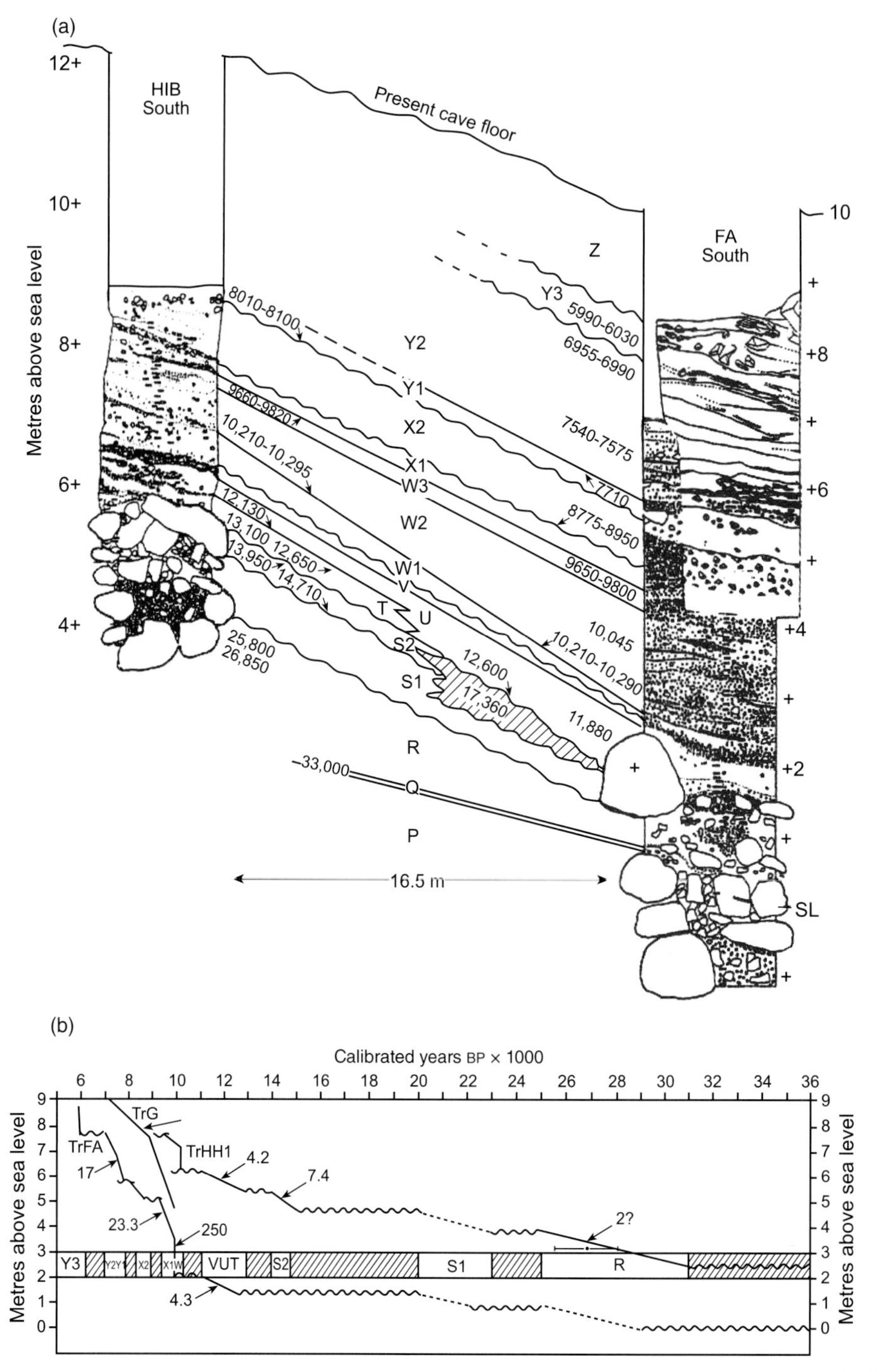

Fig. 2.33 (a) Sediment sections and (b) reconstructed sedimentation rates (base) from trenches FA and HH1, Franchthi Cave, Greece, illustrating variable sediment types and accumulation rates with significant hiatuses indicated by cross-hatching (Farrand, 2001). Very high (anthropogenic) accumulation rates are found in the Mesolithic around 10 ka. Correlation was aided by rocky layers X1 and V and volcanic tephra horizon Q. U and S1 are weathered horizons. Ages are from calibrated ^{14}C dates. Cross-hatching in (b) indicates time gap.

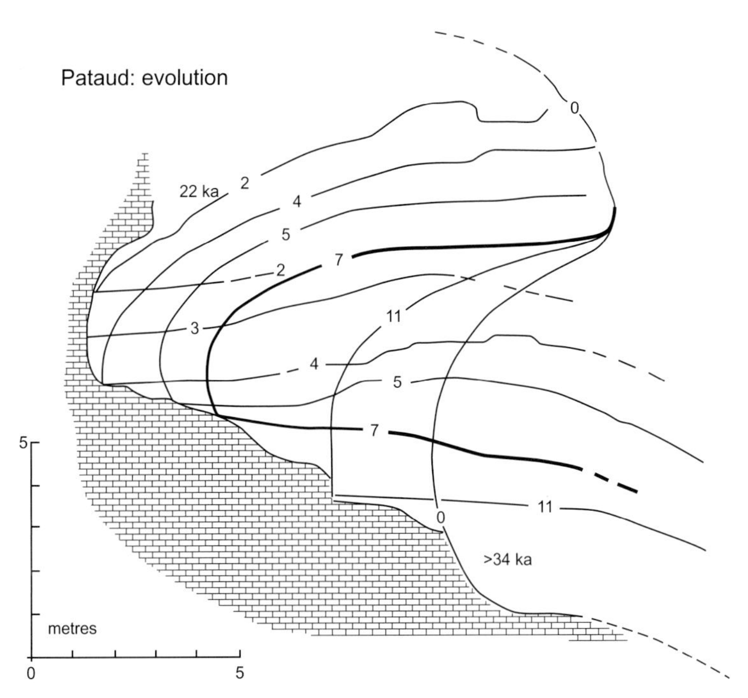

Fig. 2.34 Reconstructed profiles of the Abri Pataud rock shelter (Dordogne, France) between 34 and 22 ka. Profiles renumbered from Farrand (2001). The overhang reached a maximum around occupation '3', but subsequent collapses have reduced it to a modest size.

there are many variations on the definition (Osborne, 2000). At burial depths of up to a few kilometres, caves can resist collapse indefinitely, but they may gradually become lined with mineral cements (Fig. 2.35, right top and middle). More unstable cavities are progressively eliminated during burial from hundreds to thousands of metres depth (Plate 2.9) and Loucks (1999) describes a characteristic set of properties displayed by collapsed caves, including distinctive breccia sequences in which chaotic breccias (becoming re-brecciated by mechanical compaction) passes up to mosaic and crackle breccias, and then into undisturbed bedrock (Loucks, 1999, and Fig. 2.35 left, and right bottom). The network of discontinuities including cavities and sediment-fills created by karst processes can be influential in the subsequent burial history of the carbonate rock, including its capacity as an oil reservoir and a repository for mineral deposits. Exhumation of karstic carbonates containing palaeokarst and collapsed cave levels can give rise to complex palimpsest effects whereby ancient karstic features can guide the development of the new system, particularly for caves development by ascending water; another effect is the chemical disturbance associated with mineralized palaeokarst rocks that can lead to local changes in speleothem mineralogy (Osborne, 2000). The idea of studying ancient speleothems in palaeokarst to learn about climatic conditions deep in geological time is attractive, and examples are known (Loucks, 1999), but most palaeokarst is not associated with well-developed speleothems (see, for example, Smith et al., 1999) presumably because speleothem development occurs in the upper, vadose part of the karst system which has a greater probability of being eroded. Nevertheless this is an area in which progress is being made as discussed in Chapter 12.

2.6 Conclusion

We hope that this chapter has demonstrated answers to questions about how specific properties

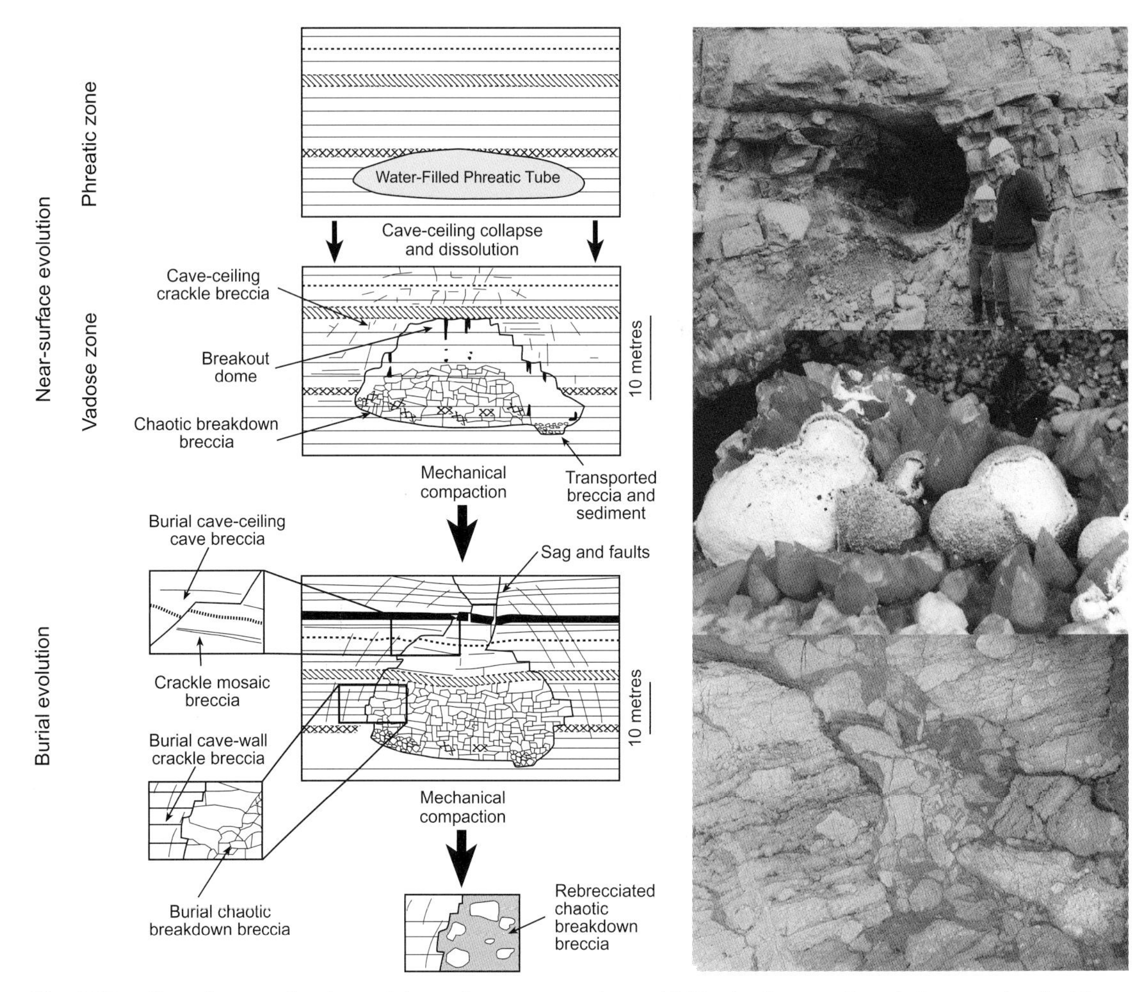

Fig. 2.35 Collapsed-cave palaeokarst. Schematic characterization of phenomena associated with cave collapse in the subsurface (Loucks, 1999). Photographs: top, open phreatic tube in Carboniferous limestone lined with crystals formed during burial, Olveston, UK; middle, as above, close-up of calcite crystals and white botryoidal barite; bottom, Brecciation associated with collapse of palaeocave exposed on wall of modern cavern, Jenolan caves, New South Wales, Australia. Relatively unmodified strata dip at low angle to left and are interrupted by disorganized breccia. Field of view approximately 3 m.

of speleothems may relate to properties of the bedrocks (e.g. matrix porosity, fracture systems, bedding orientation, faults, lithology distribution). To do this, we have covered a wide range of topics in order to demonstrate the common geological contexts in which speleothems occur. The opportunity to take such contexts into account has often not been taken in palaeoclimate research on speleothems to date. On the other hand, much work on the evolution of cave spaces and their fillings has had quite a specific focus, either on mechanisms or the local context, and has typically not been aimed at palaeoclimate workers. A more holistic approach in future, bridging different research communities, is to be encouraged, in the same way that speleothem workers have already shown the valuable part they can play in helping archaeologists to understand the significance of their discoveries (e.g. Bar-Matthews et al., 2010).

CHAPTER 3
Surface environments: climate, soil and vegetation

This chapter provides an overview of the modern surface environment, focusing on the processes that determine the climate, development of soils, and vegetation patterns. Our aim is to focus on those aspects of the surface environments that either determine or affect the speleothem palaeoenvironmental archive. For more detailed treatments of meteorology, climatology and atmospheric processes see, for example, Henderson-Sellers and Robinson (1999), Ruddiman (2008), Solomon et al. (2007), and Barry and Chorley (2009); for soils Goudie (2006) and Retallack (1998); for terrestrial vegetation and ecosystems, Campbell (1996).

For speleothems to form there has to be groundwater recharge and hence atmospheric precipitation. Therefore in section 3.1 we consider the climate system using the classic energy balance approach, leading to an explanation of the global patterns of temperature, rainfall and water excess. We also consider ocean circulation and ocean–climate interactions that lead to phenomena such as the Southern Oscillation and North Atlantic Oscillation, before concluding with an overview of modern-day climate classification and the concept of climatological 'hot spots'. Section 3.2 considers the specific case of water isotopes in the modern environment; these, in particular oxygen when incorporated in calcite, are the pre-eminent speleothem palaeoclimate proxies. Section 3.3 considers the soil system, with a particular focus on processes, leading to an understanding of the soils that one might expect to find in karst regions and their evolution over time. Reactions such as the leaching of soil minerals or the acidification of soil water through microbial and root respiration, may determine, or are the sources of, some of the speleothem palaeoenvironmental proxies to be considered in later chapters. The production and concentration of soil CO_2 is also considered in detail, being the driving force behind carbonate dissolution. In section 3.4 we consider modern vegetation patterns, again with specific reference to karst regions, with a particular focus on understanding changes over time and their sensitivity to climate variability, important concepts for speleothem records. Section 3.5 synthesizes concepts introduced in this chapter. The atmospheric and surface environments are the domains where the raw materials for the speleothem factory are processed, before being delivered to the underlying speleothem incubator (Chapters 1 and 4).

3.1 The modern climate system

3.1.1 The global energy budget

All aspects of the climate system result from the energy transfers and transformations within the Earth/atmosphere system. This process originates with solar energy which is either transferred through the atmosphere or interacts by reflection or absorption. Variations in the amount of energy

Speleothem Science: From Process to Past Environments, First Edition. Ian J. Fairchild, Andy Baker.

radiation from the Sun, and in the interaction with the Earth and atmosphere, create the spatial and temporal variations in energy exchanges that lead to our climate. Quantitatively, approximately 30% of incoming solar energy is reflected back to space (the Earth's albedo), 43% is converted to sensible heat, mostly via absorption at the Earth's surface and re-radiation in the infrared, 22% is converted to latent heat in the form of water in liquid or vapour, and 5% is stored in other forms of energy via photosynthesis, organic decay, tides and currents, convection, hot springs and thermal energy. It should be noted that there is a large temporal variation in the energy fluxes: surface advection can occur on a daily scale whereas thermal energy in the form of fossil fuel may take geological timescales to recycle. All of the above energy transfers must balance for the globe as a whole, otherwise climatic change will occur (Fig. 3.1). Today there is an imbalance, with the relatively rapid addition of significant concentrations of greenhouse gases into the atmosphere; long-term imbalances have occurred throughout Earth history over a variety of timescales.

The amount of energy received at the top of the atmosphere is called the solar constant, and has a global value of 1366 W m^{-2}, averaged parallel to the direction of radiation or 342 W m^{-2} averaged over the entire Earth's surface. However, it is not really constant and varies over several timescales, from that of solar evolution, through changes in the Earth's orbit over millions of years down to changes in solar activity over the time frame of sunspot cycles. The latter occur at approximately 11 and 22 year periods and may affect the total energy emitted by 0.2 W m^{-2}, which influence upper atmosphere temperatures and have been estimated to cause surface temperature changes of only a few tenths of a degree (Le Treut et al., 2007). Far more important are the spatial and seasonal variations in the

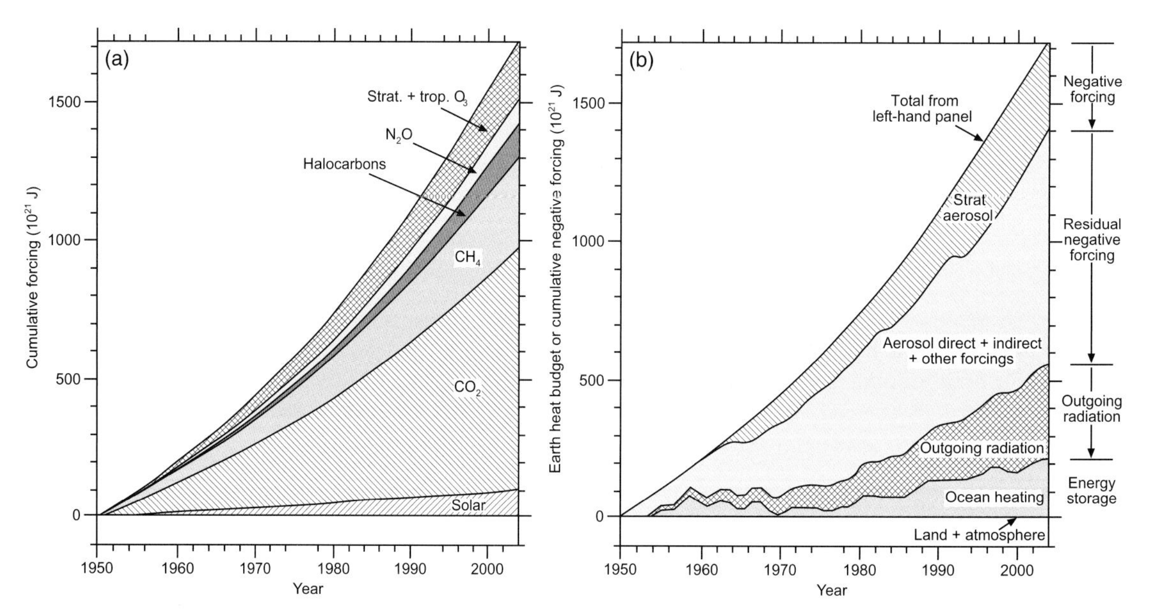

Fig. 3.1 A best estimate of the cumulative forcing of the Earth's climate since 1950 AD based on observational evidence (Murphy et al., 2009). (a) The sum of the long-lived forcing agents from 1950 to 2004 AD, which are mostly positive forcings and include increasing anthropogenic emissions. (b) The Earth's heat budget, demonstrating how the effect of the positive forcings has been balanced by stratospheric aerosols, direct and indirect aerosol forcing and an increased outgoing radiation from a warming Earth, with the amount remaining heating the Earth. The aerosol direct and indirect effects portions are not measured and are calculated as a residual after computing all other terms.

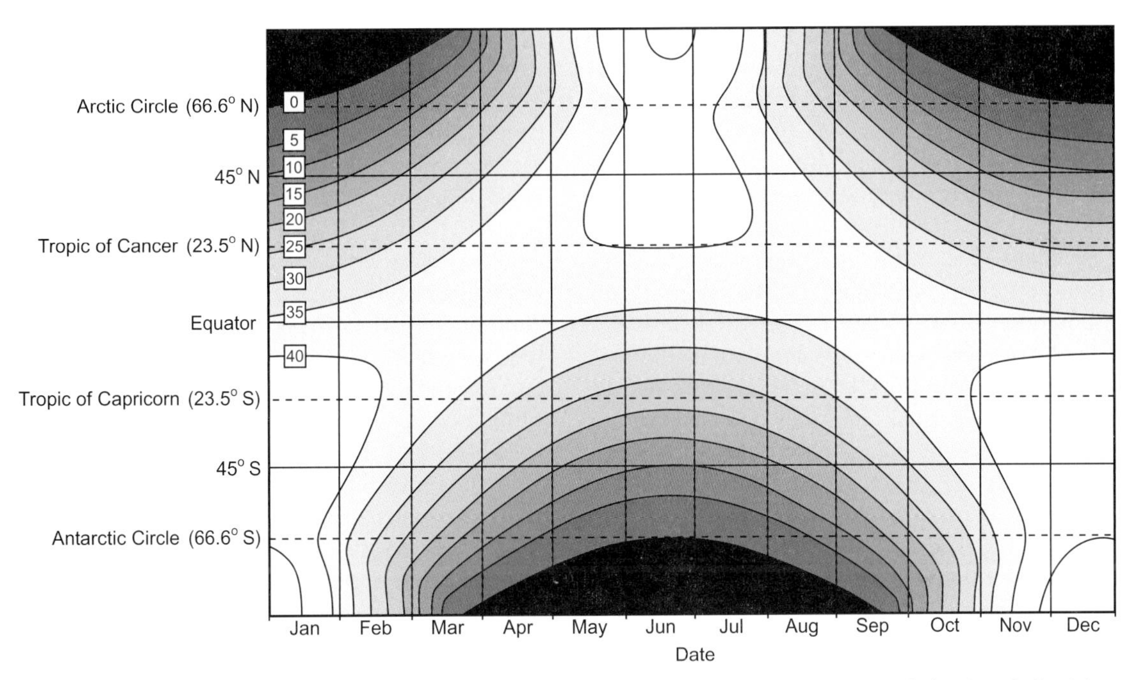

Fig. 3.2 Variation in insolation at the top of the atmosphere as a function of latitude and month for the whole globe. Image credit: NASA Earth Observatory (http://earthobservatory.nasa.gov). Contours are in megajoules per square metre.

receipt of solar radiation across the Earth's surface owing to the elliptical orbit of the Earth around the sun and its axial tilt. These cause seasonal variations in the radiation reaching the Earth in the Northern and Southern Hemispheres (Fig. 3.2). At high latitudes, it is evident that maximum insolation occurs in June (Northern Hemisphere) or December (Southern Hemisphere). Maximum insolation occurs where the sun is directly overhead, the position of which varies seasonally from 23.5° N in June to 23.5° S in December. The effect of orbital variations over 10^5–10^6 year timescales is considered in more detail in Chapter 12.

3.1.2 Global patterns of temperature, rainfall and evapotranspiration

The insolation that is absorbed and re-emitted as long-wave radiation dictates the temperature pattern over the globe, with highest temperatures near the equator and a decrease towards the poles (Plate 3.1). There is also a straightforward link between the energy at the Earth's surface and the energy available for potential evaporation (PE) or evapotranspiration (PET). PE/PET at any moment is controlled by four factors: the energy available, the humidity gradient away from the surface, the wind speed immediately above the surface and water availability. Any one of these four factors may limit PE/PET, but to the first order one can estimate the amount of PE knowing just the temperature and the amount of solar insolation as first shown by Thornthwaite (1948). Evaporation and evapotranspiration is of particular importance to the speleothem palaeoclimatologist, as groundwater recharge is the absolute fundamental necessity for speleothem growth to occur. Evaporation and evapotranspiration reduce the amount of water available for groundwater recharge.

Evaporated or evapotranspired water enters the atmosphere as water vapour. The air can only hold a limited amount of water vapour before it becomes saturated, depending on air temperature (section

4.2.2 and Fig. 4.3). When saturation occurs, the water vapour gives up energy and forms liquid droplets, which can ultimately form clouds and fall as precipitation. Precipitation duration and intensity depends on cloud type, which in turn depends on the temperature and therefore the energy budget. Thus, to the first order, we should be able to generalize the global distribution of precipitation (Plate 3.1). In the Tropics, high precipitation (>2000 mm) is related directly to temperature and convective activity, and short duration, high intensity storms occur from cumulus type clouds. In some places, seasonal monsoons occur because of land–ocean heating differentials driving convection (see section 3.1.3). In mid-latitudes, rainfall is associated more with depressions and fronts (see section 3.1.2). Rainfall rates are variable but generally of lower intensity than equatorial areas. Convective storms generally only occur in the warm summers. In high latitudes, low precipitation is associated with a lack of atmospheric moisture, as well as a lack of uplift mechanisms to cause cloud formation.

Knowing the global patterns of precipitation (P) and potential evaporation (E, more completely evapotranspiration) means that we can now plot the global distribution of water excess, or $P - E$, which is the water which is potentially available for groundwater recharge once any soil moisture deficit is overcome. Plate 3.1 shows $P - E$ for both January and July: $P - E$ is positive in equatorial regions where rainfall amounts are high, but negative in the adjacent low latitudes where evaporation exceeds precipitation and leading to desert-like conditions. Water excess is again positive in mid-latitudes, with amounts decreasing towards the poles as precipitation amounts decrease. To the first order, as the $P - E$ patterns presented in Plate 3.1 should relate to groundwater recharge, then a correlation with the abundance of speleothems in caves should be expected.

3.1.3 The general circulation of the atmosphere

An understanding of the general circulation of the atmosphere is of primary importance to any palaeoclimatologist, as the general circulation is the product of the global energy budget and has a strong influence on speleothem proxies such as $\delta^{18}O$ via the composition of precipitation. The general circulation transports moisture and energy across the globe and rebalances the spatial and temporal imbalance of energy receipt (Plate 3.2). The general circulation generates wind movement, another characteristic of the global climate. The general circulation may be considered in both the atmosphere (wind movement) and ocean (currents); both are important in transport of energy. The scale of the general circulation can also be considered: primary features are persistent, large-scale features that vary in detail but are permanent. Secondary features are short lived, such as cyclones and anticyclones. Evidence of both may be preserved in speleothem proxies.

The general circulation occurs because low latitudes absorb more energy than they emit, and thus have a positive energy balance. At latitudes greater than 30° N and S, the opposite occurs. To balance this imbalance, energy must be transported and the general circulation does this. Figure 3.3 shows the relative importance of different energy fluxes at different latitudes. The total energy flux is highest midway between poles and equator. Dynamic transport, through wind motion, makes up about 60% of total energy transfer, latent heat flux, through moisture transport, accounts for 15%, while ocean energy flux, through currents, accounts for the remaining 25%. Some constraints on the general circulation are (1) there must be a overall global water balance, (2) atmosphere mass much be maintained, (3) the Earth's angular momentum must be maintained (therefore there must be a balance between easterly and westerly winds).

Sea-level pressure exhibits variations over space, and Plate 3.1 (row 4) shows *isobars* (lines of equal pressure) and wind vectors. Pressure is low in regions of high temperature (following the ideal gas law where the product of pressure and volume divided by temperature remains constant) and low in regions of ascending air, where mass is being removed from the surface. These pressure patterns are persistent, and are thus responsible for primary circulation features. Pressure variations can occur

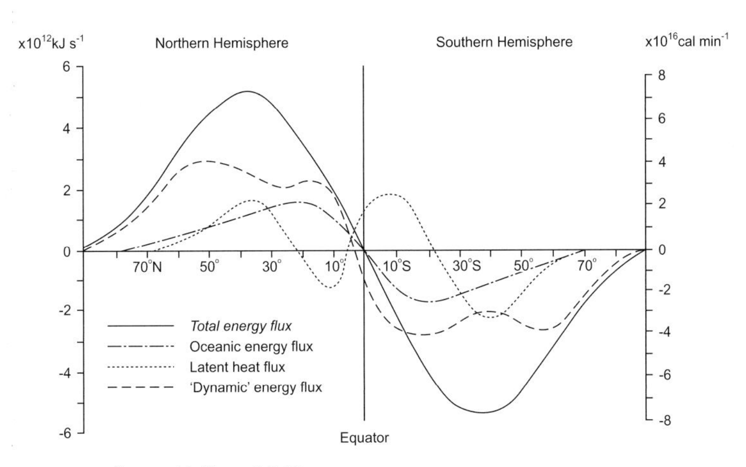

Fig. 3.3 Annual mean energy fluxes (Sellers, 1965).

higher in the atmosphere. These can be presented as either isobar maps at a fixed height or height contours at a fixed pressure. The 500 hPa (500 mb = 0.5 atm) surface is often plotted, because this is essentially the mid-point in the atmosphere; this surface is at lower altitude with low pressure. Air motion (e.g. wind) is a response to variations in atmosphere pressure along a horizontal plane. Vertical air movements can be significant, but these tend to be restricted to limited areas of the globe and to limited duration and arise normally because of temperature changes with height. The resulting general circulation is shown in Plate 3.2. Close to the equator, the *inter-tropical convergence zone (ITCZ)* is a region of convective rainfall driven by solar heating. This vertical air motion draws in the surface *trade winds* and is also often influenced by land–ocean temperature gradients to generate *monsoon* troughs. The ITCZ moves north and south seasonally with the passage of the sun and forms the ascending branch of the *Hadley cells*. These form because of the small difference between summer and winter temperatures; surface temperatures are uniform with just a poleward decrease. Under these *equivalent barotropic* conditions, rising winds at the equator transfer heat polewards. However, there are some significant complications, notably the difference between surface land and ocean heating leading to distinct motions such as the monsoons (see section 3.1.4).

In mid-latitudes a westerly circulation dominates with high velocities centred on a core or jet that changes position year by year. This forms a wave around the globe, called a *Rossby wave*. Energy transfer to the poles is completed by the waves and by the associated jet stream, cyclonic development and frontal depressions. The strength of the westerly circulation in the North Atlantic region is captured through the North Atlantic Oscillation (NAO) index and the poleward pressure gradient by the Arctic Oscillation Index (AO) (see Box 3.1 and Table 10.1). At high latitudes, there is again an equivalent barotropic zone caused by uniform and low temperatures at the surface. Complications exist, such as a jet-like velocity maximum at 75–80° N which is an extension of the mid-latitude westerlies, and thus dominates the polar circulation, but in reality with very little poleward transport.

3.1.4 Ocean circulation and land–ocean interactions

The oceans have two major characteristics that affect global circulation. The oceans can store more heat, and because the emission of energy is a function of temperature, the ocean will lose its heat more slowly by radiation than land. Hence ocean temperature changes are slower and of a smaller magnitude than land. Also, oceans transport energy in the form of deep circulation and surface currents

Box 3.1 Climate indices

Climate indices are used by climate scientists to simplify the complexities of the general circulation into a generalized description of the ocean–atmosphere system. Typically, climate parameters in local regions or points on the Earth's surface are compared to derive the index; the regions are chosen to be relevant to large-scale climate processes. A range of indices have been developed and a useful on-line resource for downloading data series can be accessed via the National Oceanic and Atmospheric Administration (NOAA) website at http://www.esrl.noaa.gov/psd/data/climateindices/. When using climate indices, one must remember that they are a simplification of the climate system and that the processes determining index values might change over time and space and might not be fully captured. Some commonly used indices of relevance to palaeoclimatology are detailed below. Many can be further explored using the Koninklijk Nederlands Meteorologisch Instituut's (KNMI's) Climate Explorer (see section 10.1).

Southern Oscillation Index (SOI). The SOI is calculated as the *surface air pressure difference* between Tahiti and Darwin, Australia. In an El Niño event, the SOI is negative, with pressure at Tahiti low compared with Darwin.

El Niño indices. The strength of the El Niño is typically measured by measuring *sea surface temperature (SST) anomalies* in two regions of the Pacific Ocean. The NINO3 index is the anomaly in the region bounded by 5° S–5° N, 90–150° W, and the NINO4 index by the region 5° S–5° N, 150° W–160° E. NINO3.4 is the central overlapping region of these indices (5° S–5° N, 170–120° W) and typically most commonly used in the literature. Most recently (Yeh et al., 2009), it has been suggested that there are two forms of El Niño: a 'typical' El Niño with Eastern Pacific Warming and a 'El Niño Modoki' with a central Pacific Warming, with the latter increasing in frequency in recent years.

Monsoon indices. The regional monsoons can be captured by indices which use the differences in *zonal wind strength* between regions (indicative of the strength of the surface land–ocean pressure gradient) or heights over land (indicative of the strength of convection) (Wang & Fan, 1999; Wang et al., 2001; Webster & Yang, 1992; Kajikawa et al., 2009). For example, the Indian Monsoon index is the difference in zonal wind strength at 850 hPa between the regions 40–80° E, 5–15° N and 70–90° E, 20–30° N; the Western Pacific Monsoon index is the difference in zonal wind strength at 850 hPa between 100–130° E, 5–15° N and 110–140° E, 20–30° N; and the Australian Monsoon index the zonal wind strength in the region 110–130° E, 15–5° S.

Pacific Decadal Oscillation (PDO). The PDO is an index that measures SSTs in the Pacific Ocean north of 20° N. During a warm (positive) PDO, the west Pacific becomes cool and the eastern Pacific warms. The PDO is highly correlated with SST in the northern California Current (CC) area and the index is correlated with salmon landings from Alaska, Washington, Oregon, and California.

North Atlantic Oscillation. The NAO measures the strength of the westerly circulation in the North Atlantic, with the NAO index defined as the difference of sea-level pressure between two stations situated close to the hot spots of Iceland and the Azores. Stykkisholmur (Iceland) is typically used as the northern station, whereas either Ponta Delgada (Azores), Lisbon (Portugal) or Gibraltar are used as the southern station. It is also possible to use only the pressure in Iceland as a proxy index for the NAO because of its negative correlation with pressure at the southern stations.

Indian Ocean Dipole (IOD). The IOD is a coupled ocean and atmosphere phenomenon in the equatorial Indian Ocean that affects the countries that surround the Indian Ocean basin. It is measured by an index that is the difference between SST in the western (50–0° E, 10° S–10° N) and eastern (90–110° E and 10–° S) equatorial Indian Ocean. A positive IOD period is characterized by cooler than normal water in the tropical eastern Indian Ocean and warmer than normal water in the tropical western Indian Ocean.

(Fig. 3.4). Thus transport of *sensible heat* (heat that can be sensed by a thermometer) may be considerable. Ocean circulation changes dynamically over annual timescales. In particular, the Pacific Ocean normally has a *Walker Cell* (Plate 3.2). At the South American coasts there is a predominant offshore breeze (owing to land heating). This blows offshore and causes water movement, which in turn causes upwelling. This deep water is cold, and stabilizes the air flowing over it by cooling it and preventing it from rising. If this were not the case, the air would rise and form the northeast trade winds: instead it flows west, gains heat, and finally rises in the Western Pacific. This flow alters every 2–7 years, when the cold ocean current off the South American coast is disrupted, the cause of which is not clear. This phenomenon is called *El Niño*, and disrupts the Walker Cell, making it weak and

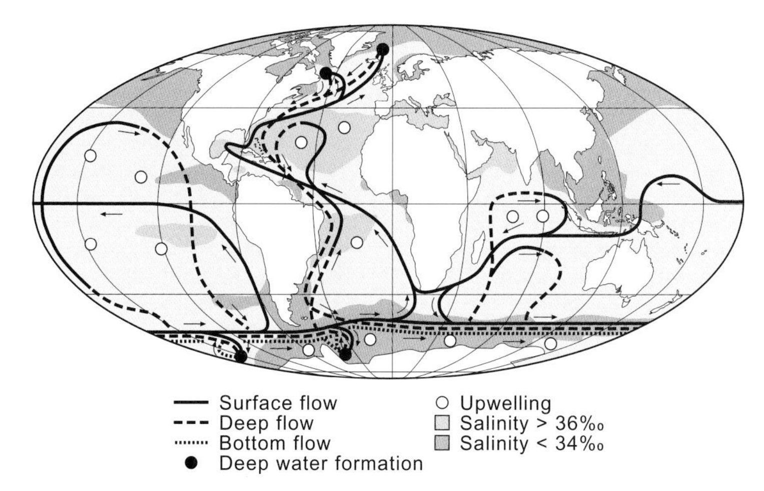

Fig. 3.4 Schematic of the global overturning ocean circulation (adapted from Kuhlbrodt et al., 2007).

reversed, and enhances the Hadley Cell. The atmospheric effect of this change in ocean circulation is measured by the Southern Oscillation Index (see Box 3.1). The implications of the combined ocean–atmosphere interaction (the *El Niño-Southern Oscillation, ENSO*) are widespread, with an exchange of air between Indonesia and the southeast Pacific that occurs at the same interval which affects regions as widespread as northern Australia and Southeast Asia. Decadal-scale variations in ocean surface temperatures and associated sea level pressure also occur in both the Pacific and Atlantic Oceans and are captured by various climate indices (see Box 3.1).

Monsoon climates are the most important example of large-scale land–ocean interactions affecting precipitation patterns. They are caused by the larger amplitude of the seasonal heating of the land compared with the ocean. This surface temperate imbalance causes air to rise more rapidly over land than over the oceans, creating an area of low pressure. This creates a steady wind blowing toward the land, bringing the moist near-surface air from the oceans. As soon as this moist air starts to rise over land, the air cools owing to expansion as pressure decreases, which in turn produces condensation and rainfall. This might be enhanced in different regions owing to to orographic uplift over mountains or convergence of air masses. Monsoons are therefore similar to the sea breezes typically experienced at a daily timeframe in coastal regions, but they are much larger in scale, stronger and seasonal. The best known monsoon climate regions are found in India, Southeast Asia and West Africa, and can be defined by various climate indices (see Box 3.1). See section 3.2.3 for a detailed appraisal of the relationship between the monsoons and rainfall isotopic composition.

The speleothem palaeoclimatologist is mostly interested in understanding the climate system and the general circulation that affects an individual point on the Earth's surface, where their cave record is located. The combination of the atmospheric and ocean circulation means that the path by which a particular air mass reaches this location could be very predictable (e.g. in a monsoon climate) or complex (in a mid-latitude location close to the position of the jet-stream). Box 3.2 details the technique of back trajectory analysis, a tool which may be of use in identifying moisture source or air mass source regions (Draxler & Rolph, 2003). The approach uses climate re-analysis products (see section 10.1), and is therefore only available for use with climate data for periods within the past approximately 100 years, but if one is interested in the source and trajectory of a 'parcel' of air which has generated precipitation over a cave site, the technique is likely to be useful.

Box 3.2 Back trajectory analysis

Trajectory analysis uses climate re-analysis products to track a parcel of air: typical applications would be to track a pollution plume or volcanic eruption plume. Back trajectory analysis is the inverse of this technique, and allows one to answer the question, 'where did a parcel or air originate from, and where did it uptake its moisture?'. It is therefore of interest to palaeoclimatologists who may be interested to know the moisture source of precipitation at a particular site.

A widely used and publicly available model is the NOAA Air Resources Laboratory's Hybrid Single-Particle Lagrangian Integrated Trajectory (HYSPLIT) Model (Draxler & Rolph, 2003). HYSPLIT computes air mass position through time using pressure, temperature, wind speed, vertical motion and solar radiation inputs from the NOAA FNL Meteorological dataset. Back-trajectories can be computed for a time period of several days before each rainfall day from the co-ordinates of the cave site in question.

HYSPLIT provides the capability of performing a cluster analysis, which can group the trajectories with similar paths, by comparing the cluster spatial variance (the sum of the squared distances between the endpoints of the clusters' component trajectories, and the mean of the trajectories in that cluster). The total spatial variance (TSV), which is the sum of all the cluster spatial variances, is then calculated. The trajectories that are grouped together are

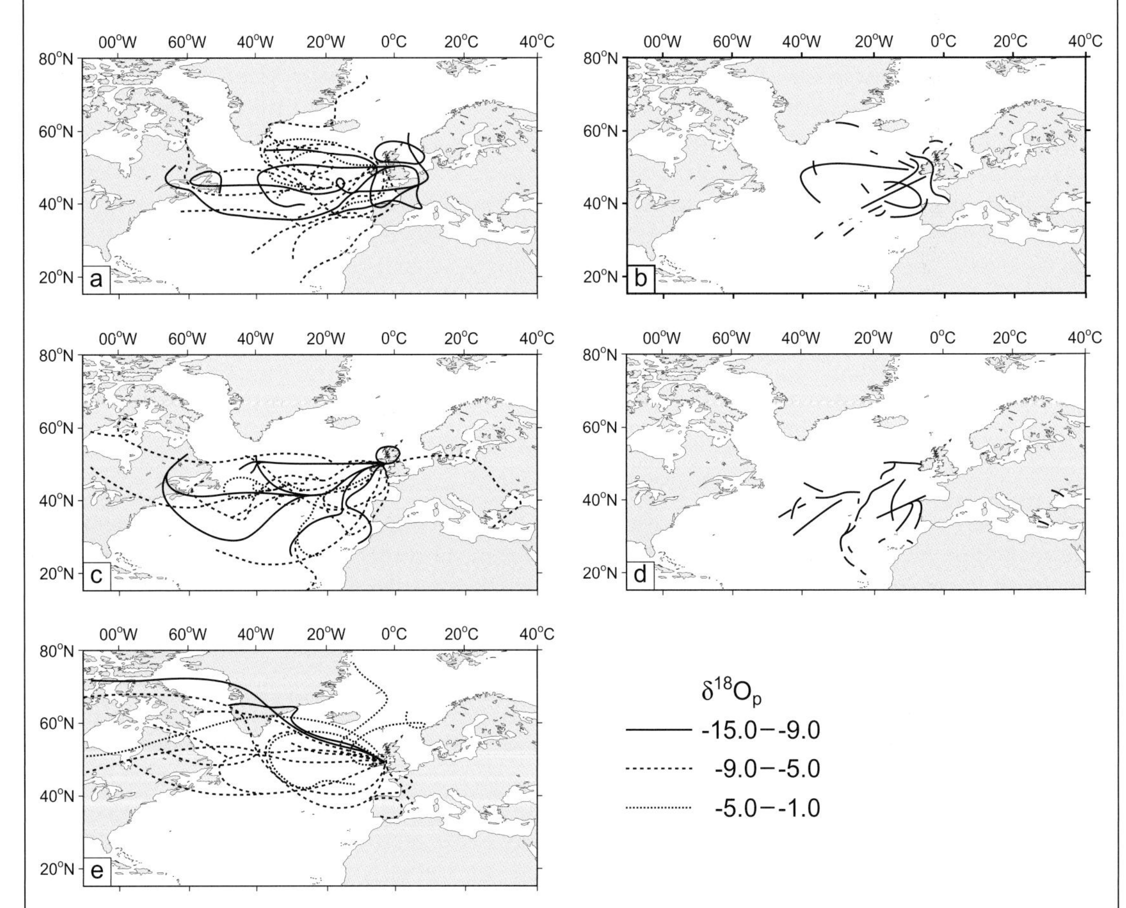

Fig. 3.B1 120-hour back-trajectories for Dublin rainfall events over the period 2003–2005 AD at (a) 850 hPa with (b) associated moisture uptake regions; (c) 700 hPa with (d) associated moisture uptake regions; and (e) 850 and 700 hPa trajectories computed for eight precipitation events for which no moisture uptake was identified (lower panel). Trajectories in (a), (c), and (e) are distinguished according to their oxygen isotope composition. (redrawn from Baldini et al., 2010.)

the ones with the lowest increase in TSV. The final number of clusters is determined by a plot of the TSV compared with the number of clusters.

Understanding the trajectory of a parcel of air is important, but the determination of the moisture source region(s) along the trajectory is also of interest, especially when investigating rainfall $\delta^{18}O$ variations. This is more complex, as to determine this, the boundary layer height (BLH) needs to be known. Some studies estimate this value as a defined pressure value, but a better approach is to obtain actual BLH data independently from re-analysis data. Knowing the BLH, plus the relative humidity, pressure and ambient temperature at each point along a trajectory, then for each time step the specific humidity can be calculated. Once calculated, changes in specific humidity can be compared with the altitude: if the altitude is lower than the boundary layer height where the specific humidity value increase is greater than 0.2 g/kg over the time period, then that region of the trajectory can be assumed to be a moisture source region (Sodemann et al., 2008).

An example of HYSPLIT output is shown below: in this case the trajectories of rainfall events at Dublin, Ireland, over a 2-year period of $\delta^{18}O_w$ rainfall sampling. For further reading, the Air Resources Laboratory HYSPLIT webpage has large amounts of practical information: http://www.arl.noaa.gov/HYSPLIT_info.php. Other recommended sources are Sodemann et al. (2008) and Baldini et al. (2010).

3.1.5 Climate classifications and 'hotspots'

The processes of the general circulation and ocean–atmosphere interactions lead to regions of similar climate characteristics, which lead to the possibility of regional climate classification. There are two general types of climate classification. A *genetic classification* is based on the origin of the observed features and thus requires knowledge of climate dynamics. This provides a good understanding of the processes causing climate regions, but are often complex and obtaining qualitative data is difficult. In the previous sections, we have attempted such a genetic classification. More common is the *empirical classification*, which is based on observed climatic conditions averaged over a set time period. All such classifications require arbitrary cut off levels, so none are perfect, and presume a stationary climate state. Input parameters include temperature, precipitation, evapotranspiration or soil moisture as well as the seasonality of each. Data have to be summarized, normally over monthly periods, and data has to be used for many years to be assured of representative data. The most commonly used climate classification is that devised by Köppen, and the recently revised classification for both the modern period and future anthropogenic global warming is presented in Fig. 3.5 (Kottek et al., 2006; Rubel & Kottek, 2010; raw data (including animations) are available to download at 0.5° resolution for the past 100 years and future scenarios until 2100 AD from http://koeppen-geiger.vu-wien.ac.at/). Input data are the monthly and annual means of temperature and precipitation, and the output is a three-letter code. The first letter separates moist from dry climates and moist climates by temperature. The second letter defines the degree of dryness for dry climate and the seasonality of rainfall for moist climates. The third letter is used to characterize seasonal variations in mid- and high latitudes.

Although spatial classifications such as Kottek et al. (2006) and Rubel and Kottek (2010) can be useful to provide a climate context for speleothem palaeoclimate research, a more useful concept is that of climate-sensitive regions, 'hot spots' (sometimes also referred to as 'sweet spots'). These regions may show the following characteristics:

1 A strong correlation with regional or global climate phenomena, such that results from one location can be extrapolated to a wider region. These may include teleconnections: climate anomalies that are related despite the locations being up to several thousand kilometres apart owing to large-scale processes related to the general circulation. Therefore, a speleothem site close to one node of the NAO (e.g. northwest Scotland) should teleconnect to one at the other node (e.g. Spain), with wetter conditions at one node occurring when there are dry conditions at the other.

2 Be sensitive to relatively small changes in the general circulation, and therefore amplify relatively small variations in climate. These might be sites close to the limit of a monsoon climate system,

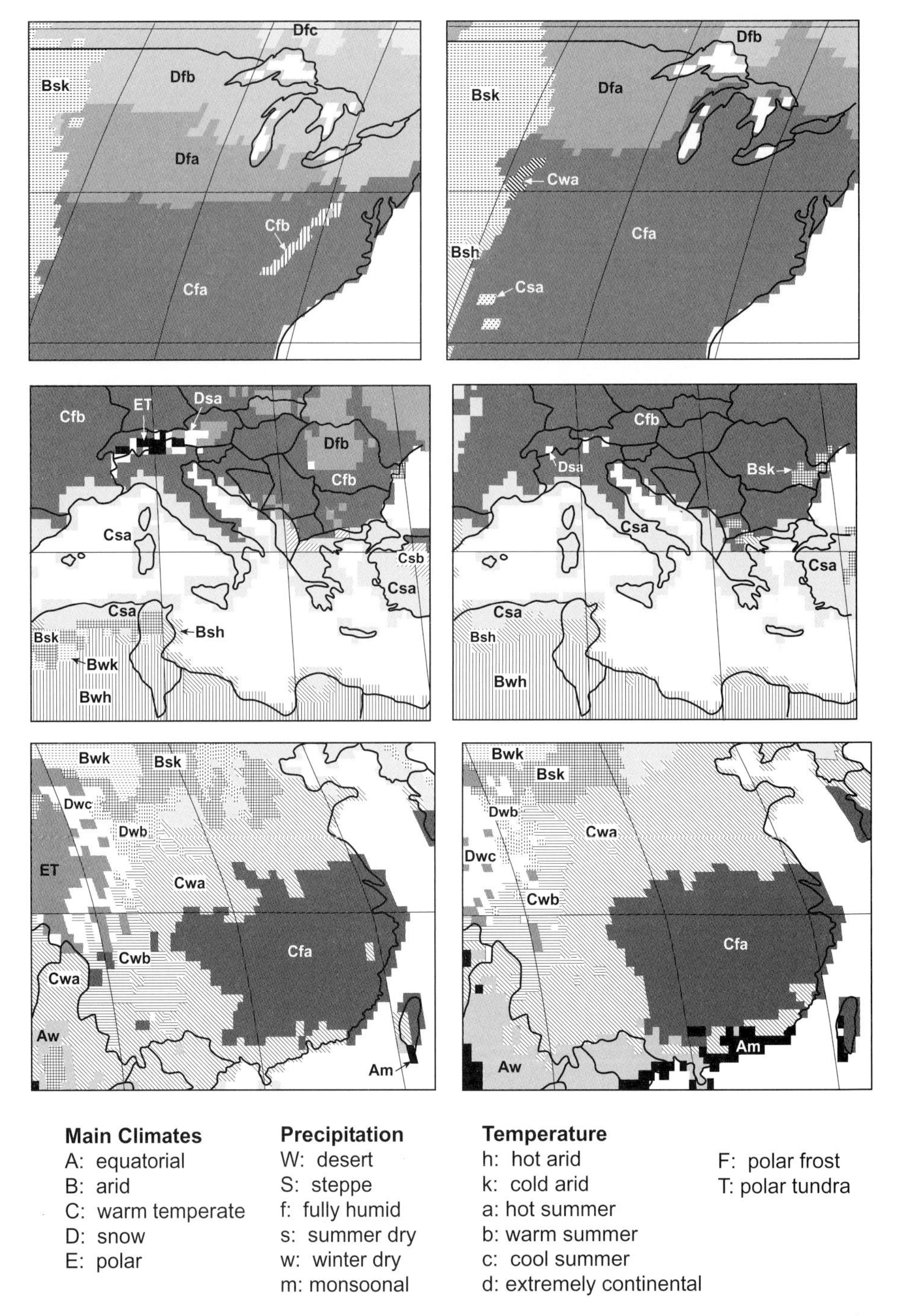

Fig. 3.5 Comparison of Koppen-Geiger climate classification for 1976–2000 (left) and 2076–2100 (right) for IPCC climate scenario B2. (top) Central and east USA; (centre) the Mediterranean; (bottom) China. Based on data archived at http://koeppen-gieger.vu-wien.ac.at (adapted from Rubel and Kottek, 2010).

under the mean position of the polar front, or sensitive to the frequency of tropical cyclones.

Figure 3.6 gives two examples of potential climate hot spots. One is the karst region of Oman, where modern total rainfall amounts can be seen to teleconnect with Pacific and East Asian Ocean sea level pressure, implying a process-based link between ocean sea surface temperatures, surface pressure and rainfall amount. The second is from the European Alps, where temperatures correlate with the strength of the NAO, particularly in winter. Here, a positive NAO brings relatively warm

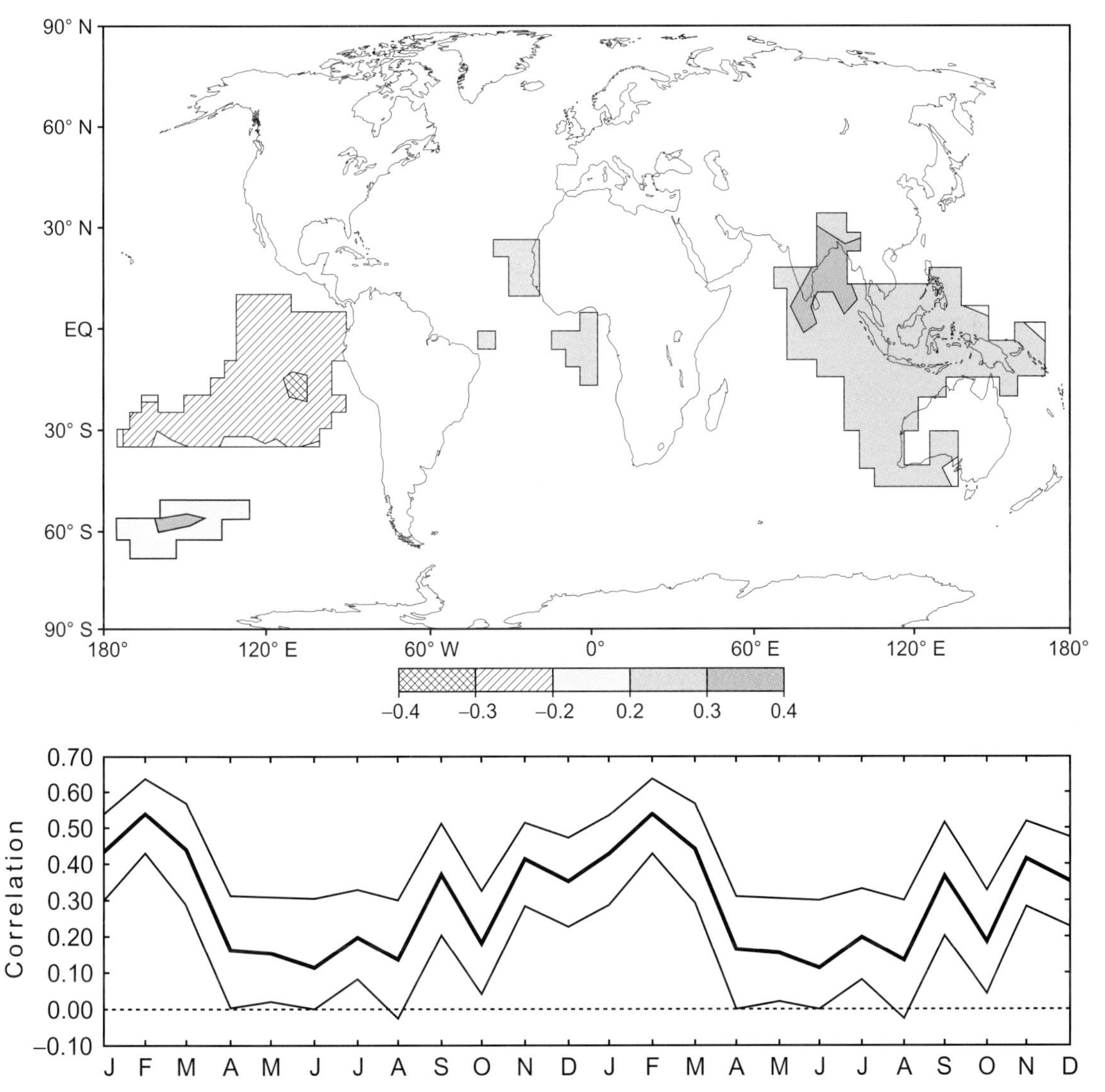

Fig. 3.6 (a) Spatial correlation between total annual precipitation in Oman and global sea level pressure. Correlation between total annual precipitation in the 1° grid 57–58° E, 22–23° N (using the CRU-TS3 precipitation dataset) and sea level pressure (using the Had-SLP2 dataset) over the period 1901–1996 AD. (b) Correlation between mean annual temperature anomaly in northern Italy and NAO. Mean monthly temperature is from the 1° grid square 11–12° E and 46–47° N using the CRUTEM3 dataset, correlated against the NAO index. Data analysed using Climate Explorer: see Chapter 10 for further details.

air masses from the south and west, whereas a negative NAO allows blocking anticyclonic conditions over central Europe and colder air masses from the north and east. Further examples of hot spots related to rainfall isotopic signatures are given in section 3.2. In the context of climate change, Giorgi (2006) identified climate hot spots that are modelled to be particularly sensitive to future climate change (Fig. 3.7), and therefore potential target regions for palaeoclimatologists interested in the relationship between past natural climate variability and future climate change. The two most prominent hot spots are the Mediterranean (MED) and North Eastern Europe (NEE), the former resulting from a forecast large decrease in mean precipitation and an increase in dry season precipitation variability, the latter from a large increase in dry (cold) season precipitation, a large warming and an increase in precipitation interannual variability. In contrast, South American regions appear to be generally less responsive to global change than other regions.

3.2 Water isotopes in the atmosphere

Our understanding of the causes of variations in the ratios of stable isotopes of oxygen ($^{18}O/^{16}O$) and hydrogen ($^{2}H/^{1}H$) in the atmosphere has developed rapidly. We can learn about these variations in the past through measurements on water inclusions found within speleothems and can often interpret past changes in atmospheric oxygen isotopes by

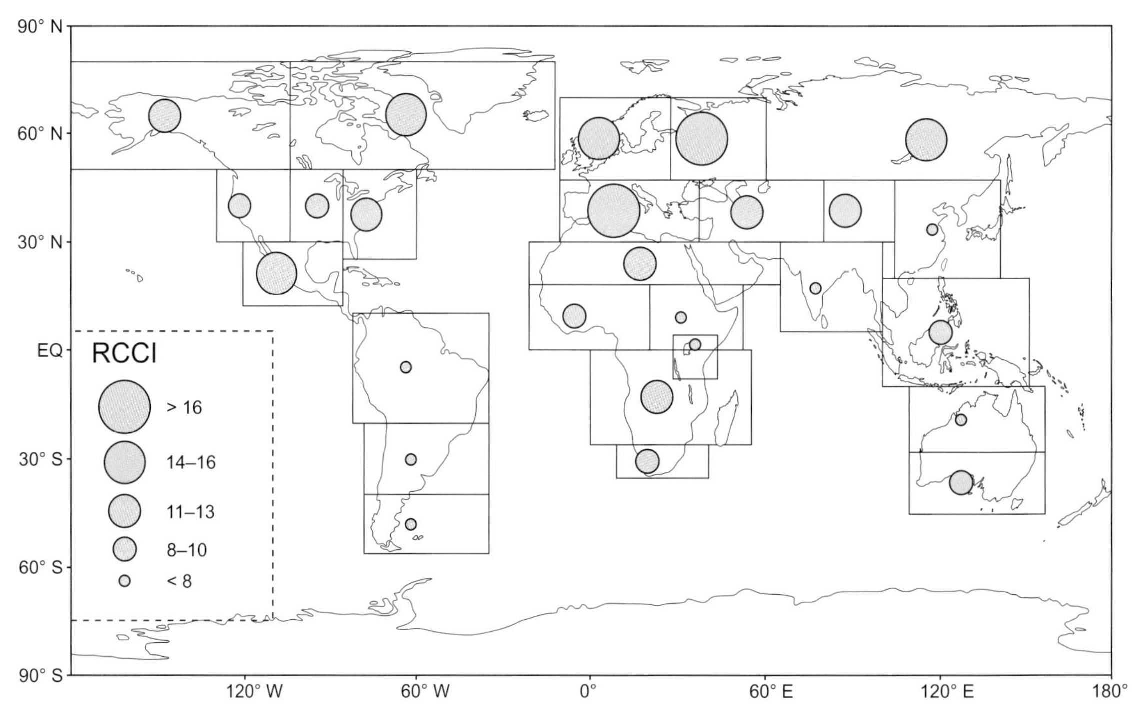

Fig. 3.7 Climate hot spots (from Giorgi 2006). The regional Climate Change Index (RCCI) is a relative index showing regions most sensitive to climate change based on a variety of climate change scenarios and is based on the change in (1) regional mean surface air temperature relative to the global average temperature change, (2) percentage change in mean regional precipitation compared with modern day, (3) percentage change in regional surface air temperature inter-annual variability compared with modern day and (4) percentage change in regional precipitation inter-annual variability compared with modern day.

analysis of oxygen in $CaCO_3$. An exciting recent development is that isotope ratios have emerged as a way of characterizing the activity of the climate system that turns out to be of independent value (Schmidt et al., 2007; Cheng et al., 2009b; Pausata et al., 2011), comparable to the pressure, temperature and precipitation fields. The state-of-the-art has evolved significantly from the 1990s when Holocene variations in temperature were interpreted from stable isotopes using either a modern calibration of $\delta^{18}O$ with temperature (Lauritzen & Lundberg, 1999), or from modern spatial gradients in isotope composition (McDermott et al., 1999): for further details see Chapter 10. The use of spatial gradients uses the *erdogic principle* of substituting space for time, and was classically used in the construction of the isotope thermometer, particularly in ice core work (Dansgaard, 1964; Holdsworth, 2008). However, we now realize that present-day spatial variations do not capture all the features of variation over time. Hence, in palaeo-studies, attempts need to be made to assess potential source areas for vapour, seasonal variability in meteorology and controls on inter-annual variability (Fricke & O'Neil, 1999; Alley & Cuffey, 2001; Sauer et al., 2002; Fairchild et al., 2006a; Baker et al., 2007). For early Holocene and older times, major changes in boundary conditions such as an expanded cryosphere and removal of source areas by sea-level variation also need to be considered (Alley & Cuffey, 2001; Griffiths et al., 2009). Identifying spatial patterns of variation in the past (e.g. Roberts et al., 2008; McDermott et al., 2011) is a key priority (Schmidt et al., 2007).

Our understanding of stable isotope variations has come from theory, from direct observations and from modelling. Several theoretical relationships have been used to code isotope modules within GCMs, starting with Jossaume et al. (1984), and these computer codes have been modified in response to further theoretical developments and observations. A major international project that has coordinated a global network of sites at which integrated monthly samples of atmospheric precipitation are collected (Global Network of Isotopes in Precipitation, GNIP) has been running since the 1960s, coordinated by the International Atomic Energy Agency (IAEA). This has been supplemented by many studies at shorter time intervals sampling individual events or even minute-by-minute variations. Systematic studies of vapour composition have been few (e.g. Strong et al., 2007; Angert et al., 2008), but now automated instrumentation for vapour and water analysis is available which can be deployed in the field (Wen et al., 2008; Gupta et al., 2009) and vapour and condensate have also been sampled from aircraft (He & Smith, 1999). Remote sensing of isotope composition is now possible from spacecraft, which is allowing important new insights into atmospheric processes over large spatial scales (Worden et al., 2007; Brown et al., 2008).

We do not deal with the essentials of isotope notation and fractionation in this chapter, but a summary for reference can be found in Box 5.1. Lachniet (2009) is recommended for a succinct overview of oxygen isotopes in speleothem science.

3.2.1 Variation in stable isotopes owing to evaporation and Rayleigh condensation

Seawater is the largest H_2O reservoir and its typical composition has been chosen as the reference standard (VSMOW, Box 5.1) to which stable isotope measurements are referred. In practice it is slightly isotopically heavy in regions such as the tropical Atlantic which are net exporters of moisture, and light in estuaries and enclosed and Arctic seas, but open ocean $\delta^{18}O$ values rarely vary by more than ±2‰ (LeGrande & Schmidt, 2006). Because water molecules containing either of the heavier isotopes ^{2}H or ^{18}O have a lower vapour pressure than $^{1}H_2{}^{16}O$ molecules, vapour is isotopically light (i.e. $\delta^{18}O$ and $\delta^{2}H$ values are negative, because seawater is zero). At equilibrium (100% (relative) humidity) the composition of vapour co-existing with seawater is shown in Fig. 3.8. However, active evaporation is not an equilibrium process and is subject to kinetic fractionations that are much stronger for oxygen than hydrogen isotopes. As a result, vapour at lower humidity is displaced to the left of that at 100% humidity. When this vapour condenses to water, it does so at equilibrium and the vapour and

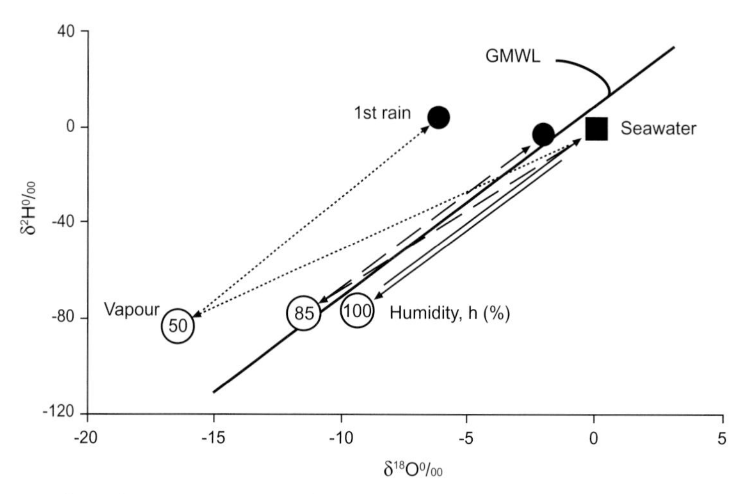

Fig. 3.8 Kinetic isotope effects during evaporation of seawater at 25 °C. Open circles refer to percentage relative humidity (h) and are linked to the first rain (filled circles) formed by equilibrium condensation. See text for discussion. (From Clark & Fritz, 1997.)

water compositions are connected by a line parallel to that at 100% humidity, but displaced to the upper left. Globally, meteoric water roughly follows the relationship first established by Craig (1961):

$$\delta^2H = 8\delta^{18}O + 10 \quad (3.1)$$

This is the equation of the global meteoric water line (GMWL). The line occupies the position expected if vapour originated in air masses with on average 85% relative humidity (Fig. 3.8). In any given region there will be a local variant of it (a local meteoric water line, LMWL) which will have a specific slope and intercept. For comparative purposes, the intercept of the GMWL is used as a standard to compare the offsets of LMWLs. Hence the deuterium excess (d) is defined for a water analysis as

$$d = \delta^2H - 8\delta^{18}O \quad (3.2)$$

A theoretical evaporation model to explain the position of the GMWL was developed by Merlivat and Jouzel (1979) and takes the form

$$1 + \delta_{vo} = (1/\alpha)(1 - k)/(1 - kh)(1 + \delta_{ocean}) \quad (3.3)$$

where δ_{vo} is the isotopic composition of the vapour forming by evaporation from sea water with local composition δ_{ocean} at a humidity (h) expressed as a fraction and related to the ocean temperature. Temperature has an effect via the isotope fractionation factor (α: see Box 5.1) and likewise wind speed through the parameter k (fractionation would be diminished in rough seas). The model predicts that d increases at lower relative humidity such that d/h is −0.43%/% and that it increases with temperature at the rate of 0.35%/°C (Johnsen et al., 1989). This approach, within the overall model of Craig and Gordon (1965) has been widely adopted within GCMs, although many authors doubt if wind speed is typically a relevant factor, and Pfahl and Wernli (2009) proposed an alternative formulation that is independent of it. Because temperature and relative humidity tend to vary inversely in the oceans, it has been difficult to verify the relationship observationally, but the data of Uemera et al. (2008) lie close to the values predicted by Merlivat and Jouzel's model (Fig. 3.9).

As mentioned in section 3.1, back-trajectory tools are now available for tracing the origin of air masses, and Pfahl and Wernli (2009) and others have added isotope systematics to such calculations, allowing a more detailed testing of how the isotopic characteristics of vapour are derived in relation to the locations where moisture in a humid air mass has been acquired.

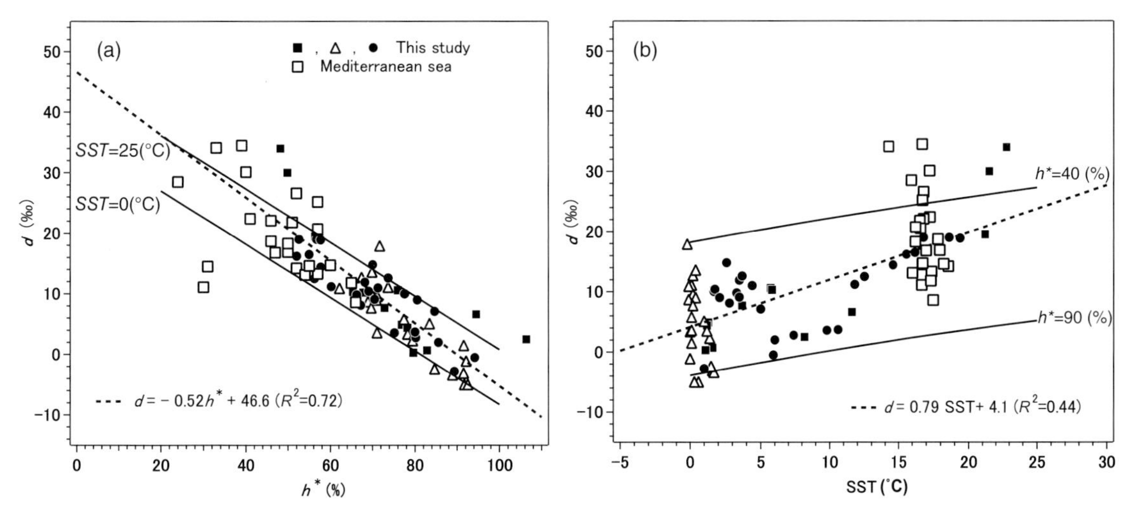

Fig. 3.9 Observations of the relationships between *d, h* and sea surface temperature (*SST*) in the southern oceans and the Mediterranean Sea. (Uemera et al., 2008).

Condensation of moisture arises because of cooling of air masses until they exceed their saturation vapour pressure (see Fig. 4.2). Water condensation takes place at equilibrium and the classic way in which it has been modelled is as a *Rayleigh fractionation* process. In a pure Rayleigh process, the condensed moisture is physically removed from interaction with the vapour, and the initial (R_o) and final (R) vapour isotope ratios are related by

$$R = R_o\, f^{(\alpha-1)} \tag{3.4}$$

where f is the fraction of remaining vapour. Each increment of water is heavier than the vapour from which it forms (Fig. 3.8) and this leads the residual vapour to become isotopically lighter, and hence successive water increments likewise become lighter. The different water compositions define a meteoric water line. Figure 3.10 illustrates calculated examples in relation to the composition of seawater at zero on both axes. The slopes of all the lines are fairly close to 8, but increase slightly at lower temperature. When vapour forms from seawater at 25 °C and 70% relative humidity and cools, moisture can condense at 19 °C (point A) and the subsequent condensates form a line (shown as far as 0 °C) which is very close to the GMWL and so the deuterium excess varies little. The vapour pressure of water continues to diminish steadily below 0 °C at all atmospheric temperatures and so the Rayleigh process can be continued. Such an analysis is important for the understanding of ice cores, but there are several additional complications such as the degree of interaction of ice or supercooled water with vapour (Ciais & Jouzel, 1994) and kinetic factors (Jouzel & Merlivat, 1984) which need to be considered. Figure 3.10 also clearly shows the strong impact of humidity in the zone of evaporation on the value of *d* and that the condensation process retains a memory of the conditions in the source area. This proves to be a vital tool in ice core research (Petit et al., 1991; Masson-Delmotte et al., 2005) and has also been used in speleothem fluid inclusion studies (McGarry et al., 2004). However, air masses from different source areas may be responsible for precipitation in any given place at different times and in exceptional cases may plot on quite distinct meteoric water lines. This is the case in Oman where Mediterranean-sourced water with high d plots quite distinctly from that derived from the Indian Ocean (Weyhenmeyer et al., 2002). Cool temperate areas lie close to the boundary of polar and tropical air masses and here successive rain events can come from condensate from a warm air mass derived

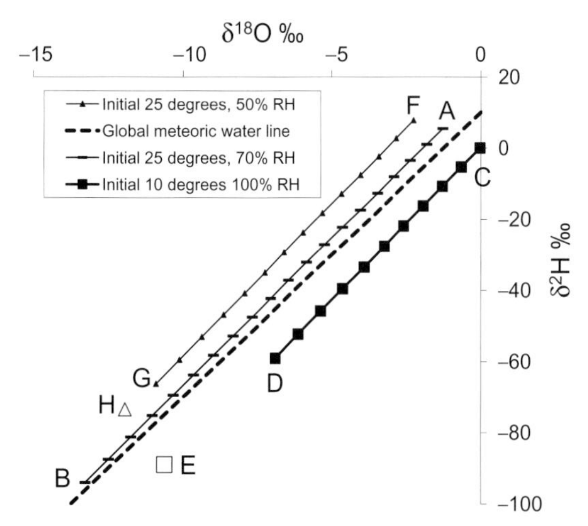

Fig. 3.10 Calculations using eqn. 3.4 and *d* values estimated from Fig. 3.9 to illustrate the progressive evolution of water compositions during Rayleigh fractionation of vapour under several different initial conditions. For example line A–B (which lies near to the GMWL) represents the composition of the condensates that progressively form from vapour that was originally generated by evaporation into air at 70% relative humidity (RH) above seawater at 25 °C. A corresponds to the temperature at which condensation first occurs (19 °C) and each point on the line corresponds to a one-degree cooling; the last point shown (B) is for 0 °C. Likewise C–D and F–G represent condensation from different initial conditions and the final condensates shown (D, G) correspond to 0 °C. E and H represent the composition of vapour in equilibrium with condensates at C and F respectively.

from tropical seas (e.g. point A) and a cool air mass (e.g. point D), the latter being much isotopically lighter. However, Baldini et al. (2010) showed that variable cooling from a single dominant source region explained variations in isotope composition in Dublin, Ireland (see also Box 3.2). The physical reality of the Rayleigh process is that falling droplets reach equilibrium by exchange with moisture near cloud base which conveniently means the temperature recorded is that of the cloud base, much closer to that at the ground surface than those of initial condensation at the cloud top (Gat, 2000). Kohn and Welker (2005) find a close agreement with Rayleigh theory for sites in the USA, but cautioned that the temperatures at the time of precipitation are not representative of those during the total climatic period under scrutiny.

There are variants of the Rayleigh model where some re-equilibration of water and vapour is permitted, leading to reduced or enhanced fractionations (Gat, 1996) which are more realistic in some situations. Nevertheless, this progressive condensation process is the dominant cause of isotopic variability in rain and snow and leads to progressively more negative isotope compositions with decreasing temperature (*temperature effect*) or increasing height (*altitude effect*), latitude (*latitude effect*) or distance from oceanic source regions (*continentality effect*); the magnitude of each of these are summarized by Mook (2001) and Lachniet (2009). In specific regions, geographic position in relation to moisture source and altitude are capable of explaining greater than 80% of the variation in isotope compositions (e.g. Central America: Lachniet & Patterson, 2006, 2009; Australia: Guan et al., 2009). In mountainous catchments, there has been a strong focus on altitudinal effects on isotopic composition, stimulated particularly by interest in reconstructing palaeo-elevations from the oxygen isotope composition of minerals forming in ancient catchments (Garzione et al., 2000; Poage & Chamberlain, 2001; Blisnuik & Stern, 2005; Rowley & Garzione, 2007).

3.2.2 Other factors responsible for variations in isotopic composition

Early syntheses (e.g. Dansgaard, 1964; Rozanski et al., 1993) focused on the geographical gradients corresponding to the effects mentioned above, but also recognized several disturbing factors. The isotopic composition of rainfall often varies strongly (by over 10‰) within events (see, for example, Celle-Jeanton et al., 2001). Another important observation, in tropical and some semi-arid regions in particular, is that the isotopic composition is inversely related to the amount of rainfall. This empirical observation, which can be observed in annual or monthly records, or individual events, is referred to as the *amount effect*. It is polygenetic and its understanding has been obscured by the lumping together of different timescales of observation into the one term. For example, Treble et al. (2005c) showed that there was a strong amount effect in individual rainfall events in Tasmania, whereas monthly means showed only a relationship to temperature. On the other hand, daily rainfall data from certain tropical islands failed to show the effect, even though it was obvious in seasonal data (Kurita et al., 2009). In some locations, the effect is non-stationary: Fuller et al. (2008) found that using monthly data from the GNIP site at Wallingford, southern England, the effect was present in certain years whereas there were covariations with temperature in others.

A clear model for development of a strong amount effect arises in the case of tropical cyclones. Here vapour can be taken to particularly high altitudes where it condenses at low temperatures to form isotopically light rain or ice. As the condensate falls, it re-equilibrates with moisture as expected and so the initial rainfall reaching the ground has lost the memory of its initial very light composition. However, because these systems extend to very high altitudes and are relatively long-lived as they contain so much moisture, after a time the low-level vapour becomes much lighter, and soon the very light signal is retained within rain reaching ground level (Lawrence & Gedelzman, 1996). More details of the processes operating, including drop re-evaporation and convection phenomena, have been studied by Risi et al. (2008) and Lee and Fung (2008). This phenomenon has stimulated research into the identification of tropical storm events within speleothems (Frappier et al., 2007).

In semi-arid areas in particular, the amount effect appears to be specifically associated with the partial re-evaporation of rain during relatively weak events, leading to isotopically heavy rain which also has a raised value of d (Ayalon et al., 1998). Such observations allow regional palaeo-precipitation records to be obtained from speleothems (Bar-Matthews et al., 1998). Within the soil profile, evaporation will generate a typical depth profile where soil surface evaporation causes deuterium and ^{18}O enrichment near the soil surface, which decreases and is smoothed with depth (Zimmerman et al., 1967; Tang & Feng, 2001; Brooks et al., 2010). The amount of surface isotope enrichment has been shown to vary with the amount of evaporation but is poorly constrained with very few monitoring studies. In karst systems, the extent to which this enriched soil water may be subsequently flushed into the groundwater and contribute to speleothem $\delta^{18}O$ is not well understood (but see Fig. 8.13).

In continental areas, significant quantities of moisture can be produced by evaporation from the ground or from lakes, or by transpiration from plants (see, for example, Ingraham & Taylor, 1991; Cui et al., 2009). In the latter case, the deuterium excess of the water vapour is unaltered, but otherwise d increases. Moisture recycling was identified as a major cause of the variability in stable isotope compositions (Koster et al., 1993), a conclusion reinforced by satellite observations (Worden et al., 2007). The presence of a secondary source of moisture in addition to a marine-derived precursor, is just a special case of the more general phenomenon of mixing of water vapour from different air masses which contributes to some degree to the history of all atmospheric precipitation, and dominates in some (Petit et al., 1991; Bhattacharya et al., 2003; Sengupta & Sarkar, 2006; Brown et al., 2008).

3.2.3 Isotopic variations in space within the annual cycle

Both observations (e.g. Dansgaard, 1964; Rozanski et al., 1993; Araguas-Araguas et al., 2000; Bowen & Wilkinson, 2002) and models (e.g. Hoffman et al., 1998) have been used to gain a global understanding of isotope distributions and their causes. A particular focus has been the extent to which geographic regions show an overall relationship between isotopic composition and either rainfall or precipitation. Figure 3.11 presents results from the first coupled atmospheric–oceanic modelling study to address this issue (Schmidt et al., 2007). As has long been known in outline (Dansgaard, 1964), isotope signatures covary with temperature (Fig. 3.11, left) at high latitudes in continental interiors in particular, whereas the amount effect at a monthly scale dominates in tropical oceans and adjoining coastal areas (Fig. 3.11, right).

A comparable analysis has been performed on the GNIP observational data (Bowen, 2008; Feng et al., 2009), but with particular emphasis on seasonal relationships. Bowen (2008) produced the first maps illustrating the degree of isotopic variability within a year (Fig. 3.12). This is an important parameter that can be revealed in suitable speleothems (Johnson et al., 2006; D. Liu et al., 2008; Mattey et al., 2008). Bowen (2008) emphasized the spatial instability of the relationships (shown also in Fig. 3.11 from Schmidt et al. 2007), by the variability of the regression slopes), which means that the ergodic assumption is liable to high errors, particularly if source area temperature changes.

Feng et al. (2009) demonstrate that there are four world zones of isotope seasonality (Fig.3.13a), bounded by three nodes of relatively little seasonal variation in isotopic composition (Fig.3.13b). The higher-latitude zones display strong gradients with temperature and hence latitude (Fig. 3.13b) and show maximum isotope signatures in the summer. The northern and southern tropical zones show isotope maxima in the spring when rainfall is weak (Fig.3.13a), contrasting with the summer period when intense rain, associated with the intertropical convergence zone, develops from vapour already depleted during its transport from the subtropical highs. Another distinctive feature is that the value of the deuterium excess is highest in the winter when humidity is relatively low in the source areas of the sub-tropical highs. This seasonal high in d is recorded in a wide variety of stations, including the Arctic and Antarctic, and reinforces the importance of the sub-tropics as export zones for moisture as well as heat.

Much interest has been stimulated in monsoonal areas by the spectacular results from speleothem

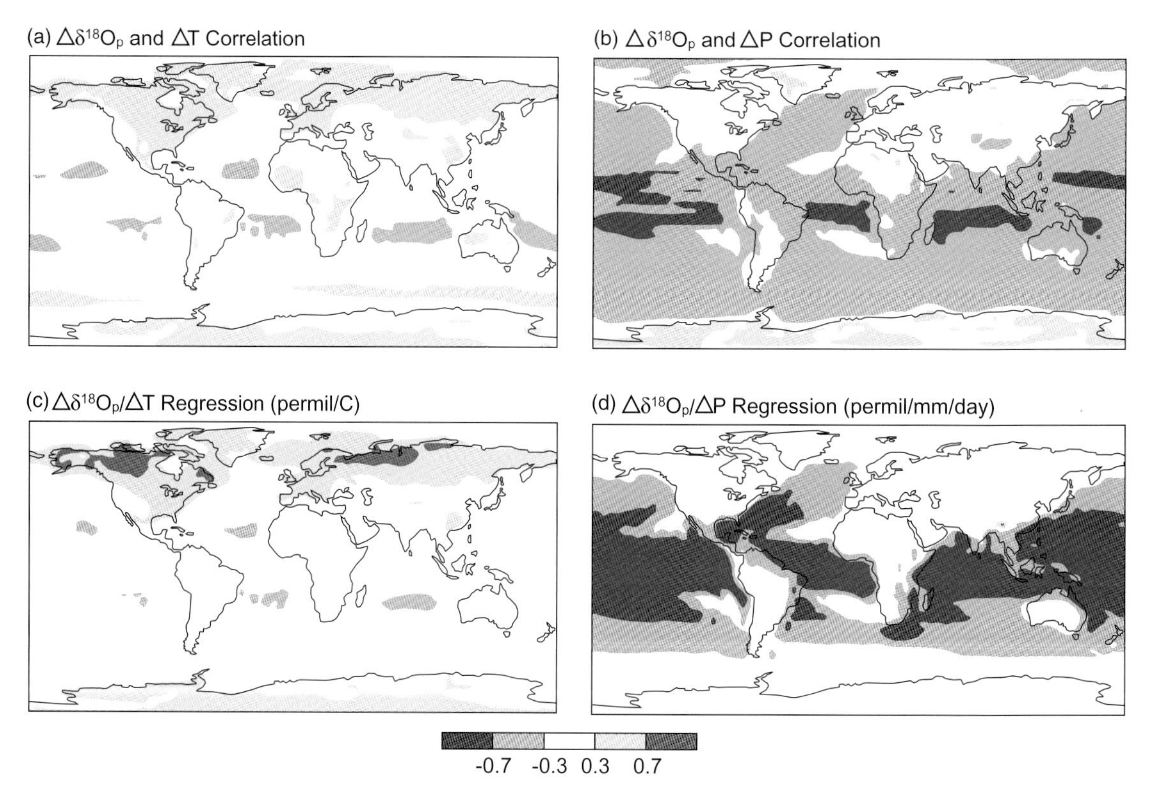

Fig. 3.11 Results from coupled atmospheric–oceanic GCM, using monthly anomalies, on the local correlations (top) and regressions (bottom) of $\delta^{18}O$ in precipitation to surface annual temperature (left) and precipitation (right) for inter-annual variability. Temperature displays positive relationships and precipitation negative ones. Simplified from Schmidt et al. (2007).

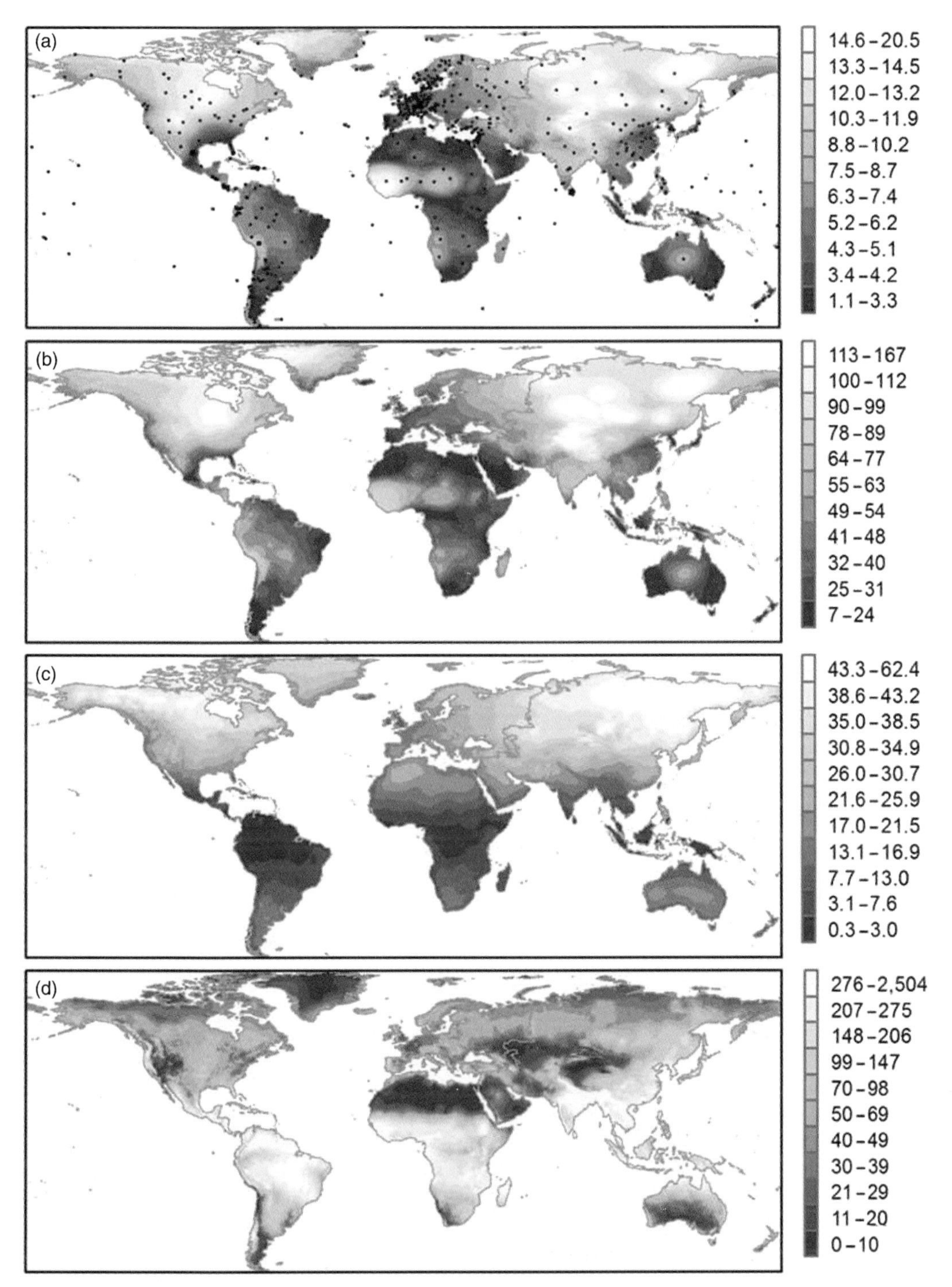

Fig. 3.12 Mapped estimates of the intra-annual range of isotopic and climatological parameters. The divisions of shade are in percentiles for (a) $\delta^{18}O$ values of long-term, monthly average precipitation (per cent); (b) $\delta^{2}H$ values of long-term, monthly average precipitation (per cent); (c) monthly mean temperature (degrees Celsius); and (d) precipitation amount (millimetres). Black dots in (a) show the distribution of isotope monitoring stations used to create the map. Bowen (2008).

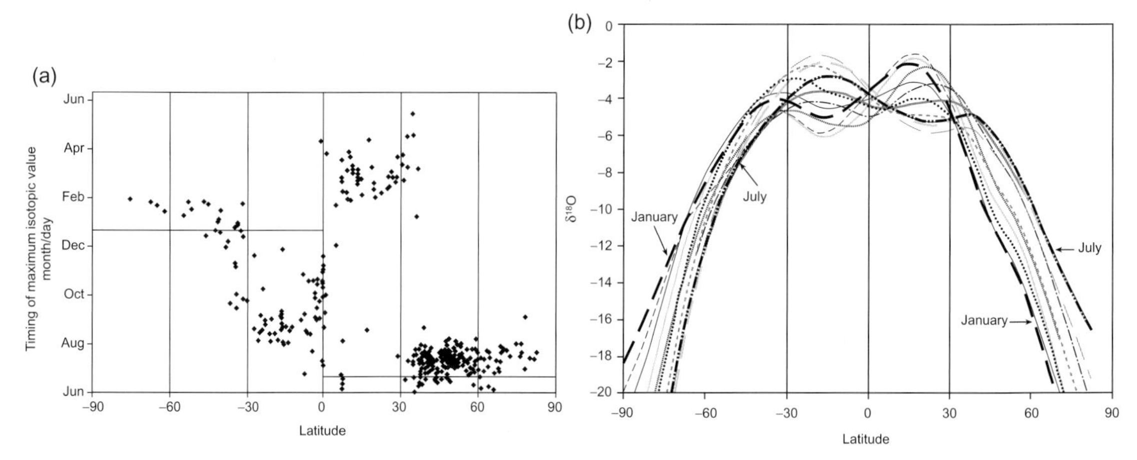

Fig. 3.13 (a) Seasonality, expressed as the timing of the maximum isotope value in the annual cycle (beginning of each month shown on the *y*-axis), as a function of latitude. The vertical axis is expressed as month/day and the horizontal lines are the solstices. (b) Smoothed monthly isotope data versus latitude showing meridional trends in $\delta^{18}O$; the seasonal maximum values represent the positions of the sub-tropical highs. Least variability is shown at the equator and just poleward of the furthest influence of the sub-tropical highs. Feng et al. (2009).

studies. The term monsoon originally referred to the seasonal reversal of winds adjacent of the Indian Ocean, but has come to refer more specifically to the associated periods of rainfall. More widely, the term is used for a summer period of high rainfall that is found in East and South Asia, southern Arabia, East Africa, northern Australia and parts of North and South America. It is linked to the strong heating and associated low pressure of these continental sub-tropical areas in summer which induces inflow of moist air from the surrounding oceans. Monsoonal areas cause confusion in terms of the types of spatial analysis shown in Fig. 3.12 because of the co-variation of temperature and rainfall (Z. Liu et al., 2008), making it difficult to distinguish separate influences (Johnson & Ingram, 2004). Satellite data show that monsoon systems are associated with intense distillation processes that tend to exceed the heavy isotope depletions expected from the Rayleigh model, although there are also local mixing and convective effects that create variability (Brown et al., 2008). Seasonal comparisons often lead to the observation of an amount effect because the intense summer rains are often the most isotopically light precipitation received during the year and this is also true of precipitation on tropical islands (e.g. Cobb et al., 2007). However, an interpretation of speleothem isotope records in China as reflecting precipitation amount (e.g. Yuan et al., 2004) is misleading (Maher, 2008) because comparison of annual mean precipitation isotope values do not necessarily show the amount effect and Holocene results disagree with other proxy evidence. Indeed, a comparison of daily mean isotope compositions and rainfall amounts of tropical stations does not display an amount effect because of lateral variations in cyclone intensity (Aggarwal et al., 2004). Instead, water isotope variability in Asia and other monsoonal areas reflects the integrated amount of landward vapour transport, rather than local precipitation (Sturm et al., 2007; Kurita et al., 2009; LeGrande & Schmidt, 2009), but it is precisely for this reason that the isotope composition bears witness to the large-scale properties of the climate system. Where estimates of rainfall amount over time have been made by comparison of isotope records along the mean vapour transport direction, they are clearly completely different from the trends from the raw $\delta^{18}O$ values (Hu et al., 2008a; Fig. 8.22). The current interpretation of long Chinese records is that they reflect the relative intensity of the summer monsoon system, reflected in the proportion of total annual precipitation that it represents (Cheng et al., 2009, see Chapter 12), although the modelling study of Pausata et al. (2011) points to an over-riding control by the south Asian monsoon on the composition of moisture exported to east Asia. Meanwhile, in Thailand, there is an excellent inverse correlation of $\delta^{18}O$ with the ratio of isotopically light rainfall from the late monsoon (August–October) with vapour sourced from the West Pacific Warm Pool and isotopically heavy rainfall in the May–July early monsoon derived from the Indian Ocean (Cai et al., 2010).

3.2.4 Inter-annual isotopic variations

The systematic relationships shown in Fig. 3.11 from model output also apply both on the inter-decadal as well as inter-annual scale (Schmidt et al., 2007). The latter relates to intrinsic variability generated by chaotic dynamics of the atmosphere and its interaction with the oceans. Additionally there are some links with specific modes of variability: for example, a strong relationship is that variations in the northern annular mode (effectively equivalent to the NAO) show patterns of isotope variability that closely match geographic anomalies in temperature. However, testing of *tropical* patterns is limited by model limitations.

The identification of climatically sensitive 'sweet spots' (section 3.1) has particular resonance in isotope science and is being used to guide research programmes (Lawrence & White, 1991). For example, a study of the spatial distribution of correlations of $\delta^{18}O$ in precipitation with the NAO index in winter in Europe identified central Europe as a suitable site for targeting Holocene samples (L. Baldini et al., 2008). Sodemann et al. (2008) report on a Langrangian study (i.e. following water isotopes from source to site of deposition of precipitation) of isotope fractionation to Greenland in relation to the NAO cycle and report that the warmer seas of the source area during negative phase of the NAO lead to significantly isotopically heavier snowfall. Vuille et al. (2005) show that $\delta^{18}O$ in East African spring and autumn rainfall is a good indicator of the state of the Indian Ocean Dipole. At Bangkok, Thailand, both the IOD and ENSO significantly influence the inter-annual variability of rainfall in individual months (He et al., 2006), which might account for the excellent relationships obtained from Thai speleothems by Cai et al. (2010) mentioned above.

ENSO variations are known to have powerful teleconnections and these include distinct stable isotope patterns. Schneider and Steig (2008) deduce that extreme positive $\delta^{18}O$ anomalies in snow accumulating in West Antarctica during the 1936–1945 decade were linked to the major 1939–1942 El Niño event via tropical deep convection. Also, a GCM-modelling study showed that ENSO-modulation of the annual summer monsoon in tropical and subtropical South America explains the dominant variation in water isotopes across a wide area of the continent (Vuille & Werner, 2005). A distinctly different, consistently Atlantic/Caribbean-sourced rainfall regime exists in the Yucatan peninsula however, where a simple relationship of $\delta^{18}O$ to rainfall amount is present and has been used to develop speleothem records of drought coinciding with retrenchment of Mayan civilization (Medina-Elizade et al., 2010).

Speleothem data are now starting to be used to extend backwards the post-war stable isotope observational period to understand the impact of atmospheric modes on water isotopes, as detailed extensively in Chapter 10. For example, Fischer & Treble (2008), working in southwest Australia where daily rainfall data showed a clear amount effect (Treble et al., 2005b). A conundrum was that relatively heavy $\delta^{18}O$ values were found not only in the late 20th century when instrumental data showed a drying trend, but also in the period 1930–1955 AD which was observed to be wet. It was found that the intercept of the amount effect could vary over time depending on the nature of atmospheric circulation. A principal components analysis of regional sea level pressure revealed three modes of variability, of which one showed a significant correlation with speleothem $\delta^{18}O$. Consequently, in the post-1970 AD period, dominant meridional circulation from the south was associated with low rainfall and hence high $\delta^{18}O$. Conversely, in the period 1930–1955, although the rainfall was higher, this was compensated for by the warm air source, so speleothem $\delta^{18}O$ was also relatively high. Although in this climatic context, $\delta^{18}O$ values are ambiguous, Mg concentrations do correctly reflect the differing rainfall regimes (Treble et al., 2008), which illustrates the utility of multi-proxy studies.

3.3 Soils of karst regions

Soils have classically been viewed as the product of five processes: climate, organisms, relief, parent material and time. Climate, in particular the balance between precipitation and evaporation, i.e. hydrologically effective precipitation, controls the water movement through the soil. Temperature is also important as it in part controls the rate of biological activity and therefore soil carbon dioxide production. It is this water, and its acidity, that drives much of the chemical weathering reactions. Organisms provide the organic input to the soil profile from both vegetation and soil flora and fauna. The relative balance of organic matter input to its decomposition and degradation, which is primarily temperature and moisture dependent, determines soil properties such as structure, and in some climates leads to the accumulation of significant carbon stores in the form of peat. Relief is a locally important factor that determines soil drainage and stability. Changes in elevation also lead to large differences in temperature and moisture under otherwise similar environmental conditions, and an increase in the amount of soil organic matter with elevation is well reported (Zimmermann et al., 2009). The parent material provides the raw material for the soil, which has undergone physical and chemical weathering. Finally, the soil forming processes are continuously acting; therefore a soil can be viewed at any one time as being a snapshot in its evolution.

Ninety-eight per cent of soil, by weight or volume, is made up of just eight elements (oxygen, silicon, aluminium, iron, calcium, sodium, potassium and magnesium) whose relative proportions reflects the elemental make-up of the underlying bedrock. Some elements are associated with stable (insoluble) phases whereas others are soluble and readily released: for an overview of properties see section 5.3.3, Railsback (2003) and the associated website http://www.gly.uga.edu/railsback/PT.html. Common minerals present in soils also reflect the elemental composition and bedrock character and are silicates, clays, carbonates, sulphates, oxides and hydroxides. Soils have typically been classified by assessing their vertical profile and variations in composition (structure, texture, grain size, organic matter decomposition and clay content). The vertical soil profile is divided into horizons, with a surface organic horizon (O) underlain by one or more soil horizons (A, B, C) which overlie the bedrock (R). Example soil horizons are shown in Fig. 3.14. Based on the profile description, global soils have been classified and named, and Plate 3.3 presents one example of a global soil classification. See also the global soil classification from the Food and Agriculture Organisation (the Harmonised World Soil Database) and which is publicly available for download (http://www.iiasa.ac.at/Research/LUC/External-World-soil-database/HTML/). Soil classifications form another example of empirical classifications, based on visual description, rather than a process- based classification related to the factors generating the soil profile, although ongoing global efforts are attempting to develop a high resolution digital soil database based on soil functional properties (Sanchez et al., 2009).

3.3.1 Processes of soil formation

Soils can be considered to be a dynamic product of chemical and physical weathering, and biological processes, altered in the modern environment by human impacts. In most circumstances, physical processes such as hydration/dehydration, fire and freeze–thaw action are less important than chemical and biological processes, but do have consequences for the latter. Soils are dynamic, open systems, and Fig. 3.15 shows the common soil-forming processes referred to in this section. Particularly of interest is the loss of material, in dissolved or colloidal form, to the groundwater during periods of hydrologically effective precipitation. Chemical weathering is an important driver, with hydrolysis, oxidation, dehydration and dissolution all important (see Chapter 5). Hydrolysis by

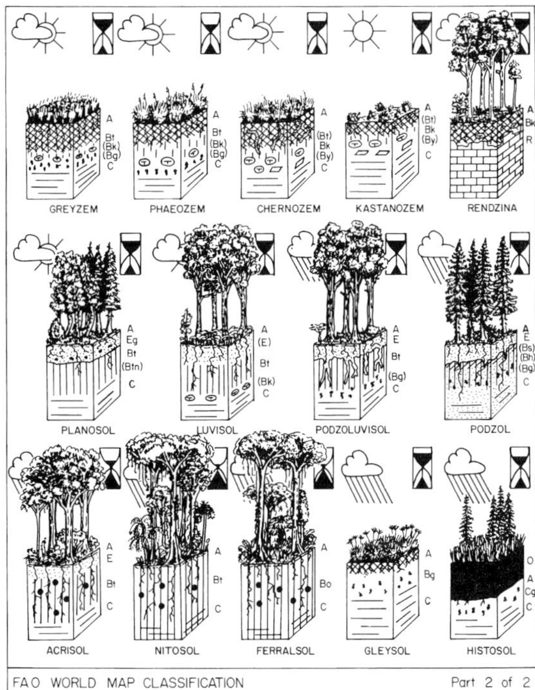

Fig. 3.14 Schematic representation of climate, vegetation and soil profile form on the major soil units of the FAO Soil Map of the World (Retallack 1998). Soil horizon nomenclature is as follows: O, organic; A, topsoil; E, eluviated soil; B, subsoil; C, lower subsoil; R, regolith. Further differentiation includes suffixes c, mineral concretions, for example: f, amorphous Fe-Al organic horizon; g, gley horizon; k, calcareous horizon; t, illuvial clay horizon; y, cryoturbated horizon

carbonic acid typically produces both new insoluble products which are not leached from the soil profile, as well as the leaching of very soluble elements such as Na and K. The pH of soil water is a good indicator of the importance of this process; in very acidic conditions the acid breakdown of clay complexes can lead to extensive loss of soluble material. In the palaeosol record, the Al_2O_3/SiO_2 ratio is indicative of hydrolysis as feldspars are more readily hydrolysed than quartz. Soils are normally aerobic and hence oxidizing except waterlogged soils in regions of very high rainfall, topographic hollows or impermeable bedrock. Iron compounds are particularly prone to oxidation and the extent of oxidation in a soil can be determined by the Fe_2O_3/FeO ratio. Measurement of the redox state (*Eh*) of a soil water gives an indication of the likelihood of oxidation occurring. Dehydration and rehydration processes particularly affect colloids and clay–humus complexes. Finally, dissolution produces soluble products which are lost to the soil. The electrical conductivity of soil water gives an indication of the likely importance of this process.

Chemical weathering processes primarily determine the loss of chemical elements from a soil, and the nature of the weathering is primarily determined (in non-waterlogged soils) by the $P - E$ balance. Figure 3.16 presents examples of modern soil relationships with mean annual rainfall amount (Retallack, 1994, 2005; Ready & Retallack, 1995). Where $P > E$, then hydrolysis and dissolution processes dominate, and H^+ ions replace metals present in minerals to form a weathered B_w horizon. Insoluble by-products re-precipitate, and soluble by-products (e.g. Ca^{2+}, Mg^{2+}, Na^+, K^+) are lost by solution to groundwater. In very acid conditions,

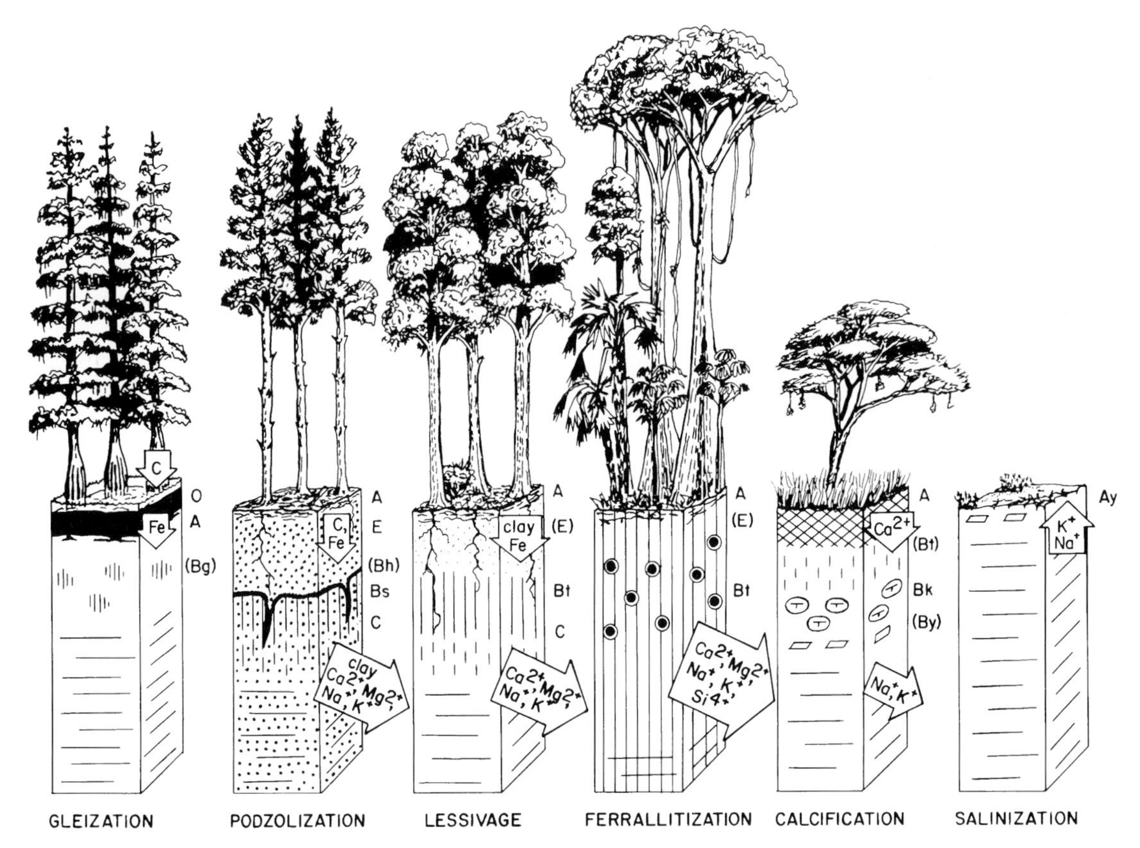

Fig. 3.15 Common soil forming processes (Retallack, 1998).

such as when >>*E*, the H^+ ions also break down the clay complexes with even greater loss of Na^+ Ca^{2+} Mg^{2+} and K^+, and re-precipitation of (alumino) silicate minerals to form a podzol. Where waterlogging occurs, iron is reduced or removed to form concretions. Typical soil categories when $P > E$ are cambic, podzols and ferallitic soils. Where $P \approx E$, then leaching is limited in extent, and the small amount that occurs leads to calcite dissolution and reprecipitation (calcification) lower in the profile, and dehydration processes start to become more important, such as of dehydration of iron oxides to haematite, giving a distinctive red colour to the soil profile. Where $E > P$ then only minimal leaching occurs, and only the most soluble products are dissolved such as Na^+, which may be reprecipitated lower in the soil profile as Na_2CO_3. Where $E >> P$, then salinization occurs as water is drawn upwards from the groundwater owing to the strength of surface evaporation. In this case, salts are drawn upwards and precipitate at the surface.

Biological processes of particular importance are those related to the breakdown of the surface vegetation or litter to form humic material. This can be classified as mor, moder, mull or peat, depending on the extent of humification, with peat being the least modified and having the greatest carbon accumulation, and mor being well humified with no identifiable organic components remaining. Well-humified organic material forms organo-mineral complexes and clay–humus complexes, which improve the soil structure, increase aeration, and help retain nutrients and metals in the soil. However, the extent to which metals are released to solution, rather than being fixed in the solid phase, varies by

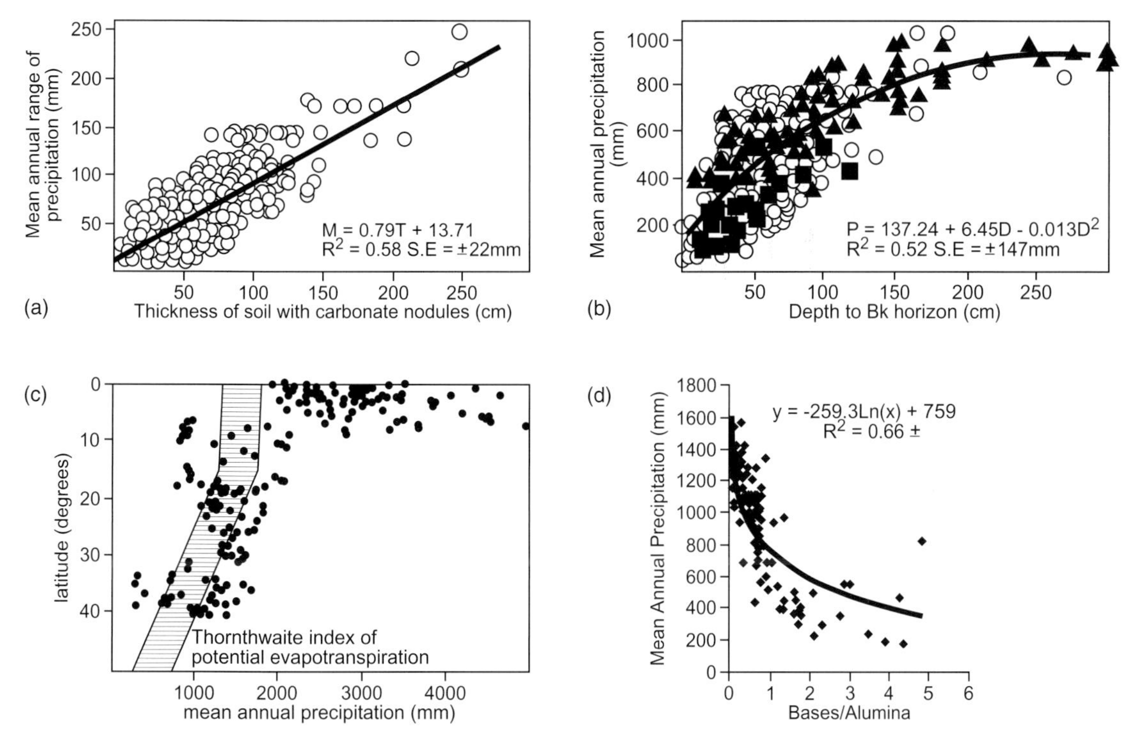

Fig. 3.16 Examples of the relationship between climate and soil properties. Relationship between mean annual precipitation and (a) the thickness of soils with carbonate nodules and (b) the depth to the Bk horizon (horizon of carbonate accumulation) (Retallack, 2005). (c) Compilation of the modern occurrence of nonfrigid peats (histic epipedons) related to precipitation and potential evaporation (Retallack, 1998). (d) Base/alumina ratio in the Bt horizon plotted against mean annual precipitation for 127 North American soils (redrawn from Sheldon et al., 2002).

orders of magnitude, mainly because of differences in pH and salt content of the soil waters, and the abundance and characteristics of the organic matter and the metals (Degryse et al., 2009). A variety of elements known to be bound to organic matter in soils are found in speleothems (Borsato et al., 2007) and in section 5.3.5, we discuss the controls on their release and transport.

The soil microbial community is also a source of soil CO_2, along with root respiration (Witkamp & Frank, 1969); this CO_2 generates the soil water acidity that enhances hydrolysis reactions, and in a karst context provides a source of acidity for the dissolution of limestone soil clasts and bedrock. One gram of soil may contain up to 10 billion bacteria, made up of 4000–7000 species (Dubey et al., 2006). The relative importance of root and microbial respiration in soil CO_2 concentrations is not well understood; the efflux of CO_2 is caused by these two factors plus carbonate dissolution and oxidation. Root respiration is variously estimated to be between 10 and 90% of total efflux (Pumpanen et al., 2003). CO_2 is transported to the atmosphere by concentration-controlled molecular diffusion: CO_2 losses are via dissolution in soil water and chemical reaction with soil minerals and are estimated to be less than 10% of that diffused. Empirically, it has been shown that soil CO_2 concentrations are a function of temperature and soil moisture, because respiration increases with temperature and is limited by soil moisture. For example, Brook et al. (1983) and Raich and Potter (1995) developed regression model approaches to predict soil CO_2 based on AET and temperature

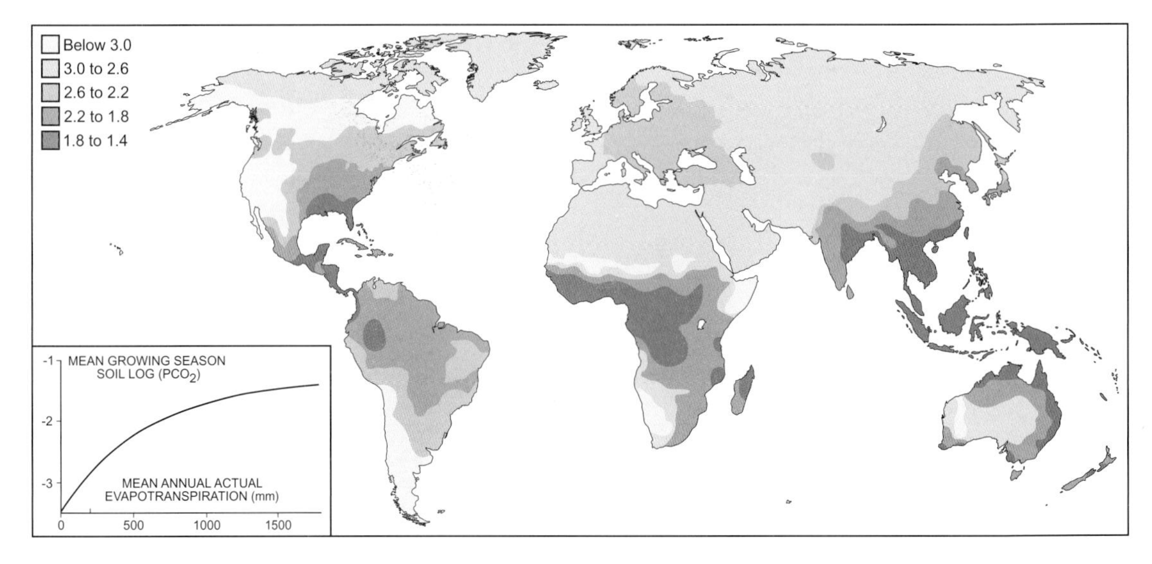

Fig. 3.17 Predicted soil log P_{CO_2} concentrations based on the observed, modern empirical relationship between P_{CO_2} and mean actual evapotranspiration (adapted from Brook et al., 1983).

and precipitation respectively. Raich and Tufekcioglu (2000) demonstrated that temperature, moisture availability and substrate properties that controlled the production and consumption of organic matter were in most cases more important in controlling soil respiration than vegetation type. Figure 3.17 presents the latitudinal and seasonal variations in modelled soil respiration, based on the empirical regression approach of Brook et al. (1983). More recently, modelling approaches (including Solomon & Cerling, 1987; Simunek & Suarez, 1993; Fang & Moncrieff, 1999; and Pumpanen et al., 2003) have been developed to predict soil efflux and concentration for a range of environmental conditions.

Important clues to soil processes and climate come from study of carbon isotopes, both the abundance of radiocarbon (^{14}C) and the $\delta^{13}C$ signature (Fig. 3.18). Primary production by higher plants follows two main alternative photosynthetic pathways (C_3 and C_4), which have quite different $\delta^{13}C$ signatures, although both are lower than that of the atmosphere, and even more so compared with typical carbonate bedrocks (Fig. 3.18a). The modern atmosphere contains ^{14}C (Fig. 3.18a) continuously created from ^{14}N by cosmic rays in the stratosphere and which exhibits radioactive decay with a half-life of 5730 years. The reference level of 100% modern ^{14}C is set at the level in 1950, before a period when atmospheric nuclear tests created an excess of ^{14}C which is slowing decay back to its natural values (Fig. 3.18b; it was 106pmc (percent modern carbon) in 2010 AD). This 'bomb spike' can be preserved in modified form in speleothems and has been used as a tool for demonstrating their modern origin (see Chapter 9) and for examining the nature of the soil carbon pools (Genty et al., 1998; Genty & Massault, 1999). Stalagmites also display a proportion of dead carbon (i.e. old carbon lacking ^{14}C), expressed as the *dead carbonate percentage* (dcp). This primarily reflects the dissolution of bedrock, but also arises whenever old carbon in the soil is metabolized (Genty & Massault 1999; Genty et al., 2001a). Figure 3.18a implies that the dcp should correlate with the $\delta^{13}C$ signature. This often occurs, but puzzles remain in interpreting the ^{14}C – $\delta^{13}C$ systematics of speleothems (Genty et al., 2001a). A way forward is to use a dynamic modelling approach. We illustrate such an approach for soils below and discuss the modifications to carbon isotope signatures owing to dynamic processes in the karst and cave environment in Chapter 5.

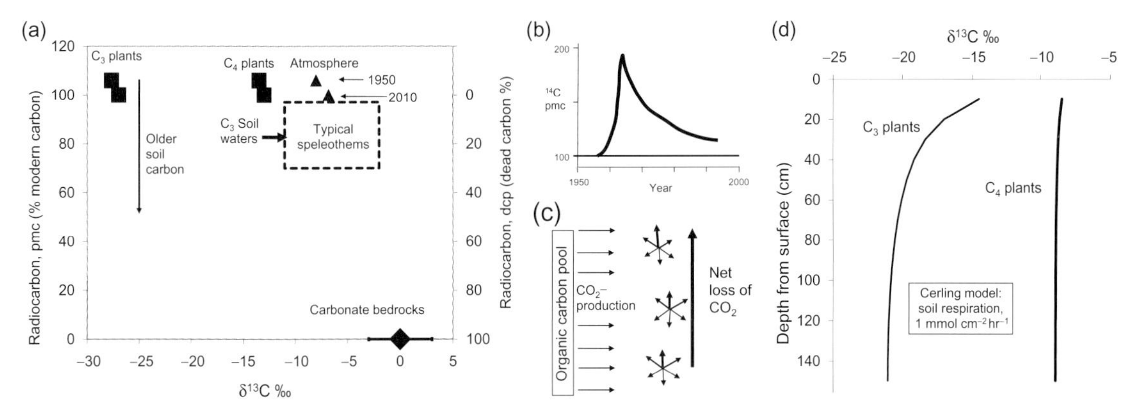

Fig. 3.18 (a) Carbon isotope composition of speleothems compared with carbon sources: organic matter in soils, carbonate bedrocks and atmospheric CO_2. (b) Generalized 'bomb spike' in radiocarbon in the late 20th century. (c) Structure of Cerling's (1984) dynamic soil carbon model. (d) Example output of depth variation in $\delta^{13}C$ composition of soils dominated by C_3 or C_4 plants using this model (see section 5.3 for a detailed analysis of carbon isotopes).

A dynamic approach so far has only been developed for a two-component (atmosphere – organic carbon) system (Cerling, 1984), but in principle a three-component system (including a $CaCO_3$ source for carbon) could be required to explain the composition of some soil carbonates (Sheldon & Tabor, 2009), and would certainly be essential for speleothems. Cerling's (1984) model is simple in principle, involving the steady production of CO_2 by microbial oxidation and plant respiration throughout a given soil interval and the net loss of carbon by diffusion of CO_2 gas through soil air into the atmosphere (Fig. 3.18c). Only a few parameters have to be assigned: principally the soil porosity and tortuosity (path length) and the CO_2 production rate. The model displays vertical gradients in $\delta^{13}C$ and CO_2 concentration reflecting the consequences of diffusion. At any point there will be an effective admixture of atmospheric and organic components, with the atmospheric air component dominating near the soil surface: the penetration of air into the soil arises because diffusion arises from random motions of molecules, although the net effect is a loss of CO_2 from the soil. Because lighter C atoms diffuse quicker, the $\delta^{13}C$ of CO_2 remaining in the soil is also higher than that which is lost (the soil-respired CO_2) by around 4.4‰ (Cerling 1984), but see Davidson (1995) for variations. The model clearly shows the difference between C_3 and C_4 vegetation (Fig. 3.18d) and the history of evolution of C_4 grasslands has been determined from soil carbonates, by assuming precipitation of soil carbonate in isotopic equilibrium with gas (Cerling et al., 1993).

A first attempt to apply such approaches has been made here (Fig. 3.19) using the data of Frisia et al. (2011), which refers to a brown forest soil above the Ernesto cave, northeast Italy (Fig. 3.20; Frisia et al., 2011). The soils show pronounced seasonal variations in soil P_{CO_2} with the deeper soil sampling site displaying a lagged response and values do not drop so low in winter. The Cerling model approximates the combined P_{CO_2}–$\delta^{13}C$ conditions under different combinations of productivity and soil porosity, but only by using a soil-respired $\delta^{13}C$ of –29‰ which compares with a –27.7‰ expected for grassland soils under early 21st century conditions. Note that the assumption of a constant CO_2 production with depth appears reasonable. This is a promising result for understanding soils under relatively dry conditions and in places where the soil volume has predominantly air rather than water in its pores. Note that there is no hint in the CO_2 data of the presence of isotopically heavy carbon from dissolution of carbonate and so it can be inferred that the radiocarbon activity of the soil CO_2 at depth

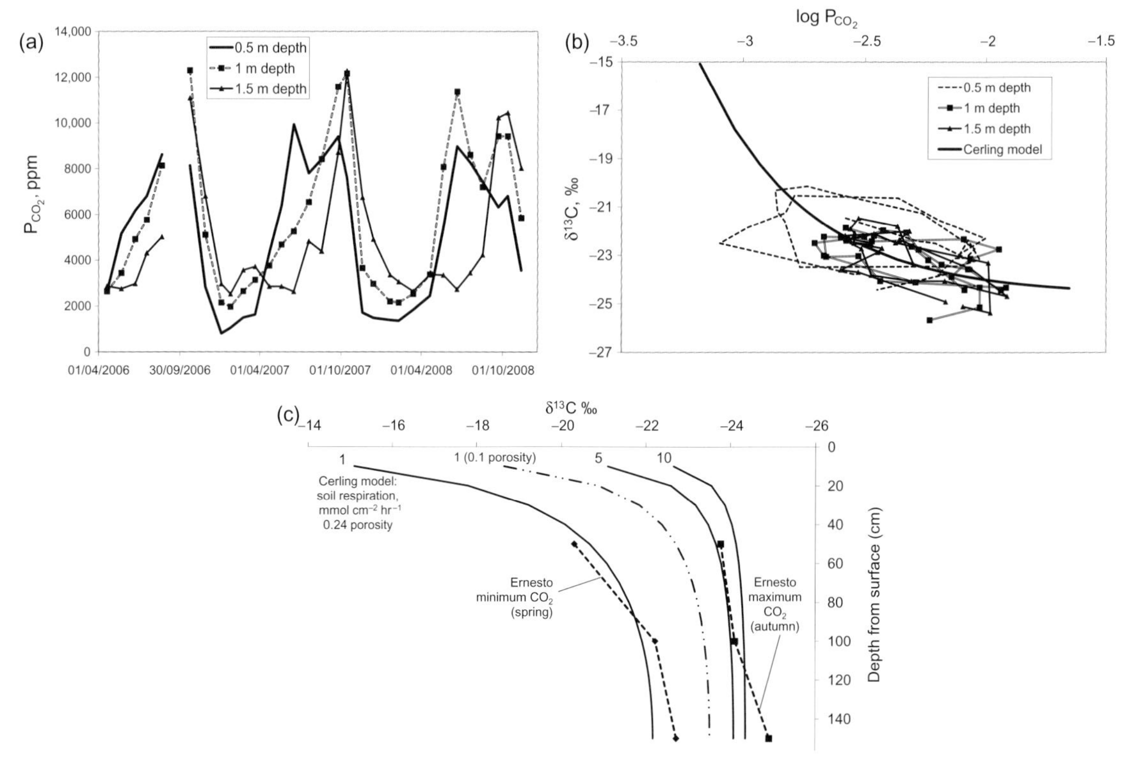

Fig. 3.19 Data from soils above the Ernesto cave (Frisia et al., 2011) compared with the Cerling model). (a) Time variation in P_{CO_2} in soil air at three different depths: note the strong seasonal variation and the tendency for values from deeper in the soil to be lagged and to reach less extreme low values in the spring. Temperatures of soil lysimeters at 1 m depth display a seasonal temperature range of 2–15 °C. (b) Cross-plot demonstrating the weak negative correlation between P_{CO_2} and $\delta^{13}C$ isotope values of the CO_2. Predictions of the Cerling model (with a soil-respired CO_2 value of −29‰ and production rate of 1 mmol m^{-2} hr^{-1}; values are irrespective of porosity and tortuosity). (c) Variation of $\delta^{13}C$ of CO_2 with depth from observations compared with models. The minimum CO_2 values can be fitted with a productivity of 1 mmol m^{-2} hr^{-1} given a porosity of 0.24 and tortuosity of 0.6 as used by Cerling (1984). However, the predictions are offset if a different porosity is assumed and the same result is obtained with a porosity of 0.1 and a production of 0.4 mmol m^{-2} hr^{-1} (this may be more realistic if there is significant pore water blocking the passage of air flow).

would be close to that of the mean soil organic carbon being converted to CO_2. However, the stalagmites at this site display evidence of a dcp raised by carbonate dissolution (Fohlmeister et al., 2011b). Hence this dynamic approach now needs to be extended to conditions during which water is infiltrating through the soil and recharging the epikarst; we will demonstrate some features that such a model could display in Chapter 5.

The above analysis has examined conditions at a specific time, but there is considerable interest in understanding how the soil CO_2 production will be influenced by changing temperatures. An important concept especially relevant to speleothem researchers is the Q_{10} value, which is the factor by which respiration is multiplied for a 10 °C increase in temperature and which often falls within the range of 2 to 4. This value is a crucial parameter for empirical and modelling studies of soil CO_2 concentration, and whose value is known to increase exponentially with temperature at low temperatures, and vary with soil moisture, organic matter

Fig. 3.20 Soil gas and water samplers installed above Ernesto Cave showing water collection by syringing from a 1-m-deep lysimeter. Methods for soil sampling are described in Tooth and Fairchild (2003), Spötl et al. (2005) and Frisia et al. (2011).

lability and drought stress (Fang and Moncrieff, 2001; Reichstein et al., 2002). Davidson et al. (2006) reviewed the processes that can affect the reported values of Q_{10}. Interestingly, the data of Fig. 3.19 imply a Q_{10} value of around 3.8, although this is complicated by the hysteresis of P_{CO_2} and temperature.

The overview of soil formation processes leads to the conclusion that it is the water balance and organic content that determines the amount of material leached from a soil profile, and the bedrock provides the raw materials for chemical and physical weathering. In karst regions, one would expect the bedrock to be relatively permeable, allowing the free draining of soil profiles and aerobic conditions. An exception would be where limestone-shale sequences are present and a surface shale bed causes waterlogging, or where a surface deposit of low permeability, glacially deposited clay forms an aquitard. Therefore a wide range of soil types are found on karst landscapes, whose characteristics are primarily dependent on the water balance and temperature, and whose CO_2 concentration depends on temperature, soil moisture and organic matter supply and composition.

3.3.2 Soil development through time

In section 3.3.1, the soils have been viewed as a modern 'snapshot', with no consideration of their development through time. In reality, the rate of formation of soils, and whether soils observed today are a final equilibrium product or in disequilibrium, are active research questions. Soils are increasingly being viewed as being dynamic, open systems, which may never reach a long-term equilibrium state. Literature estimates of soil age vary owing to the difficulties of dating soil profiles: peat accumulations are readily dated and have been shown to have accumulated since the last glaciation in arctic regions. Soil 'A' horizons have been suggested to take 10–100 years to develop; conversely, deep oxic soil horizons in tropical environments have been suggested to have formed over millions of years. Figure 3.21, generalized from observations in various regions, shows a soil profile development for limestone bedrock in climate zones sufficiently arid (<750–1000 mm yr^{-1} (Cerling, 1984)) for carbonate accumulation in the soil to occur. The development of soil over time, and the associated changes in chemical weathering processes and leached material should be expected to be visible in speleothem palaeoenvironmental proxies covering appropriate time periods. Arguably the largest influence on soil properties has been the impacts of humans on soil since the mid-Holocene. These include the following:

- Soil loss due to deforestation and intensification of agriculture, evidence for which in lake proxy archives includes an increased sedimentation rate such as that observed in Finnish varved lake records (Tijander et al., 2003).
- Agricultural 'improvements' such as liming, burning and fertilizer use. The former increases soil pH and decreases the extent of hydrolysis and

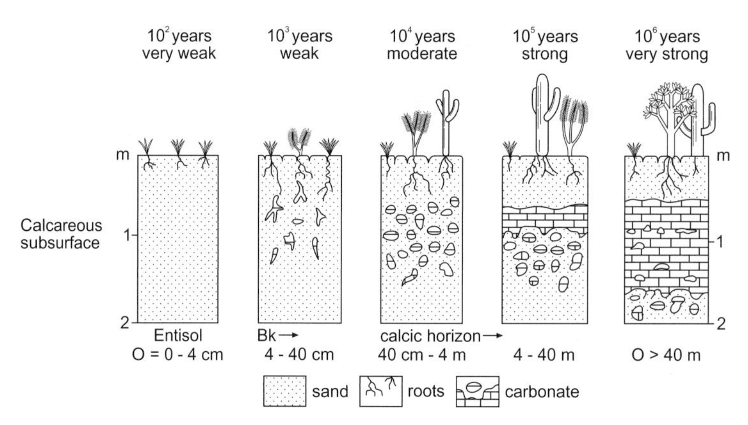

Fig. 3.21 Soil development over time for a calcareous subsurface (Retallack, 1998).

associated leaching products, whereas burning may generate polycyclic aromatic hydrocarbons (PAHs), which may be detected in speleothems (Perrette et al., 2008). The use of fertilizers increases soil nutrients, replacing those that may be lost by leaching through chemical weathering processes, but in itself may leach to provide a new groundwater metal and uranium fingerprint (Hamato et al.,1995).

• Physical changes in soil profile, for example the development of a plough horizon owing to deep ploughing, or the physical soil removal such as that caused by peat cutting for fuel.

• Changes in soil acidity, in poorly buffered soils, caused by acid rain pollution. A decrease in acid rain over recent decades in Europe and North America has led to an increase in the mobilization of soil dissolved organic carbon that had been artificially held in the soil during the previous period of acid rain (Monteith et al., 2007).

3.3.3 Concluding views on karst soils

Soils can be viewed as 'open systems', which in some environments may be in a dynamic equilibrium with climate and vegetation, but in many environments in a disequilibrium state, particularly through the impact of humans on the soil environment. Soils can also be viewed as important 'carbon stores', an increasingly important concept with a growing focus on carbon cycling and carbon sequestration in world of rising atmospheric CO_2 and global warming. In karst regions, soil types can be expected to be the same as found on other permeable bedrocks for the sample climate, with temperature, the water balance, and the quantity and character of organic matter, the key drivers of soil processes. Plate 3.4 demonstrates some of this diversity of karst soil types. Leached, soil derived, material that recharges to the groundwater may be preserved in speleothems, and soil CO_2 provides the primary driver for limestone dissolution and subsequent stalagmite formation. Soil organic matter that is leached into the groundwater provides the essential food for the groundwater and cave microbial ecosystems. Several speleothem climate and environment proxies may in part, or in whole, derive from the overlying soil, and are considered in detail in later chapters. These include stable carbon isotopes (see next section and section 5.3), trace elements (sections 5.2 and 5.5), dissolved organic matter (section 6.3) and annual growth rate (section 7.4).

3.4 Vegetation of karst regions

The inter-relationship between surface vegetation and the subsurface environment is dramatically exemplified in Fig. 3.22. At both global and regional scales, much of the variation in vegetation characteristics can be explained by the climate, in particular temperature and precipitation amounts and seasonality, which define the energy and moisture

Fig. 3.22 Deep-rooted vegetation in southwest Western Australia. Left, view of eucalypt forest from the underworld of Golgotha Cave. Right, stalagmite (15 cm high) covered with calcite-encrusted roots that have emerged from the cave floor seeking moisture, Moondyne Cave.

balances throughout the year. This has led to the definition of biomes, the world's major communities, classified according to the predominant vegetation and characterized by adaptations of organisms to that particular environment (Campbell, 1996). A biome might also be referred to as an ecosystem, and can be viewed as a major habitat type, albeit one which contains a large amount of variability. Biomes can be defined by the plant structures (e.g. trees vs. grasses), leaf type (broad leaf vs. needles), plant spacing (forest versus savanna) and climate. Different biomes will have different biodiversities, with biodiversity typically decreasing with temperature and moisture availability. Anthropogenic biomes exist because of the effects of agriculture and urbanization: contrasting anthropogenic and natural biomes in karst regions are shown in Plate 3.5.

A global terrestrial biome map is presented in Plate 3.6. Climate can be seen to control the distribution of biomes. Temperature controls the latitudinal variations in biomes from Arctic to tropical, with habitat types also varying with altitude. Humidity and the seasonality of rainfall further define the biomes. Plate 3.6 does not include anthropogenic biomes that exist today because of urbanization and agriculture. The modern-day spatial pattern of biomes must be viewed as a snapshot of a dynamic and changing vegetation pattern; changes may occur both rapidly and gradually. Rapid change is most likely to occur over relatively small regions and might be the effects of fire, floods, landslides and storms. Slow change is more likely to be due to ecological succession, the change in biome composition as individual species adapt to changes in climate and environment over longer time periods. The combination of these two rates of change, driven by two different sets of processes, namely extreme (often nonlinear) events and slow (linear) environmental change, means that within a biome there is significant heterogeneity at a small scale, with different places being at different developmental stages owing to different local histories, in particularly the times since their last major disturbance.

Significant research effort has gone in to modelling the global pattern of biomes both today and for past climates. Prentice et al. (1992) developed a global biome model of plant physiology and dominance based on input parameters of soil characteristics and monthly climate. This approach has been subsequently developed into the BIOME family of models: BIOME4 (Kaplan et al., 2003) is the most advanced and is a publicly available coupled biography and biogeochemistry model which simulates 28 biomes from the input parameters of latitude (which controls temperature and solar radiation), atmospheric CO_2, mean monthly temperature, rainfall and sunshine and soil properties. Models such as BIOME4 are of interest to speleothem researchers as they can be used to simulate climate-vegetation interactions in the past (for example, Claussen, 1997). An example of a BIOME4 simulation of modern-day biomes is presented in Plate 3.6.

Speleothem proxies might directly derive from surface biomes (such as pollen preserved in speleothems, which may derive from both regional and global biomes; see section 6.2) or may be indirectly affected (such as proxies that are influenced by the water balance and therefore the vegetation evapotranspiration). Traditionally, speleothem researchers have spent very little time investigating the surface vegetation characteristics or history, exceptions being Charman et al. (2001) who compared vegetation, peat and stalagmite environmental archives, and some modern calibration studies of cave pollen records (see section 6.2). Stable carbon isotopes have been interpreted as an indication or surface vegetation photosynthetic pathway (see Chapter 5). A new global map of C_3 and C_4 vegetation types is presented in Plate 3.7 (Edwards et al., 2010). In general, the environmental controls on C_3 and C_4 plant distribution at the biome scale relate to the CO_2 crossover leaf temperature during photosynthesis; above this temperature C_4 plants are more competitive at photosynthesis, and good correlations exist between the fractional contribution of C_4 plants and summer or growing season temperature (Wynn & Bird, 2008). The relationship between the C_3:C_4 plant ratio and climate variability is the focus of recent research using the BIOME4 model (Flores et al., 2009; Hatté et al., 2009) and changes in soil $\delta^{13}C$ with depth and over space are routinely used in the modern environment to investigate recent changes in the forest–savanna boundary (Boutton et al., 1998; Wynn et al., 2006).

An understanding of biomes and how they change over time can assist the interpretation of speleothem proxy records. In particular, vegetation 'sweet spots' such as sites at the boundary between two biomes could be the target for research projects investigating climatically forced changes in biomes through time. Conversely, researchers wishing to minimize any impact of vegetation change over time might seek to work close to the central region of a biome. Rapid changes in proxy records of vegetation need not be indicative of fast climate change, but rather of nonlinear changes in vegetation caused by local extreme events.

3.5 Synthesis: inputs to the incubator

Chapter 3 has focussed on the surface environments which determine the inputs of chemical species to the incubator. In the following chapters, the processes of speleothem formation are considered in detail, but here we summarize how the water supplying individual speleothems is closely related to the surface environment. Groundwater supply is the primary requirement for speleothem formation, which is determined by the surface water balance. Groundwater recharge is likely to vary seasonally because of variations in the relative timings of any wet seasons and maximum temperatures. The magnitude of recharge will determine the transport of soil-derived colloidal and organic material, and the relative timing of any seasonal groundwater recharge with respect to antecedent climate conditions will determine the characteristics of this material. The stable isotopic composition of this water reflects both the surface water balance and the moisture source of the precipitation (LeGrande & Schmidt, 2009). Other factors being equal, faster speleothem growth correlated with the amount of CO_2 dissolved in the groundwater. Here we have shown that soil CO_2 concentrations are primarily determined by the soil P_{CO_2} concentration, which is determined by soil microbial and root respiration. Both are primarily driven by temperature, with a secondary influence of soil moisture. Finally, soil-derived organic matter provides the essential energy to drive the within-cave and groundwater ecosystem (Simon et al., 2007). Overall, one can summarize that the inputs to the incubator are climatically determined, although predominantly a mixed signal of both temperature and precipitation, and provide essential energy inputs.

II
Transfer processes in karst

CHAPTER 4
The speleothem incubator

In this chapter we emphasize the energy and mass flows that maintain conditions within cave chambers. In appropriate circumstances, these flows allow the formation of speleothems. In this case, the cave chamber and the architecture of its surroundings define a speleothem life-support system or *speleothem incubator* as introduced in Chapter 1. Following an introduction to the basic physical parameters and transport processes, we deal separately with the issues of water movement, air circulation and heat flow and then synthesize the outcomes in terms of cave climatologies.

4.1 Introduction to speleophysiology

Speleophysiology was introduced by Fairchild et al. (2006c) as a conceptual term to draw attention to the behaviour of cave environments as functioning systems on an environmental scale, intermediate between organism physiology on the one hand and planetary physiology (Lovelock, 1988) on the other. There are many parallels between caves, organisms and planets, for example, in terms of their dynamic fluid motions and exchanges of carbon dioxide. Unlike organisms, the constant temperature of caves is a function of thermal insulation rather than of active homeostatic processes. Nevertheless, the presence of ice-accumulating caves in settings with mean annual temperatures above zero points to a specialization of function that is consistent with the physiology concept. Another simile is that their functions can be impaired by anthropogenic abuse, for example the deterioration of prehistoric cave paintings from excessive visitor numbers (Fernandez et al., 1986).

Preservation of such archaeological remains has stimulated a strong research methodology, pioneered by scientists such as Claude Andrieux (see, for example, Andrieux 1969) encompassing measurements of radon, carbon dioxide, pressure, humidity and temperature, which has produced some fine studies of cave climatology (e.g. Fernandez et al., 1986; Sánchez-Moral et al., 1999; Bourges et al., 2006b; Cuezva et al., 2011; Fernandez-Cortes et al., 2011). An analogous research driver has been concern about the health effects of radioactive progeny of radon gas in caves (Wilkening and Watkins, 1976; Fernandez-Cortes et al., 2009). Such studies include those aimed at protecting speleothems from the corrosive effects of condensation (e.g. Fernandez-Cortes et al., 2006; de Freitas & Schmekal, 2003). Conversely, most of the cave monitoring work directed at understanding the geochemistry and growth rates of speleothems has had a narrower focus on temperature measurements on the one hand, and dripwater hydrology and hydrogeochemistry on the other, sometimes supplemented by measurements of carbon dioxide. Logistical constraints have limited study at many sites: only few caves have published records over at least two annual cycles, the norm in surface hydrological studies. However, there are several multi-year studies orientated towards understanding speleothems, which are progressively reaching

Speleothem Science: From Process to Past Environments, First Edition. Ian J. Fairchild, Andy Baker.

the public domain, based on caves in Europe (e.g. Genty et al., 2001b; Spötl et al., 2005; Mattey et al., 2010; Miorandi et al., 2010; Frisia et al., 2011), China (e.g. Hu et al., 2008b) and Australia (e.g. McDonald et al., 2007; Treble et al., 2009).

An important consideration related to the regulation of the speleothem incubator is that of the fate of incoming signals of changing temperature or air or water composition. In karst studies, this approach can be traced back to the study on calcite dissolution in capillaries by Weyl (1958). The use of dimensionless numbers for such broad problems is now quite widely used in theoretical and empirical hydrogeological and geomorphological studies and was applied to cave atmospheres by Wigley and Brown (1971). Recently, Covington et al. (2011, 2012) have provided a generalized approach, using the one-dimensional case of karst conduits for simplicity (Fig. 1b). They have shown that if there is a process of change acting linearly on a fluid of constant velocity V, the change in parameter C from its input value C_0 to its final value C_∞ takes place exponentially. Hence

$$(C_{(x)} - C_\infty)/(C_0 - C_\infty) = e^{-x/\lambda_p} \qquad (4.1)$$

where x refers to a distance along the flow and λ_p is a relaxation parameter called the *process length*. They showed that it is possible to add multiple effects, for example flow from the matrix, to the specified conduit flow, and the approach can be used to describe different parameters including heat and extent of calcite dissolution. The analysis can be used in reverse to deduce characteristics of the water flow path. For example, transmission of a thermal signal to the outlet stream requires large diameter conduits or fast flows. Where the process is nonlinear, the concept still applies, but the mathematical formulation will differ. Covington et al. (2012) identified future applications to studies of, for example, cave ecology, sediment transport and dripwater characteristics, although issues of how to combine characteristics of multiple flow paths need to be examined.

We can summarize the length-scale concept in terms of the fate of an external signal from the environment. At one extreme, the signal is trans-

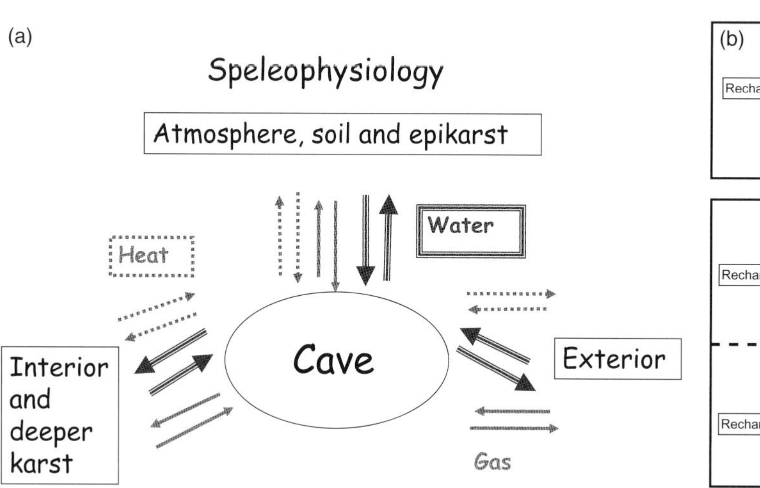

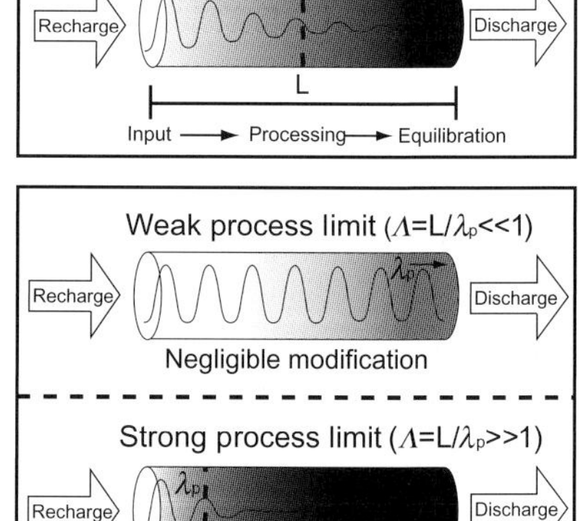

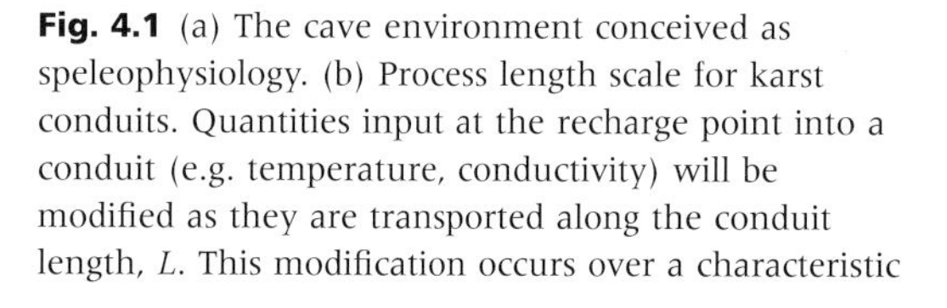

Fig. 4.1 (a) The cave environment conceived as speleophysiology. (b) Process length scale for karst conduits. Quantities input at the recharge point into a conduit (e.g. temperature, conductivity) will be modified as they are transported along the conduit length, L. This modification occurs over a characteristic length scale, λ_p, (process length). Shading and wave amplitude indicate the extent of modification, which can be quantified using the dimensionless process number, $\Lambda = L/\lambda_p$. When $\Lambda << 1$, little modification occurs. When $\Lambda >> 1$ the process reaches equilibrium before discharge. From Covington et al. (2011).

mitted to the cave interior and is expressed as deviations in the time series of the cave environment. At the other extreme, the signal is suffocated by the inertia of the karstic aquifer and merely contributes to the development of its long-term mean value.

4.2 Physical parameters and fluid behaviour

In this chapter we focus on physical parameters, together with measurements on radon and carbon dioxide concentrations of cave air, because they reveal important information on cave dynamics. Otherwise, chemical aspects are discussed in Chapter 5.

4.2.1 Measurement of parameters

We do not aim to provide a detailed guide as to how to undertake cave monitoring, but Table 4.1 lists several useful parameters, together with an indication of the potential quality of the results under the best circumstances and Fig. 4.2 illustrates some successful installations. Many difficulties are found in practice in physically locating the instrumentation, providing electrical power, and avoiding problems of equipment failure by moisture condensation. Although it is possible to make continuous logs of all of the parameters in Table 4.1, spot measurements still have considerable value, particularly as part of a traverse which defines gradients in properties. Even in the best-funded projects in the most sensitive locations, we still lack enough knowledge about the values of different parameters and how they vary in space and time. It is hoped that this chapter will help draw attention to the key issues that require specific programmes of observations to distinguish between different hypotheses of cave processes.

4.2.2 Static parameters in air

Some important behavioural attributes of air in cave and karst environments stem from the basic static physical properties, whereas for water it is the static *chemical* properties that are crucial. The ability of air to hold water vapour is strongly temperature-dependent (Fig. 4.3), condensation occurring when the saturation vapour pressure is exceeded. The *relative humidity* represents the proportion of the saturation vapour pressure of water that is present in a given case. Figure 4.3 illustrates that, for example, cooling air with 80% relative humidity at 25 °C results in saturation at around 21 °C.

Air density is the key parameter affecting air motion in caves. Predictions of air density dependent on temperature and pressure can be made using the gas law modified for water vapour:

$$\rho = (pM_a/RT)[1-((1-\varepsilon)(H_r p_{vs}/100p))] \quad (4.2)$$

(Fernandez-Cortes et al., 2009)

where p is atmospheric pressure (atm), M_a is the molecular weight of dry air (28.97 g mol^{-1}), ε is the ratio between the molecular weights of water vapour and dry air (0.6220), R is the gas constant (0.08206 L atm mol^{-1} K^{-1}), T is air temperature (K), H_r is relative humidity (%) and p_{vs} is the saturation vapour pressure (atm). An alternative formulation substitutes for temperature in the gas law a virtual temperature, which is the temperature at which a theoretical dry air parcel would have a total pressure and density equal to a real natural air parcel (Kowalczk & Froelich, 2010). Figure 4.4 (from eqn. 4.2) illustrates the large effects of atmospheric temperature and pressure on air density. Representative pressures of 1013 mb (atmospheric mean) and 980 mb (typical of a moderate depression) are used for illustrative purposes. Another effect is the reduction in pressure with altitude; for example in a local area a sea-level pressure of 1013 mb would reduce to 980 mb at around 280 m altitude. Whereas addition of water vapour decreases air density, carbon dioxide has the opposite effect (Fig. 4.4). The effect of carbon dioxide has been recognized to be sufficiently important to be factored into a revised general definition of the virtual temperature allowing for changes in trace composition in air in addition to water vapour. Such a re-definition aids calculations of atmospheric processes, particularly those involving gas exchange with the vadose zone, as well as helping predict the movement of underground air (Kowalski & Sánchez-Cañete, 2010).

Table 4.1 Techniques for monitoring cave climate and dripwater with indicative measurement errors. Information compiled from Hakl et al. (1997); Buecher (1999); Dueñas et al. (1999); Muramatsu et al. (2002); Perrier et al. (2004); Cigna (2002, 2005); Spötl et al. (2005); Bourges et al. (2006a); Fernandez-Cortes et al. (2006, 2009); Baldini et al. (2006a); Faimon et al. (2006); Collister and Mattey (2008); Mattey et al. (2010) and various manufacturers' websites.

Parameter and key references	Technique	Sensitivity/precision of measurement
Air and water temperature	a. Mercury thermometer (dry-bulb) b. Electronic thermocouple c. Optical fibre distributed temperature sensing	a. 0.05 °C b. Accuracy 0.01–0.05 °C, resolution 0.001–0.01 °C with careful calibration c. Spatially variations in temperature with precision of up to 0.01 °C
Air pressure	Electronic pressure sensor	0.1 mb (millibars = hPa) precision
Humidity	a. Chilled mirror (dewpoint) hygrometer b. Capacitative sensor c. Ventilated wet bulb versus dry bulb temperatures (whirling hygrometer)	a. 1% accuracy, 0.1% resolution; operative to 99+% humidity b. 2% accuracy, 0.1% resolution; inoperative above 98% relative humidity c. 2% precision; operative to 95% humidity. Evaporimeters could be useful close to 100% humidity (Buecher 1999).
Carbon dioxide concentration	a. Infra-red spectroscopy (e.g. Vaisala CARBOCAP® GM70 hand-held instrument, or GM353 for higher precision) b. Reagent tubes (e.g. Draeger-tubes™)	a. Resolution 1–10 ppm, accuracy of spot readings 1–10% depending on concentration and probe, improved by averaging logged data. b. 10% precision Breathing apparatus may be necessary.
Radon activity	a. Real-time sampling (e.g. alpha-counting, or continuous measurement of radon progeny by Instant Working Level Meter or spot measurements by air sampling and counting decay events on a filter) b. Cumulative sampling: e.g. by permeater or etched track detector	a. Accuracy 5–10%, but precision depends on count times (e.g. Pylon AB5 instrument has precision of 0.041 cpm $Bq^{-1} m^{-3}$) b. Precision >10%; complicated by behaviour of decay products
Air movement	a. Observation of vapour or smoke b. Hot-wire anemometer (non-directional) c. Ultrasonic/Doppler anemometer	a. <1 $cm s^{-1}$; semi-quantitative b. 0.1 $m s^{-1}$ detection limit, accuracy 5% c. 1 $cm s^{-1}$
Drip discharge	a. Collection of water in graduated container b. Measurement of drip interval	a. 1% accuracy over short time periods b. 1% if drip volume known; can be logged using acoustic triggering, or breakage of infra-red beam
Discharge of cave streams	Construction of rating curve by calibration of depth (stage) measurements (e.g. by pressure transducer) using independent technique such as salt dilution	Typical 10% accuracy

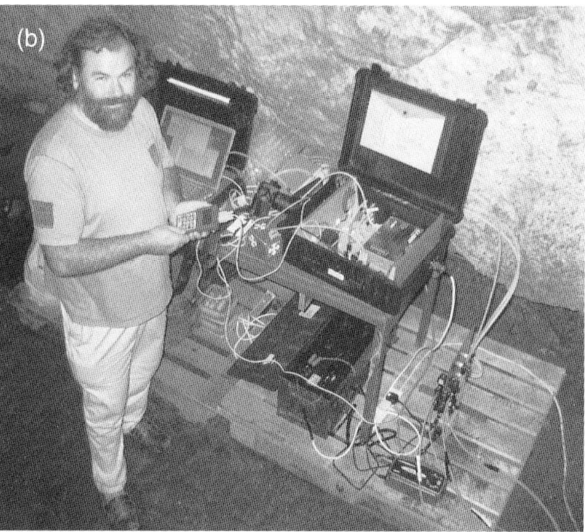

Fig. 4.2 Cave air monitoring. Left, system designed by José-Maria Calaforra in Soplao showcave (Spain) for monitoring temperature, humidity and P_{CO_2}. Right, Dave Mattey with his bespoke system for drawing air from different locations in Lower St. Michaels Cave (Mattey et al., 2008, 2010) and sequentially analysing P_{CO_2}.

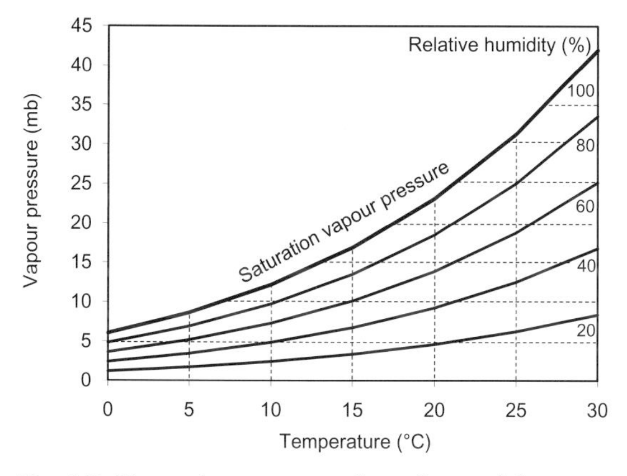

Fig. 4.3 Observed temperature-dependency of the saturation vapour pressure of water in air plotted with relative humidity values.

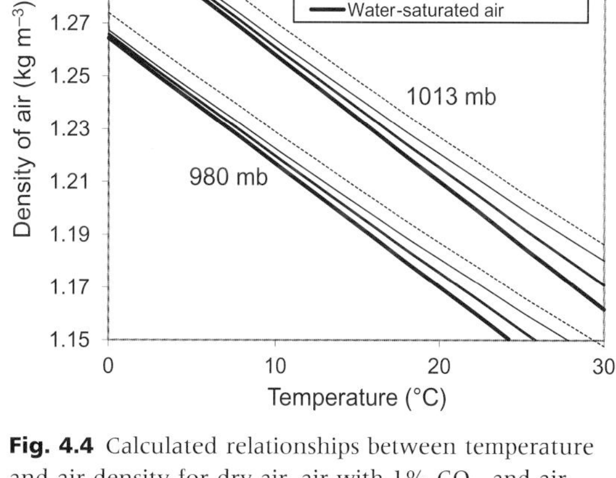

Fig. 4.4 Calculated relationships between temperature and air density for dry air, air with 1% CO_2, and air with 50% and 100% humidity, at two representative air pressures at sea level.

4.2.3 Dynamic fluid behaviour: laminar versus turbulent flow

Slowly advecting fluids are dominated by viscous forces and show linear decreases in velocity towards the edge of the flow. This *laminar flow* contrasts with *turbulent flow* characterized by chaotic eddies. Turbulent flows keep particles up to a certain size in suspension and are strongly retarded if rough flow boundaries are present. The criterion for the onset of turbulence is the exceedance of a critical value (2250) of the dimensionless Reynolds number (*Re*):

$$Re = \rho UD/\mu \qquad (4.10)$$

where ρ is the fluid density (dimensions ML^{-3} where M is mass and L is length), U is the flow velocity (LT^{-1} where T is time), D is a characteristic length (L) which for pipe or tube flow is defined as the hydraulic diameter ($4A/P$ where A is the

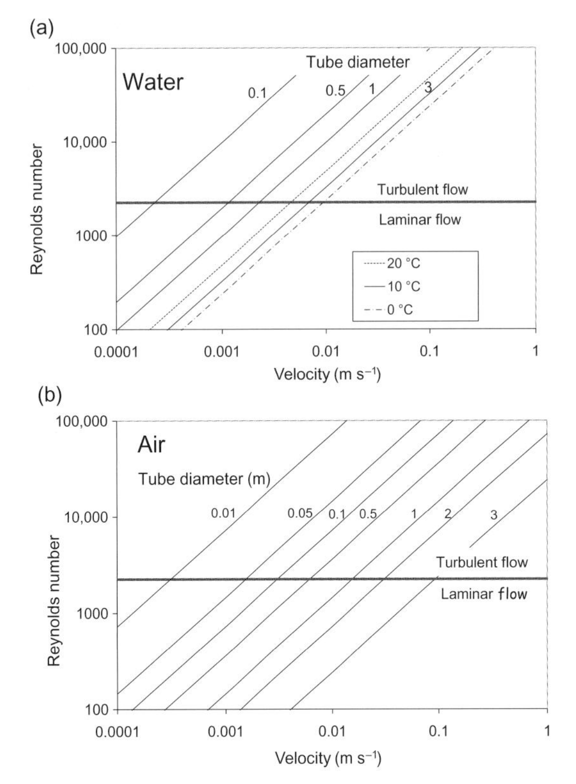

Fig. 4.5 Relationship between flow velocity and onset of turbulence for cavities of circular diameter (or other shapes with the same perimeter as the circle). The effects of temperature on water are shown for the 3 m case, but have similar offsets from the 10 °C lines for other diameters. For air, calculations are for 10 °C, but there is little variation with temperature. To become turbulent, air flows need to be much faster than do water flows, or to lie within in smaller tubes.

cross-sectional area and P is the perimeter), and μ is the dynamic viscosity ($ML^{-1}T^{-1}$).

Figure 4.5a illustrates that temperature has a moderate effect on Reynold's number, arising from the increased viscosity of colder waters (a similar effect arises from high loads of suspended sediment). Although there is a lower limit of about 1 mm diameter for turbulence to be developed (Chapter 2), generally even very slow flows in narrow tubes are turbulent and even in a 3 m diameter tube the flow rate only needs to be 1 cm s^{-1} for the onset of turbulence. By contrast, air (Fig. 4.5b) shows little temperature-dependency because both density and viscosity change by similar amounts, and flow velocities need to be around 10 times greater than in water for turbulence to occur. The terms 'static' and 'dynamic' have been applied both to the causes of air circulation in caves (Cigna, 1967) and to contrast caves with slow laminar flows, from those with turbulent flows (Wigley and Brown, 1971). In practice there is interest in speleothems forming in cave chambers with a range of airflow conditions.

Movement of water through porous rock has been given more attention in the karst literature than the movement of air. A relevant issue is that the ratio of fluid density to viscosity (around 10 times higher for water than air) controls the hydraulic conductivity of a permeable medium and hence air would require an order of magnitude higher pressure gradient to display the same flux through a dry permeable rock as would water through the same rock mass if it were saturated. Nevertheless, it is increasingly realized that such air movements need to be taken into account in understanding karst dynamics (Cuezva et al., 2011).

4.2.4 Dynamic fluid behaviour: advective versus diffusive transport

Advection refers to movement of fluid under forces, whereas *molecular diffusion* arises by the random (Brownian) motion of molecules and tends to reduce concentration gradients of solute species. Quantitative analysis of changing cave conditions requires the relative magnitude of these processes to be known. In the case of dripwaters, because they penetrate under gravity relatively rapidly, transport is normally assumed to be entirely advective. However, the necessity for considering some exchange of solutes between pores and fractures has been demonstrated by Fairchild et al. (2006b). This exchange might be related to diffusion or dispersion caused by convoluted flowpaths and awaits detailed modelling. In some geometrically simple cases, air movement is more tractable for analysis as developed below.

Diffusion between two adjacent and static air masses can be predicted by Fick's laws. Fick's first law (see, for example, Berner, 1980) states that the

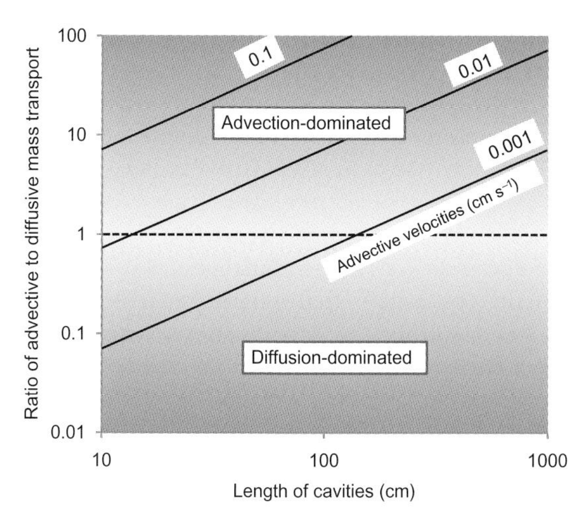

Fig. 4.6 Ratio of mass transport respectively by advection and diffusion versus length of cavities across which transport occurs, based on eqn. 4.10. Diffusion is the more important only under conditions of extremely slow advective air movement and short distances of transport.

diffusive flux (denoted F with dimensions $ML^{-3}T^{-1}$) can be calculated as

$$F = -D_t(dC/dx) \tag{4.3}$$

where the gradient of concentration with distance is denoted by dC/dx, D_t is the tracer diffusion coefficient and the negative sign is a convention. An example of the application of Fick's law to a gradient in carbon dioxide in a cave is given by Baldini et al. (2006a) who showed that, even over the large distance of 50 m, the gradient would be eliminated within 3 years; likewise gradients at the metre-scale would be eliminated within a small number of weeks. Hence, CO_2-rich air pockets, which are a danger to cavers (Smith, 1997), are not in themselves stable (Badino, 2009), but must be continuously maintained by other production or transport factors. Diffusion in water is very slow: values of D are around 1000 times lower than in air and so diffusion can be a rate-controlling process (e.g. for speleothem growth, see section 7.1). Rigorous calculations of diffusion require the use of Fick's second law to calculate changes over time and must deal with the complexities of simultaneous diffusion of different components, possibly in different directions. Where both carbon dioxide and temperature differ between two adjacent air masses, complex double diffusion phenomena can be anticipated (Bourges et al., 2006a). Up to now, cave conditions have not been defined specifically enough to allow such conditions to be quantified.

Diffusion is regarded as dominant in the loss of carbon dioxide from soils because of the small size of the pores (Cerling, 1984), but the range of pore diameters in epikarst is larger and typically unknown in detail. We present below a general treatment of the circumstances where advection is more effective at moving matter in air than diffusion. We consider the case of a trace gas such as CO_2 exchanging between a cave passage and the external atmosphere via tubular connections. We neglect frictional drag because we are only dealing in this case with slow (laminar) advective flows. The number of separate cavities is irrelevant, but we can characterize the cavities in aggregate in terms of their mean length (L) connecting cave and exterior, and total cross-sectional area (A) perpendicular to L. An advective flux (F_{adv}) can be defined in terms of the product of concentration (C), here the difference in concentrations (ΔC) between outflowing air and inflowing atmospheric air, and velocity of air flow (V):

$$F_{adv} = \Delta C \cdot V \tag{4.4}$$

A diffusive flux (F_{diff}) occurs in still air by migration of trace gas down a concentration gradient and is expressed as

$$F_{diff} = D \cdot dC/dx \tag{4.5}$$

where D is the diffusion coefficient (with dimensions L^2T^{-1}) and dC/dx is the gradient of concentration with distance. dx corresponds to the length L and dC approximates ΔC in eqn. (4.4) and so eqn. 4.5 becomes

$$F_{diff} = dC \cdot D/L \tag{4.6}$$

If the total export of CO_2 from the cave is known from trace gas studies (see later), it can be expressed in an analogous way to the above fluxes as an observed mass flux (M, with dimensions MT^{-1}) per unit area (A):

$$F_{obs} = M/A \quad (4.7)$$

Hence for advection, combining (4.4) and (4.7)

$$M_{adv} = \Delta V \cdot A\, C \quad (4.8)$$

and for diffusion, combining (4.6) and (4.7):

$$M_{adv} = D \cdot A \cdot dC/L \quad (4.9)$$

The relative importance of the mass transports are plotted in Fig. 4.6 using the relationship in eqn. 4.10 where $0.14\,cm\,s^{-1}$ is substituted for D and ΔC is taken to be equal to dC.

$$M_{adv}/M_{diff} = V \cdot L/0.14 \quad (4.10)$$

It can be seen that even at extremely slow air velocities of $10^{-3}\,cm\,s^{-1}$ ($3.6\,m\,hr^{-1}$) diffusion is only more important in short connecting passages (<1 m). This approach was applied to the Ernesto Cave (northeast Italy) as part of a carbon mass balance study (Frisia et al., 2011) and it was found that advection must dominate air exchange with the exterior even though air movement was too slow to be measured directly.

4.3 Water movement

As described in Chapter 2, karstified carbonate rock is heterogeneous, highly fractured, and with a permeability developed such that water movement occurs below the surface. Three levels of porosity can be distinguished within karstified carbonate rocks: primary, secondary and tertiary (Ford & Williams, 2007). Primary porosity is that associated with inter-granular pore space, secondary porosity is associated with joints and fractures, and tertiary porosity with solution-enhanced conduits. Note that primary and secondary porosity in this hydrogeological sense have different meanings from their geological sense as used in Chapter 2. Water movement can occur through any or all of these structures, leading to a continuum of subsurface water flow pathways. At one extreme, flow may be rapid when associated with tertiary porosity, through interconnected conduits and caves, effectively working as underground rivers. At the other extreme is matrix flow: water movement associated with primary porosity. Here, water movement can be very slow and can be considered to form part of an unsaturated water store. Speleothems are primarily fed by waters that have passed through primary and/or secondary porosity structures. The observation of speleothem deposition implies that the dissolution has occurred earlier in the flow path and by implication that at least some flow has occurred through secondary porosity structures, either in the epikarst or within the bedrock itself.

Speleothems rely on a water supply for their continued survival and growth in the incubator. The characteristics of that supply will depend on the flow pathways to the individual speleothem, and therefore the properties of the karstified carbonate bedrock (and in some cases the properties of overlying non-carbonate strata) above the cave, as well as the surface climate (see Chapter 3). The actual water flow pathway for an individual speleothem is difficult to ascertain without empirical cave measurements of water flux (drip rate or discharge), because for a constant climate input it will reflect the connectivity between structural voids within the karst, the degree to which the permeability has developed over time, and the configuration of karst water stores, whether within the matrix, micro-fractures, dissolutionally widened fractures or structural voids within the karst (Bradley et al., 2010). Some hypothetical karst water flow pathways are illustrated schematically in Fig. 4.7, showing possible mechanisms by which water might be delivered to individual speleothems, although in reality combinations of these flow pathways will always occur. Water dripping onto stalagmite A comes from the overlying limestone bed. No fissures can be seen in this limestone bed, and it is therefore likely that the stalagmite is mostly fed by diffuse flow, through either the limestone matrix or through very fine fractures. In porous limestone bedrocks, a large number of small stalactites is an indication that matrix flow is occurring (Fig. 4.8). In this case, water movement will be a function of the primary porosity of the karst, with flow rates proportional to the matrix permeability. Where diffuse flow is dominant, rates of change of water movement are likely to be slow,

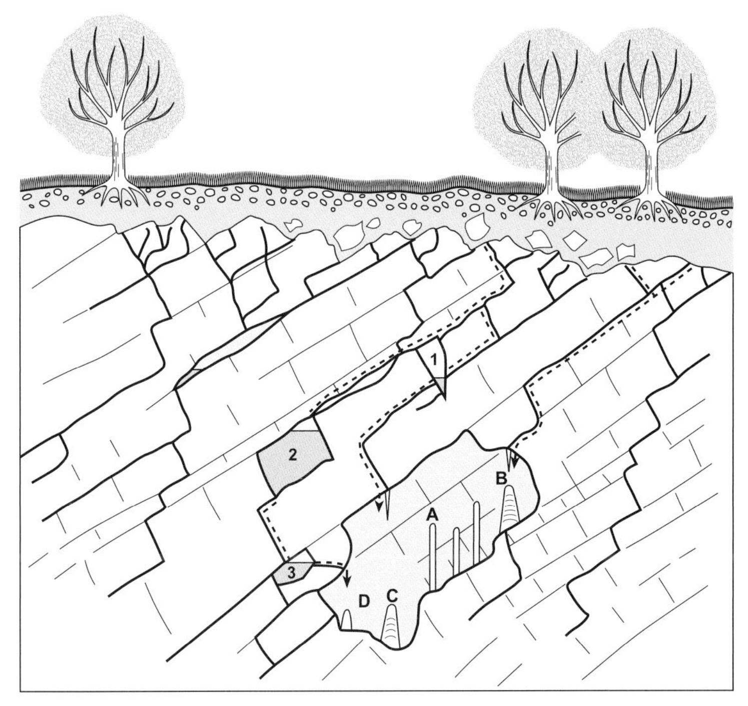

Fig. 4.7 Schematic illustration of possible karst water flow pathways to individual speleothems (after Bradley et al., 2010). See text for details of the possible flow routes to stalagmites A–D. 1, 2 and 3 are water reservoirs.

Fig. 4.8 Wide distribution of soda straws reflecting seepage flow to cave roof from overlying late Quaternary aeolian sandstone aquifer (Golgotha Cave, southwestern Western Australia).

with slow drip rates of low variability, which are likely to form stalagmites of a 'candlestick' shape (see Chapter 7). In contrast, stalagmite B is fed by a greater proportion of fracture flow. Flow is likely to be a mixture of relatively fast fracture flow, as shown by the dotted line, as well as slow diffuse flow from the overlying strata, similar to that postulated for stalagmite A. The proportion of fracture flow is greater than stalagmite A, and drip-rates are likely to vary over time, depending upon the mode of water delivery to the preferential flow system. The latter will reflect the surface water balance, including evapotranspiration rates and precipitation, in addition to potential water storage within the epikarst (Williams, 2008). Stalagmites formed by this water flow type are likely to form wider examples than those fed by matrix flow, and when fissure components of flow dominate, then flowstones, curtains and other flowing water related speleothems are likely to be formed (Figs. 4.9–4.12).

Fig. 4.9 Fracture-fed flow leading to linear cluster of speleothems (Goda Mea, Ethiopia). See also Box 7.1 for a consideration of speleothem shapes in this region.

Fig. 4.10 Fractures in Carboniferous limestone cave ceiling associated with linear groups of soda-straw stalactites (Pooles Cavern, Derbyshire, UK).

Drip-waters associated with stalagmites C and D (Fig. 4.7) also include diffuse and preferential flow through secondary porosity structures, but here a proportion of flow is routed through a reservoir, or store, in the epikarst or karst. These water stores may function in different ways: varying in their volume, their characteristics (whether air- or sediment-filled), and in the mode with which they fill and drain. Thus, the water store above stalagmite C is situated in a large fissure which drains by 'underflow': a combination of flow through discrete outlets, or diffuse seepage across the area of

Fig. 4.11 Group of stalagmites showing identical changes in morphology with height. Two have been studied and have similar histories of intermittent growth. This is inferred to relate to growth from a group of soda straws with a common source including an intermittently sealed fracture, reopened during seismic activity. (Refugio Cave, southeast Spain).

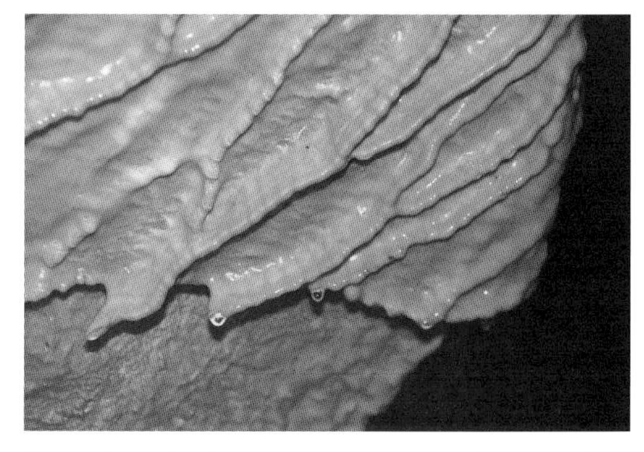

Fig. 4.12 Speleothem curtain guiding the location of dripwaters (former mine in Jurassic limestone, Nailsworth, Gloucestershire, UK).

contact between karst and water. Waters are likely to continue to drain from the underflow store for as long as water remains available. If the storage volume is large compared with the drip discharge, drip variability will be very low, and a candlestick-shaped stalagmite will result, despite having a different hydrological routing compared with stalagmite A. In contrast, stalagmite D receives waters from a 'overflow' storage reservoir, characterized by discontinuous water outflows, during periods when the water storage volume exceeds a certain threshold, and with flux rates that vary according to the characteristics of the outlet point(s). In terms of morphology, one would find it impossible to see any difference between stalagmites B and D, despite being fed by different flow pathways.

As mentioned previously, the only method of determining the nature of water movement to an individual speleothem is through cave observations of discharge. A pioneering study was that of Pitty (1966, 1968) who manually measured seasonal variations in drip-water hydrochemistry at Pooles Cavern, Derbyshire, UK, over a 12 month period. Smart and Friederich (1987) and Gunn (1974, 1983) undertook similar studies in British and New Zealand caves using a combination of manual and automated (tipping bucket gauges) approaches which helped develop conceptual models of cave dripwater hydrology. Smart and Friederich (1987) classified waters by their mean drip rate and drip variability. Later, Baker et al. (1997b) and Baker and Barnes (1998) demonstrated through additional manual measurements of drip rates that most stalagmite drip-waters are characterized by a discharge of less than 10^{-5} L s^{-1} and a CV (coefficient of variation, i.e. relative standard deviation) of between 10 and 200%. Higher CVs were associated with drip-waters that displayed a significant seasonal variation in flow regime, with an apparent increase in the proportion of seepage waters routed via preferential flow as CV increases. Figure 4.13 presents a summary of speleothem forming water fluxes after Baker et al. (1997b), showing a lack of relationship between flow variability and mean discharge, and significant inter annual variability in water flux (mean and variability) between years.

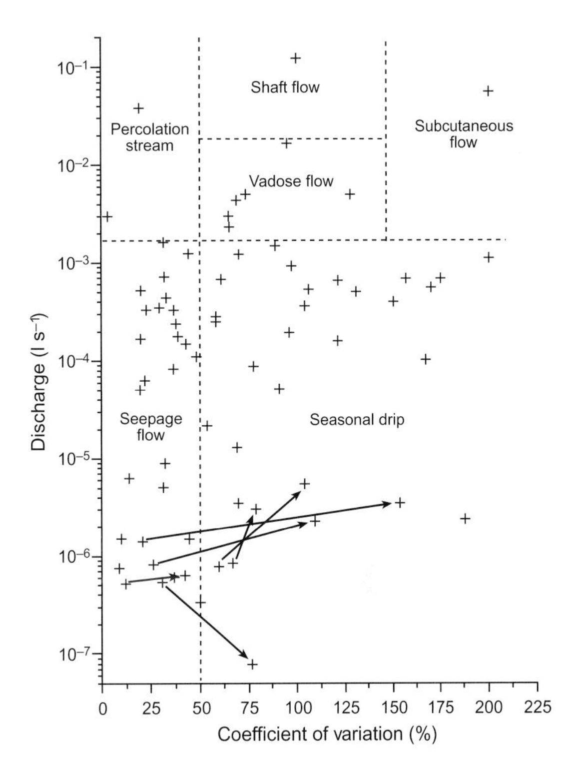

Fig. 4.13 Relationship between discharge and the intra-annual variability of discharge compiled in published studies on Carboniferous limestone cave sites in England. Intra-annual variability of Lower Cave sites measured in 1991–1992 and 1994–1995 are labelled and linked by arrows. After Baker et al. (1997b).

Huge improvements to our understanding of water movement to speleothems has come from the development of instruments to continuously measure drip rates in caves (early efforts using tipping-bucket approaches typically suffered from mechanical failures at high humidity). Technological improvements allow the continuous sensing of drip rates in caves using infra-red beam splitters or drip-sensitive pressure transducers. Pioneering work by Genty and Deflandre (1998) monitored drip-waters below a stalactite in the Père Nöel Cave, southern Belgium, for more than 5 years and identified periods of water flushing from the soil or epikarst, indicating piston, or displacement flow, with sequential water movement through linked water stores. Additional high frequency, low amplitude variations in drip-water discharge were inversely correlated with air pressure, suggesting two-phase flow. Baker and Brunsdon (2003) continuously monitored six drip sites for up to 4 years at Stump Cross Caverns, UK, and observed two-phase flow and nonlinear drip rate behaviour. Figure 4.14 presents a compilation of some of the longest logged series from Baker and Brunsdon (2003), Baker et al. (2000) and Fairchild et al. (2006b). Nonlinear behaviour of water movement is widely observed across these studies at drip rates of approximately one drip per minute or higher, and has been observed at the much greater discharge observed in karst springs (Labat et al., 2000a, b, 2002), suggesting that this behaviour was possible at a wide range of water movement rates. One might hypothesize that the possibility exists for all speleothems where water movement is affected by secondary permeability.

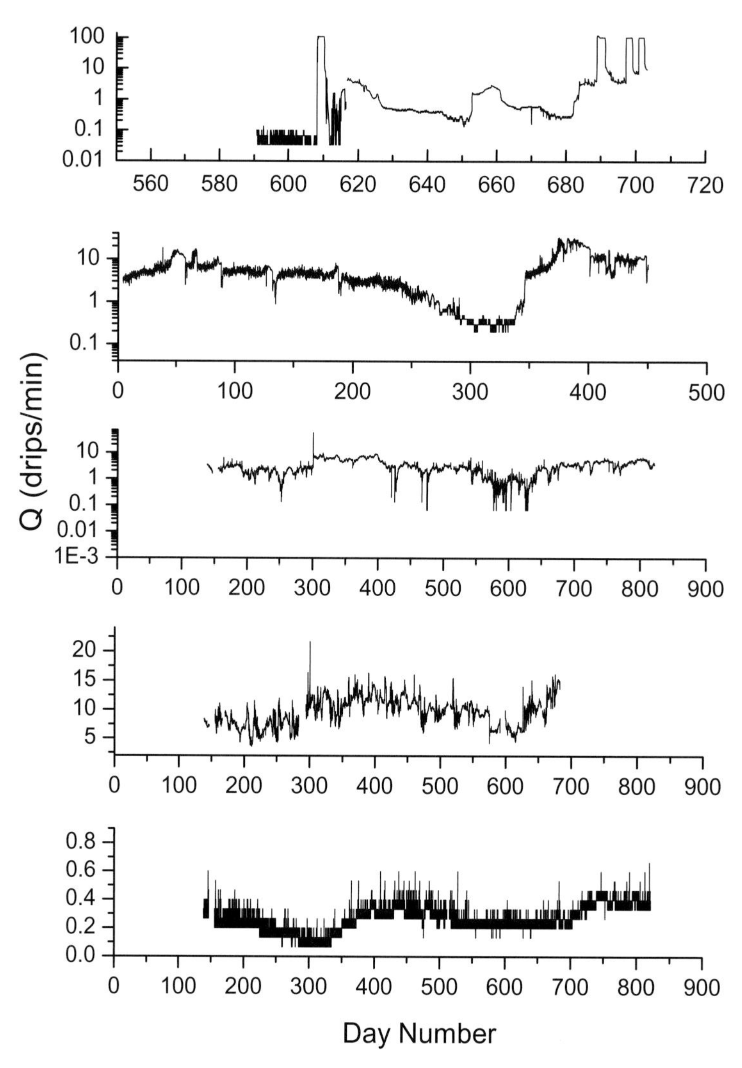

Fig. 4.14 Comparison of five long continuous records of stalagmite drip hydrology, ordered from top to bottom by decreasing maximum discharge. Note the log scale for the top three discharge records. From top to base: drip BFM-B, Brown's Folly Mine (Fairchild et al., 2006b); Vil-10a, Grotte de Villars (Baker et al., 2000); SC-1, SC-2 and SC-5 from Stumps Cross Caverns (Baker & Brunsdon 2003).

These dynamic variations in karst water routing, and their variability between individual stalagmites and over time, have important implications for speleothem research, influencing, for example, solute concentrations, Mg/Ca ratios, and the dynamics of carbonate precipitation (see section 5.5). Thus there are several implications of any change in water residence time, in flux rates, and particularly in the relative proportions of dripwaters associated with diffuse and preferential flow (Tooth & Fairchild, 2003; Fairchild et al., 2006b). Surprisingly, the mean groundwater age of speleothem dripwaters (recognizing that it is a composite of waters of different residence times) is not often known. Fluorescent dyes, which might be useful where fracture flow is thought to be significant, have been rarely applied owing to the impracticably long observation times required and loss of fluorescent tracer due to absorption in the soil and epikarst (for an exception see Bottrell and Atkinson, 1992). For example, Kluge et al. (2010) reported a total lack of detection of dye in a cave site where tritium analyses successfully yielded a mean groundwater age. Kluge et al. (2010) provided the most comprehensive analysis of tritium tracing of cave dripwaters, using tritium, ^3H–^3Be and natural ^{18}O as tracers. Analysing tritium data from three German caves, and comparing groundwater tritium with rainfall tritium concentrations, Kluge et al. (2010) observed a mean residence time of between 1 and 4 years. Previous studies have suggested that older ages are possible where diffuse flow dominates. Chapman et al. (1992) measured both dripwater $\delta^{18}O$ and tritium at Carlsbad Caverns, USA, where the tritium data confirmed a mean age in the order of decades, which was confirmed by an invariant $\delta^{18}O$ in the same dripwater. Kaufman et al. (2003) at Soreq Cave, Israel, also report dripwater residence times of decades in a study which compared a discontinuous dripwater tritium data sets collected over several decades. Kluge et al. (2010) recommended using tritium as a tracer of groundwater age only after the extent of dripwater $\delta^{18}O$ variability over time has been determined.

The best empirical evidence for the hydrogeological influence on dripwaters should therefore be seen in field campaigns where repeat measurements of dripwater $\delta^{18}O$ and precipitation $\delta^{18}O$ have been made over time, together with water discharge measurements used to characterize flow regime. However, despite the recommendations of Kluge et al. (2010), repeated measurements of dripwater $\delta^{18}O$ where the dripwater flow regime is well understood are not common, with most studies focusing on either flow regime (see above) or the isotopic composition of dripwater and precipitation (for example, Chapman et al., 1992; Williams & Fowler, 2002; van Beynan & Febbroriello, 2006; Fuller et al., 2008; Luo & Wang, 2008; Pape et al., 2010). An exception is Bar-Matthews et al. (1996), who compared the $\delta^{18}O$ of rainfall with dripwater flow types that we have renamed as seepage-flow derived (continuous all year; likely to have a dominant storage component, such as that of Fig. 4.8) and fracture-flow derived (discontinuous dripwaters from fissures, likely to be fissure-flow dominated, such as shown in Fig. 4.9). Figure 4.15 presents box plots comparing the isotopic composition of these two flow types with the $\delta^{18}O$ of pool waters and the mean $\delta^{18}O$ composition of precipitation. 'Fast-drip' regimes can be observed to have a lower $\delta^{18}O$ than 'stalactite' regime drips, with the pool waters a composite of these two hypothesized end members. The more negative $\delta^{18}O$ of the 'fast regime' drips likely represents the more negative $\delta^{18}O$ of the more intense rainfall events which preferentially recharge these fracture-flow routes. Both drip regimes exhibit a more positive $\delta^{18}O$ than average rainfall owing to the mixing of event water with older, evaporatively enriched, soil and epikarst water.

The combination of improved observational data has been coupled with a continual refinement of conceptual models of water movement, including that of Tooth and Fairchild (2003) which included a soil water flow component for the first time. This combination has led to the first attempts to model hydrologically the water movement to speleothems. These models are considered again in section 10.2.2 in the context of being able to forward model stalagmite proxy climate data; here we will consider the hydrological framework to existing dripwater hydrological models (Fig. 4.16) and implications for the interpretation of

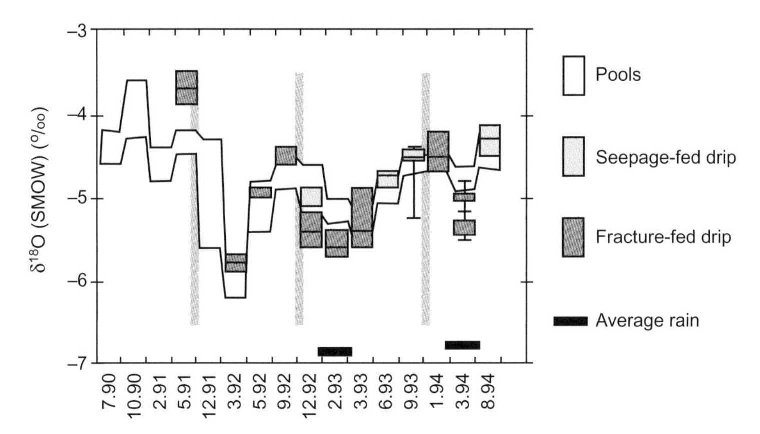

Fig. 4.15 Box plot illustrating the seasonal and annual variations in $\delta^{18}O$ of cave waters over a 4 year sampling period. The boxes enclose 50% of the data and the median is displayed as a line. Sampling date is indicated by the month and the year. Vertical lines are the boundaries between hydrological years and horizontal lines the weighted average $\delta^{18}O$ values of rainfall during 1992–3 and 1993–94. Soreq Cave (after Bar-Matthews et al., 2006).

speleothem proxies. The pioneering work of Fairchild et al. (2006b) developed a model to attempt to explain both water movement and chemistry at Brown's Folly Mine, UK. Their model (Fig. 4.16a) contained two layers; an upper layer which represented the soil and epikarst and contained two parallel reservoirs, one fed by matrix flow and one by fissure flow, reflecting primary and secondary porosity respectively. Flow switching from the matrix to the fissure-fed reservoir occurred as at predefined storage volume, and both stores fed a third reservoir where mixing was allowed. Stalagmite drip rate and chemistry was modelled from an underflow from the third reservoir. The model was demonstrated to predict drip rates and chemistry reasonably well for two sites, but for others, drip and chemical variability was not fully captured by the model at times of lowest flow, suggesting that greater model complexity is required. Baker et al. (2010) developed a model of dripwater $\delta^{18}O$ in an attempt to explain modern $\delta^{18}O$ variability in an Ethiopian stalagmite (Fig. 4.16b). Stalagmite proxy data suggested that the stalagmite in question was fed by an overflow water supply, and therefore in this case a simple single reservoir was used, fed by both diffuse and fissure flow components whose relative input was surface climate dependent. Baker and Bradley (2010) used an even simpler model to attempt to model $\delta^{18}O$ in a Gibraltar stalagmite (Mattey et al., 2008) (Fig. 4.16c). This comprised just a single reservoir with an underflow water supply to the stalagmite. Finally, Bradley et al. (2010) developed a lumped parameter model that included soil water storage and evaporative loss, an epikarst store and overflow and underflow movement. (Fig. 4.16d). Using model outputs, a continuum of stalagmite flow regimes and associated $\delta^{18}O$ composition can be modelled. In terms of implications for climate proxies preserved in speleothems, Bradley et al. (2010) demonstrated that for a single climate input series and simple conservative mixing of ground waters: (a) model-predicted stalagmite $\delta^{18}O$ can vary by approximately up to 2 ‰; this hydrological uncertainty introduced by variations in the hydrological routing from the surface to the cave on flow route; (b) that an offset of dripwater $\delta^{18}O$ from the weighted mean $\delta^{18}O$ of precipitation (typically to depleted values) can be generated owing to activation of overflow or bypass flows at high recharge; (c) that an offset of dripwater $\delta^{18}O$ from the weighted mean of precipitation to enriched values will occur if evaporation of soil or shallow groundwater occurs.

Hydrological models of increasing complexity may, in the future, be able to capture the water

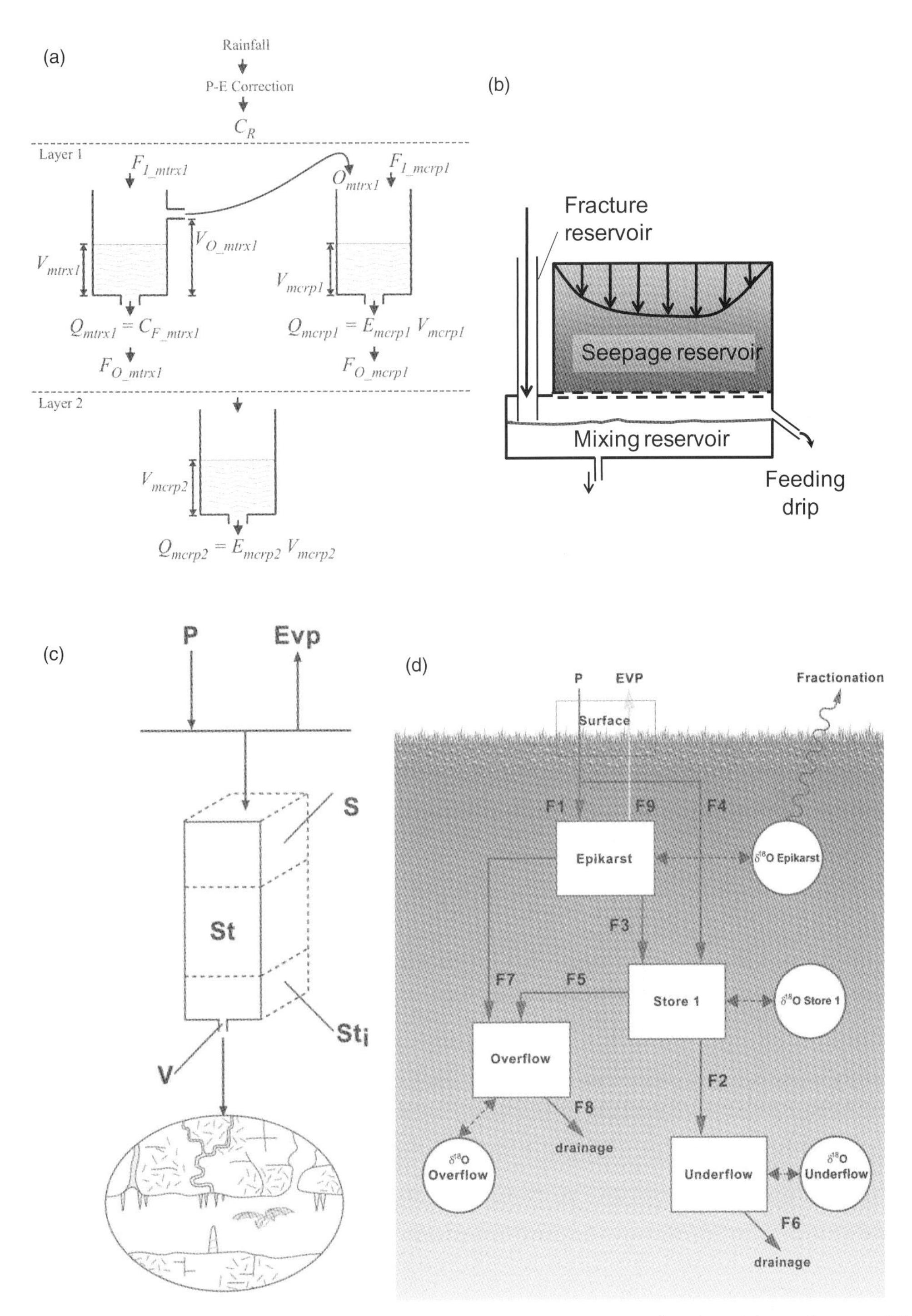

Fig. 4.16 Conceptual diagrams of numeric models of karst water movement to stalagmites. (a) Two-layer model of Fairchild et al. (2006b). (b) Single reservoir with overflow feed model of Baker et al. (2010). (c) Single reservoir model with underflow feed of Baker and Bradley (2010) and (d) lumped parameter model of Bradley et al. (2010).

movement to speleothems, and in particular the nonlinear behaviour of water flux over time. In many parts of the world, water supply over time is nonlinear owing to recharge occurring when the precipitation>evapotranspiration threshold being passed (Chapter 3). Coupled with the expectation of nonlinear water movement in the soil and karst aquifer owing to a combination of diffuse and fissure flow, as well as overflow behaviour from storage reservoirs, water movement to speleothems can be expected to be non-linear over time. In terms of speleophysiology, and the life of a speleothem in the incubator, one can hypothesize that a consistent water supply of low variability might be expected for diffuse-flow dominated stalagmites, leading to long-lived, candlestick-shaped stalagmites but which might have a very damped influence from the atmosphere, soil and epikarst (Fig. 4.7). For speleothems fed by both diffuse and fissure components, a long and simple life might be expected if fed by a significantly large reservoir compared with water outlets, but with decreasing reservoir size, nonlinear water supply is increasingly likely, ultimately leading to seasonal or episodic growth and more complex growth shapes. The latter are considered further in section 7.2.

4.4 Air circulation

Natural ventilation has been most extensively studied by building engineers, but it has been found that even simple geometrical situations are hard to model quantitatively and insights often need to be gained by the use of wind tunnels and scale models (Linden, 1999). In parallel, an understanding of the physics of air circulation in underground passages has been essential for the development of ventilation systems for mines. In the post-war period, electrical circuits were constructed to serve as analogue models (network flow analysers) for understanding how to optimize the process of ventilation (Hartman, 1961), and the parallelism of air flow with electrical current was also used as a conceptual aid to karst studies by Badino (1995). Given that mines consist primarily of large passages of well-defined geometry, the difficulties in fully understanding air movement in natural cave systems can be seen to be formidable. Nevertheless, cavers (and miners) of many nations developed an intuitive understanding of the key principles by personal experience and this was summarized within early modern synthetic accounts of speleology (Trombe, 1952; Cullingford, 1953) in terms of qualitative key influences on air circulation. Later, physical insights and syntheses were developed by among others Cigna (1967), Wigley and Brown (1976) and Choppy (1986), but the complex geometry of caves prevented analytical solutions from being developed, apart from some specific special cases (see, for example, Wigley, 1967; Atkinson et al., 1983). In recent years, there has been a wave of observational studies that constrain the patterns of air movement and allow the syntheses to be brought up to date. In this section we first explain the several different physical causes of air circulation that can now be distinguished and then discuss approaches to determining the relative magnitude of air exchange in different seasons and its significance for speleothems.

4.4.1 Physical causes

Table 4.2 classifies several types of phenomenon that differ in terms of the physical driver for air movement, and Fig. 4.17 illustrates some cave geometries that are susceptible to each category.

Cave breathing

Air flows which attempt to maintain pressure balance in response to atmospheric fluctuations (Fig. 4.17a) represent the phenomenon of *cave breathing* (Moore & Nicholas, 1964; Conn, 1966), and are distinct from very short-term resonance effects (Wigley & Brown, 1976). One of the most striking environments for demonstrating the effect of barometric pressure variations is the vast karstic Nullarbor Plain of southern Australia where a thin calcrete cap overlies karstified Cenozoic carbonates displaying cave chambers with bell-shaped roofs which have locally collapsed to connect to the exterior (Fig. 4.18). Reversing air velocities of several metres per second at the resulting narrow vertical entrances (blowholes) can be observed at subdiurnal timescales. This region inspired the study

Table 4.2 Synthesis of physical controls on air movement in caves and surrounding karstic rocks.

Phenomenon	Physical driver	Magnitude	Preferred geometry	Duration
Cave breathing Diurnal or synoptic atmospheric pressure variations (or anthropogenic disturbance, e.g. opening cave door)	Pressure equalization	Up to tens of millibars (i.e. thousands of pascals), but strongly dependent on ratio of cross-sectional area of entrance to cave volume	Pressure wave (and variable mixing) with strong effects near entrance where entrance passage to cave volume ratio is high.	Typically hours to days
Wind-induced flow High wind speeds	Wind forcing of air movement into cave, or sucking air out when blowing across the mouth (Venturi effect)	Up to 60 Pa for 10 $m s^{-1}$; 15 Pa for 5 $m s^{-1}$ wind.	Cave entrance oblique to wind or Venturi effect from wind blowing across entrance	Hours to days (can be diurnal)
Chimney circulation or stack effect Sustained cooling (winter/summer)	Sustained difference in density between interior and exterior	$\Delta\rho$ can be estimated dropfrom Fig. 4.4. Pressure across the two entrances corresponds to 44 Pa per degree per 100 m of height difference between entrances. For initially descending caves, is augmented by gravity flow in winter.	At least two open or diffuse entrances at different altitudes	Up to seasonal
Convection Cooling or warming of exterior or interior air compared with other parts of cave interior	a. Forced convection induced by gravity flow of fluid with differing density to ambient cave air forcing convective circulation b. Free (thermal) convection	$\Delta\rho$ can be estimated from Fig. 4.4. Velocity proportional to the square root of $\Delta\rho$ and height difference over which the flow moves.	a. Descending cave (flows of cold air at base of passage). Ascending cave (flows of warm air at top of passage) b. Favoured by tall, wide chambers.	Varies from short-term burst (hours or diurnal) to seasonal
Water-induced flow	a. Cave stream forces air circulation b. Water induces airflow in biphasic karstic fissures	Localized physical drag	a. Descending cave passage b. Biphasic, fractured aquifer	Event-based, seasonal or perennial

of Wigley (1967) who obtained, from consideration of air flow in slits or cylindrical voids in impermeable bedrock, the relationship between velocity and time-varying air pressure (dP/dt) at a given point:

$$PUA = V\, dP/dt \tag{4.11}$$

where P is mean air pressure, U is the mean (inward) air velocity over cross-sectional area A, and V is the volume of the cavity beyond the point where U is measured. Pressure changes and velocity are in phase, whereas an out-of-phase relationship applies if the matrix voids also contribute: an interpretation favoured by Wigley (1967) for preliminary field data. However, more detailed recent observations of flow characteristics coupled with gravity observations demonstrates the presence of

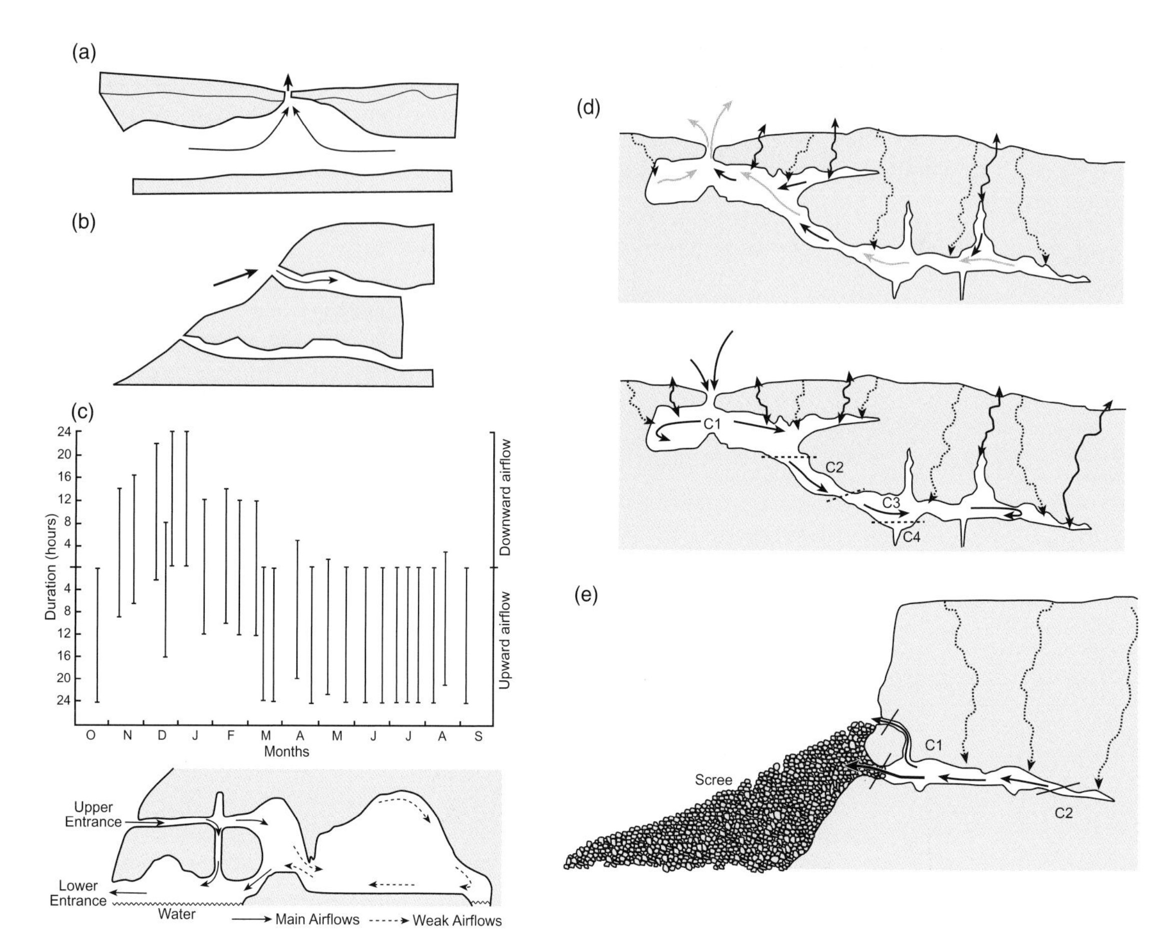

Fig. 4.17 Cartoons of air circulation in different cave systems. (a) Geometry favouring strong cave breathing; (b) Geometry favouring wind-induced flow into the cave. (c) Chimney-effect ventilation from Glowworm Cave, New Zealand (de Freitas et al., 1982) showing duration of airflows in a particular direction during 24 hour monitoring periods and cartoon of prevailing downward summer circulation (winter flows are exactly reversed). Height difference between entrances is 14 m. (d) Cave dominated by convective effects (Aven D'Orgnac, France, based on Bourges et al. (2001, 2006a, b)). (e) Constant circulation forced by biphasic epikarst flow (Chauvet Cave, France (based on Bourges et al., 2006b)).

large unexplored cave systems; quantitative analysis of the data reveals that non-cave rock porosity does not contribute significantly to the air mass (Doerr et al., 2006; S. Doerr, personal communication 2009). Another famous example of a breathing cave system is Lechugilla, (New Mexico) which has 86 km of passages, although convection is more important for air exchange in some more remote areas of the system (Cunningham & LaRock, 1991). Breathing responses have been monitored in detail in a Rn study at a shaft accessing a limestone mine near Paris (Perrier et al., 2004) where there is evidence for Rn transfer to the mine from small pore spaces during low-pressure impulses; the effects are complicated by mixing processes. Conversely, pressure fluctuations have minimal impact where the cross-sectional area of voids exposed to the atmosphere is relatively large compared with the internal volume, and at sites or at times of the year when other circulation mechanisms are operating. Where

Fig. 4.18 Thampanna cave and the base of its entrance shaft, Nullarbor, Australia. Many Nullarbor caves have even narrower entrances to extensive passage networks, stimulating strong cave breathing.

circulation is very limited, external pressure effects (including responses to diurnal temperature variations), transmitted as a wave, are felt deep in the cave interior, and under these conditions is shown a weak semi-diurnal pressure pulsing arising from Earth (i.e. solid earth) tides (Richon et al., 2009). The most obvious implication of cave breathing for speleothems is that their formation will be complicated by regions of high airflow and that means that P_{CO_2} levels, and hence calcite saturation state, may depend more on distance from the entrance rather than seasonal changes.

Wind-induced flow

Moving wind exerts a pressure (P_{wind}) that depends, by Bernouilli's principle, on the square of the velocity (U):

$$P_{wind} = 0.5\rho U^2 \tag{4.12}$$

The resulting *wind-induced flow* (Fig. 4.17b) is a very complex phenomenon to model as its magnitude is always less than in eqn. (4.12), dependent on the exact configuration, even for the simple geometries of buildings (Linden 1999). In the case of caves, the factors are the external wind direction, landscape surface morphology and geometry of the cave opening. For example, both inward and outward directed forcing of air flow, depending on external wind direction, was demonstrated from Sainte-Catherine caves in the French Pyrenees by Andrieux (1969) and the impact on trace gas levels of the suction (venturi) effects of wind blowing across an entrance during meteorological events are well illustrated by Kowalczk and Froelich (2010). Such events disturb seasonal patterns of air composition and associated speleothem properties.

Chimney circulation

The *chimney (or stack) effect* (Fig. 4.17c) depends fundamentally on air density differences (Wigley & Brown, 1976) and temperature is the main influence on density (Fig. 4.4). During the summer, external temperatures are higher than in the cave and air density is correspondingly lower; the reverse applies in the winter and this leads to reversing airflows within the cave network. The principle of hydrostatic balance, following from Newton's Laws of motion, requires that for a still fluid, the gravitational force is balanced by the pressure gradient force related to the upward reduction in atmospheric pressure (dP) with height (dh):

$$dP/dh = -\rho g \tag{4.13}$$

where ρ is the air density and g is the acceleration due to gravity. In the case of a cave with upper and lower entrances:

$$P_{extL} = P_{extU} + \rho_{ext} gh \tag{4.14}$$

where P_{extL} and P_{extU} are the external pressures respectively at lower and upper cave entrances, h is the elevation difference between entrances, and ρ_{ext} is the mean density of the external air column of height h above the lower entrance. The internal cave air exerts a pressure (P_{intL}) at the lower entrance of

$$P_{intL} = P_{extU} + \rho_{int} gh \tag{4.15}$$

The mean internal density (ρ_{int}) is averaged in the vertical dimension. Air flow at the lower entrance results when $\rho_{ext} \neq \rho_{int}$ and a circuit is set up connecting airflow between upper and lower entrances. Fig. 4.17c illustrates the example of the Waitomo Glowworm cave in New Zealand resulting from a carefully planned observational campaign including release of tracer gases (de Freitas et al., 1982). Here the chimney effect, although controlling the predominant sense of airflow, is relatively weak, because the difference in elevation of the entrances is only 14 m; as a result there is significant diurnal variation in flow intensity resulting from external temperature changes affecting density. A key point is that the 'entrances' could be a network of fractures rather than a macroscopic cave; this applies to the upper entrance of the Eisriesenwelt of Austria (Wigley and Brown, 1976) and other cases (Hakl et al., 1997; Buecher, 1999; Spötl et al., 2005). In practice, many caves have voids that connect to the surface but are too small to be accessed by speleologists, indeed some notable caves, such as Wind Cave in the USA, have been discovered by cavers who have excavated small openings that were noticed because of the draft through them. As long as the cave interior has the thermal capacity to modify the temperature of the incoming air, the larger the area of opening, the larger the discharge will result from this effect.

Waters (2003) gives an expression for the pressure difference which arises in the chimney effect:

$$\Delta P_{\text{ext-int}} = \rho g * 273(h)\,[1/(T_{\text{ext}} + 273) - 1/(T_{\text{int}} + 273)] \quad (4.16)$$

where T is in degrees Celsius.

This can be simplified at typical values to:

$$\Delta P_{\text{ext-int}} \approx 0.043(h)\,(T_{\text{int}} \quad T_{\text{ext}}) \quad (4.17)$$

In the simplest case, the flow velocity will be directly proportional to this calculated pressure difference. In reality, if air is moving there is a gradual change in pressure within the cave and the total magnitude of pressure differences drives air movement; there will also be frictional resistance to flow. Atkinson et al. (1983) showed that in Castleguard Cave, a large open system in the Canadian Rocky Mountains, this mechanism generated seasonally reversing velocities of 1–2 m s^{-1} over a lateral distance of 8 km and with h = 350 m. These authors modelled the manner in which the turbulent air movement is increasingly resisted by frictional forces at higher velocities, in order to understand the effects of roughness of passage geometry. In systems displaying strongly reversing circulation, but consisting of conduits linking chambers, air movement may only be felt at the constrictions between chambers as at the Obir Cave, Austria (Spötl et al., 2005). Fairchild et al. (2010) speculated as to whether speleothems within such a chamber captured trace chemical species via atmospheric aerosols in addition to directly from dripwater as normally assumed. Some types of speleothem can develop asymmetry in response to prevailing air currents and these can be used to help map the dominant current paths (e.g. Buecher, 1999).

Convection

Convective motion can arise either as a forced effect when a temporary or permanent density current exists (i.e. a flow of differing density from the main body of air) or as free thermal convection (Fig. 4.17d). Mixing by thermal convection is inevitable in any substantial air mass, unless in a narrow shaft with a stable upwards-decreasing density and is particularly important in caves with large entrances (Hakl et al., 1997). In the case of the Aven D'Orgnac, southeast France (Bourges et al., 2001, 2006a, b), the inferred upward transport of air through epikarst fissures is facilitated by convection and provides a means of loss of cave air to the exterior. Thermal convection is, however, inhibited by a steady flow produced by another mechanism, such as the chimney effect. Indeed, the relative lack of mixing is a primary feature of displacement ventilation of rooms in buildings (inflowing area at low level and outflowing air at high level (Linden, 1999)) equivalent to chimney ventilation, whereas rooms ventilated by a single entrance do so less efficiently because of mixing effects.

Very commonly, cave passages descend from a surface entrance. In this case, in the winter, convection can facilitate the entry of cold, dense air, as in the Aven d'Orgnac (Fig. 4.17d) where nocturnal flows of cold air are generated close to the cave floor. Generally, for a tubular downward-sloping passage the flow will take the form of a low-level discrete gravity current, balanced by a return flow of warmer cave air at a higher level (and possibly within the same cave passage). This was, for example, directly observed in Carlsbad Caverns one January where the boundary between cold incoming air and warmer outgoing air enriched in radon was found to be between 5 and 9 m above the passage floor (Wilkening and Watkins, 1976). Geologists will be familiar with the sedimentological version of the gravity current—the turbidity current. The velocity of a transient gravity flow can be given as (Leeder, 1999):

$$U = k\sqrt{[gh(\Delta\rho / \rho)]} \quad (4.18)$$

where k is a constant for turbulent flows, but varies inversely with Reynolds number for laminar flows, h is the height difference over the region in which the head flows, and $\Delta\rho$ is the difference in density between the flow and the ambient (still) medium. A crucial point is that, irrespective of other effects, total flow in winter can be significantly augmented by gravity flows or thermal convection in cases where there is a descent from the lower entrance. This can be linked to the lower winter levels of trace gases commonly exhibited by caves. Conversely in summer, it is possible for warm air to penetrate at high level at cave entrances (e.g. Smithson, 1991) and this is a possible mechanism to explain the lower trace gas composition found at several sites in this season (e.g. Hoyos et al., 1998). Preferential circulation in relation to cave geometry can give rise to permanent temperature disturbances: upward, blind passages may be warm-traps and likewise descending blind passages can be cold traps (e.g. Andrieux, 1969).

Water-induced flow

Where significant water flows are present, they are likely to have a profound effect. Air is entrained by its boundary with a stream and achieves a velocity proportional to that of the stream (Cigna, 1967) which can result in complex circulation effects, including the possibility of a return airflow on the upper wall of the entrance passage. Flooding streams lead to stronger circulation effects and isolated areas of pressurized air may result at the top of flooded passages. Bourges et al. (2006b) also drew attention to the biphasic (air-water) nature of the epikarst and the role that downward-percolating water plays in entraining air. In the case of the Chauvet Cave, France, these authors believe that this accounts for its consistent direction of air circulation through the year (Fig. 4.17e).

4.4.2 Radon studies as indicators of rates of air-exchange

The mechanisms for air movement described above clearly vary on different timescales, with annual variability expected to be dominant. In order to quantify the extent of air-exchange with the external atmosphere, the ideal tracer would be a trace gas phase that is generated at a constant rate throughout the cave volume. The closest approximation to this ideal is radon, published time-series measurements of which have been particularly extensive in central Europe because of the use of caves as sanitoria, and in southwest Europe as part of environmental monitoring of caves with prehistoric paintings.

Radon is a radiogenic and radioactive gas whose most abundant form (^{222}Rn) is created within a decay chain involving U and Th isotopes (see Chapter 9) and ejected (recoiled) into fluids where it can be transported into the cave. In turn ^{222}Rn decays, with a half-life of 3.8 days, to continue the chain of radioactive progeny. Although radon is a noble (unreactive) gas, its progeny can attach to particles and become lodged in the lung, where they release harmful alpha radiation. This health hazard has led to a considerable effort to monitor the concentrations of Rn and its progeny in indoor environments and prescribe safe limits (Cigna, 2005). Radon concentrations in cave air can be expressed in terms of (radio)activities which are given in becquerels (Bq, one nucleus disintegrating per second), expressed per unit volume. Progeny concentrations are commonly expressed in working levels (WL) over an interval of time (e.g. WLh, WLM) and the radiation dose accrued from exposure to radon is measured in millisieverts (mSv).

Uranium and thorium tend to be more strongly associated with silicates and phosphates rather than carbonates and so sites of radon generation are strongly associated with shale horizons in carbonate rocks. Varying bedrock composition accounts for most of the four orders of magnitude range in concentrations found in caves (Fig. 4.19a from Hakl et al., 1997). In addition, an outlying high value of 155,000 Bq m^{-3} was found in Giant's Hole in Derbyshire, UK (Gunn et al., 1991). Hakl et al. (1997) presented a variety of long-term monitoring

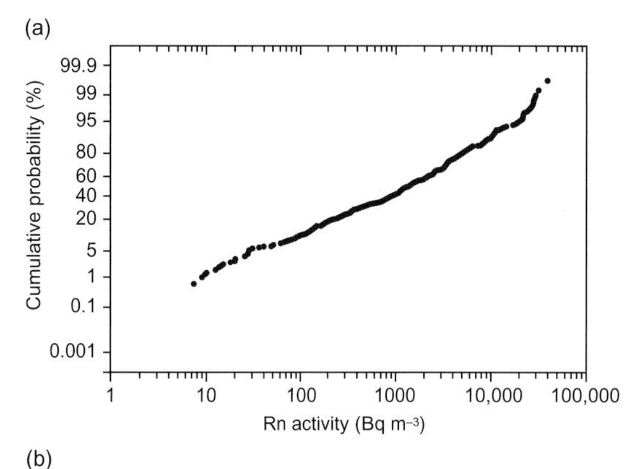

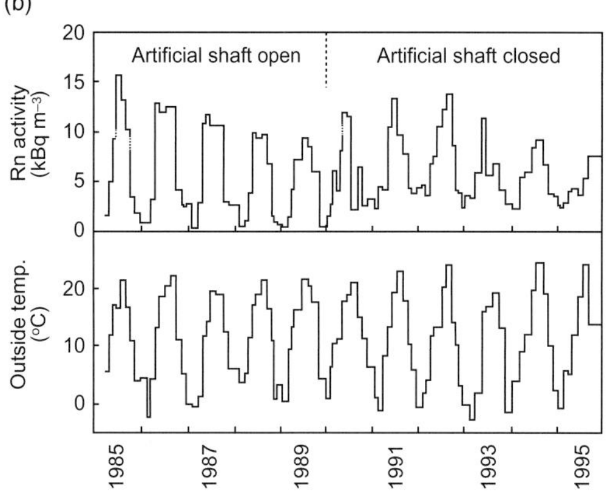

Fig. 4.19 (a) Distribution of Rn activity measurements from 220 caves. (b) Radon levels at the General Assembly Room of the Szemlő-hegy cave, Buda Mountains, central Hungary. The ratio of summer to winter activities reduced from more than ten to less than five after sealing of an artificial shaft in 1990. (after Hakl et al., 1997)

results, many sites displaying a pronounced seasonal fluctuation in Rn concentrations paralleling that of external temperature (Fig. 4.19b) which they explain primarily in terms of variations in cave ventilation, with lower Rn concentrations in winter because of the increased exchange with external air with negligible Rn content. In the example from Szemlő-hegy cave, for example, the insertion of a cave door in winter 1989–1990 increased the winter minimum level of Rn because of reduced circulation (Fig. 4.19b). An example of direct evidence of higher winter exhalation rates are the high rates of Rn detected above a slit over Hajnóczy Cave in Hungary (Hakl et al., 1997). Sites varied as to their susceptibility to atmospheric pressure variations or external temperature as anticipated from the discussion in the previous section.

The quantitative interpretation of Rn levels follows the pioneering study of Wilkening and Watkins (1976) from Carlsbad Caverns (New Mexico, USA) at a time when the need to monitor the health hazards of Rn progeny was becoming known. They constructed a mass balance model of the form shown in Fig. 4.20, which leads to the following relationship:

$$dC/dt = ES/V - \lambda C - (Q/V)(C - C_{ext}) \quad (4.19)$$

where C is the cave radon concentration (or activity), E is the rate of radon exhalation from rock and sediment surfaces, S is the rock surface area and V the volume of the cave, λ is the radioactive decay

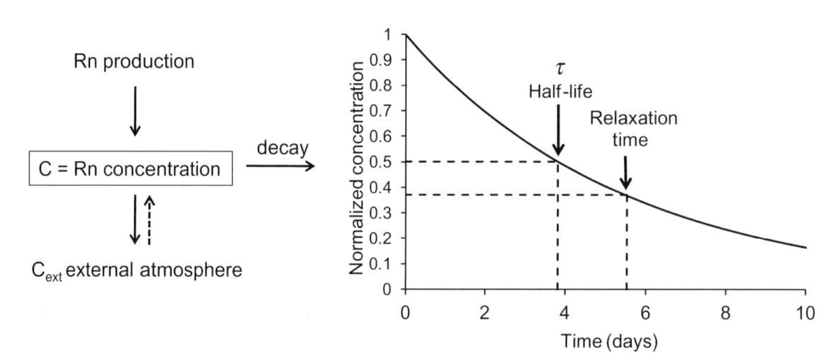

Fig. 4.20 Mass balance for radon as in eqn. 4.19. $C_{(t)} = e^{-\lambda t}$. Conventionally radioactive decay is expressed in terms of a half-life; alternatively the relaxation time is $1/\lambda$.

constant (0.181 day^{-1}), Q is the rate of air-exchange with the exterior (ventilation) and C_{ext} is the radon concentration in the air outside the cave (normally very low). When the change of concentration in time (dC/dt) is zero, and in the usual condition where $C >> C_{ext}$, eqn. 4.20 applies:

$$ES/V = C(\lambda + Q/V) \quad (4.20)$$

Two assumptions can be made to interpret time variations. If the production of Rn is assumed constant, Rn activities are inversely proportional to the rate of air-exchange and so the ratio of Rn maximum and minimum activities (at times a and b) is a function of the ratio of air exchange times (Fig. 4.21), approaching a 1:1 relationship for short exchange times during which little Rn decay occurs.

An alternative assumption (e.g. Duenas et al., 1999; Hoyos et al., 1998) is that the rate of air exchange is minimal at the time of year when Rn levels are highest. If both assumptions are made then the maximum (winter) ventilation rate (Q_{wint}) is

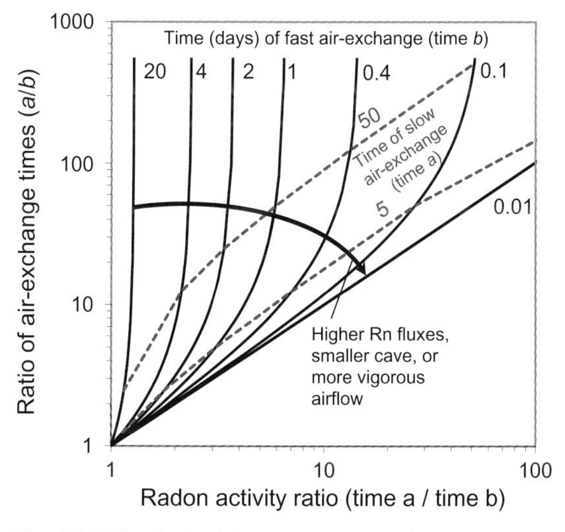

Fig. 4.21 Plot derived from eqn. 4.19 under an assumption of constant Rn production rate showing the relationship between the ratio of maximum to minimum Rn concentrations to the calculated ratio of corresponding air-exchange times. Equation 4.21 (Perrier et al., 2004) corresponds to the case at the top of the diagram where a particular radon activity ratio is associated with a unique exchange time.

$$Q_{wint} = \lambda(C_{summer}/C_{winter} - 1) \quad (4.21)$$

(Perrier et al., 2004)

Normally, it would be an oversimplification to assume no air-exchange took place when Rn was at a maximum, but Perrier et al. (2004) had independent evidence to show that it was reasonable in their case. More generally, it is likely that changes in production rates do occur as they are sensitive to the degree of water saturation of the fissures close to the cave (Ball et al., 1991; Muramatsu et al., 2002), but it is not usually possible to measure production directly.

A recent excellent case study of a dynamic cave system (Hollow Ridge Cave, Florida (Kowalczk & Froelich, 2010)) used data from a 15-month monitoring period. Cave ventilation was constrained by measurements of radon activity calibrated by direct determination of its production from the cave walls. Ventilation rates calculated using the relationships in eqn. 4.19 show interesting seasonal patterns (Fig. 4.22). In the cave interior (station 2), air movement was low in the summer and particularly high during the autumn when diurnal temperature variations created density instabilities. Such disturbances were also common near the entrance (station 1) in summer. Kowalczk and Froelich (2010) also documented several responses to synoptic meteorological variability; the sensitivity of the site relates to the presence of multiple upper entrances above the main cave passages.

4.4.3 Carbon dioxide and its variability

Carbon dioxide in caves has several sources (James, 1977; Baldini, 2010): (1) diffusion within epikarst air ultimately derived from root respiration and organic matter decay, (2) degassing from cave waters with enhanced CO_2 levels inherited from soils, (3) biological productivity in the cave (mainly from micro-organisms feeding on organic material brought into the cave by allogenic streams and, in some cases, on bat guano) and (4) deep-seated (thermal). Both (1) and (3) are associated with an equivalent reduction in oxygen levels and are a widespread cause of foul air in caves (James et al., 1975). Geothermal sources may have a distinctive

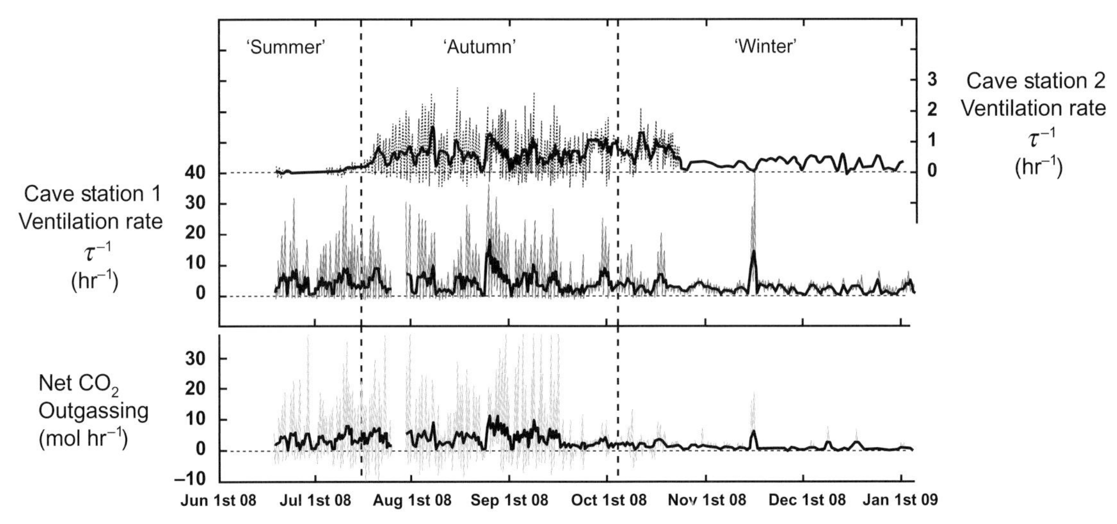

Fig. 4.22 Calculated cave ventilation (from Rn data) and CO_2 degassing, Hollow Ridge Cave, Florida (modified from Kowalczk & Froelich, 2010). Cave station 1 is close to the entrance and station 2 is further into the cave interior. Ventilation rates vary inversely with radon concentrations and are given as the reciprocal of the turnover time (τ) in hours. Net CO_2 degassing from the cave is calculated from the turnover time and CO_2 profiles.

fingerprint, such as $\delta^{13}C$ and He-isotope signatures (Bourges et al., 2001).

The concentration of CO_2 in a cave passage is a function both of production and ventilation processes and so can be modelled in the same way as radon. A simplification is the lack of radioactive decay, but a complication is that the pattern of CO_2 production, from multiple sources, is more heterogeneous than radon. Pioneering studies by Gewelt and Ek (1983) and Ek and Gewelt (1985) using breathing apparatus in two Belgian caves established relationships such as linear increase from the cave entrance and sources of CO_2 both from a cave stream and from passage ceilings. Mapping studies continue to be valuable in identifying sources and understanding heterogeneities (Baldini et al., 2006b; Fernandez-Cortes et al., 2006), particularly in poorly ventilated caves.

An alternative approach to modelling CO_2 abundances has been provided by J. Baldini et al. (2008). The production of CO_2 is modelled as a temperature-dependent modified Arrhenius equation while ventilation is expressed as a multiplicative modifying factor to derive predictions of cave air P_{CO_2}. All necessary parameters are empirically fit to datasets. Baldini (2010) demonstrated an overall consistency with P_{CO_2} in a variety of caves of different temperatures, consistent with other work on variation of soil productivity with altitude (Borsato & Miorandi, 2003), but the within-season variability of different caves is large in relation to the difference in mean temperature between sites.

There are several potential reasons for seasonal differences in P_{CO_2}. Troester and White (1984) presented a clear example of a case (Tytoona cave, Pennsylvania) where the CO_2 content of cave air was closely linked to that of a stream flowing through the system (Fig. 4.24). The higher CO_2 contents of streams during the summer relate to the higher soil CO_2 productivity (Chapter 3) and so this is likely to be a factor influencing cave air CO_2 whenever there is significant stream flow (e.g. J. Baldini et al., 2008).

A second explanation for seasonality in P_{CO_2} can be proposed: a direct link between seasonally varying soil P_{CO_2} contents and its transport and release into cave air. In very shallow caves, particularly those with extensive penetration by roots

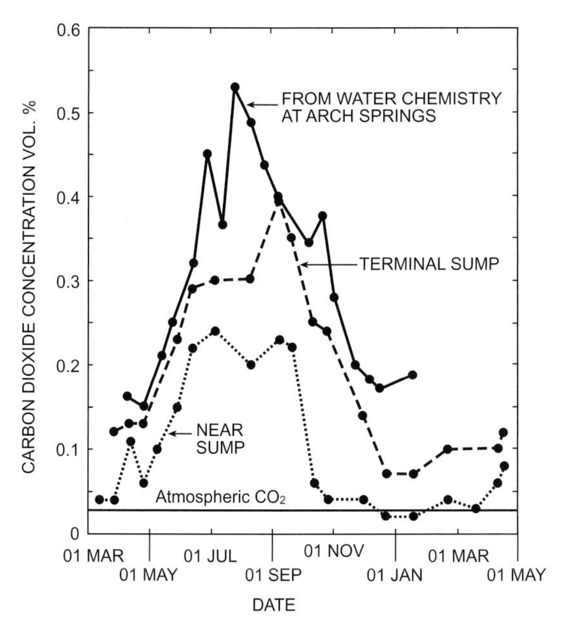

Fig. 4.23 Parallelism of P_{CO_2} of cave air and the calculated P_{CO_2} of a cave stream indicating the control of the former by the latter (Troester & White 1984).

such a connection may exist, but more commonly there is usually a significant epikarst region between the two which may have an independent character in terms of CO_2 concentrations (e.g. Atkinson 1977a; Benavente et al., 2010). Indeed epikarst (ground) air may be rather higher in P_{CO_2} than the overlying soils because of decay of inwashed organic matter. Nevertheless, there is quantitative evidence to indicate that the flow of CO_2 from the epikarst can be lower in winter (e.g. Fig. 4.22 (Kowalczk & Froelich, 2010)). Finding the correct explanation for this needs care, however. At the Ernesto cave, northeast Italy, a constant P_{CO_2} of the epikarst feeding particular cave chambers was argued on the basis of the constancy of Ca content of drips (Fairchild et al., 2000). A calculated reduction in flux of CO_2 in air from the epikarst in winter was interpreted by Frisia et al. (2011) to be associated with the filling of the aquifer by water, cutting off sources of epikarst air, as has been demonstrated at Castañar de Ibor in Spain (Fernandez-Cortes et al., 2009).

A third explanation is that the variations are primarily associated with seasonal changes in ventilation. Tests for this include the similarity of P_{CO_2} and Rn profiles, and the nature of the spatial structure of P_{CO_2}. For example, winter and summer circulation regimes are well-contrasted in Srednja Bikambarska Cave, Bosnia-Herzegovina (Milano & Garovšek., 2009) by CO_2 profiles. This is a descending cave with only one entrance which in winter is well-ventilated, except that in the worst-ventilated region, towards the back of the cave carbon dioxide rises linearly with distance. The Aven d'Orgnac has been the subject of intensive studies and a good understanding has been gained of CO_2 fluxes (Bourges et al., 2001; 2006a, b). Despite the sub-horizontal surface landscape, a significant air-exchange through the wide entrance can be measured directly and this increases significantly in winter to rates of $1\times10^5-5\times10^5\,m^3\,day^{-1}$. The presence of nocturnal temperature inversions in the first chamber in winter directly indicates the presence of gravity flows of cold air (Fig. 4.24). This winter regime develops progressively through the cave system, several emptying events being observed, characterized by a sudden drop of CO_2 concentration, beginning near the entrance of the cave (Fig. 4.25) and propagating stepwise toward its deeper parts, but not reaching the furthest chamber 1 km from the entrance. In the summer regime, when the system is stabilized by a thermal inversion, sluggish air exchange takes place with the epikarst whose rate can be influenced by high levels of water transfer or strong variations in atmospheric pressure. Here, small ventilation events allow the flow of epikarst air into the cave to be calculated resulting in values that are consistent with the direct measurements at the natural cave openings. When carbon dioxide is high in summer, air movements of $0.1\,m\,s^{-1}$ are observed in large fissures, and with this flow rate, air-exchange and CO_2 removal through only $0.78\,m^2$ of fissure area is needed for compositional balance.

It is generally recognized that cave chambers with high P_{CO_2} values (of the order of several thousands of ppm) must reflect very poor air circulation. In the study of Bourges et al. (2006a, b) the

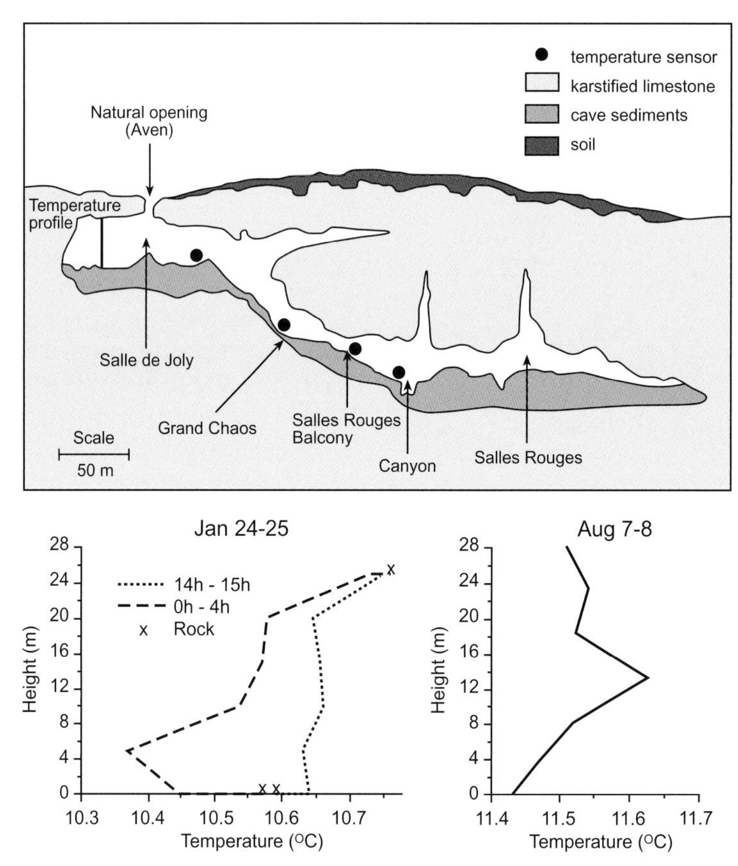

Fig. 4.24 Cross-section giving locations within Aven d'Orgnac and mean vertical temperature profiles through the Salle de Joly for 2 days each in January and August. A nocturnal temperature inversion in January promotes convective airflow. The winter circulation regime was established in Salle de Joly during the period 25 October to 6 November (Fig. 4.24) then extended to Grand Chaos on 14 November, Salles Rouges Balcony on 22 November and Salles Rouges after 9 December (Bourges et al., 2006a).

most distant chamber displays low amplitude (0.03 °C) temperature changes, with major diurnal and semi-diurnal components, strongly correlated with the rate of change of pressure and they proposed as a criterion for a confined system. More generally, air exchange with the epikarst can be tightly coupled to pressure changes and rainfall events (e.g. Bourges et al., 2006a; Fernandez-Cortes et al., 2009). At Altamira, Spain, Sánchez-Moral et al. (1999) even attributed a prominent and temporary reversion from the summer to the winter circulation regime (signified by Rn and CO_2) levels to a strong period of rainfall. This draws attention to the need for better understanding of the relationships between water content of the aquifer and karst air movement (Cuezva et al., 2011), particularly as this will be relatively much more important in tropical climates where seasonal temperature variations are small.

Because air circulation is such an important phenomenon, it is fortunate that at many caves there exists an opportunity to use carbon dioxide as a natural tracer to quantify it. Disturbance of the cave system by visitors (Fig. 4.26) provides a short-lived CO_2-pulse whose build-up is inversely proportional to the effective cave volume while its

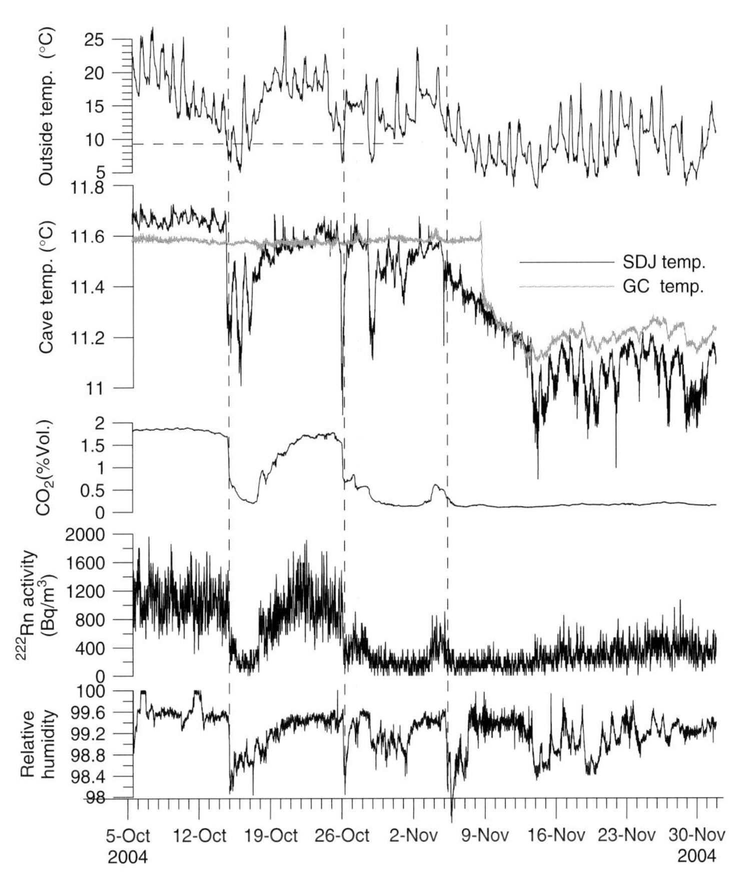

Fig. 4.25 Monitoring data from Aven d'Orgnac (Bourges et al., 2006a). Switching toward the winter regime in the Salle de Joly, as recorded by the temperature, CO_2 concentration, ^{222}Rn activity and relative humidity. The first change on 16 October is marked by a drop in temperature, ^{222}Rn activity, CO_2 concentration, and relative humidity, followed by a total recovery of these variables. The second switch, during the night of 25 October is followed by only a partial recovery. The definite switch to the winter regime occurs during the night of 6 November.

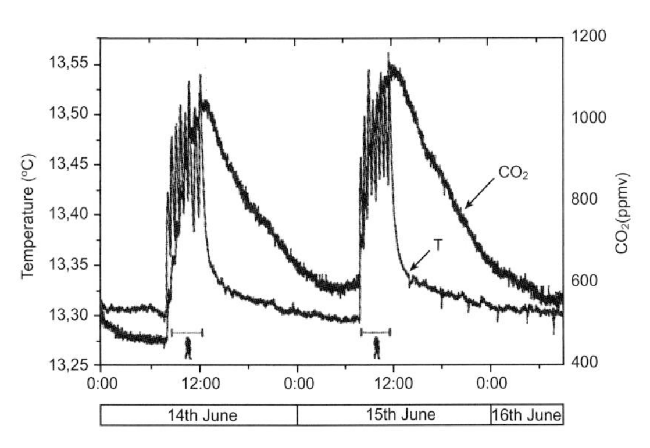

Fig. 4.26 Typical examples of disturbances to temperature and CO_2 concentration caused by visitors (Polychromes Hall of Altimira Cave (Sánchez-Moral et al., 1999)).

exponential decay reveals the rate of cave ventilation. Simply stated, the relaxation time corresponds to the air-exchange time (around 8 hours for Fig. 4.26). Faimon et al. (2006) provided a formal mathematical treatment as follows. Cave ventilation is associated with a characteristic airflow discharge (Q with dimensions L^3T^{-1}) between a cave (chamber) of volume V and the exterior, concentrations of incoming (external) and outgoing air having dimensionless CO_2 concentration (i.e. a mixing ratio) of respectively of C_e and C_c. The total CO_2 flux (J) into the cave is the sum of input fluxes (j_1, j_2, etc., which are products of a volume flux multiplied by a mixing ratio) from epikarst air, degassing of water, human input and microbial decay) plus the air-exchange flux terms:

$$J = VC_c / dt = j_1 + j_2 + QC_c - QC_e \quad (4.22)$$

when each of the fluxes is constant, integration of eqn. (4.22) gives:

$$C_c = J_1 / Q(1 - e^{-Qt/V}) + J_2 / Q(1 - e^{-Qt/V}) + C_e(1 - e^{-Qt/V}) + C_c^i e^{-Qt/V} \quad (4.23)$$

where C_c^i is the initial CO_2 concentration in the cave. If C_c is constant (steady-state) then

$$C_c = (j_1 + j_2)/Q + C_e \quad (4.24)$$

Care needs to be taken only to use this approach where the origin of the carbon dioxide variations is clearly known and the system is in approximate steady-state over the period of observation. For example, Milano and Gabrovšek (2009) attempted to use these relationships to calculate epikarst gas fluxes, but this was done on natural CO_2 variations whose origins are unknown, so the calculations are unsafe.

Deduction of air-mixing times by observing the decay of anthropogenic perturbations was a key part of the strategy used by Frisia et al. (2011) in constructing a CO_2 budget for a small, well-mixed Italian cave (Grotta di Ernesto), known to display significant annual CO_2 variations. Three different methods of quantifying CO_2 release from dripwaters showed that it was an insignificant part of the budget (there is also no cave stream). The budget was dominated by flow of CO_2 from the epikarst, and removal by cave ventilation. A combination of faster ventilation and lower epikarst flows were responsible for low P_{CO_2} in winter.

4.4.4 Generalizing seasonality and its implications for speleothems

Information on the absolute or relative seasonal magnitude of air circulation at a total of 19 underground sites have been identified from the literature (Table 4.3). A variety of approaches, as detailed in the above sections on radon and carbon dioxide have been used. It should be noted that the ratio of carbon dioxide concentrations in different seasons is insufficient to make this comparison because of the likelihood that supply factors varied seasonally. In 17 of the sites, the comparison is made from Rn concentrations and the template of Fig. 4.21 has been used to plot them (Fig. 4.27). A wide range of degrees of seasonality is revealed, but in no case is the ratio of seasonal high to low Rn activities significantly less than two.

Figure 4.28 relates the seasonality to the volume of the cave network where this is known. The lack of a strong correlation is an interesting finding. Figure 4.28 is contoured in terms of efficiency where it can be seen that despite its sluggish fastest air–exchange time of around two days, the huge Carlsbad Caves network exchanges just as efficiently as the small Obir site (Fig. 4.29). In the Obir cave system (Fig. 4.29) several different cave chambers display simultaneous CO_2 variations and in the winter air-exchange of individual chambers takes of the order of 0.01 days.

More extreme seasonality is more common amongst sites with winter-dominated ventilation, presumably because of the greater power of convective mixing in descending caves in winter. However, significant seasonality is always present in the studied caves which are all from cool to warm temperate environments. The seasonality of circulation in tropical caves awaits description, although it is predicted that rainfall variations may influence circulation here. The seasonality in P_{CO_2} will have a strong influence on the saturation state for carbonate minerals (Fig. 4.30), causing significant variations in growth rate or allowing corrosion to occur during a particular season. Again, this is shown to be independent of the size of the cave system (Fig. 4.30). There are also implications for external CO_2-flux measurements because the air-exchange in karstic environments is likely to be significant in relation to CO_2 fluxes as a whole (Kowalski et al., 2008; Cuezva et al., 2011). The study of Cuezva et al. (2011) at the shallow Altimira cave site, northern Spain, is particularly significant in showing the importance of seasonal dryness in controlling this air exchange. Whereas in winter, the aquifer is water-filled and is impermeable to gas-exchange, in summer there is an important

Table 4.3 Summary information concerning caves plotted in Figs. 4.27, 4.28 and 4.30.

Code	Location	Dominant season for ventilation	References
1	Gyokusen-do, Okinawa, Japan	Winter	Tanahara et al. (1997)
2	Phulchoki tunnel, Kathmandu, Nepal	Winter	Perrier et al. (2007)
3	Niedźwiedzia, Poland	Winter	Przylibski (1999)
4	Létrási-Vizes, Hungary	Winter	Hakl et al. (1997)
5	Szemlő-hegy, Hungary	Winter	Hakl et al. (1997)
6	Radochowska, Poland	Winter	Przylibski (1999)
7	Altimira, N Spain	Summer	Sánchez-Moral et al. (1999)
8	Candamo, northwest Spain	Summer	Hoyos et al. (1998)
9	Ernesto, northeast Italy	Winter	Frisia et al. (2005a, 2011)
10	Obir, southeast Austria	Winter	Spötl et al. (2005); Fairchild et al. (2010)
11	Srednja Bijambarska, Bosnia and Herzegovina	Winter	Milano & Gabrovšek (2009)
12	Hajnóczy, Hungary	Winter	Hakl et al. (1997)
13	Aven d'Orgnac	Winter	Bourges et al. (2001, 2006a)
14	Nerja, southeast Spain	Winter	Dueñas et al. (1999), Liñan et al. (2008)
15	Carlsbad, New Mexico, USA	Winter	Wilkening and Watkins (1976)
16	Karchtner, Arizona, USA	Summer	Buecher (1999)
17	Císařská, Czech Republic	Summer	Faimon et al. (2006)
18	Castaña de Íbor, central Spain	Summer	Lario et al. (2006); Fernandez-Cortes et al. (2009, 2010)
19	Hollow Ridge Cave, Florida (station 2)	Autumn, winter	Kowalczk and Froelich (2010)

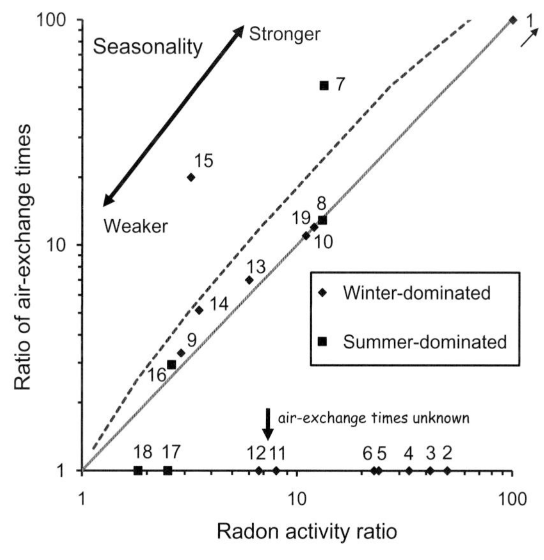

Fig. 4.27 Diagram as Fig. 4.21 with data from the literature on cave air circulation (numbers are keyed in Table 4.3). Five of the 18 sites are dominated by summer ventilation, the rest by winter ventilation and on average the winter-dominated sites show a more extreme seasonality. Two sites (Ernesto and Obir) have no radon data, but relative seasonal Rn variations are predicted using independent knowledge of air-exchange times in summer and winter.

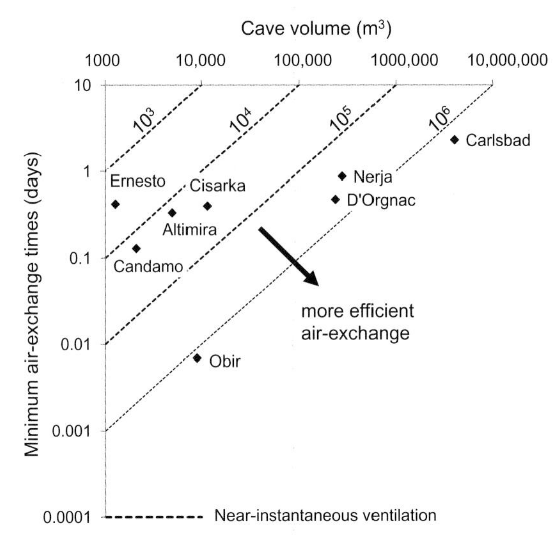

Fig. 4.28 Crossplot of air-exchange times and cave volume to illustrate the concept of efficiency of air-exchange (contoured in units of $10^6 m^3 day^{-1}$). Both very large and very small caves can show efficient ventilation.

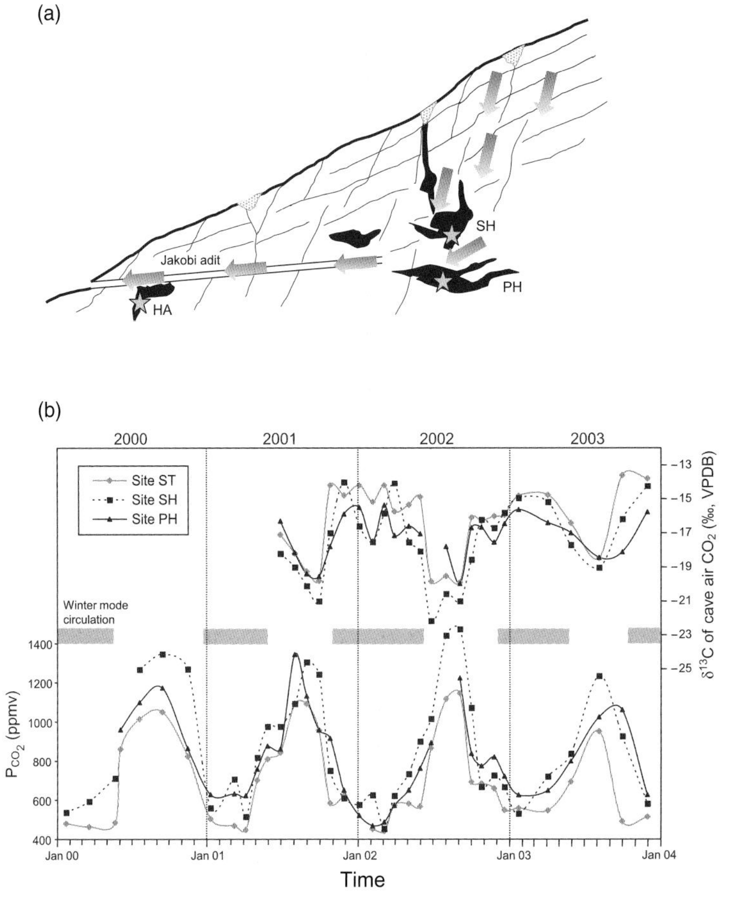

Fig. 4.29 Obir Cave (Spötl et al., 2005). This cave has a prominent chimney circulation. (a) Summer circulation (arrows reverse in winter); scale of sketch is approximately 250 m long. (b) Seasonal pattern of P_{CO_2} variation illustrating similar patterns in all cave chambers. This strong winter circulation is presumably augmented by convective flows.

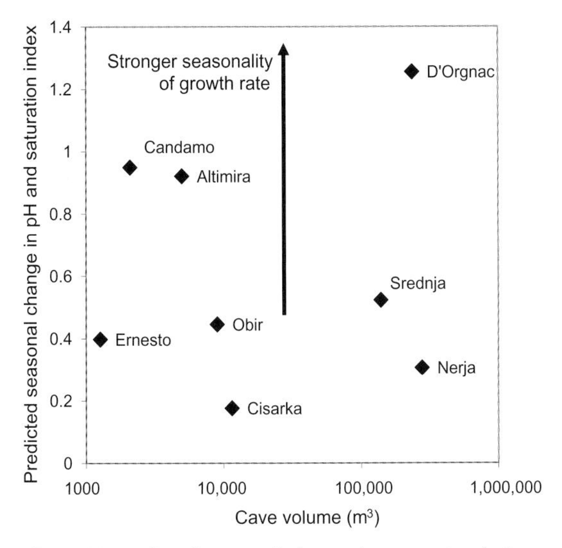

Fig. 4.30 Predicted seasonal change in saturation index versus cave volume, illustrating the presence of strong seasonal variations in caves of all sizes.

diurnal exchange of gas from the cave across the soil-epikarst membrane to the atmosphere.

4.5 Heat flux (authored by David Domínguez-Villar)

Underground temperature is known to be rather stable during the year below several metres depth. For this reason, cellars, abandoned mines or tunnels, and of course caves, have been traditionally used as homes, for food storage, fermentation of cheese or wine, among other uses (e.g.Silvia & Ignacio, 2005). However, underground galleries represent a complex system in which temperature is controlled by an energy balance. For example, the potential or kinetic energy of an air mass entering the cave is partially transformed into thermal energy or heat, affecting the temperature in the system. Therefore, to understand cave temperature it is necessary to evaluate heat fluxes.

Caves show an extraordinary variety of behaviours, with temperature being controlled by one or other factor or several in combination; they include: external climate, cave depth, morphology and size of cave galleries, presence/absence of a river or lake in the cave, and the existence of geothermal activity in the region.

4.5.1 Sources and mechanisms of heat transfer into caves

Two main sources of heat are available in a cave system: the heat transferred from the Earth's interior and the heat transferred from the atmosphere to the underground. The mechanisms for heat transfer in caves are mostly related to advection and conduction. Owing to the high humidity in cave environments, the processes of evaporation and condensation play a major role and heat can be transferred as sensible or latent heat. Sensible heat is the thermal energy that is transformed in a change of temperature, whereas latent heat is the heat needed for a phase change which is isothermal (e.g. from gas to liquid). Thus, during a phase change, the energy supplied is not transformed in terms of changes in temperature, but in evaporation or condensation processes.

Geothermal heat flux

Temperature profiles from the upper crust are available from deep caves, mines and wells (e.g. Badino, 1995, 2005; Pollack & Huang, 2000). In regions where no major water flow exists in the phreatic zone, temperature profiles show a clear division between a surficial or heterothermic zone and a geothermal or homothermic zone (Fig. 4.31). In the geothermal zone the temperature increases at a constant rate with depth. This rate is known as the *geothermal gradient* and is the result of the heat flow from below (dependent on the particular geo-tectonic history of the region) and the thermal conductivity of the rocks transferring such heat. The geothermal gradient varies regionally, although common values range from 1.8–3.5 °C/100 m (Luetscher & Jeannin, 2004). The temperature profiles can be disrupted by advection of waters, because of a local phreatic flow, or by differential water flows in the geothermal zone as a result of an alternation of rock types or fault zones with different permeability (Lismonde, 2004; Luetscher & Jeannin, 2004; Anderson, 2005). The surficial zone shows wider range of temperature during the year because the proximity of the surface

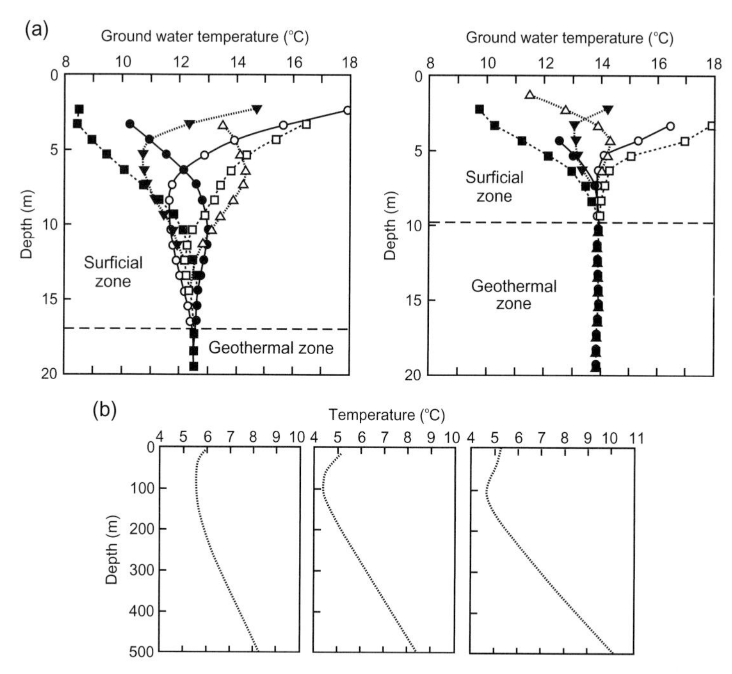

Fig. 4.31 (a) Thermal profiles showing the surficial and geothermal zones in two different wells in Nagaoka (Japan) at different times of the year. Note that in recharge areas the geothermal zone is deeper than in discharge areas and in both cases temperature in the surficial zone change seasonally (modified from Anderson, 2005). (b) Three deep cores from eastern Canada showing a constant increase of temperature with depth in the geothermal zone. Note that the geothermal gradient differs from site to site and that these are non-karstic sites (after Pollack & Huang, 2000).

represents another source of heat. The limit between the zones is variable but is normally shallower than 150 m (Stevens et al., 2008). Additionally, the depth of such zone is much shallower if the region represents a discharge zone instead of a recharge zone (Anderson 2005). In most cases the geothermal heat flux in shallow caves (e.g. <20 m beneath the surface) is not considered, because the influence of heat provided from above is responsible for most of the signal and its variability. This is not applicable to geothermal regions where even shallow caves can be dominated by the geothermal heat flux.

Surface heat flux

The mechanisms for transferring heat from the surface to the cave system are variable, and highly dependent on the presence of moving air or water that could advect heat. In the case of absence of significant heat advection, the heat flux transferred by conduction through the bedrock dominates the thermal signal of the cave. Typically several factors are integrated to produce distinct thermal regimes to different sectors of the same cave system.

Heat transferred by the atmosphere

The most likely influence of external air in all caves is in their entrance sectors (Fig. 4.32). If internal and external air temperature and humidity are known, the distance into the cave in which external temperature is equilibrated with cave temperatures can be calculated according to eqns. 4.25 and 4.26 (Wigley & Brown, 1971):

$$T = T_a + (T_0 - T_a)e^{-X} + \frac{L}{c}(q_0 - q_a)Xe^{-X} \tag{4.25}$$

$$(1+A)\ln\left[\frac{T_1 - T_a}{T - T_a}\right] + AB(T_1 - T) = X - X_s \tag{4.26}$$

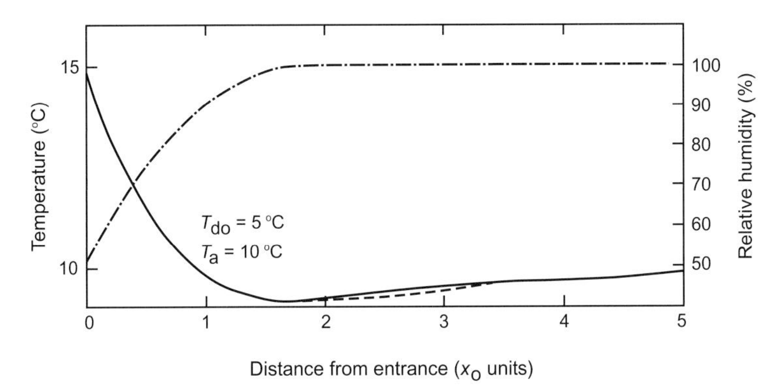

Fig. 4.32 Modelled temperature and humidity gradients from a cave entrance. The distance from the entrance is measured in relaxation length units (after Wigley & Brown, 1971).

where T is the temperature in the entrance sector, T_a is the stable cave temperature, T_0 is the temperature of air entering the cave, X is the ratio of distance from cave entrance and the relaxation length (equivalent to the process length of section 4.1 of this chapter), X_s is X at saturation, L is the latent heat of vaporization, c is the specific heat of air, q_0 the specific humidity of air entering the cave and q_a specific humidity in the corresponding stable cave interior, T_1 is the temperature at which the cave air becomes saturated with water vapour and A and B are constants. On the right-hand side of eqn. 4.25 the middle term accounts for the difference between internal and external temperature, whereas the right-hand term considers the latent heat and the modifications caused by evaporation or condensation processes. As the cave atmosphere cannot be supersaturated, during periods when cave atmosphere is saturated at some sector of the entrance, that is $X \geq Xs$ (predominantly in the summer season), the eqn. 4.26 provides the solution to the temperature transition between cave interior and the surface. In both cases the temperature describes a roughly exponential profile that asymptotically reaches the T_a value. The relaxation length of external temperature depends not only in the difference in temperature and humidity, but also on the diameter of cave passages, the wind speed, and the Reynolds number of the flow (Wigley & Brown, 1971, 1976; de Freitas & Littlejohn, 1987). The diameter of cave passages is critical in comparison with wind speed, and caves with narrow entrances are expected to have a limited influence from external temperatures owing to advection, unless forced ventilation occurs. The external air influence in cave entrances range from just some metres to several kilometres, and the distance varies with external and cave temperature and humidity.

Changes in cave temperature (e.g. from groups of humans) affect the density of air, driving pressure differences in the cave system causing air currents inside the cave (Pflitsch & Piasecki, 2003). Effects of atmospheric pressure can also drive changes in cave temperature in confined rooms by adiabatic heating (Perrier et al., 2001), although more commonly the rate of pressure change correlates with shifts in cave temperature (Perrier et al., 2001; Bourges et al., 2006a). Rates of temperature change with pressure gradients have been reported from less than 0.001 to 0.042 °C/hPa (Bourges et al., 2006a) depending on the mechanisms involved, although some systems have demonstrated to be thermally invariant to pressure modifications (Salve et al., 2008). Pressure changes can be forced by water inflow into cave passages, as in the case of floods or tides in coastal caves, affecting their cave temperature (Gamble et al., 2000).

In deep caves, the pressure changes owing to the weight of air column; deeper sectors of the cave are warmer. The thermal gradient with height or adiabatic gradient depends on the atmosphere humidity, ranging from 0.4 °C/100 m in saturated air to 1 °C/100 m in dry air, although this range is highly modified by the water flow along the karst massifs that is described in the next section. Additionally, although air has lower mass and heat capacity than water, in caves with strong winds, airflow can have an importance in the heat balance of the cave owing to friction with cave walls (Atkinson et al., 1983; Jernigan & Swift, 2001).

As discussed in section 4.4, air density represents a boundary condition for atmosphere interchanges and air stratification in caves is common, with warmer air circulating in the upper sectors of galleries and cooler and denser air circulating at the bottom (Bourges et al., 2006a; Milanolo & Gabrovšek, 2009). Morphology of caves represents a critical factor in isolating air masses. Thus, stagnant cold air masses can be isolated from the rest of the cave in the so-called 'cold traps' (Luetscher et al., 2008), and chimneys or vaults can keep warm air masses in 'warm pockets' (Pflitsch & Piasecki, 2003). The existence of cold traps is of particular interest because of the preservation of ice during the whole year in some ice caves (Silvestru, 1999; Luetscher et al., 2008).

Heat transferred from water

Water has a high specific heat and it is able to modify cave temperature by advection. This is the case of galleries in caves through which a large river passes. The stream temperature normally dominates the cave atmosphere temperature, although other sectors of the cave may not be affected (Kranjc & Opara, 2002; Fuller, 2006). Galleries that are subject to occasional stream flow also change their air temperature during flood events (Lismonde, 2004). Lakes may also affect the temperature of their galleries (Pulido-Bosch et al., 1997; Sarbu & Lascu, 1997).

Infiltration water, too, has a substantial effect on the temperature in karst areas, having a major impact in Alpine tunnels in limestone which are much cooler than in basement rocks. Figure 4.33 illustrates the case of an Alpine karstic massif, where the movement of water dominates temperature up to hundreds of metres below the surface owing to the transformation of potential energy into heat, owing to friction of water against the conduit walls. This hydrologic gradient which account for a cooling of 0.234 °C/100 m counteracts in part the adiabatic gradient (0.4 to 1 °C/100 m) being responsible for net warming gradients ranging from 0.18 to 0.78 °C/100 m (Luetscher & Jeannin, 2004). Since the resulting thermal gradient is lower than the adiabatic gradient in the external atmosphere, mean annual temperature of karst waters is cooler than that of surficial waters at the same altitude (Badino, 2005). Despite the progressive warming of waters along the karst massif, such flow towards the springs entirely removes the original geothermal gradient signal and cools down karst massifs. This hydraulic gradient is important for deep cave atmosphere temperatures, because they will tend to be equilibrated with host rock temperature, whereas it can be mostly neglected for shallow caves.

Humidity in caves is frequently close to saturation and under these circumstances evaporation and condensation process are common. Latent heat due to water change of phase requires energy from the system and significantly affects cave temperature. Evaporation and condensation can be the result of air inflow at the cave entrance (Wigley & Brown, 1971), air advection in the cave (de Freitas & Schmekal, 2003; Dreybrodt et al., 2005; Salve et al., 2008), temperature difference due to a local geothermal source (Sarbu & Lascu, 1997), tourists entering the cave (Pulido-Bosch et al., 1997; Hoyos et al., 1998) or barometric pressure changes in the system (Perrier et al., 2001). In all cases the latent heat effect reduces the thermal differences between end-members and tends to mitigate anomalies. Thus, evaporation causes cooling of the cave atmosphere whereas condensation occurs during cave air warming at relative humidity of 100%. In L'Aven d'Orgnac in France, a thermal difference of 7 °C between air masses caused only a thermal anomaly of 2 °C because of the condensation of 350 litres of water from the $10^5\,m^3$ of air entering the cave daily (Bourges et al., 2006a). In Glowworm Cave in New Zealand, evaporation rates of $2.41\,g\,m^{-2}\,h^{-1}$ have been reported as the result of a 14.8 °C temperature drop in the cave atmosphere, with a non-equilibrium

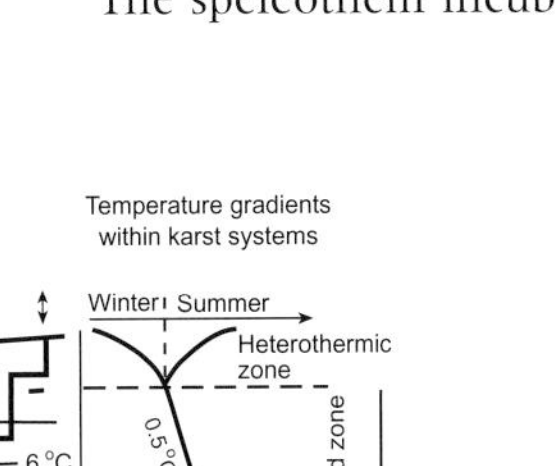

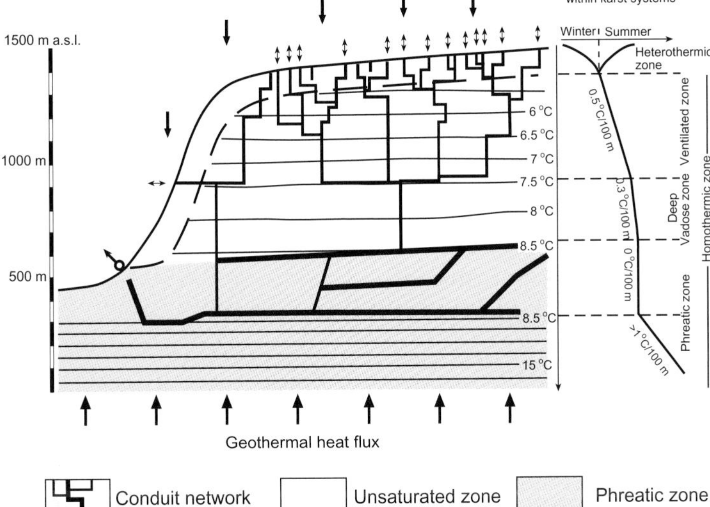

Fig. 4.33 Conceptual model of controls of temperature with depth in an Alpine limestone massif. The interior of limestone massifs shows only a small increase in temperature depth, contrasting with mines and tunnels in rocks lacking cavernous porosity. The near-surface heterothermic zone of up to a few tens of metres is affected by atmospheric temperature disturbances (e.g. seasonal). Below this, temperatures are stable (homothermic) with small increases in temperature with depth caused by loss of gravitational energy: from moist air in the vadose zone and from water in the phreatic zone. Rapid lateral flow of water to springs effectively flushes out geothermal heat so that steeper rises of temperature with depth corresponding to geothermal gradient are only found beneath the karstified zone. In a lowland environment, the intermediate zones are compressed and may be absent in geothermal areas.

system in which measured condensation (10.3 g m^{-2}) is higher than the 9.9 g m^{-2} of evaporation of condensate (de Freitas & Schmekal, 2003).

Heat transferred from the rock

Rock has a large thermal inertia and unless there is significant heat input or output due to water or air advection, rock temperature will dominate the cave temperature. This is the case for most caves in the surficial zone thoroughly isolated from direct external influences: cave temperature is very stable and close to that of the mean annual surface atmospheric temperatures where geothermal flux can be neglected. The transfer mechanism is simple and implies heat conduction. The atmospheric temperature is recorded by the soil and then transmitted to the bedrock (Pollack & Huang, 2000; Beltrami & Kellman, 2003; Pollack et al., 2005; Smerdon et al., 2006). The thermal signal propagates by conduction through the rock until reaches the cave. The effective transmission of temperature by conduction in a substance is dependent on the thermal diffusivity, which is specific for each lithology. Rock thermal diffusivity values of 10^{-6} m^{2}s^{-1} are commonly used in calculating the propagation of thermal signatures in deep profiles (e.g. Beltrami, 2001), although owing to the shallow nature of most caves and anisotropy of karst massifs, errors are significant and it is recommended to calculate the local thermal diffusivity value for each cave site (Fig. 4.34a).

The absolute temperature recorded in the cave will be equivalent to the soil temperature, but two main effects are involved: attenuation of thermal anomalies and phase lag with distance (Figs. 4.34 and 4.35). The thermal signal in depth is diffused and external anomalies are attenuated at depth. Thus, daily temperature oscillations are completely attenuated before reaching the first half metre,

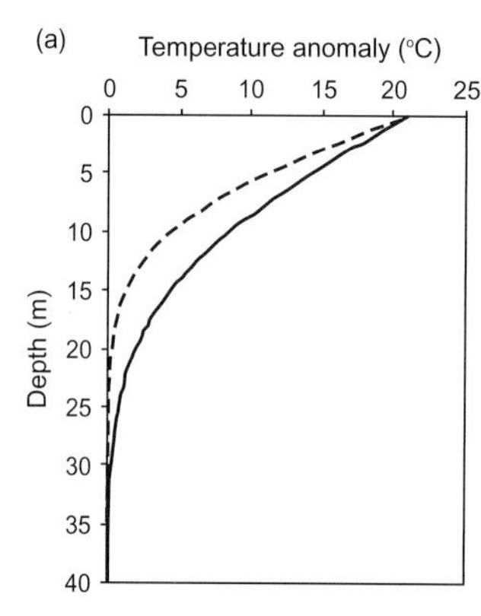

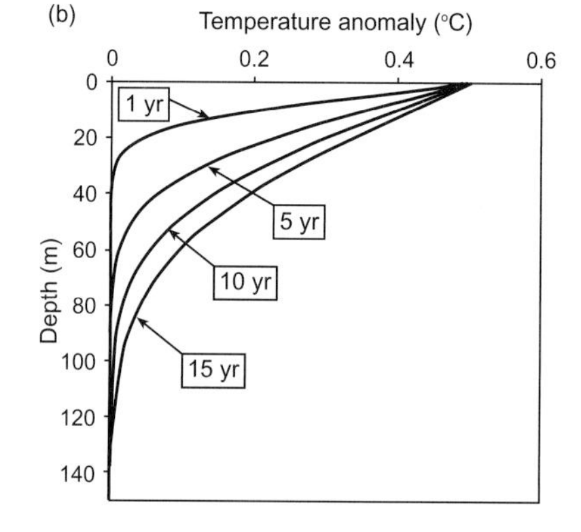

Fig. 4.34 Temperature profiles for Postojna Cave (Slovenia). (a) Difference between expected seasonal thermal anomaly in depth calculated using an specific thermal diffusivity value for this site (solid line) in comparison with the standard used rock value (dashed line) of $10^{-6}\,m^2s^{-1}$ in order to show the importance of calculating local thermal diffusivities. (b) Propagation of thermal changes versus depth expected for a hypothetical 0.5 °C surface temperature anomaly with persisting durations ranging from 1 to 15 years, showing the importance of the duration of the anomaly in order to be recorded in the cave.

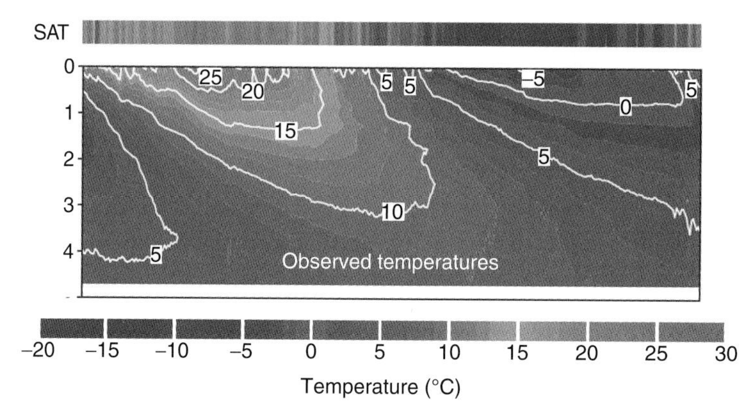

Fig. 4.35 Temperature profile in Fargo (USA) showing thermal evolution in the zone within 5 m of the ground surface during a year in relation to the surface atmosphere temperature (SAT). Note that with depth the temperature anomaly decreases and the thermal signal has an increasing lag time up to several months (Smerdon et al., 2006).

whereas seasonal oscillations could reach depths of 15–45 m depending on the thermal diffusivity and thermal seasonal amplitude. The heat conduction equation, which describes the thermal energy propagation through solid media by pure conduction, is given by eqn. 4.29 (Carslaw & Jaeger, 1959), and its solution in terms of temperature is given in eqn. 4.30:

$$\frac{\partial T}{\partial t} = \kappa \frac{\partial^2 T}{\partial z^2} \qquad (4.29)$$

$$T = T_0 erfc\left(\frac{z}{2\sqrt{\kappa t}}\right) \qquad (4.30)$$

where T is temperature, t is the time, κ is thermal diffusivity, z is the depth and *erfc* is the complementary error function. Hence at greater depths the thermal seasonal amplitude is progressively smaller until it is completely lost. In considering conduction as the main mechanism of heat transport, the thermal amplitude of the anomaly is not as important as the duration of the anomaly (Fig. 4.34). Thus, a fire lasting for one day will be damped within the first centimetres of soil even if the thermal anomaly could approach a thousand degrees, whereas temperature variations of the Little Ige Age in the order of 0.1 °C are recordable

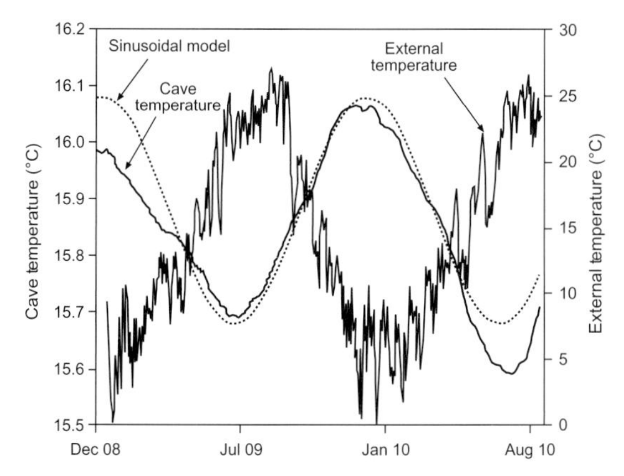

Fig. 4.36 Sinusoidal temperature record from Eagle Cave (Spain) dominated by heat conduction through the rock massif. A simple theoretical sinusoidal signal (dotted line) has been added to illustrate the symmetry and harmony of the temperature signal (solid line) dominated by conduction. Major deviations from the sinusoidal signal are caused by inter annual trends transferred by conduction. Cave temperature is at a maximum in winter, but the lag time in cave temperature is thought to be several years, not 6 months.

in deep thermal profiles because that period lasted over a century (Beltrami, 2001).

On the other hand, the thermal signal will take some time to be transmitted by conduction through the rock massif (Figs. 4.35 and 4.36). Thus, the temperature recorded in the cave will experience a phase lag in relation to the soil temperature. In soil experiments with depths between 3 and 7 m, lag times ranged from 1 to 6 months (Pollack et al., 2005; Smerdon et al., 2006; Mazarrón & Cañas, 2009). Rock thermal diffusivities are normally smaller than the soil thermal diffusivities found in these experiments, and delay times are expected to be larger. So, delay times larger than a year are expected for many of the caves unless the galleries are very shallow (e.g. less than 5 m in depth) or present anomalously large thermal diffusivities. Therefore, in the case of Eagle Cave in Spain (Fig. 4.36) where the site is approximately 20 m under bedrock, it is prudent to say that there is a phase lag of approximately $n2\pi$ years instead of a delay of half a year. In this case, the delay in thermal signal is thought to be between 5 and 10 years (Domínguez-Villar et al., 2009a).

It is more typical that the cave atmosphere will be influenced by rock temperature rather than vice versa, because considering the same mass the heat capacity in rock is approximately 1800 times larger than in air. Thus, cave wall temperatures are normally stable (de Freitas & Schmekal, 2003; Bourges et al., 2006a) or show variations in the order of approximately 0.1 °C (Crouzeix et al., 2003), although if cave temperature has large amplitude oscillations it will affect cave wall temperatures (Luetscher et al., 2008). The cave temperature in relation with rock conduction normally shows a quasi-perfect sinusoidal signal if the annual cycle has been not completely attenuated (Fig. 4.36), and asymmetrical records would indicate multiple sources of heat. Owing to seasonal ventilation of many caves, it is relatively common to find cases in which the rock signature dominates the cave temperature when the ventilation is attenuated, whereas during the enhanced ventilation season heat advection will provide additional noise to the cave temperature record (Fig. 4.37). This is typical of sites near the entrance where the transition zone between isolated and external influenced cave sectors fluctuates seasonally or even at higher frequencies depending on external weather conditions (Mazarrón & Cañas, 2009).

4.5.2 Thermal equilibrium in caves

Temperature in caves is thought to be equivalent to mean annual surface atmosphere temperature (e.g. Wigley & Brown, 1976; Moore & Sullivan, 1978). In the case of cave systems dominated by advection, there is a multitude of variables to disrupt the thermal equilibrium with the cave, although caves with temperatures similar to those recorded on the surface are not uncommon (de Freitas & Littlejohn, 1987; Smithson, 1991; Kranjc & Opara, 2002). However, thermal shifts of several degrees can be found owing to advection or geothermal fluxes (Petkovšek, 1968; Atkinson et al., 1983; Buecher, 1999; Milanolo & Gabrovšek, 2009). Caves where the temperature is dominated by the heat transferred by conduction from the surface are *a priori* better candidates to show thermal equilibrium with surface temperature (e.g. Amar & de Freitas, 2005; Domínguez-Villar et al., 2010). In these systems, although cave temperature is

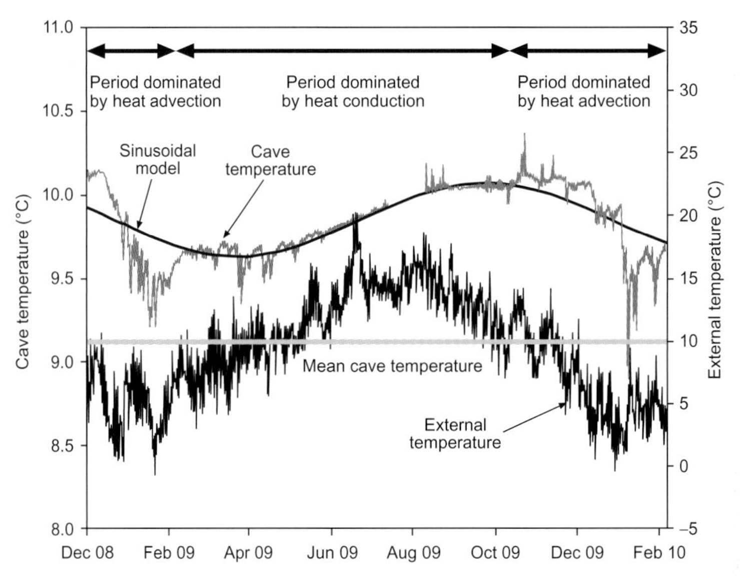

Fig. 4.37 Asymmetrical temperature record in Lower Balls Mine (Nailsworth, UK), whose signal is seasonally dominated by conduction through bedrock or advection by air flow. A simple theoretical sinusoidal signal is shown to illustrate the good fit of the mine temperature, when the thermal signal is dominated by conduction in comparison with the periods where advection is dominant. When external temperature is above the mean cave temperature, ventilation is less effective and conduction dominates the thermal signal in the monitoring site, whereas when there is enhanced ventilation the thermal instability caused by advection goes deeper into the galleries.

considered stable, temperature changes at the surface can modify cave temperature if thermal anomalies last sufficiently long, although a certain delay for the signal to reach the cave depth is expected (Perrier et al., 2005; Genty, 2008; Domínguez-Villar et al., 2009b). Although thermal changes are less abrupt than in the surface they finally equilibrate, and inter-annual rates of 0.04 °C/year were measured in Villars Cave in France (Genty, 2008) and minimum rates of 0.07 °C were recorded in Eagle Cave in Spain (Domínguez-Villar et al., 2009a). However, caves dominated by conduction can be out of equilibrium with surface air temperature by more than a degree (e.g., Fernandez-Cortés et al., 2006, 2009). Additionally, caves with galleries at different depths show different temperatures (Vokal, 1999; Genty, 2008). Although some aspects of these thermal disequilibria can be explained by the mechanisms previously reported in this section, other causes can be responsible for uncoupled surface and cave temperatures.

The cave receives heat transferred from the soil by conduction from the surface. Although the source of soil temperature is from the atmosphere, there are several causes that can be responsible for thermal shifts in this interface, including natural and anthropogenic factors. Seasonal snow cover is an effective isolating mechanism, causing soil temperatures to maintain at approximately 0 °C during the freezing period when the snow persist over the soil (e.g., Vieira et al., 2003). In higher-latitude regions with a continental climate where snow cover is persistent during several months in winter, soil temperatures are higher than atmosphere temperatures (Smerdon et al., 2006), thus soil and not atmosphere temperature should be expected to be recorded in the cave. This explains, at least in part, the existence of temperatures some degrees above zero in caves where mean surface atmosphere temperature is below zero, as in Castleguard Cave in Canada (Atkinson et al., 1983). Soil temperature is also dependent on factors such as hydrological changes in the region (Beltrami & Kellman, 2003); hence the development of agricultural regions in Southwest United States has modified the deep temperature records of the region (Stevens et al., 2008).

Another important factor affecting soil temperature is the forest cover over the cave. Soil temperature experiments in forested and non-forested regions have reported temperature differences between them of 1–3 °C (Nitoiu & Beltrami, 2005) and ground temperature records demonstrate a warming of 1–2 °C in different North American regions owing to the effects of deforestation in past decades (Lewis & Wang, 1998). On the contrary, the recovery of forest induces a cooling of soil temperatures, which can be recorded in caves after conduction through bedrock. This is found in the case of Eagle Cave in Spain, where much of the 2 °C drop in temperature over the past 30 years is explained by forest recovery (Domínguez-Villar et al., 2009b). Density of forest canopy is important in protecting soil from the direct effect of sun; for example, a dense beech forest is thought to be more effective than acacia woodlands. Additionally, latitude is an important factor, because lower latitude soils are more sensitive to deforestation, owing to higher solar radiation. Except in uninhabited regions, the changes in forest cover over the past centuries generally have an anthropogenic origin, and temperature in associated caves is expected to have been anthropogenically modified. On the other hand, changes in forest canopy occur owing to natural causes, including fire, or ecotone shifts due to major climatic changes, or species migration. Thus, caves located today in forested regions experienced a negative feedback on falling temperature during past glacial periods where they became overlain by steppe vegetation.

4.6 Synthesis: cave climatologies

We can now return to the concept of process length (section 4.1) to help summarize the way that the speleothem incubator functions. Process length refers to the maximum distance within a cave system that short-term external signals penetrate. For this purpose we can think of annual variations as the upper limit of 'short-term' because they are such a dominant feature of the climate system. Hence, beyond the process length, annual and longer-term variability just contribute to the mean value of the climate parameter.

Consider for example, the predominantly vertical movement of water feeding speleothems. This is a climatological parameter analogous to rain and snowfall in the external environment. Rate of water flow primarily depends on the contrast between seepage and fracture-fed flow. The proportion of dripwater whose infiltration time is primarily limited by seepage flow increases with depth, whereas that of fracture-fed flow diminishes (Fig. 4.38b), although the exact depth depends on aquifer structure. In contrast, if we consider the likely distribution of process lengths with depth, they are much shallower for seepage flow than for fracture-fed flow (Fig. 4.38a). This means that the annual variability of infiltration is damped out at shallow depths in the matrix porosity of the aquifer, whereas fracture-fed flow can propagate distinct impulses much deeper in the system (e.g. Genty & Deflandre, 1998; Fairchild et al., 2010). Hence any features of a speleothem which depend on changes in the annual quantity of water or a time-limited chemical property of the water must depend on fracture-fed flow deeper in the system, but less so at shallow depths.

Now consider the propagation of temperature anomalies into a cave network. Figure 4.38d shows in principle how distinct fluctuations in temperature die out in the interior. These fluctuations could be daily or annual, so there is no absolute scale on the diagram. Although many parts of a cave system may be at the mean annual external temperature, the geometry of the cave system can give rise to cold or warm traps (blind upward or downward passages). In section 4.5, we have also seen how the surface vegetation can influence temperature, and that anomalies arise where geothermal heat is able to propagate into shallow environments. When we consider the process lengths associated with temperature, we find that they tend to be shorter than those associated with CO_2 (Fig. 4.38c). For example, during chimney circulation, incoming

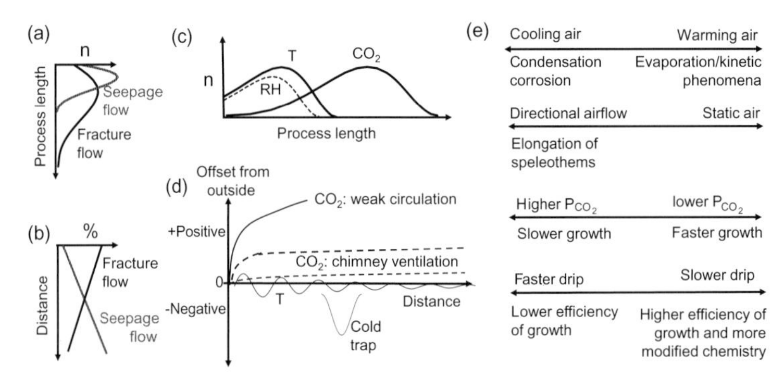

Fig. 4.38 Schematic synthesis of properties of the speleothem incubator. (a) and (c) Schematic probability density functions of process lengths (where *n* is the number of observations) for different parameters. (b) and (d) Schematic variations of parameters with distance. (e) Some key links between cave climate and speleothem properties. *RH*, relative humidity; *T*, temperature. See text for discussion.

air is warmed or cooled as appropriate to a constant temperature, but can still retain a distinctive carbon dioxide content deeper into the interior of the system.

Relative humidity changes with distance are largely controlled by temperature and to a lesser extent by the availability of moisture. In Figure 4.38c, the process lengths of relative humidity are shown as being slightly shorter than those of temperature: this is depicted thus simply to reflect that temperature variations can be measured more precisely. Relative humidity is difficult to measure, but when care is taken, quantitative results can be obtained for condensation or evaporation (e.g. de Freitas & Schmekal, 2003).

The composition of the trace gases carbon dioxide and radon are invariably higher than the external atmosphere and are controlled primarily by the rates of air circulation, and secondarily by variations in their rates of production within caves, transported in air and water. In Fig. 4.38d, an example of a weakly circulating cave is depicted as one where the P_{CO_2} rises to high levels relatively close to the exterior; in this case, this could be a seasonal pattern, or one that is year-round. This is contrasted with the highly effective chimney circulation where steady air movement deep within a cave network can be generated (e.g. Spötl et al., 2005; Mattey et al., 2010), but where variations in its intensity or flowpath give rise to seasonal differences in P_{CO_2}. Accordingly, Fig. 4.38c depicts a wide range in process lengths for signals (e.g. annual) of P_{CO_2} variability.

In this chapter, we have seen several consequences of distinct external climatologies. A contrast between a warm and a cold season typically has a major effect on air circulation and hence on P_{CO_2}. Incoming cold air is warmed and therefore exhibits a fall in relative humidity; the opposite applies to warm air in summer (de Freitas & Littlejohn, 1987). In entrance passages, there may be a significant direct annual temperature change. On the other hand, a contrast between a wet and a dry season can result in notable changes in the amount of dripwater, its $\delta^{18}O$ or trace element composition, and the relative humidity of the cave. Degassing of CO_2 from cave streams will be much more pronounced in wet seasons as will be air circulation driven by water movement.

The consequences of differences in climatological parameters are summarized in Fig. 4.38e. The condensation of moisture from cooling air causes corrosion (e.g. Dreybrodt et al., 2005), particularly on less active growth surfaces. This can be important in summer near to inlets for chimney circulation, or in the upper parts of systems, for example geothermal caves, with significant unforced convection (Palmer, 2007). The converse from warming air is evaporation which is typically associated with enhanced kinetic effects on speleothem growth. A directional airflow can influence speleothem morphology, although this is normally most pronounced for helictites. Changes in P_{CO_2} of cave air have a direct impact on the degree to which drip-water can degas and hence low P_{CO_2} results in higher supersaturations and faster speleothem growth. A faster drip rate, other things being equal, will tend to result in faster growth, although a plateau is reached at higher drip rates (see Chapter 7). However, faster drips often have lower super-saturations and a seasonal corrosion of growth may even occur. Conversely slower dripping can result in a more efficient removal of Ca from each increment of water. Because this is also often associated with earlier degassing of water, the result may also be a more evolved chemistry via prior calcite precipitation (see sections 5.5 and 8.5). In some cases, dripping may cease which is another route to forming a seasonal hiatus.

The simple ideal of speleothem growth in a cave environment under essential constant equilibrium conditions, which was the initial driver for palaeo-climatological work, is unusual in practice. Significant seasonal variations of cave conditions modulate growth and indicate that a rich variety of information is recordable at high resolution. The speleothem incubator turns out to be a most lively environment indeed.

CHAPTER 5
Inorganic water chemistry

The chemistry of speleothems largely reflects drip-water chemistry, the topic of this chapter. The processes that determine the composition of oxygen and hydrogen isotopes in water are discussed in Chapters 3 (atmosphere and soil) and 4 (karst). The fractionation and partitioning effects related to stalagmite formation (crystallization processes beneath a thin and dynamic water film) are discussed in Chapter 8.

We start by outlining the parameters to be studied and optimal analytical methods (5.1). A brief treatment of classical carbonate system chemistry (5.2) is used to explain how to predict dissolution or growth of carbonates. Section 5.3 explains how the solute chemistry arises (5.3), firstly by focusing on calcite weathering at a fundamental level, then dealing more generally with the weathering of other minerals, and finally discussing two increasingly important special topics: metal isotopic signatures, and colloidal trace element transport. Carbon isotope chemistry (5.4) requires an understanding both of soil and cave processes, and much progress is being made currently on these topics. Finally in section 5.5 we show how qualitative and quantitative forwards and backwards modelling can help understand the origin of cave waters and how they reflect climate. Box 5.1 summarizes general aspects of solution chemistry that are needed for this chapter.

5.1 Sampling protocols for water chemistry

In recent years, rapid development of analytical instrumentation has significantly reduced the sample sizes required for multi-parameter analysis (Table 5.1) and there have also been advances in automated logging, which can be coupled with sampling. Several bespoke sampling solutions have been developed by individual researchers for cave deployments and there is some small-scale commercial production.

A valuable parameter is the conductance of electricity by the water (electroconductivity, EC) which is expressed here in units of microsiemens per centimetre ($\mu S\, cm^{-1}$). This parameter is a measure of solution charge, with a greater contribution by ions with larger charge, and so EC correlates with *ionic strength, alkalinity* and the cation load resulting from carbonate dissolution (Crowther, 1989; Krawczyk & Ford, 2006), moderated by re-precipitation. Although EC is often replaced by ion analyses by the conclusion of a study, it is important for three reasons. Firstly, robust algorithms by Rossum (1975) and Hughes et al. (1994) allow the conductivity to be calculated from ion analyses and, because EC is readily measured in the field, so drip sites can be immediately compared. Secondly, if both ion analyses and EC are measured and compared with calculated conductivity (e.g. by using the spreadsheet in the website linked to this book), errors in the analysis can be identified (excess positive or negative charge; EC too low or high) and attributed to an error in either the cation or anion analyses. Thirdly, EC can be continuously logged more readily than other parameters of solute composition which provides valuable data. The EC increases by a factor of around two from 0 to 25 °C and modern digital instruments automatically

Speleothem Science: From Process to Past Environments, First Edition. Ian J. Fairchild, Andy Baker.

Box 5.1 Aqueous chemistry definitions

Concentrations in solids: mg kg^{-1} (ppm), $\mu g\,kg^{-1}$ (ppb).

Concentrations in solution: the following definitions are simplifications for dilute solutions of density ≈ $1\,kg\,L^{-1}$.

Weight concentrations (g, mg or $\mu g\,L^{-1}$ solution) ≈ ‰, ppm and ppb respectively. These units are often convenient for expressing results of chemical analyses.

1. *Molar concentrations* (e.g. mmol or $\mu mol\,L^{-1}$). To convert from weight to molar units divide by the formula weight. Molar units are needed for equilibrium calculations.
2. *Equivalent concentrations* (e.g. meq or $\mu eq\,L^{-1}$). To convert from molar to equivalent units multiply by the charge of the ion. Equivalent concentrations are used in titrations or to check the balance of electrical charges in a solution. They are also used to express the *cation exchange capacity* in meq per 100 g of dry substance, representing the availability of cations that can be readily displaced into solution.
3. *Mixing ratios*. The mixing ratio of a gas X (equivalently called the mole fraction) is defined as the number of moles of X per mole of air, with units of mol/mol (or ppm if multiplied by 10^6) which is equivalent to v/v (volume of gas per volume of air) because the volume occupied by an ideal gas is proportional to the number of molecules.

The effective amount of species i that is available to take part in a reaction is the *activity*, here denoted by (i). It is defined in such a way as to be numerically equal to concentration in mol L^{-1} in very dilute solutions: $(i) = \gamma_i m_i$ where γ_i is the activity coefficient (L mol^{-1}) and m_i is the molar concentration. Activity is dimensionless and is equal to one for pure substances at one atmosphere pressure and 25 °C. Gases behave sufficiently ideally that the mixing ratio (mol/mol) and the pressure of CO_2, P_{CO_2}, are both equivalent to activity, but the value should be corrected to 1 atmosphere (1013 bar or hPa) total pressure for calculations.

The *ionic strength* (I) of the solution is

$$I = 0.5\sum m_i (z_i)^2, \qquad \text{(I)}$$

where z_i is the charge of ion i. Except in brines, the higher the value of I, the stronger the electrostatic attractions between ions (γ_i drops below one and more strongly so for highly charged ions) and the greater the abundance of ion pairs forming between specific ions.

The relative stability of different species at chemical equilibrium can be expressed thermodynamically in terms of free energy change for a reaction or as an *equilibrium constant* (K), where

$$K = (D)^d (C)^c / (A)^a (B)^b \text{ for the equilibrium: } aA + bB \leftrightarrow cC + dD \qquad \text{(II)}$$

(i.e. the forward and back reaction, the forward version of which is that *a* molecules of species A react with *b* molecules of species B to form *c* molecules of C and *d* of D).

In the special case of the dissociation of a salt BC:

$$B_bC_c \leftrightarrow bB^+ + cC^-, K = (B^+)(C^-)/(BC) \qquad \text{(III)}$$

but because BC is a pure substance, its activity is one and the equilibrium constant becomes a *solubility product*:

$$K_s = (B^+)(C^-) \qquad \text{(IV)}$$

In any given solution, the *ionic activity product (IAP)* is calculated as $(B^+)(C^-)$ for that particular solution (which may not be at equilibrium with salt BC). The departure from equilibrium can be expressed by the *saturation index* (Ω), which in karst literature is normally defined as

$$\Omega = \log_{10}(IAP/K_s). \qquad \text{(V)}$$

Positive values of Ω represent *supersaturated* or *oversaturated* solutions, which will tend to precipitate the salt, whereas negative values are *undersaturated* and will tend to dissolve it.

Reactions involving entirely dissolved species can be considered to be so fast that they are at equilibrium all the time. An exception is the reaction $CO_{2(aq)} + H_2O \rightarrow H_2CO_3$, which Dreybrodt et al. (1997) showed can limit the rate of calcite precipitation in thin films. However, reactions involving solids are normally out of equilibrium with fluids, and gases may also not have had time to equilibrate with the solution.

The dissociation of water can be represented as $H_2O \longleftrightarrow H^+ + OH^-$ with $K_w = (H^+)(OH^-) = 10^{-14}$ at 25 °C. Where the activities of H^+ and OH^- are equal, the solution is described as *neutral*. This corresponds to a value of 7 on the pH scale where pH = $-\log(H^+)$. Lower pH solutions are acid and higher pH solutions alkaline.

Alkalinity is a different property from alkaline pH and is defined as the capacity of a solution to react with acid. In freshwaters the total alkalinity is normally equal to the carbonate alkalinity, which is defined as the sum of (CO_3^{2-}) and (HCO_3^-) in equivalent units and is also a measure of the 'hardness' of the water. Alkalinity is determined by titration with a strong mineral acid which successively reacts with CO_3^{2-} and HCO_3^- ions until the end-point of the titration is reached at a pH of around 4.5; note that a CO_3^{2-} ion reacts twice with H^+ during the titration and so contributes twice to the alkalinity. At mid-range pH values, alkalinity can also be expressed accurately in units of mg/l HCO_3^- (=61× meq alkalinity) or mg/l $CaCO_3$ (=50× meq alkalinity). The total dissolved inorganic carbon (DIC) is normally calculated from pH and alkalinity using the equations given in the text, but can also be measured by the volume of CO_2 produced when a solution is strongly acidified.

(Continued)

Parameters such as saturation state and P_{CO_2} can be calculated directly, but for convenience, and to account for the various ion pairs correctly, geochemical speciation software is normally used and the definitive thermodynamic parameters used are those of Nordstrom et al. (1990), for example as in Stumm and Morgan (1996). One or other version of PHREEQ, developed at the US Geological Survey, provides a comprehensive solution and Appelo and Postma (2005) provide a structured guide to its use. For carbonate systems, the simplified software MIX2 was also developed by the US Geological Survey (Plummer et al., 1975) and is available on the website associated with this book (www.speleothemscience.info), in the version MIX4 with an improved interface. It models typical karst waters accurately (Fairchild et al., 2000), and can deal with the addition of strong acids such as that derived from pyrite oxidation (Fairchild et al., 1994), and concentration of ions by freezing (Killawee et al., 1998) or evaporation.

Table 5.1 Water chemical parameters (see also section 5.3.4 for other isotope species). See Spötl et al. (2005) as an exemplar of methods for a field measurement campaign.

Parameter	Critical issues	Method of determination/sample size/precision
$\delta^{18}O$, $\delta^{2}H$	Absolute avoidance of evaporation after collection	Isotope ratio mass spectrometry/0.2 mL (but larger samples are more conveniently handled)/ typically 0.05‰ and 0.5‰ respectively. New alternative: wavelength-scanned cavity ring-down spectroscopy (CRMS) (Gupta et al., 2009)/2 μL aliquots from 2 mL autosampler/ precision <0.1 and <1‰ respectively).
$\delta^{13}C$	Avoidance of degassing or microbial activity after collection	Isotope ratio mass spectrometry/0.1–0.2 mL if field-acidified; otherwise several mL/0.1‰ CRMS: $\delta^{13}C$ on gas, inorganic and organic C in water to 0.3‰ (Zare et al., 2009).
Electro-conductivity (EC)	Need sufficient volume to cover the probe; avoidance of damp	Electrode/usually several millilitres, although single-drop instruments available/1% for larger samples.
pH	Quality of electrode; instrumental drift; avoidance of damp	Electrode: glass is ideal, but all electrodes deteriorate over time; calibrate against pH buffers daily; ideally check with other instruments or electrodes regularly/0.1 to several millilitres/±0.05–0.1 units under field conditions.
Cations and silica	Acidify to prevent $CaCO_3$ precipitation and loss of colloids	Inductively coupled plasma mass spectrometry (ICP-MS) or for more abundant species; inductively coupled plasma atomic emission spectrometry (ICP-AES); atomic absorption analysis or ion chromatography/typically several millilitres (samples can be diluted)/<5% precision which can be improved by matching with standards for majors, but usually limited by varying sample composition.
Anions	Phosphate is a nutrient and is modified or lost within 24 hours; nitrate may also be modified by microbial action	Ion chromatography; molybdate reaction for phosphate/<1 to several millilitres/typically <5%.
Alkalinity	Ideally perform before any $CaCO_3$ has had a chance to precipitate	Acid titration/25 or 50 ml convenient for sample handling; small samples can be diluted with de-ionized water for analysis/<5%.

correct data to a set temperature (typically 25 °C). This correction is made using a notional salt composition which has a different temperature dependency to that of typical karst waters. Hence, for a precise calibration, it can be useful to experimentally calibrate the temperature dependency of EC readings by observing the values obtained during laboratory warming of a representative sample.

To determine carbonate system parameters (section 5.2), ideally one needs analyses of major cations and anions, together with *pH* and *alkalinity* determinations. However, given logistical problems in sampling cave waters, some short-cuts may be justifiable. For example, once the ion content of a sub-set of samples have been determined to check for the presence of any unusual species, alkalinity can be estimated by charge balance. It can also be set equal to the concentration of (Ca + Mg), that is, it can be assumed that the alkalinity arises through dissolution of Ca–Mg carbonates (this should not be done if Ca or Mg-salts may be present). If typical contents of Na^+, K^+, Cl^- and SO_4^{2-} are known for the site, their contribution to EC can be subtracted. Carbonate system parameters can then be estimated purely on field measurements of pH and EC as is discussed in the next section.

It is commonly impractical to visit cave sites frequently, so there is a strong incentive to develop continuous logging systems. Commercial multiparameter probes are widely used for monitoring river water chemistry, but some standard parameters (turbidity, and pressure as proxy for flow depth) are not useful for dripwaters. The practical problems of collecting sufficient water for analyses while also measuring the discharge (e.g. drip rate), have been solved in various ways by researchers, but there are difficulties in designing an apparatus to function usefully at widely varying drip rates, in addition to the usual problems of water condensation on electrical components. Perhaps the most useful parameter is EC because of its correlation with alkalinity, (Ca + Mg), and the ionic strength of the solution. Temperature-logging can also pick up discharge disturbances, although its signal will be damped and be affected by anthropogenic disturbances to air temperature (Borsato, 1997). pH can be measured, but the response of the electrode is likely to change after a few days and will need recalibration; similar problems arise with ion-specific electrodes.

It is not normally required to filter dripwaters to remove suspended sediment, as this is typically present in very low quantities compared with surface waters, and the process of filtration itself may introduce impurities (filters should always be pre-rinsed and will need acid-cleaning for ultra-trace element work). For study of colloidal fractions, filtration at 1 µm, 0.1 µm, and possibly ultrafiltration can be performed, but it is not practical to do such operations within the cave environment. These and other specialized separation techniques must be done within 48 hours of collection, because of the instability of colloids (Lead & Wilkinson, 2006).

Samples for organic carbon or fluorescence analysis should be collected in dark glass bottles, but other samples are normally collected in high-density polyethylene (HDPE) vials which are gas-tight and from which evaporation does not occur; some workers prefer also to cover the top with Parafilm ® which provides an additional seal and protects from dirt. For several analytical purposes (especially cation analysis) it is necessary to acidify the solution. It can be convenient to pre-acidify cation bottles, marking them appropriately, with a drop (0.1 ml) of 20% v/v ultrapure hydrochloric acid per 10 ml solution. This prevents $CaCO_3$ precipitation in the bottles and adsorption of colloidally bound ions to the vessel. Soil waters are often sampled from suction lysimeters, but these processes cannot be prevented in this case. Sample for anions and water isotopes should not be acidified. Aqueous carbon isotopes can either be converted to CO_2 by injection of a small volume into an argon-filled tube containing a small amount of phosphoric acid (Spötl, 2005), or taken in a bottle lacking airspace, for laboratory equilibration. Alkalinity is determined by acid titration. Because alkalinity is a conservative property uninfluenced by gas-exchange, the measurement can be done either in the field or the laboratory, but in the latter case care must be taken to re-dissolve any $CaCO_3$ precipitates.

5.2 The carbonate system

The carbonate system in freshwater, although simplified compared with seawater (Morse & Mackenzie, 1990), offers plenty of research challenges which include routes to interpret past conditions from speleothem chemistry. Aqueous carbonate chemistry is a core topic in all aqueous geochemistry texts and several accessible, thorough treatments are available (e.g. Garrels & Christ, 1965; Drever, 1982; Langmuir, 1997; Appelo & Postma, 2005). Following the pioneering approach of Bögli (1980), there are also very good summaries in karst texts by White (1988) and Ford and Williams (2007). Here we will focus on the treatment of freshwater environments, with particular emphasis on explaining the possible pathways that determine dripwater chemistry.

Considering first the gas phase, carbon dioxide is a trace species with a mixing ratio that has risen from 180 ppm during the last glacial period to 280 ppm in the pre-industrial period; at the beginning of 2010 it was around 388 ppm ($P_{CO_2} = 10^{-3.4}$) at Hawaii, climbing at around 5 ppm per year (NOAA, 2009). As discussed in Chapter 3, in soils CO_2 mixing ratios are typically 0.1–10% ($P_{CO_2} = 10^{-3}$–10^{-1}); and in Chapter 4 it was found that P_{CO_2} values in different cave chambers and in different seasons span the entire range from soil to atmospheric values.

Carbon dioxide gas ($CO_{2(g)}$) dissolves in water to form the species ($CO_{2(aq)}$) which then reacts with H_2O to yield carbonic acid H_2CO_3 (Fig. 5.1). Although $CO_{2(aq)}$ is 800 times more abundant than H_2CO_3 molecules (Adamczyk et al., 2009), it is conventional to combine the two forms as $H_2CO_3^*$ and the overall reaction becomes

$$CO_{2(g)} + H_2O \rightarrow H_2CO_3^* \qquad (5.1)$$
$$K_H = (H_2CO_3^*)/P_{CO_2} = 10^{-1.28} \text{ at } 10\,°C$$

Carbonic acid is a 'weak' acid which progressively dissociates (de-protonates) at higher pH:

$$H_2CO_3^* \rightarrow HCO_3^- + H^+ \qquad (5.2)$$
$$K_I = (H^+)(HCO_3^-)/(H_2CO_3^*) = 10^{-6.46} \text{ at } 10\,°C$$

$$HCO_3^- \rightarrow CO_3^{2-} + H^+ \qquad (5.3)$$
$$K_{II} = (H^+)(CO_3^{2-})/(HCO_3^-) = 10^{-10.49} \text{ at } 10\,°C$$

At near-neutral pH values, HCO_3^- is the dominant species of the total *dissolved inorganic carbon (DIC)* (Fig. 5.1c), but the small proportion of CO_3^{2-} is vital for determining the stability of carbonate minerals:

$$CaCO_3 \rightarrow Ca^{2+} + CO_3^{2-} \qquad (5.4)$$

The *solubility product* value for calcite, $K_{cc} = (Ca^{2+})(CO_3^{2-}) = 10^{-8.41}$ at 10 °C, whereas aragonite is more soluble and $K_{arag} = 10^{-8.25}$. The temperature effects on these various equilibria are shown in Fig. 5.1b; the overall consequence is that the solubility of $CaCO_3$ in water, in equilibrium with the atmosphere, decreases from 75 to 55 mg L^{-1} from 0 to 25 °C. However, the indirect effect of temperature on organic productivity and hence soil P_{CO_2} is far more important (see Chapter 3), as are effects such as air circulation (Chapter 4), that influence P_{CO_2} and hence solution pH.

The role of P_{CO_2} is central to understanding the behaviour of the carbonate system. Whereas P_{CO_2} of the gas phase is a simple concentration or activity, the P_{CO_2} of a water sample refers to the gas concentration of CO_2 with which it is at equilibrium, i.e. as specified by the CO_2 dissolution (K_H) equilibrium constant. Figure 5.2 depicts a spatial framework for the carbonate system demonstrating the relationship of P_{CO_2} and K_{cc}. In this section we consider the forward evolution of water parcels, whereas in section 5.5 we examine the extent to which the history of a water parcel can be determined.

Figure 5.2c illustrates example dissolution trajectories that, for a given initial P_{CO_2} in the soil/epikarst, follow paths to calcite saturation. For the open system lines, the same P_{CO_2} is maintained throughout the calcite dissolution process, whereas in the closed system dissolution purely uses the carbon that is dissolved in the solution in the initial phase, i.e. it is not open to re-supply of CO_2 from the gas phase. In soils, the open and closed systems have also been referred to respectively as the coincident and sequential systems (Drake, 1983) meaning that dissolution of $CaCO_3$ can coincide with, or follow, a period when CO_2 is available to dissolve in the water. The sequential system might be expected to arise where a limestone regolith is covered with thick young volcanic ashes lacking carbonate minerals (Gunn, 1981); complete loss of soil carbonate

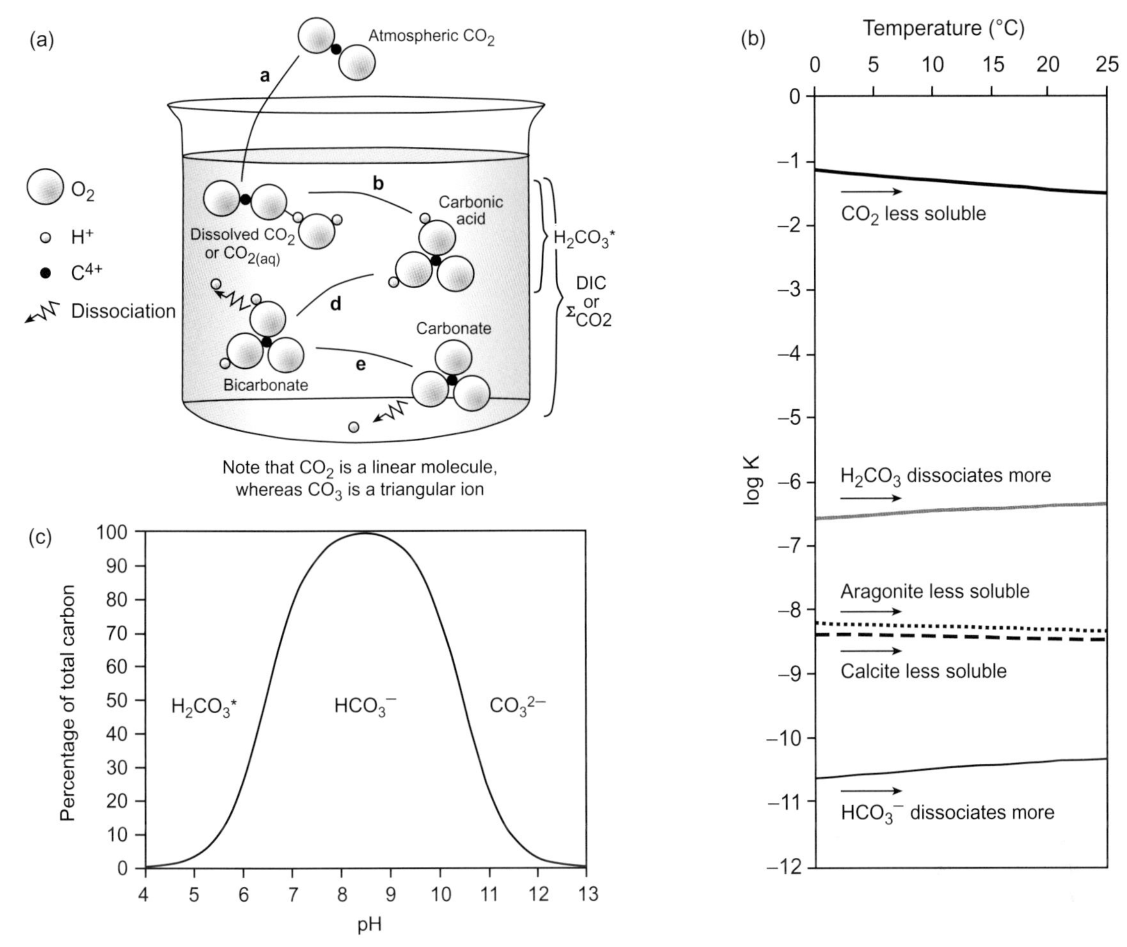

Fig. 5.1 (a) Graphic view of carbonate equilibria (Railsback, 2008). (b) Relative importance of different inorganic carbon species in solution as dictated by pH (values plotted are for 10 °C). (c) Temperature dependency of equilibrium constants in the carbonate system.

is also likely generally in the tropics. In such cases, P_{CO_2} is expected to decrease downwards below the soil horizon. On the other hand, the P_{CO_2} is often higher in the carbonate epikarst than the soil either because of oxidation of inwashed plant matter (Atkinson, 1977a), or more commonly when soil-CO_2 production is low in winter because the epikarst reservoir of water and gas will vary much less seasonally in P_{CO_2} than the soil. Both temperature and water saturation are strong controls on CO_2 production in soils, optimal CO_2 production arising in warm conditions at intermediate water saturations (Chapter 3). Hence another case when closed system conditions arise is when soils become temporarily waterlogged; then microbial activity is inhibited and may be insufficient to keep pace with CO_2 lost by carbonate dissolution. However, an open system during dissolution to calcite saturation is a common situation in the soil–regolith system (more common than in groundwaters generally), because it is normal to find that soils are not completely water-saturated and that there is a large surface area of $CaCO_3$ as fragmented bedrock available for dissolution, often supplemented by wind-blown dust, which almost invariably contains carbonate. In some cases, it can indeed be

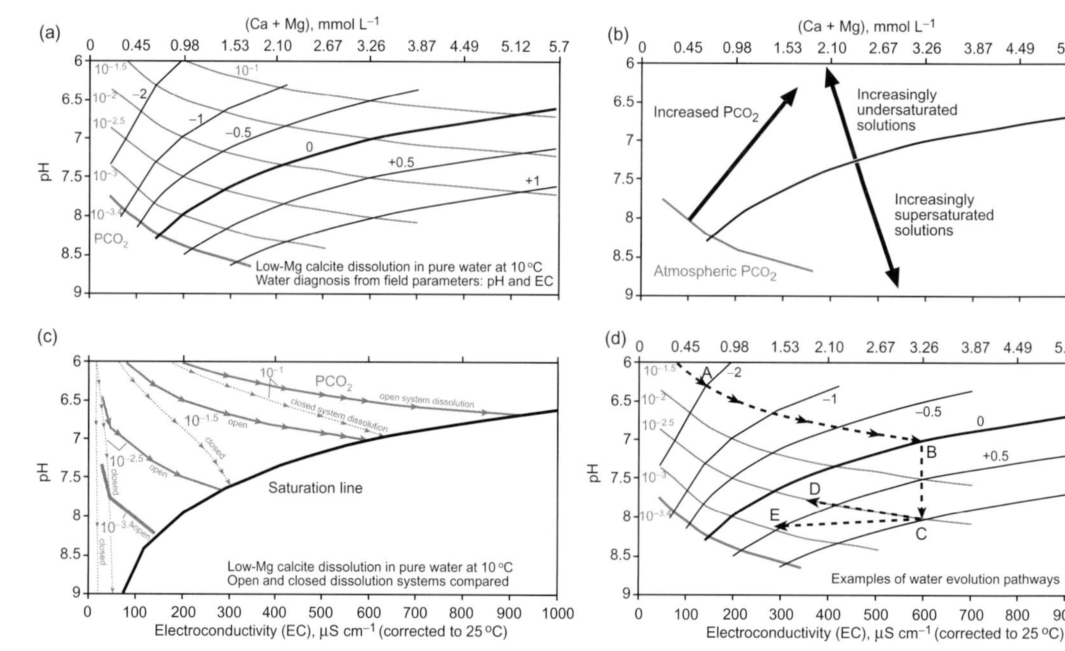

Fig. 5.2 The carbonate system set in a field-relevant reference frame. The diagrams are calculated using software tools that are available on the website associated with this book. The diagrams apply to a pure limestone karst environment where other salts contribute minimally ($11\,\mu S\,cm^{-1}$) to electroconductivity, where the Mg/Ca ratio in solution is 1/20, and at a temperature of 10 °C. The molar (Ca + Mg) is also given on a second *x*-axis. Addition of (non-Ca) salts shifts the lines to the right. (a) Grid from which field measurements of pH and electroconductivity can be used to answer questions requiring a knowledge of saturation state (will this solution precipitate $CaCO_3$?) or P_{CO_2} (will this solution tend to degas?). (b) Essential elements of the reference frame illustrating the atmospheric P_{CO_2} condition and the regions of undersaturation and supersaturation. (c) Vectors of limestone dissolution under several different initial P_{CO_2} conditions, illustrating the much smaller amounts of dissolution that occur in closed compared with open systems. (d) Chemical evolution of a typical dripwater in which dissolution occurs (AB) under open system conditions at P_{CO_2} of $10^{-1.5}$, a high value for 10 °C, but within observations (Baldini et al., 2008). Then follows rapid degassing along vertical line BC in order to reach a strong supersaturation (e.g. as found in Texan caves by Banner et al. (2007)); afterwards $CaCO_3$ precipitation occurs. Line CD depicts precipitation following a line of constant P_{CO_2}: this is most likely to arise if this was the P_{CO_2} of the cave air. CE represents a situation where degassing continues towards the lower P_{CO_2} of cave air at the same time as $CaCO_3$ precipitation. This could arise, for example, if the water–air interface did not present a large surface area at that time, limiting the rate of degassing.

demonstrated that the mean (Ca + Mg) content of cave dripwaters matches that expected from dissolution in an open system (i.e. at P_{CO_2} values that are close to the annual mean of those found in the soil), although this relationship can be obscured where $CaCO_3$ has already precipitated before the water is sampled (Fairchild et al., 2000). A more specific test of dissolutional conditions is provided by ^{14}C activity, as discussed in section 5.4.

Figure 5.2d illustrates the evolutionary pathway of a specific water parcel which dissolves to calcite saturation at a constant P_{CO_2} of $10^{-1.5}$. At point B, the water encounters a gas phase of lower P_{CO_2} and so degasses CO_2 to move towards point C. This process may only takes seconds to minutes, depending on the exposed surface area of $CaCO_3$. Hence, the water rapidly becomes supersaturated and $CaCO_3$ precipitation occurs (lines CD and CE). The amount of degassing can also be limited by the cave air P_{CO_2} and line CD illustrates a case where $CaCO_3$ precipitation follows a line of constant P_{CO_2}. Normally, approximately equal amounts of CO_2 and $CaCO_3$ are generated according to the usual summary reaction:

$$Ca^{2+} + 2HCO_3^- \rightarrow CaCO_3 + H_2O + CO_2 \qquad (5.5)$$

The excess CO_2 will be expected to degas so that near-constant P_{CO_2} conditions are maintained. Line CE, however, depicts a case where P_{CO_2} is lowering during $CaCO_3$ precipitation, presumably because the cave air P_{CO_2} falls, or because degassing was initially (line BC) incomplete.

There has been a lot of confusion in the literature about the interplay between degassing of water and precipitation of $CaCO_3$ from it. This seems to arise because of implicit, but often incorrect, assumptions that (1) the degassing starts when a drop emerges on the cave ceiling or at a stalactite tip; and (2) $CaCO_3$ precipitation has only occurred at sites visible to the observer. However, prior calcite precipitation (PCP) is common (Fairchild et al., 2000) as discussed in section 5.5. On the other hand, modelling studies of the tops of stalagmites often assuming the drip has already lost its excess CO_2 (e.g. Romanov et al., 2008a), although Dreybrodt and Scholz (2011) allow for degassing to occur at a soda-straw tip. Apparently the only previous textbook treatment, by Bögli (1980), follows work by Roques (1969) which assumes that water within a soda-straw stalactite cannot degas. This assumption was challenged by Maltsev (1998) who suggested that water runs down the outside of longer soda straws and is only drawn inside near the tip. Nevertheless water in soda straws is likely to be less degassed than that on solid stalagmites. Roques (1969) used experiments and a first-order decay model to illustrate the progress of degassing and he summarized results in Fig. 5.3a. Figure 5.3b illustrates field data from a limestone mine site in western England, taken using a pH probe capable of measuring pH on single drops. The data are placed over a model line for degassing; in such a

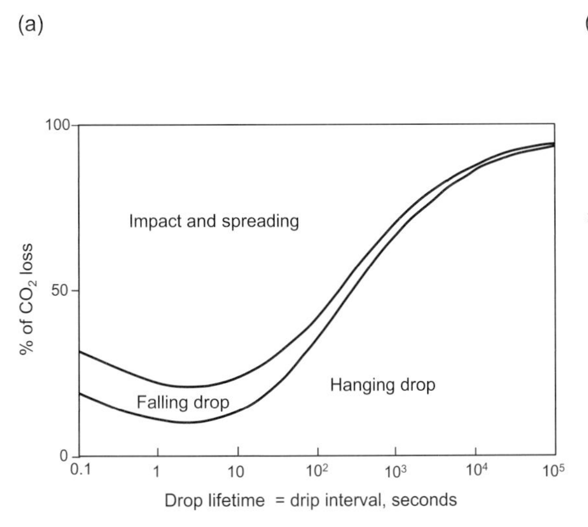

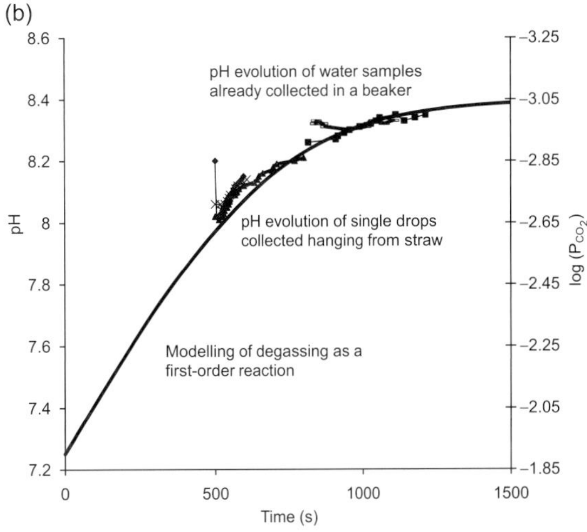

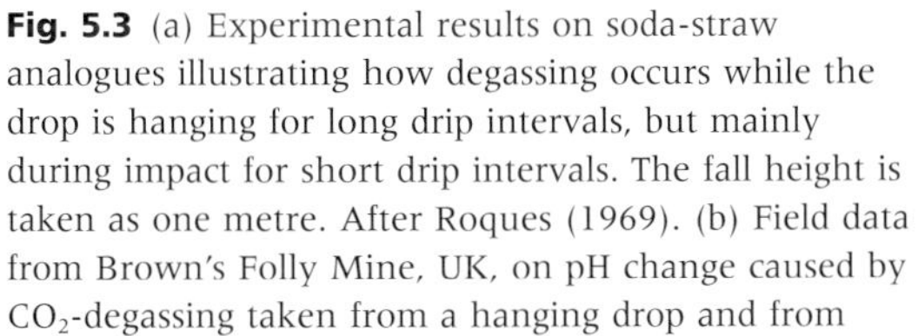

Fig. 5.3 (a) Experimental results on soda-straw analogues illustrating how degassing occurs while the drop is hanging for long drip intervals, but mainly during impact for short drip intervals. The fall height is taken as one metre. After Roques (1969). (b) Field data from Brown's Folly Mine, UK, on pH change caused by CO_2-degassing taken from a hanging drop and from water in a beaker compared with a first-order degassing model. The collected drops had evidently already significantly degassed before the start of measurement. The model constrained by data on (Ca + Mg) allows one deduce that this water was at equilibrium with calcite at a pH of 7.25 corresponding to a P_{CO_2} of $10^{-1.9}$.

model the P_{CO_2} excess (in ppm) over that found in the cave air at time t is

$$P_{CO_2}(t) = P_{CO_2(0)}^{e^{-\lambda t}} \quad (5.6)$$

where $P_{CO_2(0)}$ is the original P_{CO_2} at the onset of degassing and λ is the decay constant of the degassing of excess CO_2. From the (Ca + Mg) content of the solution it can be deduced from Fig. 5.3 that the water was at equilibrium with calcite at a pH of 7.25, corresponding to a P_{CO_2} of $10^{-1.9}$. Hence it is clear that the water at the soda straw has already extensively degassed. A model fit to a time frame mutually consistent with the data (Fig. 5.3b), reveals that the water has had of the order of 500 seconds of degassing under similar conditions to those pertaining during the observations. In fact, degassing may have started and stopped at different times, even above the cave chamber, because the rate of CO_2 loss is absolutely dependent on the surface area of water exposed to the atmosphere; likewise degassing will be enhanced at the point of impact of a falling drop as shown in Fig. 5.3a. Bögli (1980) attempted to generalize the extent to which water on any stalagmite tip will be already degassed, depending on its drip rate. This is not feasible because degassing depends not only on the drip time, but on the surface area to volume ratio (Roques, 1969) and hence on the pore system geometry and air-filling, which cannot be directly observed. Complete degassing, shown by a congruence of P_{CO_2} values measured in air and calculated from waters, has been shown in aggregate monthly samples from Obir Cave, Austria (Spötl et al., 2005). However, the dominant modulating process on calcite supersaturation and speleothem growth rate is likely to be the seasonal variation in cave air P_{CO_2} (see Chapter 4), and whose effects have been documented by Genty et al. (2001b), Banner et al. (2007), Mattey et al. (2010) and Miorandi et al. (2010) among others.

Other mechanisms of $CaCO_3$ precipitation can occur in caves. In caves undergoing significant ventilation with relative humidity below 100%, evaporation can be a significant factor (Thrailkill, 1971; Hill & Forti, 1997). In caves subject to seasonal freezing, or with cold-traps freezing can stimulate precipitation, typically in the form of a powder (Zak et al., 2008). Such phenomena can be recognized in terms of the water chemistry by increases in concentration of solute species down flow in addition to creation of characteristic speleothem morphologies and/or isotope compositions. The processes can be modelled using speciation software (Killawee, et al., 1998).

A completely different mode of precipitation also occurs in highly alkaline solutions. Above a pH of around 9.5–10.5, the decarboxylation reaction

$$H^+ + HCO_3^- \rightarrow CO_2 + H_2O \quad (5.7)$$

becomes less important than the dehydration reactions

$$HCO_3 + OH^- \rightarrow CO_3^{2-} + H_2O \quad (5.8)$$

$$H_2CO_3^* + OH^- \rightarrow HCO_3^- + H_2O \quad (5.9)$$

This can arise, for example, when burnt lime or concrete is found in a soil, where ultrabasic igneous rocks are weathered (Clark et al., 1992), or in basements of buildings by reaction of water with mortar. We can represent a summary reaction as:

$$Ca^{2+} + 2OH^- + CO_2 \rightarrow CaCO_3 + H_2O \quad (5.10)$$

This represents a different case from Fig. 5.2 because $CaCO_3$ is not being dissolved and so Ca concentrations can be high, not being limited by the saturation line; also pH is very high (values of up to 13 are known in such environments) and P_{CO_2} is very small. Water emerging from such a soil or aquifer to contact the atmosphere is extremely strongly saturated with $CaCO_3$ and will precipitate as quickly as CO_2 can be sequestered from the air; stalagmites in such situations can grow at rates of up to 1 cm yr^{-1} (Fig. 5.4, Baker et al., 1999; Sundqvist et al., 2005; Hartland et al., 2010a).

5.3 Weathering, trace elements and isotopes

5.3.1 Overview of element sources and sinks

Section 5.3 is concerned with some important raw ingredients for our speleothem factory analogy: calcium and trace element species, and the envi-

Fig. 5.4 Highly decorated abandoned mine tunnel (Derbyshire, UK) in which speleothems form quickly from hyperalkaline water.

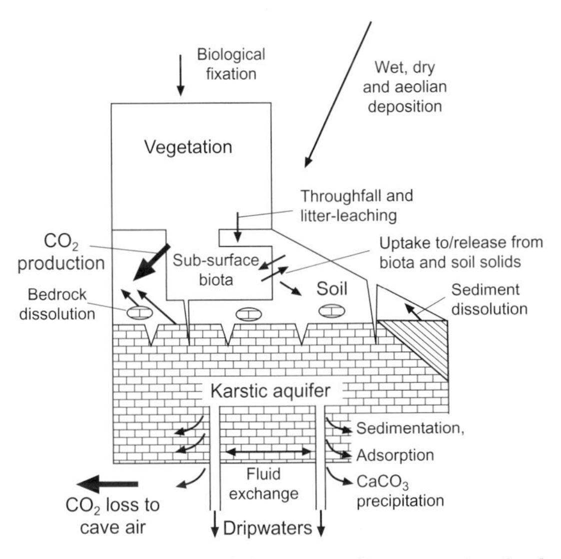

Fig. 5.5 The sources of elements and processes involved in their transport to caves (from Fairchild & Treble, 2009). Arrows indicate element fluxes as particulates, colloids or solutes in aqueous solutions.

ronmental context is summarized in Fig. 5.5. The atmosphere provides a direct source of solutes in rainfall, supplemented by dry deposition and changes over time of relatively immobile elements have been monitored by mosses (Rühling & Tyler, 2004): decreases in these elements and other forms of anthropogenic pollution occurred towards the end of the 20th century in developed countries. There may be addition of wind-blown (aeolian)

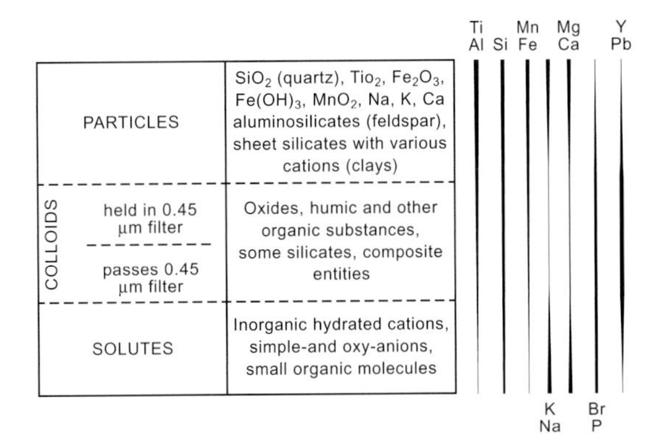

Fig. 5.6 Modes of occurrence of trace elements in karst waters (from Fairchild & Treble, 2009). Colloids are defined as entities with at least one dimension in the size range 1 nm to 1 µm (Lead & Wilkinson, 2006). The tendency of selected elements to be transported in a given mode is shown schematically on the right.

particles, typically silt-sized, and likewise solid material can be removed by wind. Encouraged by the acid generated by organic decomposition and plant respiration, bedrock surfaces and fragments dissolve to provide solutes, and an insoluble residue accumulates as part of the soil mass. Clay minerals contain important reservoirs of exchangeable ions. Plants assimilate various elements as nutrients, which are re-released and cycled by microbes and other soil organisms, with many elements being characteristically bound to organic molecules. Transpiration removes water and solutes from the soil, while evaporation leads to re-precipitation of $CaCO_3$ and more soluble salts. Precipitation events tend to cause movement down the soil profile of fine particles and colloids, and leaching of elements into solution. The tendency of elements to be associated with each of these three modes is illustrated in Fig. 5.6. Once in the karstic aquifer, the flow is partitioned into different hydrological pathways (Chapter 4) and slow reactions with the bedrock may continue. Sediment can become physically trapped within the aquifer and colloids will tend to adsorb to aquifer walls. Typically, the water will encounter open spaces at depth at lower P_{CO_2} than the soil and $CaCO_3$ may accumulate before the water enters a cave site being studied (Fairchild et al., 2000). The dripwaters will tend to be depleted in particles and colloids compared with karstic waters in general (McCarthy & Shevenell, 1998; Mavrocordatos et al., 2000).

When seeking to understand the trace-element composition of speleothems, one should look first to the composition of cave waters. The changing chemistry of cave waters over time must relate either to the supply (e.g. rates of weathering, atmospheric fluxes, change in steady-state of soil–vegetation system, see Chapter 3) or the removal rates en route (e.g. by adsorption or mineral precipitation). One key factor will be the residence time of water in the soil–epikarst system. Long residence times allow slower weathering reactions to have more impact, and less permeable parts of carbonate aquifers are often impure (clay-rich) and a wider variety of solutes may be available (Fairchild et al., 2006b). Hence we need to deal with the rates of weathering processes.

5.3.2 Calcite dissolution as an exemplar of weathering processes

A focus on calcite dissolution serves several purposes:

(i) it introduces the crystal chemistry of calcite, which is also needed to understand speleothem growth and geochemistry (Chapters 7 and 8);

(ii) it accounts for the bulk of the ions that are found in dripwaters;

(iii) it can be used to illustrate the key processes that occur in mineral weathering in general, including the molecular mechanisms of dissolution revealed by modern microscopic techniques.

Much of the literature has focused on marine systems (Morse & McKenzie, 1990; Morse et al., 2007), which are more difficult to handle than typical karstic waters because of their salinity and complex composition. Empirically, rates of dissolution are a power function (= reaction order) of the degree of undersaturation (expressed as IAP/K_s) of the solution, and parallel expressions can be formulated in terms of free energy change (Morse & Arvidson, 2002). A different approach was adopted in the influential study of Plummer et al. (1978) who used carbonate system parameters to describe the dissolution rate, inferring possible reactions that may have occurred at the crystal surface. Thus, in this Plummer, Wigley and Parkhurst (PWP) model, the reaction kinetics are described by

$$R = k_1 \times (\mathrm{H^+}) + k_2 \times a(\mathrm{H_2CO_3^*}) + k_3 \times a(\mathrm{H_2O}) - k_4 \times a(\mathrm{Ca^{2+}}) \times a(\mathrm{HCO_3^-}) \quad (5.11)$$

where R is the reaction rate per unit area per unit time, k_1–k_4 are rate constants, $a(\ )$ are the activities of chemical species at the surface of the dissolving mineral, and H_2CO_3* refers to the sum of H_2CO_3 and dissolved CO_2. Both R and the rate constants have units of mass (dissolved) per unit surface area (of mineral) per unit time. The rate constants k_1, k_2 and k_3 refer respectively to the following three putative forward reactions:

$$(k_1)\ \mathrm{CaCO_3 + H^+ \rightarrow Ca^{2+} + HCO_3^-} \quad (5.12)$$

$$(k_2)\ \mathrm{CaCO_3 + H_2CO_3 \rightarrow Ca^{2+} + 2HCO_3^-} \quad (5.13)$$

$$(k_3)\ \mathrm{CaCO_3 + H_2O \rightarrow Ca\text{-}^{2+} + HCO_3^- + OH^-} \quad (5.14)$$

The rate constant k_4 refers to back-reactions, which have in common calcium and bicarbonate as reactants, and which become increasingly important as equilibrium is approached. The PWP model for calcite dissolution has been applied to a variety of situations in natural environments (Fairchild et al., 1999a), and has been extended by Dreybrodt and co-workers for particular application in karstic environments (e.g. Dreybrodt, 1988; Dreybrodt et al., 1996). These authors quantified the importance of CO_2-hydration and mass transfer as potential rate-limiting factors. Where mass transfer is rate-limiting, bulk solution chemistry is not equal to that at the crystal surface, unlike the situation in the experiments from which the PWP theory was formulated. Thus, when the ratio of solution volume to calcite surface area is low, CO_2-demand is high and dissolution rates can be significantly slower than predicted by the PWP model.

The PWP model also fails to predict the very low rates of dissolution close to saturation (see also section 2.4), and so a convenient simplification is to examine dissolution only up to 95% of the saturation solute activities (or concentrations). Appelo and Postma (2005, p. 215) calculate that this degree of saturation is reached after around 30,000 seconds (8.3 hours) under conditions where transport of ions in solution is not a limiting factor. Hence, one would normally expect waters percolating into caves to have approached saturation for calcite, unless they were transmitted very rapidly following a rainfall event. Decreases in conductivity (as a surrogate for Ca) following rainfall can be identified in fast drips (e.g. Borsato, 1997). At Crag Cave (Ireland), Baldini et al. (2006b) found that Ca systematically decreased with discharge only above discharges of 2 ml min^{-1}; for slower discharge rates at the same site Tooth and Fairchild (2003) found that dilution effects were only found in species like SO_4^{2-} which originate from more slowly dissolving phases.

The reasons for the failure of the PWP model close to calcite saturation relate to the properties of the crystal surface and so we now turn to these aspects. The crystal structure of calcite is illustrated in Plate 5.1A/B and Fig. 5.7. Layers of calcium ions and carbonate ions alternate parallel to the vertical axis (*c*-axis of Fig. 5.7), with a distance of 1.42 between like layers, and the oxygen atoms surround each carbon atom in a plane perpendicular to *c* (Plate 5.1B). The Miller–Bravais notation is used to describe characteristic orientations within the lattice, including the sets of crystal faces that develop when the mineral grows freely. As explained in the legend to Fig. 5.7, the most characteristic form also parallels the cleavage (breakage) surfaces developed in calcite; this rhombohedron is referred to as $(10\bar{1}4)$. Over the past 15 years, a detailed knowledge of calcite dissolution behaviour has been obtained using the technique of atomic force microscopy (AFM), which requires very flat surfaces handily provided by calcite cleavage.

Figure 5.8 shows that the $(10\bar{1}4)$ surfaces are parallel to planes of carbonate ions, which makes them particularly mechanically stable as well as electrically neutral. True atomic-scale imaging is now possible (Fig. 5.9; Rode et al., 2009), which reveals defects in the crystal lattice. A point defect (Fig. 5.9a) represents the case where an ion is missing, or the site is occupied by an ion of different size (and possibly charge). Steps on the surface (Fig. 5.9b) arise in relation to other types of defect as well as occurring on the edges of crystals. One-dimensional defects are referred to as dislocations

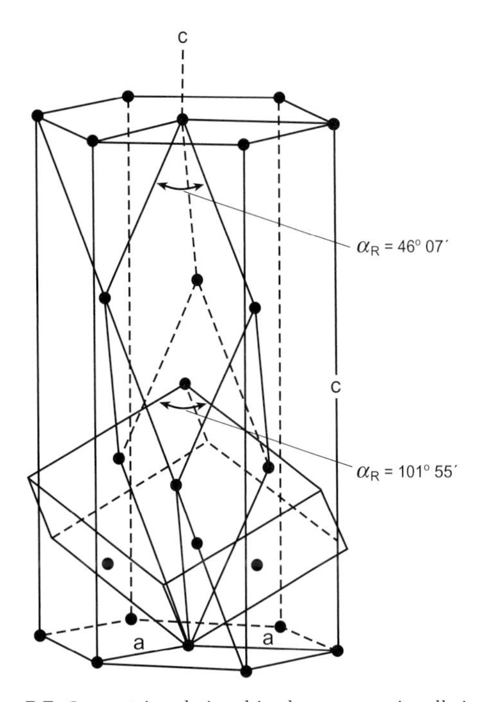

Fig. 5.7 Geometric relationships between unit cells in calcite (Reeder, 1993). The hexagonal unit cell is shown as the vertically sided figure with a rhombic cross-section (three of which make up the hexagon shape on the top and base of the figure). The steep rhombohedron (with interfacial angle of 46°) is the true (simplest) rhombohedral unit cell and contrasts with the cleavage rhombohedron with interfacial angle close to 102°. Crystal faces are indexed using the Miller–Bravais system with reference to the large hexagonal prism in terms of three axes (*a*) of equal length and vertical axis *c*. The Miller–Bravais indices consist of four positive or negative numbers, the first three of which sum to zero. The numbers are the simplified reciprocal of the intercept of the faces with each of the four crystallographic axes. The cleavage rhombohedron occurs in a form of six faces referred to as (10–14), which can also be written as $10\bar{1}4$ or (10.4). The steep rhombohedron is (10–11). Some literature uses a different convention of the height of the cleavage rhombohedron as the reference distance for the *c*-axis (Dickson, 1983).

that may represent the local termination of a lattice layer (edge dislocation) or a spiral ramp (screw dislocation), and are also important sites for impurities to be incorporated in calcite. Two-dimensional defects include both geometrically regular and

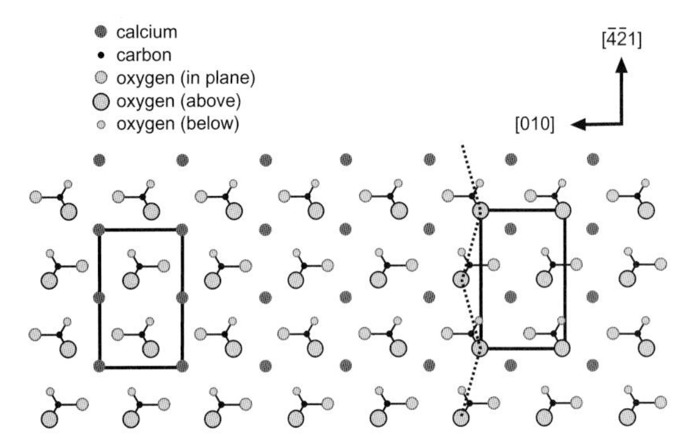

Fig. 5.8 Calcite unit cell and cleavage plane. Truncated bulk surface structure of the (10–14) cleavage plane (Rode et al., 2009). Two different unit cells are shown, defined by the calcium ions (left) or the protruding oxygen atoms (right).

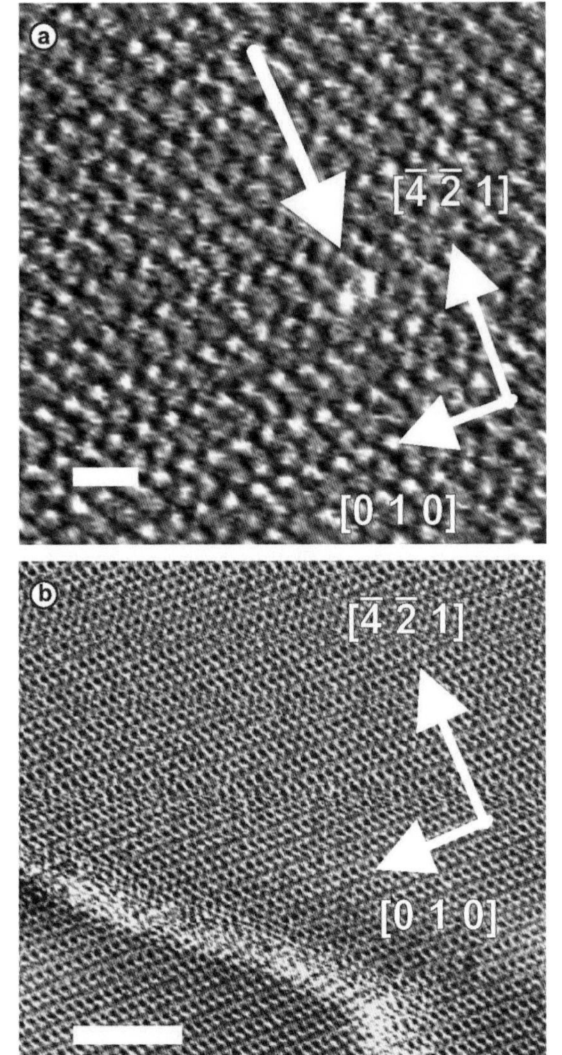

Fig. 5.9 Atomic force microscope images in frequency modulation mode to achieve true atomic-scale imaging of calcite cleavage surfaces (cf. Fig. 5.7) (Rode et al., 2009). (a) An atomic point defect (arrowed); scale bar 1 nm. (b) An atomic-scale growth step (; scale-bar 4 nm.

irregular boundaries between lattice domains. Irregular boundaries arise during growth whereas in calcite regular boundaries (twins) occur in repeated sets and arise by deformation; twins are ubiquitous in larger crystals in limestones and can also be generated when limestone or speleothem samples are mechanically cut.

Real-time dissolution can be observed by AFM and, disconcertingly, it was discovered that under humid air the surface becomes retextured by small amounts of dissolution and re-precipitation (Stipp et al., 1996). Theoretically, a strongly bound and ordered monolayer of water molecules will develop on the calcite surface from undersaturated vapour (Lardge et al., 2009) and AFM observations show that above a relative humidity of 55%, a layer several water molecules thick forms (Kendall & Martin, 2005). Hence migration of ions along this film can occur.

A fundamental concept in chemical kinetics is that there are several steps to the process of dissolution and that the slowest of these will limit the dissolution rate (Morse & Arvidson, 2002). Reactants will need to diffuse through the solution to the mineral surface, adsorb onto it, perhaps migrate to an active dissolution site, the chemical reaction will need to occur, and the products will need to be removed from the surface and diffuse away from it. Berner (1980) generalized that sparingly soluble substances will tend to dissolve very slowly and are limited by surface reactions, not by transport. This will apply to calcite and all less soluble substances in karstic environments.

Crystal surfaces, subject to surface reaction controlled dissolution, tend to have irregular surfaces

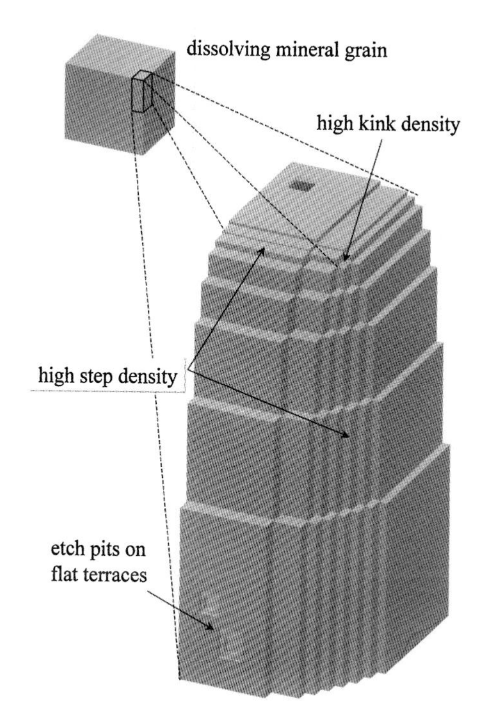

Fig. 5.10 Distribution of steps and kinks on mineral particles showing higher densities at edges and corners versus flat faces (Arvidson et al., 2003).

because the rate of dissolution is very locally variable (Berner, 1980). Dissolution is easiest at growth steps or isolated islands on the crystal surface, because only some of the ionic bonding is satisfied in the lattice. Kink (niche) sites and holes (pits) on the surface are also favoured (Fig. 5.10). Direct observations by AFM (Teng, 2004) showed that above an undersaturation (defined as log IAP/K_s) of −0.27, dissolution primarily took place at existing steps; between −0.27 and −0.39, pits started appearing; and below −2.15, there was a wholesale rapid development of numerous pits. Close to saturation therefore, the surface had a particularly strong control because lattice defects controlled many locations where dissolution was occurring, but at higher undersaturation, more widespread dissolution occurs. Use of vertical scanning interferometry allows the rate of dissolution of entire faces to be measured, which demonstrates that dissolution rates increase as expected at more extreme undersaturations, but that dissolution rates on crystal faces are much lower than rates obtained from bulk experimental powders (e.g. Plummer et al., 1978), indicating that the powders must contain a high proportion of edge and corner sites (Arvidson et al., 2003). Similar studies of dolomite dissolution show that dissolution below a critical undersaturation occurs as a series of propagating waves (macrosteps) away from a pit (Lasaga & Luttge, 2001). These studies are important in demonstrating that a change to a slower, more specific dissolution regime occurs at moderate undersaturations: something that had not been adequately predicted from theory or bulk experiments.

These techniques have also been applied in pioneering studies to assess the impact of microbial biofilms colonizing carbonate surfaces. Davis et al. (2007) allowed a strain of the bacterium *Shewanella*, originally isolated as a Fe- and Mn-reducing microbe, to colonize calcite cleavage surfaces. Microbial cells dissolved calcite by attaching to the surface, and subsequently also secreted extracellular polysaccharides. Overall the effect on calcite was to reduce dissolution by around a factor of two by inhibition of dissolution at growth pits (Fig. 5.11), but on dolomite, which dissolves more slowly, the additional dissolution related to cell attachment was the more important process.

Another mechanism that can contribute to reduced dissolution rates and low undersaturations is inhibition, either by mineral coatings or by attachment of specific ions that limit dissolution. For example, calcite commonly contains significant Fe^{2+} and Mn^{2+} originating from reducing conditions during limestone diagenesis. These ions will tend to oxidize rapidly and oxide films may coat mineral surfaces. The process is rather more important for dolomite, which tends to contain substantially more of these ions. If reducing conditions are locally maintained at the crystal surface, dissolved Mn^{2+} (Arvidson et al., 2003) can act as an effective growth inhibitor by strongly adsorbing to defect sites. In oxidizing solutions, the orthophosphate ion displays this property particularly strongly and several other species can also act as inhibitors (Table 1 in Morse and Arvidson, 2002). By contrast, AFM has also been used to study the effects of

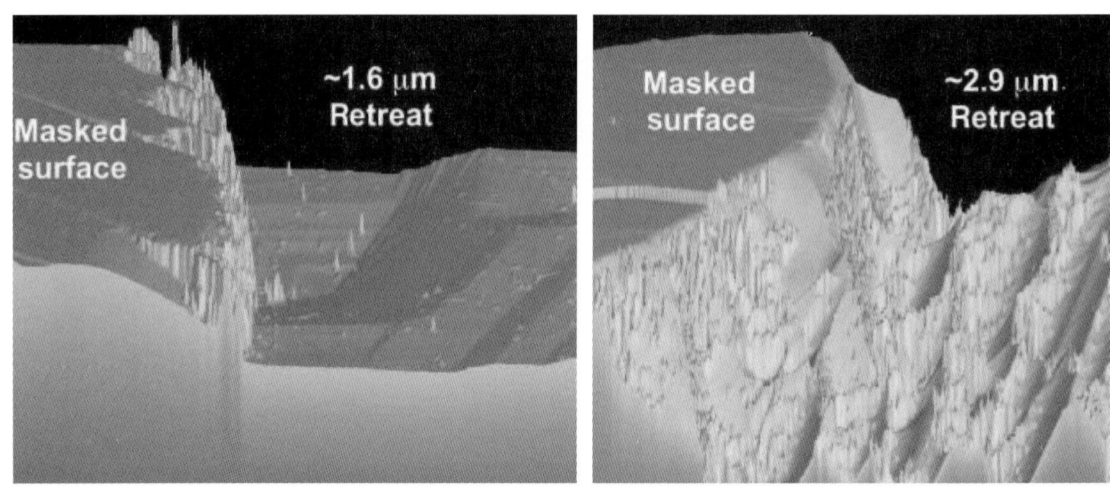

Fig. 5.11 Effects on dissolution of microbial colonization by a culture of Shewanella (left, field of view 650 μm) compared with a cell-free control (right, field of view 165 μm) (Davis et al., 2007). Each map has been made by vertical scanning interferometry after 33.5 hours of dissolution and the original surface, protected during the experiment by a mask, is shown to the left. The images are vertically exaggerated.

high salt concentrations in enhancing weathering (Ruiz-Agudo et al., 2009). Mg^{2+} is particularly effective and in theoretical simulations it tends to attract water molecules very strongly in order to reach full hydration; the loss of surface water molecules from calcium ions encourages their detachment from the surface. SO_4^{2-} and Na^+ also encourage dissolution and the effects overall may help to explain the enhanced weathering of limestone in coastal and desert areas, and in building stones affected by pollution, along with the expansion effects associated with salt crystallization (Ruiz-Agudo et al., 2007).

Apart from the oxidation of transition metals, mentioned above, it has been assumed that dissolution of low-Mg calcite was congruent, i.e. that the composition of the solution directly mirrored that of the calcite. However, observations made in glacial areas (Fairchild et al., 1994, 1999b) combined with experiments (Fairchild & Killawee, 1995; McGillen & Fairchild, 2005) demonstrated that freshly broken surfaces yielded excess Sr^{2+} and Mg^{2+}. McGillen & Fairchild (2005) argued that this was a genuine selective leaching rather than a dissolution–re-precipitation reaction and calculated that a surface layer must be generated, preferentially leached in these species, and that it must be at least two or three lattice layers thick (around 0.2% of the bulk). More extensive dissolution of the powders was congruent, implying a model as shown in Fig. 5.12. Evidence for this style of dissolution, in the form of enhanced Mg/Ca and Sr/Ca in relation to bedrock, but with no evidence of PCP, was found at the Ernesto Cave, northeast Italy

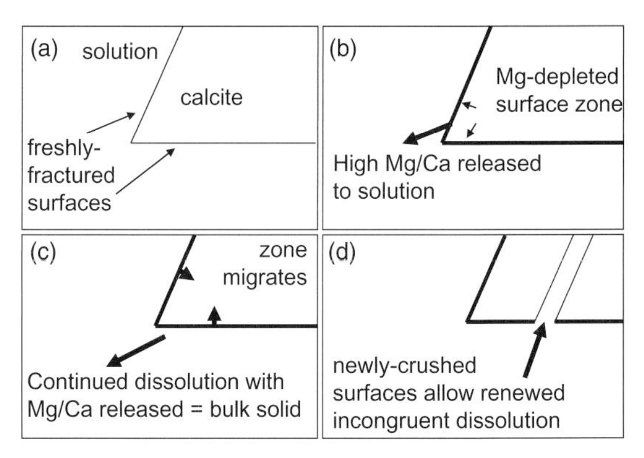

Fig. 5.12 Conceptual model for the creation of Mg-depleted surface layers on fresh calcite surfaces; the surface layers are maintained in steady-state during further dissolution McGillen and Fairchild (2005).

by Fairchild et al. (2000). This may relate to new calcite surfaces created by winter frost action, or possibly the same effect could arise by re-dissolution of soil-zone salts. Sinclair (2011) has mathematically modelled processes of incongruent calcite dissolution under a range of kinetic conditions and found that the increase in the cation ratios was essentially identical to those found by PCP.

The observations described above are consistent with important new observations in the physical chemical literature of uptake of protons by fresh calcite surfaces (Villegas-Jiménez et al., 2009a). These authors performed careful titrations at circum-neutral to alkaline conditions, demonstrating that the uptake of H^+ ions by the crystal surface far exceeded available surface sites. They interpreted their observations as reflecting fast proton-calcium exchange equilibrium between the solution and cations in lattice positions, down to several lattice layers below the surface. This stable Ca-deficient, proton-enriched layer then migrates during dissolution in the manner shown in Fig. 5.12. One can infer that this should be the same phenomenon that gives rise to incongruent dissolution of trace ions and therefore that they are selectively expelled by exchange with H^+ ions.

Such observations have led to a new model of the fundamental chemistry of the calcite surface: the one-site model of surface complexation on cleavage rhombohedral surfaces (Villegas-Jiménez et al., 2009b). This is simpler than the dominant two-site model summarized by Schott et al. (2009). The generic surface site has an adsorbed water molecule and has no charge, so is denoted as $\equiv(CaCO_3)\cdot H_2O^0$. Surface reactions involving metal or hydrogen anions, or carbonate system anions, may occur, generating negative, positive or no charge. This accounts for why it has been difficult to determine a clear pH-dependent point of zero charge for such carbonates. The task now is to test whether this model provides the best fit to experimental dissolution data.

5.3.3 Mineral weathering

We now deal with the wider scene of mineral weathering within which calcite dissolution is set. One point of comparison of different minerals is their solubility (Fig. 5.13b); carbonates have mid-range solubility with a significant control by P_{CO_2}. By contrast, highly soluble salts simply dissolve in water, nearly irrespective of acidity, to yield their constituent ions as solutes. Another means by which ions are rapidly introduced to solution is from exchangeable sites on clay minerals, such as smectites and illites, where they can be displaced, for example, by H^+ from infiltrating rain water.

The solubility of a few minerals with metals in a reduced state depends on the rate of microbially catalysed oxidation reactions (Sharp et al., 1999). Pyrite is particularly important because it is common as a dispersed phase in rocks and its oxidation yields sulphuric acid:

$$2FeS_2\ (\text{pyrite}) + 7.5O_2 + 5H_2O \rightarrow 2FeO.OH\ (\text{colloid or precipitate}) + 4SO_4^{2-} + 8H^+ \quad (5.15)$$

This acid can help drive the dissolution of both carbonates and silicates, but the main source of H^+ ions is normally carbonic acid.

To understand the relevance to karstic environments of the huge body of experimental literature on the kinetics of mineral dissolution, it is necessary only to concern ourselves with a narrow range of pH values that are typical of carbonate-buffered soils. We have plotted results in Fig. 5.13a of the changing dissolution rates of a range of minerals during the progressive open-system dissolution of calcite in a soil with a P_{CO_2} of 10^{-2} atmospheres at 25 °C, because this is the temperature where most experiments have been done. There are significant decreases in dissolution rate at lower temperatures, but these will impact little on the relative solubility of minerals. For carbonates the key factor is the P_{CO_2} which will be lower, other things being equal in colder soils (Chapter 3). However, again the main features of Fig. 5.13a will change little if a different controlling P_{CO_2} is used. Note that for minerals plotted on both diagrams, there is an overall similarity in the hierarchy of reactivity: this is because reactions tend to occur quicker if there is a stronger thermodynamic drive for them to occur (Morse & Arvidson, 2002).

Normally it is only possible to deal with relative rates, partly because the reactive surface area of all

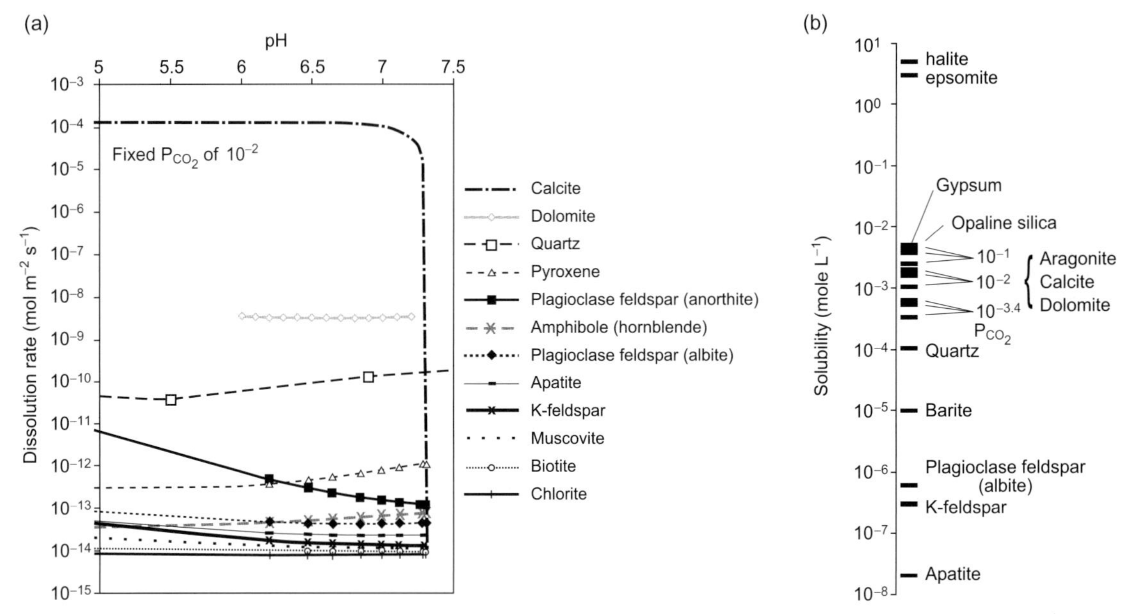

Fig. 5.13 (a) Comparative dissolution rates of common rock-forming minerals during progressive pH rise accompanying calcite dissolution at a fixed P_{CO_2} of 10^{-2} at 25 °C (the initial pH of pure water is just below 5 at this P_{CO_2} value). Dissolution rate is conventionally expressed as mass per unit area of reactive mineral surface per unit time. The experimental data used are from Chou et al. (1989) for carbonates, Brady and Walther (1990) for quartz, and for other minerals we have used the algorithms derived from experimental data by Sverdrup and Warfvinge (1995), as summarized by Appelo and Postma (2005). The dissolution equation terms related to activities of H^+ and OH^-, and P_{CO_2} of these authors are used, but not the term related to organic radicals The dissolution kinetics of dolomite are not well known close to its saturation under higher pH conditions. (b) Solubility of some common minerals at 25 °C using data from Berner (1980), Nordstrom et al. (1990) and calculations for aragonite, calcite and dolomite using K_s values of $10^{-8.28}$, $10^{-8.48}$ and $10^{-8.5}$ respectively at various P_{CO_2} values as shown.

the mineral species is very difficult to characterize and as discussed in the previous section there are effects of other ions and colonization by microbes to consider. Also, there are other reasons for mismatches between experimental and field data. Sverdrup and Warfvinge (1995) account for the commonly observed much faster laboratory dissolution rates compared with field conditions both in terms of both field and laboratory factors. Field factors make a range of 16–500 times difference in terms of issues such as partial wetting of minerals in the field, differences in temperature, and the inhibition owing to lack of removal of solid reaction products. In the laboratory, contributing factors include inappropriate chemical bounding conditions and the use of freshly ground solids for experimentation. The results of long-term laboratory weathering studies by White and Brantley (2003) are consistent with these conclusions.

In catchment studies, considerable attention is paid to host rock compositions in relation to solute yield (e.g. Bickle et al., 2005; Anderson, 2007). Only limited work has been done to date using this approach to interpret cave water and speleothem chemistry, but when this is done significant insights are gained (Fairchild et al., 2000; 2006b). Initially, the composition of marine-derived aerosol, whose composition is given in Table 5.2, must be removed (Fairchild et al., 2000). This is important as it can be the main source of some ions (e.g. for Mg in the Villars Cave, southern France (Baker et al., 2000)). Then the ratios of elements to Ca can be assessed

Table 5.2 Weathering behaviour of phases in soils and associated host rocks. Compiled from various sources including Appelo and Postma (2005). Note that silicate weathering is usually incongruent, with clay minerals and oxides as solid products. The clay minerals include illites, smectites and kaolinites and the chemical reactions are classically written in such a way as to conserve Al in the solid phase (Appelo & Postma, 2005), but alumina is commonly released in colloidal form and may also bind and transport other elements.

Mineral or solid	Formula	Solubility or dissolution rate	Significant solutes sourced
Atmospherically derived substances including cyclic salts (marine aerosol)	Complex mixture, variably enhanced by pollution (e.g. SO_4^{2-}, NO_3^-) or dust dissolution (Ca^{2+}, HCO_3^-)	High or already solutes	SO_4^{2-} (polluted areas); marine aerosol (Chester, 1990) has the composition (molar ratios): Cl^- (1), Na^+ (0.857), Mg^{2+} (0.0974), SO_4^{2-} (0.0517), Ca^{2+} (0.0189), K^+ (0.0187), Sr^{2+} (0.000165)
Halides and sulphates (either in host rocks or as secondary soil precipitates resulting from evaporation)	NaCl (common salt); $CaSO_4$ (anhydrite); $CaSO_4.2H_2O$ (gypsum); $BaSO_4$ (barite), and various others	High to very high Medium for barite	Na^+, K^+, Cl^-, Ca^{2+}, Ba^{2+}, Sr^{2+}, SO_4^{2-} and other alkali elements and halides
Calcium carbonate (calcite, high-Mg calcite, aragonite); normally host rocks but can form in soils	$CaCO_3$ (or e.g. $Ca_{0.9}Mg_{0.1}CO_3$ for high-Mg calcite)	Medium	Ca^{2+}, Mg^{2+} (calcite); Sr^{2+}, Ba^{2+} (especially aragonite); SO_4^{2-} ; locally others, including UO_2^{2+}
Other carbonates (host rocks)	$CaMg(CO_3)_2$ dolomite; $Ca(Mg,Fe)(CO_3)_2$ ankerite; $FeCO_3$ siderite; $MgCO_3$ magnesite, and others in mineralized rocks	Medium	HCO_3^- , Ca^{2+}, Mg^{2+} plus Fe^{2+} and Mn^{2+} in reducing conditions; locally others
Phosphates (host rock apatite or complex biogenic precipitates)	Apatite ($Ca_5(PO_4)_3OH$)	Very low	PO_4^{3-}, trace impurities including UO_2^{2+}
Quartz, opaline silica (mostly host rocks; occasionally formed in soils)	SiO_2 (hydrated for opal)	Medium	$Si(OH)_4$
Feldspars from host rocks	Plagioclase (solid solutions of albite ($NaAlSi_3O_8$) to anorthite ($CaAl_2Si_2O8$); alkali feldspar ($KAlSi_3O_8$)	Low	HCO_3^- Ca^{2+}, Sr^{2+}, N^+, K^+, UO_2^{2+}
Igneous ferromagnesian silicates (olivine, pyroxenes, amphiboles, biotite mica)	Includes Fe^{2+}, Mg^{2+}, Si and O, and often Al, Ca^{2+}, Na^+, K^+	Low	HCO_3^-, Mg^{2+}, sometimes a variety of other cations; H_4SiO_4, UO_2^{2+}
Other igneous and metamorphic silicates	Variable cations; also includes Si and O, and usually Al	Low to very low	HCO_3^-, alkali and alkali earth metals; sometimes distinctive source of rarer ions
Muscovite mica (high-temperature mineral)/illite (clay mineral formed in soils or sedimentary rocks)	$K_{1-x}[Si_{3+x}Al_{1-x}][Al_2]O_{10}(OH)_2$ where the atoms in [ ] are in particular lattice sites	Low	Exchangeable ions, K^+, Rb^+, Sr^{2+}, UO_2^{2+}

(Continued)

Table 5.2 (Continued)

Mineral or solid	Formula	Solubility or dissolution rate	Significant solutes sourced
Smectite group of clay minerals, formed by soil or diagenetic alteration of primary silicates	e.g. montmorillonite, $Na_{0.5}(Al_{1.5}Mg_{0.5})Si_4O_{10}(OH)_2$	Low	Exchangeable ions, Na^+, Mg^{2+}, Ca^{2+}, Sr^{2+}, UO_2^{2+}
Kaolinite, formed by soil, diagenetic or hydrothermal alteration of primary silicates	$Al_2Si_2O_5(OH)_4$	Low	$Si(OH)_4$ in extreme leaching
Iron and manganese oxides, variably hydrated from host rock or precipitated in soil	FeO.OH (brown goethite or amorphous oxides) to Fe_2O_3 (red, haematite); trace Fe_3O_4 (biogenic magnetite); MnO_2 or hydrated forms	Very low	Fe^{2+}, Mn^{2+} under reducing conditions; otherwise can be mobilized as colloidal oxides transporting a variety of elements
Aluminium oxide (alumina), normally a soil product	$Al(OH)_3$ gibbsite	Very low	May be mobilized as colloids transporting a variety of elements
Sulphides of iron (or, where host rock is mineralized, base metals such as lead or zinc)	e.g. FeS_2 (pyrite), ZnS (sphalerite), PbS (galena)	Medium	Fe^{2+}, HS^-/S^{2-} where reducing conditions; otherwise SO_4^{2-} and organically bound base metals where present
Humic substances generated from plant matter		Can be mobilized as molecules with a range of sizes from sub-colloidal to coarse colloids	Wide variety of elements transported (e.g. Y, REE, Pb, Zn, Cu, Co, Ni, P, F, Br, I): see Borsato et al. (2007), Fairchild et al. (2010), Fairchild & Hartland (2010)
Titanium dioxide from rock	TiO_2	Extremely low	Ti content is often used as a tracer for detritus as it is so insoluble

in relation to bedrock composition. It should be apparent from the above that the relatively fast dissolution and abundance of carbonate phases should make them the first target. Fairchild et al. (2000) demonstrated the consequences of competitive dissolution of calcite and the more slowly dissolving dolomite. Fig. 5.14 illustrates how the bedrock composition can be estimated from dripwater composition. For example, where the dripwater has a Mg/Ca composition of 80×10^{-3}, and dissolution has taken place at P_{CO_2} of $10^{-2.6}$, the original host rock composition was a 50:50 mix of dolomite and calcite. Where two components are analysed, additional tests can be made. In both the Ernesto and Clamouse Caves, a plot of Sr/Ca versus Mg/Ca (Fairchild et al., 2000) revealed that the bedrock compositions define clear mixing lines on which cave water should lie if dissolution was congruent. Instead, waters were enriched in both Sr and Mg demonstrating that additional processes had occurred (e.g. incongruent dissolution, PCP, etc.). These issues are discussed further in section 5.5.

More generally, it is an advantage for a specific element to be derived from a slower dissolving phase than Ca, because higher quantities of the

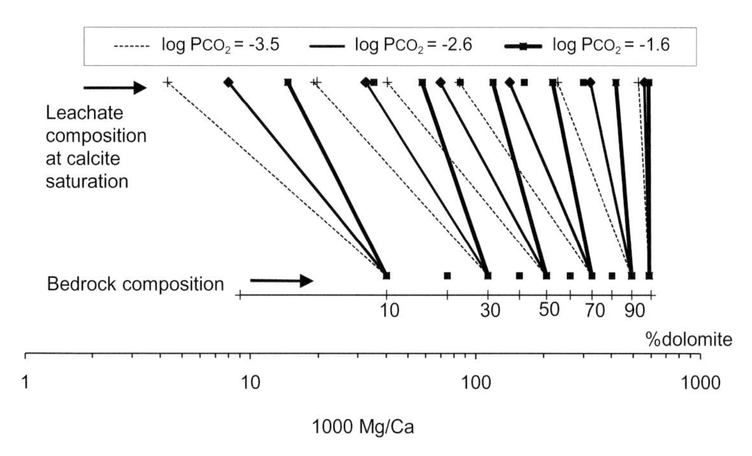

Fig. 5.14 Diagram to illustrate that dripwaters tend to have lower Mg/Ca weight ratios than mixed dolomite–limestone bedrocks under conditions of congruent dissolution to calcite saturation (at different P_{CO_2} values) (Fairchild and Treble (2009) after Fairchild et al. (2000)).

trace element imply longer water residence time and hence a drier climate. Table 5.2 offers a menu of possibilities as to which minerals have sourced which dissolved species, and this can be narrowed down by study of soil and bedrock composition at specific sites. However, to date, comparatively little mineralogical analysis of bedrocks has been made, and isotopes have been preferred as markers of specific phases and changing environments, as is discussed in the next section.

5.3.4 Isotope studies

Table 5.3 summarizes the species that are discussed in this section (S is discussed in Chapter 6) and summarizes the isotope notation which is used in the text. The most extensively used isotope system is the $^{87}Sr/^{86}Sr$ ratio, which was briefly introduced in Chapter 1. ^{87}Sr is radiogenic: it forms from ^{87}Rb with a half-life of 4.9×10^9 years (Banner, 2004), whereas the original Rb is typically substituted for K in feldspars, micas and clay minerals. As a result, K-rich rocks (e.g. granites) have high $^{87}Sr/^{86}Sr$ ratios (>0.71), whereas K-poor rocks such as basalts have low ratios (0.703–0.705). Seawater $^{87}Sr/^{86}Sr$ is intermediate in composition (Fig. 2.1) and helpfully there is no fractionation of the isotopes during precipitation of $CaCO_3$, i.e. $^{87}Sr/^{86}Sr$ ratio of the solution is the same as the precipitating carbonate. The ratio can thus be used to indicate the sources of Sr and it should be easier to do this in the speleothem context than on the larger catchment to mountain-belt scale, on which this approach is often tackled (e.g. Bickle et al., 2005). Figure 5.15 shows the relationship between cave dripwater compositions at Crag Cave, Ireland, and potential Sr sources (bedrock carbonate, overlying (silicate) till, rainwater and fertilizer), demonstrating how the proportion of different end-members can be estimated by systematic leaching studies (Tooth, 2000). Changing stalagmite $^{87}Sr/^{86}Sr$ ratios at this site over time should reflect the residence time of water in the till. In the literature it is established that changing Sr isotope ratios over time is caused in different cases by changing aeolian input in response to sea level changes (Goede et al., 1998), or atmospheric circulation (Ayalon et al., 1999; Frumkin and Stein, 2004), or changing intensity of weathering in response to rainfall (Banner et al., 1996; Verheyden et al., 2000; Li et al., 2005).

A reason to pay attention to source materials (rocks or soil particles) in Sr-isotope studies is that it clarifies which end-members are actually present. Schettler et al. (2009) made a careful study of the relationship between grain size and sources of Sr isotopes in loess which is a good exemplar. Another issue is that leachates may have enrichment in ^{87}Sr

Table 5.3 Isotopes and isotope fractionations of Mg, Ca, Sr and sulphate. Compiled from various sources; generalized isotope ratios from http://www.sisweb.com/referenc/source/exactmaa.htm. Because the analysis is rather specialized, the interested reader should consult the literature cited in the text for methodologies.

	Sources of ions	Isotope ratios and notations	Isotope fractionations
Sr^{2+}	Calcite, dolomite, silicates (especially feldspars and micas)	$^{87}Sr/^{86}Sr$; absolute ratio used.	^{87}Sr generated by decay of ^{87}Rb. No fractionation during precipitation of $CaCO_3$.
		$\delta^{88}Sr$ referred to standard NBS987 with $^{88}Sr/^{86}Sr = 8.37$	Small fractionation in soil of unknown origin found for $^{88}Sr/^{86}Sr$.
Mg^{2+}	Dolomite, calcite, silicates	$\delta^{26}Mg$ referred to standard DSM (Dead Sea Magnesium) with $^{26}Mg/^{24}Mg = 0.140$.	Light isotopes preferred during $CaCO_3$ precipitation with also a weak temperature effect and fractionations diminish at fast growth rate; fractionations also noted during uptake by wheat.
Ca^{2+}	Calcite, dolomite, feldspar	$\delta^{44}Ca$ referred to standard National Institute of Standards and Technology Standard Reference Material (NIST SRM) 915a with $^{44}Ca/^{40}Ca = 0.0014$	Depletion in ^{44}Ca found during plant uptake and in certain growth conditions for $CaCO_3$.
SO_4^{2-}	Atmosphere, calcite, dolomite, pyrite, sulphate minerals (see Chapter 6 for discussion)	$\delta^{34}S$ refers to CDT standard with $^{34}S/^{32}S = 0.0045$. $\delta^{18}O$ referred to VSMOW (Box 5.3)	None during $CaCO_3$ precipitation as CAS. Minor $\delta^{34}S$ fractionation during plant assimilation and subsequent re-oxidation; major changes in $\delta^{34}S$ in microbial oxidation/reduction. $\delta^{18}O$ is re-set if reduction and re-oxidation to sulphate occurs.
UO_2^{2+}	Silicates, carbonates and phosphates	$^{234}U/^{238}U$; absolute ratio used. For speleothems, normally the original concentration of ^{238}U at the time of deposition ($^{238}U_o$) is used.	None during $CaCO_3$ precipitation. Fractionation as a consequence of radioactive decay before or during weathering (α-recoil and other processes).

compared with the bulk composition (Bullen et al., 1997; Brantley et al., 1998). Study of soil chronosequences demonstrates that the older soils have lost more ^{87}Sr (Bullen et al., 1997), and the feldspars studied in the laboratory by Brantley et al. (1998) were inferred to have a leached surface layer (probably similar to that shown on calcite in Fig. 5.12), from which preferential release of Sr, and especially ^{87}Sr, occurred. Alternatively for micas, inter-layer cations may be lost during glacial grinding (Anderson, 2007). Conversely, Banner et al. (1996) found that Sr in soil leachates in Barbados was relatively unradiogenic compared with the bulk soil composition, and attributed this to leaching of an exchangeable pool of Sr derived by gradual weathering of volcanic ash; speleothem studies show the varying influence of this end-member over time. In summary, $^{87}Sr/^{86}Sr$ ratios have proved effective in source distinction and palaeoenvironmental analysis, and a significant expansion in their use is warranted. Work is also in progress on the use of $^{143}Nd/^{144}Nd$ in speleothems as a tracer of dust sources.

Improvement in analytical techniques has recently allowed determination of stable isotope ratios of alkaline earth cations (Johnson et al., 2004) which has revealed some interesting variations, but the field is still young and the ground-rules are still being laid down. For example, measurements have been made of the ratios of the

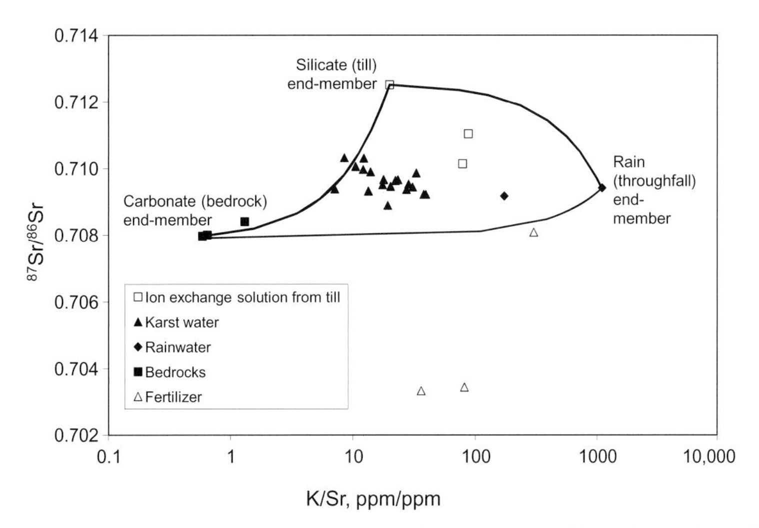

Fig. 5.15 Data on Sr sources at Crag Cave, western Ireland replotted from Tooth (2000). This site is notable for the presence of a thick carbonate-free till overlying carbonate bedrock (Tooth & Fairchild, 2003) and provides an example of determination of the relative contribution of different potential sources of Sr to dripwaters. Rain under vegetation (throughfall) was thought to be more representative of the rainfall end-member. The composition of exchangeable ions in silicate sources (till) overlying the carbonate bedrock, were determined by ammonium acetate extraction buffered to pH 8. Consideration also of similar plots with Na/Sr and Ca/Sr leads to the conclusion that carbonate and silicate sources are sub-equal in importance in present-day dripwaters. Early Holocene speleothem Sr isotopes are more radiogenic, implying changes in hydrology, perhaps anthropogenically related.

stable isotopes ^{88}Sr and ^{86}Sr; a speleothem and a terra rossa soil in Israel were depleted in ^{88}Sr by around 0.4‰ compared with all known marine carbonates, but the origin of these fractionations is currently unknown (Halicz et al., 2008). Initial Mg-isotope studies on speleothems also revealed that there was depletion in heavier stable isotopes during precipitation, implying that PCP might be identifiable. There was also a weak temperature effect on fractionation and changes with time within speleothems were observed (Galy et al., 2002; Buhl et al., 2007). Immenhauser et al. (2010) found quite a significant growth rate effect, with a reduction in fractionation at high growth rate. However, the influence of biospheric modifications remains to be fully understood and a study has already found that wheat seeds and their exudates preferentially uptake heavy Mg isotopes (Black et al., 2008). Vegetation is established as a significant influence on Ca isotope patterns in terrestrial catchments with depletion in heavier isotopes owing to preferential ^{40}Ca take-up by plant root systems (Cenki-Tok et al., 2009). Although Lemarchand et al. (2004) found experimentally that calcite displayed lighter values than its precipitating solution, and that this effect diminished with increasing precipitation rate, Fantle and DePaolo (2007) demonstrated that there is no fractionation of Ca isotopes during very slow calcite precipitation during burial of deep-sea sediments. Reynard et al. (2011) found increased fractionation associated with presumed kinetic effects in cave-analogue dripwater growth experiments. Coupling of Ca isotope, oxygen isotope and trace element analyses may help with understanding growth mechanisms and kinetic influences (Tang et al., 2008b; Reynard et al., 2011), a topic covered in section 8.3. Each of these isotope systems will undoubtedly contribute to our understanding of speleothems, but the immediate challenge is to find settings where the range of potential processes is sufficiently constrained to allow the potential variable factors to be isolated.

It is well known that there are fractionation processes affecting U isotopes during weathering of silicate minerals. The typical pattern is to find enrichment in ^{234}U which is related to the process of α-recoil (Kigoshi, 1971). Solubility of decay products can be increased either owing to the kinetic energy imparted during α-decay from ^{238}U to ^{234}Th, and subsequent rapid β-decay to ^{234}Pa and ^{234}U, or to enhanced solubility of U following lattice damage from such decay. $^{234}U/^{238}U_0$ ratios are generated as part of the process of U-Th dating (Chapter 9) and several studies have observed correlations between these isotope ratios and other speleothem proxies, and their authors have inferred that the U isotope ratios can retain palaeohydrological information. For example, a positive correlation with $\delta^{18}O$ from Soreq Cave, Israel implies that $^{234}U/^{238}U_0$ is a proxy for low rainfall in that setting (Kaufman et al., 1998). Zhou et al. (2005) developed a complex model to account for low $^{234}U/^{238}U_0$ ratios, and generalized also from other studies that fast speleothem growth (thought to be reflective of high fluid flow rates) was associated with a tendency towards a ratio of unity. These U-isotope studies are interesting, but need to be placed within a broader multi-proxy process understanding to be confirmed.

5.3.5 Colloidally bound elements

Box 5.2 defines technical terms used in this section. Here we develop some of the soil-related phenomena that were introduced in Chapter 3 and which will be linked to other biogeochemical phenomena in Chapter 6. Distinctive laminae in stalagmites containing fluorescent organic substances (Baker et al., 1993, 2008a) are now known to be enriched in a variety of trace elements (Fairchild et al., 2001; Richter et al., 2004; Borsato et al., 2007), distinct from the alkaline earth cations that have been the focus of most speleothem studies. This has drawn attention to the importance of understanding element transport processes in karstic systems.

In the steady-state, a significant reservoir of elements, many of them trace nutrients, is held in the soil and karst ecosystem. Losses by leaching to the underlying karst will balance elements derived either from mineral sources (bedrock, weathering residues and wind-blown sediment) or dry and wet solute deposition. As in the case of sulphate (Box 6.3), we are currently *not* in steady-state which makes it difficult to reconstruct the natural balance, for example in terms of residence times. For example, studies of moss chemistry from southern Sweden show that atmospheric deposition from pollution sources of a large number of trace elements has decreased significantly between 1975 and 2000 (Rühling & Tyler, 2004).

The case of phosphorus is particularly pertinent because it is a key element for marking annual laminae in speleothems (Fairchild et al., 2001; Huang et al., 2001; Baldini et al., 2002; Borsato et al., 2007), but some speleothem sites underlie arable land where direct application of fertilizers has transformed P budgets (Sharpley & Rekolainen, 1997). The natural source of soil P is likely to be slow leaching from trace mineral apatite, but it is tightly cycled by plants, animals and microbes in the soil, as well as being extensively adsorbed on soil materials. Variable release of P during an annual cycle could be mediated by microbial activity (Stewart & Tiessen, 1987), because microbes may act as a reservoir for plant growth in the summer, reducing P outputs at that time (Sharpley & Rekolainen, 1997). A significant proportion of bio-available soil P can be leached from soil, but the amount depends on soil hydrology (Pote et al., 1999). In agricultural areas, the first autumnal storms yield high dissolved P concentrations (80–1700 ppb (Heathwaite, 1997)). Then in the autumn and winter, freezing and thawing of plant residues may cause leaching of dissolved P, which is commonly transported in snowmelt in Nordic countries (Rekolainen et al., 1997). Pionke et al. (1997) stress the role of leaching from vegetation during storms as the main source of dissolved P. These data suggest that P supply to karst settings should be related to the timing of vegetation die-back, in climates where this is seasonal. A complication, however, is that P exists in several forms (Chapman et al., 1997), of which nominally dissolved P is present both as free phosphate ions and both weakly and strongly bound to organic substances. Specifically in soil humic substances it occurs both as phosphate and phosphonate ($C\text{-}PO(OR)_2$ where R is H or an alkyl

Box 5.2 Ion behaviour and complexation

A *complex* is a dissolved species representing a combination of a cation with a *ligand*, that is an anion or a neutral species. Cations in solution are surrounded by (typically four to six) water molecules which tend to orientate themselves with the oxygen atom facing the cation, as a consequence of the polar nature of the water molecule (Fig. 5.B1). Weak interactions (*outer-sphere complexes*) exist between anions and such hydrated cations to form *ion pairs* as mentioned in Box 5.1 when discussing the concept of activity. Strong interactions (*inner-sphere complexes*) arise when the cation accepts a pair of electrons from the ligand which makes it a *Lewis acid* and the ligand a *Lewis base*. The water of hydration is such an inner-sphere complex. A water molecule must be displaced whenever another ligand bonds with a metal to form an inner-sphere complex. The behaviour or elements in relation to water molecules is related to the *ionic potential*, the ratio of the charge (z) to the radius (r) of the cation. Small ions are more strongly hydrated in solution. Where the charge is high, hydrogen tends to be lost from water molecules so that hydroxy complexes are formed (or, at the highest z/r values, oxy-anions such as carbonate and sulphate). Generally, high values of z/r are associated with bonding which is more covalent in character in complexes. The degree of covalency of bonding is also related to the

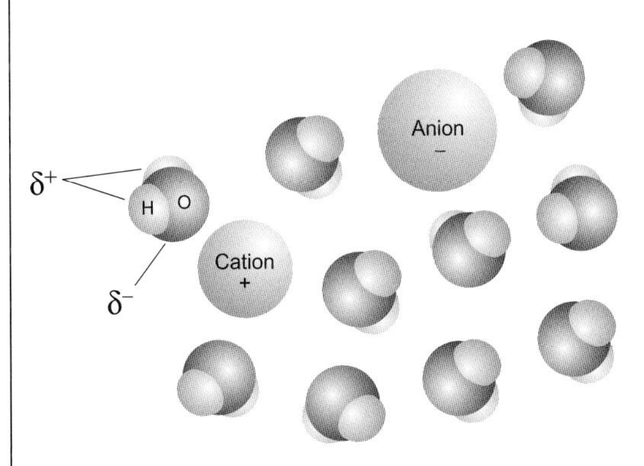

Fig. 5.B1 Illustration of how the charge polarity of water molecules influences the configuration of molecules in solution: O faces cations and H faces anions. This enables water to be a highly effective agent for dissolution.

electronegativity (the tendency for an atom to attract electrons): similar electronegativity values between components of a complex result in more covalent bonds. A further understanding of the stability of complexes can be gained by classification of ions as *hard* and *soft* Lewis acids and bases; this is developed in the earth science context by Railsback (2003) and http://www.gly.uga.edu/railsback/FundamentalsIndex.html.

Because cations are relatively small compared with ligands, a cation may form multiple bonds. If one pair of electrons is accepted by the cation, the bond is *monodentate*, whereas organic molecules may have many close-spaced ligands and so multiple (*polydentate*) bonds may result. Where the cations are held inside the organic molecule in a cage-like structure, this is known as a *chelate* and is important for humic substances.

Humic substances (HS) make up the bulk of non-particulate organic C in solutions derived from soils, and are important agents for transportation of elements in soils. Using operational definitions based on solubilities, fulvic acids are soluble over a wider range of conditions than humic acids and are traditionally thought to be smaller molecules. However, recent work has shown that in modern soils, individual molecules are smaller in mass than previously thought, and that soil organic matter represents a mixture of plant biopolymers and microbial breakdown products (Kelleher & Simpson, 2006; Lehmann et al., 2008) rather than macromolecules. Generically the dissolved organic species are acidic molecules, contain a high surface charge (becoming more negative at higher pH) with a high cation exchange capacity, and can be sorbed onto base-ion exchange resins. An example composition of fulvic acid is $C_{135}H_{182}O_{95}N_5S_2$ (Sposito, 1989). The functional ligands in humic substances are oxygen-containing carboxylic (–C(=O)OH), phenolic (OH attached to aromatic ring) and more complex groups containing N and/or S (Kinniburgh et al., 1999). The displacement of hydrogen in water by cations (*hydrolysis*) to form inner-sphere complexes becomes increasingly strong at higher pH values and this also applies to the analogous reaction with organic ligands. Above a critical pH region (the adsorption edge), binding of specific cations to HS becomes the dominant of occurrence of metals (Bradl, 2004) and the issue of metal transport then becomes one of transport of the host organic molecule.

group) (Myneni, 2002). At the Ernesto Cave site, Huang et al. (2001) found 7 and 3 $\mu g\,L^{-1}$ P in soil and cave dripwater respectively: these levels are difficult to detect with standard techniques and illustrate that P in speleothems is scavenged from solution (see Chapter 8).

Borsato et al. (2007) and Fairchild et al. (2010) reviewed the behaviour in terrestrial aquatic solutions of several other elements which had been found associated with fluorescent laminae in speleothems (Pb, Zn, Cu, Y, Br, F and I) and found that despite their varied properties, they displayed a strong affinity for organic substances. Generic studies of elemental behaviour in soil ecosystems and in relation to humic substances, have focused on cations and some generalizations have emerged that allow predictions to be made about element associations to be expected in karst waters, and by extension speleothems.

Tyler (2004) found that the *ionic potential* was a good predictor of the extent to which cations were taken up by plant roots, regardless of whether or not the elements were known to be essential nutrients. The more readily soluble ions were taken up in higher abundance, and to fulfil charge balance this should also apply to anions. A converse process, the release of ions from decomposing leaf litter, revealed a hierarchy that bears some similarity with the uptake process. Larger ions, and those that had higher electronegativities, were preferentially retained, presumably by *chelation* in the residual organic material (Tyler, 2005). Study of the effect of adding lime to natural topsoil shows a strong increase in the yield of dissolved organic carbon above pH 7, a pattern also followed by several elements (Tyler & Olsson, 2001).

In terms of the expected behaviour in karstic environments, it is important to appreciate that the 'dissolved' organic carbon is largely colloidal and consists dominantly of humic substances, with some less well-known polysaccharides (Lead & Wilkinson, 2006). The complexation of metals to humic substances has been extensively studied, initially by two competing models (NIC(C)A-Donnan model of Benedetti et al. (1996) and Kinniburgh et al. (1999); WHAM/Model VI of Tipping (1994, 1998, 2002); see also Appelo and Postma (2005, p. 348 ff.) and latterly by combining the insights from the models (Milne et al., 2001, 2003). These models deal both with the specific binding of protons and cations and the weakly complexed (non-specific) binding of ions, especially alkali and alkali earths. Milne et al. (2001) demonstrated the similarity of a variety of humic substances, which display a typical pattern of pH-dependence of availability of proton-binding sites (fulvic acids have slightly higher quantities per unit mass than humic acids). Direct evidence that ions bond strongly to humic substances is shown using X-ray absorption

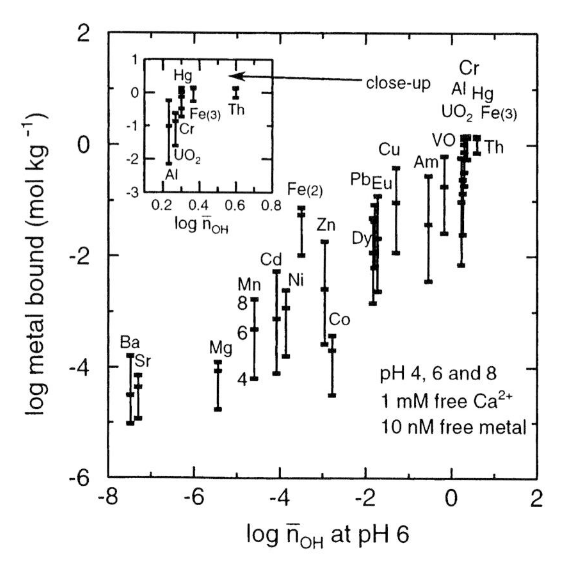

Fig. 5.16 Calculated variation in metal-ion binding by the generic humic acid versus average number of hydroxyl ligands attached to the metal ion at pH 6 (this is a function of the tendency of the metal ion to hydrolyse) (Milne et al., 2003). The three tick marks on the vertical lines indicate the binding at pH 4 (lower), 6 (middle) and 8 (upper).

techniques (e.g. for Co, Ni, Cu, Zn and Pb ions (Xia et al., 1997a, b)). Milne et al. (2003) calculated the binding behaviour of a variety of cations under varying conditions of pH, ionic strength and ion concentration, and established a clear hierarchy (Fig. 5.16) which paralleled the tendency of the metal ion to hydrolyse (which in turn is correlated with the ionic potential). Among the ions most strongly adsorbed to humic substances are U (as UO_2) and Th, which could account for some difficulties encountered in fulfilling the requirement for U-series dating of speleothems (Chapter 9), of a clean separation of the two elements, and lack of Th incorporation during speleothem growth (Kaufman et al., 1998).

The colloidal substances themselves are typically heterogeneous aggregates (Fig. 5.17) containing both organic molecules and crystalline or amorphous solids such as the Fe, Mn or Al oxides (Buffle et al., 1998; Lead & Wilkinson, 2006). Some elements show preferential binding to oxides rather than humic substances; this can be predicted from

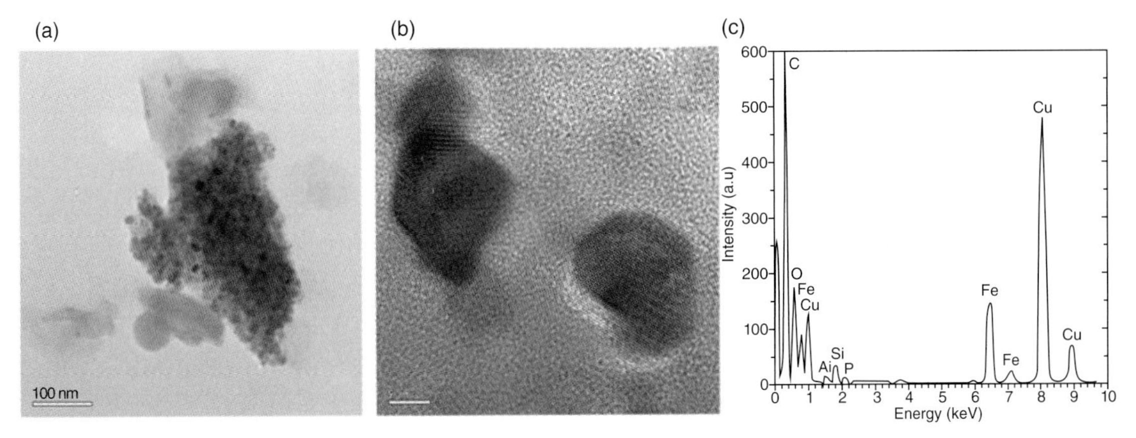

Fig. 5.17 Transmission electron microscope images of colloidal materials in dripwater samples from Pooles Cavern (UK) and corresponding energy-dispersive X-ray spectrum (Fairchild & Hartland, 2010). Left to right: aggregate (organic matter and iron oxide with very minor aluminosilicates) with scale bar 100 nm, enlarged image (scale bar 5 nm) showing iron oxide particles (parallel lines are diffraction from atomic planes). The energy-dispersive spectrum includes Cu and C peaks originating from supporting grid.

models such as WHAM and can also be revealed by examining the patterns of distribution of elements and colloidal carriers with colloidal size (Pédrot et al., 2008). Weakly bound species are expected to be the dominant cations in solution (Atteia et al., 1998); hence Ca and Mg in karst waters are more abundant (McCarthy & Shevenell, 1998).

Colloids can readily be destabilized by aggregation in solution and by adsorption onto solid substrates. Thus their concentration in cave water is much lower than in the overlying soils, and sampling methods that rely on collection times of more than 1–2 days will fail to sample their natural abundance. Analyses of elements likely to be strongly bound to colloids in solution in cave dripwaters were presented by Borsato et al. (2007) and Fairchild et al. (2010), but the frequency of analysis was insufficient to establish seasonal variations. Unusually high-pH cave environments provide opportunities to work with higher-than-normal concentrations of colloids and to demonstrate directly the effects of pH variations (Fairchild & Hartland, 2010). The distribution of strong- and weakly bound cations fits with the general expectations outlined in the previous paragraph.

In summary, a huge variety of elements with characteristic soil complexation behaviour can be mobilized during infiltration events along with their binding ligands in organic-dominated colloids. A proportion will escape adsorption to the karstic aquifer allowing them to be incorporated in speleothems (see also Box 6.2).

5.4 Carbon isotopes

Varied interpretations of $\delta^{13}C$ signatures and dead-carbon percentages of speleothems (see Chapter 3) have been offered in the literature (Baker et al., 1997a). Although some are carefully argued and are supported by a range of ancillary data (e.g. Dorale et al., 1992; Genty et al., 2003), most $\delta^{13}C$ data lack a convincing interpretation, and indeed are often unreported. Given the universal availability of $\delta^{13}C$ data alongside $\delta^{18}O$ during analysis, this is clearly an unsatisfactory situation, but ultimately a multi-proxy approach will be needed, as is discussed in the next section.

Because there is only a small fractionation between HCO_3^- and $CaCO_3$ during $CaCO_3$ precipitation (Box 5.3), the variability in the carbon isotope compositions must arise because of the generation of signatures in the soil and their modification in the cave system. Such processes were a major

Box 5.3 Stable isotopes and their fractionation

For further reading and quantitative exercises, Clark & Fritz (1997) and Sharp (2007) are recommended.

The term *isotope* is often (as here) used synonymously with the word *nuclide*, which is an atom of specified mass, but strictly *isotopes* are a set of nuclides, differing in mass, for a given element. For example, hydrogen is defined as an element with one proton in the nucleus, but can exist as the radioactive species ^{3}H (tritium) with two neutrons: the stable nuclide ^{2}H (deuterium) with one neutron, or common stable hydrogen ^{1}H with none. For light elements, there commonly exist one or more nuclides that are heavier than the dominant one and significant variations in the ratios of heavy to light isotopes occur in nature. The table below summarizes measurements of the relative abundance of the routinely measured isotope pairs.

Element	Lighter isotope: mass; abundance	Heavier isotope: mass, abundance	Standard	Absolute ratio (*R*) in standard
Hydrogen	1; 99.985%	2; 0.015%	VSMOW (Vienna Standard Mean Ocean Water)	0.00015575
Carbon	12; 98.9%	12; 1.1%	VPDB (Vienna PeeDee Belemnite) ($CaCO_3$)	0.01237
Oxygen	16; 99.76%	18; 0.204%	VSMOW	0.0020672
			VPDB	0.0020052
Sulphur	32; 95.02%	34; 4.21%	CDT (Canyon Diablo Troilite) (meteorite)	0.045005

The relative difference in isotope ratio between the VSMOW (water) and VPDB ($CaCO_3$) standards is a good illustration of the degree of discrimination or *fractionation*, between isotopes that occurs between one phase and another. The *fractionation factor* (alpha) is equivalent to the equilibrium constant for the exchange reaction of the heavier isotope between the substances and is defined as

$$\alpha_{B-A} = R_B/R_A \qquad \text{(I)}$$

where R_B and R_A refer to the ratios of the heavier to the lighter isotope in substances B and A respectively. Numerically, the fractionations are subtle and so values of α are close to 1. However, isotopic ratios can be measured with mass spectrometers highly precisely relative to standards, whereas measurement of absolute ratios is far more exacting. It is convenient to define a *delta value* that expresses the extent to which the ratio is larger or smaller than a standard:

$$\delta = (R_{sample} - R_{standard})/R_{standard}, \text{ which is mathematically equivalent to} \qquad \text{(II)}$$

$$\delta_B = (R_{sample}/R_{standard}) - 1 \qquad \text{(III)}$$

Normally values are expressed in parts per thousand (‰) after division by 0.001 (Coplen, 2008) because fractionations are typically of the order of several thousandths. Delta values are simply numbers so the terms 'enriched' and 'depleted' should not be applied to them (Sharp, 2007); they lie along a number line from negative to positive numbers and so a 'low' value is always smaller than a 'high' value.

Figure 5.B2 illustrates three substances and a standard with isotope ratios $X > Y >$ standard $> Z$. When comparing two substances, the difference in δ values (the isotope separation Δ) is convenient, but for theoretical calculations the *enrichment factor* ($\varepsilon = \alpha - 1$), which is also commonly expressed in‰ and approximates Δ, is used. Another useful relationship, linking α and‰ δ values, is

$$\alpha_{X-Y} = (1000 + \delta_X)/(1000 + \delta_Y) \qquad \text{(IV)}$$

Example: fractionation between water vapour and standard sea water at 20 °C. The heavier molecule has a lower vapour pressure and so is depleted in the vapour phase. For the exchange reaction:

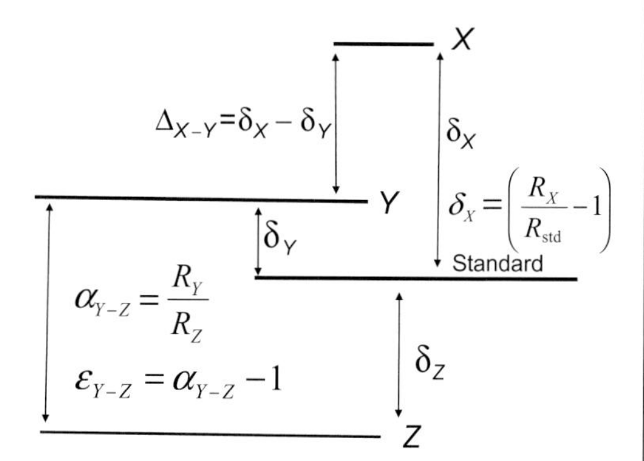

Fig. 5.B2 Diagrammatic illustration of the definitions of α, δ and ε. Differences from a standard are described using δ values, whereas ε is more generally is used for fractionations between substances.

$^2H^1HO_{liquid} + {}^1H_2O_{vapour} \leftrightarrow {}^2H^1HO_{vapour} + {}^1H_2O_{liquid}$

$\alpha_{vapour-liquid} = 0.9216$

Applying eqn. (IV), we note that δ_Y is zero as Y is the VSMOW standard.

$0.9216 = (1000 + \delta_{vapour})/1000.$

Hence $\delta_{vapour} = 921.6 - 1000 = -78.40‰$

And similarly for $^{18}O/^{16}O$: $\alpha_{vapour\text{-}liquid} = 0.99029$

$\delta_{vapour} = -9.710‰$

Note that the hydrogen isotopes are fractionated more by a factor of 78.4/9.71 = 8.1 compared with oxygen isotopes. It is this relative fractionation during condensation of water vapour that gives rise to the characteristic slope of 8 of the *meteoric water line*, when water data of these two isotopes are plotted against other (see Chapter 3). The effect of the arithmetic of δ values is to cause the ratio (the meteoric water line slope) to become lower when isotope values are lighter, but this is counteracted by the effect of temperature on α values.

Temperature effects. Isotope fractionation depends fundamentally on differences in mass, but becomes insignificant at high temperatures where all atoms are more energetic (Urey, 1947). Hence, it is important to characterize the temperature-dependence of fractionation.

Experimental data typically take the form

$$\varepsilon \text{ (or } 10^3 \ln\alpha) = a(10^6/T^2) + b(10^3/T) + c \qquad \text{(V)}$$

where a, b and c are constants and T is the absolute temperature. Data of this type on the carbonate system is plotted below for $\delta^{13}C$ and for $\delta^{18}O$.

Figure 5.B3 shows the composition of carbon isotopes with reference to a HCO_3^- $\delta^{13}C$ value of zero. The most important fractionation during $CaCO_3$ precipitation from HCO_3^--rich solutions is the generation of ^{13}C-depleted CO_2 gas. Because there are typically no other carbon sources during precipitation, the result is an increase in the $\delta^{13}C$ value of the solution. The fractionation between the HCO_3^- and $CaCO_3$ is much smaller and the temperature effect is small in relation to variations in aqueous $\delta^{13}C$ during progressive precipitation.

By contrast, the temperature variations of oxygen isotopes give rise to significant effects under equilibrium conditions, because the oxygen isotope composition of all phases is buffered by the semi-infinite reservoir of H_2O molecules. Figure 5.B4 shows the composition of phases on the VSMOW scale with respect to a H_2O composition of zero. The large fractionation between heavy carbonate and light water diminishes at higher temperatures as predicted by theory. Figure 5.B4 also shows the equivalent

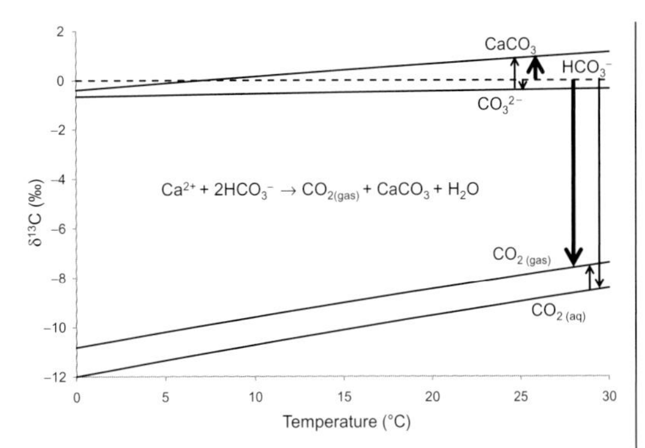

Fig. 5.B3 Carbon isotope fractionations and their variation with temperature, based on experimental data of Vogel et al. (1970), Mook et al. (1974) and Mook and de Vries (2000). The thick arrows refer to the overall fractionations from HCO_3^- to $CaCO_3$ (calcite) and gaseous CO_2 as in the chemical equation.

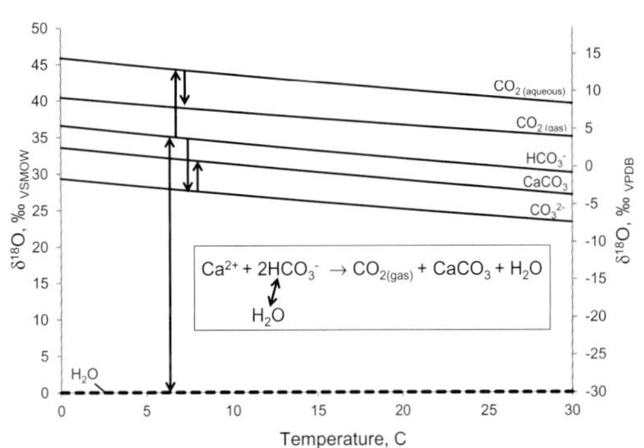

Fig. 5.B4 Oxygen isotope fractionations and their variation with temperature. Based on experimental data by Kim and O'Neil (1997), Thorstenson and Parkhurst (2004) and Beck et al. (2005).

values on the VPDB scale to which carbonate analyses are typically referred. The conversion equations are:

$$\text{VPDB} = 0.97002\text{VSMOW} - 29.98 \qquad \text{(VI)}$$

$$\text{VSMOW} = 1.03091\text{VPDB} + 30.91 \qquad \text{(VII)}$$

Starting with water of composition zero on the VSMOW scale, the diagram shows $CaCO_3$ compositions just above and below zero on the VPDB scale, lower temperatures leading to higher $\delta^{18}O$ values and vice versa. At constant temperature, the approximate effect of changing water composition by x‰ is to change the calcite composition by the same value.

(Continued)

The diagram also shows the consequences of $CaCO_3$ precipitation, accompanying CO_2-outgassing faster than the typical exchange time of several minutes between HCO_3^- and H_2O (Scholz et al., 2009). For example at 10 °C, using the values plotted on the diagram, let us calculate the effects of removal of 10% of the HCO_3^- reservoir:

the reaction is $20\ HCO^-_{3(initial)} \rightarrow CO_2 + CaCO_3 + H_2O + 18\ HCO^-_{3(residual)}$

$\delta^{18}O$ compositions: $60(34.19) = 2(38.41) + 3(31.28) + 0 + 54(\delta^{18}OHCO^-_{3(residual)})$

$\delta^{18}O_{HCO_3^-(residual)} = 34.80‰$ (VSMOW)

Hence, $\delta^{18}O$ is perturbed upwards (by around 0.6‰), because isotopically light H_2O is one of the products. Hendy (1971) accordingly argued that covarying increases in $\delta^{18}O$ and $\delta^{13}C$ along speleothem laminae were an indication of kinetically disturbed fractionation.

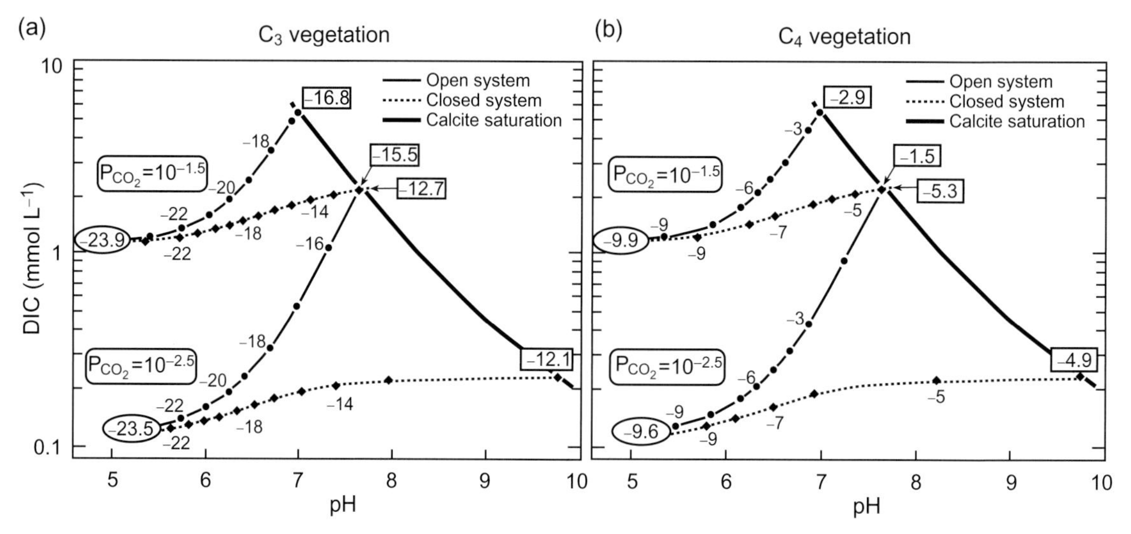

Fig. 5.18 Systematic changes in the bulk carbon isotope composition of the dissolved inorganic carbon (DIC) of water during progressive dissolution of $CaCO_3$ where the source of CO_2 is from decay of C_3 vegetation (a) or C_4 vegetation (b). Open and closed systems are discussed in the text. After Clark and Fritz (1997).

theme of the extraordinarily ambitious paper by Hendy (1971), although this article is more commonly cited for the prediction that covarying carbon and oxygen isotope fractionations would result from kinetic effects during rapid calcite precipitation (section 8.4). Hendy (1971) laid out a detailed basis for consideration of open and closed systems (Garrels & Christ, 1965), the necessity for consideration of kinetic effects, distinguishing between slow and fast CO_2-outgassing, and explicitly treating equilibration between gas and solution. He was also aware of the implications of the different isotope compositions of C_3 versus C_4 plants, as discussed in Chapter 3, and the issues that would arise where evaporation was a significant factor. To clarify issues in the nascent science of speleothem palaeoclimatology, he clearly pointed out the small influence on carbon isotope systematics of temperature (from physicochemical considerations) compared with the larger effects related to changing P_{CO_2}.

One area that was soon followed up, because of its importance for groundwaters, was the detailed prediction of carbon isotope evolution of open and closed system end-members (Deines et al., 1974; Clark & Fritz, 1997; Fig. 5.18). The open system was considered to represent a complete equilibration with a gas phase of infinite extent, so that not only would P_{CO_2} be maintained throughout dissolution, but also that the $\delta^{13}C$ composition of DIC would be controlled by the gas phase. The closed system envisages a solution with an initial relatively high P_{CO_2} set by the soil, followed by isolation from the gas phase and carbonate dissolution. The increase in the $\delta^{13}C$ signature as dissolution proceeds results from two factors. Firstly, the bulk DIC changes its composition from one dominated by aqueous CO_2 below pH 6.4, to one dominated by HCO_3^- at higher pH. Because HCO_3^- is considerably enriched in ^{13}C compared with gaseous and aqueous CO_2 (Box 5.3), this is the key factor in the open system. For the closed system however, the important issue is the addition of $CaCO_3$ to the water with a much higher $\delta^{13}C$ composition and even higher $\delta^{13}C$ values are attained. For example, in Fig. 5.18a closed system dissolution from a starting P_{CO_2} of $10^{-1.5}$ leads to a $\delta^{13}C$ at calcite equilibrium of −12.7‰, whereas the figure is −16.8 for the open system. Note that the former has exactly the same chemical composition as water resulting from open system dissolution at P_{CO_2} of $10^{-2.5}$, but the $\delta^{13}C$ composition of the open system water is −15.5‰. This is a useful property in groundwater studies where *equifinality* (many paths leading to one particular composition) is often a problem. Figure 5.18b illustrates the much higher $\delta^{13}C$ compositions expected where C4 vegetation dominates in the soil.

A clear weakness in this classical approach is that it does not explicitly deal with the kinetics of the different processes that operate, and so underplays the full extent of potential variability in processes in the soil and epikarst zone before degassing starts. Accordingly, an attempt at a new systems analysis is shown in Fig. 5.19. Here, infiltrating water is carried through up to four stages represented by depth zones during the process of equilibration with carbon dioxide and carbonate minerals. In zone 1, which might represent the topsoil, or the whole soil profile if carbonate-free, the water is primed with carbon dioxide and completely equilibrates with soil gas both chemically and isotopically. Once carbonate minerals are encountered (depicted as $CaCO_3$), zone 2 is reached. It is possible that at this point that water is moving sufficiently rapidly that there is insufficient time to reach calcite saturation; if so then input of CO_2 may be kinetically limited as may the process of equilibration with soil gas. Such rapidly flowing undersaturated waters may pass through macropores to the

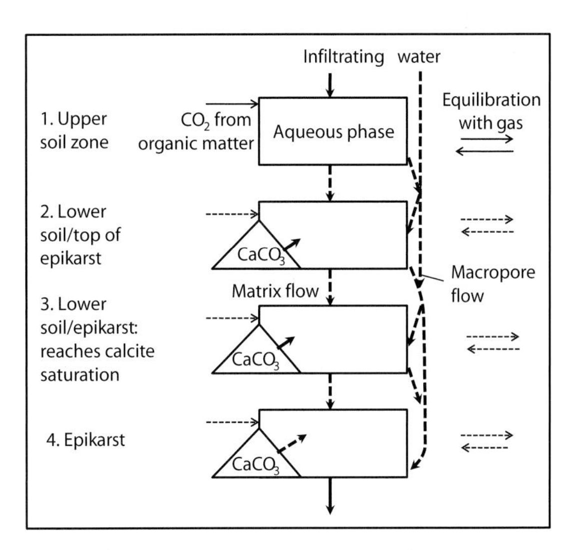

Fig. 5.19 Systems analysis of processes affecting carbon isotope evolution in the soil and epikarst; evaporation is neglected. Dashed lines denote optional processes. The rectangles indicate water percolating downwards through the system through zone 1 (upper soil) to zone 4 (epikarst). The water either seeps from box to box, or by-passes certain stages by macropore flow. CO_2 from decomposition of organic matter is added directly to the aqueous phase in zone 1 and optionally in the lower zones. Equilibration with soil gas takes place in zone 1 and optionally/partially in the lower zones. Dissolution of calcium carbonate takes place in zones 2 and 3 until saturation is reached and may also continue in zone 4 if additional CO_2 is generated from inwashed organic matter.

epikarst as shown in Fig. 5.19, and may even reach an underlying cave. On the other hand, more slowly percolating water will be expected to reach calcite saturation (zone 3), and this would also make it more likely that the soil or epikarst solution will retain the soil P_{CO_2} (moderated by diffusive mixing as discussed in Chapter 3) and have exchanged more thoroughly with a gas phase. At this point, we note that $\delta^{13}C$ signatures should stabilize much faster than ^{14}C activity because the distance from equilibrium is much smaller in the former case than the latter (up to 10 parts per thousand versus up to tens of percent). We also draw attention to the possibility that the carbon dioxide is added directly to the solution phase (as invariably happens in waterlogged environments).

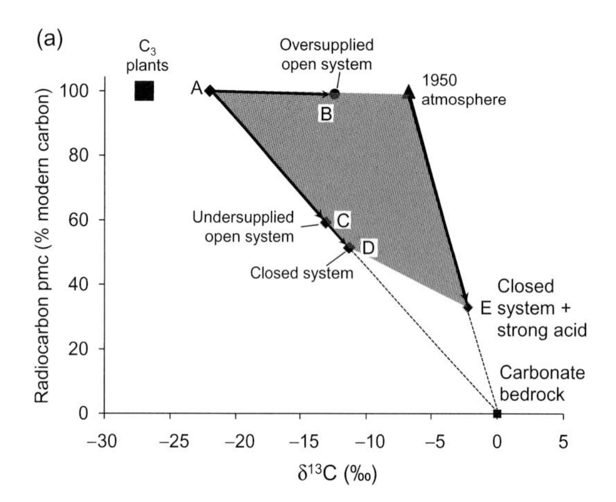

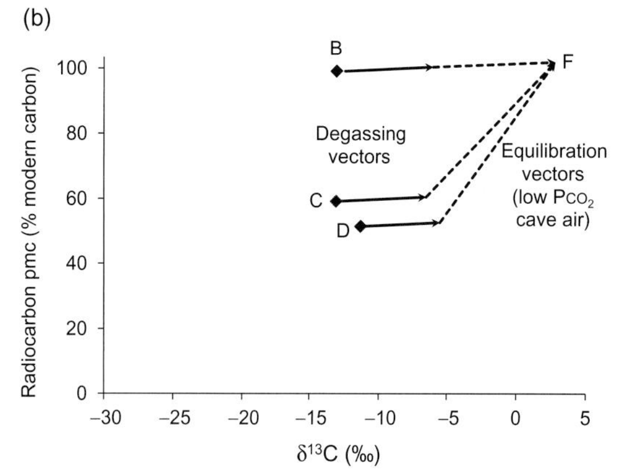

Fig. 5.20 Potential paths of evolution of carbon isotope systematics in karstic environments. Soil P_{CO_2} is set at 10^{-2}, all vegetation is C3, and ^{14}C activities are set at 1950 (pre-bomb) values for simplicity. Diagram (a) focuses on the dissolution to calcite saturation whereas (b) deals refers to degassing and equilibration. In a realistic case, the lines in (b) would be curved as degassing merges into equilibration. Shaded area in (a) covers the expected range of data. See text for discussion.

Sometimes, as in stage 4, a higher P_{CO_2} is encountered in the epikarst than in the overlying soil (Atkinson, 1977a), for example because of inwashed organic matter, allowing further $CaCO_3$ dissolution and gas exchange.

Figure 5.20 now looks at the consequences of these issues and some others for carbon isotope systematics. In zone 1, the water composition will lie close to point A (Fig. 5.20a). The exact position depends on fractionation during respiration or decomposition of organic matter, and the small fractionation between gaseous and aqueous CO_2, but the main effect is likely to be some diffusive admixing with atmospheric air, leading to somewhat heavier $\delta^{13}C$ compositions compared with organic matter as discussed in Chapter 3.

Closed system dissolution simply involves addition of carbon from $CaCO_3$ and so follows line AD which is co-linear with the bedrock composition (Fig. 5.20a). At calcite saturation the percent modern carbon (pmc) is close to 50%, but this can vary somewhat depending on the initial P_{CO_2}. As discussed earlier in this chapter, some speleothems form in the absence of soils; in such cases P_{CO_2} is raised by the production of strong acid from pyrite oxidation (e.g. Atkinson, 1983). The stoichiometry of these reactions is such that the calcite-equilibrated solution lies on a mixing line between atmospheric CO_2 and carbonate bedrock, with pmc values capable of lying rather lower than 50% (e.g. at point E, Fig. 5.20a). Intermediate cases with some contribution from soil CO_2 can be envisaged (e.g. between D and E).

An open system with complete isotopic equilibration would result in cave waters with nearly the same radiocarbon activities as CO_2 generated by organic matter decomposition in the soil, which for simplicity we will refer to as 100% modern carbon, assuming the date is 1950 and that the soil carbon is all modern (Fig. 5.20a). However, this is virtually never found in speleothems. Conversely, the closed system would generate quite high dead carbon percentages (dcp = 100 − pmc). Dead carbon percentage values approaching 50% are quite unusual and, in the absence of strong acids, the best documented examples represent growth beneath old peat soils (Genty et al., 2001a). It is a matter of observation that higher temperature soils with higher P_{CO_2} are associated with low dcp (high pmc) values (Genty et al., 2001a). It therefore seems that one can envisage a range of systems which are open systems as far as chemical equilibrium are concerned, but which vary in their degree of isotopic equilibrium, with $\delta^{13}C$ equilibration being much more readily established than ^{14}C. The fully equilibrated system is termed the *oversupplied open system* (point B, Fig. 5.20a). This term is used because complete equilibration is promoted by over-production of carbon dioxide. Higher levels of CO_2, whether introduced directly to the solution or in an adjacent gas phase, will promote equilibration. One could envisage a situation in waterlogged soils where the aqueous composition tends towards the carbon isotope composition of the metabolic products (i.e. approaching point A, or even lighter $\delta^{13}C$ compositions if methane is in the system), but it is not known whether this applies to speleothem-forming settings. Conversely, if carbon dioxide equilibration between gas and solution is barely sufficient to maintain the P_{CO_2} of the water to be in equilibrium with the gas phase, and does not equilibrate isotopically, we have an undersupplied open system (point C on Fig. 5.20a) which lies on the closed system line because it effectively represents a simple mixture of CO_2 dissolved from soil gas and $CaCO_3$. The $\delta^{13}C$ composition of water equilibrated with soil gas lies close to point C in any case, but clearly waters with such compositions have not equilibrated with ^{14}C.

Once in the cave environment the degassing process is liable to begin, and because the expelled CO_2 is isotopically light (Box 5.3), the solution increases in $\delta^{13}C$ value. Hendy (1971) modelled the consequences both of degassing and of isotope exchange with cave air. Figure 5.20b illustrates consequences of rapid degassing followed by slower equilibration with cave air, which for this purpose is taken as that of the external atmosphere (i.e. a strongly ventilated cave). Equilibrated waters converge on point F. In practice lines CF and DF would be smooth curves because the processes of degassing and equilibration can overlap. Hendy (1971) was the first to attempt to model this, although noting the necessity for more field and experimental data to constrain the system. The same caveat applies even now (Dreybrodt & Scholz, 2011). Comparable processes occur in rivers downstream of springs where isotope exchange via equilibration is argued to proceed only once chemical equilibrium is approached (Doctor et al., 2008).

One limitation in Hendy's analysis was the assumption that $CaCO_3$ precipitation would start as soon as the solution was saturated whereas it is now clear that a critical supersaturation has to be reached first. Dulinski and Rozanski (1990) were the first to attempt to model the impact on carbon isotope systematics of a phase of CO_2-degassing before the onset of calcite precipitation, and recognized that the rapidity of this reaction implies that the back-reaction of dissolution of CO_2 (i.e. equilibration) should be neglected in constructing a kinetic model. Their analysis assumed equilibrium fractionation between the DIC and the released CO_2, which is now known to underestimate fractionations, at least in dynamically ventilated caves (Spötl et al., 2005; Frisia et al., 2011; Lambert & Aharon, 2011). The results of their modelling indicate that an increase in the $\delta^{13}C$ signature of 2–4‰ would occur from degassing from initial P_{CO_2} values of $10^{-1.5}$–10^{-2}, to an end-solution with CO_3^{2-} of $2.2 \times 10^{-5}\,mol\,L^{-1}$ taken to be sufficient for $CaCO_3$ precipitation. In a second stage, the authors then used the same approach to model continued outgassing and $CaCO_3$ precipitation, but the attempt to put timescales on the isotope changes was speculative in the absence of firm data on equilibration rates.

A series of evidences for kinetic effects on $\delta^{13}C$ via degassing have emerged. Baker et al. (2011) presented a set of data on instantaneous $\delta^{13}C$ composition of dripwater from soda straws at Tartair Cave, Scotland and a wide range from −16 to −10‰ was encountered, with some drips showing a clear relationship of higher $\delta^{13}C$ with slower drip rate. Modelling of the data using eqn. 5.5 was successful provided that a kinetically enhanced fractionation applied. Also, mass balance studies at Ernesto Cave, where clearly degassing precedes $CaCO_3$ precipitation (Frisia et al., 2011), has established that there is a strong kinetic fractionation effect on ^{13}C during the fast process of degassing, and slower equilibration between dripwater and cave air (Fig. 5.21). Degassing of carbon dioxide at isotopic equilibrium with DIC would result in only a small increase in $\delta^{13}C$ values as shown. A mass-balance comparison of the composition of initial soil/epikarst waters with stalagmite tops, which represent the first point for precipitation of $CaCO_3$, demonstrates that the released CO_2 has an isotopic composition of around −70‰; similar results have been obtained by

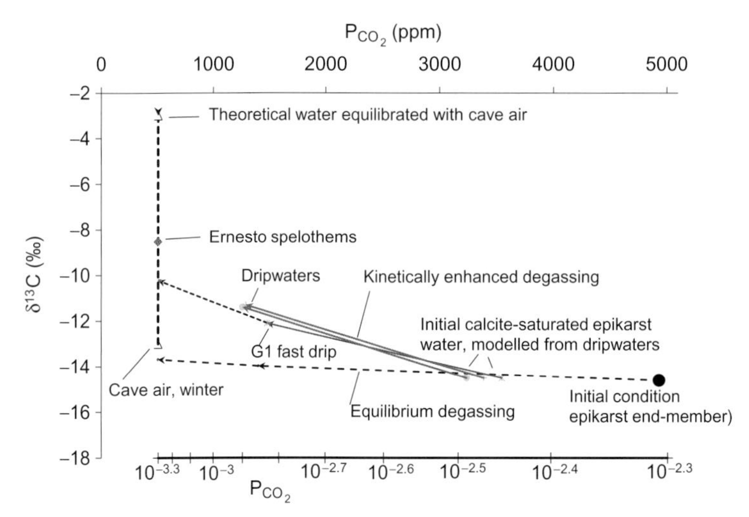

Fig. 5.21 Data and modelled trajectories of waters from Ernesto Cave, northeast Italy (after Frisia et al., 2011). Comparison of elemental concentrations, $\delta^{13}C$ and ^{14}C isotope systematics of soil and cave dripwaters, indicates that there was a stage of closed-system dissolution below the soil followed by kinetically enhanced ^{13}C-fractionation in the cave before dripwater collection. Because stalagmite tops have much heavier compositions (and there is virtually no PCP), further kinetically enhanced degassing together with some equilibration is inferred. The end-product of equilibration between the atmosphere and water running over and beyond the stalagmites is a value of around −3‰ under low-P_{CO_2} winter conditions.

re-analysis of data of Spötl et al. (2005) from Obir Cave. Equilibration should be particularly effective in caves with dynamic air circulation because the cave air is then effectively an infinite reservoir allowing equilibration to be complete. Complete equilibration can be observed in a dripwater site (SH-2) from Obir Cave that had been left sampling for a month. Such an equilibration is the most obvious explanation for the synchronous evolution of stalagmite and atmospheric ^{14}C values during and following the 1960s bomb spike at this site (Smith et al., 2009), rather than the lagged effect normally encountered.

5.5 Evolution of cave water chemistry: modelling sources and environmental signals

5.5.1 Forward modelling

Figure 5.5 provides a conceptual summary of various modifiers for water composition which can be summarized as atmospheric input, interactions with soil organic matter, mineral leaching, and aquifer modification. The composition of resulting solutions can be calculated by mass balance equations. For example

$$M = X_1M_1 + X_2M_2 + X_3M_3 \ldots \quad (5.16)$$

where M is the mean composition and M_1, M_2 . . . are the compositions of sources with mole fractions X_1, X_2 . . . that add up to 1. Compositions may be concentrations or isotope signatures (see, for example, Figs. 5.15 and 5.20a), but where evaporation occurs, concentrations need to be modified by the degree of evaporation. Where there are losses as well as gains, as in the case of carbon dioxide in soils, terms need to be added for these, taking into account processes leading to isotope fractionation.

Commonly, one aspect of the chemistry may be uniquely determined whereas others are not. An equation of the form (5.16) can often be written for Sr isotopes (Banner et al., 1996) or for a combination of the Mg/Ca and Sr/Ca composition (Fairchild et al., 2000). This then results in an understanding of the mass balance of Ca whose origins we currently cannot deduce from Ca isotope studies. A distinctive case is the contribution from marine aerosol (Table 5.2), where the ratio of an element to Cl is fixed, and chloride is usually not lost or gained by soil or aquifer interactions (excepting seasonal salt precipitation in soil).

Wherever an isotope fractionation arises by selective removal of a component from the solution, the process can be described in terms of a Rayleigh fractionation equation, as was discussed in section 3.2.1 in the case of condensation of water vapour. Taking carbon isotopes as an example, with the fractionation equation expressed in terms of δ values:

$$\log(1+\delta^{13}C/1000) = \log(1+\delta^{13}C_o/1000) + \varepsilon \log f \tag{5.17}$$

Here $\delta^{13}C_o$ refers to the initial $\delta^{13}C$ composition, f is the fraction of carbon remaining in the system and $\varepsilon = \alpha - 1$ as in Box 5.3. The appropriate value of ε is the mean value related to all the fractionation reactions. For example, where CO_2-degassing and calcite precipitation are both occurring, the mean value of ε is used according to the mass balance (Bar-Matthews et al., 1996; Spötl et al., 2005). Because the speciation of aqueous carbon can change during such a calculation, it is most accurate to perform such a calculation iteratively in conjunction with a speciation programme such as PHREEQC (Appelo & Postma, 2005), as has been done by Johnson et al. (2006) and Mattey et al. (2010). An example of the resulting trend in $\delta^{13}C$ is shown in Fig. 5.22b.

For trace elements it is conventional to express the relationship between solution composition, and that of the calcite removed from it, in terms of a distribution or partition coefficient (Fairchild & Treble, 2009). Using Mg as an example:

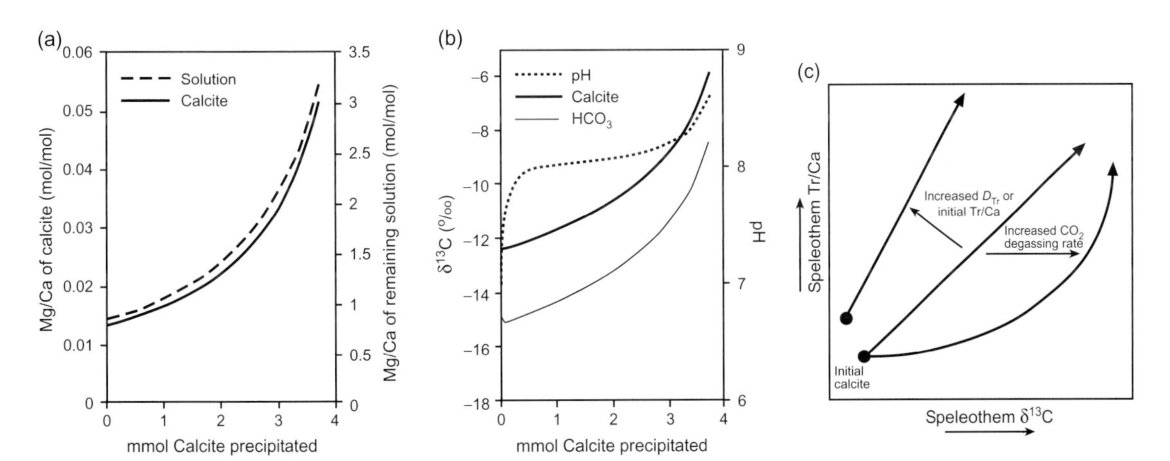

Fig. 5.22 Summary of changes in solution chemistry accompanying degassing and calcite precipitation from solution at constant supersaturation. (a, b) Calculations based on data from Heshang Cave, China, of the increase in Mg/Ca and $\delta^{13}C$ as calcite is precipitated (in the Heshang case up to 1.5 mmol L^{-1} of calcite would be required to explain observed changes). The calcite compositions shown are those at each increment of precipitation. The calculations used the software PHREEQC (Appelo & Postma, 2005) which allows for changes in carbonate speciation in solution. Modelled conditions are temperature 18 °C initial solution Mg/Ca = 0.84, $D_{Mg} = 0.016$, $\delta^{13}C_{DIC} = -16.75‰$. (c) Generalizations of covariation of Tr/Ca (where Tr is a trace element such as Mg or Sr) and $\delta^{13}C$. Steeper covarying slopes are expected where the partition coefficient (D) is higher or where the initial Tr/Ca is higher. A higher initial $\delta^{13}C$ will shift the line to the right. If CO_2 degassing exceeds that required to maintain constant supersaturation, the covariation will show a concave-up curvature because degassing to reach a critical supersaturation for calcite growth increases $\delta^{13}C$ (Fig. 5.21), but does not change Mg/Ca. Modified from Johnson et al. (2006).

$$D_{Mg}(Mg/Ca)_{water} = (Mg/Ca)_{calcite} \quad (5.18)$$

For Sr and Mg in calcite the value of K is << 1 (see also Chapter 8), so there is a characteristic increase in the ratios of Sr/Ca and Mg/Ca as more calcite is precipitated. An iterative application of this equation takes this form:

$$(Mg/Ca)_{water} = (Mg/Ca)_{water_0} - (Ca_0 - Ca)_{water} D_{Mg} \quad (5.19)$$

where the suffix 0 refers to the original (or previous) condition. An example of the resulting trend is shown in Fig. 5.22a. Sinclair (2011) referred to a rigorous mathematical proof showing that the slope of a crossplot of ln(Sr/Ca) versus ln(Mg/Ca) is given by $(D_{Sr} - 1)/(D_{Mg} - 1)$ and is approximately equal to 0.88 ± 0.13 using typical D values.

Calcite precipitation clearly results in covariations of trace elements such as Sr and Mg, the evolving $\delta^{13}C$ composition of the solution, and hence the composition of calcite further down the water flowline. The exact form of these lines will depend on different factors as shown in Fig. 5.22c. In cases where initially only degassing occurs, $\delta^{13}C$ rises until a critical supersaturation is reached at which calcite precipitation keeps pace with degassing. Steady degassing accompanying calcite precipitation at constant supersaturation gives rise to linear covarying trends but the slope depends on values of D and the absolute initial value of Tr/Ca (Fig. 5.22c). More complex but partly conceptual forward models, directly involving climatic parameters are now being generated (e.g. Wong et al., 2011) and this will be an important avenue for future research.

5.5.2 Backward modelling

It became apparent to early investigators that cave environments could provide model systems for study of carbonate geochemistry. Pioneering works such as Trombe (1952) and Murray (1954) were followed by a study by Holland et al. (1964) that provided a benchmark in showing the deductive power of carbonate water chemistry in cave environments, making effective use of trace elements for the first time. Studying three caves in Pennsylvania and Virginia, they estimated the minimum necessary P_{CO_2} to reach the maximum (Ca + Mg) concentrations observed. They also established that the cave waters typically became supersaturated by degassing of CO_2, rather than evaporation. They used the evidence that Mg was constant in different parts of a cave, but Ca varied and reasoned that evaporation would lead to an increase in Mg concentration, whereas degassing followed by calcite precipitation would lead to little change in Mg. They also overcame, for the time, significant analytical difficulties to explore how the Sr/Ca ratio in dripwater evolved down flow owing to precipitation of aragonite and/or calcite. Although degassing has remained the dominant paradigm, Thrailkill (1971) demonstrated that both degassing and evaporation were required to explain the range of solution compositions and speleothem mineralogies (including Mg-carbonates) within parts of Carlsbad Caverns, New Mexico, and Gonzalez and Lohmann (1988) extended such studies in these caves to include stable isotope patterns of the precipitates.

In groundwater geochemistry in general, there has been extensive use of reaction-transport models to deduce the history of waters and the geochemical processes to which they have been subjected (Appelo & Postma, 2005). Such transport models assume large-scale aquifer properties and cannot cope with the case of flow to a unique dripwater point; hence linear mixing models have been developed as a first attempt at an alternative (e.g. Fairchild et al., 2006b; Fig. 5.23). It became immediately obvious that although there is qualitative agreement between such mixing models and real data, there is quantitative disagreement (Fig. 5.23b) pointing to additional processes such as exchange between fractures and matrix. Nevertheless, it is encouraging that particular drips do display distinctively different chemistries at low flow. A convenient test for discovering the reasons for this is a comparison with PCP lines calculated from eqn. 5.19 (Fig. 5.23c, d). In some cases high and low-flow waters differ in the amount of PCP, whereas in others there are higher trace element concentrations at low flow implying that low-transmissivity portions of the aquifers yield enriched trace element compositions (in this case of Sr and Mg). In both

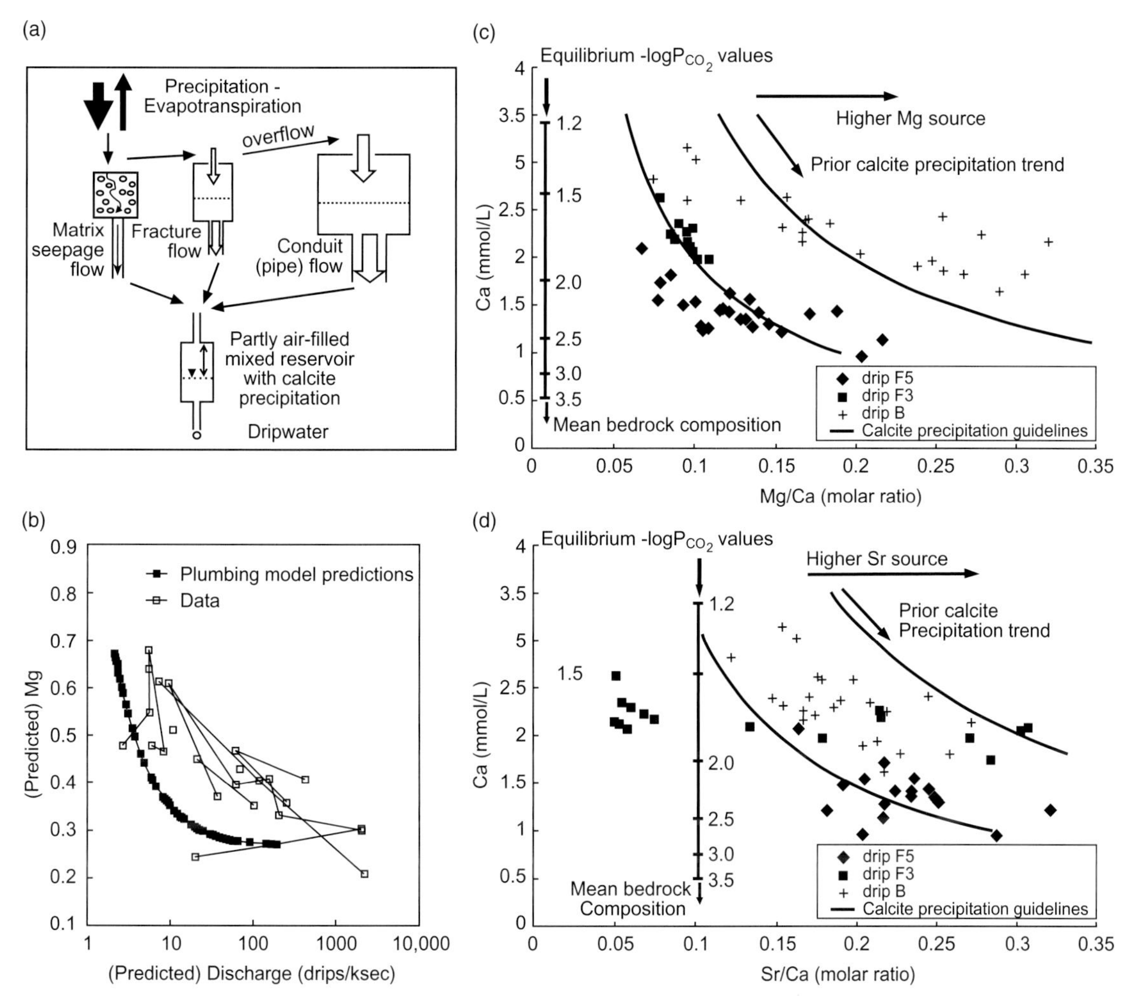

Fig. 5.23 Geochemical relationships of bedrocks and dripwaters at a Jurassic limestone mine site (Brown's Folly Mine, Bath, UK) from Fairchild and Treble (2009) after Fairchild et al. (2006b). (a) Linear systems hydrological model illustrating the role of processes such as fluid mixing and PCP (b) Comparison of model predictions with data from a drip site; deviations are due to nonlinear mixing effects. (c, d) Diagrams to determine the nature of processes causing chemical changes at low flow. Low Ca is found at low flows; in some cases this relationship can simply be explained by PCP (e.g. Mg/Ca data at site F3 and Sr/Ca data at sites B and F5). In the other cases, there must also be a high trace element content of dripwaters at low flow, drawing on low transmissivity regions of the aquifer enriched in Sr (probably from relic aragonite) or Mg (from clay or dolomite).

cases, such enrichments in an associated speleothem could be interpreted as evidence for lower flows (and lower atmospheric precipitation). However, in some caves, the reason for higher PCP is the greater rate of calcite precipitation when P_{CO_2} is seasonally low owing to increased ventilation (Spötl et al., 2005; Boch et al., 2011; Wong et al., 2011). PCP has been identified as a widespread process of major importance in modifying dripwater chemistry (Tooth & Fairchild, 2003; Musgrove & Banner, 2004; McDonald et al., 2004, 2007; Johnson et al., 2006, Karmann et al., 2007;

Verheyden et al., 2008b; Griffiths et al., 2010a; Oster et al., 2010) and analogies are found for lake waters in surface environments (Kober et al., 2007; Jin et al., 2010).

Because the movement of water is sub-vertical to caves, it may be possible to constrain the components that contribute to water chemistry by study of soil profiles and bedrock exposures. The most specific results for water flowpaths can be obtained by the use of artificial tracers (Bottrell & Atkinson, 1992; Mattey et al., 2010) or anthropogenic contaminants (Jiménez-Sánchez et al., 2008). Results are complicated by retention or transformation by bacterial reactions in the aquifer as well as mixing processes. Fairchild et al. (2000) specifically used experimental data on dissolution kinetics (Fig. 5.14) to constrain the evolution of cave water chemistry. By determining end-member compositions aspects of the water chemistry can be parsed into different sources, correcting first for marine aerosol, and allowing for PCP (Banner et al., 1996; Fairchild et al., 2000, 2006b). For complex changes, PHREEQC modelling can also be instructive (e.g. Turin & Plummer, 2000).

Carbon isotope patterns offer a great challenge to interpretation as is clear from the discussion in section 5.4 but back-modelling approaches are always instructive. Bar-Matthews et al. (1996) demonstrated from Soreq Cave that $\delta^{13}C$ variations could be accounted for quantitatively by simultaneous balanced $CaCO_3$ precipitation accompanied by CO_2 degassing and Johnson et al. (2006) inferred likewise for annual $\delta^{13}C$–Mg covariations in stalagmites. However, in some dynamically ventilated caves, there is clear evidence for a phase of degassing and a rise in $\delta^{13}C$ before calcite precipitation starts (Spötl et al., 2005; McDermott et al., 2005; Fairchild & McMillan, 2007; Mattey et al., 2010; Frisia et al., 2011). This process may result in kinetic fractionation (Fig. 5.21) and an inverse variation of $\delta^{13}C$ with drip rate may be present (Baker et al., 2011). Also in the case of slow drips, there can be equilibration with cave air. Even the distinctively large control by C_3 or C_4 vegetation could be masked if such in-cave processes vary in space or time.

Several modern monitoring studies are aimed at providing a deep understanding of processes which in turn inform interpretation of speleothem records. For example, the combination of Sr isotope measurements with trace elements in Natural Bridge Caverns, Texas, has proved effective in demonstrating the extent to which water chemistry changes reflect PCP, changing amounts of water–rock interaction or some combination (Wong et al., 2011). Mattey et al. (2010) have provided one of the most comprehensive programmes to date and Figure 5.24 illustrates a range of data from three dripsites and a lake, in Lower St Michael's Cave in Gibraltar, demonstrating the strong seasonality in hydrology, cave air P_{CO_2} and hydrochemistry. Backward modelling demonstrates that PCP is a key driver and it is very pronounced in the summer when the two potential forcing factors reinforce each other. Summer is the season when there is a strong soil moisture deficit and drip rates decline. The increased air space in the aquifer encourages degassing and PCP and higher PCP relate to seasonal dryness has been invoked at several other sites (Fairchild et al., 2000; Baker et al., 2000; McDonald et al., 2007). However, at this site, P_{CO_2} of cave air is also low in summer which independently tends to increase PCP; seasonally higher rates of calcite precipitation in relation to low P_{CO_2} has also been widely recognized (Frisia et al., 2000; Liñán Baena et al., 2000; Spötl et al., 2005; Banner et al., 2007; see also Fig. 4.30). There are many sites where these two effects are likely to be in competition (summer dry and winter low P_{CO_2}), but too few caves displaying summer dryness have also been monitored for P_{CO_2} to be able to generalize about controls.

Characterization of seasonal variations in dripwater chemistry allows comparison with annual spatial scales of observation in speleothems. In this way, the seasonal system behaviour in the past can be investigated (Johnson et al., 2006; Mattey et al., 2008, 2010). A further step is possible when it can be argued that multi-annual variability is likely to be caused by the same factors responsible for annual changes. This was argued by McMillan et al. (2005) in the case of Clamouse Cave, where

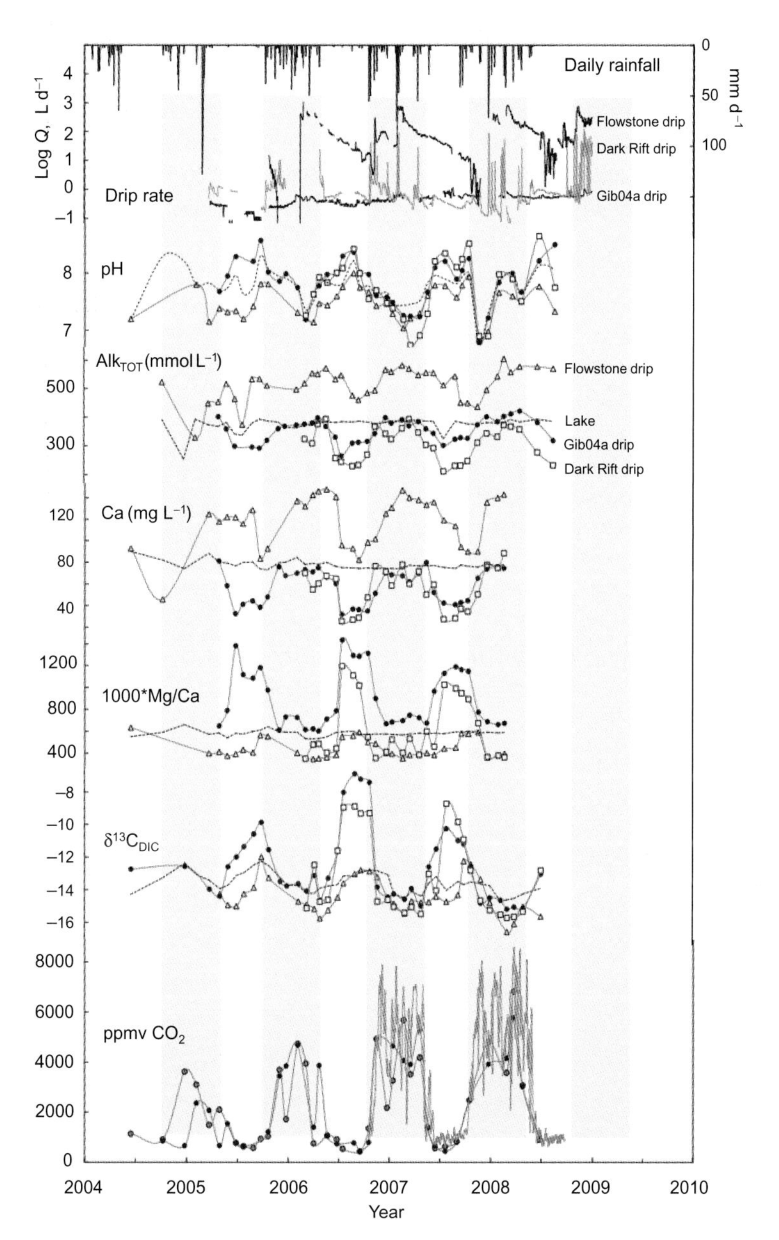

Fig. 5.24 Data from New St. Michaels Cave, Gibraltar (Mattey et al., 2010). Relationships between rainfall amount, drip discharge, dripwater composition and cave air P_{CO_2} measured between 2004 and 2009. From top to bottom: daily rainfall, drip discharge on log scale, monthly dripwater pH, total alkalinity, Ca concentration, 1000Mg/Ca and $\delta^{13}C_{DIC}$. Lower plot is the concentration of CO_2 in cave air, measured on a combination of monthly spot samples at the Gib04a (black circles) and Lake (grey circles) sites and continuous logging at the Gib04a site. Strong seasonality is present; the similarities in aqueous chemistry of sites with very different hydrology implies an in-cave control (varying amounts of PCP limited by seasonal changes in P_{CO_2} of cave air).

increases in both Sr/Ca and Mg/Ca owing to enhanced PCP during summer dryness, were seen in past records to be modulated by longer-term variability with the same geochemical signature, and interpreted as an aridity signal.

Our understanding of the distinct processes responsible for changes in cave water chemistry clearly indicates limitations of single-proxy studies in speleothems. They also point to the advantages in understanding the seasonal responses of the speleothem factories and incubators in which they form. We will return to the use of multiple proxies and seasonally resolved chemistry when discussing speleothem chemistry in Chapter 8.

Plate 1.1 Stalagmites as individuals. Bespoke vases compared with adjoining stalagmites from Yonderup Cave, Perth, Western Australia. The left stalagmite is a composite of two conjoined columnar ('candle-shaped') individuals which are currently inactive and coated with white moonmilk. The central stalagmite is active, but its feeding drip shows inter-annual variation in discharge (P. Treble, personal communication, 2008); ripples on its surface attest to periodically high discharge. The right stalagmite also displays conjoined individuals whose shape indivates periodically high flows. Locally there is a white deposit that appears younger than the rest of the surface. Vase images reproduced by permission of the Wedgwood Museum, Barlaston, Staffordshire, UK.

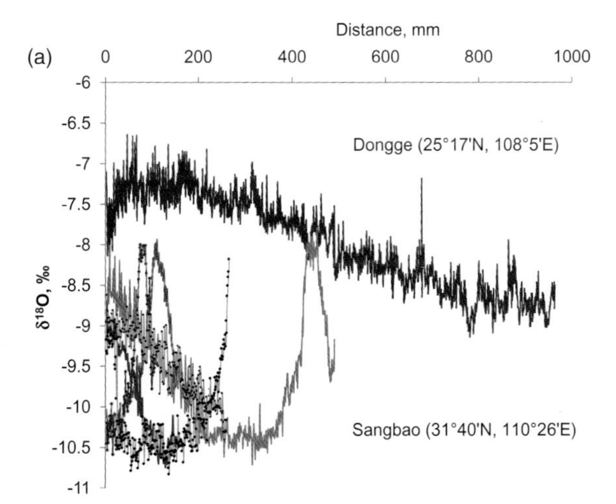

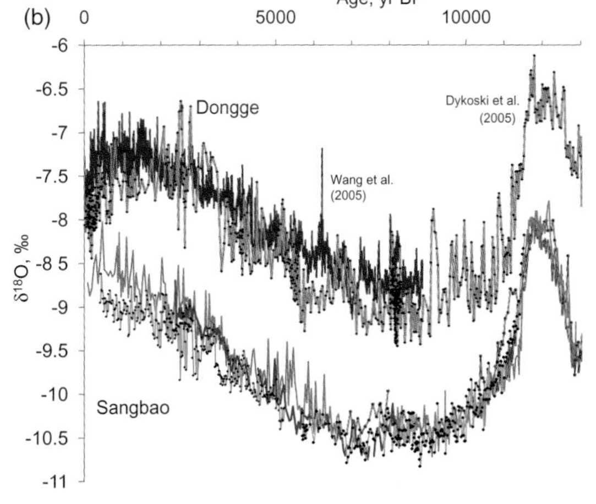

Plate 1.2 Plots of data from the studies of Wang et al. (2005) and Dong et al. (2010) lodged at the NOAA Data Center. (a) Oxygen isotope time series from a sample from the Dongge cave site and several samples from Sangbao cave, the latter lying in southern China 750 km south–southwest of the former. (b) Same oxygen isotope data as (a) (supplemented by a second speleothem from Dongge from Dykoski et al. 2005) plotted against sample ages from U–Th dating. More negative oxygen isotope values correspond with a relatively stronger summer monsoon compared with winter rainfall.

Speleothem Science: From Process to Past Environments, First Edition. Ian J. Fairchild, Andy Baker.

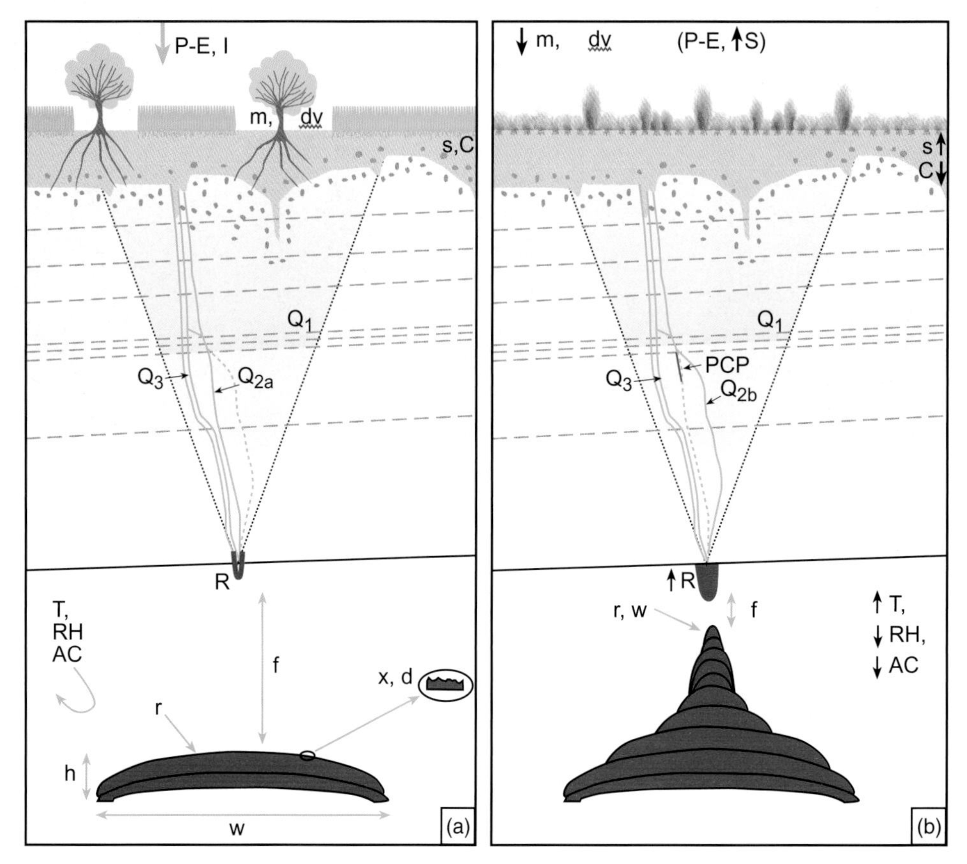

Plate 1.3 Conceptual model of factors influencing speleothem properties. Scale increases downwards. Multi-decadal to multi-millennial-scale variations in system parameters for the speleothem factory. See Table 1.1 for summary of the parameters and the controls, and the text for discussion. (a) A more humid environment with colder winters compared with the later semi-arid environment in (b).

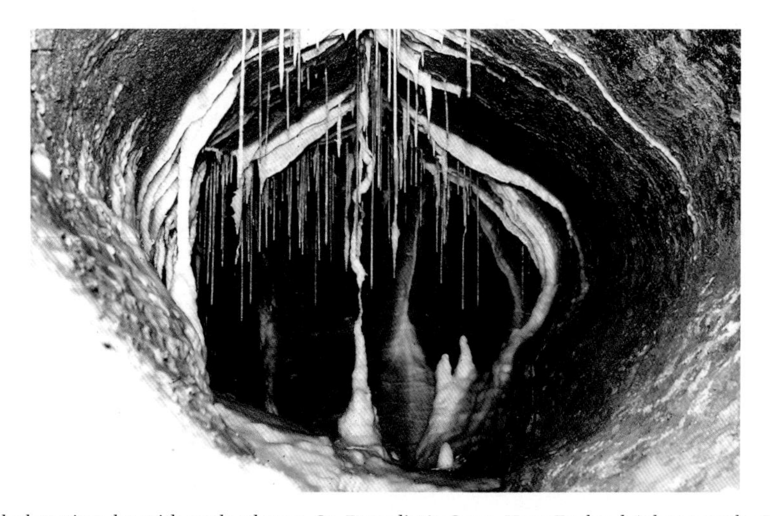

Plate 2.1 Drained phreatic tube with speleothems, St. Benedict's Cave, New Zealand (photograph: John Gunn).

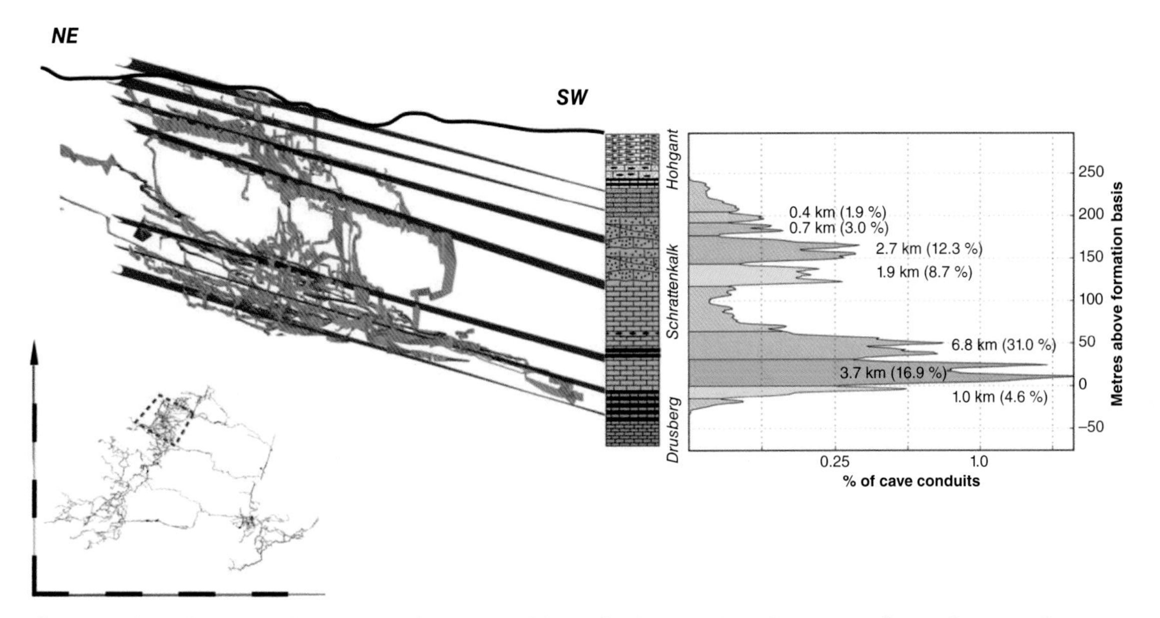

Plate 2.2 Three-dimensional projection of one part of the Siebenhengste Cave System, southwest Germany. Seven potential inception horizons can be identified within the Schrattenkalk Formation. From Filipponi et al. (2009).

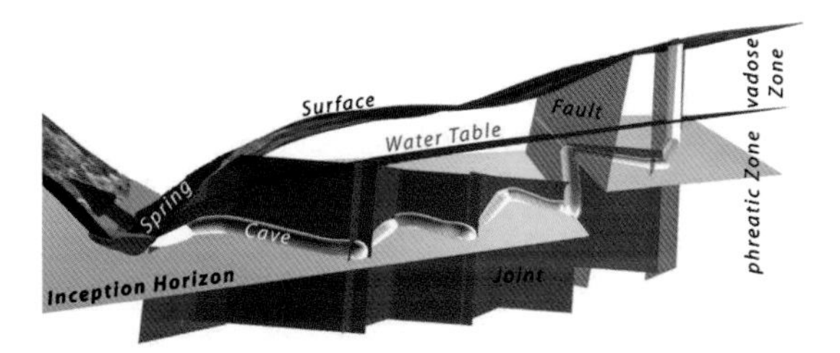

Plate 2.3 The preferred directions of phreatic cave conduit development parallel to an inception horizon, roughly following the hydraulic gradient, but following the intersection with joint sets, and offset along faults. From Filipponi et al. (2009).

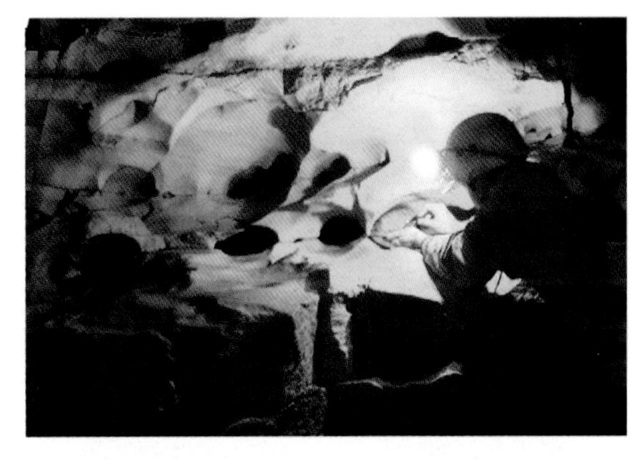

Plate 2.4 Nidlenloch cave system (Switzerland). Anastomoses (small openings) are present, features commonly found in cave passages located along an inception horizon. (Photograph: M. Widmer). From Filipponi et al. (2009).

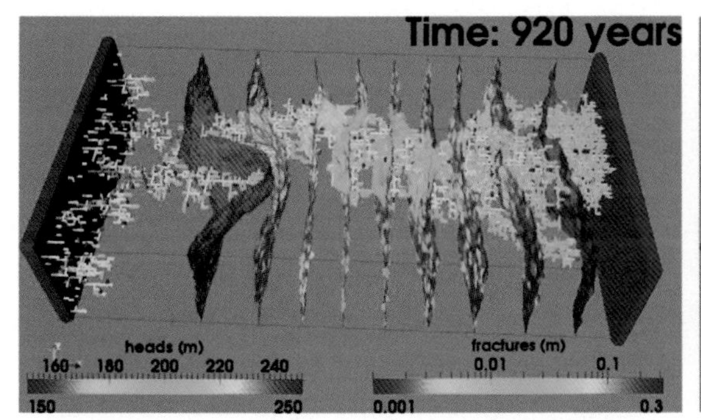

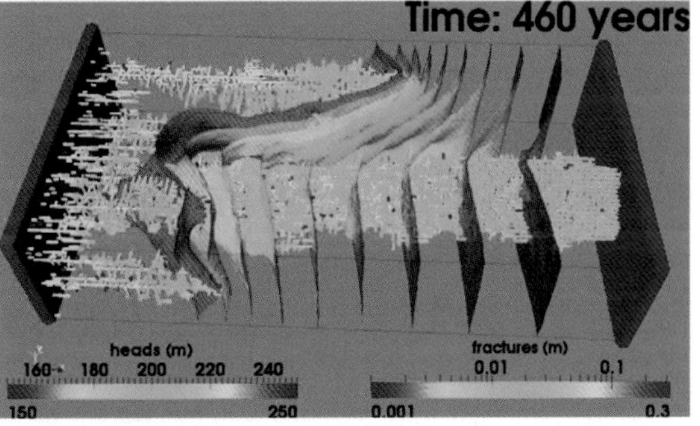

Plate 2.5 Three-dimensional computer model of the development of a conduit network in a limestone with microscopic fractures having a distribution of initial diameters of 0.15±0.02 mm. In both diagrams the driving pressure heads (lower left key) fall from left to right in the model and the fracture width is given in lower right key. Breakthrough has occurred of a central wide conduit surrounded by a network of enlarging fractures. In the right-hand model result, the central conduit is much straighter than in the left-hand image where its sinusoidal nature reflects a lower density of initial fractures. From Kaufmann et al. (2010).

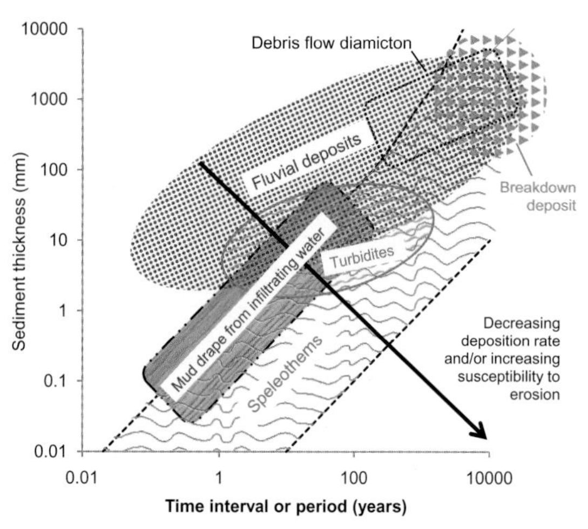

Plate 2.6 Comparison of approximate rates of deposition of clastic sediments (usually episodic) and speleothems (continuous to episodic). Appropriate references are in text discussing of each type of sedimentation. The time axis represents recurrence interval for episodic deposition of a certain magnitude or (for mud drapes and speleothems) the total period of deposition.

Plate 2.8 Archaeological investigations into complex slope and anthropogenic deposits at cave entrance (10 m wide view), La Garma, northern Spain.

Plate 2.7 Eroded deposits in the upper passage of Crag Cave, Ireland. Flood gravels are overlain by a bench of flowstone and stalagmite.

Plate 2.9 Palaeo-cave infills (2 m high), Phosphate Mine, Wellington Caves, New South Wales, Australia. Dark grey bedrocks representing the floor underlying the palaeo-cave. A thick sediment fill with blocks of bedrock was generated, cemented and then eroded; the erosion surface has been lined subaqueously with a white mineral cement, then filled with more sediment including the prominent dark red layer in the centre of the photograph . Following later burial, some small sub-vertical white tension gash veins cut the sediment.

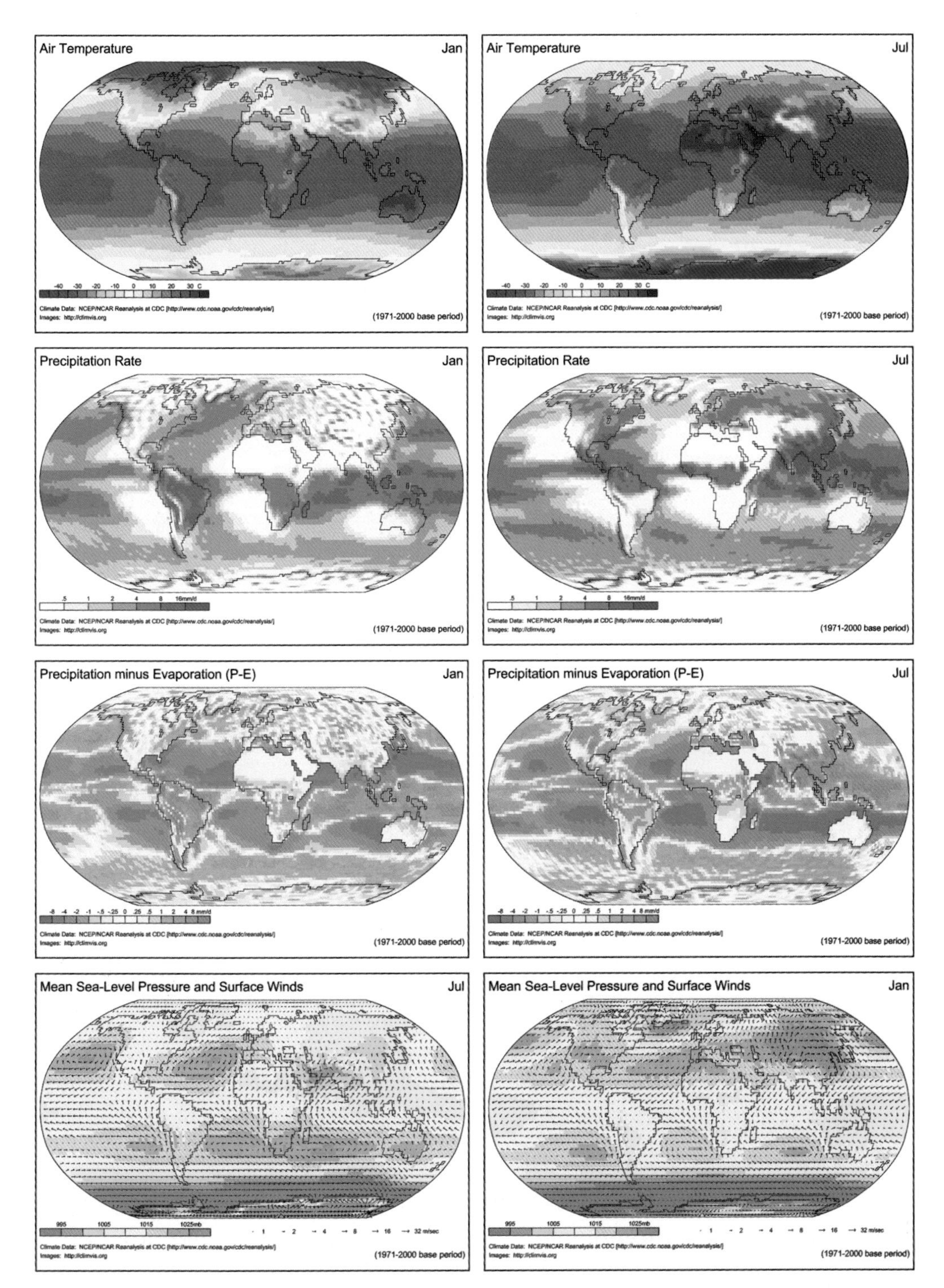

Plate 3.1 Global patterns of air temperature (row 1), daily mean precipitation rate (row 2), precipitation minus evaporation (row 3) and global sea-level pressure and winds (row 4) in January (left column) and July (right column). From Shinker (2007).

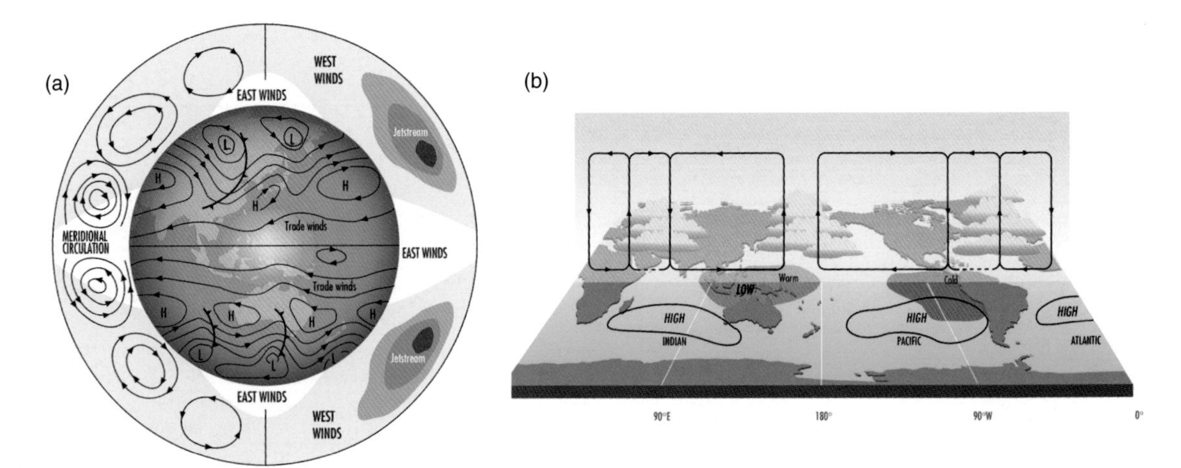

Plate 3.2 Left, essential features of the general circulation of the atmosphere showing a typical daily pattern of surface pressure systems and (in greatly exaggerated vertical scale) the zonally averaged meridional (left of globe) and zonal circulation (right of globe). Right, schematic representation of the east–west Walker Circulation of the tropics. In normal seasons air rises over the warm western Pacific and flows eastward in the upper troposphere to subside in the eastern Pacific high pressure system and then flows westward (i.e. from high to low pressure) in the surface layers across the tropical Pacific. Weaker cells also exist over the Indian and Atlantic Oceans. In El Niño years, this circulation is weakened, the central and eastern Pacific Ocean warms and the main area of ascent moves to the central Pacific. Source: Australian Government Bureau of Meteorology (2010). Copyright Commonwealth of Australia reproduced by permission.

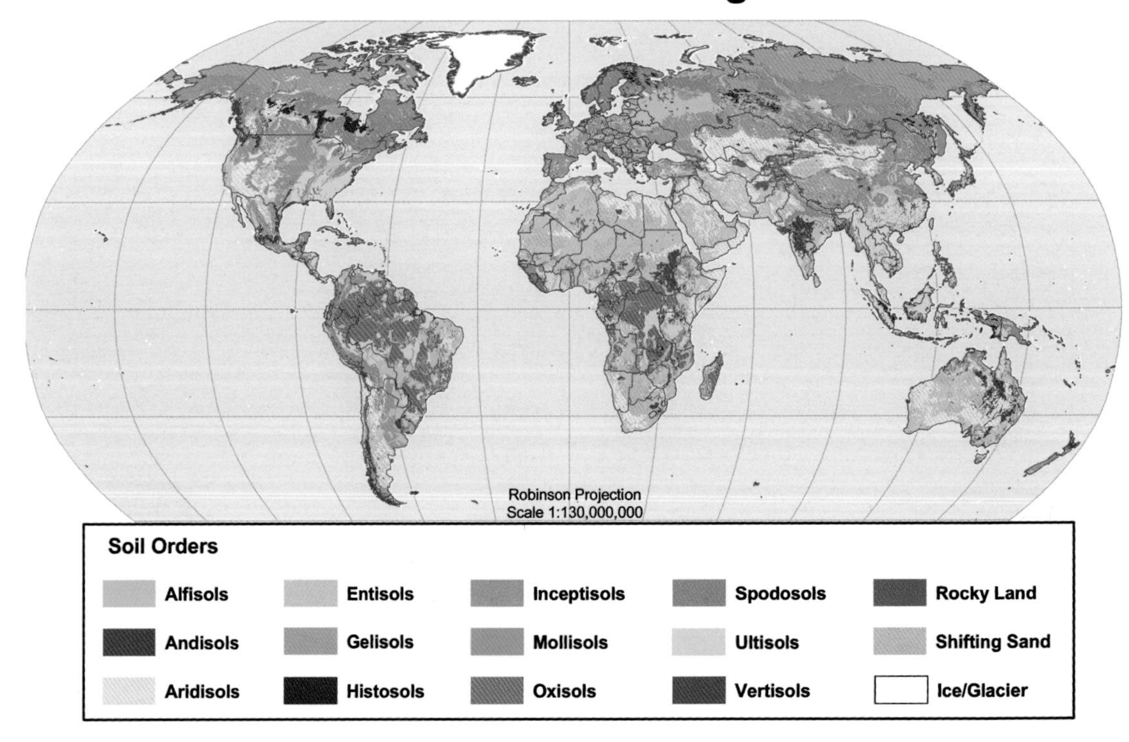

Plate 3.3 Global soil regions. Source: United States Department of Agriculture (2005): http://soils.usda.gov/use/worldsoils/mapindex/order.html.

Plate 3.4 Top left, peat core (50 cm) resting on heather-*Sphagnum* moss peatland above Tartair cave, NW Scotland. Bottom left. Quarry in steeply dipping Jurassic limestones, PS of Lisbon, Portugal showing penetration of red (terra-rossa) soil into the epikarst (photo approximately 20 m high). Right, trench (80 cm) in forest soil above Ernesto Cave, Italy. Humic A horizon, coarse to medium textured B-horizon and a limestone gravel-rich C horizon.

Plate 3.5 Biomes. Left, an example of an anthropogenic biome: Ethiopian agricultural landscape, Mechara. Right, Heshang Cave, central China (temperate deciduous broadleaf forest biome of Plate 3.6).

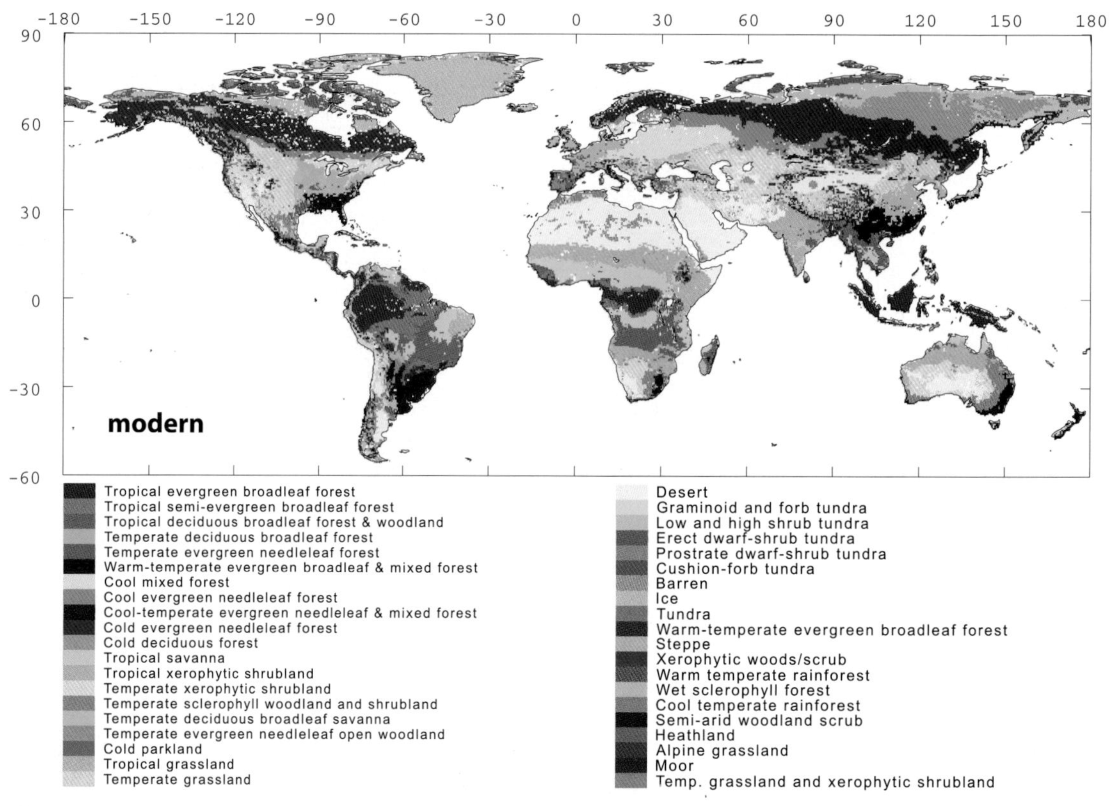

Plate 3.6 Example of a biome model output (BIOME4; Kaplan et al., 2003).

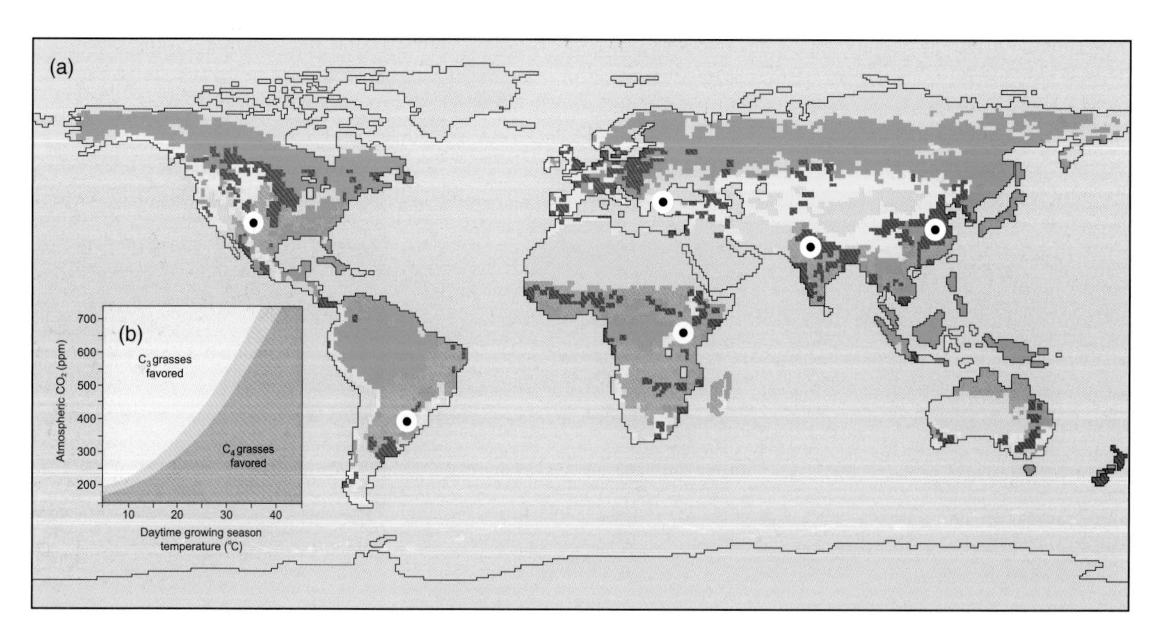

Plate 3.7 (a) Global distribution of forests (green) and woodlands, savannas, and grasslands with a ground cover dominated by either C_4 (orange) or C_3 (yellow) grasses. Cropland (red) and shrubs, desert, bare ground, and ice (beige/brown) are also shown. Dots show regions where geological history of C_4 plants is best known. (b) The predicted atmospheric CO_2 and growing-season temperature conditions that favour the growth of C_3 or C_4 grasses. Adapted from Edwards et al (2010).

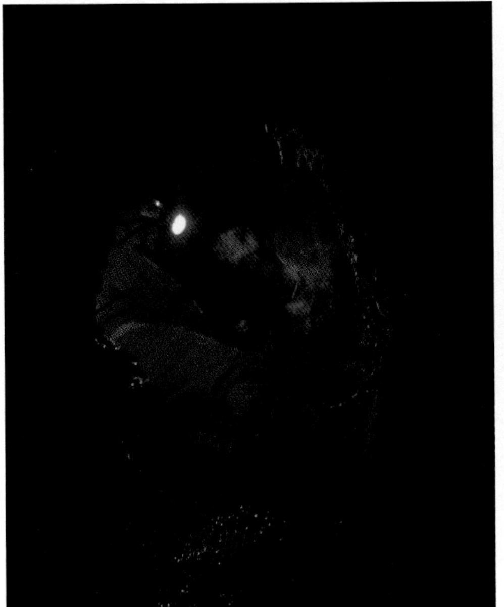

Plate 4.1 Left, slow-moving artificial smoke in Lower St. Michael's Cave , Gibraltar. Right, mist formed by condensation of moisture from summer ambient air onto cold air stream from artificial entrance to Obir cave. Air movement that is fast enough to be felt is uncommon and characteristic only of restricted or wind-susceptible passages even in dynamically ventilated caves such as these (Spotl et al., 2005; Mattey et al., 2010; Fairchild et al., 2010).

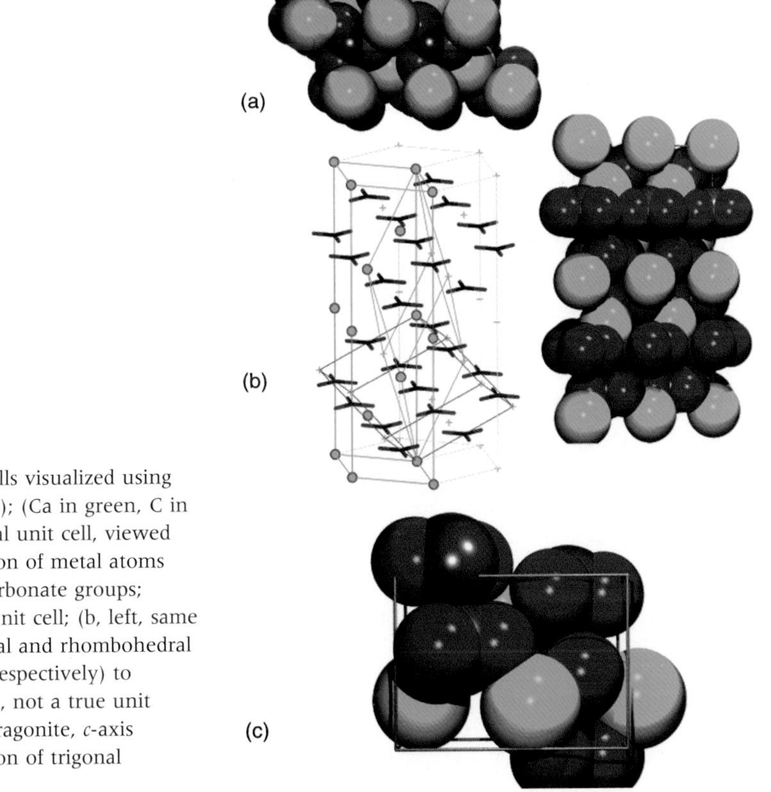

Plate 5.1 Carbonate mineral unit cells visualized using PLATON software (Morse et al. 2007); (Ca in green, C in black, O in red): (a) calcite hexagonal unit cell, viewed down the *c*-axis, showing coordination of metal atoms by oxygens belonging to different carbonate groups; (b, right, *c*-axis vertical) hexagonal unit cell; (b, left, same orientation) relationship of hexagonal and rhombohedral unit cells (green and blue outlines, respectively) to cleavage rhombohedron (red outline, not a true unit cell); (c) orthorhombic unit cell of aragonite, *c*-axis vertical, showing staggered orientation of trigonal carbonate groups.

Plate 7.1 Examples of reduction in diameter of a stalactite-stalagmite pair as they grow together and the drip height reduces. Left:,Orient Cave, Jenolan, New South Wales, Australia. Photograph: Stan Robinson. Right, Christmas Cave, near Tham Lod, Thailand. Photograph: Otto de Voogd.

Plate 7.2 False-coloured electron backscatter diffraction image of aragonite rays (blue) in calcite, Clamouse Cave (see study of McMillan et al. (2005) and compare with backscattered electron image in Fig. 7.11c). Black areas are where the crystal structure was not resolved.

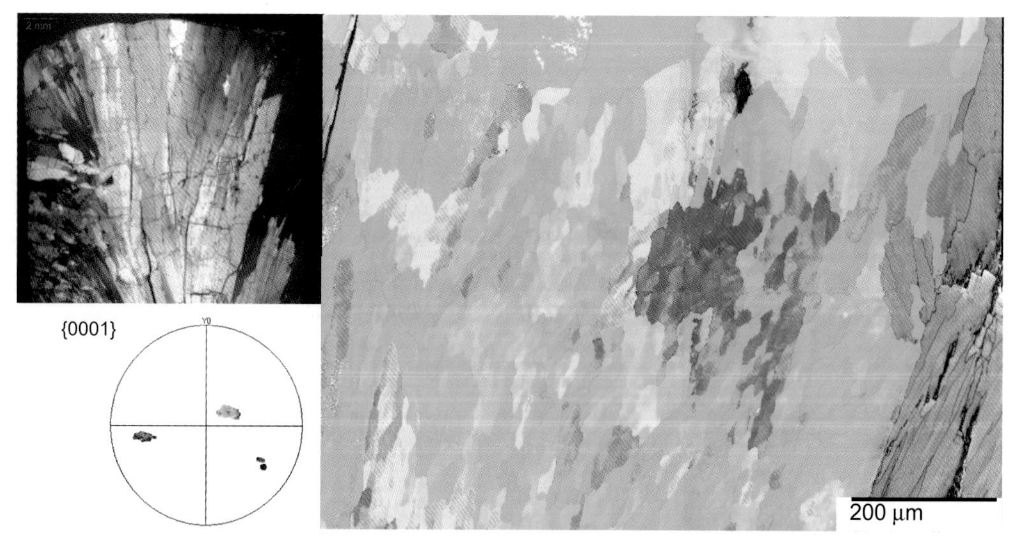

Plate 7.3 Microcrystalline columnar fabrics of stalagmite Obi84 (Obir Cave, Austria); sample top also shown in Fig. 1.2b. Top left, crossed-polars image showing sweeping radiaxial extinction; red box locates large image. Right, false-coloured electron backscatter diffraction image. Main crystals vary in the orientation of their *c*-axes by around 8°, with the colour sequence blue–green–yellow–red representing increasing divergence from the mean orientation. Lower left, Wulff stereonet of orientation of poles to {0001}. Grey crystals have a completely different optical orientation. After Fairchild et al. (2010).

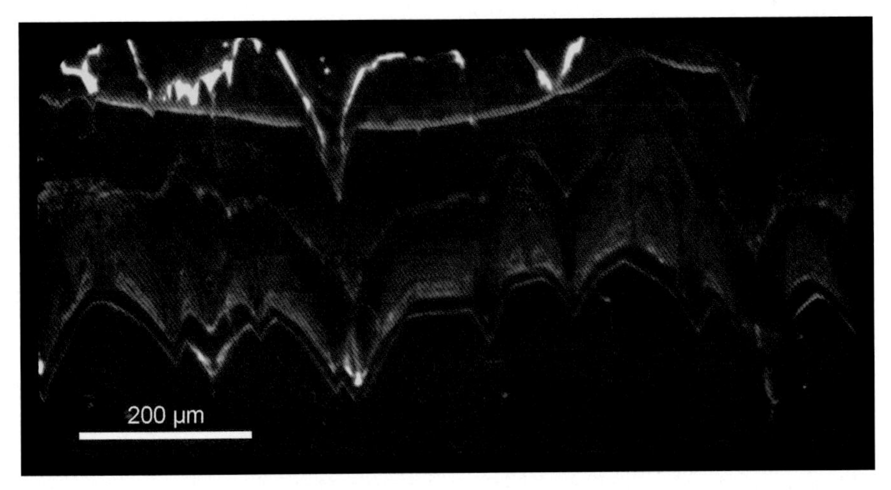

Plate 7.4 Changing growth morphology of a stalagmite surface revealed by Mn-activated cathodoluminescent zones, sample OE21,Ostenberg Cave, Germany. Richter et al. (2004).

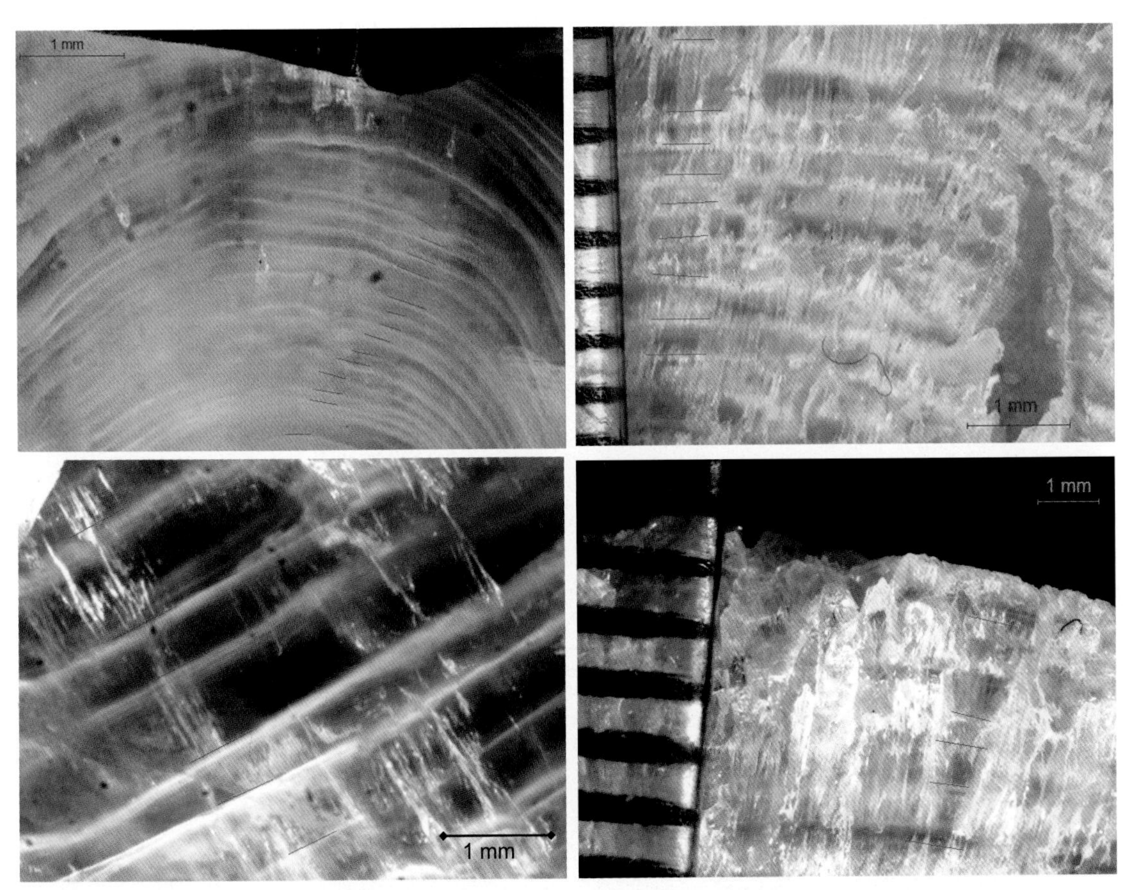

Plate 7.5 Examples of visible annual laminae in two stalagmites from Akçakale Cave, Gümüşhane province, northeast Turkey. The top two images are from stalagmite 'Colin', the bottom two images from stalagmite '2pac', both stalagmites were actively growing when sampled. Left, visible laminae in generally compact calcite deposited more than 500 years BP. Right, visible laminae in porous calcite fabric deposited less than 100 years BP. From Baker et al. (2008a).

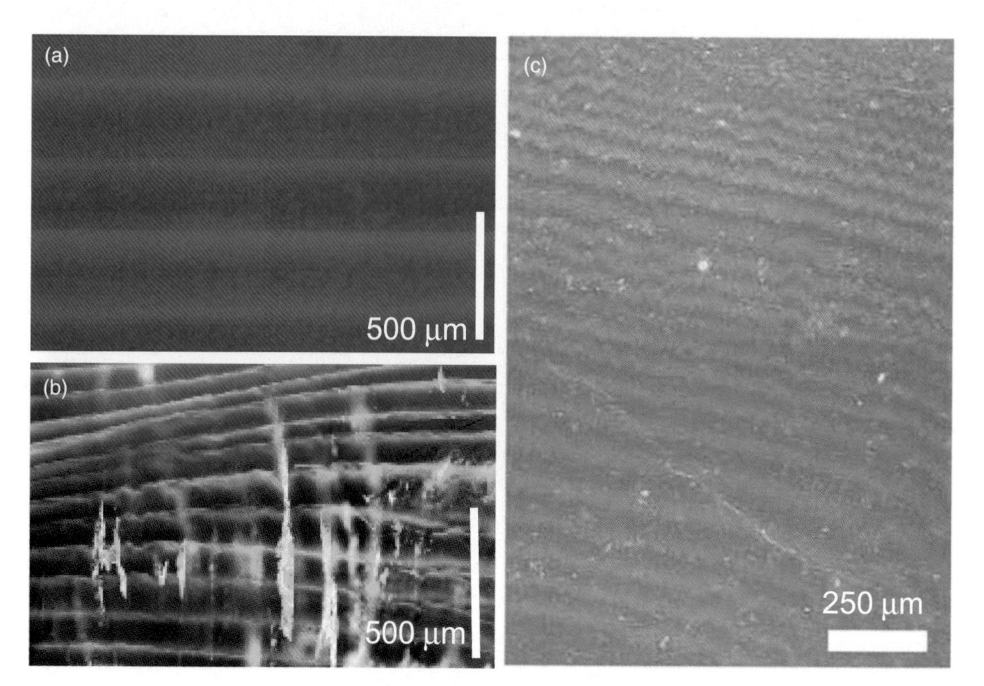

Plate 7.6 Examples of annual fluorescent laminae. (a) Shihua Cave, Beijing. (b) Stalagmite ER-77, Ernesto Cave, Italy (note fluorescence below sample surface causes some apparent peak broadening. (c) Flowstone WM1, dated using U–Pb to 2 Ma (Meyer et al. 2009); see also Fig. 12.6.

Plate 7.7 Stalagmites formed during times of glacial low sea level, now submerged by eustatic sea-level rise, Argentarola Cave, Italy. Right photograph illustrates cross-section showing a laminated stalagmite core encrusted with serpulid worms and cut by borings. Photographs courtesy of Fabrizio Antonioli.

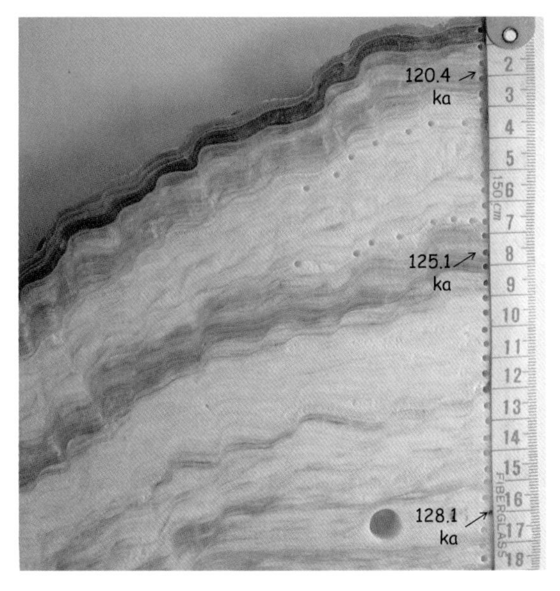

Plate 7.8 Flowstone dating from the last interglacial (isotope stage 5e) from Cascade Cave, New Zealand, illustrating characteristic rippled growth morphology.

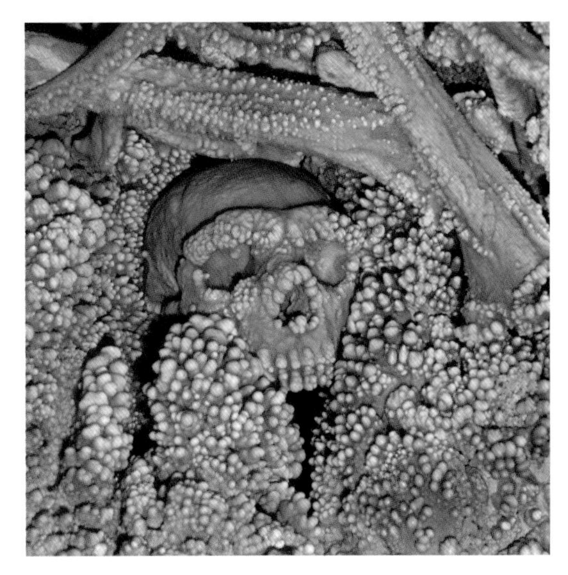

Plate 7.9 The Altamura skull, Italy: remains of an archaic Neanderthal covered in cave popcorn (splash deposits). The material has been left undisturbed *in situ* so its age is not known, but thought to be 150–250 ka (Vacca & Delfino, 2004).

Plate 7.10 Stalagmite backlit to highlight young growth phase of translucent calcite overlying older opaque stalagmite, Wombeyan Caves, New South Wales, Australia.

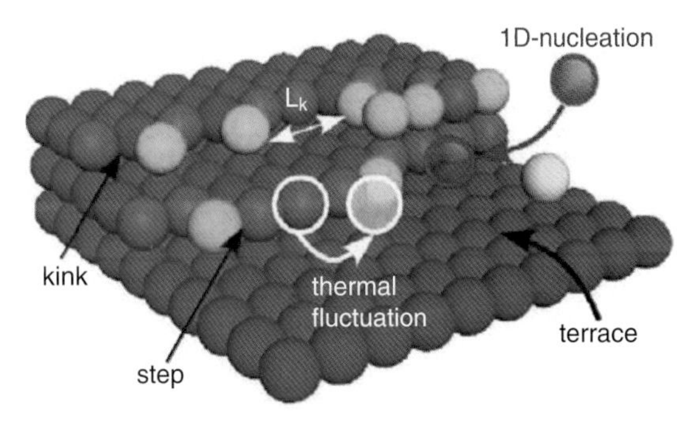

Plate 8.1 Terrace, step, and kinks on a crystal surface. Kinks are created either via movements of molecules on the step edge (thermal fluctuations) or attachment of new solute molecules from solution (one-dimensional nucleation). The step advances because addition of molecules to right- and left-facing kinks leads to their lateral movement and eventual annihilation. L_k is the kink spacing. The different colours show the different degrees of bonding that are satisfied from gold (none) through yellow, green, pale blue and dark blue to purple (complete). From de Yoreo et al. (2009).

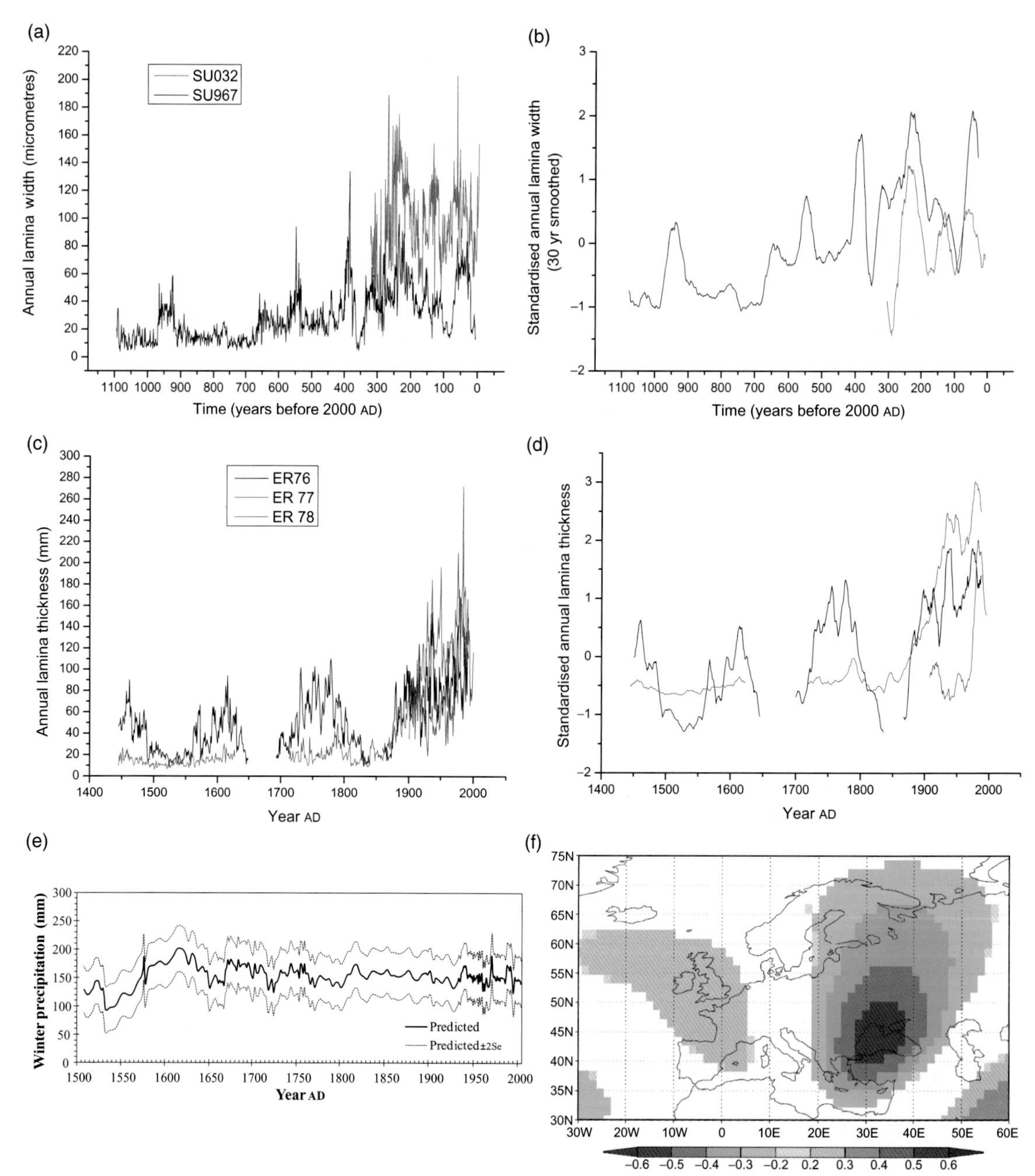

Plate 11.1 Reconstructions. (a) Annual lamina thickness series from northwest Scotland. SU967, SU961 and SU962 published in Proctor et al. (2000, 2002) and SU032 in Fuller (2006 and Baker et al. (2008a). SU967 and SU032 have correlations with decadal average total annual precipitation / mean annual temperature of 0.79 and 0.60 respectively. (b) (derived from (a)) Standardized (Z) 30-year smoothed series of northwest Scotland lamina thickness series. (c) Annual lamina thickness series from Ernesto, Italy. ER76, ER77 and ER78 published in Frisia et al. (2003). (d) Decadal smoothed and standardized Ernesto annual lamina thickness series. (e) Reconstructed late autumn–winter precipitation in northern Turkey based on annually resolved stalagmite $\delta^{18}O$ (Jex et al., 2011). Lines illustrate an uncertainty of ±40 mm (2 standard errors of the regression). f. Late autumn–winter precipitation at the cave site versus sea level pressure field (HADSLP, AD 1932–2004). Correlation map shows regions where $p < 1\%$; data were obtained from, and the maps drawn in, KNMI Climate Explorer.

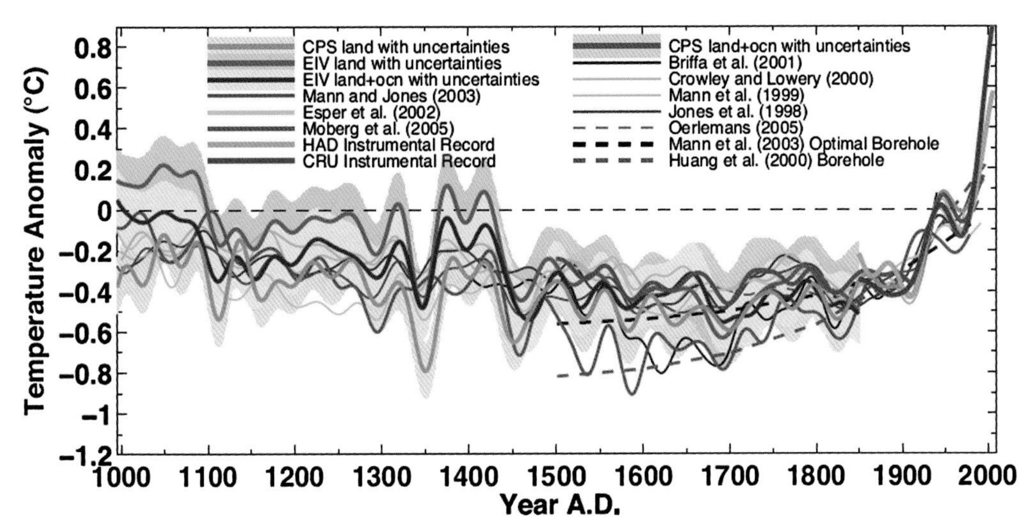

Plate 11.2 Comparison of multiproxy reconstructions for the past Millennium (from Mann et al., 2008). Reconstructions are shown as 'composite plus scale' (CPS) and 'total least squares' (error in variables, EIV) for Northern Hemisphere (NH) land, and land plus ocean, temperature reconstructions and estimated 95% confidence intervals. Shown for comparison are published Northern Hemisphere reconstructions (see references), centred to have the same mean as the overlapping segment of the Climate Research Unit (CRU) instrumental temperature record 1850–2006 that, with the exception of the borehole-based reconstructions, have been scaled to have the same decadal variance as the CRU series during the overlap interval. All series have been smoothed with a 40-year low-pass filter. Confidence intervals have been reduced to account for smoothing. Speleothem series feature in Moberg et al. (2005) and Mann et al. (2008).

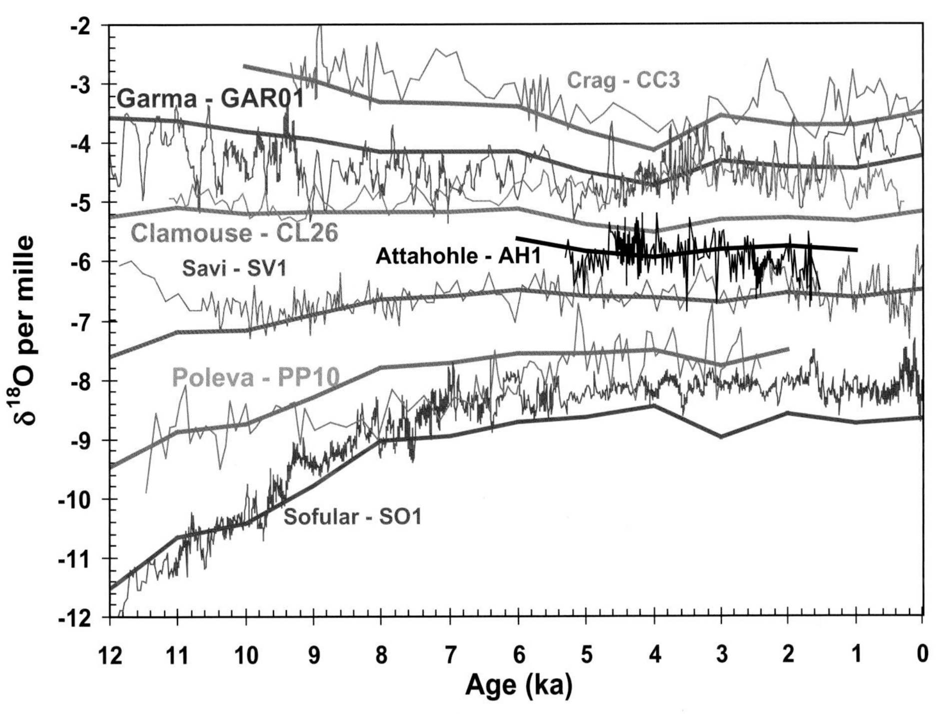

Plate 11.3 Comparison of European Holocene low-frequency stalagmite $\delta^{18}O$ trends. Selected stalagmite $\delta^{18}O$ series are shown on a gradient from west to east from top to bottom compared with $\delta^{18}O$ values predicted from regression models. From McDermott et al. (2011).

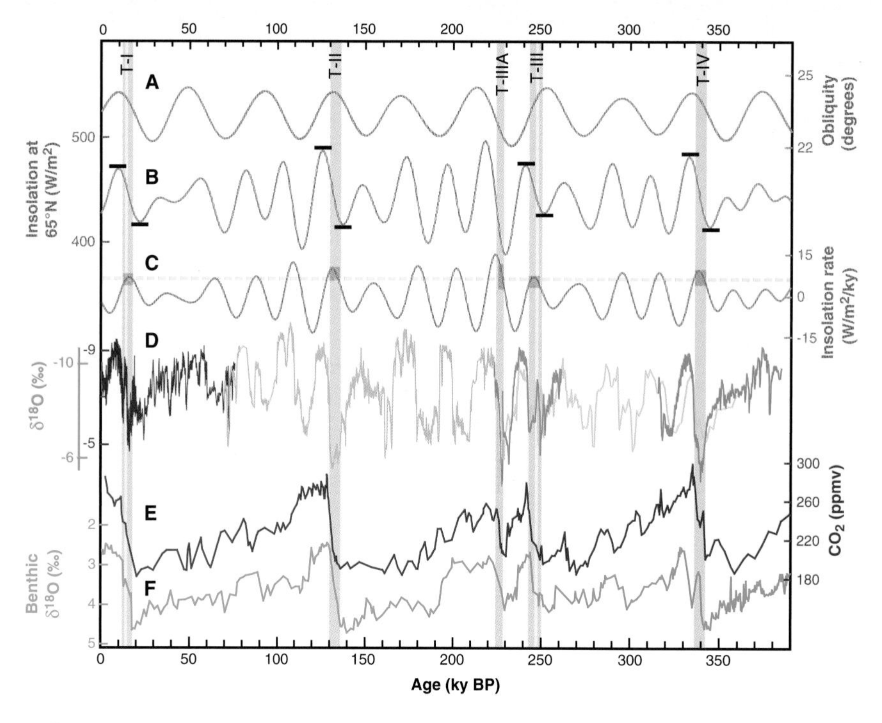

Plate 12.1 Comparison of Chinese stalagmite $\delta^{18}O$ records (D, from Wang et al., 2001; Dykoski et al., 2005; Wang et al., 2008; Cheng et al., 2009b) to orbital forcing (A—obliquity, B—July insolation, C—rate of change of July insolation at 65° N), Vostok ice core CO_2 and benthic $\delta^{18}O$ from ODP 980. The last two records are tied to the speleothem $\delta^{18}O$ at the glacial terminations. Diagram from Cheng et al. (2009b).

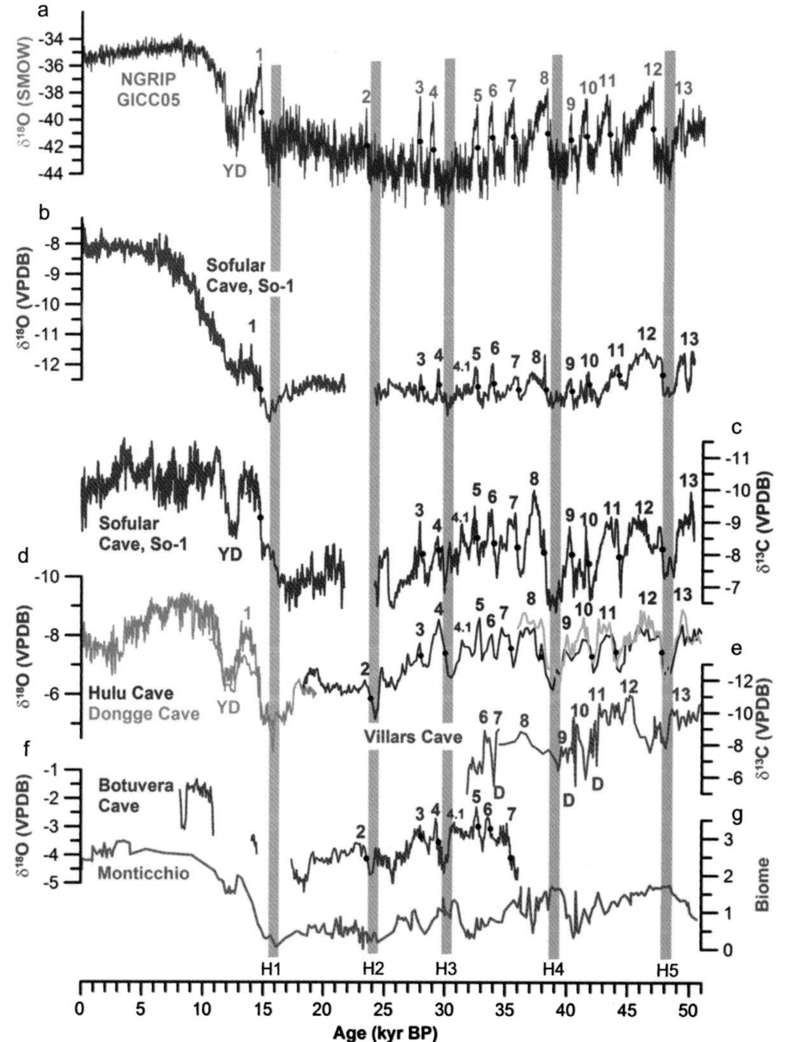

Plate 12.2 Adapted from Fleitmann et al. (2009). (a) NGRIP $\delta^{18}O$-profile from Greenland (Svensson et al., 2008). (b), (c) The $\delta^{18}O$ and $\delta^{13}C$ time series of stalagmite So-1 from Sofular Cave. (d) Hulu and Dongge caves records from China (Wang et al., 2001; Dykoski et al., 2005). (e) Villars Cave $\delta^{13}C$ record, southwestern France (Genty et al., 2003); D indicates discontinuity. (f) Botuvera Cave $\delta^{18}O$ record, Brazil (Wang et al., 2006). (g) Pollen record from Lago Grande di Monticchio, southern Italy (Allen et al., 1999). Numbers denote Greenland interstadials. Grey shaded bars denote Heinrich (H) events 1–5.

CHAPTER 6
Biogeochemistry of karstic environments

6.1 Introduction

An understanding of the biogeochemistry of karstic environments is essential for the full appreciation of the palaeoenvironmental record contained in speleothems. The historical focus on climate proxies such as oxygen and carbon isotopes, primarily owing to the ease of analysis of their elemental composition, has lead to an emphasis on the inorganic aspects of speleothem palaeoenvironmental analysis in the literature. Historically, organic material, although recognized as a minor component of speleothems, was mainly of interest as a potential contributor to the carbon isotope record (see section 5.4) and as a source of speleothem colour (Hill & Forti, 1997). However, the biogeochemistry of karst environments has several important effects on transfer processes. Firstly, *primary* organic material may be sourced from the overlying soil and vegetation, from within the aquifer and from within the cave. Secondly, these biological materials may be degraded or biologically processed during the transfer process to form *secondary* materials. The rate and extent of degradation will be affected by a range of processes that include, for example, climate, air movement and rate of water movement as introduced in Chapter 2. Finally, both the primary and secondary organic materials may interact with inorganic elements and compounds (as introduced in Chapters 3 and 5).

The earliest speleothem biogeochemical research focussed on fluorescent organic matter, primarily owing to its ease of analysis. Researchers in the 1950s observed that stalagmites appeared more 'coloured' in photographs, owing to the effect of ultraviolet-excitation of organic matter, thanks to the presence of ultraviolet light in traditional flash guns. Early work by White in the 1970s identified the spectrum of emitted fluorescence, and by the 1990s variations in speleothem fluorescence intensity and wavelength were being reported and interpreted as palaeoenvironmental proxies. Interestingly, this work was undertaken in isolation from similar studies investigating organic matter fluorescence in marine (Coble, 1996), freshwater (McKnight et al., 2001; Baker, 2001) and engineered systems (Reynolds & Ahmad, 1997). Around the same time that the optical properties of speleothems were being investigated, pioneering researchers such as Brook and Bastin were extracting pollen grains from speleothems. A large body of work exists, but has not been widely used by the wider palaeoclimate community: partly owing to a large amount of work published in non-English language journals, and partly owing to a lack of understanding of the pollen transfer process. Most recently, molecular biomarkers such as lipids and amino acids have been analysed in speleothems. Compared with other fields of palaeoenvironmental analysis, speleothem research in this area is

Speleothem Science: From Process to Past Environments, First Edition. Ian J. Fairchild, Andy Baker.

nascent but now rapidly developing owing to improvements in technology allowing smaller sample sizes and improved detection limits. We view the analysis of organic molecular biomarkers as a major research gap in speleothem palaeoenvironmental analysis (see section 6.2).

In this chapter, the biogeochemistry of karstic environments will be considered by the source of potential macromolecular or molecular biomarkers. For each source zone, the potential for degradation or transformation of the biomarker is considered. An overview of the sources and processes is given in Fig. 6.1. Pollen, spores, lipids, lignin, fluorescence organic matter, amino acid and genetic material, sourced from vegetation and/or fauna are discussed in turn. Transformations of these materials during groundwater transport and in the cave environment is discussed. Interactions between the organic and inorganic realms are discussed in Boxes 6.1 and 6.2, where the role of storage and cycling of sulphur and the interactions between dissolved organic matter and trace elements and colloids are discussed. The chapter concludes with a review of research gaps (section 6.4). Figure 6.2 gives an example of how the preservation of gastropod fossils within a stalagmite sample can be understood when taken in the wider context of the cave environment.

6.2 Organic macromolecules

6.2.1 Fluorescent organic matter

The use of speleothem organic matter as an environmental proxy has been dominated by the observation that speleothems fluoresce. The observation that cave speleothems are fluorescent was first made by cavers experimenting with underground flash-photography, who noticed an 'after-glow' effect (O'Brien, 1956). This fluorescence is generated by the dissolved organic matter preserved in the speleothems, and which has derived from the overlying soil (White & Brennan, 1989). Organic matter fluorescence relies upon the presence of a loosely held electron in the outer orbit of a molecule: this situation occurs in humic and fulvic substances (see Box 6.1) which have aromatic carbon ring structures which allow fluorescence (Senesi et al., 1989, 1991). When the electron is excited to a higher energy level by the absorption of energy (e.g. a photon), fluorescence occurs when energy is lost as light, as the electron returns to its original energy level (ground state). There are three measurable parameters: the fluorescence intensity, which should reflect the amount of organic matter present per unit area of illumination; and the excitation and emission wavelengths, which are specific to the molecule, and therefore can be used to differentiate organic matter source or structure. Fluorescence spectrophotometry has been widely applied to speleothems at a variety of timescales from annual (e.g. Baker et al., 1993) to multidecadal (e.g. Proctor et al., 2000). Modern calibration studies have been widely undertaken, all of which confirm a seasonal flux of organic matter that is related to groundwater recharge from the soil zone and stalagmite dripwaters (e.g. Baker & Genty, 1999; van Beynen et al., 2000). For a detailed review of speleothem organic matter fluorescence, refer to McGarry & Baker (2000) and Blyth et al. (2008), and for a general review of aquatic organic matter fluorescence refer to Baker et al. (2012).

Three techniques are typically used to observe organic matter fluorescence. *Lasers* have the advantage of high energy, which can excite a greater number of fluorescent molecules, and high resolution depending on laser spot size. However, the laser also excites fluorescence within a stalagmite sample, not just the surface, and is limited to a fixed wavelength of excitation which might not match the optimal wavelength absorbed by organic molecules. Shopov et al. (2004) used thick slabs of speleothem and measured the transmitted fluorescence intensity through the slab; Baker et al. (1996) measured fluorescence intensity emitted from the speleothem surfaces; Perette et al. (2005) used a laser and measure emitted fluorescence at two wavelengths to measure both intensity and wavelength of fluorescence. *Spectrophometers* equipped with fibre optic probes use a xenon light source and diffraction gratings to allow multiple excitation energies to be applied to samples, but have a lower power than lasers and just excite the near surface.

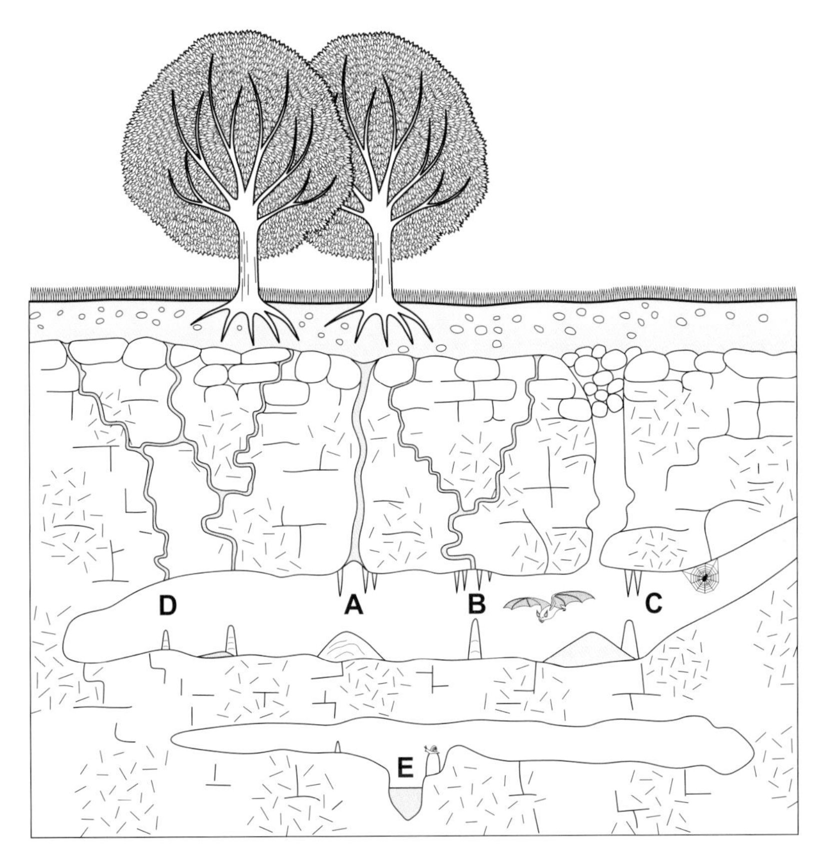

Fig. 6.1 Sources, pathways and sinks of speleothem organic material. *The surface* provides a unique source of vegetation-derived lignin macromolecules. Lipid macromolecules may be derived from vegetation, animal, or soil microbial sources. Soil microbes provide a potential source of microbial ribosomal DNA (rDNA) and the soil is the source of dissolved organic matter. Plants generate pollen and spores; spores are also generated by fungi. Degradation and damage to organic materials will occur depending on various inter-related factors such as soil residence time, soil type, moisture and temperature, vegetation type and microbial activity. *In the aquifer*, organic matter is transported to the speleothems. Potential losses of organic materials occur in the aquifer owing to adsorption, prior speleothem precipitation, degradation, and damage. Potential gain of lipid and rDNA macromolecules may occur from aquifer biofilms. *In the cave*, stalagmites form from the degassing of ground waters. Rapid preservation of organic materials into stalagmites is expected. In the cave, there is a potential addition source of lipid and rDNA macromolecules from cave microbes, fungi and animals. The relative importance of faunal sources can be hypothesized to decrease from entrance (C) to deep cave (D). Location E illustrates the exchange of organisms along a streamway (see also Fig. 6.2).

Resolution also depends on the size and focussing of the fibre optic. Spectrophotometers have the advantage of measuring both intensity and wavelength of fluorescence and can be purchased 'off the shelf'. Finally, *microscopes* can be used to image fluorescence when equipped with a *mercury source* (e.g. Baker et al., 1993). Observed fluorescence is dependent on filter sets used (an ultraviolet or violet excitation filter and emitted fluorescence detected in the visible) to pass both excitation light and emitted fluorescence. A vastly under-used imaging technique in speleothem science is the use of *confocal optical microscopy*, which allows the selection of fluorescence from one point at a specified

Box 6.1 Organic macromolecules in speleothems

A wide variety of organic macromolecules can potentially be detected in speleothems, but three groups are becoming routinely analysed. A brief overview of humic substances, lipids and lignins is provided here.

Humic substances (HS) are a major component of natural organic matter, with the literature typically quoting a figure of around 50% of all dissolved organic matter comprising HS. HS chemistry is highly variable, comprising heterogeneous mixtures of polydisperse materials (having a wide range of size, shape and mass characteristics) formed by biochemical and chemical reactions during the decay and transformation of plant and microbial remains (humification). Therefore HS comprise plant lignin and its degradation products (see Fig. 6.4), as well as polysaccharides, melanin, cutin, proteins, lipids and nucleic acids, etc.). HS have traditionally been subdivided into humic and fulvic acids and humin, based on their solubility, and in karst environments the predominant HS would therefore be classified as fulvic acids (soluble at pH >2). Figure 6.B1 shows a model fulvic acid monomer, which in the aquatic environment would be expected to aggregate into large structures depending on concentration, ionic strength, pH etc. It is these HS which are observed using fluorescence (section 6.2.1).

Lipids. Lipids are formally defined as a substance that dissolves in alcohol but not in water. Depending on their chemistry, they may be described as being hydrophobic (water hating molecules) and lipophilic (fat-loving), that is, apolar molecules with the inability to form hydrogen bonds with other compounds. Some compounds, such as cholesterol (Fig. 6.B1) are amphiphilic, containing both water-loving and fat-loving properties. It is these proper-

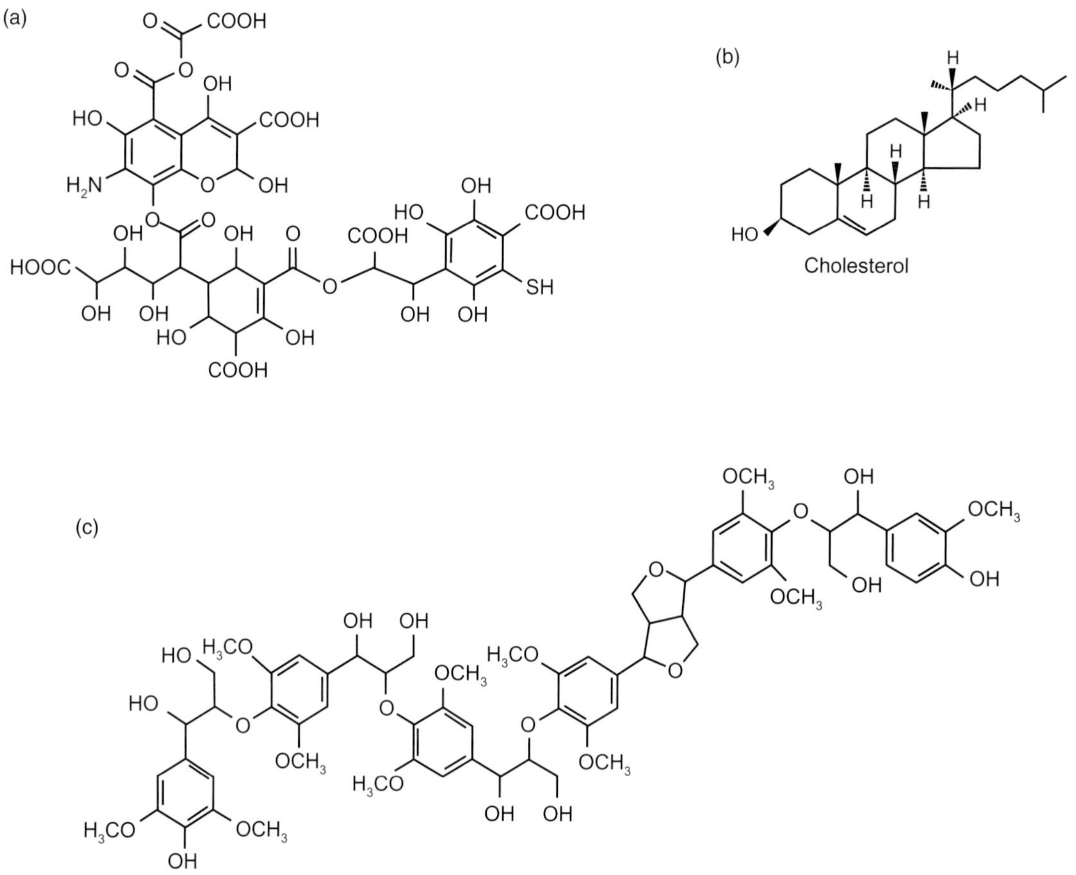

(a)

(b)

Cholesterol

(c)

Fig. 6.B1 (a) An example fulvic-acid monomer, from Alvarez-Puebla et al. (2006). (b) Cholesterol, an example of an amphiphilic lipid biomarker. (c) A putative primary sequence structure of a lignin fragment (Davin & Lewis, 2005).

ties which allow their separation by chromatography. Lipids are sourced from a wide range of materials including bacterial cell membranes but with subtle variations in the length of carbon chains (e.g. a C_{18}-alkane is an 18 carbon chain alkane) and position of functional groups. Empirical investigation of modern and fossil lipid extracts of soils and sediments has enabled the interpretation of differences in relative abundance, chain length and function group position in terms of environmental change (section 6.2.2).

Lignin. Lignin is a polymer that is produced by all vascular plants. It is most commonly derived from wood and is also an integral part of the cell walls of plants. It is the most abundant organic polymer after cellulose, constituting from a quarter to a third of the dry mass of wood, and is notably heterogeneous. Lignin is one of the most slowly decomposing components of dead vegetation, and therefore contributes significantly to soil humus. Geochemically, lignin is thought to be a large, relatively hydrophobic and aromatic macromolecule. A putative chemical structure of a lignin fragment is presented in Fig. 6.B1. Pyrolysis of lignin yields lignin phenols, which have been used in other disciplines to obtain records of vegetation structure. Initial results on speleothems are presented in section 6.2.2.

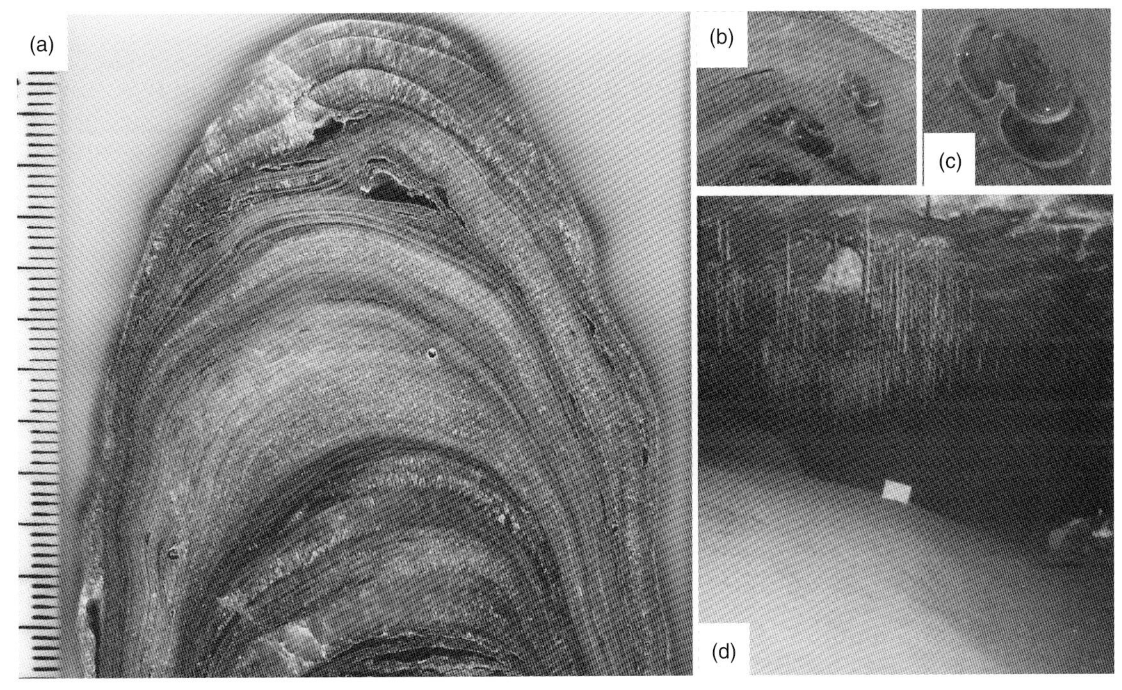

Fig. 6.2 (a) Stalagmite (Crag Cave, Ireland) showing multiple growth laminae. Lensing black area are voids where mud layers washed out during sample polishing; dark layers are calcite with mud impurities whereas layers with radiating crystals are pure calcite. Gastropod fossils are frequently found within the sample. (b) Detail of reverse of slab showing mud lenses and gastropod which is enlarged in (c). (d) Context of sample showing sloping mud bank on flank of streamway (stalagmite grew on opposite mudbank and was fed by a soda straw like those visible in the photograph).

depth in a sample, therefore increasing the contrast in the image. It is ideal for speleothems where the imaged point might be within a thick sample, or surrounded by other bright points (Webb, 1996). Confocal microscopy typically has a laser light source and operates in scanning mode to produce high-resolution two- or three-dimensional images (for example, Menéndez et al., 2001). Although extensively used in biomedical science, only few speleothem studies have been published so far (e.g.

Ribes et al., 2000; Orland et al., 2009; Dasgupta et al.; 2010); see also Figs. 8.2 and 12.1.

Stalagmite organic matter fluorescence is dominated by excitation wavelengths between 300 and 420 nm and emission between 400 and 480 nm. Fluorescence in this region has been shown in modern marine, soil, river and groundwater samples to be derived from 'fulvic-like' material. In the fluorescence literature this is often referred to as 'peak C' (Coble, 1996); surface and ground water studies have shown that the intensity of this fluorescence correlates with total dissolved carbon (Baker & Spencer, 2004; Hudson et al., 2007) and that variations in the emission of this fluorescence to longer wavelengths reflect an increase in molecular weight, hydrophobicity and/or aromaticity in the signal (Baker et al., 2008b). Figure 6.2 shows the relationship between emission wavelength and dissolved organic matter hydrophilicity, and metal and organic pollutant binding for a selection of surface waters (Baker et al., 2008b). No equivalent study has been undertaken on karstic groundwaters. However, the chemical function of dissolved organic matter is important as it determines both the extent to which it is transformed within the karst groundwater system as well as its interactions with trace elements and colloids (see Box 6.2). Less well understood are the extent of any transformations in dissolved organic matter (DOM) structure during transport from the soil to the speleothem; although limited observations of paired soil water and dripwater confirms that dripwater DOM fluorescence is of a lower emission wavelength than that observed in the soil (Baker & Genty, 1999), confirming the loss of the relatively hydrophobic fraction, probably owing to absorption to sediment and bedrock surfaces. In general, fluorescence characterization of dripwater indicates that what reaches the cave is of much lower concentration than observed in surface rivers and lakes, and which is relatively hydrophilic. DOM might also be both produced and degraded by biological action in the epikarst and groundwater: for example biofilms are likely to be present in the karstified zone, with microbial communities able to decompose organic matter and generate CO_2; the observation of 'ground air CO_2' was first made in the 1970s (Atkinson, 1977a).

There is a relatively poor understanding of the extent of DOM transformation during transport and incorporation within the stalagmite. For example, for the latter, one might expect a competition between a surface charge effect and a 'chromatographic' effect on the stalagmite cap, with both more hydrophobic material and appropriately charged material incorporating more rapidly into the stalagmite calcite. However, both modern process studies and stalagmite records confirm an empirical relationship between fluorescence wavelength and the organic matter composition. For example, it has been utilized to determine the extent of humification of overlying peat soils (Proctor et al., 2000), long-term changes in soil from peat to non-peat soils (Baker et al., 1998b), and to determine the relative proportions of surface and storage components feeding stalagmites (Asrat et al., 2007). Speleothem fluorescence intensity also provides a rapid and non-destructive approach for determining organic matter relative concentration. Fluorescence intensity variations, reflecting variations in concentration, have been shown to occur seasonally in stalagmites, with annual fluxes of fluorescent material occurring in some regions in winter owing to snowmelt (Linge et al., 2009a), in other regions during monsoon seasons (Tan et al., 2006), and in others owing to seasonal variations in hydrologically effective precipitation (Proctor et al., 2000). These can create annual lamination of the stalagmite, which can then be used to constrain dating through the counting of annual bands, and for palaeoclimatic purposes as the rate of annual growth can be determined and related to climate and environmental factors. A review of the processes forming annual fluorescence laminae can be found in Tan et al. (2006) and this is considered in more detail in section 7.4, with examples of climate reconstructions from stalagmites containing annual fluorescent laminae in sections 10.3 and 11.2. Interestingly, direct measurement of organic carbon concentration in stalagmites has rarely been undertaken, despite its relative ease. If the growth rate is known, the total organic carbon per gram calcite can be converted to a flux, and in general the information is essential as a screening method

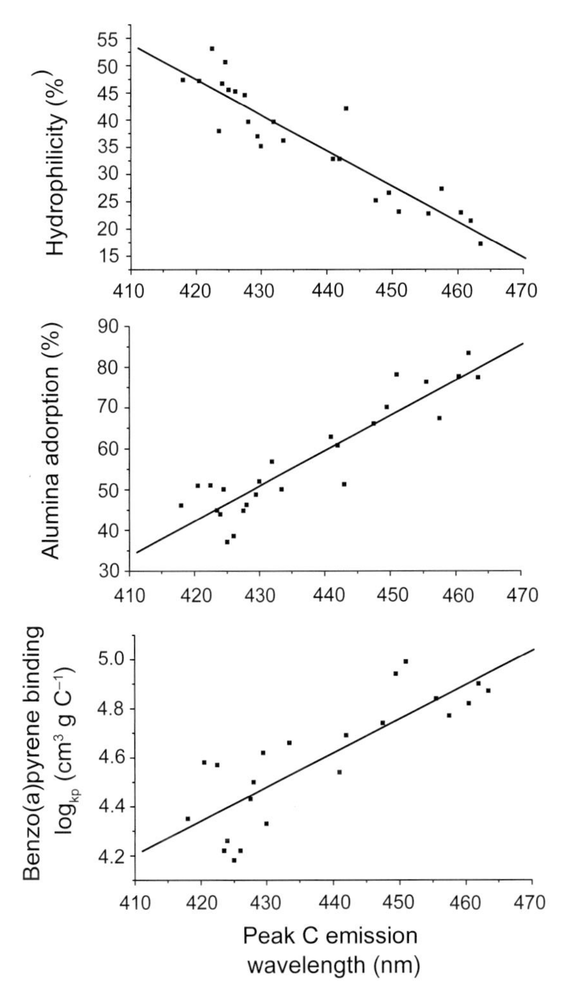

Fig. 6.3 Relationship between the emission wavelength of fluorescence and chemical function for a selection of UK surface waters (Baker et al., 2008b). For comparison, cave dripwaters are typically in the range 400–420 nm, indicating highly hydrophilic material with relatively low metal and organic binding capacity.

before undertaking biomarker analyses. These are considered in the next section.

6.2.2 Lipid and lignin macromolecules

Lipid macromolecules are biologically derived fatty molecules such as fatty acids, alcohols and sterols. They are common across all environments, but different compounds may be specific to particular parts of the ecosystem (e.g. vegetation, bacteria, fungi). Accordingly, by measuring the relative quantities of lipids present in an environmental record, it is possible to identify how the contributions of these different parts of the ecosystem have changed through time. The use of lipid biomarkers in climatic and environmental research in marine sediments has become well established over the past 20 years (Brassell et al., 1986). More recently, terrestrial records (e.g. lake sediments, peat and soils) have also been studied, with success in characterizing changes in vegetation and climate (for example, Jacob et al., 2007). Despite a long history of research in other palaeoenvironmental archives, biomarkers in general have been rarely analysed in speleothems. In part this is due to the large sample size needed with early instrumentation, but also it is due to the general lack of consideration of the organic component in speleothems.

The first lipid macromolecular analyses on speleothems were those of Rousseau et al. (1992, 1995), who showed that lipid biomarkers are preserved unaltered in speleothems at a 100,000 yr timescale, and Cox et al. (1989) in the analysis of a stromatolitic stalagmite from southeast Australia. Following these pioneering studies, no research was attempted until the studies of Xie et al. (2003) and Blyth (Blyth et al., 2007, 2008; Huang et al., 2008; Blyth & Frisia, 2008). Lipids have been the most frequently investigated (examples are shown in Fig. 6.B1), and challenges to the analysis and interpretation of lipids biomarkers in speleothems are many. Firstly, speleothems need to have an organic matter concentration that is high enough to allow detection of lipids: work to date has primarily focused on material from either warm climates where organic matter productivity in the soil is high (such as the Chinese material analysed by Xie et al. (2003) and Huang et al. (2008) and an Ethiopian stalagmite by Blyth et al. (2008)), or cool temperate climates (Rushdi et al., 2011), often with thick overlying peat soil cover (Blyth et al., 2011). Interpretation of the biomarker signal is also complex. Biomarker data reported in speleothems have to date focussed on compounds such as high-molecular weight (HMW) *n*-alkanes: straight-chain hydrocarbons with a chain length of 25 or more carbon atoms. These are frequently derived from higher plant waxes, a source indicated by an *n*-alkane distribution with a strong odd-over-even carbon number predominance (Bray & Evans, 1961; Eglinton & Hamilton, 1967). Research has shown that different carbon chain lengths are dominant in different plant types, providing potential for identifying vegetation change from the molecular signature preserved (Rieley et al., 1991; Marseille et al., 1999; Pancost et al., 2002). HMW *n*-alkanols and HMW *n*-fatty acids also relate to the type of vegetation present and the amount of plant-derived input to the soil (Bull et al., 2000; Wiesenberg et al., 2004). However, these studies all relate to non-karstic environments (for exceptions see Cui et al., 2010 and X. Li et al., 2011), and although some lipids are uniquely derived from surface plant matter, others have a mixed signal and may be sourced from microbes, plants or animals in the soil, aquifer or cave. Observation of one cave dripwater over 2 years at Heshang Cave, China, tentatively indicates the presence an additional temperature-dependent signal in the ratio of monosaturated to unsaturated low molecular weight lipids (X. Li et al., 2011). Finally, the soil- and vegetation-derived lipid biomarker signal is likely to undergo transformations within the groundwa-

Box 6.2 Colloids and gels: interactions between organic matter and inorganic stalagmite proxies (lead author Adam Hartland)

Organic matter, particularly the dissolved organic matter pool, can both chemically and physically interact with inorganic ions at all stages of the pathway from soil through to stalagmite. Our understanding of the importance of these interactions is relatively immature. In particular, physical interactions between organic and inorganic matter are very poorly studied. As well as the process of mineralization, organic matter can form polymers with inorganic positively charged metals (Verdugo et al., 2008). Studies in the laboratory with ocean and lakes waters demonstrated that these gels can self-assemble to form self-assembly microgels (SAG) (SAG, Fig. 6.B2), macromolecules formed with solvated porous networks interconnected by chemical or physical cross-links. Ca^{2+} in particular helps promote polymer assembly, which can comprise a mixture of organic matter, proteins, polysaccharides, DNA material, etc. held together by Ca^{2+} crosslinks and tangles (Verdugo et al., 2008). No experimental observations of polymer and gel properties have yet been made on cave waters, and this is a gap in the research that requires addressing. Observations based on ocean and alkaline lake water (de Vicente et al., 2009, 2010; Ortega-Retuerta et al., 2009, 2010) suggest that at the low organic matter concentrations found in natural environments, physical processes such as gel formation dominate over chemical reactions.

Recent applications of specialized colloidal techniques (e.g. atomic force microscopy and transmission electron microscopy X-ray energy dispersive spectroscopy (TEM-X-EDS)) have begun to expand our understanding of the conformation and composition of organic matter in a range of sizes in karst dripwaters (Fig. 6.B3). As in other freshwater systems, natural organic matter (NOM) in dripwaters encompasses the particulate (>1 μm), coarse colloidal (1 μm – 100 nm), fine colloidal/nanoparticulate (100 nm – 1 nm) and nominally dissolved (<1 nm) phases (Hartland et al., 2010a, b). As expected, organic entities above the

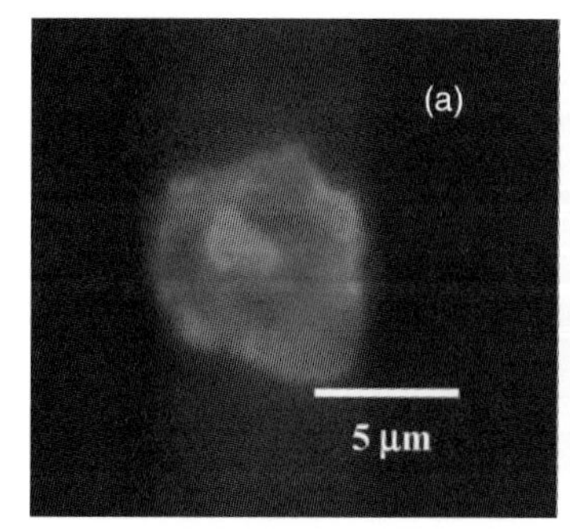

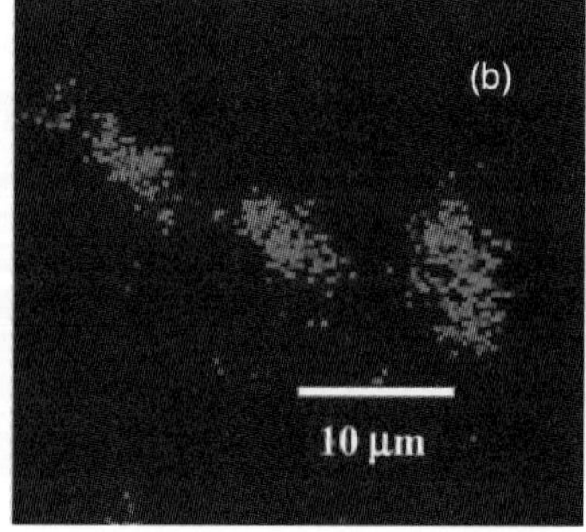

Fig. 6.B2 (a) Ca^{2+} bound to SAG shown by fluorescence staining. (b) Microbes stained showing their presence in three SAG. From Verdugo et al. (2008).

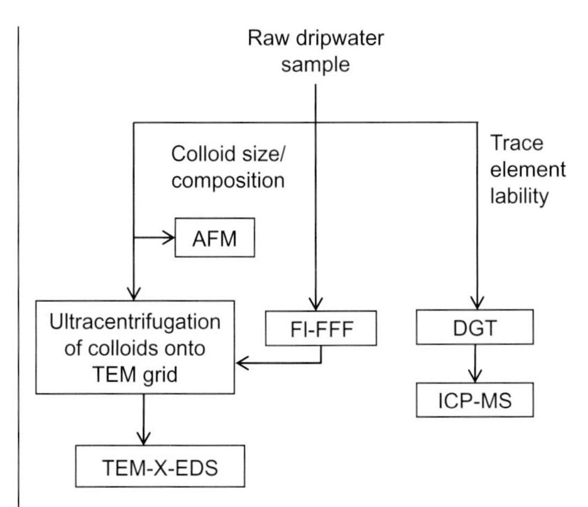

Fig. 6.B3 Multi-methodological approach to the characterization of colloid-trace metal complexes.

level of macromolecular nanoparticles (ca. 1 – 10 nm) are mainly composition of NOM and mineral phases (Hartland et al., 2010a), the formation of which may be enhanced by divalent cation bridging reactions (Hartland et al., 2010a). TEM-X-EDS analyses of dripwaters demonstrate that colloids and particulates possess heterogeneous compositions consistent with organo-mineral complexes of NOM and mineral colloids (e.g. aluminosilicates, Fe oxides) (Hartland et al., 2010a). As in other environments (Wells, 1998), colloidal aggregates may be important vectors for lipids, DNA material and other organic components. Preliminary data on the *n*-alkane content of freeze-dried dripwater samples indicates that lipids may be directly measured in dripwaters (A. Thompson, personal communication), leading the way to their physical speciation being determined.

Diffusive gradients in thin films (DGT) works by ion-exchange with Chelex resin embedded in a hydrogel which immobilizes free metal ions and prevents reformation of NOM-metal complexes in solution. Diffusion coefficients of metal ions in the hydrogel are similar to those in water but colloids are restricted in their diffusion. Variation of gel thicknesses and pore sizes enables properties such as colloid-size and dissociation kinetics to be examined. Fl-FFF separates colloids and particles based on their diffusion coefficients; smaller, more-diffusive particles pass into the highest velocity flow and are eluted first.

Natural aquatic colloids are important complexants of trace metals in freshwaters (Lead & Wilkinson, 2006), with the fine organic fraction (<10 nm) being the most ubiquitous (Buffle et al., 1998) and important fraction for trace element binding (Tipping & Hurley, 1992). Size fractionation of NOM in hyperalkaline cave dripwater samples by flow-field flow fractionation (Fl-FFF), in conjunction with TEM, has identified an abundant class of fluorescent, globular nanoparticles with hydrodynamic diameters between 1 and 10 nm, consistent with humic and fulvic macromolecules (Hartland et al., 2011). Furthermore, in-situ trace element speciation by the DGT technique has been applied in the same dripwaters and has demonstrated that a suite of surface-reactive trace metals (e.g. Cu, Ni, Co) are coordinated with functional groups in both colloidal and nominally dissolved NOM. These metals exhibit varying tendencies to dissociate from complexes with NOM in the studied samples, with important implications for the understanding of trace metal composition of speleothems (Hartland et al., 2011).

Traditional conceptualizations of trace element partitioning between solution and speleothem are generally based on the assumption that metal ions in solution are 'free' (i.e. not bound in complexes) and partition based on the relative affinities for the $CaCO_3$ crystal lattice (Chapter 8). Although this is the case for elements which tend to form non-specific ion pairs in solution (e.g. Sr, Mg), the partitioning behaviour of metals which form complexes with colloids and DOM (e.g. Pb, Cu, Y) does not conform to predictions based on free metal ions at equivalent concentration (Fairchild et al., 2010). The insight that the availability of trace metals for incorporation in speleothems is restricted by kinetic limitations on dissociation (i.e. are complexed by colloids and DOM) opens up a new area of speleothem-based research. Experimental studies have demonstrated that macromolecular organics (such as fulvic acids) readily adsorb to the calcite surface (Lee et al., 2005), and in speleothems variations in the delivery of organic species gives rise to annual, to seasonal variations in fluorescence attributes (Baker et al., 1993, 1999). Variations in a suite of trace elements in dripwaters and speleothems are also linked to the abundance of fluorescent organic entities (Fairchild et al., 2001; Borsato et al., 2007). Deviations in the ratio of OC to trace metal between solution and speleothem reflect the interaction between the lability of the trace-metal–NOM complex, the stability of the $CaCO_3$–NOM surface complex, and the relative affinity for each metal competing for available sites in the $CaCO_3$ crystal lattice (Hartland, 2011). This process is likely to be subverted by the effects of surface-adsorbed NOM, i.e. through adsorption of metal ions in defect sites. However, the crucial observation is that the abundance over time of surface-reactive metals in speleothems is linked to variations in colloid– and DOM–metal transport in dripwaters. Because colloids and particulates are hydrologically dynamic, variations in complexed metals may be linked to hydrologic processes, such as infiltration events (Jo et al., 2010; Hartland, 2011). Further work is needed to characterize colloid–metal interactions in a wide range of cave environments, but the initial data indicates that the transition metals are commonly complexed and transported by NOM in a range of sizes in cave waters, the behaviour of which contrasts markedly with that of the alkaline earth elements (Hartland, 2011).

ter and cave as lipids can degrade over time with oxidation or bacterial reworking in either the aquifer or the soil (van Bergen et al., 1998; Bull et al., 2000). Blyth et al. (2007) demonstrated a rapid response of lipid biomarkers preserved in an Ethiopian stalagmite to surface vegetation change, and equate this to a preferential preservation of lipids that have been rapidly transported to the sample via fissure flow routes. Finally, additional lipid sources might be present from biofilms in the overlying karst and within the cave, both from any *troglodytes* (cave-dwellers), as well as potential chromatographic fractionation on the stalagmite surface. Therefore the interpretation of any speleothem lipid signals is likely to be cave-specific.

Other biomarkers are likely to be present in speleothems and are yet to be thoroughly researched. For example, in non-karstic systems, lignin biomarkers have been used in situations where tracers are required that are ubiquitous, indicative of surface organic carbon and stable over long periods. Figure 6.4 presents some lignin phenol structures identified in the pyrolysis of a stalagmite from northwest Scotland (Blyth & Watson, 2009). Modern calibration studies using fresh plant materials, forest floor litter, soils, and aquatic sediments have demonstrated that lignin phenols are indicative of certain plant taxa and tissue types (Hedges & Mann 1979). In particular, vascular plants can be distinguished from non-vascular plants because only vascular plants produce vanillyl phenols. In modern soils, sediments and river systems, the syringyl to vaninyl (S:V) ratio has been shown to increase with the proportion of angiosperm woody material, and the cinnamyl to vaninyl (C:V) ratio to the proportion of non-woody angiosperms. In addition, the acid-to-aldehyde ratio of vanillyl phenols may indicate the extent of microbial degradation of organic matter (Hedges et al., 1988). These properties have been used to determine the amounts, types and diagenetic histories of vascular-plant tissues in several biogeochemical analyses of sediments and soils. Only recently (Blyth & Watson, 2009; Blyth et al., 2010) have lignin macromolecules been used as a macromolecular biomarker in speleothems, and this can be seen as another area of future research potential.

6.2.3 Ribosomal DNA

Direct information on both soil and cave microbial communities can be determined through molecular methods of microbial community analysis, in particular the analysis of rDNA. In environmental samples, microbial community structure, such as bacterial and archaeal diversity and species richness, has been determined from amplified 16S rDNA genes extracted from environmental samples, amplified and cloned into libraries for sequencing. This has recently been applied to a wide range of environmental samples including streams, ground and marine waters, thermal springs, sediments and soils (Willerslev et al., 2003). Ancient biomolecules have been extracted from permafrost, cave sediments and deep ice cores, demonstrating that preservation is possible for at least 100,000 years, and issues relating to possible DNA contamination with modern material are now firmly understood and controlled (Binladen et al., 2007). However, despite reviews of the potential of molecular techniques to subterranean biogeography, and active research into geomicrobiology (microbe–mineral interactions in caves), there are few previous studies of speleothem microbial community genomics. The typical focus of these studies is on sulphur systems or the culture of microbes found on modern speleothems to investigate their use in bacterially induced carbonate mineralization to restore carbonate historic buildings (Porter, 2007; Barton et al., 2001; Cacchio et al., 2004). The ubiquitous presence of other soil-derived biomarkers in speleothems, plus the successful analysis of ancient DNA from cave sediment, leads one to predict that sufficient DNA is present to be able to extract and amplify this genetic material. In speleothems, one would anticipate that 16S rDNA would reveal information about both community structure as well as the relative importance of soil- and cave-derived microbes. Swabs of speleothem surfaces are starting to yield intriguing results. Swabs taken from two adjacent speleothems in Kartchner Caverns revealed that the microbial communities were similar in structure, but that both bacterial and archaeal communities are speleothem specific, and that they contained numerous phylotypes which have no close relationship to known organ-

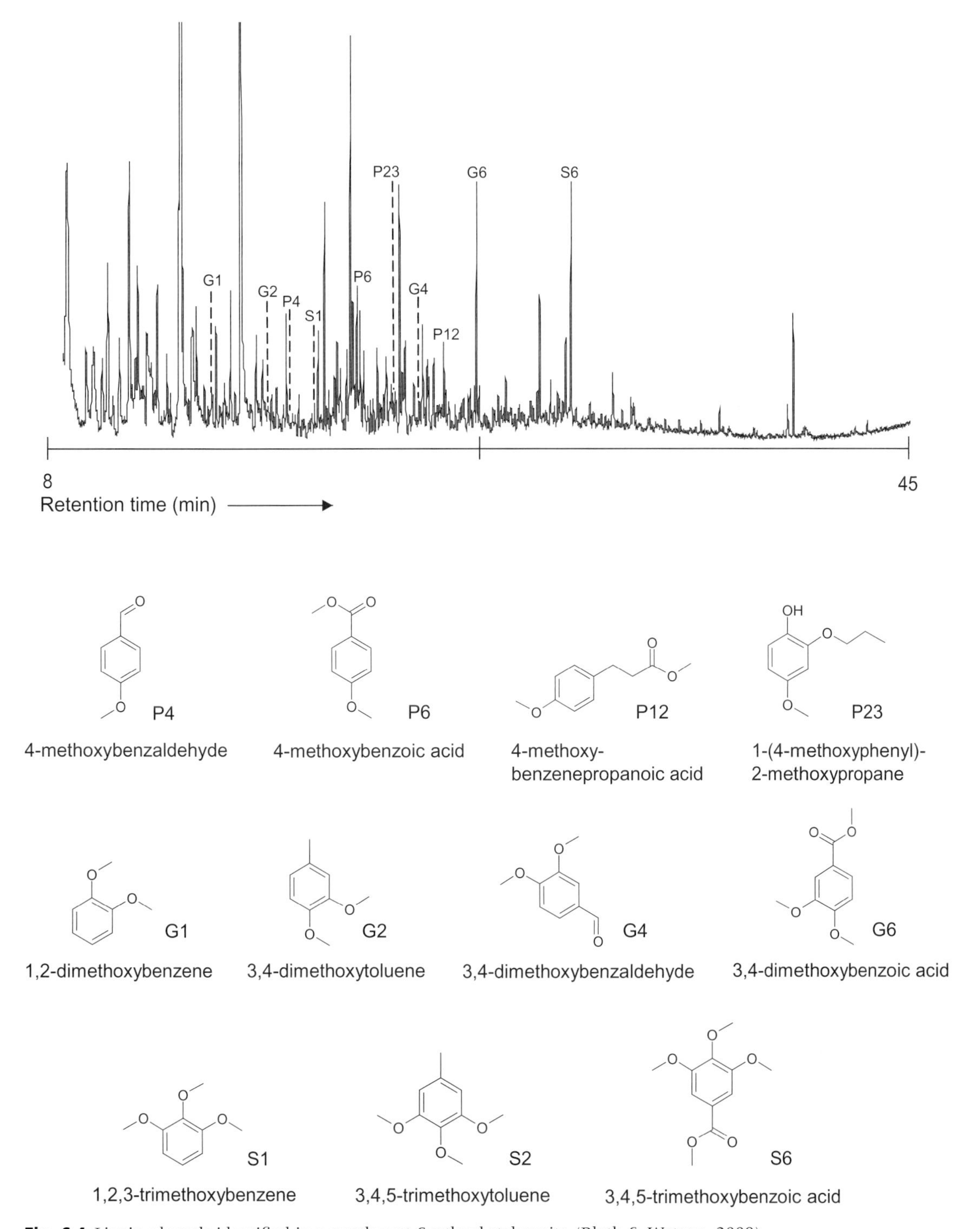

Fig. 6.4 Lignin phenols identified in a northwest Scotland stalagmite (Blyth & Watson, 2009).

isms (Legatzki et al., 2011). Interestingly, in the same caverns, Vaughan et al. (2011) identified 21 genera of fungi based on 18S rDNA analysis of swabs taken from speleothem surfaces. Wang et al. (2008) presented a short abstract with details of the first stalagmite 16S rDNA results from a Holocene stalagmite from Heshang Cave, China, demonstrating the validity of this approach. Banks et al. (2010) isolated 51 culturable bacteria from coralloid speleothems (cave popcorn) in a Kentucky cave in an investigation of the role of microbes in calcite precipitation.

6.3 Pollen and spores

6.3.1 Pollen

Speleothem pollen derives from both air-blown and water-transported pollen grains and typically shows a taphonomic bias towards species that grow close to the cave entrances, although pollen transported a long distance can also be observed; a full review of pollen transport and preservation in speleothems can be found in McGarry & Caseldine (2004). In summary, pollen is of mixed preservation, depending on its source and therefore the speed of transport and preservation. Preservation state can be a good indicator of pollen source: corroded pollen is indicative of poor preservation in the overlying soil, and is indicative of a dripwater source and potential reworking on the soil. However, most pollen observed in stalagmites is of excellent preservation, suggesting very rapid transport from the plant to the stalagmite: such preservation is strongly indicative of a wind-blown transport mechanism and rapid calcification into the speleothem.

Pollen in speleothems can provide a direct measurement of vegetation cover. However, concentrations of pollen in speleothems are typically one grain every 1–10 g, preventing high resolution analysis except in fast growing samples or material such as flowstones, which allow sampling of a large lateral section. To achieve a statistically significant sample of pollen, 200 grains would normally be identified and hence samples sizes need to be of the order of hundreds of grams. Therefore, even with fast growing samples, these sample sizes result in significant time averaging of the pollen record, typically of the resolution of several hundreds of years at best. Pollen-rich samples are typically found close to cave entrances, where wind-blown material is most likely to enter the cave. This pollen is therefore more likely to be biased towards material derived from close to the cave entrance. Quantifying the relative proportions of local and distal pollen, and air and water transport mechanisms, remain significant issues in interpreting speleothem pollen, noting that pollen might also be sourced from animals accessing a cave. A final problem is that the near-entrance samples which are likely to contain the most pollen, are also most likely to contain detrital material which could hamper radiometric dating. A taphonomic model for speleothem pollen is presented in McGarry and Caseldine (2004).

Despite these potential problems, pollen has been extracted from speleothems in several successful studies. Table 6.1 summarizes the literature, including both modern calibration studies and speleothem records that date back to the mid-Quaternary. A common thread is a focus on archaeological sites and research undertaken by French-language researchers, and a use of flowstone deposits to provide the necessary sample for adequate pollen recovery. Despite the problems of quantifying taphonomic biases and the need for large samples sizes, speleothem pollen can provide essential palaeoenvironmental information when these issues can be negated.

6.3.2 Spores

Fungal hyphae and spores are typically present in the soil in greater abundance than pollen grains, but have yet to be analysed in speleothems, despite observations of their presence in speleothem pollen extractions (McGarry, 2000). Primarily through the work of van Geel (1986, 2001) and van Geel et al. (1995), several hundred fungal spore 'types' have been identified from lake, soil and peat archives, and the taxonomy, ecology and therefore their use

Table 6.1 Examples of speleothem pollen research.

Author	Subject matter	Location
Bastin (1978, 1990)	Extraction methodology	
Coles et al. (1989)	Review of transport mechanisms	
McGarry & Caseldine (2004)	Review of speleothem palynology	
Navarro et al. (2000)	Transport mechanism experiments	
Bui-Thi-Mai & Girard (1988)	Pollen trapped in modern calcite	France
Burney and Burney (1993)	Modern speleothems, moss, pollen traps	USA
Genty et al. (2001c), Genty (2008)	Modern speleothems and water	France
van Campo & Leroi-Gourham (1956)	Modern pollen trapped in caves	France
Bastin et al. (1982)	Holocene multi-proxy comparisons	Belgium
Bastin & Gewelt (1986)	Holocene multi-proxy comparisons	Belgium
Bastin (1990)	Holocene multi-proxy comparisons	Belgium
Burney et al. (1994)	Holocene	Botswana
Bastin et al. (1986)	Late Quaternary	Belgium
Bastin et al. (1988)	late Quaternary	Belgium
Brook et al. (1990)	Late Quaternary	Zaire
Brook & Nickmann (1996)	Holocene and late Quaternary	USA
Carrion & Scott (1999)	Travertines	South Africa
Lauritzen et al. (1990)	Flowstone	Norway
Baker et al. (1997c)	Mid-Quaternary stalagmite	UK
Caseldine et al. (2008)	Late Quaternary speleothems	UK

as a palaeoenvironmental indicator ascertained. Potential interpretations includes soil wetness, the presence of standing water, or even surface fire occurrences. Vaughan et al. (2011) recently demonstrated the presence of fungi growing on speleothem surfaces, indicating that this within-cave community is yet another potential source of spores preserved in speleothems. Further research from speleothems would undoubtedly yield novel palaeoenvironmental data.

6.4 Cave faunal remains

Despite the observation of a variety of cave fauna in modern cave systems (for example, see many contributions to Culver and White (2005)), and widespread analysis of faunal remains in cave sediments, there has been almost no systematic study of cave faunal macrofossils preserved in stalagmites. The exception is the study of mites preserved in two Holocene stalagmites from New Mexico (Polyak et al., 2001): an example of one is shown in Fig. 6.5. Polyak et al. (2001) used pitfall traps, swabs of stalagmite surfaces and vacuum cleaners to determine the modern cave mite population, and compared this with 12 genera of mites preserved in the stalagmites, to infer changes in cave environmental conditions over the late Holocene. Polyak et al. (2001) also reported the presence of spider leg segments, moth wing fragments and arthropod faecal pellets preserved with the stalagmites; McGarry (2000) similarly reported insect fragments in her pollen preparations and gastropods were illustrated in Fig. 6.2. There is undoubtedly further work to be undertaken in the analysis

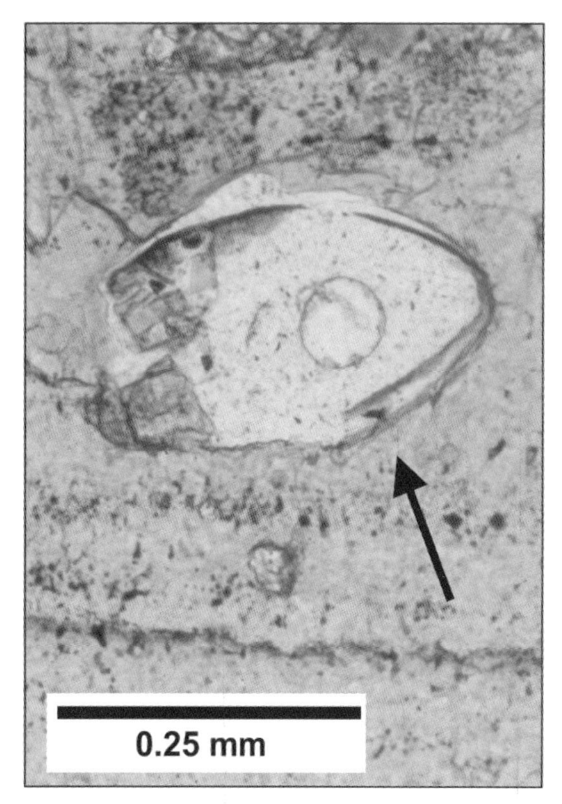

Fig. 6.5 Mite encased in a Holocene stalagmite (Polyak et al., 2001).

Fig. 6.6 Stromatolitic stalagmites at Jenolan, southeast Australia. See Cox et al. (1989) for organic geochemical analyses of these speleothems.

of cave faunal remains found in speleothems and their environmental interpretation.

6.5 Synthesis and research gaps

The biogeochemistry of karstic environments is a relatively immature subject area compared with other aspects of the karst system such as the understanding of karst hydrogeology and inorganic geochemistry. Mostly, this has been due to technological factors, which have historically limited the detection of what is a trace component of speleothem calcite. However, recent advances in mass spectrometric, optical and genomics-based technologies are likely to revolutionize this aspect of speleothem palaeoenvironmental analysis. In particular, spectrophotometers that can be used to measure organic matter fluorescence are becoming faster, with the next generation of instrument equipped with charge-coupled device (CCD) imaging systems that allow the near-instantaneous imaging of emitted fluorescence at all wavelengths. Within mass spectrometry, developments include the use of pyrolysis-MS (see, for example, Blyth & Watson, 2009), which allows the analysis of smaller sample sizes and with less potential for contamination during extensive extraction procedures. Within genomics, next-generation sequencing now allows the use of smaller sample sizes (Schuster, 2008).

Despite these rapid advances in our ability to measure organic macromolecules in speleothems, our understanding of the source, transport and fate of many macromolecular and fossil organic materials found in speleothems remains relatively weak. The understanding of transformations during transport and deposition is relatively poor, and modern calibration studies are few (exceptions being Blyth et al. (2007), detailed in Section 10.3; and Polyak et al. (2001), detailed in section 6.5) owing to the relatively large sample sizes needed until recently. Some biochemical proxies, such as amino acids, have remained uninvestigated subsequent to lone pioneering papers (Lauritzen et al., 1994), as have some biologically mediated growth morphologies such as stromatolitic stalagmites (Fig. 6.6; Cox et al., 1989).

Box 6.3 Vegetation and soil cycling of inorganic proxies: evidence from sulphur isotopes

Sulphur analyses on speleothems were first made by Frisia et al. (2005b) who used micro-X-ray fluorescence stimulated by synchrotron radiation to establish a rising trend during the late 20th century in a sample from Ernesto Cave, northeast Italy. These authors also demonstrated that the S was present as expected in its oxidized form as carbonate-associated sulphate (CAS), as has been found to be normal for carbonates in the geological record (Gill et al., 2008).

Sulphur sources (Fig. 6.B4) can be distinguished effectively using $\delta^{34}S$ signatures because pollution and volcanic sources have much lower values than marine aerosol or other sulphate sources. Hence, Wynn et al. (2008) developed an optimal method for extraction of sulphate from speleothem carbonate for measurement of its isotopic composition. They presented data from a site in southern England which showed the expected isotopically light pollution signal in 20th century growth, whereas a sample from Crag Cave, western Ireland, demonstrated that sulphate reduction had occurred; such reducing conditions above the cave had not previously been suggested. Wynn et al. (2008) also demonstrated that the value of $\delta^{18}O$ determinations in sulphate independently recorded the redox history at Crag.

In more slow-growing samples from Alpine regions, physical extraction of the sulphate with good time-resolution is not possible, but Wynn et al. (2010) have performed the first determinations of CAS by ion microprobe to resolve this variation. At both the Ernesto site, and the Obir Cave in southeast Austria (Fig. 6.B5), rising sulphate concentrations in the 20th century were associated with falling $\delta^{34}S$ values. Although the atmospheric pollution in these regions declined after 1980, sulphate concentrations in dripwaters did not decline until the 2000s. In the meantime, S was being stored in the soil-ecosystem. Fairchild et al. (2009) demonstrated that conifers can act as S archives and found that at the Ernesto site, the S rise in the wood of these trees slightly pre-dated that of the speleothems. Unpublished $\delta^{18}O$ data on sulphate in cave dripwater by Wynn show that it differs from that in

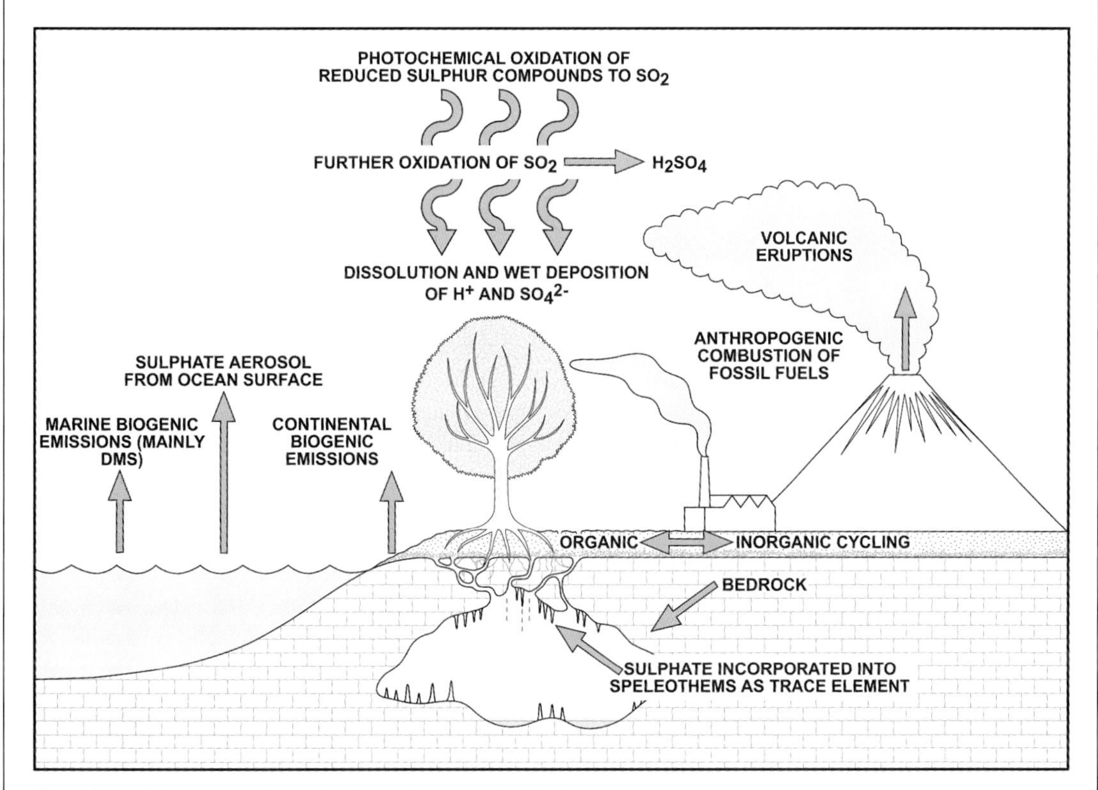

Fig. 6.B4 Sulphur sources for speleothems (Wynn et al., 2008).

(Continued)

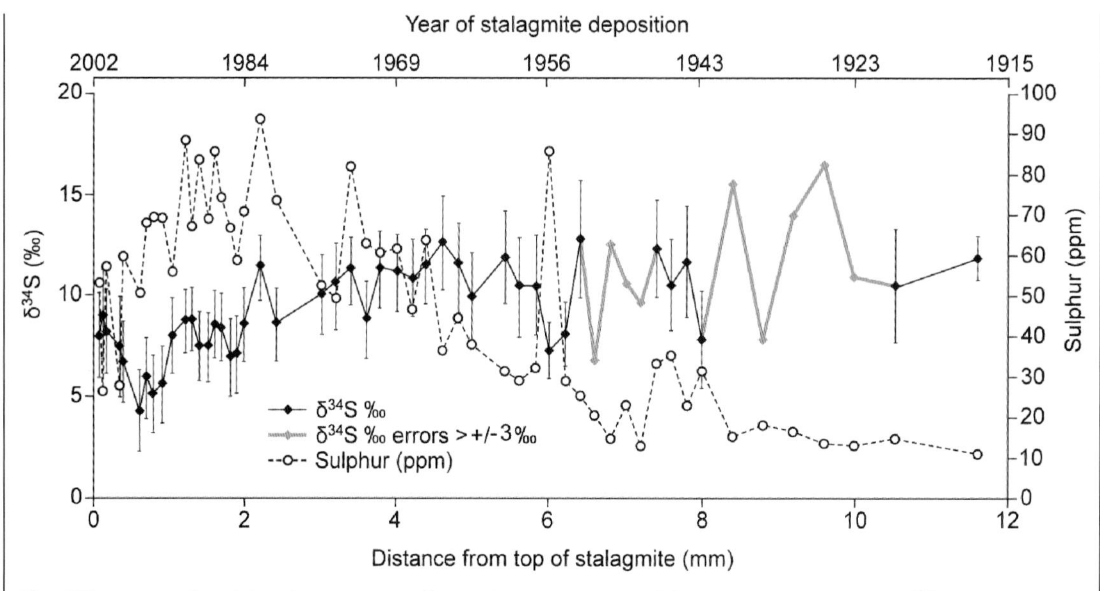

Fig. 6.B5 Microanalytical data demonstrating a late 20th century rise in sulphate content and decline in $\delta^{34}S$ composition in a stalagmite from Obir Cave, southeast Austria, related to regional pollution. A mixing model between natural and pollution end-members indicates that the pollution end-member became heavier over time (Wynn et al., 2010).

rainfall and has been reduced (presumably by assimilation into organic matter) and re-oxidized, during which process the $\delta^{18}O$ signal is re-set as has been shown in karst water elsewhere (Einseidl & Mayer, 2005). This detailed process understanding gives a clear context for the search for ancient volcanic S signals in speleothems (Frisia et al., 2008).

In some areas, speleothem biogeochemical research falls behind the use of biomarkers that have been successfully used in other archives. One example is in the field of lipid analyses which, as well as their use as vegetation biomarkers described earlier, can be developed as a temperature proxies based on the relative distribution of glycerol dialkyl glycerol tetraethers (GDGTs) (Schouten et al., 2007). In marine and lacustrine waters, these have been used to generate an index that has been called the TEX86 index. This is the relative abundance of cyclopentyl-containing isoprenoid GDGTs derived from Crenarchaeota, and has been found to be highly correlated to the temperature at their time of growth. Developing on this research, the relative proportions of branched and isoprenoid tetraether lipids has been developed as the BIT index, based on the relative abundance of terrestrially derived tetraether lipids versus marine crenarchaeol (Hopmans et al., 2004). Recently, branched GDGTs that have been measured in terrestrial peat bogs and soils have been calibrated against surface temperature. In a novel analysis of over one hundred globally distributed soil samples, it has been shown that the relative amount of cyclopentyl moieties, expressed in the cyclization ratio of branched tetraethers (CBT), and the relative amount of methyl branches, expressed in the methylation index of branched tetraethers (MBT), can be used to derive mean annual temperature (Weijers et al., 2007). A technological advance using high-performance liquid chromatography/atmospheric pressure chemical ionization-mass spectrometry (HPLC/APCI-MS) has recently led to radically improved analytical sensitivity (Schouten et al., 2007). Novel lipid GDGT analysis to obtain past temperatures has only just started to be undertaken in speleothems (Yang et al., 2011).

Another avenue of potential research is the analysis of stable hydrogen and carbon isotope ratios in individual organic compounds. This is a growing technique in organic geochemistry and of particular use in using carbon isotopes to identify changes between C_3 and C_4 dominant vegetation (see, for example, Wiesenberg et al., 2004) and oxygen and hydrogen isotopes as archives of the composition of rainfall at the time of vegetation growth (see, for example, Brenninkmeijer et al., 1982; Daley et al., 2010). However, these approaches have rarely been applied to either specific organic compounds or bulk organic matter preserved in speleothems, an exception being the analysis of lipids in a Chinese karst soil (Cui et al., 2010).

Future research efforts should aim to generate multiproxy, multi-site calibrations of organic biochemical markers in stalagmites. This should include the analysis of the complete transportation pathway of biomarkers from source to stalagmite, as the few previous studies have investigated individual historic or ancient samples, without detailed analysis of the modern-day transfer process. Further efforts to analyse replicate material for the same time period and from several caves of contrasting vegetation and temperature would allow the quantification of the uncertainty in biochemical proxies and demonstrate their utility to the wider palaeoenvironmental community.

III
Speleothem properties

CHAPTER 7
The architecture of speleothems

7.1 Introduction

This is the first of three chapters that focus on speleothem properties. It considers the architecture of speleothems, and in particular focuses on the morphology and rate of deposition of stalagmites, stalactites and flowstones, because these are the types dominantly used in palaeoenvironmental reconstruction. It demonstrates how both morphology and rate of deposition of these speleothems are related to the transfer processes in respect of, for example, water movement (section 4.3), air circulation (section 4.4) and alkaline earth composition (section 5.2).

This chapter is organized as follows: section 7.2 scrutinizes the controls on the rate of deposition, as well as the form, of speleothems, starting with a theoretical consideration of the factors determining speleothem growth rate. These are compared with empirical observations, before being combined with geometric considerations in the development of models of stalagmite and stalactite shapes. Section 7.3 presents a broader geometric classification of speleothems. It specifically examines a link between form (architecture) and process (hydrology) for classifications of stalagmites, stalactites and flowstones. We then move on to the internal structure and petrology (section 7.4), including the processes determining mineralogy, the nature and causes of growth phases and hiatuses, carbonate petrology and cave environments, and the types of speleothem laminae and the causes of their development. Links to process, in particular water chemistry, cave atmospheric composition, and hydrology, are emphasized. The chapter concludes (section 7.5) with a synthesis and overview of areas for further research.

7.2 Theoretical models of stalagmite growth and of stalagmite and stalactite shapes

7.2.1 Theories of speleothem growth rate

The first attempt to develop a kinetic model of speleothem growth was made by Dreybrodt (1980). The model considers stalagmite formation from a constant supply of water dripping onto a plane stalagmite surface, precipitation occurring from a stagnant thin film of water remaining on the stalagmite cap. The water drops are saturated with respect to calcite, and slowly degas until the CO_2 concentration of the solution equals that of the cave air. With degassing, calcite precipitation may occur by the reaction:

$$Ca^{2+} + 2HCO_3^- \rightarrow CaCO_3 + CO_2 + H_2O \qquad (7.1)$$

Four processes were shown potentially to limit precipitation rates. These were the following.

1 The diffusion of CO_2 molecules within the solution.

2 The diffusion of Ca^{2+} and CO_3^{2-} within the solution.

3 Deposition of $CaCO_3$ at the solid surface.

4 Production of CO_2 at the solid surface.

Speleothem Science: From Process to Past Environments, First Edition. Ian J. Fairchild, Andy Baker.

Dreybrodt (1980) established the kinetics of each reaction using the calcite dissolution experiments of Reddy and Nancollas (1970, 1971). They established that from the four potentially limiting reactions, the production of CO_2 at the solid surface was rate-determining. This in turn depended on the calcium concentration of the water, water film thickness and temperature. Dreybrodt (1981) developed the kinetic theory further by using the empirical data from more recent dissolution experiments performed by Plummer et al. (1978, 1979), reporting an approximate doubling in the predicted growth rates compared with those using the Reddy and Nancollas equation. Later work by Buhmann and Dreybrodt (1985) incorporated an additional rate-determining step and tested the equation experimentally. In this formulation, the four potential rate-determining processes listed above, together with an additional reaction, the kinetics of the CO_2 to $H_2CO_3^0$ conversion, were calculated simultaneously using an iterative procedure. Results showed that both precipitation and dissolution rates (R) could be approximated by the equation

$$R = \alpha([Ca^{2+}]_{eq} - [Ca^{2+}])\,(\text{mmol cm}^{-2}\,\text{s}^{-1}) \qquad (7.2)$$

where the calcium concentration, $[Ca^{2+}] > 0.2\,[Ca^{2+}]_{eq}$ (the aqueous calcium concentration at equilibrium with calcite at the specified P_{CO_2} rate), and α is the kinetic constant, a function of film thickness, cave air P_{CO_2}, temperature and flow regime. Temperature was shown to affect growth rate, because the CO_2 to $H_2CO_3^0$ reaction is strongly temperature-dependent. Film-thickness variations were also important: under especially thin films, CO_2 conversion is rate-determining, whereas in thicker films diffusion-limited reactions become important (only thin films had been considered by Dreybrodt (1980, 1981). Cave P_{CO_2} levels are also an important factor as they control the diffusion gradient across the water film. Finally, dynamic flow conditions were also incorporated into the model for the first time: under a turbulent flow regime, the diffusion coefficient is 10,000 times higher than for molecular diffusion which occurs under laminar flow. Figure 7.1 (Baker et al., 1998a) shows the relationship between theoretical growth rate and calcium ion concentration for a range of

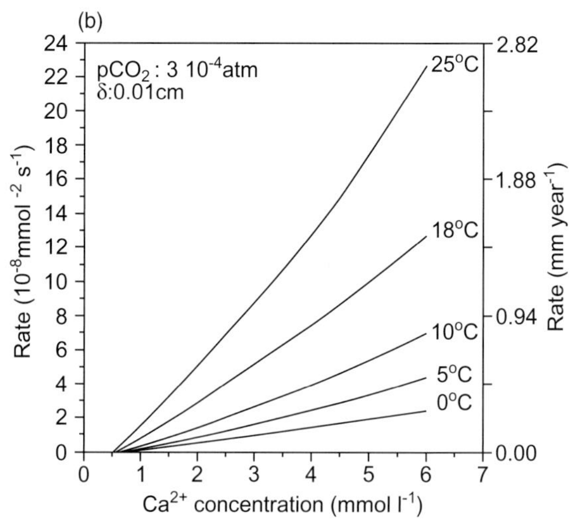

Fig. 7.1 (a) Calcite precipitation rates on a stalagmite top at various water film thicknesses as a function of the average calcium concentrations in the film (A: 0.005 cm (50 μm), B: 0.0075 cm, C: 0.01 cm, D: 0.02 cm, E: 0.04 cm). (b) Precipitation rate with water film thickness = 0.01 cm for various temperatures between 0 and 25 C. In both cases $P_{CO_2} = 3 \times 10^{-4}$ atm (i.e. atmospheric composition). These rates assume laminar flow conditions and a constant water supply, and are therefore most relevant to fast drip rate stalagmites or flowstones deposited under laminar flow conditions. From Baker et al. (1998a).

film thicknesses and temperatures under laminar flow conditions.

Equation 7.2 assumes a continuous water supply, such as that typically found for flowstones. For stalagmites, the time between drips might mean that that the rate of supply of saturated water could limit the rate of stalagmite growth. In this case, Dreybrodt (1988) and Baker et al. (1998a), after Curl (1973), demonstrated that the instantaneous accumulation rate R at any one time is given by:

$$R = \delta\phi c_o[1 - e^{(-T/\tau)}]/(T[1-(1-\phi)e^{(-T/\tau)}]) \quad (7.3)$$

where τ is a time constant defined as δ/α, and represents the relaxation time of excess saturation, where α is as defined in eqn. (7.2) and δ is the water film thickness, T is the time between drips, c_o is the maximum saturation excess of the water, and ϕ the mixing coefficient between new dripwater and the existing water film. Figure 7.2 presents the relationship between c, ϕ, T and τ. Where $T << \tau$ and $\phi = 1$ (high drip rates and complete replacement of water), then water supply does not limit dripwater calcium concentration, and eqn. (7.2) applies. For low drip rates, $T >> \tau$, the supersaturation excess approaches zero before a new drip falls. Dreybrodt (1988) demonstrated that when $T > 100$ s, then eqn. (7.3) has to be applied (Fig. 7.3), but for $T < 100$ s the calcium concentration in the water remains relatively constant and eqn. (7.2) is an appropriate approximation.

The growth rate theory was tested experimentally by Buhmann and Dreybrodt (1985). Precipitation experiments were performed by dripping water drops onto a calcite slab of known weight embedded into a stalagmite cap, which was later removed and reweighed. Agreement between experiment and theory is good, although growth rate determined with laboratory experiments was systematically higher than that predicted from theory. However, when compared with actual growth rates measured using modern annually laminated stalagmites (see section 7.4.3), or speleothems formed over known recent time periods, a good agreement was noted (Baker & Smart, 1995; Baker et al., 1998a; Genty et al., 2001b). Figure 7.4a presents the most comprehensive test to date: theoretically predicted, versus observed growth rates,

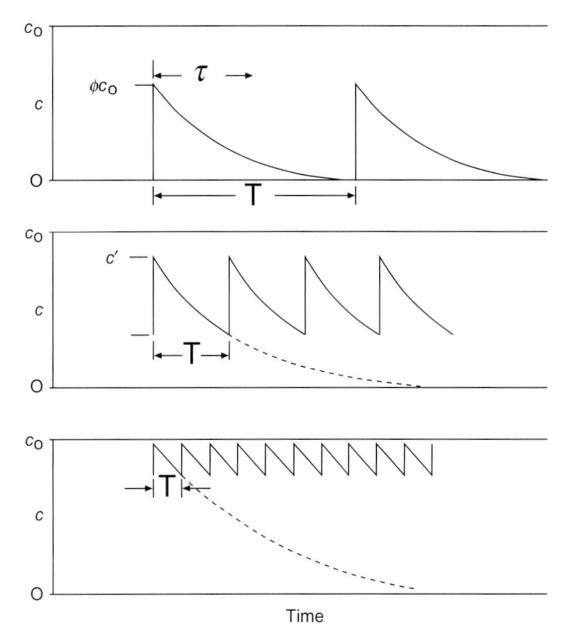

Fig. 7.2 Time histories of the supersaturation of a water solution on a stalagmite cap for different drip intervals T. Top, $T >> \tau$ (condition of low flow). A drip of supersaturation c mixes incompletely (here the mixing ratio, ϕ, is set at 0.6). Time between drips is greater than the time constant τ and there are periods of no deposition. Middle, $T \sim \tau$. The dashed line shows the composition of the water film if there were no further drips. Initial supersaturation c' is therefore higher than the low flow example. Bottom, $T << \tau$ (conditions of high flow). Here c' approaches the maximum c_o, and hence growth rate is close to the maximum possible. Both supersaturation and time are on arbitrary scales; c_o is a function of $[Ca^{2+}]_{eq}$ and $[Ca^{2+}]$ as in eqn. 7.2. Adapted from Curl (1973).

for stalagmites from five European sites as in Genty et al. (2001b). When compared with the 1:1 fit line between theoretical and actual growth rates, a reasonable agreement is observed (with one outlying sample with greater than predicted growth rate from a site where turbulent flow was likely). For the same sites, the observed growth rate (R, mm yr^{-1}) also demonstrated a strong correlation with mean annual surface temperature (T):

$$R = 0.193T - 1.67 \quad (R^2 = 0.63) \quad (7.4)$$

where T ranged from 7 to 14 °C. Therefore, despite the fact that for slow drip-rate samples growth rates

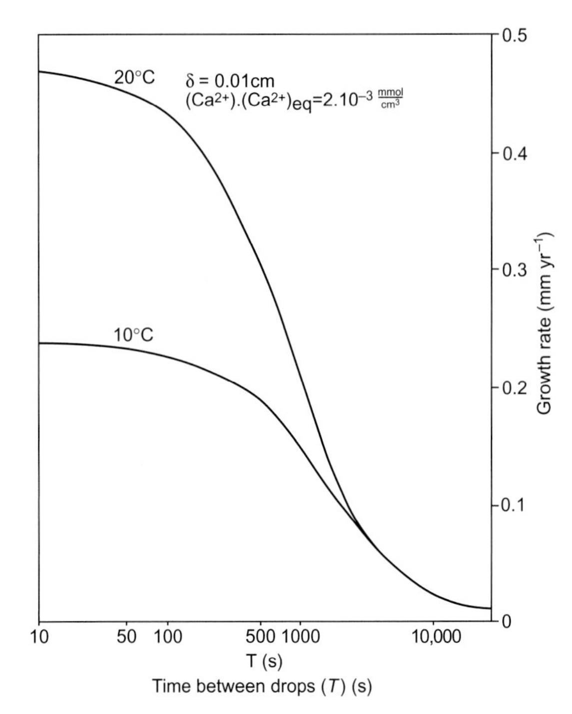

Fig. 7.3 Growth rate as a function of time between drips (T) for constant film thickness (0.01 cm) and water supersaturation (c_o; in this case $c_o = 2 \times 10^{-3}$ mmol cm^{-3}). Adapted from Dreybrodt and Franke (1987).

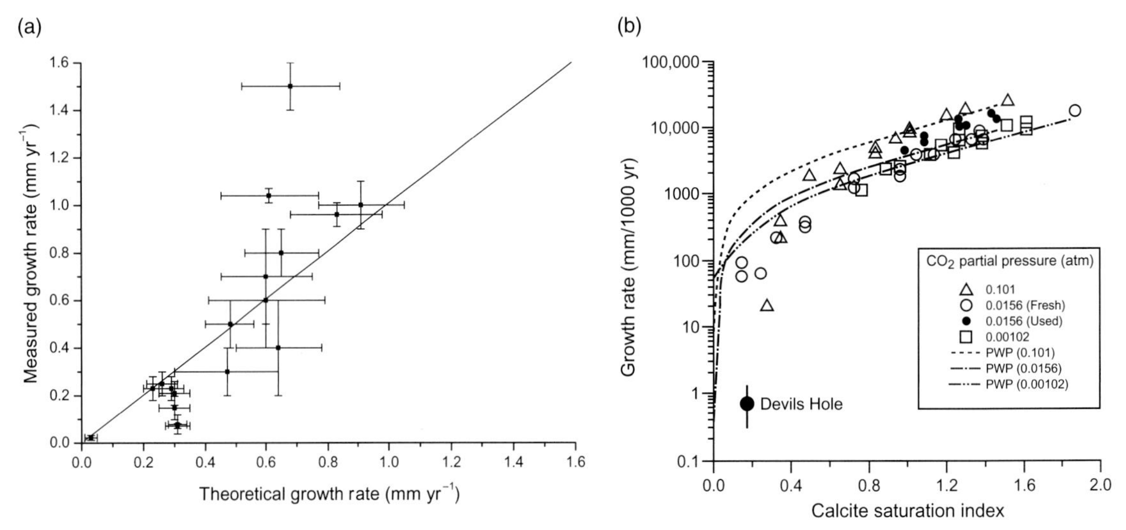

Fig. 7.4 Tests of growth rate theory. (a) comparison of theoretical and actual growth rates for 18 modern stalagmite samples from five European caves. For each sample, air temperature, water drip rate, and Ca^{2+} concentration were measured to calculate theoretical growth rates using eqns. (7.2) and (7.3). Actual growth rates were determined using annual lamina thicknesses (see section 7.4.3) or total growth since a known event horizon. Note that cave air P_{CO_2} was not known in all cases. Compiled from data in Genty et al. (2001b). (b) comparison of growth rate predictions from the PWP theory, experimental data, and Devil's Hole waters. There is a kinetic inhibition effect at low supersaturations. Devil's Hole subaqueous speleothems are discussed in section 12.1.1.

would be less than the maximum possible, growth is dominated by the greater control of temperature on growth rate. For these soil and vegetation covered sites, this was demonstrated to be through its influence on dripwater supersaturation owing to the relationship between temperature and soil CO_2 productivity. The relationship would not hold for sites where vegetation cover was sparse, or soil CO_2 limited, by summer drought, or where much prior calcite precipitation had occurred (Fairchild et al., 2007).

Another test of growth rate theory, but this time for subaqueous speleothems where issues of film thickness do not arise, was provided by Plummer et al. (2000). They used predictions of growth rate using Plummer, Wigley and Parkhurst (PWP) theory (Plummer et al., 1978) which was introduced in section 5.3.2 in the context of calcite dissolution in comparison with laboratory experiments, and found agreement was poorer at slow growth rates. The disparity with modern groundwater from which very long records of slow growth have been obtained (see Chapter 12) is marked (Fig. 7.4b). These discrepancies are attributed to kinetic inhibition effects due to impurities in solution which can adsorb onto calcite surfaces and such phenomena may also explain the tendency for growth to be less than predicted near the origin in Fig. 7.4a.

7.2.2 Models of stalagmite shapes

Given an understanding of theoretical stalagmite growth rates, it is then possible to develop models of stalagmite shape by accumulating appropriate amounts of calcite per year. Building on eqns. (7.2) and (7.3), a theoretical diameter of a stalagmite can be determined neglecting growth from water flowing down the flanks of the hypothetical stalagmite. In that situation, Curl (1973), Dreybrodt and Franke (1987) and Dreybrodt (1988) show that, under high drip rates, by multiplying eqn. (7.3) throughout by the surface area of the stalagmite, and then solving for the stalagmite diameter (d):

$$d = (4V / \pi \alpha T)^{0.5} \tag{7.5}$$

where V is the drip volume. For a slow drip rate, such that all the growth occurs on the stalagmite top surface, Curl (1973) deduced that the growth rate equation simplifies, such that diameter is a function of just drip volume (V) and water film thickness (δ):

$$d = 2(V / \pi \delta)^{0.5} \tag{7.6}$$

Curl made preliminary estimates of film thickness and drip volume, to derive solutions which were similar to observations of stalagmites suggesting a minimum diameter of around 3 cm.

Dreybrodt and Lambrecht (1981) presented the first computer visualizations of stalagmite growth. Following the concepts of stalagmite growth first postulated by Franke (1965), they assumed a drop of solution which spreads out radially, a monotonic decrease in growth rate with radius, and growth perpendicular to the existing surface (Fig. 7.5a). Growth rates were kept constant, but realistic 'candlestick' shapes were generated, independent of the nature of the deposition surface. This approach was later built upon by Kaufmann (2003), Kaufmann and Dreybrodt (2004), Dreybrodt and Romanov (2008) and Romanov et al. (2008a). Kaufmann (2003) and Kaufmann and Dreybrodt (2004) assumed a stagnant film that was being completely replaced, and decline of precipitation rate was a modelled as an exponential (latterly typified as an EXP-model). This approach also generated realistic 'candlestick' shapes (Fig. 7.5c, d). However, numerical solution for the precipitation rate was not unique, and was not able to realistically represent the actual Ca^{2+} concentration profile on the stalagmite surface. This factor becomes crucial when linking growth kinetics and the stable isotopic composition of stalagmites (see section 8.4). Mühlinghaus et al. (2007) attempted one numerical solution by allowing the mixing of water in the surface film, and Romanov et al. (2008a) removed the assumptions of both a stagnant water film and the requirement that deposition rates decrease with distance from the centre (Fig. 7.5b). Their 'FLOW' model simulated laminar flow spreading radially across a stalagmite surface, using gravity-driving equations (Short et al., 2005); the decline in precipitation rate with distance follows an approximately Gaussian shape. Figure 7.5c shows a comparison of the modelled stalagmite shapes with varying rates of change of drip rate, using the EXP and FLOW models. The FLOW model was shown to give stalagmite shapes

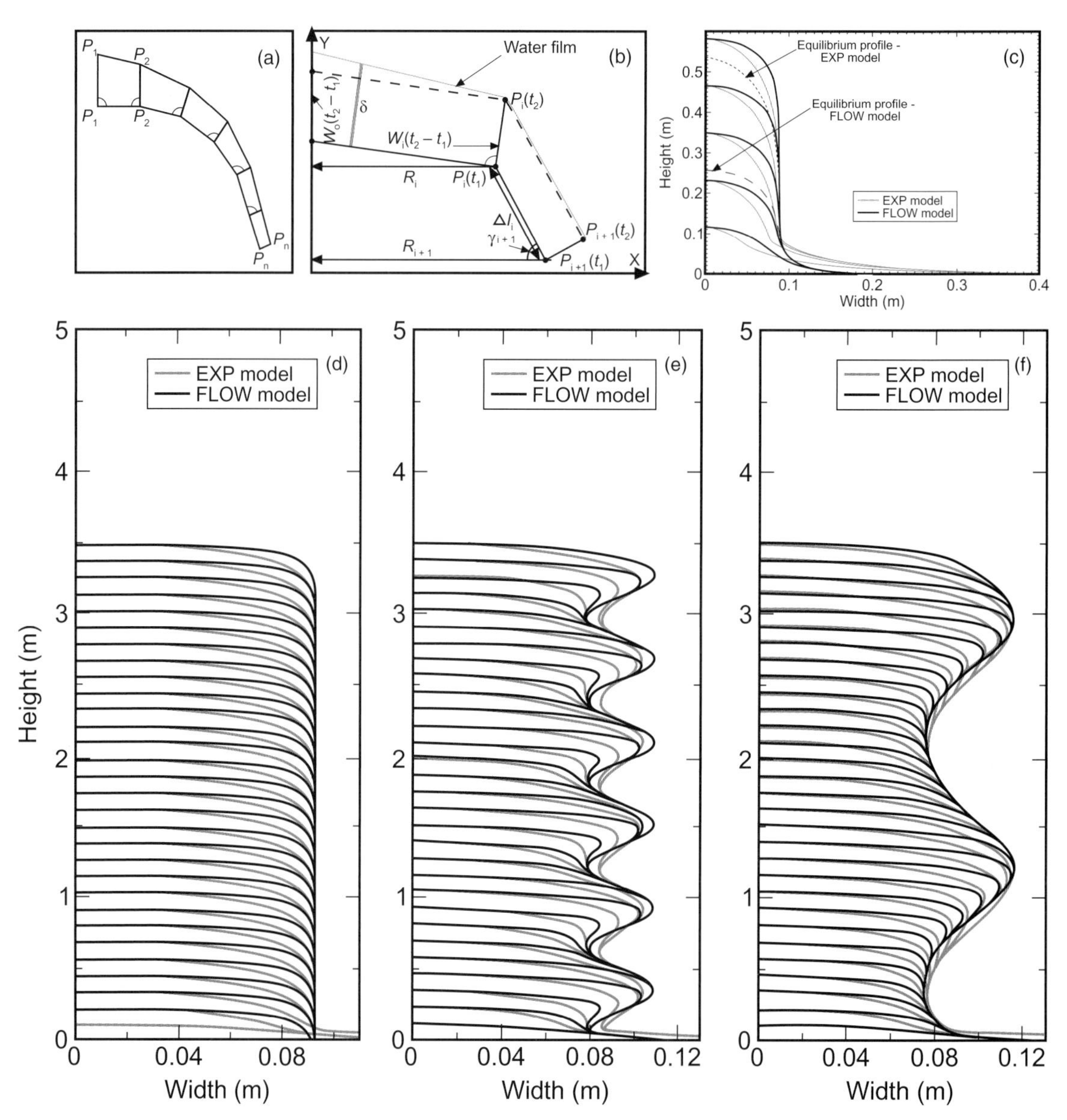

Fig. 7.5 Models of stalagmite shape and growth rate. (a) Geometric construction of stalagmite growth from Dreybrodt and Lamprecht (1981). Deposition is modelled between point P_1 and P_n, assuming a monotonic decrease from the central growth axis P_1' to P_n'. Deposition rate $(R) = R_0 \exp(-P_1P_n/\lambda)$ where λ and R_0 represent lateral spread and initial growth rate terms. (b) Geometric construction of stalagmite growth from Romanov et al. (2008a). Growth occurs between times t_1 and t_2 within the water film. The water film decreases with increasing distance from the centre R owing to radial spreading and radial flow is calculated, from which dissolved calcium concentrations and calcite deposition at points P_i, P_{i+1}, etc. are determined. Growth between time periods is shown (W_0, W_i). (c)–(f) Simplified from Romanov et al. (2008a). (c) Modelled stalagmite shapes for EXP and FLOW models under constant boundary conditions (T = 30 s, $\alpha = 1.3 \times 10^{-7}$ mol s^{-1}). The EXP model gives a less realistic shape and reaches equilibrium later than the FLOW model. (d)–(f) Comparison of stalagmite shapes under by varying drip rates from 13 to 30 s per drip over time periods of 500, 5000 and 15,000 yr respectively. Growth layers are shown every 1000 yr.

comparable with actual samples, and an equilibrium shape was calculated to be reached when height is three times the diameter. Figure 7.6a illustrates a modern stalagmite whose diameter is in the process of narrowing as a consequence of reducing drip rates.

Modelled stalagmite shapes in Fig. 7.5c, d do approximate those observed for 'candlestick'-shaped natural stalagmites, where precipitation is focussed on the stalagmite cap and flow does not occur down the flanks of samples. However, stalagmites comprise a much wider range of morphologies, where the impacts of splash and precipitation on the flanks of the specimen occur, and these other forms are widely used for palaeoenvironmental research. Further research is necessary to model the wider variety of stalagmite forms: several factors which need to be considered are described in section 7.3 as part of a broader classification of speleothem form.

7.2.3 Models of stalactite shapes

Curl (1972) addressed the morphology of the slender hollow stalactites known as 'soda straws', showing that their size was controlled by surface tension effects of the water droplet hanging from the tip. The analysis draws on the identification of a dimensionless ratio (Bond number, *Bo*) of gravitational to surface tension forces, which controls when a drop breaks and falls from the tip:

$$Bo = d^2 \rho g / \sigma \tag{7.7}$$

where d is the stalactite diameter, ρ the water density, g is the acceleration due to gravity and σ is the surface tension (g cm^{-2}). From experiments on glass capillary tubes, *Bo* was found to be 3.5, corresponding to a minimum diameter of 5.1 mm.

Short et al. (2005) modelled growth of solid stalactites as a free-boundary problem, by assuming that the diffusion time across the thin film is not rate-limiting. They showed that the Stokes approximation for gravity-driven laminar flow applies even to sub-vertical films. This relationship is

$$\partial^2 u / \partial y^2 = g \sin\theta \tag{7.8}$$

where the left-hand side of the equation refers to the rate of change of the gradient of velocity u with distance y, g is the acceleration due to gravity and θ is the tangential angle of the surface from the horizontal. They defined a characteristic length l_Q (i.e. a convenient representative length scale) as

$$l_Q = 3v_w Q / (2\pi g)^{0.25} \tag{7.9}$$

where v_w is the kinematic viscosity of water (0.01 cm^2 s^{-1}) and Q is the volumetric water flux.

Likewise, v_c, a characteristic (growth) velocity is given by

$$v_c = v_m l_Q(X) \tag{7.10}$$

where v_m is the molar volume of calcite and X is a complex chemical term which relates to the supersaturation of the solution.

By assuming that the time the water remains in contact with the stalactite is much longer than the diffusion time, and that the growth time is much longer than the contact time, they derived the growth equation

$$V_n = v_c[l_Q / r \sin\theta]^{0.33} \tag{7.11}$$

where V_n is the growth velocity normal to the interface, and r the radius of the stalactite. An approximation of this equation led to simulated stalactite shapes which compared closely with a small dataset of actual stalactites (Fig. 7.6b), and a wider intercomparison is undoubtedly warranted. Further research into models of stalactite shape are also required that remove the assumption that diffusion time is limiting (for very slow flowing waters) and which can model the widely observed centimetre-scale ripples on stalactite surfaces.

7.3 Geometrical classification of speleothems

The shape and form of speleothems has long fascinated researchers, well before recent attempts to model stalagmite growth rate and form, and there is a broader interest in the uniformity of geological patterns formed by the combination of laminar and turbulent flow conditions and precipitation/dissolution reactions (Meakin & Jamtveit, 2010). Allison (1923) presented a remarkable early attempt to classify speleothem shapes, examining a

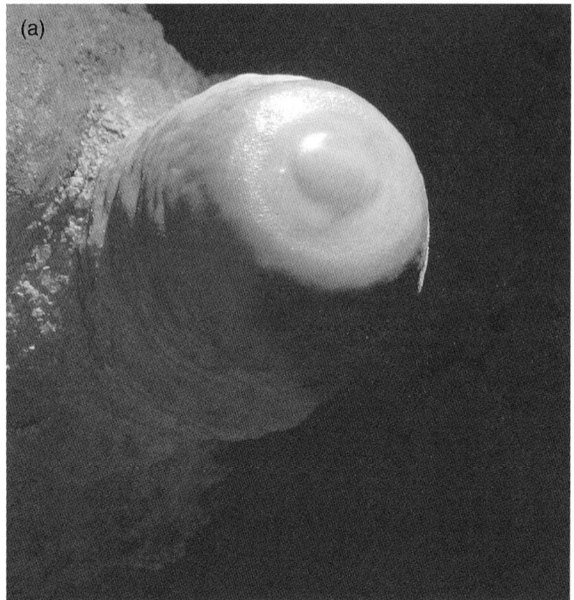

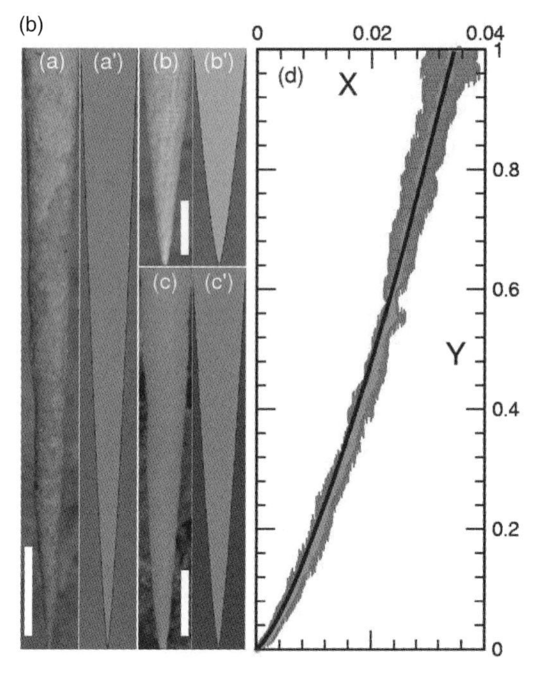

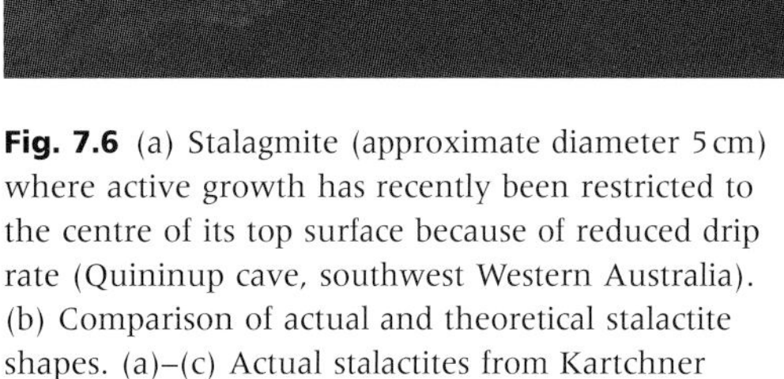
Fig. 7.6 (a) Stalagmite (approximate diameter 5 cm) where active growth has recently been restricted to the centre of its top surface because of reduced drip rate (Quininup cave, southwest Western Australia). (b) Comparison of actual and theoretical stalactite shapes. (a)–(c) Actual stalactites from Kartchner Caverns, USA, and their scaled shapes a′–c′. d. Average shapes of 20 stalactites (inner grey) with uncertainties (outer grey) and theoretical shape in black. *X* and *Y* are stalactite width and height respectively. From Short et al. (2005).

series of growth factors leading to 32 separate types. He deduced that stalagmite growth would be vertical and symmetrical upward growth unless conditions change, and identified cup morphologies formed as a result of splashing. In addition, he made what were rare early observations of growth rates, including the accelerated growth characteristic of those forming from lime (section 5.2). The work was flawed, however, by a mistaken emphasis on evaporation, rather than degassing, as the key geochemical process. Several decades later, the first explanation of stalagmite growth as a function of dripwater CO_2 and drip rate were provided by Franke as summarized in his 1965 paper. He considered the concentration of CO_2 in both air and water, and its diffusion in thin films. Curl's work (1972, 1973) provided the remainder of the foundation of speleothem morphology studies as discussed in the previous section. Gams (1981) considered the additional influences of drip fall height and splash effects on stalagmite shape, but many of his ideas remain to be followed up, later workers focusing largely on applying improved growth rate theory to stalagmites as considered in section 7.2. Here we consider the broader geometric classification of speleothems of relevance to palaeoclimate and palaeoenvironmental research.

7.3.1 Soda-straw stalactites

These sub-vertical tubes, with walls typically 100–300 μm thick made of few calcite crystals, are very common phenomena. Growth occurs at their tips where it initially can either be of feathery (dendritic) morphology, or more solid rhombohedra (Fig. 7.7). Although soda straws exceptionally can reach several metres in length, their fragility typically leads to breakage while they are less than tens of cm long; they may also be damaged by

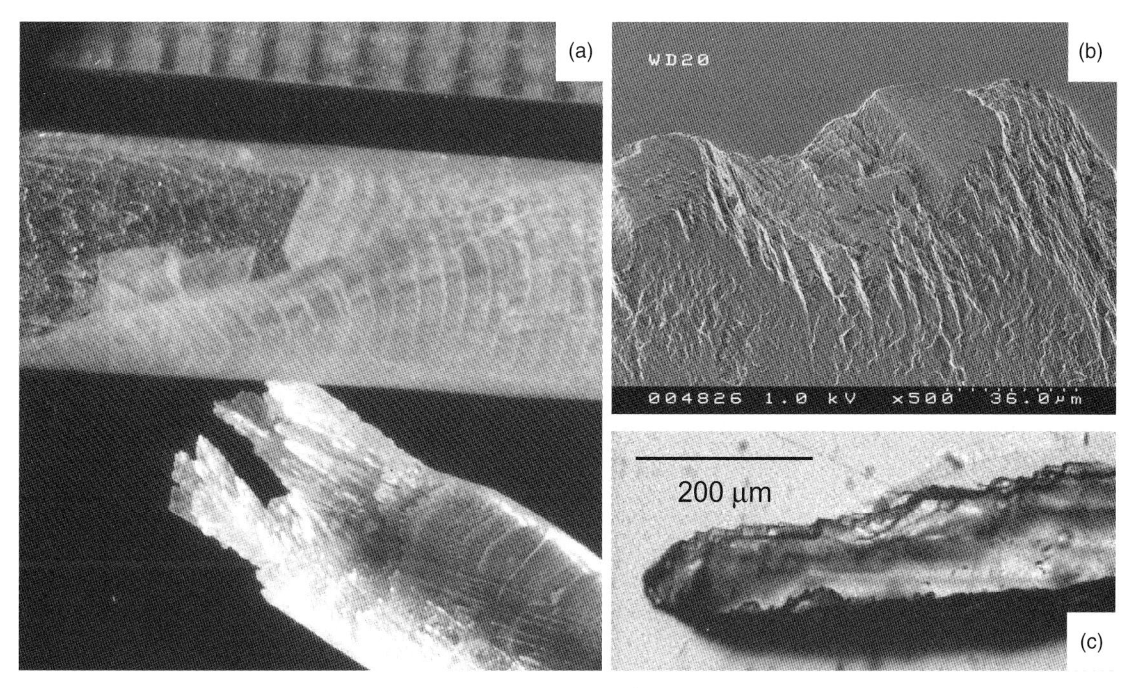

Fig. 7.7 Soda-straw stalactites. (a) Pieces of broken soda straw with annual growth bands (millimetre scale at top) and feathery terminations showing at growth tip in lower fragment (Ernesto Cave). (b) Growing tip of soda straw, Ernesto cave (secondary electron image) (c) View of approximately 1 year's growth at soda-straw tip in section showing smooth outer wall and crystallite faces on inner wall (Crag Cave (Fairchild et al., 2001)).

flooding. It is often assumed that water flows entirely internally, but Maltsev (1998) has disputed this, arguing that straw redevelopment on broken stalactite samples does not follow previous channels, but that straws arise automatically by degassing-induced crystallization at the margins of a drop where it detaches (and hence are related to the size of drops, as dictated by surface tension, as discussed in the previous section). Water can flow down the outside of straws, but is sucked inside by negative pressure in longer straws through cleavage defects.

Soda straws provide the most obvious evidence for seasonal cyclicity of speleothem growth in the form of annual growth laminae. These represent alternate thickenings and thinning of the walls (Fig. 7.7a), usually reflecting growth on the inside, but sometimes on the outside. First noted by Moore (1962), the phenomenon appears to have been rediscovered more recently, where it has been related to annual changes in chemistry (Fairchild et al., 2001; Desmarchelier et al., 2006). Huang et al. (2001) found evidence in one case of an internal annual layering (similar to that found in stalagmites) coinciding in spacing with that of the external morphology. The key point is that these variations in wall thickness directly demonstrate seasonal variations in growth rate as can be predicted from changes in air and water composition as discussed in Chapters 4 and 5. Data from the Ernesto Cave on soda-straw extension rates over time support this (Fig. 7.8). The absolute extension rate differs between straws from two different chambers, correlating with the Ca content of dripwater at this site (Fairchild et al., 2000). A pronounced increase in extension rate coincides with the ending of the Little Ice Age regionally as is found too in stalagmites from the same cave (Frisia et al., 2003). The faster-growing straw responds to deforestation in World War I, which is known to

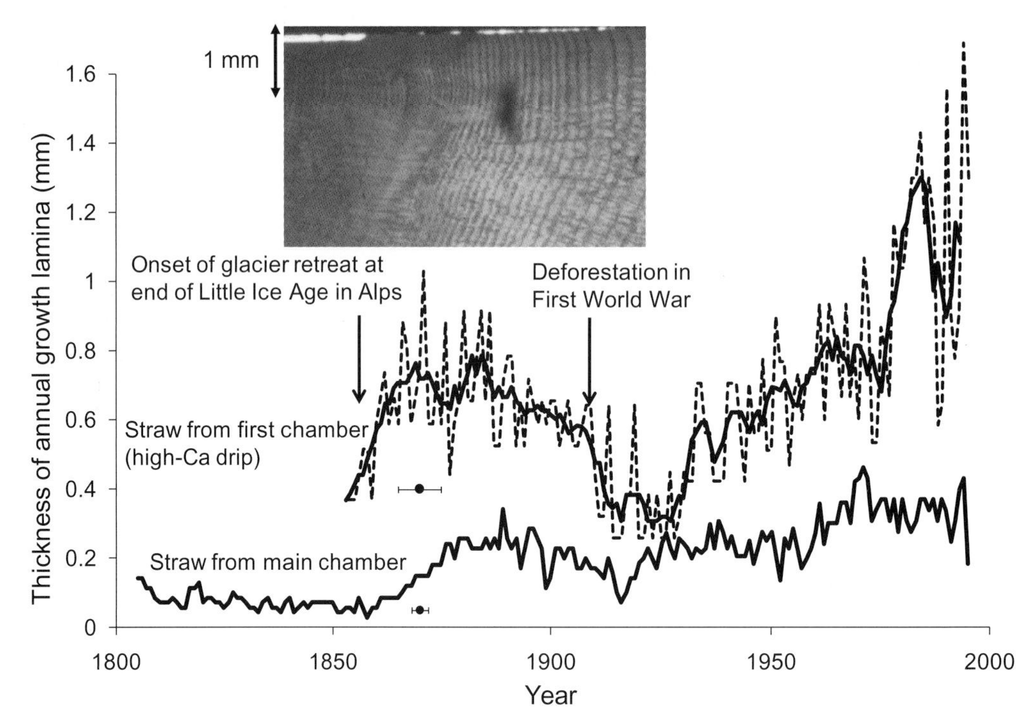

Fig. 7.8 Parallel changes in extension rate of two soda-straw stalactites from Ernesto Cave, Italy, inferred to be caused by variations in soil P_{CO_2} production linked to climatic temperature and forestry influences. Inset photograph illustrates increase in extension of the slower-growing straw around 1860. Age-uncertainties shown as error bars.

influence stalagmite chemistry at this time (Borsato et al., 2007) as well as late 20th century warming. Complementary stalagmite lamina thickness records are discussed in section 11.2.1.

The total annual accumulation rate of calcite on straws is small. Taking Fig. 7.7a as a typical example, this straw extended by around 0.5 mm per year, but the precipitation of such a 5 mm ring of calcite 200 μm in thickness corresponds to only around 1 μm per year growth on the top of a minimum diameter stalagmite.

Many key questions remain about the formation and duration of soda straws. Straws tend to have a relatively slow and invariant drip rate, although this has not been quantified. Also unknown is the maximum and mean 'lifetime' of a soda-straw stalactite (presuming both ideal and varying growth conditions of the 'incubator' and inputs). Both factors are of importance if a stalagmite under investigation has an associated soda-straw, as this might imply a relatively constant and slow drip rate, significant prior calcite precipitation on the soda-straw, and a possible change in stalagmite morphology if the associated soda-straw is blocked by sediment or precipitate, damaged or destroyed. The spatial pattern of soda-straw stalactites within a cave under investigation can also provide useful information on the flow regime and relationship to aquifer properties such as fracture networks (Perrette & Jailett, 2010). There is undoubtedly more scope for use of soda straws in modern calibration studies, although use of replicated samples is particularly important as it can be difficult to identify broken and re-healed sections.

7.3.2 Non-'soda-straw' stalactites

If soda-straw stalactites are the low discharge and low discharge variability end-member of speleothems that are associated with water ingress to a cave roof, then at increasing discharge and discharge variability typical forms are 'icicle shaped' stalactites and 'curtains'. In these morphologies, the stalactites thicken outwards and downwards with time, and often a surface rippling is observed with a characteristic wavelength of approximately 1 cm. Recent mathematic investigations into the development of similarly shaped morphologies in other environments (icicles, etc.) suggest that a similarity in process (gravity and laminar water flow) results in this similarity of form (Meakin & Jamtveit, 2010). However, as is the case for soda-straw stalactites, the relationship between stalactite form and discharge amount and variability is still not completely understood. Baldini (2001) used a disused limestone mine with modern stalagmite and stalactite growth to relate stalactite and stalagmite volumes with drip rate, although hydrological sampling was undertaken during a very short field season and a full range of discharge variability was unlikely to be captured. Despite this weakness, the study showed a relative decrease in stalactite volume to stalagmite volume with increasing discharge: slower discharge allowing more time for degassing on the stalactite. Kim and Sanderson (2010) demonstrate that in fractured limestone the size and distribution of stalactites can relate to water flow via the fractures, and is strongly controlled by the fracture aperture, the intersection of fractures and the development of damage zones around a fault. Further work is undoubtedly necessary, linked to the palaeoenvironmental proxies preserved in speleothems (see Chapter 8). Although stalactites have been used for palaeoclimatic studies along with stalagmites in specific cases for conservation reasons (e.g. Bar-Matthews et al., 1999, 2003), typically the spatial resolution is much inferior to stalagmites.

7.3.3 'Minimum-diameter' stalagmites

It is extremely uncommon to see a stalagmite with a diameter of less than 3–4 cm, and Curl (1973) addressed this observation as discussed in section 7.2.3. This work depended on estimates of film thickness and drip volume, both of which have since become better known. Baker and Smart (1995) used a vernier caliper to measure the film thickness on stalagmite caps and obtained values of 0.045 ± 0.019 mm ($n = 72$), with no relationship with drip rate observed. On the other hand, Dreybrodt (1999) continued to use a wide variety of values (0.05–0.4 mm) to bracket model results, and J. Baldini et al. (2008) calculated a value of 0.32 mm from a relatively fast-growing stalagmite, pointing out that this relatively high value was consistent with the micro-topography on the stalagmite surface (cf. Fig. 1.5b).

Drip volume is increasing well measured (Genty & Deflandre, 1998; Collister & Mattey, 2008), with a typical drip volume of 0.14 ml measured by Genty and Deflandre (1998), with higher volumes observed at drip rates of less than 10 s per drip. Collister and Mattey (2008) demonstrated a maximum drip volume of 0.25 ml from a 'non soda-straw' stalactite tip. Using a range of drip volumes of 0.14–0.25 ml and film thickness of 0.02–0.07 mm, 'minimum diameter' stalagmites can be calculated to be in the range 1.6–4 cm, matching the original observations of Curl (1973) which predated measurements of water film thickness and underestimated the drip volume. On the other hand, a film of 0.32 mm thickness would imply a minimum diameter of 0.7–1 cm, which is not found in practice.

Key gaps in Curl's (1973) analysis are the effects of drip splash, or the effects of changing drip fall height on the amount of splash, factors considered further in the next section. Furthermore, as is the case for stalactites, the relationship between diameter and changing drip hydrology is not completely understood, especially the discharge threshold at which dripwater starts to flow down the stalagmite flanks to form wider morphologies. An understanding of this parameter would allow speleothem researchers to know the discharge amount and variability of minimum-diameter stalagmites.

7.3.4 Non-'minimum-diameter' stalagmites

Minimum-diameter stalagmites represent one unique end-member of stalagmite forms. At higher drip rates such that flow occurs down the flanks of the samples, the stalagmites thicken to form a more typical conical shape. Centimetre-scale rippling can also be observed on the flanks of some samples (Plate 1.1), with a similar wavelength to that observed on stalactites. The numerical models of stalagmite morphology and growth rate (section 7.2.2) fail to model flow down the flanks of stalagmite, giving over-simplistic morphologies. Additionally, many other factors can affect stalagmite shape: Gams (1981) presented a series on hypothetical environments where, for a constant hydrological input, a wide range of forms results (Fig. 7.9). Hydrological factors include the splash of the water drop on the stalagmite cap, broadening the deposition area, and the drip fall height, which will affect the drip velocity and therefore the amount of splash (Fig. 7.9, example 1; Plate 7.1). This is quantified in the literature on soil erosion (Brandt, 1990) and splash physics (Mutchler & Larson, 1971). Water drops of 6 mm diameter (equivalent to a drop volume of approximately 0.11 cm^3, a typical value for cave dripwaters) have a terminal velocity of approximately 9 $m s^{-1}$, which is approached within the first few metres of descent from a cave roof. In caves where the fall height is greater than a few metres, drips will have reached terminal velocity: Mutchler and Larson (1971) calculate that approximately 20% of a water drop will be splashed at terminal velocity. This loss of splashed water implies that the observed vertical growth rate of stalagmites should be less than that predicted. Gams (1981) also theorizes that the migration of the drip point in the cave may occur during the formation of a non soda-straw stalactite, or owing to the calcification of the feed source(s), and either the stalagmite or stalactite themselves might move because of tectonic activity (Fig. 7.9, examples 2–4) or movement of unconsolidated sediments. Tectonic activity can yield a unique population of speleothem morphologies, as described in Box 7.1.

7.3.5 Flowstones

Flowstones (Fig. 1.2a) precipitate at the base of flowing rivulets or streams of water, and therefore represent higher mean discharges than stalagmites. A higher discharge typically requires a fissure or conduit flow component to the groundwater flow and discharge is therefore typically more variable than for stalagmite dripwaters. It is not uncommon, therefore, for flowstones formed from water sources entering a cave above floor height to contain both horizontal flow components (flowstones, precipitated predominantly at high flow) and vertical flow components (stalactites and stalagmites, formed at low flow). At a smaller scale, centimetre-scale micro-terraces are again observed, whose wavelength is inversely proportional to slope, leading to wavy morphologies in cut sections (Plate 7.8). Dreybrodt and Gabrovsek (2009) suggest that hydrochemical modelling is required that links water flow, water depth and turbulence to calcite precipitation rate. The higher discharges and water depth on flowstones means that turbulent flow is possible, leading to faster growth rates (see section 7.2). Chan and Goldenfeld (2007) have shown that on inclined planes with turbulent flow, linear instability on all scales is likely to be the cause of the micro-terraces. Nearer to cave entrances, microbial activity disrupts crystal growth, and a 'calcareous tufa' form of flowstone is often observed (Frisia & Borsato, 2010).

Flowstone deposition therefore occurs in the direction of the surface of the water film (typically vertically), but centimetre-scale micro-terraces and the deposition of calcite means that 'self-damming' is likely to occur with flow migrating from one part of flowstone to another, even with a constant input discharge. Therefore, care has to be taken that any one vertical section of flowstone sampled for palaeoenvironmental analysis contains a complete time sequence.

7.3.6 Other speleothem forms

The presence of other speleothem forms, although not in themselves best suited to geochemical analysis, may provide useful contextual information about the cave environment. Bulbous coralloid forms (also known as 'cave popcorn', Plate 7.9) typically grow in splash zones. A suite of forms occur which are typical of cave environments where evaporation may occur. Eccentrically shaped stalactites such as *helictites* (slender irregular forms with fluid supplied along a central canal by a combination of hydrostatic pressure and capillarity) typically evince evaporative conditions, and may be aligned to give evidence of air flow (Fig. 7.10). The presence of soft white microcrystalline *'moonmilk'* is often indicative of biological activity. Mammillary and botryoidal forms of flowstone demonstrate

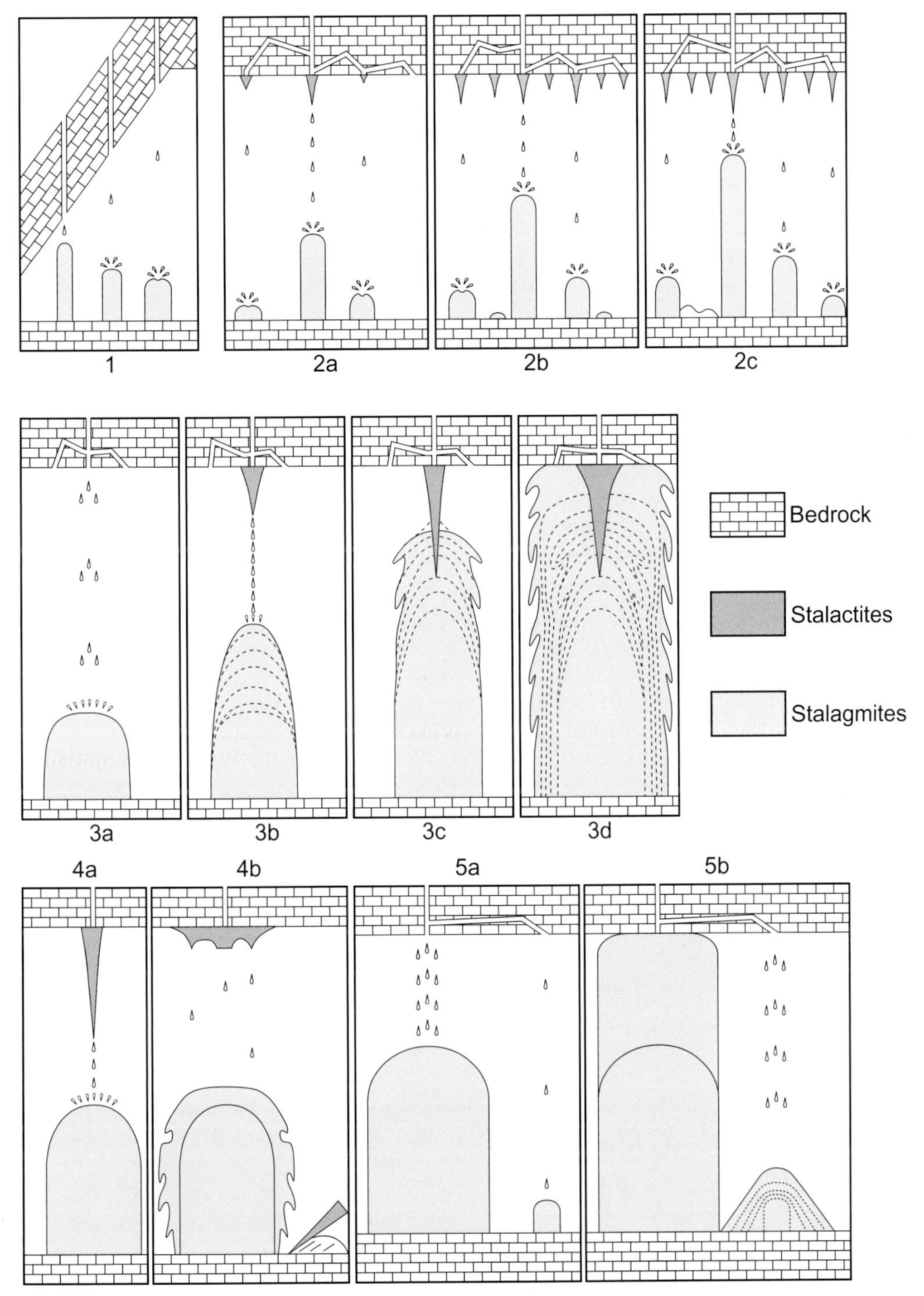

Fig. 7.9 Modifiers of stalagmite form (Gams, 1981). Example 1 shows the effect of drip fall height on splash and therefore stalagmite diameter for three samples. Example 2 indicates how the calcification of the roof drip sources would lead to increasing complexity in stalagmite growth. Example 3 demonstrates the effect stalactite growth may have on focussing drip supply and therefore stalagmite shape; example 4 the effect of the breakage of the stalactite (e.g. by tectonic activity). Example 5 shows the effect of the formation of a column on an adjacent stalagmite. All examples have a constant water input to the system e.g. no external climate forcing.

Box 7.1 Speleoseismicity in the Mechara karst, southeastern Ethiopia (authored by Asfawossen Asrat)

In karst caves, broken speleothems and those with non-vertical growth axes are very widespread (Becker et al., 2006 and references therein). In many cases it is very difficult to determine the real reason for this because several interacting mechanisms can be responsible (Becker et al., 2006; Šebela, 2008): instability of the ground owing to its composition (loose sand or clay), removal of ground owing to water flow, collapse of cave floor, gravitational deformation close to valleys, ice growth or melting, dissolutional loosening of ceiling deposits, earthquakes, and anthropogenic and faunal impacts. Among these, earthquakes and associated tectonic activities have been singled out in many karst caves as the prominent mechanisms responsible for broken and non-ideal speleothems (see, for example, Gilli, 1986, 1992, 1999; Postpischl et al., 1991; Bini et al., 1992; Gilli & Serface, 1999; Delaby, 2001; Forti, 2001; Angelova et al., 2003; Becker et al., 2005). Broken speleothems have also been used to identify unrecorded, or pre-historic earthquakes and/or to confirm recorded historic earthquakes in many parts of the world (see, for example, Lacave & Koller, 2004; Kagan et al., 2005; Szeidovitz et al., 2008; Panno et al., 2009).

The Mechara karst caves (Fig. 7.B1) located at the shoulder of the Main Ethiopian Rift (MER), in southeastern Ethiopia, and in close proximity to the Afar Depression, the most tectonically and volcanically active zone in the whole African continent, contain many speleothems with characteristic episodic growth (with numerous hiatuses among growth phases) and/or non-ideal growth. Asrat et al. (2008) considered that these non-growth periods might be initiated by major earthquake events, which created instability in dripwater supply, or instability in the bedrock or the ceiling. The MER is a Miocene–Quaternary, northeast-trending extensional tectonic feature that crosses the northwest and southeast Ethiopian plateaus. The Mechara cave system developed within Mesozoic limestones along

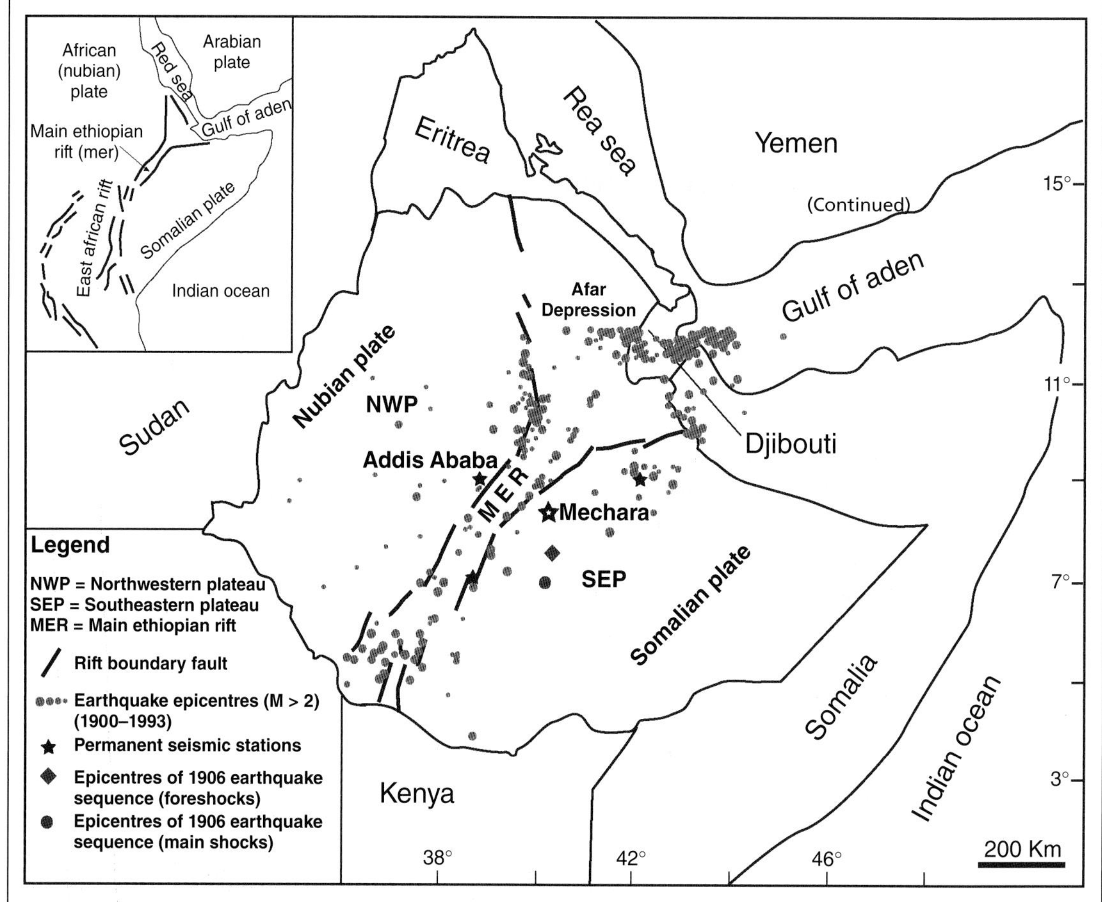

(Continued)

Fig. 7.B1 Seismicity of the Main Ethiopian Rift for the period 1900–1993. Seismicity of the northern part of the Afar Depression is not included (Gouin, 1979; Ayele and Kulhánek, 2000; Asrat et al., 2008).

major northeast–southwest aligned rift passages, parallel to the general orientation of the MER, suggesting karstification triggered by rift-related extensional tectonics (Asrat et al., 2008).

Seismic activity over the past century (Fig. 7B1) demonstrates that both the MER and surroundings are very active, related to extension between the Nubian and Somalian plates (e.g. Gouin, 1979; Ayele & Kulhánek, 2000; Ebinger & Cassey, 2001). In fact, the epicentre of the largest earthquake in the past 100 years or so (magnitude 6.5, 25 August 1906) lies only 50 km south of the Mechara karst (Ayele & Kulhánek, 2000) and more than 110 earthquake episodes have been documented at epicentres located within the MER, the Afar Depression and on the southeast Ethiopian Plateau (Gouin, 1979; Ayele & Kulhánek, 2000).

Three stalagmites from two caves in the Mechara karst are shown in Fig. 7.B2: Asfa-3 is a modern stalagmite from Rukiessa Cave (Baker et al., 2007), Merc-2 is an undated stalagmite from Rukiessa Cave, and Achere-4 from Achere Cave, with a basal age of 20 ka (unpublished data). These stalagmites display one or more tectonically influenced or controlled features such as anomalous laminae, deviations from vertical growth axis, abrupt changes in stalagmite morphology, and numerous hiatuses among growth phases.

Both Asfa-3 and Merc-2 have continuous visible annual laminae with alternating growth phases dominated by light- and dark-coloured calcite, which were used as proxy for palaeoclimate reconstruction (Baker et al., 2007). Asfa-3, which was deposited within a stable chamber 30 m below the surface, displays distinctive black annual bands dated

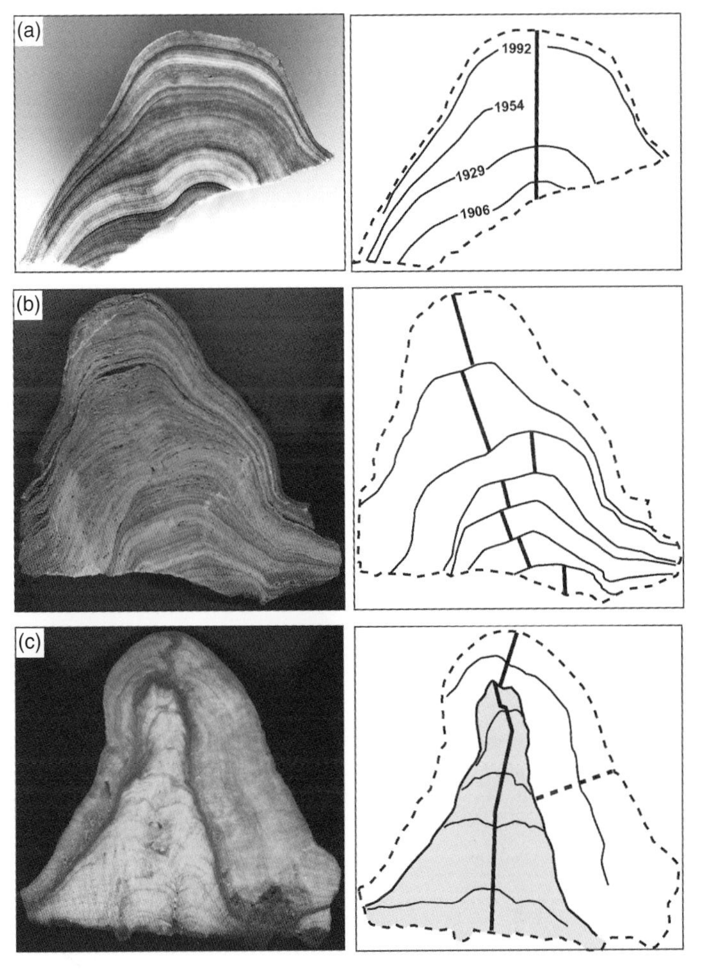

Fig. 7.B2 Polished and scanned images of (a) Asfa-3, (b) Merc-2 and (c) Achere-4 stalagmites from the Mechara caves. Corresponding sketches to the right of each stalagmite show the manifestations of earthquake activities in their growth: Asfa-3 by anomalous growth bands, Merc-2 by deviation of growth axis from vertical, and Achere-4 by a prominent hiatus with distinctly differing stalagmite shapes on either side.

(Continued)

to 1906, 1929, 1954 and 1992 (Fig. 7.B2a), containing more impurities and are thicker than those above and below. These growth events correspond to the major earthquake events of the past century: magnitude 6 earthquakes occurred in 1929–30, 1952–54, and 1989–92 with epicentres near the Mechara karst. The 1906 event triggered the same effect in many stalagmites deposited in various caves in the region (Asrat et al., 2008). Asrat et al. (2007) showed that stalagmite morphologies, laminae thickness and colour are strongly controlled by hydrological variations above the cave. Similar studies from Italy (e.g. Postpischl et al., 1991) indicated that colour anomalies in speleothem bands resulted from changes in the physical or chemical character of percolation water input, owing to earthquake activity. In particular, the significant input of impurities in these bands is considered to result from a change from a matrix flow to fracture flow of the percolating water triggered by seismic activity. The anomalous laminae represent isolated years in-between periods of continuous matrix flow, implying short-term events like earthquake sequences.

Merc-2 (Fig. 7.B2b) shows a distinct non-ideal growth, where the growth axis has been deviated from vertical growth six times, implying either a movement of the supporting block or shifting of the drip location, owing to local factors (e.g. permanent air currents deflecting the drip, movement of the drip source along a fracture on the ceiling, or gravity sliding of the stalagmite) or tectonic events and earthquakes (Forti, 2001). In the case of Merc-2, the local factors can be easily ruled out: Merc-2 was deposited on a stable limestone bed below a drip from a straw stalactite, hanging from a non-fractured ceiling, far away from the entrance to the cave. The growth axis deviation has been reversed (fifth growth phase, Fig. 7.B2b) implying a tectonic movement rather than local sliding. The growth phases in Merc-2 can easily be explained by an episodic shaking of the cave owing to earthquakes, which led to slight deviation of the drip location, or the location of the drip impact surface on the growing stalagmite. Moreover, the deviations occurred at more or less regular intervals, which are comparable to the intervals among the major documented earthquakes in the region during the last century (25–40 years).

Achere-4 (Fig. 7.B2c) has a peculiar shape, including an important growth hiatus above a candle-shaped stalagmite with regular laminae, followed by a change to a flowstone form indicating fracture-dominated flow (shaded in Fig. 7.B2c). This can be explained by a change in the flow regime above the cave, which is likely to have been triggered by an important seismic or tectonic event.

In conclusion, the studied stalagmites show that the Mechara karst, which is located in a tectonically and seismically active zone, can be an ideal site for conducting a systematic field and experimental investigation of speleothems, supported by high-resolution U–Th dating of growth phases, in order to identify undocumented prehistoric earthquakes in the region.

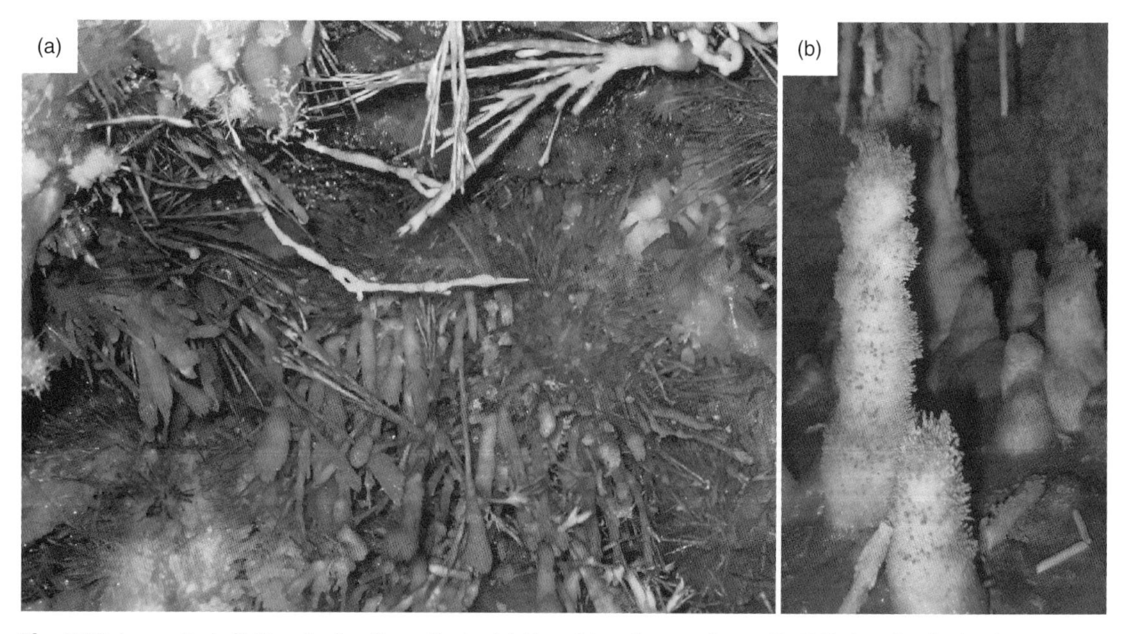

Fig. 7.10 Aragonite helictites, Soplao Cave, Spain. (a) Branching elongate forms. B. Helictites developing is a decoration on previously established stalagmites. Such morphologies can be preferentially developed in one direction in relation to air flow.

underwater formation. These forms are all described in detail in Hill and Forti (1997) and some examples given later in section 7.4.4.

7.4 Mineralogy and petrology

7.4.1 Mineralogy: aragonite versus calcite

The occurrence of the carbonate minerals aragonite, calcite and dolomite in carbonate sediments and rocks was introduced in Chapter 2, with more insights into calcite crystal structure in section 5.3.2. Here we focus only on calcite and aragonite, because speleothems used for palaeoenvironmental analysis are almost always composed of one or both of these phases. A variety of other carbonate mineral phases do sometimes occur, associated with speleothems forming from Mg-rich waters and in evaporative settings (Hill & Forti 1997, pp. 142–146). However, although dolomite has been described interlaminated with high- and low-Mg calcite (Bar-Matthews et al., 1991), other phases (e.g. huntite $CaMg_3(CO_3)_4$, hydromagnesite $Mg_5(CO_3)_4(OH)_2 4H_2O$) are typically characteristic of more delicate, evaporation-induced forms such as helictites that are not used in palaeo-studies.

Kinetic factors can often lead to the initial precipitation of metastable or very finely crystalline phases, which tend to stabilize to larger crystals (Ostwald's ripening), or be transformed to more stable phases (Ostwald's step rule) over time (Morse & Casey 1988). At Earth surface temperatures and pressures, calcite is the most stable (and hence least soluble) form of $CaCO_3$ (Chapter 5). However, aragonite commonly grows when calcite growth is inhibited, and short-lived phases such as the third polymorph vaterite, or even amorphous calcium carbonate, may sometimes have a role to play in the initial crystallization. Although vaterite and amorphous $CaCO_3$ are rarely found preserved, aragonite has been described in speleothems as old as Pliocene (Hopley et al., 2009), but on the other hand it can also be replaced by calcite soon after deposition.

Aragonite develops distinctive morphologies (Fig. 7.11) which can be diagnostic. Aragonite and calcite can precipitate simultaneously on the same

Fig. 7.11 Scanning electron microscope images of aragonite in speleothems from Grotte de Clamouse, southern France. (a) Aragonite rays on stalagmite surface with blunt pseudo-hexagonal terminations. (b) Aragonite needles (note narrowing towards crystal tips) on subaerial frostwork. (c) Backscatter electron image of stalagmite horizon of nucleation and growth of aragonite rays. (The aragonite is paler than the surrounding calcite because it contains more Sr and less Mg and so has a higher mean atomic number.) (a) and (b) from Frisia et al. (2002); (c) from McMillan et al. (2005).

growth surface (Fig. 7.11c), but more usually when they co-exist in the same speleothem, they reflect either separate episodes of growth at the speleothem surface under changing conditions (Railsback et al., 1994), or partial secondary alteration of aragonite to calcite. Aragonite differs from co-existing primary calcite dramatically in trace element content and more subtly in isotope composition (Chapter 8). Secondary calcite is unsuitable for palaeoenvironmental study because it is typically significantly modified in chemistry from the original aragonite, including disturbance to the U-series systematics (Ortega et al., 2005; Hopley et al., 2009). Hence it is very important to know if speleothems originally grew as aragonite. Thankfully secondary alteration of aragonite can be readily recognized petrographically (section 7.4.2, Railsback et al., 2002; Martín-García et al., 2009). It is also useful to determine mineralogy directly by X-ray diffraction of powdered samples and/or electron backscatter diffraction (Table 7.1).

Hill and Forti (1997, pp. 237–239) summarized factors proposed to influence the crystallization of aragonite rather than calcite, although several these are only of historical interest, such as the pioneering attempt by Moore (1956) to relate mineralogy to cave temperature. We now know that the two key issues are the solution Mg/Ca ratio (Murray, 1954) and the mineralogical substrate. Aragonite requires a Mg/Ca ratio in aqueous solution greater than one in order to form and normally this reflects significant dolomite content in the host rock (Cabrol & Coudray 1982). If such a water precipitates calcite, this mineral necessarily includes much Mg (section 8.3), but such Mg-calcites are more soluble than pure calcite. Hence Mg-calcites grow more slowly at a given supersaturation, such that aragonite crystals may out-compete them. Thus, experiments in dilute solutions resembling karst waters demonstrate that a progressively higher supersaturation level is required for calcite growth as Mg/Ca ratios in solution increase (de Choudens-Sánchez & González 2009; Fig. 7.12). Studies on caves in southern France have shown that aragonite is also associated with drier conditions (Cabrol & Coudray 1982) and indeed slow drip rate can correlate with reduced supersaturation, higher solution Mg/Ca and the occurrence of aragonite (Frisia et al., 2002). Crystal growth that continues an existing lattice structure of the same mineral (known as *syntaxial* growth in the carbonate literature) is energetically advantageous, and so a crystal of a particular mineral may continue growing even if solution composition changes to favour the other polymorph. In practice, changes in mineralogy over time during continuous growth can be either abrupt or transitional via a period of simultaneous growth of aragonite and calcite (McMillan et al., 2005; Fig. 7.11c; Plate 7.2).

7.4.2 Crystal fabrics

In carbonate literature, the term *fabric* (*texture* is an alternative) is used to encompass both the nature of the rock components, and their orientation and arrangement. In the case of stalagmites and other regularly growing speleothems, the property emerges within a single growth layer as the result of the distinctive geometry and spatial arrangement of simultaneously accreting crystals (Stepanov 1997; Self & Hill 2003). In turn this property depends on the style of crystal nucleation and the crystal morphology. An accessible introduction to the fundamental literature on crystal growth is provided by Sunagawa (2005) from which parts of the following discussion are drawn.

Nucleation

The driving force for crystal growth can be represented as:

$$\ln S = \Delta\mu / kT \qquad (7.12)$$

where S is a supersaturation ratio, $\Delta\mu$ is a change in chemical potential associated with the phase change, k is the Boltzmann constant and T is the absolute temperature. Note that ($\ln S$) is similar to Ω as defined in Chapter 5. The biggest hurdle in crystal growth is to nucleate the crystal, because a critical size has to be exceeded for the crystal lattice to be stable. At low driving forces, crystals nucleate on pre-existing surfaces (heterogeneous nucleation) and most of them continue the growth of an existing crystal lattice of the same mineral: hence crystals grow to be relatively large. At high driving

Table 7.1 Petrographic techniques for study of speleothems.

Technique	Principle and spatial scale	Types of observation	References
A. Optical sources			
Observation of polished hand specimens	**a.** Reflected light with images captured by computer scanning or photomicroscopy	**a.** Inclusion-rich or finely crystalline precipitates are bright; larger translucent crystals are paler; some impurities have distinctive colour (e.g. red Fe_2O_3); identification of laminae and growth phases; can be quantified by image-processing methods	**a.** Genty & Quinif (1996)
	b. Hyperspectral imaging 0.05 mm to decimetre-scale	**b.** Enhanced distinction of laminae using specific wavelength ranges, including near infra-red for fluid inclusions	**b.** Jex et al. (2008)
Thin section petrography	Transmitted light optics. 1–5 μm to 3 cm	Confirmation of crystal size, identification of mineral and fluid inclusions, laminae and hiatuses	Frisia et al. (2000), Fairchild et al. (2010)
B. Secondary images excited by electromagnetic radiation			
Infra-red imaging	Fourier transform infra-red spectroscopy. Using synchrotron source (5 μm spot) or benchtop source (50 μm spot)	Distribution of organic matter; distinction of calcite and aragonite	Bertaux et al. (2002), Frisia & Borsato (2010)
Fluorescence	Excitation by a laser, or in more usually a mercury ultraviolet lamp with excitation, emission and barrier filters 5 μm to 2 mm	Observation of luminescent areas such as laminae by light typically at 405–420 nm wavelength	McGarry & Baker (2000), Shopov (2006), Baker et al. (2008a)
X-ray imaging	Degree of transmission of X-rays, e.g. in an X-ray microscope, scanner, or by focused micro-beam in a synchrotron 1–20 μm or 1 mm to several cm	Presence of inhomogeneities or layering	Frisia et al. (2005b)
C. Electron imaging			
Cathodo-luminescence	Excitation by incident electrons 5 μm to 2 cm	Displays distribution of luminescence centres (normally trace Mn^{2+} or REE), but rare in speleothems	Richter et al. (2004)
Scanning electron microscopy (SEM)	**a.** Secondary electron imaging (intensity of secondary electrons)	**a.** 3-D images of crystal growth or breakage surfaces	**a.** Frisia et al. (2000)
	b. BSEI: backscattered electron imaging (intensity of reflected incident electrons)	**b.** Brightness of image depends on mean atomic number	**b.** McMillan et al. (2005)
	c. EBSD: electron backscatter diffraction (diffraction image of backscattered electrons) 10 nm – 1 mm	**c.** Mineralogy and crystal orientation; can be mapped at micron-scale resolution	**c.** Neuser & Richter (2007), Fairchild et al. (2010)
Transmission electron microscopy (TEM)	Imaging of samples thinned so that they are transparent to electrons 0.1–1000 nm	Images of electron transmission or electron diffraction images of points	Frisia et al. (2000, 2002)

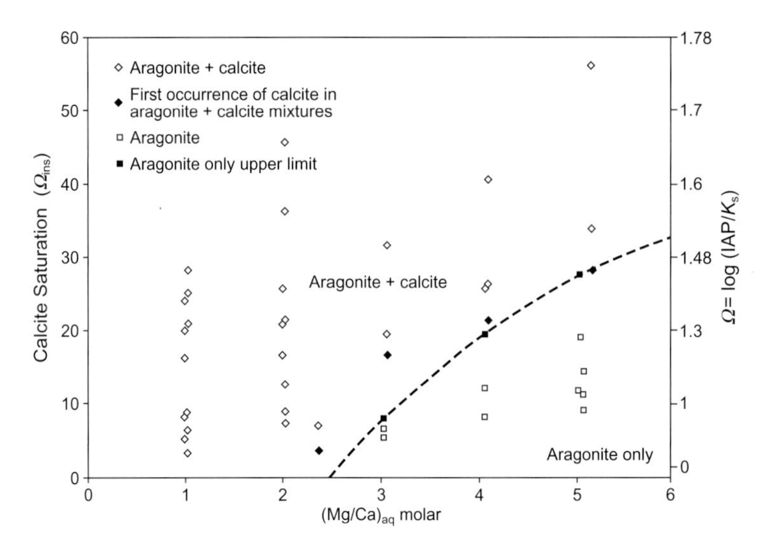

Fig. 7.12 Conditions for precipitation of aragonite versus calcite in laboratory experiments in dilute solutions. Only aragonite precipitates in the lower right area. With increasing solution Mg/Ca ratio, higher supersaturations are required for calcite to precipitate. Simplified from de Choudens-Sanchez and González (2009) who use Ω_{ins} as the ratio of ionic activity product to solubility product. Ω as defined in Chapter 5 is shown on the right *y*-axis.

forces, or where impurities of solid or gas-liquid interfaces are introduced into the system, nucleation occurs more readily. Hence crystal size is at least temporarily reduced, and the new crystals that form may have diverse lattice orientations. A nucleation event that only represents a disturbance of a few days may nevertheless have a prominent effect on the subsequent speleothem growth (Fig. 1.5b).

Although nucleation of crystals directly from solution (homogeneous nucleation) is normally not thought to apply in sediments, one case may be the nucleation of crystals around CO_2-rich gas bubbles which explode by degassing into the solution. A range of types of $CaCO_3$ occur in such situations, as is shown by experiment, including hollow calcite rhombs (Killawee et al., 1998; Aquilano et al., 2003) which may equate to bubble structures found in cave pools by Warwick (1950).

The most complex and variable speleothem fabrics are those associated with sporadic nucleation events, at complexly changing air–water interfaces, including helictites (Hill & Forti, 1997) and raft deposits (Jones, 1989), whereas some features such as the growth of feathery dendritic crystals (e.g. from soda-straw tips (Maltsev 1998)) are simplified by later overgrowth (Jones & Kahle, 1993). Biological influences on crystal nucleation can also be prominent, particularly in the twilight zone of caves (Jones, 2010; Chapter 6). Self and Hill (2003) provided a systematic summary of growth morphologies of all types of speleothem crystal aggregate, whereas our focus here is more restricted to those that provide longer records.

Crystal morphology

Under low driving forces, and where crystals do not incorporate many impurities, stable crystal faces develop that have a characteristic geometric relationship to the crystal lattice. As introduced in Chapter 5, faces occur in sets called crystallographic forms that have a unique geometrical relationship to crystal symmetry. Well-developed forms are those where growth is slow perpendicular to the faces, whereas fast-growing forms disappear during crystal enlargement. There are several theoretical approaches to predicting the most prominent 'equilibrium' crystallographic forms. The single most successful and well-known of these is the Periodic

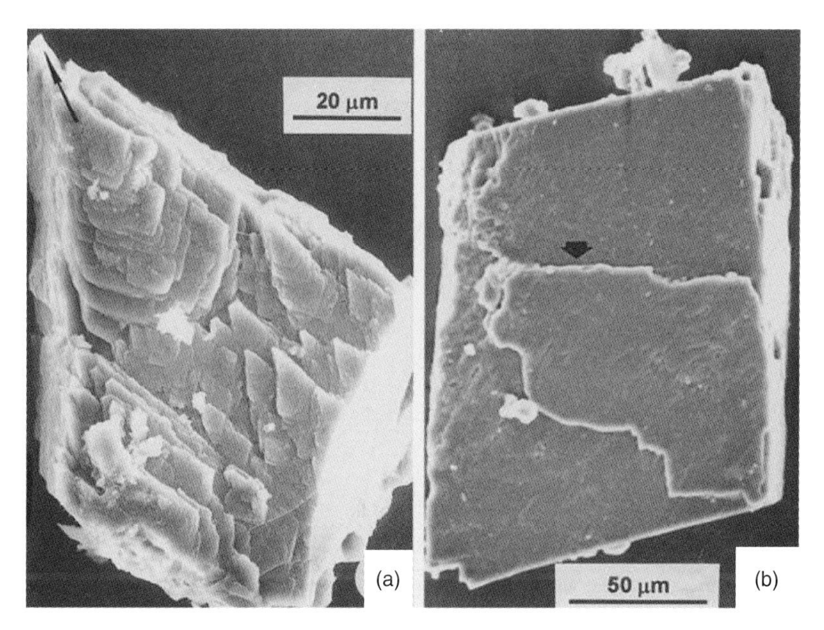

Fig. 7.13 Contrasting calcite rhombohedra that grew on a glass slide under dripwaters during different seasons. (a), Stepped (rough) surface; (b), smooth surface. (Ernesto Cave; Frisia et al., 2000).

Bond Chain (PBC) analysis (Hartman 1987) which identifies atoms and ions connected by strong bonds. The most stable faces (F or flat faces) contain two or more PBCs, whereas S (stepped) faces have one and K (kinked) faces have none. Calcite has only one F-form, the unit rhombohedron, which also corresponds to the planes along which the mineral cleaves (Fig. 7.13). The equilibrium morphology for aragonite is more complex, with several F-forms calculated (Aquilano et al., 1997; Frisia et al., 2002). In practice highly elongated prisms develop with pseudo-hexagonal cross-sections (Fig. 7.14); single or repeated twins are common (Fig. 7.14). Twin planes are surfaces bounding two lattices in different orientations, but which line up exactly and hence are energetically favoured. For both minerals, the presence of specific impurities (see, for example, Braybrook et al., 2002), or the precise growth conditions, may change the favoured crystal form, but the range of potential variables is so large that is it often difficult to identify a specific reason for development of particular faces. Different growth forms may be characteristically associated with different trace element contents: the phenomenon of sector zoning (Reeder & Grams, 1987).

The characteristic mode of growth at low driving forces, as established by Burton et al. (1951), is by accretion around a growth spiral (Fig. 7.15) centred on a dislocation, which is an offset in the lattice layers. Dislocations have a typical density of 10^8–10^{10} lines per square centimetre (Sunagawa, 2005), equivalent to a spacing of 0.1–1 μm, and are more abundant where crystal growth recommences over an etched surface, or at high driving forces. Dislocation spirals offer a favoured site for growth because a new calcium or carbonate ion can have more of their bonding satisfied than if they were starting a new flat growth layer. Two-dimensional layer spreading becomes more important at higher driving forces (Figs. 7.15 and 8.12).

At high driving forces, or where impurities are more abundant, several types of phenomenon can occur, but a common factor is a decrease in the ideal symmetry and shape of individual crystals. Figure 7.15 illustrates that there is a transition from smooth to rough surfaces (e.g. compare the rhombs

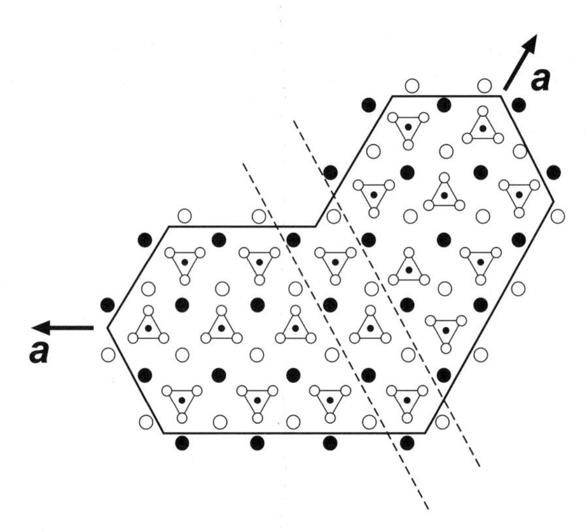

Fig. 7.14 Structure of aragonite in relation to common faces and twin. Triangles denote carbonate ions and the open and closed circles represent calcium ions. (Sunagawa, 2005.)

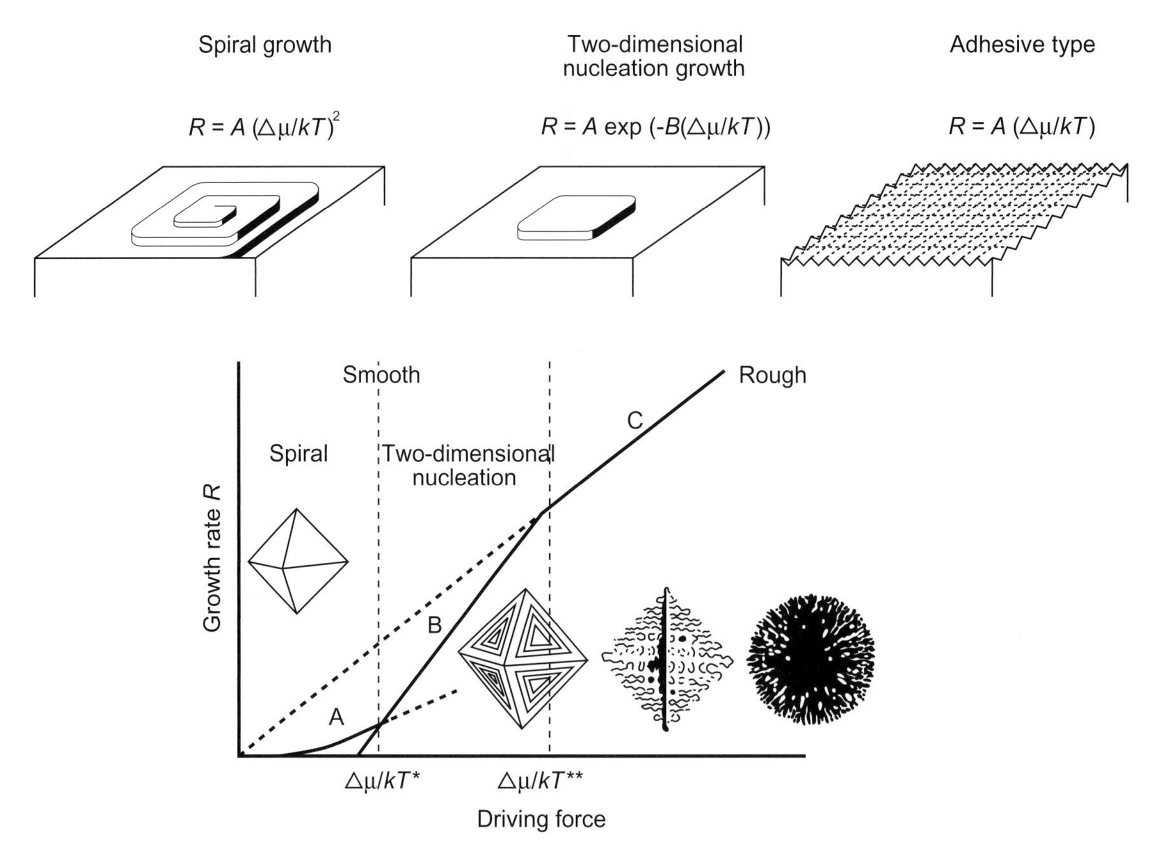

Fig. 7.15 General concepts of crystal growth (after Sunagawa 2005). The top three diagrams illustrate three mechanisms of growth of crystal faces where the growth rate (R) is a different function of the driving force $\Delta\mu/kT$ for growth. The lower diagram illustrates how, with increasing driving force (and growth rate) firstly spiral growth, then layer growth occurs. At high driving forces, dendritic and ultimately spherulitic growth result.

in Fig. 7.13) which are more like S or K faces. At relatively high growth rates, distinct differences in fluid composition may exist in a diffusive boundary layer around the crystal, and between the centres and edge of faces, and skeletal crystals are responses to diffusion-limited growth. The enlarging crystal develops a high number of defects which is a contributory factor to the development of rough surfaces. There is also a tendency to split into several units with slightly diverging lattice orientation (Plate 7.3). Where such split growth is systematic it is manifested as sweeping extinction on a polarizing microscope. The ultimate form is a spherulite (Fig. 7.16, base).

Impingement growth

Buckley (1951) demonstrated the rules of impingement growth whereby crystals grow into aggregates, forming compromise crystal boundaries. The most systematic treatment of calcite growth is by Dickson (1983, 1993) who manually constructed growth templates to compare with natural calcite cements (Fig. 7.17). Comparison of three rhombohedral forms (Fig. 7.17e) illustrates that the fastest growth vector (the greatest growth vector) may occur either parallel to the *c*-axis, in the case of an elongated rhombic form, or at high angle to the *c*-axis in the case of obtuse forms. When such crystals grow away from an interface, they compete for space, and the number of crystals diminishes (Fig. 7.17f). The crystals, with their greatest growth vectors orientated more nearly perpendicular to the substrate, will be favoured and so a crystal fabric with a preferred optical orientation develops and is observed in natural calcite cements (Fig. 7.17a–d). Optically, the calcite c-axis also corresponds to the faster of two light rays produced by crystal polarization. The observation that both length-slow (slow ray parallel to crystal length) and length-fast (fast ray parallel to crystal length) mosaics occur in speleothems (Folk & Assereto 1976; Fig. 7.15 A versus C), can thus be explained

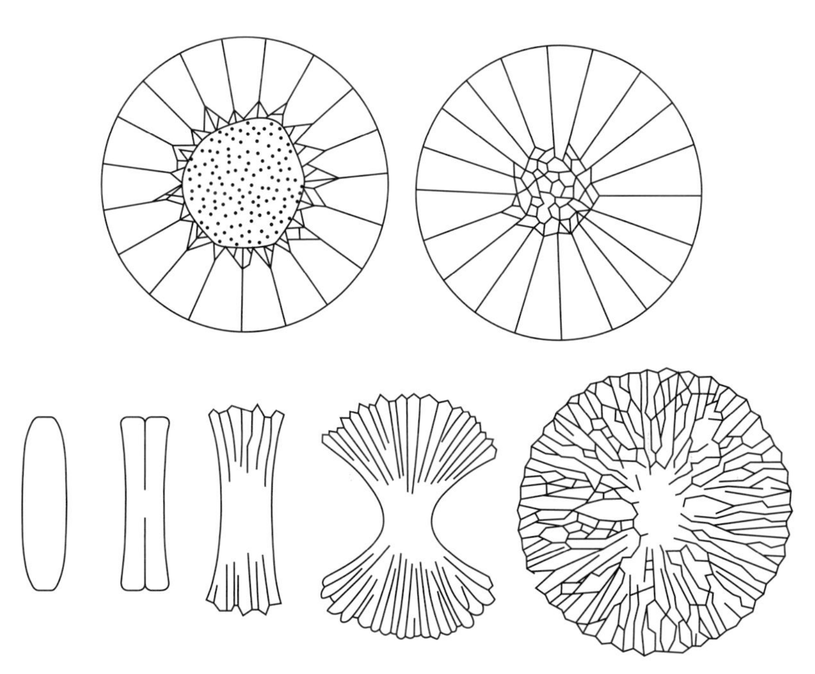

Fig. 7.16 Top, spherulitic growth representing the results of competitive growth away from a substrate. Bottom, spherulitic growth as the end-product of progressive crystal splitting. Redrawn from Grigor'ev (1965).

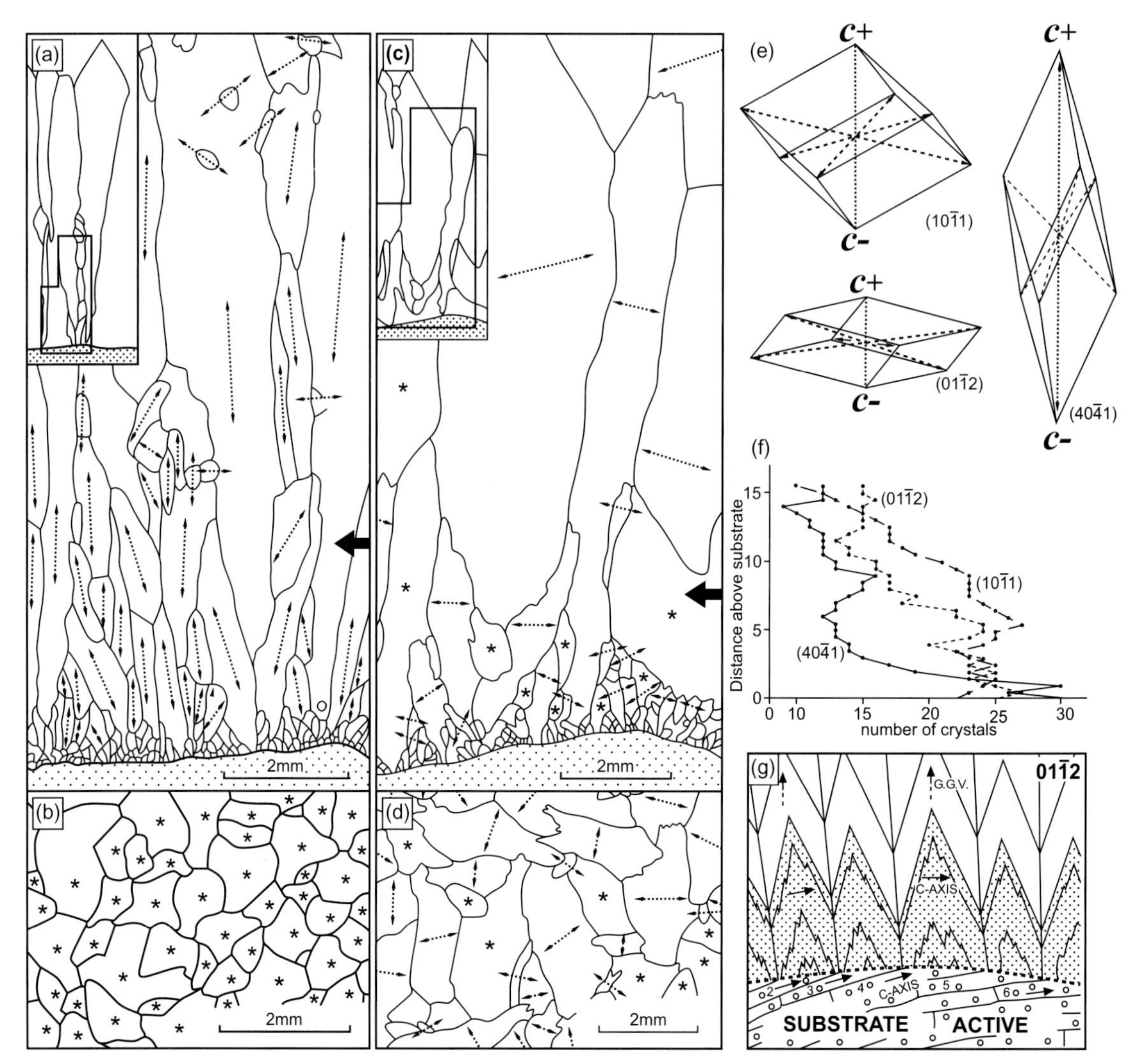

Fig. 7.17 (a, b) Longitudinal and basal sections of natural cavity-lining calcite cement with crystals that are optically length-fast (double-headed arrows denote orientation of *c*-axes within the plane of section; the full extent of the mosaic is shown in the inset). (c, d) Likewise a mosaic which is length-slow. (e) Morphology of three common rhombohedra with the orientations of their greatest growth vectors illustrated. Only the steep rhombohedron, with its greatest growth vector parallel to the c axis will produce length-fast mosaics. (f) Results of a geometrical crystal growth experiment showing how the number of crystals diminish away from the substrate for each of the three forms illustrated in (e). (g) A geometrical crystal growth experiment illustrating how nucleation on an orientated substrate can immediately result in the generation of a columnar mosaic. Within each crystal is a zig-zag pattern illustrating the presence of sideways-merged crystallites. In dripstones, the thin solution film limits the height of crystal terminations (see also Plate 7.4). Note that the nomenclature of crystal faces used in this figure uses the cleavage rhombohedron as the unit cell which is not the most common convention (see captions to Figure 5.7 and Plate 5.1).

by the impingement growth of obtuse or acute crystallographic forms respectively.

Mineralogists find such competitive growth fabrics developed very commonly in vein deposits (Grigor'ev 1965) and in palaeo-caves. Cave pearls display the phenomenon radially (Self & Hill, 2003), rather like that shown Fig. 7.16, top. Although no detailed petrographic observations have been published, it is apparent that this type of mineral deposit that forms the mammillary crusts forming from regional groundwater flows at Devil's Hole (Szabo et al., 1994) from which long palaeo-records have been obtained (Winograd et al., 1992). This growth style would also be expected for the more conventional subaqueous speleothem growth at Corchia Cave, where long-term records (up to 1 Myr (Piccini et al., 2008)) are being obtained. Other, faster growing subaqueous deposits, have more complex laminated structures (see, for example, Babić et al., 1996). Whereas in subaqueous environments, the competition of crystals can proceed freely until interrupted by nucleation or sedimentation events, on stalagmite and stalactite surfaces the growth competition has to be contained within a thin solution film. The consequences of this for fabric evolution have not been quantitatively modelled. However, it would be inappropriate to assume a continuous process of competition, because observations of growth zones in stalagmites (Plate 7.4 and Fig. 1.5b) demonstrate that from year-to-year the surface can change from prominent 100- to 200-μm-high crystal terminations to subdued, near-flat surfaces. This reflects the inherent short-term variability of growth surfaces associated with dripping water.

Stalagmite fabrics

Stalagmite fabrics have been most comprehensively studied and classified by Frisia and Borsato (2010), updating Frisia et al. (2000); a simplified version of their fabric classification is shown in Table 7.2. This synthesis is based on optical and mineralogical determinations and electron microscopy, including much TEM work, as well as related information from dripwater chemistry for modern precipitates. Some of the common calcite fabrics are illustrated in Fig. 7.18 and an example of complex microstructures under TEM is shown in Fig. 7.19.

A key feature of calcite crystals in speleothems is the observation that the macroscopic crystals comprised numerous *crystallites* of similar optical orientation (Kendall & Broughton 1978; Broughton, 1983a, b). Growth surfaces commonly developed a micro-topography consisting of micrometre- to decimicrometre-scale crystals (Fig. 7.20) which preserve the crystal shapes as former growth surfaces previously noted (Plates 7.3 and 7.4). Very similar phenomena arise when calcite crystals nucleate on crystal substrate with preferred crystal orientations (Fig. 7.17g), or a new phase of growth on a large crystal commences (Sunagawa 2005), but a crystallite arises by the development of complex growth domains during the development of a large crystal, i.e. crystal splitting.

Kendall and Broughton (1978) observed that the lateral margins of crystallites tended to growth together (coalesce) below the speleothem surface. Where this process was very incomplete, *fluid inclusions*, elongate in the direction of growth developed (Figs. 7.20a and 7.21c); these are often thorn-shaped. Lack of coalescence leads to the open columnar fabric described in Table 7.2 (Fig. 7.20b), whereas the normal columnar and acicular calcite and aragonite fabrics form continuous $CaCO_3$ crystals with or without inclusions. Inclusions constitute a major feature of the fabric-type named white porous calcite by Genty and Quinif (1996), which alternates with inclusion-poor dark porous calcite in annual couplets. However, the formation of couplets of white porous calcite and dark compact calcite in response to hydrological variations in such studies in France and Belgium is apparently genetically different from those in a perennial wet cave (Katerloch, Austria) where seasonal ventilation is a key control and inclusion-rich laminae form at seasonally low rather than higher supersaturations (Boch et al., 2011). Fluid inclusions assume major significance in the search for primary water chemistries (see Chapter 8), but they remain poorly described in most general speleothem studies. Key observations are to distinguish air-filled (Fig. 1.5b)

Table 7.2 Speleothem crystal fabrics (simplified from Frisia & Borsato, 2010).

A. Columnar fabric: regular stacking of crystallites (typically 50–100 μm wide and more than 100 μm long), typically translucent (low in inclusions); all contain dislocation microstructures.		
Open columnar calcite (parallel crystals with open pores between them)	Free growth of calcite crystals elongated along *c*-axis, typically length: width ratio of 6:1.	Elongation of calcite attributed to high Mg content; may contain many inclusions.
Columnar calcite or aragonite	Compact mosaics; length: width ratio of 6:1 separates short columnar and elongated columnar categories. Crystals have fairly straight boundaries.	Calcite (where elongated and high in Mg may exhibit modulated microstructure) Aragonite forms elongated columnar mosaics with twinned microstructures
Microcrystalline columnar calcite	Crystallites are small (4 μm side and up to 50 μm long), but with similar optical orientation. Crystals have irregular boundaries.	Inclusions abundant; microstructures include lamellae, twins and subgrain boundaries.
B. Fans and fibrous structures		
Aragonite rays	Discrete elongate crystals with square terminations, elongated along *c*-axis; uniform to patchy extinction	Microstructures include dislocations loops, twin, modulated type and subgrain boundaries.
Acicular aragonite or calcite	Compact mosaics of highly elongated crystals (length:width ratio >> 6:1) displaying sweeping extinction.	Dislocation loops; twins for aragonite
Calcite fibres (pliable) and whiskers (rigid) (e.g. found in moonmilk)	Elongated and isolated crystals (up to a few micrometres long and length:width ratio >> 6:1).	Defect-free.
C. Other types		
Dendritic calcite	Branching crystals with interfingering boundaries composed of stacked rhombohedral crystallites (4–10 μm) arranged in rods typically 10 by >100 μm.	High defect density: dislocations, lamellae, twins, subgrain boundaries.
Micrite	Randomly orientated crystals of calcite, <4 μm in size; opaque in hand specimen.	Microstructures unknown.
Calcite mosaics (possibly with aragonite relics)	Calcite spar (>15 μm) mosaics replacing aragonite.	Might show modulated microstructure.

from water-filled inclusions (Fig. 7.21), observe gas bubbles in water-filled inclusions (Fig. 7.21b), and to distinguish the shape and orientation of inclusions and whether they occur in bands parallel to growth (Fig. 7.21c). The precise micro-topography of the surface will bear a strong influence on whether inclusions are created but there is little understanding of the variability in inclusions apart from the generalized observation that they may be more abundant in more complex crystal fabrics which tend to be less compact.

It is common to find minor mismatches in orientation between adjacent crystallites, such that a sweeping extinction develops giving a fascicular-optic or radiaxial fabric (cf. Fig. 2.16). Although in some cases, this relates to growth on a curved substrate, more generally it appears to be an example of split growth (Grigor'ev, 1965). Onac (1997) distinguished dendritic crystals as forming from split growth, in relation to capillary films from skeletal crystals formed by diffusion-limited growth, for example, in feathery tips to soda straws (Maltsev,

Fig. 7.18 Stalagmite calcite fabrics (all thin-section photographs in crossed polars; scale bars = 2 mm). (a) Columnar fabric in CC4 from Crag Cave (Ireland). The straight features within crystals are cleavage planes. The upper box is an SEM micrograph showing the active stalagmite top, where the tips of crystallites composing the larger columnar individuals emerge. (b) Columnar fabric, from AL1, Aladino Cave (Italy). Note the irregular crystal boundaries and the cleavage planes. (c) Microcrystalline columnar fabric from ER78, Ernesto Cave (Italy). Note the interfingered boundaries, intercrystalline porosity, growth laminae, nucleation sites, large voids (black in the photograph), and the absence of cleavage planes. (d) Dendritic fabric in CC3 from Crag Cave (Ireland). The upper box is an SEM micrograph showing the scaffold-like arrangement of crystallites. From Fairchild et al. (2007).

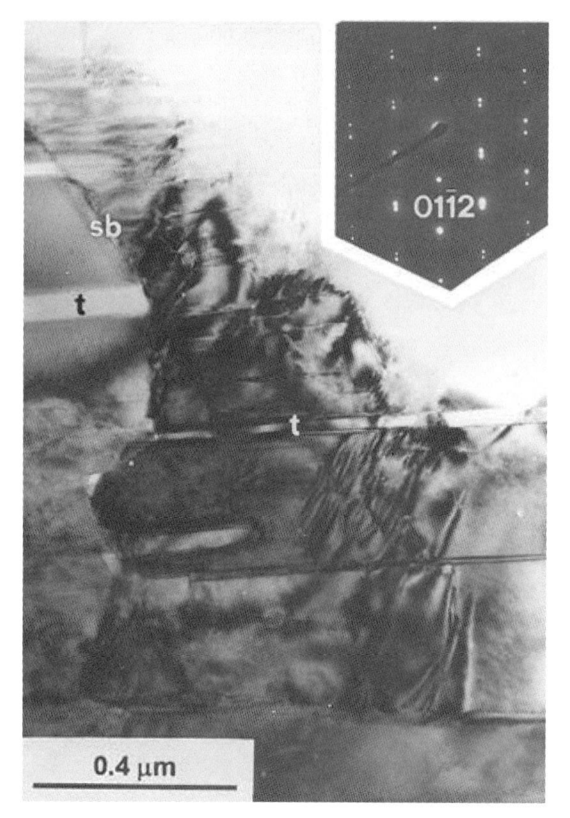

Fig. 7.19 Example of microstructures made visible by TEM in microcrystalline columnar calcite. Mottled dark areas are dislocations. t, twins; sb, sub-grain boundary adjoining dislocation-free crystallite. Inset illustrates the diffraction pattern. From Frisia et al. (2000).

1998), on rimstone dams, or in pools. In stalagmites, the modern technique of electron backscatter diffraction allows split-growth phenomena to be mapped (Neuser & Richter, 2007; Fairchild et al., 2010; Plate 7.3). Side-by-side crystals with similar orientation can also develop as fibrous mosaics through impingement growth.

The primary speleothem fabrics of Table 7.2 can be interpreted in the light of the general principles of crystal growth outlined by Sunagawa (2005), but there are additional factors in the natural environment that need to be borne in mind (Frisia et al., 2000). The columnar calcite fabrics represent growth under relatively low forcing, but in detail their properties depend also on the variability of the environment and the addition of non-carbonate solid impurities and ionic substituents from dripwater. All these factors lead to growth disruptions that are associated with more complex microstructures as seen by TEM, and enhanced crystal splitting or crystal re-nucleation as viewed in optical microscopy, which leads to the more complex microcrystalline columnar variety. Open columnar fabrics reveal the crystallographic forms stable under the growth conditions. González et al. (1992) argued that the form of modern crystals tended to correlate with aspects of water chemistry and hydrodynamics, although the role of the latter is disputed. They observed a wider range of calcite

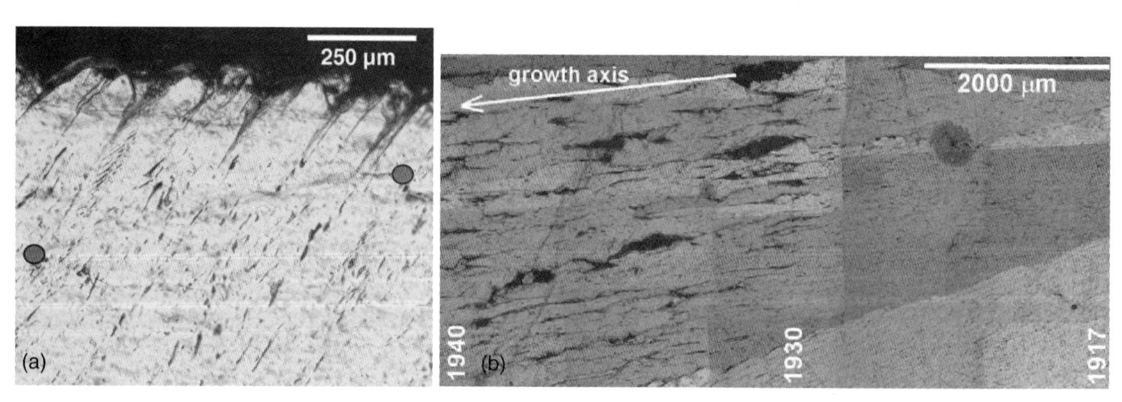

Fig. 7.20 Crystallites and inclusions. (a) Top of modern stalagmite, Moondyne Cave, southwest Australia, showing crystallite terminations and intervening pores. Below the surface the crystallites coalesce, leaving occasional fluid inclusions elongate parallel to the crystallite boundaries (Treble et al., 2005a). (b) Open and closed columnar structures (Treble et al., 2005b).

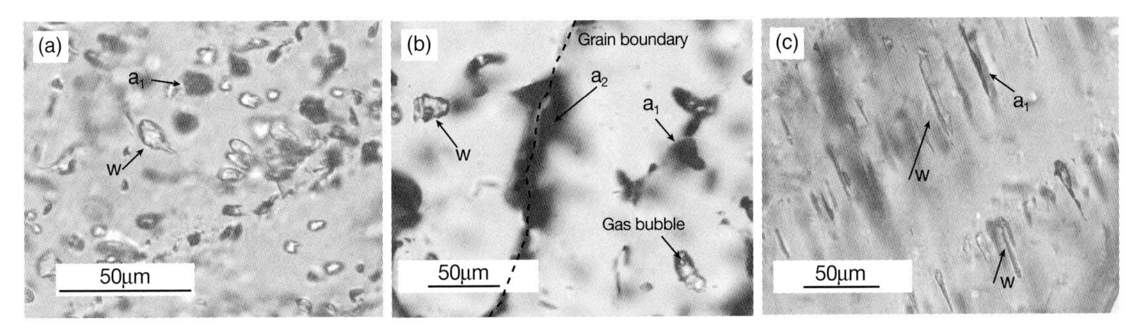

Fig. 7.21 Fluid inclusions from three German stalagmites (Scheidegger et al., 2010). Water (w) and air (a) inclusions, with the latter distinguished as intracrystalline (a_1) and intercrystalline (a_2). The shapes of water inclusion range from elliptical (in a) to strongly elongated (in c) and sometimes contain a gas bubble as indicated in (b). In (a) and (c) groups of inclusions define laminae parallel to the growth surface.

habits in more highly supersaturated solutions, with curved crystal surfaces at the highest values, observations that are consistent with general theory (Sunagawa, 2005). However, the degree of elongation of crystals with closed fabrics depends more on the growth history of the aggregate. Columnar fabric does not in itself imply any isotopic disequilibrium, but the same cannot be said for dendritic fabric which forms under conditions that depart further from equilibrium. A kinetically disturbed chemistry (e.g. with higher or lower $\delta^{13}C$ and perhaps different $\delta^{18}O$ (Frisia et al., 2000, Onac et al., 2002)) has been observed in connection with dendritic fabric, related to growth under higher or more varied supersaturation, or because of infill of dendrites at different times. Aragonite crystal fabrics tend to be more uniform in the case of acicular types, or event-related in the case of rays, but clear guidelines as to the approach to equilibrium growth are not available, except it should be noted that aragonite tends to form from more evolved Mg-rich cave solutions. In the case of isolated elongated calcite crystals in moonmilk, Frisia and Borsato (2010) distinguish pliable fibres, possibly bacteriogenically mediated, from rigid whiskers whose formation may relate to fluid–air interface phenomena. Micritic calcite can occur at nucleation horizons and is particularly common in flowstones; in some cases a bacterial influence is suspected, but interpretation needs to be made in the local context.

Calcite fibres/whiskers are widely known from settings in the vadose zone (Jones & Kahle, 1993). In the speleothem context, whiskers are particularly characteristic of moonmilk deposits (Fig. 7.22) where they appear to form from very slowly dripping, weakly supersaturated solutions (Borsato et al., 2000). Many crystalline materials can develop such a morphology, and one possible origin that also explains its defect-free character is by the rapid growth of one face around a single screw dislocation (Sunagawa 2005), while organic agencies may also promote this morphology in some cases, particularly where the pliable fibre morphology is found (Fig. 7.22b). The microstructures of cave tufas are highly variable as befits an origin under highly variable flows, with some sediment input and enhanced biological activity close to cave entrances (Fig. 7.23). Rapidly changing nucleation conditions can give rise to laminae of micrite alternating with slightly larger crystals (say 4–30 μm) for which the term *microspar* is used (Fig. 7.23b).

Documentation of speleothem fabrics is highly desirable as part of the development of a speleothem time series, particularly when short-term variability, $\delta^{13}C$ and trace element proxies are being used. For the future, the systematic study of fabrics using the technique of electron backscatter diffraction is attractive as it offers the possibility of a quantitative approach and could also be coupled with focused ion beam thinning to allow TEM study at locations whose context is precisely known.

Fig. 7.22 Moonmilk from the Italian Alps. (a) Modern occurrence (white coatings including short stalactites). (b) SEM micrograph of microwhiskers and nanowhiskers. (c) TEM images of nanowhiskers (shades reflect thickness variations; crystals are defect-free). From Borsato et al. (2000).

It will also be helpful to have more experimental and observational data relating fabrics to fluid chemistry and variability in growth conditions.

Although it is commonly assumed that secondary alteration in calcitic speleothems is negligible because of stable mineralogy and young age, some exceptions apply (Frisia, 1996). The presence of macroscopic porosity in the axial zone of a speleothem (Fig. 7.24), where a periodically unsaturated drip impinges, is a common phenomenon and there are other cases where interior calcite cementation appears to have occurred. Aggrading recrystallization may also occur locally (Frisia, 1996), although complex textures in calcite mosaics (e.g. radiaxial calcite) originally interpreted as secondary (Kendall & Broughton, 1978) were later reinterpreted as primary (Kendall, 1985). Where speleothems are flooded, secondary changes are much more likely, particularly if the water is strongly undersaturated or has a different origin from the

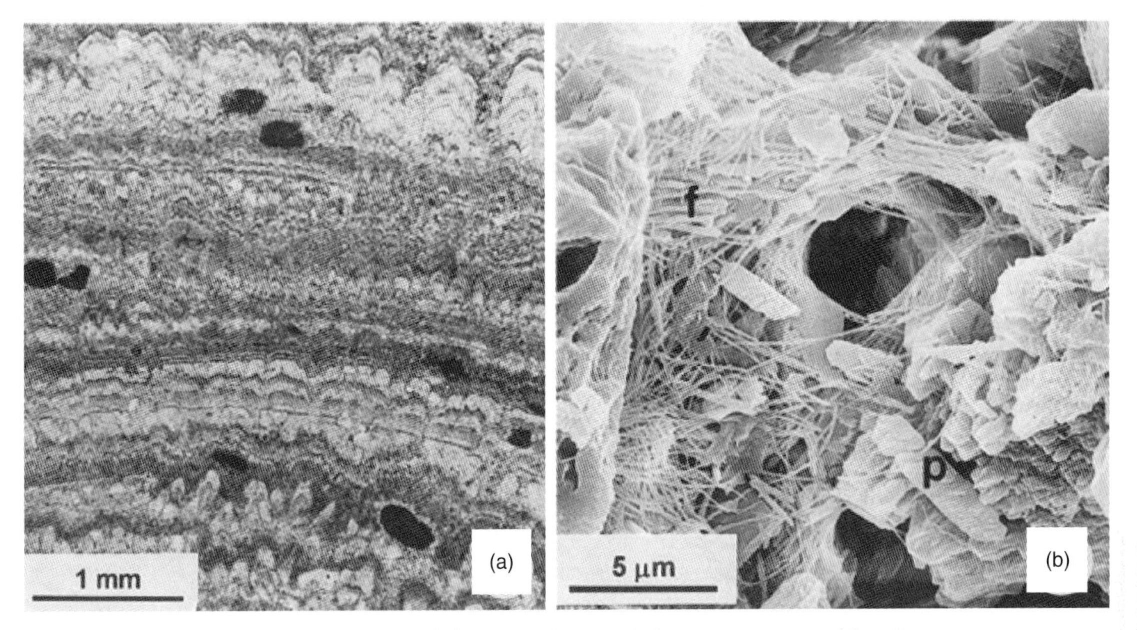

Fig. 7.23 Microstructures of cave tufa, Bus de la Spia, Italy. (a) Micrite aggregates coated by microspar cements. (b) SEM image of porous portion displaying fibres (f) and polycrystalline crystal chains (p). Frisia et al. (2000).

to calcite. Figures 7.25 and 7.26 illustrate a range of associated phenomena. Figure 7.25 shows a late Holocene speleothem where aragonite has been largely replaced by a calcite mosaic, with faces developed on the edge of the calcite crystals. This fabric is diagnostic of the role of crystallization force during the replacement (Maliva & Siever, 1988). Even in such a case, studies of aragonite-to-calcite transitions in meteoric diagenesis more generally has demonstrated that there is a fluid film between the two minerals along which chemical species migrate, explaining the changes in chemistry that are associated with the replacement (Hopley et al., 2009; Fig. 7.26b). However, it is common to find aragonite dissolution outpacing calcite precipitation such that etched aragonite fabrics can be found, sometimes coated with micritic calcite precipitates (Fig. 7.26d). On the other hand, relic aragonite needles can often be found inside secondary calcite (Fig. 7.26b). The distribution of calcitized zones is often complex and currently cannot be predicted. The essential problem is that of understanding the incidence and consequences of

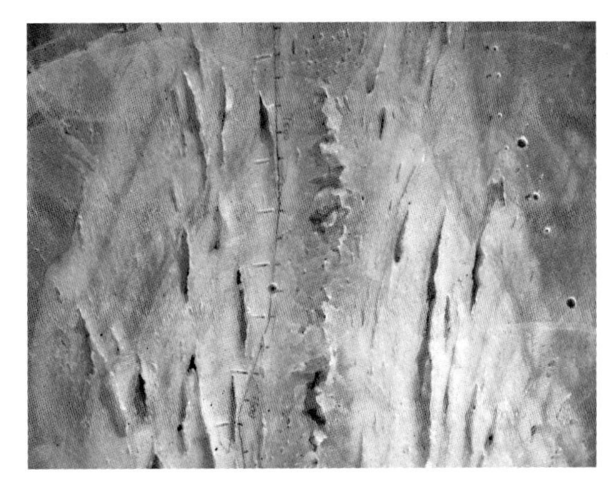

Fig. 7.24 Dissolution porosity within large stalagmite from quarry site, New Zealand (collection of Paul Williams). Tick marks are 1 cm apart.

original meteoric water (see also section 7.3.4). Finally, fungal and bacterial modification of crystals producing micrite is well-documented in some vadose settings (Jones 1987, 2010).

The most important secondary change is that of dissolution of primary aragonite or its conversion

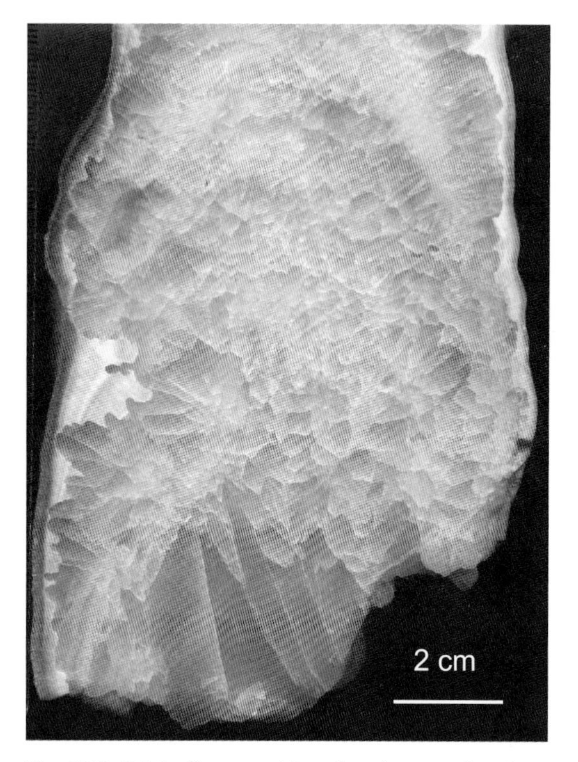

Fig. 7.25 Originally aragonitic stalagmite, now largely replaced by calcite spar with crystal faces against aragonite. Relic aragonite on left and right flanks of stalagmite (Holocene, Valporquero Cave, Spain).

fluid penetration into the interior of speleothems. This is a more complex matter than that of continuous porous media, discussed in Chapter 2.

7.4.3 Laminae

Rhythmic variations in speleothem fabric or mineralogy may occur, forming laminae that are visible in hand section or microscopically, depending on their frequency and the growth rate of the sample. Variations in dripwater chemistry and/or quantity, or cave atmospheric conditions (carbon dioxide concentration, humidity, air flow) may all occur at regular intervals, the most frequent being annual periodicity driven by surface seasonal climate variations (for example, water excess, temperature, soil CO_2 concentration). For speleothems that are forming close to a threshold condition for a particular fabric or mineralogy, small changes in hydrochemistry or cave environment can lead to rhythmic changes in fabric or mineralogy. Section 7.4.2 demonstrated that there are increasingly well understood relationships between fabric and the hydrological and climatological context of speleothem formation. Different speleothems within a cave will have different sensitivities to changing hydrochemistry or cave environment, primarily depending on their hydrological connectivity and physical location in the cave, which will determine the ability of an individual specimen to preserve laminae. 'Minimum diameter' stalagmites (section 7.3.3) are likely not to experience significant seasonal flow variations, for example, and so might be less likely to exhibit annual variations in fabric compared than broader stalagmite forms. Likewise, speleothems found in sections of a cave with little change in cave atmospheric conditions are likewise less likely to contain rhythmic variations in fabric or mineralogy, than those located in more ventilated sections. Some examples of annual laminae observed in rhythmic variations in stalagmite fabric are presented in Plate 7.5.

Laminae found within speleothems may therefore occur at a variety of timescales, depending on the frequency that critical thresholds for the formation of specific fabrics or mineralogies are crossed. For regions of the world where there are significant annual variations in climate processes that are likely to affect speleothem formation (such as hydrologically effective precipitation and soil CO_2 production), laminae of annual frequency are the most likely (Tan et al., 2006). Broecker et al. (1960) were the first to demonstrate regular annual variations in speleothem fabric, demonstrated through the ^{14}C analysis of a flowstone containing annual fabric variations. This early study was not rediscovered until the work of Genty (1992, 1993) who demonstrated annual fabric variations in stalagmites in a disused tunnel at Godarville, followed by a wider range of European materials (Genty & Quinif, 1996; Genty et al., 1997). Railsback et al. (1994) demonstrated that annual variations from calcite to aragonite occurred in a stalagmite from Madagascar. Sub- annual and super-annual laminae may also occur, and these are less well understood or reported. The visual inspection of U–Th-dated

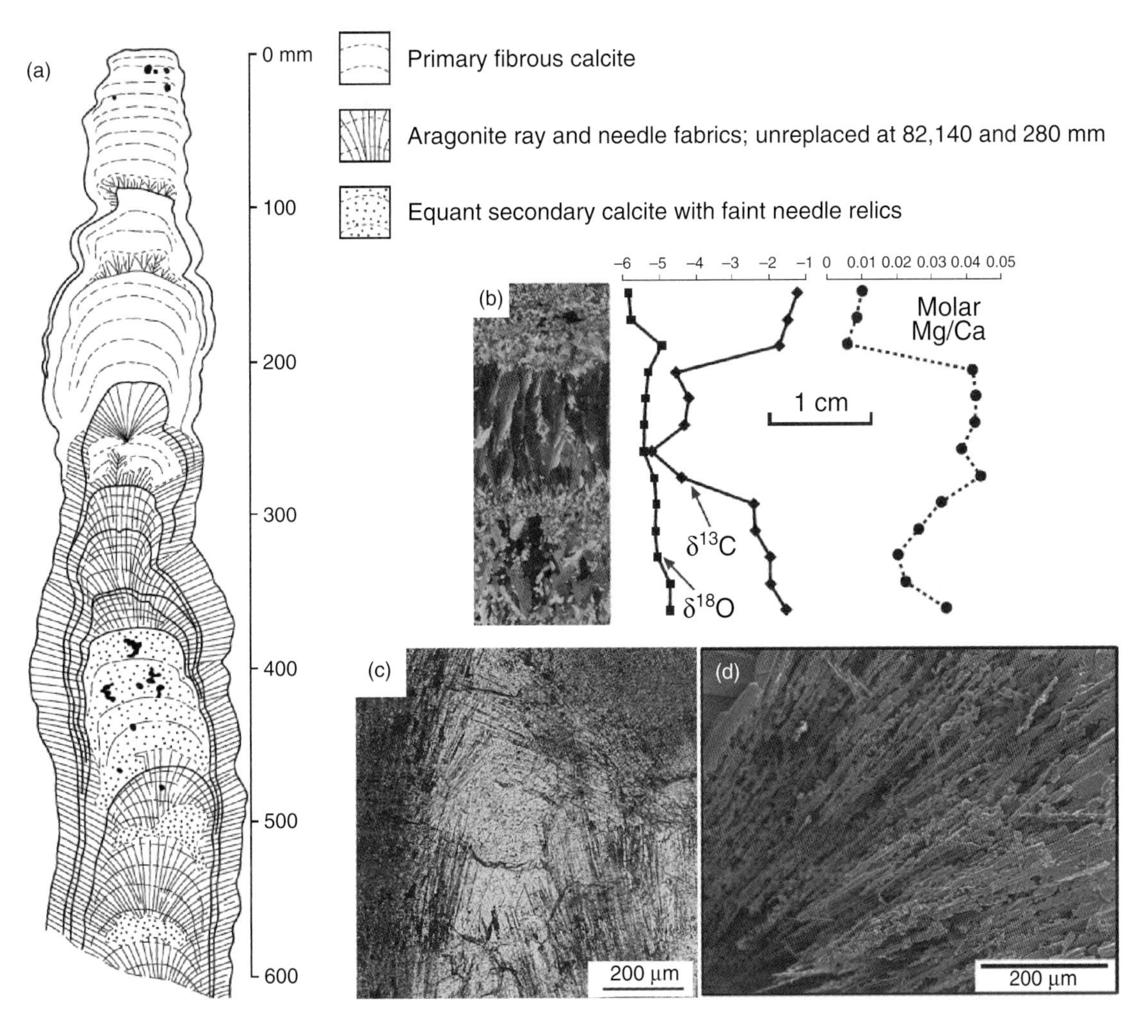

Fig. 7.26 Mixed-mineralogy speleothems.
(a) Stalagmite CL27 with primary calcite, aragonite and calcite-after-aragonite (Frisia et al., 2002). (b) Flowstone with a primary columnar calcite layer separating two calcitized aragonite layers with secondary modification of chemistry (early Pleistocene flowstone, Swartkrans Cave South Africa (Hopley et al., 2009)). (c) Relic aragonite fibres in secondary calcite, same sample as (a). (d) Partly leached aragonite fibres with partial covering of secondary calcite micrite, Castañar Cave, Spain (Martín-García et al., 2009)).

stalagmites demonstrates that super-annual variations in fabric are common, but they are less simple and regular than annual laminae. Baker et al. (2008a) demonstrated that changes in fabric observed in late-glacial stalagmites from northern Turkey occurred at centennial frequency, but did not ascertain a cause for the changes in fabric. Sub-annual variability in fabric or mineralogy is also possible in speleothems that have a rapid hydrological connectivity to the surface; samples have to be those where individual events may be preserved in the speleothem while still maintaining water supersaturation with calcite. Event-scale variations in trace constituents have been demonstrated: for example, in fluorescent organic matter (Baker et al., 1999), where sub-annual variability was related to sub-annual variations in hydrologically effective precipitation.

Speleothem laminae may also be observed in variations in trace constituents such as trace metals and organic matter; examples of the latter are presented in Plate 7.6. The extent to which these form laminae also depends on the geochemistry of speleothem formation, discussed in detail in Chapter 8. Annual variations in Mg, Sr and Ba concentrations were demonstrated by Roberts et al. (1998) and annual variations in fluorescent organic matter by Baker et al. (1993). In both cases these were observed in a speleothem comprised of columnar calcite, demonstrating that annual variations in these constituents can occur without an associated change in fabric or mineralogy. In other cases, annual variations in both fabric and trace constituents are observed, allowing one to elucidate the relationships between them (see, for example, Genty et al. (1997) and Tan et al. (1999). Several extensive reviews of trace element and organic matter annual laminae are available, Tan et al. (2006) focusing on annual laminae formed by changes in fluorescent organic matter and fabric, the relationship between lamina thickness and climate, and a comparison between the climate information likely to be contained in stalagmites in comparison with tree rings. Baker et al. (2008a) additionally considered super- and sub-annual laminae, and Fairchild and Treble (2009) review annual trace element variations (see also section 8.6.3). Distinct detrital flood layers in stalagmites from Spring Valley Caverns, Minnesota, were clearly distinguished petrographically and by Al analyses from annual fluorescent laminae by Dasgupta et al. (2010); the floods had a recurrence interval of 10–50 years and so represent one type of super-annual laminae. The ability to extract high-resolution climate information from variations in annual lamina thickness is considered in Chapters 10 and 11.

7.4.4 Growth phases and hiatuses

The use of the internal structure of speleothems to demonstrate and date environmental changes over periods of time greater than a year is under-utilized at present. Some of the best examples are provided by speleothems that document changes in water level over time (Quaternary archives are considered further in section 12.1.5). Figure 7.27 illustrates an example from the Balearic Islands demonstrating a clear contrast in morphology and internal structure of dripstones, and phreatic precipitates in meteoric waters, whose water table is controlled by sea level.

Long-term variations in water level have also been deduced from the changing structure of speleothems on the walls of a sub-vertical fissure carrying regional groundwater flow at Devil's Hole, Nevada. Szabo et al. (1994) distinguish the occurrence of flowstone consisting of length-fast fibres arranged in growth bundles between interruption surfaces, as evidence for groundwater level below the deposit, porous folia-textured deposits (composed of radiating length-fast crystals) representing deposition close to water level (Hill & Forti, 1997), and dense mammillary deposits (composed of 4 mm × 1 mm length-slow calcite spar crystals with sweeping extinction) representing submerged growth. Changes over time are illustrated in Fig. 7.28.

Particularly clear examples of growth phases are shown by speleothems that are currently in a submarine setting, but which grew during Quaternary sea-level lowstands, sometimes with calcareous serpulid worm growths characterizing the intervening highstands (Plate 7.7; Antonioli et al., 2004, Dutton et al., 2009). Near-surface columnar speleothem calcite was partly dissolved and recrystallized during marine flooding, and an oxide layer also formed in some cases (Fig. 7.29).

More typically, vadose conditions persist throughout stalagmite growth, and interruption surfaces and growth phases are less clear-cut petrographically than in the above examples. However, a prolonged gap in deposition must be owing to either undersaturation of the dripwater or cessation of dripping. Railsback et al. (2011) usefully described and illustrated dissolutional hiatuses, with truncation of underlying layers and presence of clastic impurities from aridity hiatuses, preceded by reduction in stalagmite width and modification in chem-

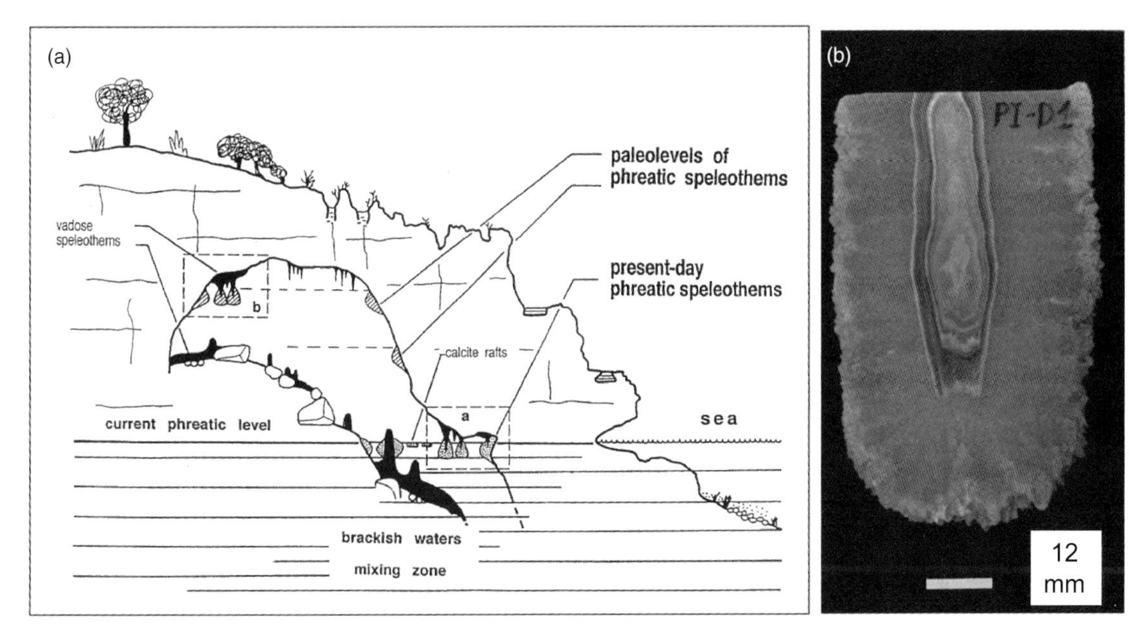

Fig. 7.27 (a) Schematic of speleothems at coastal locations in Mallorca, Spain, illustrating the occurrence of modern phreatic overgrowths on stalagmites and stalactites, and fossil phreatic precipitates representing former highstands of sea level. (b) Stalactite with smooth growth layers overgrown by subaqueous precipitate (scale bar 12 mm). Vesica et al. (2000).

istry. However, it should also be borne in mind that a non-carbonate (e.g. metal oxide) residue can accumulate on the surface by aerosol activity in the absence of dripwater. Other examples of hiatuses have been illustrated, for example, by Onac et al. (2002), Drysdale et al. (2004, supplementary material), Spötl et al. (2008), Bernal et al. (2011) and Luetscher et al. (2011). In a high-alpine example, Spötl et al. (2008) noted that ends of interglacial growth periods were marked by hiatuses displayed as bands of micrite (interpreted as dehydrated moonmilk). Both these and less significant zones of scattered opaque fluorescent inclusions tended to become more prominent on the flanks of the speleothems. Indeed a comparison of flanks and axial zones can be quite instructive in understanding growth histories (Fairchild et al., 2010).

Sometimes repetitive motifs occur in speleothems, implying repeated analogous environmental changes. The intermittent growth of flowstones, for example, is a case where fabrics might be expected to reflect hydrology. Frisia et al. (1993) illustrated several calcitic flowstones from Italy displaying fascicular-optic fabrics where interruption surfaces provide re-nucleation horizons for competitive growth fabrics. A possibly analogous case was provided by Bertaux et al. (2002) where growth-interruption surfaces followed by re-nucleation occur in an aragonitic stalagmite at 2–4 cm intervals (Fig. 7.30) compared with the 0.1 mm thickness of annual laminae where visible. In such cases, a multiproxy geochemical approach matched to the petrology seems the way forward.

7.5 Synthesis

Given the long fascination with the morphology and rate of formation of stalagmites and stalactites, research into their architecture is not as complete

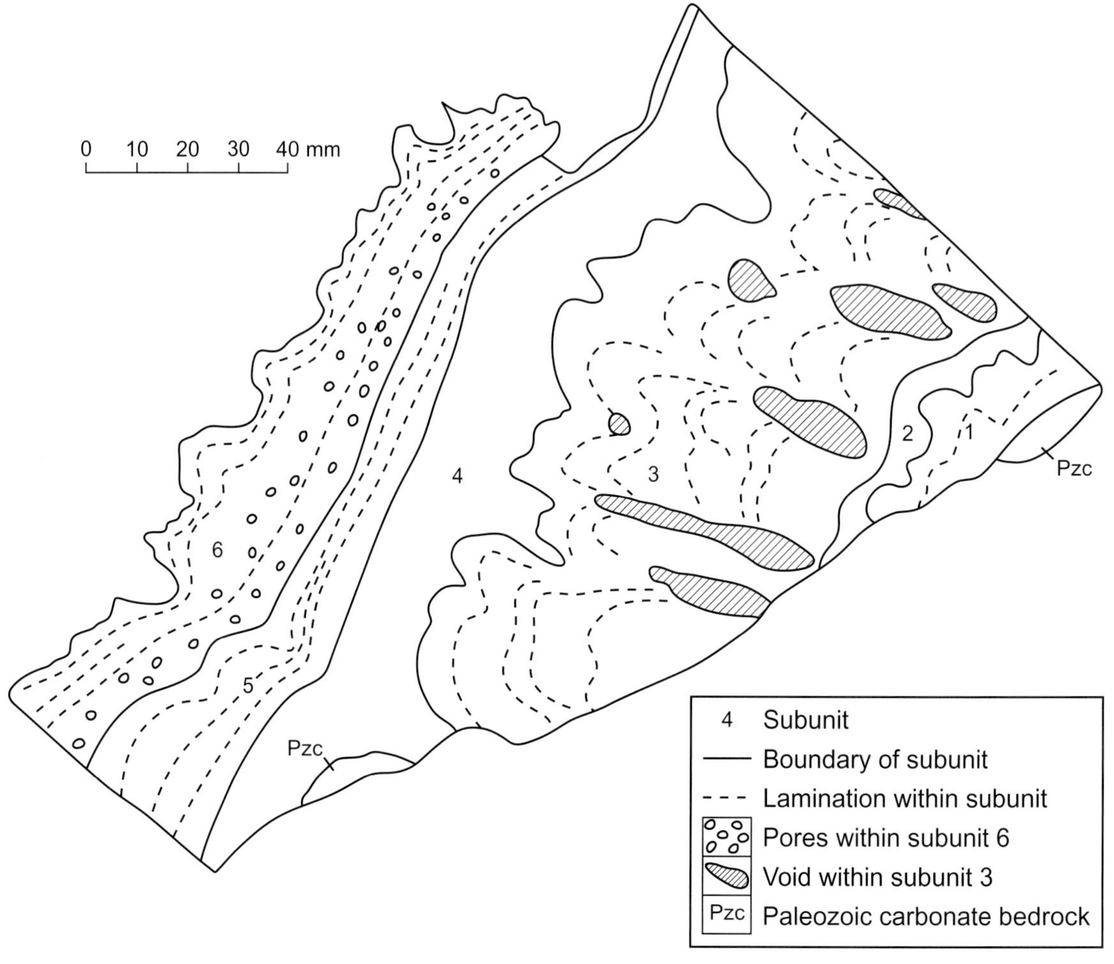

Fig. 7.28 Vertical cross-section through calcite sample DH-BR2 from Brown's Room, Devil's Hole, from 6 m above the present-day water table and orientated as in the field. The sample grew from around 116 to 17 ka. Variable deposits 1 and 2 are succeeded by growth foliae around the water table level in unit 3, mammillary calcite of subaqueous growth in unit 4 and flowstone of units 5 to 6 indicating a lowered water level. Slightly modified from Szabo et al. (1994).

as one might expect. Our knowledge of the processes determining stalagmite growth rates has greatly increased in the last 30 years (section 7.2.1), to the extent that growth rates are a routinely used palaeoclimate proxy, as considered further in Chapters 10–12. In part, this has been coupled with the relatively recent recognition that speleothems often contain laminae and that these are most typically annual (section 7.4.3). The recognition of annual laminae has revitalized interest in the processes determining the internal structure of speleothems, such as mineralogy and crystal fabric (sections 7.4.1. and 7.4.2). However, only a few researchers make use of this information, as well as the characteristics of growth phases (section 7.4.4), when interpreting speleothem geochemical proxies. Similarly, the external morphology is one of the few measurements that can be easily observed and interpreted before sampling (section 7.3), yet is rarely analysed and interpreted in

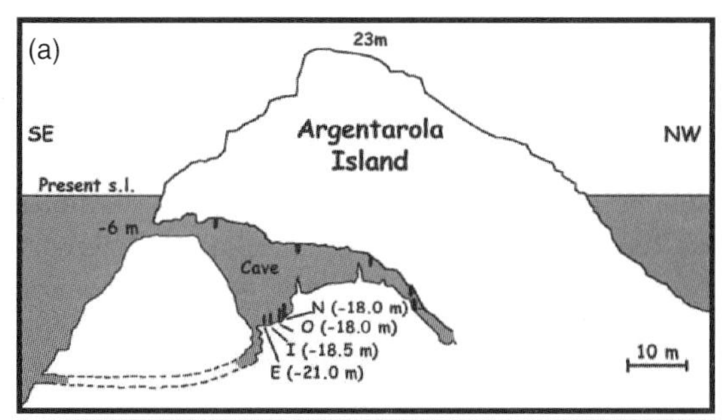

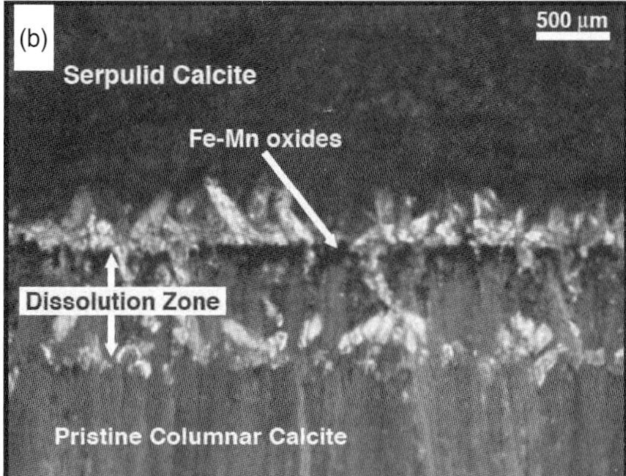

Fig. 7.29 Submerged speleothems from the coastal Argentarola Cave, Italy, where marine serpulid worm growths represent sea-level highstands. (a) Sketch cross-section through Argentarola Cave. (b) Transition from Marine Isotope Stage (MIS) 6 spelean calcite (below) to MIS 5 serpulid layer (above) in sample ASN2 (from location N) shows minor dissolution that has broken off the ends of the crystal terminations; also visible is a thin layer of oxides. Crossed polars image. Dutton et al. (2009, supplementary information).

research publications. Avenues for future research are many and varied. Recent efforts to model stalagmite shape are interesting (section 7.2.2), but assume a constant drip height and ignore the effects of drip splash, which could be incorporated into future models of both morphometry and isotope geochemistry. Models of stalactite shape are less well developed (section 7.2.3), and no attempt has yet been made to numerically model flowstone deposition over time, which would be of great usefulness in understanding the representativeness of individual flowstone sub-samples

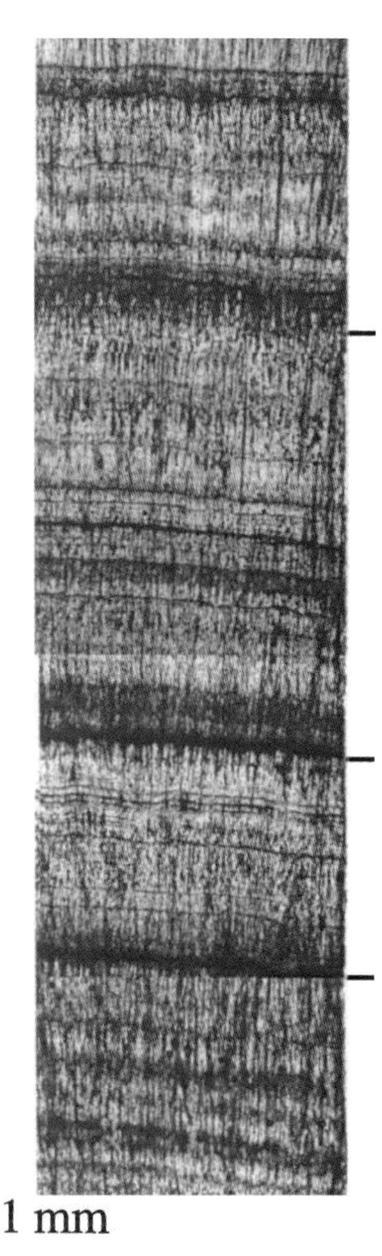

Fig. 7.30 Series of growth phases in aragonitic stalagmite initiating as a dark horizon (interpret as a growth interruption during a dry climatic spell) with fan-shaped bundles of aragonite crystals that evolve into arrays of sub-vertical aragonite fibres with some zones of fluid inclusions. Growth laminae, thought to be annual, are typically around 0.1 mm thick where visible. Mato Grosso do Sul state, Brazil (Bertaux et al., 2002).

typically used for palaeoenvironmental reconstruction. At a more detailed level, research into the depth of the water film on speleothems would be useful to improve quantification of growth rates; a better understanding of micro-scale morphologies observed on stalagmites and stalactites would be valuable, including centimetre-scale ripples and stalagmite 'splash cups'; and a better understanding of the relationship between hydrology and speleothem shape and internal fabric is still necessary, including the critical discharges at which stalagmite and stalactite morphology and fabric change.

CHAPTER 8
Geochemistry of speleothems

The changing chemical composition of speleothems over time forms a major part of their use in palaeoenvironmental analysis. Our treatment here complements recent succinct reviews on carbon and oxygen isotopes (McDermott et al., 2005; Lachniet, 2009) and trace elements (Fairchild & Treble, 2009). We first discuss chemical analysis, focusing on the key issue of spatial scale (section 8.1), and then introduce the nature of the crystal–water interface in relation to chemical composition, discuss the adsorption and incorporation of organic molecules, and assess the likelihood of active involvement of biofilms (section 8.2). We build on the treatment of water chemistry in Chapter 5 by examining the principles and practice of trace element partitioning into $CaCO_3$ in section 8.3. Section 8.4 examines the light-isotope chemistry of speleothems, starting with fluid inclusions and followed by an in-depth treatment of 'equilibrium' and kinetic effects on oxygen and carbon isotope fractionation. In previous chapters we have covered the controls on the composition of water isotopes $\delta^{18}O$ and $\delta^{2}H$ in the atmosphere (Chapter 3), soil processes affecting O and C isotopes (Chapter 3), effects of groundwater hydrology on $\delta^{18}O$ of dripwater (Chapter 4) and karst system controls on $\delta^{13}C$ in dripwater (Chapter 5). We have also examined the controls on U, Sr, Mg and Ca isotopes in Chapter 5, and organic constituents in speleothems and biogeochemistry affecting S isotopes in Chapter 6. In the final sections, we elucidate the expected spatial variability in speleothem chemistry along a water flowline in relation to in-cave processes (section 8.5) and examine the patterns of chemical variation over time in terms of processes on millennial to seasonal timescales (section 8.6).

8.1 Analysis and the sources of uncertainty

Recent rapid advances in instrumentation have outpaced available time and resources to take full advantage of its capabilities. There is increased pressure on speleothem scientists to find cost-effective, well-posed problems, justified from general principles, and appropriate for the field context. Successful execution of a project also requires careful attention to analytical detail. A systematic tabulation of analytical techniques has been presented elsewhere (Fairchild et al., 2006a) and here we focus on new developments in the wider context of justifying a particular research methodology.

8.1.1 What's the research question?

Forty years ago, the current question was how to develop the use of oxygen isotopes as a palaeothermometer (Hendy & Wilson, 1968; Duplessy et al., 1970); hopes were then raised that routine analysis of water isotopes in fluid inclusions would enable solution of the palaeotemperature equation (Schwarcz et al., 1976). However, this analytical issue has taken 30 years further to be resolved. In the meantime, the many potential causes of oxygen isotope variability have become apparent, as has the wide variety of potential analytes in speleothems, so that an extensive range of possible research questions can now be addressed, as is discussed in section 8.6.

Speleothem Science: From Process to Past Environments, First Edition. Ian J. Fairchild, Andy Baker.

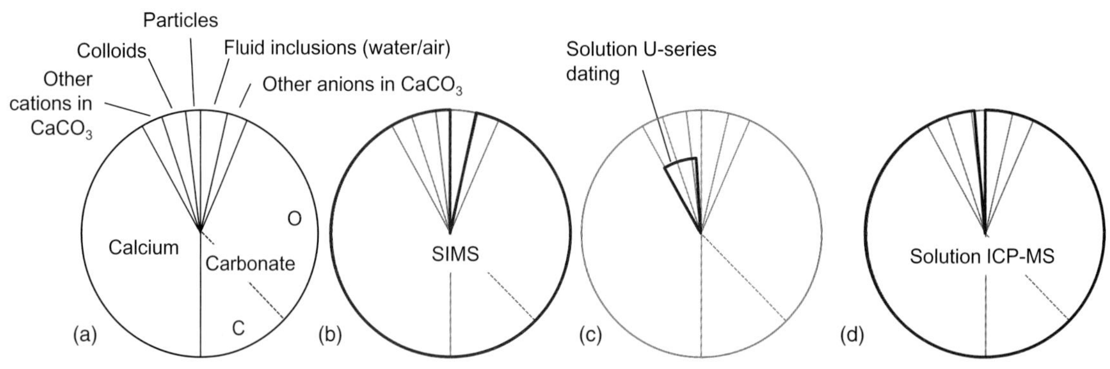

Fig. 8.1 (a) Conceptual illustration of the different genetic components and groups of analytes within speleothems. Particles (>1 μm) are likely to be mainly silicates, oxides or microfossils, whereas colloids will probably be large organic molecules or composite oxide-organic entities. (b) Analysis by secondary ionization mass spectrometry (SIMS) (secondary ionization mass spectrometry) involves complete volatilization of a small pit or narrow trench, but fluid inclusions dissipate instantaneously and would be outliers in steady-state analysis. (c) Leaching and concentration of U and Th for dating purposes primarily recovers ions dispersed in $CaCO_3$ but also includes colloids and some Th and U leached from detritus. (d) Solution-based inductively coupled plasma mass spectrometry (ICP-MS) analysis is not likely to include acid-insoluble particulates which sink to the bottom of the sample tube, whereas they would be aspirated into the instrument by laser ICP-MS analysis.

Choosing the right analytical methodology for sedimentary materials requires thought, not least because they are made of several, genetically distinct materials (Fairchild et al., 1988). Speleothems appear to be relatively simple, being fairly pure chemical precipitates, but they also contain fluid, and particulate and colloidal material (Fig. 8.1). Most of these chemical components are highly inhomogeneous. Hence, both the method and the spatial scale of analysis must fit the research question.

8.1.2 Analytical specificity

A specific analytical method may be quite selective for the different genetic components of Fig. 8.1. For example, some fraction of the colloidal component may be detected by fluorescence, part may be resolved as biomarkers, while an overlapping portion might have a characteristic association with a metal (Box 6.2). Each of these three measurements may represent imperfect proxies for total organic carbon content. On the other hand, some analytical methods do not discriminate between components when ideally we would like them to. Stable isotope analysis will determine both calcite and aragonite where they are mixed. Partial leaching of U and Th from particulate and colloidal detritus leads to significant uncertainty in U-series chronologies in impure speleothems (Fig. 8.1c and Chapter 9). Another example is provided by phosphorus. Analysis of total P by laser-ablation ICP-MS will include organic-bound phosphate, individual phosphate ions and co-precipitated calcium phosphate (Mason et al., 2007). It is very likely that other trace elements also form discrete mineral phases in particular speleothems. A further case is silica. Determination of lattice-bound SiO_2 by colorimetric methods has led to development of a proxy for rainfall, in which drier conditions correspond to high Si (more deposition of reactive aeolian silica (Hu et al., 2005)). Conversely determination of total Si by laser-ablation ICP-MS is likely to include significant siliciclastic detritus, and the rainfall proxy would be expected to work in the opposite sense.

8.1.3 The geometry of the growth surface and spatial precision

Speleothem analysis almost always involves hanging results on an age-model, which is established parallel to growth (Fig. 1.2b and section 9.3).

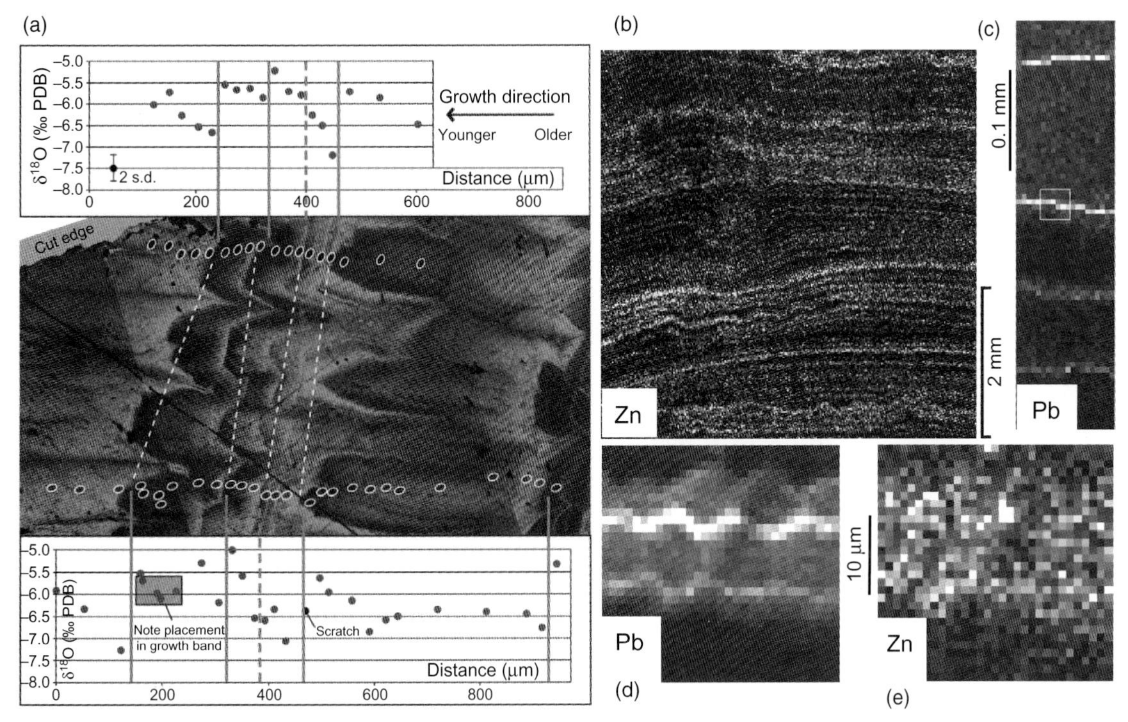

Fig. 8.2 (a) Location of SIMS analytical spots for $\delta^{18}O$ compared with growth banding (upwards to left) revealed by confocal luminescence microscopy from a stalagmite from Soreq Cave, Israel (Orland et al., 2009). Luminescence horizons are interpreted as annual periods of enhanced infiltration and are associated with sudden falls in $\delta^{18}O$ that gradually recover. (b–e) X-ray maps from benchtop X-ray fluorescence (XRF) (b) and synchrotron micro-XRF (c–e) of stalagmite (Obir cave) which has prominent annual impulse laminae, rich in Zn and Pb (as described in Fairchild et al., 2010). Box in (c) shows location of (d) and (e). (d) Individual events, probably only a few days in duration within approximate month of Pb-Zn-enrichment.

The visibility of successive growth surfaces limits the precision with which the age model can be established and hence the time intervals of investigation. Many types of imaging are available to help with this (see summary in Table 7.1), but usually optical views of a polished surface suffice for work on long-term trends. Large-area X-radiographs or elemental maps can be produced using a core scanner (Weltje & Tjallingii, 2008) or XRF instrument (Fig. 8.2b; Koshikawa et al., 2003) but concentrations of target elements need to be of the order of thousands of parts per million for chemical mapping.

The assertion of Fairchild et al. (2001) that annual-scale chemical heterogeneity is normally present in speleothems has been borne out by subsequent work. However, a problem in data interpretation may be presented by irregularities of the growth surface. A range of micromorphologies, from smooth to jagged growth relief up to 200 μm high (Figs. 1.5 and 7.30), bound the solution film during growth of dripstones. Figure 8.2a illustrates an example from a stalagmite from Soreq Cave, showing that it is only by having an image of growth relief that the interpretation of the presence of a regular annual pattern of $\delta^{18}O$ variation can be established in this sample. The confocal method is the latest and most useful mode of viewing UV-excited fluorescence, a phenomenon of great importance in mapping (see section 7.4.3). Other effective mapping tools for annual-scale phenomena, roughly in order of increasing sensitivity, are benchtop XRF (Fig. 8.2b), backscattered electrons (grey shade reflects mean atomic number, McMillan

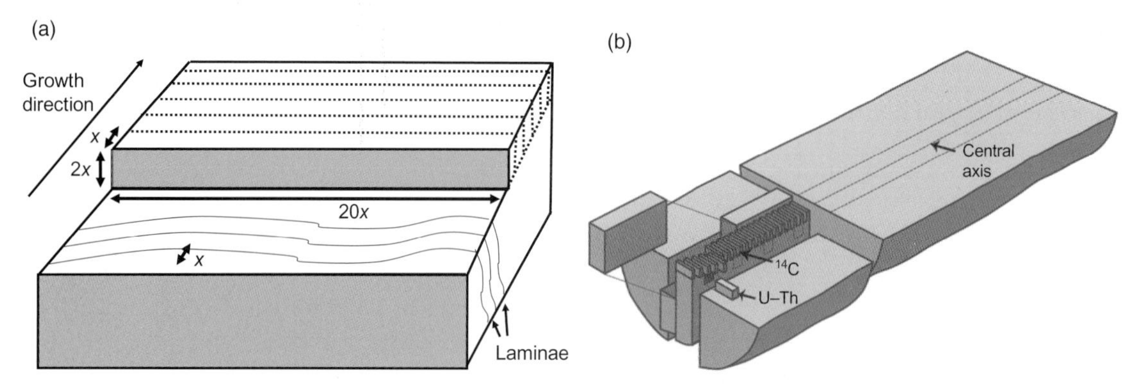

Fig. 8.3 Spatial aspects of sampling. (a) Recommended sample geometry to optimize the time-resolution of sampling (Fairchild et al., 2006a). Actual volumes and masses corresponding to this shape are shown in Table 8.1. (b) Example of sampling strategy when significant sample is need for several forms of analysis (after Hoffmann et al., 2010).

et al., 2005), laser-induced breakdown spectroscopy (Ma et al., 2010), electron microprobe, cathodoluminescence (Richter et al., 2004), laser-ablation inductively coupled plasma mass spectrometry (ICP-MS, Treble et al., 2003), proton microprobe (Ortega et al., 2005) and synchrotron X-ray fluorescence (Frisia et al., 2005b; Fairchild et al., 2010; Fig. 8.2c). An alternative strategy, where there is a clear annual chemical pulse, is to count annual laminae by statistical analysis of chemical line scans (Smith et al., 2009). Nevertheless, mapping studies do sometimes demonstrate significant lateral variations so that single traverses are not necessarily representative of mean chemistry, particularly for trace elements, even where such lines correctly record annuality.

Even where the optimal orientation of sampling parallel to growth is clear, various sampling issues need to be considered. For purposes such as age-dating and continuous sampling at the highest time-resolution, the optimal shape to maximize the time resolution of samples is an obloid parallel to lamination as shown in Fig. 8.3a and discussed in Fairchild et al. (2006a). Table 8.1 gives corresponding masses and volumes for such samples. Where several analytes requiring significant sample volume are required, thought needs to be given to a combined strategy, an example of which is shown in Fig. 8.3b. Figure 8.4 summarizes the typical sample sizes required for different analytical techniques in relation to sample growth rate and time-resolution.

Where a sample is inhomogeneous on the annual scale, and where the length *x* of the sampling obloid (Fig. 8.3a) is equivalent to between 0.3 and 2 years of growth, there are significant dangers of producing an aliased record (Fairchild et al., 2006a; Spötl & Mattey, 2006), in other words a misleading representation of the nature of chemical variability. The same issue arises where sampling is discontinuous (e.g. drilled 0.5 mm pits every 2 mm). Although for an ideal sine wave, four samples per year would approach the annual range of values (approximately 70–85% of the true range), year-to-year variations in annual thickness make a denser sampling of an average of at least six samples per year necessary. This minimum is similar to the sampling density in Fig. 8.2a.

8.1.4 Analytical precision and accuracy

Considerable attention is paid in all isotope laboratories to appropriate standardization methodologies to maximize reproducibility of results. Recent examples include sample presentation issues in (SIMS, Kita et al., 2009), design of laser-ablation ICP-MS systems (Müller et al., 2009), and a comparison of laser-ablation and micromill sampling techniques for U-series dating (Hoffmann et al., 2009). For comparison of samples within a single analytical session, an internal standardization using a reference material of known composition is sufficient to characterize precision. However, more generally, inter-laboratory (or even inter-technique)

Table 8.1 Relationships between sample dimensions and mass of $CaCO_3$ and trace species (from Fairchild et al., 2006a).

Parameter of sampled volume	Units	Examples of different sample sizes						
Distance in growth direction (*x*), obloid	mm	0.05	0.1	0.2	0.5	1	2	5
Length of trench (20*x*), obloid	mm	1	2	4	10	20	40	100
Depth into specimen (2*x*), obloid	mm	0.1	0.2	0.4	1	2	4	10
Side of equivalent sample cube	mm	0.17	0.34	0.68	1.7	3.4	6.8	17
Volume of sample $CaCO_3$	mm^3	0.005	0.04	0.32	5	40	320	5,000
Mass of sample $CaCO_3$	mg	0.0125	0.1	0.8	12.5	100	800	1,2500
Mass of trace species at concentration 1 ppm (µg/g)	µg	0.0000125	0.0001	0.0008	0.0125	0.1	0.8	12.5
Mass of trace species at 10 ppm	µg	0.000125	0.001	0.008	0.125	1	8	125
Mass of trace species at 100 ppm	µg	0.00125	0.01	0.08	1.25	10	80	1,250
Mass of trace species at 1000 ppm	µg	0.0125	0.1	0.8	12.5	100	800	1,2500

comparisons are needed using international standards of defined or exceptionally well-known compositions to gain confidence in the degree of accuracy, that is to ensure that there is no systematic or bias in the results. Analytical uncertainty ideally ought to be reported in terms of the variability over a series of analytical sessions. For a single determination, uncertainty can be expressed in terms of one or two standard deviations (s.d.) of the standard or reference material, 2 s.d. being normal in geochronology. Following repeated analysis, the uncertainty can also be expressed as the standard error (standard deviation divided by square root of number of determinations). Although every analyst makes occasional mistakes, and sporadic analytical artefacts can be difficult to discover (e.g. McDermott et al., 2001; cf. Fairchild et al., 2006a), the uncertainty in speleothem analyses is mostly related to sample inhomogeneity rather than analytical issues.

The precision of trace element analyses of speleothems can be minimized by careful standardization procedures (e.g. for SIMS, Fairchild et al., 2001; for laser ablation ICP-MS, Tanaka et al., 2007; Sylvester, 2008; Strnad et al., 2009) and is typically 1–5% when well above the detection limit (detection limit is defined generally as three times the variability of the background or blank). This contrasts with the more typical errors of measurement of 5–10% for ions in dripwaters. This higher error is because their variable composition makes standardization difficult, as does the issue of variable sample storage times. Oxygen isotopes in dripwaters in contrast can be analysed just as precisely as in speleothems. However, for carbon isotopes, a consistent procedure to minimize degassing-variability before collection is required to compare dripwaters. In speleothems, issues with $\delta^{13}C$ and $\delta^{18}O$ mainly relate to lateral variability because of kinetic effects, as discussed in section 8.4.

A special case is presented by the analytical errors associated with U-series dating in comparison with sampling errors (Fig. 8.4). For example, if the sample size required to obtain sufficient U for dating in a particular sample was 100 mg, this corresponds to a sampling obloid 1 mm in length in the growth direction (Fig. 8.3a). If growth rate was 20 µm yr^{-1}, then the spatial resolution error (1 mm = 50 yr) is equivalent to a 0.5% age error for sample 10 thousand years old (point A, Fig. 8.4). To achieve 0.5% dating errors in a younger sample, either a faster growth (e.g. point B, Fig. 8.4), or a higher U content and hence smaller required sample size (e.g. point C) would be needed.

8.2 The growth interface

The crystal surface displays irregularities on a variety of scales. An individual crystallite displays

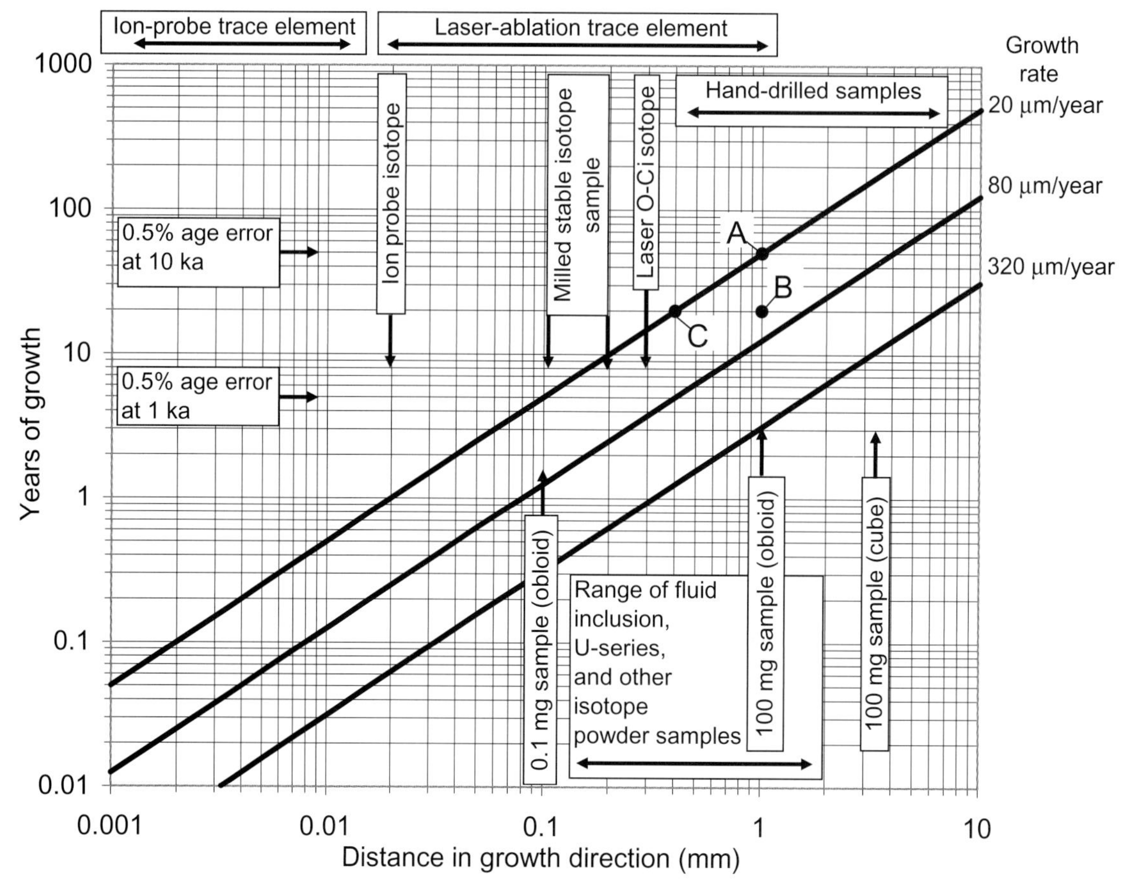

Fig. 8.4 Plot of time versus distance to illustrate sampling issues in relation to temporal and spatial variation. Diagonal lines indicate a range of growth rates inclusive of most speleothems. Table 8.1 gives the unit length of the 'ideal' sampling obloid (with sides 20:2:1, Fig. 8.2b) corresponding to different sample masses. Unit lengths corresponding to 0.1 mg and 100 mg obloid samples are shown in comparison with a cubic volume of mass 100 mg which requires a unit length 3.4 times larger. The spatial resolution of different techniques is indicated, ranging from ion-probe trace element spots of 0.001 to 0.01 mm to hand-drilled sample powders at resolution of 0.5 mm or more. Adjacent to the *y*-axis, the boxes indicate the resolutional limits of a 0.5% age error (e.g. 50 yr for a 10 ka sample). See text for discussion of points A, B and C. Modified from Fairchild et al. (2006a).

typically micron to decimicron-scale relief, and associated inter-crystalline depressions can sometimes develop into fluid inclusions (Fig. 1.5b and section 8.4). The surface of crystallites may be flat or they may display microscopic to nano-scale roughness. At the atomic scale, there are different sites to which ions can attach (Plate 8.1). In classical crystal growth theory (Burton et al., 1951), thermally generated *kinks*, representing niches along growth steps (Plate 8.1), are favoured sites. The steps could be generated as spirals around screw dislocations or, at higher supersaturations, by layer spreading (Fig. 7.15). In this section we span the atomic to microscopic spatial scales in focusing on two aspects: the relationship of morphology to chemistry and the viability of biofilm development on growing surfaces.

8.2.1 Nanostructure of the growth surface

Whereas the scanning electron microscope revolutionized microstructural studies of crystal surfaces, atomic force microscopy (AFM) has brought to life their nanostructures, allowing real-time observation of a variety of phenomena including dissolution behaviour (section 5.3.2), specific crystal growth mechanisms (as discussed in principle in section 7.4.2) and the effects of adsorption of ions and colloids from solution. AFM studies require virtually atomically flat surfaces. Hence the $\{10\bar{1}4\}$ cleavage surface of large calcite crystals has been an important target, imaged in contact with moist air or beneath a moving water film. A hallmark phenomenon at low levels of supersaturation is the development of growth hummocks of four *vicinal* (low-angle) faces around screw dislocations. Figure 8.5 (left) illustrates how each of these faces consists of a staircase of steps with a distinct orientation: the two (+) steps are geometrical equivalents, symmetrical across a glide plane, as are the pair of (−) steps. Since the work of Paquette and Reeder (1990, 1995) it has been known that ion incorporation differs at kinks (Plate 8.1 and Fig. 8.6) on hummock steps of opposite sign, with larger ions being more readily incorporated in the (+) sites. The work of Davis et al. (2000, 2004) demonstrated a change in hummock morphology with increased Mg incorporation in crystals growing from solutions with higher Mg/Ca ratios (e.g. typical of those found in caves in dolomitic host-rock). The hummocks become elongated (Fig. 8.5, centre), possibly because of slow-growth at the boundary of + and − steps where lattice strain arises related to differential Mg incorporation. In turn this leads to the creation of curved pseudo-faces parallel to *c* which explains the observed elongation of Mg-rich calcite crystals (Fig. 8.5, right).

It has been known for some time that a wide range of impurities in solution have the effect of reducing calcite growth rate. The study of Meyer (1984) found that there was a 50% reduction produced by molar concentrations as small as 2×10^{-8} for Fe^{2+}, 10^{-7} for Zn^{2+}, 10^{-6} for PO_4^{3-} and 2×10^{-5} for Ba^{2+}. In these experiments, Mg^{2+} was less sensitive an inhibitor than the other species studied, but it is normally present at much higher concentrations. The mechanism for slowing growth in the case of Mg has been argued to be the enhanced solubility of Mg-calcite, on the basis of an observed linear relationship between step velocities and supersaturation (Davis et al., 2000), whereas if the mechanism had been adsorption at growth steps, this should have had a more pronounced effect at low supersaturations. On the other hand, adsorption is the usual explanation for other growth retardants. For example, Dove and Hochella (1993) demonstrated that addition of phosphate (up to 10^{-6} molar) during growth led to the generation of jagged growth steps rather than smooth ones, implying adsorption of phosphate at kink sites. A similar observation was made for Sr by Wasylenki et al. (2005a), but only at high concentrations ($>10^{-4}$ molar). Teng et al. (1999) demonstrated the likelihood that small quantities of a variety of impurities strongly influence growth step migration even in experiments utilizing relatively pure experimental reagents.

These studies provide some useful background for speleothem chemistry, but it should be borne in mind that the growth hillocks are relatively widely spaced in many experimental studies where relatively defect-poor large seed crystals are used. Natural speleothem surfaces are too irregular to be studied by AFM and the crystals also contain a wider variety of impurities than are present in experimental solutions. Nevertheless, in section 8.3.2, we return to insights from AFM work in relation to understanding growth rate effects on trace element incorporation.

8.2.2 Organic molecules

The calcite surface is a favourable site for adsorption of organic molecules, and bonding between carboxyl (COO^-) groups of macromolecules and cations is a generally important mechanism, which is also implicated in biomineralization (Gilbert et al., 2005). Although high concentrations of carboxylate anions of certain amino acids (e.g. aspartate $HOOCCH(NH_2)CH_2COO-$) are known to promote calcite growth (Stephenson et al., 2008), the more general phenomenon is inhibition. Inskeep and Bloom (1986) demonstrated experimentally that moderate concentrations (0.03–0.15 $mM L^{-1}$ carbon) of 'soluble' (colloidal) organic ligands in the form of aqueous soil extracts or fulvic acid solutions, significantly reduce the rate of calcite precipitation. Complete inhibition corresponded to adsorption of between 30 and 90 C atoms per square nanometre.

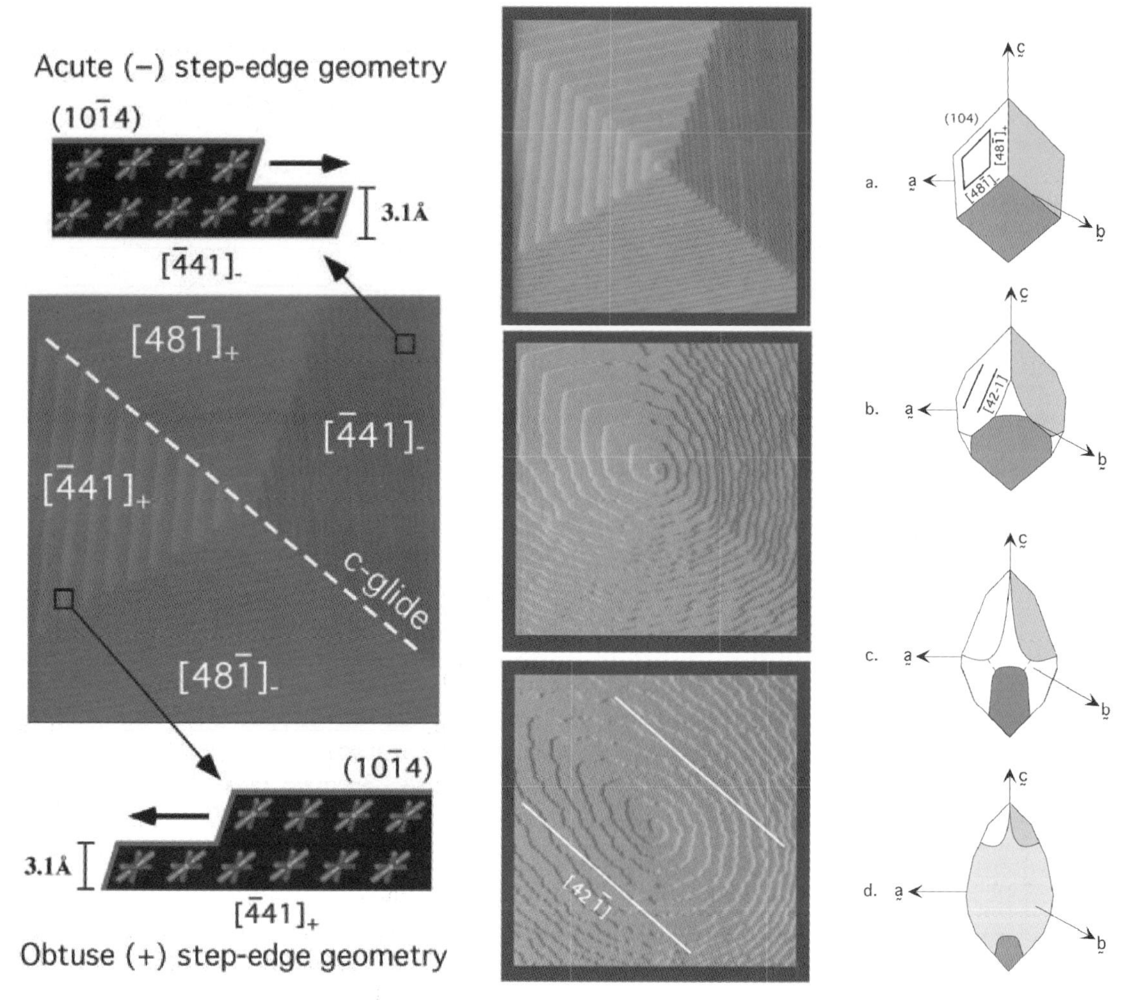

Fig. 8.5 Consequences of high Mg incorporation in calcite: changing growth step morphology leading to more elongated crystal habit (Davis et al., 2004). Left, growth hillock on (10$\bar{1}$4) face produced by spiral growth around screw dislocation, illustrating the development of two pairs of vicinal (low-angle) faces. Note two types of growth step: wide (+) obtuse and more close-spaced (−) acute which correspond to differing types of growth site. Centre. Progressive elongation of growth hillocks (images are 3 by 3 μm) with increased Mg incorporation: top image is Mg-free, centre has aqueous molar Mg/Ca = 0.6 and base = 1.2. Direction of 'illumination' is different in lower image which shows a new step direction parallel to the vector illustrated. Right, illustration of macro-morphological consequences of increased Mg incorporation and the change in the step-edge orientation. Curved growth faces sub-parallel to *c* occupy a progressively large proportion of the surface area.

Preferential adsorption of proteins on particular crystal faces can lead to changes in crystal habit (Jimenez-Lopez et al., 2003; Rautaray et al., 2003). AFM studies illustrate that subtle effects such as adsorption on growth steps in particular orientations are exploited in biomineralization (Yang et al., 2008).

In speleothems, there is evidence both from fluorescence and molecular analyses of incorporation of organic molecules (Chapter 6). Ramseyer et al.

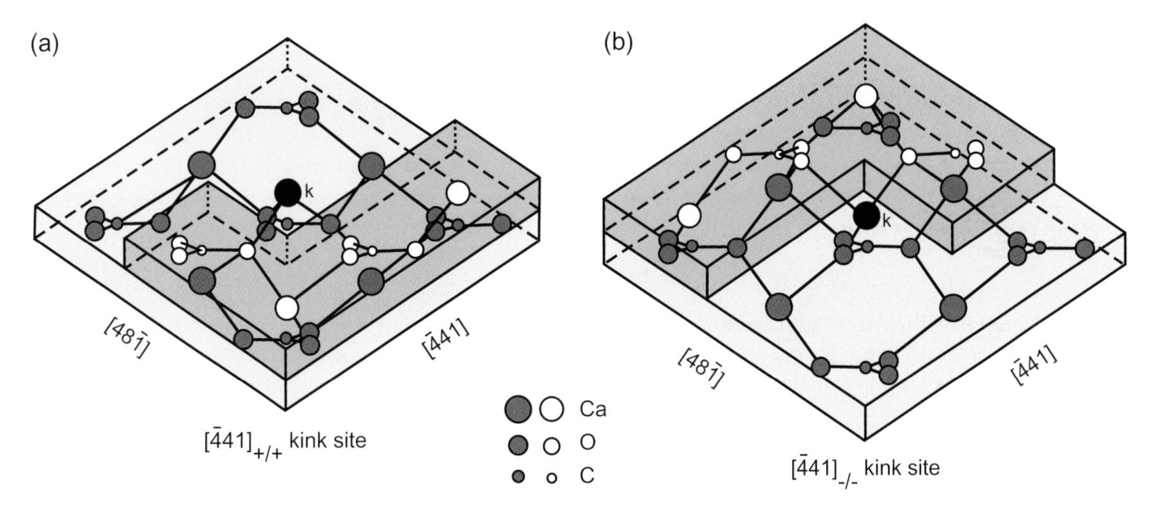

Fig. 8.6 Cartoons to illustrate the different spatial environments of (+) and (−) growth kinks (Paquette & Reeder, 1995) (a and b respectively). Larger ions can be incorporated more readily at the (+) site (diagram a) where the nearest O atoms to the incorporating cation (k) are all in the same plane. White ions are located within the upper (dark-shaded) zone in each diagram, whereas dark grey ions are located in the lower (pale grey shaded) plane.

(1997) examined SEM polished surfaces of fluorescent calcite from which organic matter had been removed. They inferred that the organic matter occurred in 30–150 nm pores between calcite crystallites and could represent an initially adsorbed phase. On the other hand, experimental studies show that co-precipitation (within-lattice incorporation) of organic species (e.g. citrate) can readily occur (e.g. Phillips et al., 2005). Because fluorescent bands can also be enriched in colloid-transported trace elements (Borsato et al., 2007), this raises the issue of whether the two are incorporated while still bound together (Ganor et al., 2009; Fairchild & Hartland, 2010): we return to this in section 8.3.

8.2.3 Biological activity at the growth interface

Jones (2010) described the common tacit assumption (e.g. Kendall & Broughton, 1978) that calcitic speleothems are abiogenic, as 'dangerous'. Certainly there is a wealth of information to be gleaned from them that is of biogenic origin as described in Chapter 6. However, is such an assumption justifiable in relation to their inorganic chemical composition? We conclude that, although there is certainly plausible evidence for bacterial involvement in speleothem genesis, particularly at hiatuses, in general their chemical composition lie within the data arrays of analogous inorganic experiments. Nevertheless, there are several interesting avenues for investigation.

Firstly, let us examine the evidence for $CaCO_3$ precipitation in connection with biological activity. The specific biological agent of interest is a surface *biofilm*, an assemblage of bacteria and fungi which bind to any moist surface and to each other through the agency of mucilaginous excreted polysaccharides known as extracellular polymeric substance (*EPS*). Biofilms in caves have been extensively studied in areas where light is present (including showcaves), and in sulphidic caves, because in both cases they are relatively thick, biologically diverse, and involved in many biogeochemical activities (Barton & Northrup, 2007; Cuezva et al., 2009). However, our main interest is in dark, aerobic environments. In this case, *autotrophs*, primary producers dependent on light or reduced chemical species to fuel photosynthesis, should be absent. However, dripwaters in dark caves do contain genetically similar microbes to those found in showcaves (Q. Liu et al., 2010) and capable of facilitating the precipitation of $CaCO_3$ in relation to sterile controls (Danielli & Edington, 1984; Wang et al., 2010).

Although the chemical mechanisms in natural environments are very difficult to elucidate (Portillo et al., 2009), it seems that both *heterotrophic* bacteria, those that metabolize existing organic matter, and fungi can precipitate $CaCO_3$. This occurs passively, by processes linked to the nitrogen and sulphur cycles causing a rise in pH or removal of growth inhibitors in solution (Barton & Northup, 2007), or actively, associated with rapid growth triggered by enrichment of the environment in organic matter (Castanier et al., 1999). Portillo et al. (2009) demonstrated that cultivated strains of heterotrophic bacteria from aerobic sites showed an ability either to precipitate or dissolve $CaCO_3$, dependent on the nutrient supply. Consumption of acetate, the simplest organic acid breakdown product from humic substances, can lead to raised pH and carbonate precipitation, whereas the converse applies to glucose. Recent research using carbon-isotope labelling suggests that $CaCO_3$ precipitation can serve a physiological function in reducing toxic high levels of calcium and thereby could generate crystal nuclei to contribute to speleothems (Banks et al., 2010). Tourney & Ngwenya (2009) showed that calcite nucleation is favoured within EPS, perhaps because of Ca^{2+}-binding. Petrographically, distinct zones of micrite imply high rates of nucleation, which could be linked to bacterial mineralization, and such phenomena are found in some pool deposits, cave pearls, and occasionally stalactites and stalagmites (Barton & Northrup, 2007; Jones, 2010).

Secondly, we review evidence for an influence on the chemical composition of speleothems: this is more restricted. The biological role in tufa construction is far more obvious than in speleothems and yet Andrews (2006) and Shiraishi et al. (2008) found little evidence of any disturbance to physico-chemical predictions of carbon or oxygen isotope composition. Under stagnant conditions there can be a slight enrichment in ^{13}C, presumably reflective of removal of light carbon through photosynthesis (Andrews, 2006). In the only study of its kind, Cacchio et al. (2004) cultured strains of Gram-positive bacteria extracted from ground-up stalactites from Cervo Cave in Italy and found that most strains yielded $CaCO_3$ with slightly higher $\delta^{13}C$ values than the mean stalactite composition, but a few were lower, by up to 20‰, but the biochemical pathways responsible for such differences are unknown. Likewise, when considering trace element (alkaline earth) chemistry of tufas, Ihlenfeld et al. (2003) found a reasonable agreement with that expected inorganically. Nevertheless, Rogerson et al. (2008) demonstrated in mesocosm experiments that tufa biofilms sequestered alkaline earth ions within EPS, with a preference for the larger ions. They suggested that this could explain some chemical properties of the inorganic precipitates within EPS, e.g. Sr–Ba covariations in tufas, and could modify the observed Mg response to temperature studied by Ihlenfeld et al. (2003). It is difficult to know how much account to take of comparisons with tufa studies because biofilms are much better developed in streams and there is significant autotrophic activity. By contrast, there is little information (e.g. mean thickness, three-dimensional structure or presence of EPS) on biofilms on dripstones in dark aerobic caves. However, even if a biofilm, as opposed to a few scattered microbial cells, is present, its mere presence does not necessarily influence solution composition at the growing crystal surface. Generally, biofilms usually have complex shapes rather than having a simple two-dimensional morphology; this facilitates the access of host solution to the substrate (Wimpenny et al., 2000). The biofilm as an agent for mediating $CaCO_3$ precipitation differs from many invertebrate groups where there is a distinct vital effect on carbonate chemistry generated by specific physiological regulatory processes either outside or within cells, including control of Ca and Mg concentrations (Weiner & Dove, 2003). Another potentially relevant fact is that enzymes responsible for CO_2-uptake (carbonic anhydrases), and which eliminate kinetic effects for $\delta^{18}O$ during $CaCO_3$ precipitation, are common in bacterial species (Barton & Northrup, 2007).

An appropriate way forward, in the spirit of this book, would be to take more account of mass balance. Biofilm growth rate is limited by supply of nutrients and the doubling time for the biomass in a biofilm has been shown experimentally to be at least 70 hours for mature biofilms, if sufficient nutrients are supplied (Davey & O'Toole, 2000). The ratio of organic C input to $CaCO_3$ produced during bacterial calcification in cultures is around 0.6 (Castanier et al., 1999). A typical stalagmite, growing at 0.3 mm yr^{-1}, might grow upwards by around 3 μm during a 3-day period. It is difficult to see how more than a tiny proportion of this calcite could relate to bacterial activity when we consider that dripstone environments are usually *oligotrophic* (nutrient-poor, oxygen-rich) environments and that modern stalagmites are compact with very low concentrations of organic matter (Huang et al., 2001). Concentrations of organic C in dripwater are of the order of 1 mg L^{-1}, some 1.5–2 orders of magnitude lower than that of Ca, although potentially significantly higher in a particular season (Baker et al., 1997b; van Beynen et al., 2000; Hartland et al., 2010b). These carbon-limited systems represent very different conditions from culture media.

The above considerations imply that normal continuous speleothem growth of sparry calcite with clearly defined chemical growth zones, and often with well-developed crystallite faces, is clearly triggered by physicochemically induced supersaturation. Conversely, when growth is intermittent or absent, it becomes more likely that microbial activity has a significant role to play, for example in generating new crystal nuclei at hiatuses. An example given by Jones (2009) is the occurrence of phosphate mineralization on corrosion surfaces on speleothems in a cave on Grand Cayman Island where moisture condensation was occurring, leading to good preservation of phosphatized microbes. Whereas Jones (2009) used such an occurrence to cast doubt on the use of phosphorus as a proxy for seasonal changes, we would argue that the context is completely different from the observed annual enrichments in P in continuously laminated stalagmites (Fairchild et al., 2001). Instead, the lesson seems to be that the microbial influence is most likely to be found in the chemistry of hiatus surfaces. Modern molecular biological techniques, particularly the 16S ribosomal RNA (rRNA) signature (section 6.1.3), has demonstrated that, just as in soils, there is a huge diversity of microbial life present at any point in a cave, most of which cannot be cultured (Barton, 2006). Depending on the supply of nutrient elements, different organisms will prosper and might be associated, for example, with Mn-oxidation, phosphatization, calcification or etching, depending on the local context.

8.3 Trace element partitioning

It seems entirely possible that calcite is capable of incorporating any element in the periodic table that presents itself in an appropriate form. More specifically, the high capacity of the calcite structure to substitute trace metal ions is thought to be because of the corner-sharing topology of the octahedra of oxygen atoms to which cations are bound, which allows tilting and bending (Reeder et al., 1999). The literature of crystal growth and incorporation of trace impurities into calcite and aragonite is enormous because there are many applications, not least in crystal design for industrial purposes, and in marine science. In the karst context, the low ionic strength of cave waters makes the physical chemistry relatively simple, but this is compensated for by their unsteady composition in respect of both P_{CO_2} and trace species, and the necessity to consider transport of many elements by colloids. Compilations of experimental data on partitioning were summarized by Rimstidt et al. (1998), Curti (1999) and Böttcher and Dietzel (2010), although experimental conditions varied significantly. From the earliest studies on trace elements in cave carbonates, there was a desire to reconstruct solution compositions from speleothem chemistry (Holland et al., 1964; Gascoyne, 1983). We show below how this can readily be done semi-quantitatively, but that exact solutions are impractical in the cave environment.

8.3.1 Thermodynamic and mixed empirical-thermodynamic approaches

Trace metals with valence two are normally regarded as substituting in Ca^{2+} lattice sites to form a solid solution. Following the approach of Lakshtanov and Stipp (2007), and taking Mg^{2+} as an example, we can consider the relative stability of the end-members of the solid solution:

$$(Mg^{2+})(CO_3^{2-}) = K_{MgCO_3} X_{MgCO_3} \gamma_{MgCO_3} \qquad (8.1)$$

$$(Ca^{2+})(CO_3^{2-}) = K_{CaCO_3} X_{CaCO_3} \gamma_{CaCO_3} \quad (8.2)$$

where parentheses denote activities in solution, K are equilibrium constants (solubility products), X are mole fractions and γ are activity coefficients in the solid phase for the appropriate end-members. In a dilute solid solution, γ_{CaCO_3} is approximately one and γ_{MgCO_3} is a constant. If we also argue that the ratio of activities in solution is equal to ratio of molalities (m) and divide eqns. 8.1 and 8.2, we obtain:

$$[X_{MgCO_3}/X_{CaCO_3}][m_{Ca^{2+}}/m_{Mg^{2+}}] = K_{CaCO_3}/[K_{MgCO_3}\gamma_{MgCO_3}] = \text{constant} \quad (8.3)$$

If the aqueous ions are not complexed by other ions, the term on the left-hand side of eqn. 8.3 has the same form as the Henderson–Kraček partition coefficient (McIntyre, 1963) and hence the constant in eqn. 8.3 can also be regarded as a partition coefficient. This relationship was introduced in an earlier chapter simply as an empirical equation (eqn. 5.18) where the partition coefficient (D_{Tr}) indicates the extent to which a trace element to calcium ratio is increased ($D_{Tr} > 1$) or decreased ($D_{Tr} < 1$) in the $CaCO_3$ compared with aqueous solution.

It can be seen from the central part of eqn. 8.3, that a key determinant of the partition coefficient is the ratio of solubility products K_{CaCO_3}/K_{MgCO_3}. For this reason, we can predict that because calcite is much less soluble than magnesite ($MgCO_3$), but more soluble than siderite ($FeCO_3$), D_{Mg} will be less than one and D_{Fe} will be more than one. However, where the mixing within the solid solution is strongly non-ideal, γ values may be much less than one and differ from element to element. Another limitation of this approach is that for many elements, there is no trigonal end-member for the trace element in question (e.g. Sr where the end-member $SrCO_3$, strontianite, has an aragonite-like orthorhombic structure); or the stability of such an end-member may not be known.

Several studies have mixed theory and empirical approaches to partitioning behaviour in order to progress these issues. The paper of Rimstidt et al. (1998) is widely cited, although Prieto (2010) has identified some theoretical inconsistencies in this work. The approach of Wang and Xu (2001), building on the model of Sverjensky (1984), appears to be the most consistent to date. Wang and Xu (2001) posit that there are three factors that control partitioning from a thermodynamic point of view:

1 the chemical bonding energies of the different cations within $CaCO_3$;

2 the excess energies created by the size difference between the substituting ion and Ca^{2+};

3 the difference in *chemical potential* (the thermodynamic energy per molecule) between the substituting ion and Ca^{2+}.

The *Gibbs free energy* is a thermodynamic measure of the work obtainable from a system. In relation to a particular chemical reaction, it has the following relationship to the equilibrium constant (K):

$$\Delta G^{o} = -RT \ln K = -2.303RT \log K \quad (8.4)$$

where ΔG^{o} is the standard change in free energy associated with a reaction, R is the gas constant, and T the absolute temperature. It is found that ΔG^{o} values for a mineral of a particular composition in a given isostructural series (e.g. $CaCO_3$–$MgCO_3$) can be expressed in terms of three linear terms:

$$\Delta G^{o} = a\Delta G^{o}_{M^{2+}} + \beta r_{M^{2+}} + b \quad (8.5)$$

where $\Delta G^{o}_{M^{2+}}$ is the free energy of formation of cation M^{2+} (in the lattice, whereas $\Delta G^{o}_{\text{solvated } M^{2+}}$ is the equivalent for the aqueous solution), $r_{M^{2+}}$ is its radius, and a, β and b are constants specific to the isostructural series. Figure 8.7a (Wang & Xu, 2001) shows how experimentally determined partition coefficients for divalent ions in calcite correlate with a combination of ion properties developed from this relationship. Wang and Xu (2001) successfully used a similar approach to explain the high values of $\log D_{Tr}$ (2–4) for triply charged rare earth elements (REE) in sea water where charge is balanced by the co-substitution of Na^{+}. Figure 8.7b shows a comparison of experimentally determined values with those calculated in three separate studies, illustrating their semi-quantitative agreement.

It has been observed since the pioneering study of Lorens (1981) that partition coefficients become closer to unity at faster growth rates (Figs. 8.7a and 8.8). One explanation for this could be that the incorporation of the ions was controlled by diffu-

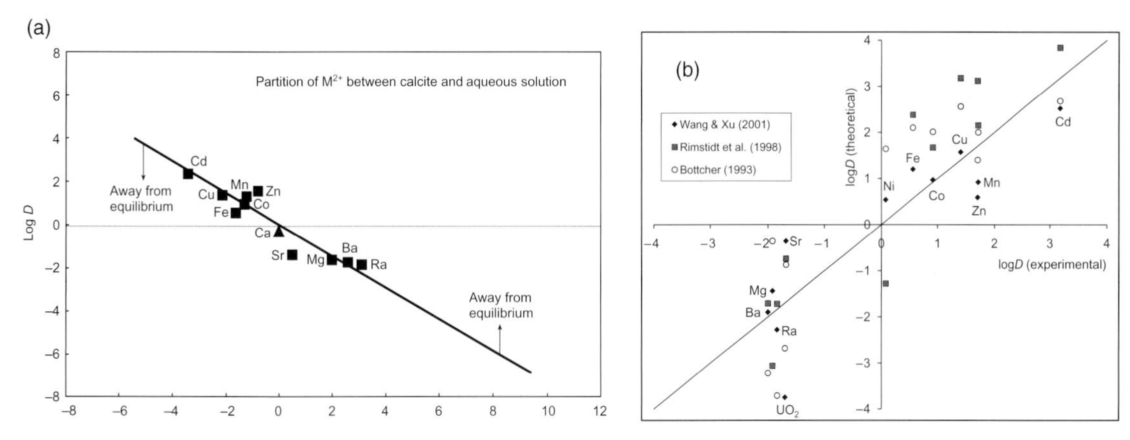

Fig. 8.7 (a) Correlation of log partition coefficients in calcite with metal cation properties (Wang & Xu, 2001). The x-axis corresponds to $0.968(\Delta G^{\text{o}}_{M^{2+}} - \Delta G^{\text{o}}_{Ca^{2+}}) + 75.17(r_{M^{2+}} - r_{Ca^{2+}}) - (\Delta G^{\text{o}}_{\text{solvated } M^{2+}} - \Delta G^{\text{o}}_{\text{solvated } Ca^{2+}})$ with units of kcal mol^{-1}. At faster growth rates ('away from equilibrium'), log$D$ values converge on zero, i.e. discrimination between substituting ions and calcium diminishes. (b) Comparison of predicted partition coefficients (log$D$) by Böttcher (1993), Rimstidt et al. (1998) and Wang and Xu (2001) with experimentally determined values at 25 °C and one atmosphere pressure. Data compiled by Böttcher and Dietzel (2010) with experimental Ni value from Lakshtanov and Stipp (2007).

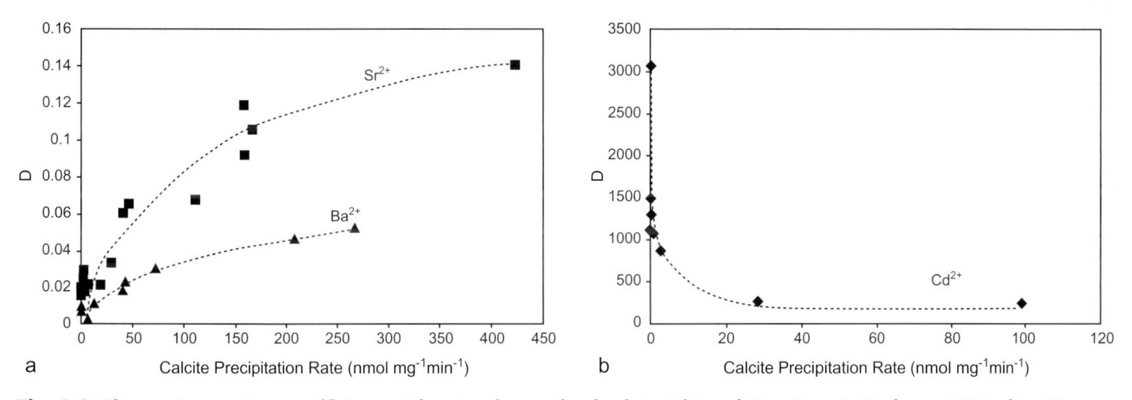

Fig. 8.8 Change in partition coefficient with rate of growth of calcite (data of Tesoriero & Pankow, 1996 after Wang & Xu, 2001).

sion in a boundary layer around the crystal (Tesoriero & Pankow, 1996). However, Wang and Xu (2001) pointed out that the values do not reach one, which would be required by this hypothesis. They articulated a broader conceptual model (Fig. 8.10). At slow growth rates, a characteristic impurity to calcium ratio develops, allowing for interdiffusion between the adsorbed layer and the bulk crystal and for adsorption and desorption at different rates. At higher growth rates, diffusion becomes more important and there is insufficient time for exchange with the interior; hence the asymptotic values at high growth rates correspond to the composition of the adsorbed layer. The lack of evidence for a growth rate effect on REE incorporation could be accounted for by their known very strong adsorption, implying that the adsorbed layer has a similar composition to the bulk calcite, whatever the rate of growth. However, this concept sits uncomfortably with the observations of Zachara et al. (1991) who found the relative order of surface affinity of divalent cations to be (Cd > Zn ≈ Mn > Co > Ni ≈ Ba = Sr), which is more or less the exact opposite to that predicted by this narrative.

The surface entrapment model of Watson (2004) develops the issue of ion diffusion within near-surface lattice layers further, and in a formal way. However, although experimental data on Sr can be made to fit the model (e.g. Gabitov & Watson, 2006, Tang et al., 2008b), this could be fortuitous because diffusion coefficients in the solid phase are not known. Indeed, entirely different mechanisms for growth rate effects have been proposed (De Yoreo et al., 2009), as we will see in the next section.

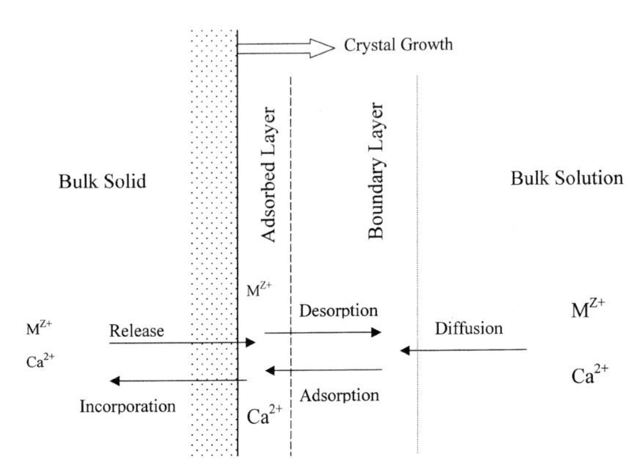

Fig. 8.9 Conceptual model of the kinetic controls on element partitioning into calcite (Wang & Xu, 2001). At faster growth rates, the composition of the bulk solid becomes closer to the composition of adsorbed ions. However, such phenomena as controls on trace element content may be outweighed by changes in kink site distribution (see next section).

8.3.2 Limitations of the partition coefficient concept

The partition coefficient concept is undermined by a series of observations, some of which we introduced in section 8.2.

1 The classic thermodynamic approach does not consider the specific growth sites on the crystal surface which can vary greatly in morphology and energetics. A clear illustration of this is the development on the common rhombohedral crystallographic form of different kink sites around spiral growth hummocks (Fig. 8.6) which display characteristically different trace element patterns (Paquette & Reeder, 1990). Larger ions (Ba^{2+}, Sr^{2+}, SO_4^{2-}, SeO_4^{-}) are enriched at the (+) kink sites while smaller ions (Mn^{2+}, Co^{2+}, Cd^{2+}, Mg^{2+}) are enriched at the (−) kink sites (Staudt et al., 1994; Paquette & Reeder, 1995; Reeder, 1996). Exceptions are the enrichment at (+) sites by Zn^{2+} and the (−) sites by AsO_4^{3-},

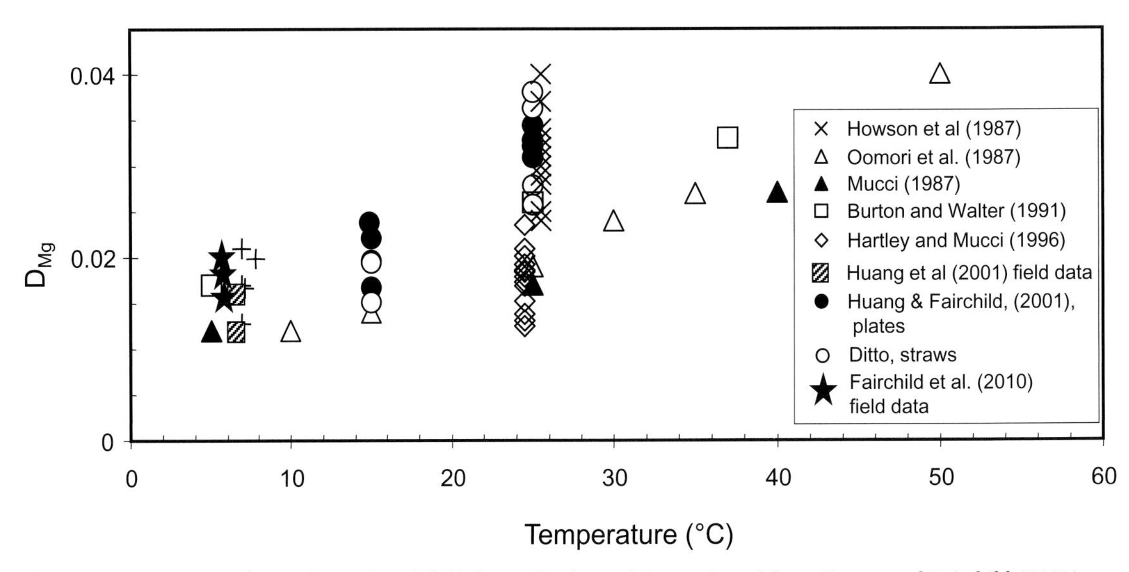

Fig. 8.10 Comparison of experimental and field determinations of D_{Mg} replotted from Huang and Fairchild (2001) with the removal of poorly constrained field data and addition of well-constrained data from three stalagmites at Obir Cave (Fairchild et al., 2010).

$B(OH)_3$ / $B(OH)_4^-$ and CrO_4^{2-} (Alexandratos et al., 2007). Enrichment factors are typically of the order of two and the phenomenon is also known as *intra-sectoral zoning*. These site-specific partitioning behaviours are a special case of the more general phenomenon that ion attachment sites differ on non-equivalent crystal faces, so that different crystallographic forms have different trace element compositions (Reeder & Grams, 1987). Such *sector zoning* is particularly an issue for aragonite, which may be the reason why trace elements form complex patterns when mapped at high resolution in this mineral (Finch et al., 2001, 2003).

2 Partitioning theory assumes the development of a solid solution with substitution of trace ion for major ion in the same lattice position. However, trace elements may be incorporated in different locations in the crystal. The incorporation of trace ions of different size to calcium generates lattice strain and tends to lead to a higher proportion of crystal defects. Incorporation of an *altervalent* species (higher or lower than the charge two of Ca^{2+} and CO_3^{2-}) leads to incorporation of ions of complementary charge (e.g. Na^+ accompanies REE^{3+} in marine calcites, Zhong & Mucci, 1995). Planar defects could have significant capacity for incorporating ions of different size or charge than Ca^{2+} or CO_3^{2-}. Work using X-ray absorption techniques in calcite has concluded that trace metals quantitatively occupy Ca^{2+} sites (Sr^{2+}, Pingitore et al., 1992; Co^{2+}, Zn^{2+}, Pb^{2+} and Ba^{2+}, Reeder et al., 1999; REE, Elzinga et al., 2002), but with complexity where altervalent substitution occurs. However, Pingitore and Eastman (1986) postulated that some Sr in calcite may reside in defect sites in additional to the dominant siting substituting for Ca in regular lattice positions, based on the lowering of the Sr distribution coefficient by addition of Na^+ to the solution, and there is still no simple alternative explanation for this relationship. This idea inspired the suggestion of Borsato et al. (2007), based on high-resolution analyses of Ernesto cave stalagmites, that annual dips in Sr content were caused by preferential incorporation of a variety of other ions at defect sites during hydrologically active periods. Another example is that the large ion UO_2^{2+} partitions preferentially into (–) kink sites in calcite, but displays significant disruption to the local structure, unlike in aragonite (Reeder et al., 2000, 2001).

3 Trace elements may be present in solution or in $CaCO_3$ at ratios to Ca much higher than the trace amount (Tr/Ca ratio < 0.01) demanded by the normal simplifications to thermodynamic theory (McIntyre, 1963). Although much progress has been made in understanding the thermodynamics of solid solutions (Prieto, 2010; Putnis, 2010; Kulik, 2010), their partitioning behaviour is generally complex—quite different from a simple constant partition coefficient. AFM studies also show that where a trace element suddenly becomes strongly enriched in solution, complex growth effects ensue which are dependent on sample history (Astilleros et al., 2006, 2010).

4 Other aspects of solution composition can influence partitioning. A clear case of pH influence is shown by SO_4^{2-}, which substitutes for CO_3^{2-} (Pingitore et al., 1995) and so is preferentially incorporated at lower pH when HCO_3^-/CO_3^{2-} and hence SO_4^{2-}/CO_3^{2-} are relatively high (Busenberg & Plummer, 1985). Solution pH also has a strong influence on ion desolvation (removal of water), this being stronger at high pH. Water-bonding varies significantly between alkaline earth ions, with the small Mg^{2+} ion forming particularly strong bonds with water molecules in solution, whereas Sr^{2+} bonding to water molecules is weaker than Ca^{2+}. Stephenson et al. (2008) demonstrated that the presence of peptide in solution led to higher Mg incorporation; one possible contributing factor could be the ability of the hydrophilic molecules to facilitate the *desolvation* from Mg ions as they approach the crystal surface. Wasylenki et al. (2005b) found changing the experimental geometry and hence the growth environment altered the relative preference for Mg at different growth sites. These studies imply that there is a kinetic aspect to Mg incorporation. Finally, as discussed in Chapters 5 and 6, it is clear that many trace elements are transported bound to colloids and their effective concentrations in solution at the calcite surface would need characterization by specialized techniques such as diffusive gradients in thin films (Hartland et al., 2011). Such studies show that some trace elements do remain tightly bound to colloidal species, while others show a greater freedom to dissociate, adsorb and be co-precipitated as independent ions. Such a combination of processes was invoked by Zhou et al. (2011) to explain the covariation of As with its presumed colloidal carrier Mn, but with excess As at low Mn concentrations.

So we see that it is quite easy to demolish the concept of a unique value of the partition coefficient in practice, but much trickier to tease out the range of circumstances in which the use of such a parameter is appropriate. If we take the influence of temperature for example, we find that, in theory and in nature (Bottcher & Dietzel, 2010), D_{Tr} changes with temperature for several trace species in calcite and aragonite, and this has been the source of great debate and confusion in the literature on marine carbonates (see overviews by Henderson, 2002, and Cohen & Gaetani, 2010). However, cave temperatures are unlikely to vary by more than a few degrees over any timescale in which other conditions remain constant. Mg in calcite is the obvious example to test a temperature response: D_{Mg} changes by around 5% per degree within individual experimental studies, although different studies show some variation in its absolute value (Fig. 8.10). Ihlenfeld et al. (2003) found a reasonable agreement between seasonal changes in Mg content of a stream tufa and the observed 11° temperature change. However, Fairchild and Hartland (2010) found no seasonality in Mg content of a stalagmite from Tartair Cave (Scotland) fed by a drip with constant Mg/Ca and annual cave air temperature variability of 4.5°. This could simply reflect the uncertainty of ±15% in experimental values determined by Huang and Fairchild (2001) under identical conditions, or might, in part, reflect lattice distortions in relation to the unusual composition of the Tartair speleothems (around 6000 ppm Mg and 7000 ppm Sr). More encouragingly, consistent seasonal differences in modern precipitates were seen at Heshang Cave in a location with a 7.5° seasonal range (Henderson et al., 2008; Hu et al., 2008a, b). Another observation that we can make about Mg partitioning is that it proves surprisingly robust to wide variations in Mg/Ca considering that some of the experimental data points in Fig. 8.10 are for marine-analogue solutions with high salinities and Mg/Ca of 5 rather than the <0.01 required by theory!

Another useful example is that of Sr partitioning, because the role of growth rate has been shown to be important (Lorens, 1981; Tesoriero & Pankow, 1996; Fig. 8.8). However, the controls on Sr incorporation are multi-faceted and it is better to refer to the influence of growth kinetics (e.g. supersaturation, growth mechanisms, role of other impurities) rather than growth rate *per se* (Huang & Fairchild, 2001).

Figure 8.11 contrasts a range of studies on Sr partitioning in relation to growth rate. The accretion is expressed both as a mass per unit area per unit time, which is standard for experimental studies, and linear extension (growth) rate for comparison with speleothems. Only the study by Huang and Fairchild (2001) set out to approach the growth geometry and water chemistry of natural caves, including growth in glass tubes (straws) and on plates (stalagmites), whereas most studies used rather different conditions. The outlying points of high D_{Sr} at low growth rate from Mucci and Morse (1983) refer to experiments to grow marine-analogue Mg-calcites which are known to incorporate more Sr. The effects of temperature are most clearly shown by the experimental work of Tang et al. (2008b) in which lower temperatures resulted in higher Sr incorporation, the exact value of which was particularly sensitive to growth rate. However, even relatively high growth rates can result in low D_{Sr} values, as shown by a group of experiments within the study of Gabitov and Watson (2006) which used similar techniques to that of Huang and Fairchild (2001), but with high ionic strength, Na-rich solutions, echoing the earlier findings of Pingitore and Eastman (1986). Huang and Fairchild's (2001) results were comparable to mean values of speleothems from Ernesto Cave reported by Huang et al. (2001), but at this site, and at Obir Cave (Fairchild et al., 2010), speleothems show annual variations of 50 to 100% in Sr content despite minimal change in solution Ca, Mg or Sr composition. The source of these variations is debatable.

Important insight into atomic-scale processes affecting Sr incorporation has come from AFM

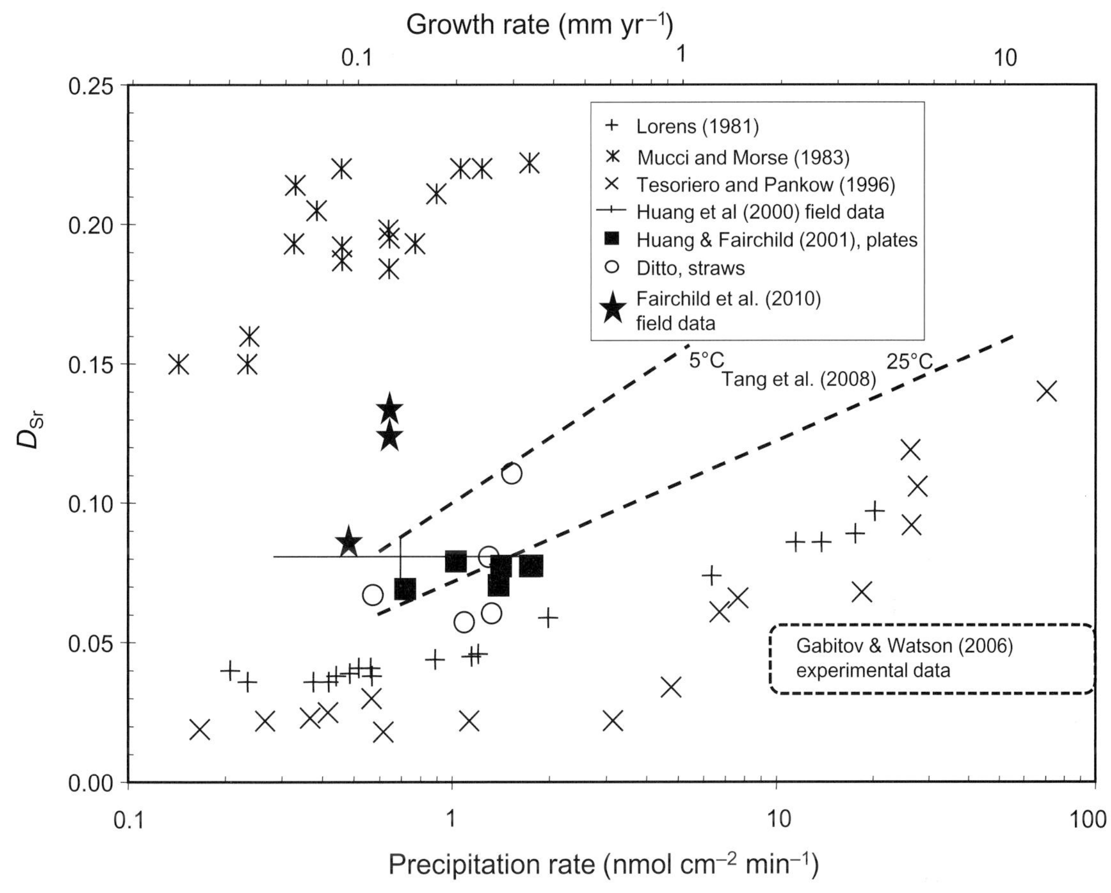

Fig. 8.11 Experimental and field data (mean cave dripwaters compared with speleothem tops) for Sr partitioning replotted from Huang and Fairchild (2001) with additional data. Note that most experiments (except Huang and Fairchild, 2001) used high salinity solutions and the those of Mucci and Morse (1983) were for marine analogues (Mg-rich). Experimental growth temperatures are mostly 25 °C except that some of the experiments of Huang and Fairchild (2001) were conducted at 15 °C. Tang et al.'s (2008b) experimental data form good fits to log-linear relationships, which are shown for 5 °C and 25 °C. The field data are from caves around 6 °C.

studies. Wasylenki et al. (2005b) demonstrated that (i) at concentrations up to 0.5×10^{-4} molar likely to be found in caves, Sr accelerates growth (either because of increased mineral stability or because of creation of kink sites for step advancement); (ii) at progressively higher concentrations than this, more and more Sr is co-precipitated; adsorbed Sr inhibits growth and leads to elongation of growth hillocks at right-angles to the effect of Mg mentioned earlier; (iii) Sr partitioning varies by a factor of two between growth segments; at low growth rates (3.5 nm s^{-1} step advancement, equivalent to around 0.1 mm annual growth according to Teng et al. (2000)), D_{Sr} is around the minimum values found in Fig. 8.11. Wasylenki et al. (2005b) found that growth hillocks of similar morphology contained more Sr at higher supersaturations, but emphasized also that layer spreading may have allowed more Sr incorporation. Teng et al. (2000) found that layer spreading became the dominant growth mode at a supersaturation (converted to a Ω value) of 0.7 (Fig. 8.12).

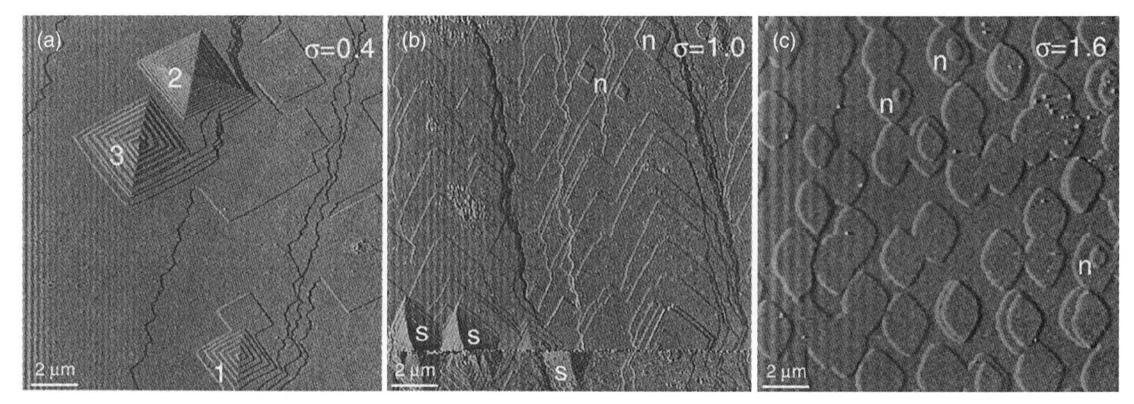

Fig. 8.12 Surface growth morphologies on experimental Iceland spar crystals visualized by AFM at different saturation states (σ) (Teng et al., 2000). Saturation index Ω as defined in Chapter 5 = $\sigma/2.3$. (a) Three spirals are observed in the imaging area: spirals 1 and 3 are single ones (step is one atomic layer high), and spiral 2 is a convolution of two double ones. In the area where dislocations are absent, growth occurred by the advancement of existing mono-molecular layers. (b) Co-existence of spiral growth (denoted by s) and homogeneous surface nucleation growth (denoted by n). Spirals occurred in the bottom part of the image where dislocations were linearly distributed. Two-dimensional nuclei appear in the upper right portion of the images. Growth by advancement of existing steps dominated in the rest imaging area. (c) Dominance of growth by two-dimensional surface nucleation. Continuous surface nucleation is also observed at several locations (denoted by n).

This work has inspired a revised concept of growth kinetics of sparingly soluble minerals such as calcite (De Yoreo et al., 2009) which display a different form of growth inhibition behaviour compared with the predictions of the original Burton–Cabrera–Frank (BCF) model of Burton et al. (1951). Vapour-deposited phases and soluble salts show highly irregular morphologies of growth steps at high resolution because of the spontaneous (thermal) formation of kinks. In contrast for sparingly soluble salts, De Yoreo et al. (2009) considered that kink formation is likely to be growth-limiting, occurring by one-dimensional nucleation (green spheres in Plate 8.1). De Yoreo et al. (2009) successfully modelled the observed growth inhibition effects by sorption of Sr, with the simplifying assumption that desorption is minimal. They implicated the higher rates of one-dimensional nucleation from more supersaturated solutions as allowing more opportunities for binding and incorporation. This model, emphasizing the kinetics of attachment, for which there is experimental evidence, is attractive, considering the difficulties of models such as Watson (2004), which emphasize solid diffusion. De Yoreo et al. (2009) did not address the observation of other authors that ions with D_{Tr} values greater than 1 display lower values of D_{Tr} at higher growth rate. However, if their concept is expressed in terms of high growth rates leading to a wider variety of opportunities for generating kinks and new lattice defects, it could make sense that fast growth will discriminate less between ions that present themselves at the crystal interface. Hence values of D_{Tr} converge towards one, the exact values depending on the rates of other processes such as desorption and the mixture of types of atomic growth morphology (Teng et al., 2000; Fig. 8.12) on the crystal surface.

Despite the progress made in mechanistic studies, there is still a significant gap to its application to speleothem studies. Lateral changes in speleothem chemistry that correspond to ion partitioning at different kink sites in calcite, has not yet been documented, perhaps because growth hillocks in natural samples are too small to be analysed. This is encouraging in that averaging may have occurred, but does not allow mechanistic insights to be made in specific samples. Nevertheless, crystallite morphology and speleothem textures can be indicative of growth character (Frisia et al., 2000).

In current speleothem-orientated work, progress is being made in several areas to complement more general studies of calcite behaviour:
(i) new types of experimental study are being performed using waters identical to cave waters, generated by degassing of solutes from limestone;
(ii) studies of the fate of colloidally transported elements. Unpublished data suggests that a significant proportion of bound metals desorb at the growth surface and become significantly enriched relative to their carrier colloids within calcite;
(iii) several multi-element studies by microanalytical techniques, particularly at cave sites where waters have been intensively monitored.

The overall lesson from this section is that currently unspecifiable kinetic growth factors can lead to significant variation in trace element partitioning, of the order of up to 100% for Sr and 50% for Mg, within the envelope of broadly consistent driving conditions. Nevertheless, it is remarkable that even where the solution Sr/Ca does not appear to change through the year, Sr in speleothems can still shows consistent annual changes in Sr (Smith et al., 2009) as discussed above and in section 8.6. In that section we also draw attention to the fact that we can obtain first-order explanations of trace element patterns where fluid composition clearly changes significantly over time, because, in this case, uncertainties in partition coefficients are unimportant.

8.4 Oxygen and carbon isotope fractionation

Oxygen isotopes are the mostly widely reported parameter in speleothems. It is now recognized that variations in input water composition are the single biggest cause of variation in $\delta^{18}O$ in speleothems over time, and the root causes for this are have been discussed in Chapters 3 and 4. Nevertheless, as shown in Box 5.3, the fractionation during $CaCO_3$ precipitation is large and temperature-dependent. It is clearly necessary to be able to invert speleothem $\delta^{18}O$ values to obtain depositional conditions. Cave temperature can be estimated if water composition can be constrained by fluid inclusion evidence (section 8.4.1). Alternatively, dripwater $\delta^{18}O$ can be estimated if temperatures can be independently determined. Either approach requires a robust transfer function, but the original idea of specifying an equilibrium composition through experiments fails to provide a highly precise result. As discussed in section 8.4.2, an empirical approach based on observed speleothem and dripwater compositions is the pragmatic way forward.

The carbon isotope signature is co-measured with $\delta^{18}O$, but as discussed in Chapter 5, is liable to have evolved significantly within the cave by degassing and it changes further during $CaCO_3$ precipitation. Its covariation with other parameters in speleothems provides good constraints on such processes. In section 8.4.3 we discuss kinetic effects on speleothem $\delta^{18}O$ and $\delta^{13}C$ composition arising during precipitation. New work on 'clumped isotope' systematics may prove to provide an alternative route to determination of temperatures of formation, but there are formidable analytical difficulties and evidence of kinetic effects here too (section 8.4.4).

8.4.1 Fluid inclusions

The rationale to use fluid inclusions to solve the $\delta^{18}O$ palaeotemperature equation was set out clearly by Schwarcz et al. (1976) and updated by McDermott et al. (2005). Because it was feared that oxygen isotopes might exchange between the water and the calcite walls of inclusions, attention was focused on δ^2H (δD) analysis. However, analytical difficulties led to inconsistent results. One strand of research focused on the thermal decrepitation method by which a sample is heated under vacuum to very high temperatures (e.g. 900 °C) such that the calcite starts to break down to CaO. Although this method gives high water yields, the δ^2H compositions have a sample-dependent offset to low values.

The most internally consistent datasets were generated on speleothem samples with low water contents from Israel (Matthews et al., 2000; McGarry et al., 2004) where the offset was found on modern samples to be consistently around 30‰. This enabled it to be shown that the deuterium excess differed significantly between glacial and interglacial periods. However, stepped heating experiments

(McDermott et al., 2005; Verheyden et al., 2008a) showed that water is released in different phases, and that which is released during decrepitation is not only isotopically light, but may have a different origin, i.e. it could represent structural water rather than distinct inclusions. Indeed annual peaks in H are commonly found in speleothems (Fairchild et al., 2001), which might represent molecular or nanoinclusion water (McDermott et al., 2005), hydroxyl ion or HCO_3^-. Because the annual H increase parallels that of P, another possible contributor is $CaHPO_4^0$, because this species was calculated by Lin and Singer (2006) to be the favoured form of adsorption of orthophosphate from solution.

The state-of-the-art is now regarded as crushing with mild heating to ensure that only water from visible inclusions is analysed. Dennis et al. (2001) solved some analytical problems by carefully avoiding adsorption of water within the preparation line and successfully obtained both $\delta^{18}O$ and δ^2H data. Vonhof et al. (2006) developed an on-line continuous-flow system analysis which allowed sample size to be significantly reduced. Dublyansky and Spotl (2009) have optimized the continuous-flow approach by maintaining all preparation lines at 120°C, concentrating the water by freezing, followed by flash heating. Based on calcite samples of less than 0.5 g, they obtained analytical precision of better than 1.5‰ for δ^2H and 0.5‰ for $\delta^{18}O$ with water samples of 0.1–0.2 μL.

Several studies on fluid inclusions have now provided crucial data for palaeoclimate analysis. Among these, Fleitmann et al. (2003b) established through δ^2H analysis of speleothems from Hoti Cave in northern Oman, that they grew rapidly in pluvial intervals within the past 330 kyr from an isotopically light, south-derived (monsoonal) moisture source quite distinct from modern conditions. van Breukelen et al. (2008) demonstrated conclusively through $\delta^{18}O$ and δ^2H analyses of inclusions in a Peruvian sample, that the sigmoidal pattern of variation of calcite $\delta^{18}O$ in the past 14 kyr, parallels the evolution of water chemistry (as controlled by the amount effect), ultimately driven by insolation variations (Fig. 8.13a). A similar parallelism over this time period is found in a study from southern Indonesia (Fig. 8.13b; Griffiths et al., 2010b), but one feature of this dataset is that both modern cave waters and fluid inclusions are consistently offset from rainfall compositions, indicating a consistent degree of evaporative modification of infiltrating water in the surface karst.

A technically advanced development has been the recent attempts to use the noble gas contents in water as a palaeothermometer, following the successful use of this technique to assess the recharge of groundwaters containing small amounts of trapped air (Aeschbach-Hertig et al., 2000). Although the solubility of each of the rare gases bears a simple relationship to temperature in dilute waters and atmospheric pressure, their relative and absolute abundances are quite different in air. Kluge et al. (2008) identified the high and variable proportion of air-filled inclusions in speleothems as being the main obstacle to obtaining precise and accurate palaeotemperatures, although they did obtain reasonable results from a speleothem of milky appearance, rich in inclusions from Bunker Cave in Germany. Scheidegger et al. (2010) noted that air inclusions tend to be inter-crystalline, whereas water inclusions are within crystals. Hence they developed a modified method of sample preparation, using 3 g sample, in which initial crushing released air from inter-crystalline inclusions. Even so, only 40% of samples of known temperature gave accurate results, and they also identified an excess of the light gases He and Ne, presumably trapped as individual atoms within the calcite lattice.

In the study of hydrothermal and burial diagenetic mineral phases, temperatures of formation are routinely determined by heating vapour bubbles, which formed within aqueous inclusion during natural cooling of the sample, until the inclusions homogenize to a single liquid phase. Such vapour bubbles (as opposed to primary air inclusions) do not occur in speleothems because of their low formation temperature, but Krüger et al. (2007) demonstrated a new method for creation of vapour bubbles in aqueous inclusions in minerals by the use of a brief femtosecond (10^{-15}) laser pulse, and forecasted the application to speleothem palaeotemperature analysis. A thorough investigation

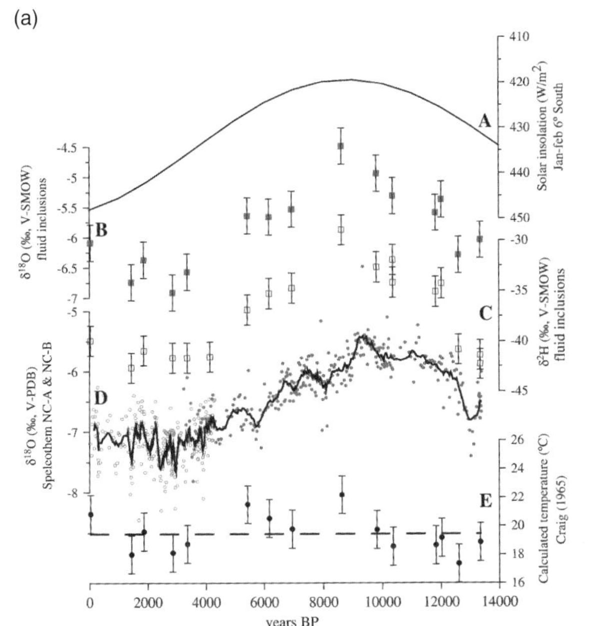

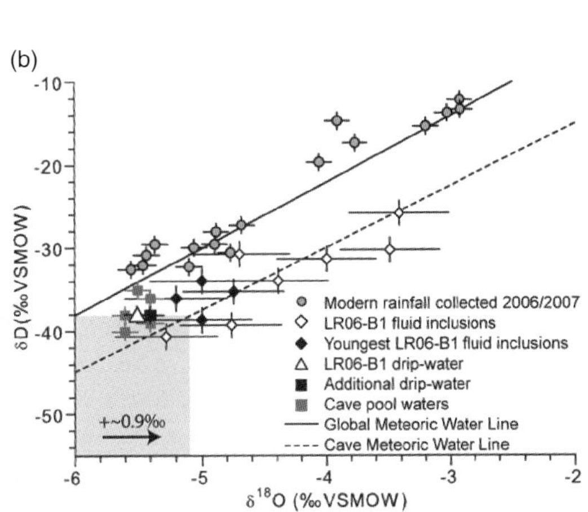

Fig. 8.13 (a) Holocene evolution of speleothem chemistry from the Tigre Perdido Cave (Peru) in which both $\delta^{18}O$ (B) and $\delta^{2}H$ (C) in fluid inclusions and speleothem calcite $\delta^{18}O$ (D) parallel the decrease in solar insolation (A, note inverted scale) clearly indicating the dominance of a change in water composition over time on the calcite signal, related to the changing position of the Inter-Tropical Convergence Zone. The calculated palaeotemperatures (E) show no long-term trends. From van Breukelen et al. (2008). (b): water isotope data from Liang Luar Cave in southern Indonesia illustrating a consistent offset between rainfall, lying on the global meteoric water line, and cave water and speleothem inclusion waters which are argued to have been enriched by approximately 0.9‰ by evaporative processes in the surface karstic environment. From Griffiths et al. (2010b).

of the technique on modern Turkish speleothems, transported without heating from their site of growth to the laboratory, has been completed by Krüger et al. (2011). The sample is cooled to minimize fluid density, then a vapour bubble is nucleated by laser, followed by re-heating of the sample to observe its homogenization temperature. The crystal structure limits the method to intracrystalline (rather than inter-crystalline) inclusions and the physics of the system allows results to be obtained only when the formation temperature exceeds 9–11 °C (dependent on inclusion size). Although the raw results showed a wide range of homogenization temperatures, the upper part of the distribution reflect density changes in inclusions relate to cracks or plastic deformation in surrounding calcite and so can be disregarded, whereas the lower part of the range reflects inclusion volume-related surface tension effects. Krüger et al. (2011) developed a model for correction for the surface tension effects leading to a final set of results within 0.2 °C of the measured mean cave temperatures. This method shows great promise for future palaeotemperature work.

8.4.2 Can an equilibrium composition be defined?

In developing a palaeoclimatic interpretation, authors often seek to justify the assumption that a particular speleothem formed at isotopic equilibrium, but is this realistic? Chemical equilibrium is by definition a reversible process, whereas sustained mineral growth implies a departure from equilibrium. We saw in section 8.3 how difficult it is to define unique stable trace element compositions of calcite co-existing with a solution of defined composition. The situation is still less secure for aragonite as it is only thermodynamically stable at high pressure and only grows by default under some circumstances where calcite fails to form (section 7.4.1).

The literature suggests that growth mechanisms should not influence the equilibrium fractionation of the stable isotopes in $CaCO_3$. Although there is some evidence of differences in $\delta^{18}O$ and $\delta^{13}C$ composition between growth sectors of calcite samples, the evidence overall is inconclusive (Dickson, 1997). Reeder et al. (1997) found that there was no fractionation between different intrasectoral growth sites that differed in trace element content. They noted that ion size, charge or electronic configuration influenced site-specific trace element incorporation, but these considerations would not apply to isotope fractionations.

Pioneers of carbonate isotope geochemistry (McCrea, 1950; Epstein et al., 1953) focused their experimental work on marine waters, both inorganic and biogenic precipitation, in order to derive palaeotemperature equations. Fractionation was revisited by O'Neil et al. (1969) (and re-corrected for phosphoric acid fractionation by Friedman & O'Neil (1977)) in a widely cited set of experiments on a range of divalent carbonates at a range of temperatures. $CaCO_3$ was dissolved in water by bubbling pure CO_2 gas and the solution filtered, before slow degassing by controlled bubbling of nitrogen gas through the solution (see summary best-fit line on Fig. 8.14). Unfortunately it was subsequently realized that a mixture of vaterite and calcite had precipitated, but Kim and O'Neil (1997) found more reliability in starting with a mixture of $NaHCO_3$ and $CaCl_2$ solutions. Their results showed some important differences with ion concentrations: their favoured 'equilibrium' results (Fig. 8.14) with 5 mM reagents showed lighter values than those of O'Neil et al. (1969), whereas rather heavier precipitates were found at higher concentrations of 25 mM ('kinetic' line on Fig. 8.14). The justification for choosing the former set of results as those of true equilibrium has been challenged as being arbitrary (Coplen, 2007).

The technique used by Kim and O'Neil (1997) did not allow growth rate and pH to be explored. This was done by Jiménez-López et al. (2001) using a different set-up (constant pH, downward-drifting saturation); they found a complete absence of growth rate effects, but experiments were complicated by the initial formation of monohydrocalcite. Dietzel et al. (2009) grew pure calcite precipitates by a CO_2-diffusion technique (the same as in Tang et al. 2008a, b), a method which enables tight control on supersaturation (and hence growth rate) and on solution pH. Crystals grew from a 0.01 molar $CaCl_2$ solution into which CO_2 diffused from an inner reservoir and whose alkalinity was maintained by titration with NaOH; however, this set-up, differs in key respects from growth in caves. Figure 8.15 illustrates some of the results from the study of Dietzel et al. (2009), clearly illustrating the strong influence that temperature has on the fractionation (1000lnα values). Figure 8.15a, from experiments at a constant pH at around 8.3, close to that of many low-P_{CO_2} cave environments, shows a reduction in fractionation (isotopically lighter precipitates) at high growth rates, although this effect is much less than the experimental scatter for growth rates equivalent to less than 1 mm yr^{-1}. Figure 8.15b shows that at pH values of 10–11 precipitates are very light, but this is predictable given that at very high pH, OH^- is a reactant (eqn. 5.10) which has a very light signature that is transferred to calcite (Clark et al., 1992). Figure 8.15b also shows a decline in fractionation from pH 8.3 to 9. This might reflect the expected change in the $\delta^{18}O$ composition of the dissolved inorganic carbon (DIC) at higher pH because of the lower $\delta^{18}O$ composition of the carbonate ion (Beck et al., 2005). Nevertheless there is a high degree of scatter at the latter value, which is typical of the variability of experimental products in general. A new set of more realistic experiments by Day & Henderson (2011) robustly confirms the reduced fractionations at higher growth rate (forced by higher rate of degassing of drips onto glass plates).

Equilibrium can also be estimated by theoretical methods and recent workers have either used developments of traditional statistical mechanics methods making use of experimental data on the vibration modes of the carbonate ion in the lattice (Chacko & Deines, 2008) or first-principles (*ab initio*) lattice

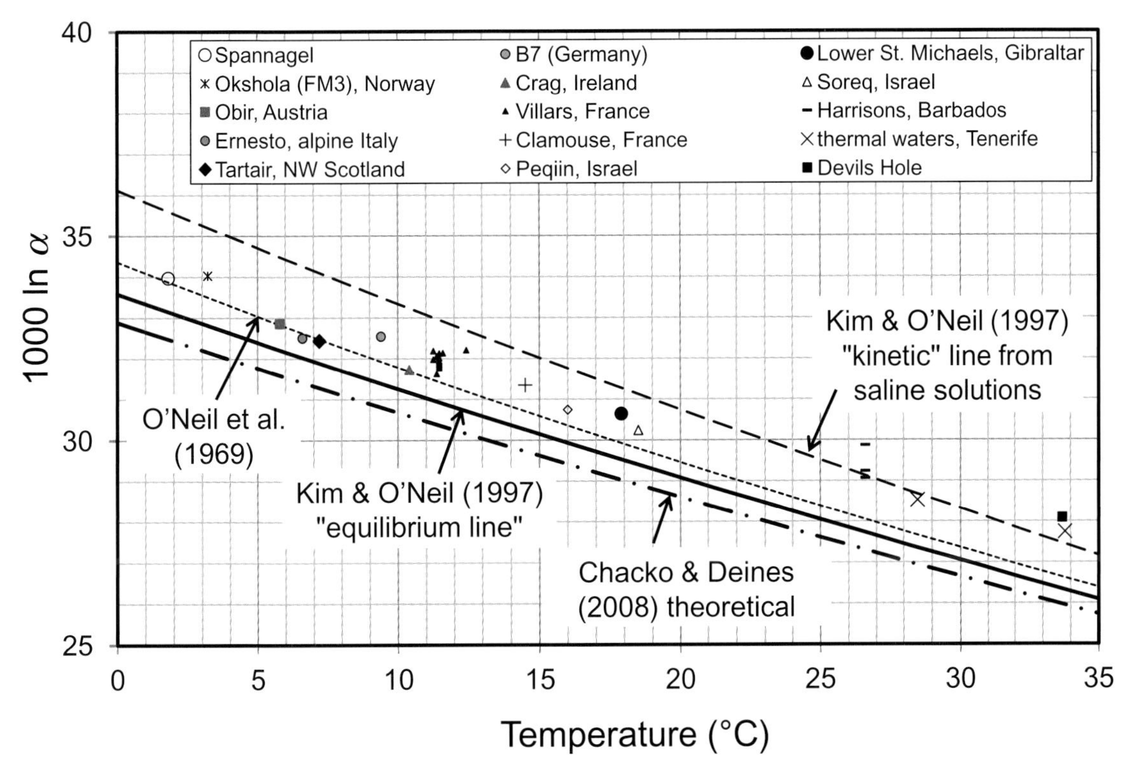

Fig. 8.14 The variation with temperature of oxygen isotope fractionation (expressed as 1000lnα) between calcite and water. The reader is reminded (Box 5.3) that the difference in δ values between calcite and water, when both are expressed on the Vienna Standard Mean Ocean Water (VSMOW) scale, is a quantity very close to the 1000lnα value. Theoretical calculations of Chacko and Deines (2008) are compared with generalized lines from experimental work by O'Neil et al. (1969) (but using the equation in Friedman and O'Neil (1977)), Kim and O'Neil (1997) and a selection of data from field sites based on monitoring dripwater composition and analysis of modern precipitates. Quoted uncertainties in mean values are typically ±0.2 on both axes; in some cases where larger several data from one site are shown. Data sources: Spannagel (Mangini et al., 2005), Okshola (Linge et al., 2009b), Obir (Fairchild et al., 2010), Ernesto, Crag and Clamouse (McDermott et al., 1999), Tartair (Fuller et al., 2008), B7 (Niggemann et al., 2003), Villars (Genty, 2008), Peqiin and Soreq (Bar-Matthews et al., 2003), Lower St. Michaels (Mattey et al., 2008), Harrisons, least-modified precipitates on glass plate (Mickler et al., 2006), Tenerife (Demény et al., 2010) and Devils Hole (Coplen, 2007).

dynamics calculations (Schauble et al., 2006). The two independent methods agree closely for calcite, and predict fractionations (Fig 8.14) closer to those observed in the experiments of Dietzel et al. (2009) than those of Kim and O'Neil (1997).

A third approach is to use the natural occurrences of speleothems to derive an empirical palaeothermometer which we call the *speleothem T function* where T can stand for temperature and/or transfer. In Fig. 8.14 we have plotted 1000lnα values derived from pairing water and modern calcite precipitates at several well-studied sites where the variability of dripwater conditions is known, and the heterogeneity of the calcite has been constrained. The data display a distinctive trend, oblique to, and consistently heavier than, theoretical values. Similar observations have been made by Demény et al. (2010) and Dublyansky et al. (2010), and compare also with the experiments of Day & Henderson (2011). The tight grouping of the natural data lends itself to regression analysis (Fig.8.16a) and good linear fit results, with

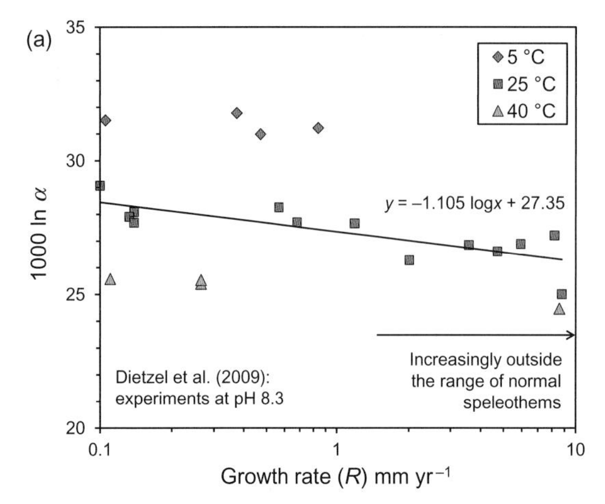

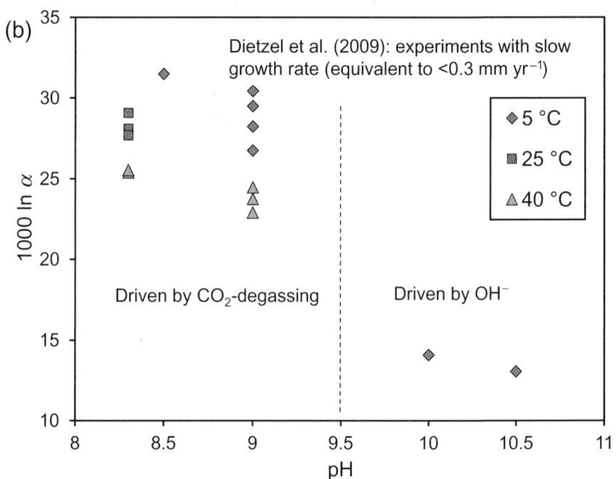

Fig. 8.15 Experimental determination of fractionation of oxygen isotopes between calcite and water, plotted from data in Dietzel et al. (2009). The fractionation is expressed as $1000\ln\alpha$, which is approximately equivalent to the difference in $\delta^{18}O$ values of calcite and water (both expressed on the VSMOW scale, see Box 5.3). Note the significant scatter in results at similar growth conditions. (a) Results of experiments at similar pH illustrate a pronounced effect of temperate and a weaker link to growth rate. Growth rate is converted from millimoles per square metre per minute (in Dietzel et al., 2009) to mm extension per year by division by 1849. (b) At very high pH, there is a profound change in fractionation (see text). Other data imply a fall in $1000\ln\alpha$ between pH 8.3 and 9 (only data from slow growth rate experiments are shown).

a gradient of 0.185‰ °C^{-1}. It must be emphasized that this is not claimed to be an equilibrium relationship because both theory and experiments over a wide temperature range show that an equilibrium fit should be nonlinear. The empirical relationship can be re-cast into the useful form of a palaeotemperature relationship for speleothems (Fig.8.16b) which is expressed here, for convenience, as the difference between δ values of calcite (on the VPDB scale) and water (VSMOW scale).

The speleothem T function, visualized in Figs. 8.14 and 8.16, can be summarized in the following equations (see also Box 5.3 for a reminder of definitions):

$$1000\ln\alpha = -0.185T + 34.1 \quad (8.6)$$

uncertainty is ±0.006 on the slope and ±0.1 on the intercept.

$$T = -5.25(1000\ln\alpha) + 179 \quad (8.7)$$

uncertainty is ±0.16 on the slope and ±5 on the intercept

$$\alpha = (1000 + \delta^{18}O_{ccVSMOW})/(1000 + \delta^{18}O_{water}) \quad (8.8)$$

$$T = 19.4 - 5.30(\delta^{18}O_{ccVPDB} - \delta^{18}O_{water}) \quad (8.9)$$

uncertainty is ±0.3 on the intercept and ±0.16 on the slope

where α refers to the fractionation between calcite and water, T is in degrees celsius, $\delta^{18}O_{ccVPDB}$ and $\delta^{18}O_{ccVSMOW}$ refer to the oxygen isotope composition of calcite on the VPDB and VSMOW scales respectively and $\delta^{18}O_{water}$ is the oxygen isotope composition of water on the VSMOW scale.

In Fig. 8.16b, the speleothem T function of Equation 8.9 is compared with a palaeotemperature function for meteoric water calcites presented by Hays and Grossman (1991) and which Andrews (2006) considered to be approximately valid for most tufa deposits. It approaches the speleothem T function at low temperatures. Hays and Grossman's (1991) equation, which simply recasts a sub-set of O'Neil et al.'s (1969) experimental data in the 0-60 °C temperature range, is:

$$T = 15.7 - 4.36(^{18}O_{ccVPDB} - \delta^{18}O_{water}) + 0.12(^{18}O_{ccVPDB} - \delta^{18}O_{water})^2 \quad (8.10)$$

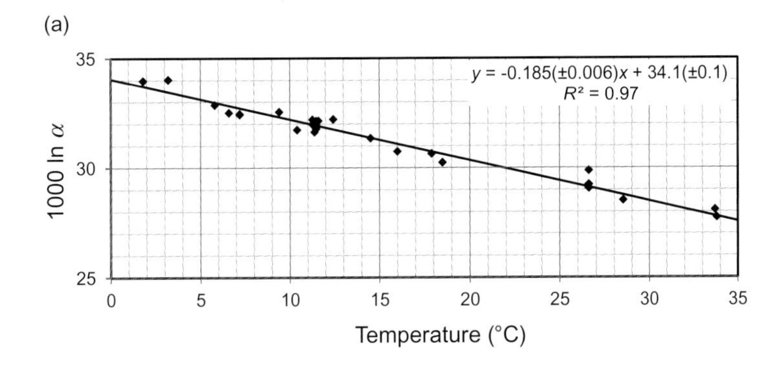

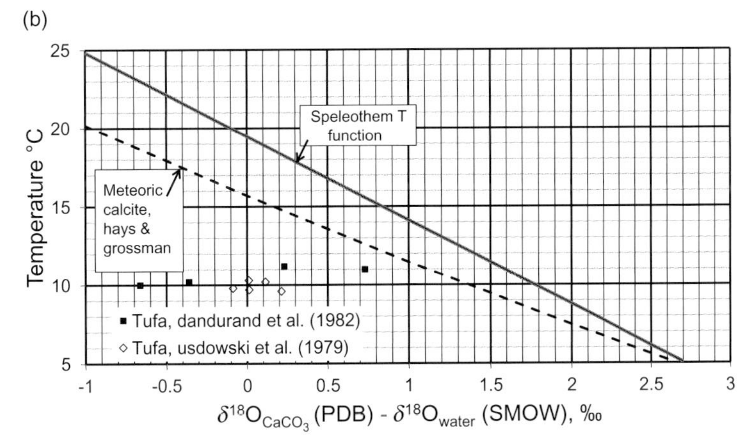

Fig. 8.16 (a) Linear regression of the site-by-site data points of Fig. 8.14. A polynomial fit does not improve the R^2 value and hence the temperature dependency is consistently 0.184‰ °C^{-1} temperature change (i.e 5.3 degrees per mil). (b) Palaeotemperature relationships as a function of the difference between $\delta^{18}O$ compositions of calcite (on the familiar VPDB scale) and water. Speleothem T function (recast from (a)) compared with the relationship of Hays and Grossman (1991) and some data from kinetically influenced tufas (Usdowski et al., 1979; Dandurand et al., 1982). The position of the lines is marginally affected by the chosen value of $\delta^{18}O_{water}$, but is only visible at low temperatures; accordingly a water composition of −10‰ was used for the calculation of speleothem T function line because it is a typical value for low temperature systems.

The analysis of environmental sensitivity of speleothem $\delta^{18}O$ has been expressed most completely mathematically by Lauritzen and Lundberg (1999) as their *speleothem delta function*, but here we consider just the sensitivity to temperature, as in Mangini et al. (2005) (see also Fig. 11.5). Difficulties arise in deriving valid palaeotemperature relationships because of the complex controls and variability in dripwater composition even under constant climatic driving conditions, contrasting with the limited variability of seawater. Classically the overall relationship of speleothem $\delta^{18}O$ to temperature can be expressed as the sum of water and cave precipitation terms:

$$[d(\delta^{18}O_{cc})/dT]_{total} = [d(\delta^{18}O_{cc})/dT]_{atm} + [d(\delta^{18}O_{cc})/dT]_{caveppt} \quad (8.11)$$

The term $[d(\delta^{18}O_{cc})/dT]_{total}$ can be either positive or negative, dependent on the sign and magnitude of $[d(\delta^{18}O_{cc})/dT]_{atm}$, the local gradient of oxygen isotopes in atmospheric precipitation with temperature. Equation 8.6 implies that the cave term, $[d(\delta^{18}O_{cc})/dT]_{caveppt}$, is −0.19‰/°. Speleothem workers generally use a figure of −0.22 to −0.24 (although Lachniet (2009) does give a wider range), as used for marine precipitates. The effects of temperature on $\delta^{18}O$ of atmospheric precipitation are sufficiently complex that expressing them as a single term may not be meaningful (see Chapters 3 and 11); hence use of eqn. 8.11 over an extended timescale may not be valid. A less controversial way to use eqns. 8.6 to 8.9 is as a ready-reckoner to estimate one of temperature, speleothem composition, or water composition, given two of the three.

The question of where equilibrium lies in relation to the speleothem T function remains unresolved because both apparent equilibrium and non-equilibrium precipitates lie on the line at the higher end of the temperature range (Fig. 8.14). Although the Devil's Hole data point (Nevada, USA) is the only one that refers to a precipitate that is not modern (the speleothem formed subaqueously at 4 ka), Coplen (2007) has made a closely argued case that the water composition should not have changed since deposition (see also section 12.1.1). He further advocated this site as one where true equilibrium would be expected, based on the slow long-term mean growth rate of less than $1\,\mu m\,yr^{-1}$. In contrast, precipitates at Harrison's Cave, Barbados display covarying kinetic ^{13}C and ^{18}O-enrichments away from the drip point (Mickler et al., 2004, 2006). This type of kinetic fractionation is the one described by Hendy (1971) (Box 5.3). The other data from warmer sites do not necessarily show kinetic effects directly, but do grow relatively quickly (0.3–2 mm yr^{-1}),which fits with the larger gradient in P_{CO_2} between soil and cave in warm climates as discussed in Chapter 3.

There are few data on carbon and oxygen isotopes on modern aragonites in caves where dripwater has been monitored. Some recent careful studies have inferred equilibrium (Lambert & Aharon, 2011; Li et al., 2011) but the problem is which of a range of experimental studies (reviewed by Lachniet, 2009) to use for comparison. However, there is a close agreement of experimental data on aragonite growth by Kim et al. (2007) and the theoretical calculations of Chacko and Deines (2008), the latter being only 0.05–0.27 higher for $1000\ln\alpha$. If one compares this with the empirical speleothem function of Fig. 8.14 for calcite, aragonite is predicted to be around 0.5‰ heavier than calcite at lower temperatures, but lighter by similar amounts at higher temperatures, over the 0-30 °C range. This contrasts with earlier work: Chacko and Deines (2008) calculated that aragonite is theoretically 1.5–1.7 higher in $1000\ln\alpha$ than calcite whereas the difference in experimental results between Kim et al. (2007) for aragonite and Kim and O'Neil (1997) for calcite is 0.7–0.8. In practice, natural aragonite appears to be slightly heavier than co-existing calcite, both in terms of $\delta^{13}C$ and $\delta^{18}O$ (Frisia et al., 2002), but it also demonstrates seasonal variations in $\delta^{13}C$ due to degassing (Lambert & Aharon, 2011) or noticeable kinetic fractionation,for needle morphologies at least, at very slow drip rates (Frisia et al., 2011). This emphasizes the point that for aragonite, equilibrium is an even more elusive concept in practice than for calcite.

8.4.3 Kinetic effects during $CaCO_3$ precipitation

The previous section has raised a series of issues about kinetic factors that may influence isotope composition. For the purposes of gaining a deep understanding of how speleothems form, there is much to do in terms of documenting and understanding these effects, not least because external environmental factors such as aridity could accentuate rates of deposition and give rise to more pervasive development of more evolved chemistries in speleothems (see, for example, McMillan et al., 2005). This research aim will require going well beyond what has been standard practice to date.

pH and growth rate effects

Much attention has been focused on the so-called 'vital effects' which lead to variations in composition of $CaCO_3$ of different species of organisms from each other and from putative equilibrium. Spero et al. (1997) found a correlation between decreasing $\delta^{18}O$ and $\delta^{13}C$, and increased aqueous carbonate ion concentration (and pH) in cultured planktonic foraminifera. Although decreased $\delta^{13}C$ is likely to relate to fluxes of organically derived carbon that are not relevant to speleothems, the $\delta^{18}O$ observations were explained by Zeebe (1999) as reflecting the changing composition of the fluid with pH. The idea is that during calcification both HCO_3^- and CO_3^{2-} attach to calcite surfaces, and in the same ratio as in solution. Because CO_3^{2-} is isotopically light (Beck et al., 2005) and becomes more important at higher pH, so $\delta^{18}O$ values fall at high pH. Adkins et al. (2003) used data from deep-sea corals to demonstrate more clearly that the mechanisms for $\delta^{13}C$ and $\delta^{18}O$ change were different and these authors supported the high-pH model. This model was also argued by Zeebe (1999) to explain experimental results for calcite precipitates of Kim and O'Neil (1997), and in particular the differences between solutions of different concentrations, based on an estimate of the mean pH at which the precipitates formed. However, unlike these experimental conditions, pH is typically around 8 on speleothem surfaces, and certainly not the values of greater than 9 that would be needed (Zeebe, 2007) to show a dominance of CO_3^{2-} and hence lighter isotope composition.

Tufa-depositing streams offer another environment in which to study kinetic effects. However, although seasonally resolved variations in $\delta^{18}O$ are found, it has proved quite difficult to establish the variability of temperature and $\delta^{18}O$ sufficiently accurately in order to constrain fractionations closely. Usdowski et al. (1979) and Dandurand et al. (1982) found evidence for isotopic disequilibrium during rapid precipitation of tufa in streams near their source, whereas kinetic effects seem to diminish downstream (Andrews, 2006). The former two sets of authors observed $\delta^{13}C$ values close to that of the DIC, rather than around 2‰ higher as expected (cf. Fig. 5B3), and $\delta^{18}O$ values are 0.7–2.5‰ lower than predicted by the speleothem T function (Fig 8.16b). The kinetic effects on $^{13}C/^{12}C$ have been attributed by Michaelis et al. (1985) to rapid precipitation at high supersaturation just keeping pace with continued degassing. On the other hand, because the streams do not have pH values above 9, the light $\delta^{18}O$ values, if interpreted as caused by incomplete equilibration, cannot be simply related to the composition of CO_3^{2-}. An intriguing idea (Tripati et al., 2010) is that increased lattice defects at relatively fast growth is associated with increased incorporation of HCO_3^- directly into the lattice: this would imply a correlation with altervalent cations.

Another possibility is that there may have been the initial formation of a metastable carbonate such as amorphous calcium carbonate, vaterite or monohydrocalcite which then transformed to calcite. Such a mechanism has been noted above in relation to some sets of experiments (e.g. Jiménez-López et al., 2001) and is also advocated as the explanation for light $\delta^{18}O$ in deep-water corals by Rollion-Bard et al. (2010), who used boron isotope evidence to show that the high-pH model was inapplicable. Such an explanation could apply to these fast-growing proximal tufas and should be borne in mind as a possibility for fine-grained (micritic) speleothem fabrics where new crystal nuclei are abundant. Where powdery $CaCO_3$ forms in relation to freezing in caves, variably heavy isotopic values tend to occur, attributed to physicochemical processes by Lacelle (2007) and experimental freezing does produce spherical (originally amorphous) forms (Killawee et al., 1998). However, normal speleothems show continuous growth of calcite fabrics and so the formation of metastable carbonates is not favoured energetically.

Day and Henderson's (2011) experiments demonstrate increased δ values where degassing or evaporation were enhanced (in their case, at higher temperatures, and the same generalization may apply to caves). They inferred a counter-tendency for a kinetic effect, leading to additional preferential ^{16}O incorporation at higher growth rates, and in quantitative agreement with that observed by Dietzel et al. (2009). However, as discussed earlier in the trace element section, it is not clear that the surface entrapment model discussed by these authors provides a meaningful explanation of the process involved.

In summary, if Coplen (2007) is correct, a general kinetic mechanism is needed to explain why speleothems are lighter than expected from the Devils Hole datum. However, speleothems normally neither grow fast enough or at sufficiently high pH for any well-documented specific mechanism to be responsible.

The Hendy test

The best documented kinetic effects are related to the fractionations accompanying degassing and calcite precipitation along a growth surface. Hendy

(1971) predicted progressively increasing and covarying $\delta^{13}C$ and $\delta^{18}O$ signatures accompanying relatively fast calcite precipitation under a solution film, and warned that if covarying data were found (he implied axially, as well as along laminae) that this would demonstrate a breach of the requirement of equilibrium deposition for palaeotemperature analysis. In later literature, the search for covariations along laminae became known as the Hendy test and became something of a totem. A few laminae are selected for analysis at various positions away from the axial zone, together with an assessment of the extent of covariation in the axial zone, in order to make a judgement about equilibrium for the sample as a whole. The approach has been increasingly criticized in recent years, and Dorale and Liu (2009) declared, 'We reach the long overdue conclusion that the Hendy Test criteria for judging speleothems (Hendy, 1971) are not reliably effective at screening stalagmites for paleoclimatic suitability'. They made the following arguments:

1 *Equilibrium might occur in the axial zone, but not on the flanks.* It is true that the rate of change of isotopes is greater further from the axis as is apparent in modelling studies as discussed below.

2 *It is very difficult to sample the same lamina precisely, and impossible where it thins on the flanks.* An alternative strategy is to sample an entire growth interval both axially and laterally, as has been done in some studies (Fairchild et al., 2006a; Baker et al., 2011), but this is very analytically intensive.

3 *There is good evidence that environmental variations can lead to axially covarying $\delta^{13}C$ and $\delta^{18}O$.* Dorale and Liu (2009) listed some 20 published studies where this has been argued to be the case, using the studies of Dorale et al. (1998), Hellstrom et al. (1998) and Baldini et al. (2005) as exemplars to show that processes leading to changes in vegetation abundance or type (and hence changes in $\delta^{13}C$) are commonly associated with changes in climate or hydrological response that cause parallel shifts in $\delta^{18}O$.

4 *Analytical effort would be better directed at replicating speleothem records (i.e. analysing different speleothems across the same time intervals).* Undoubtedly this replication test, if successful for speleothems with different feeding hydrologies, provides robust evidence that the isotope data can be interpreted directly in terms of external forcing factors. The problem arises where it is unsuccessful.

Modelling fractionation along speleothem surfaces

Recent observational, experimental and modelling works have made much progress in consolidating our understanding of this issue. The implemented models assume an absence of processes that were discussed in Chapter 5. In other words they assume that the water has already approached the P_{CO_2} of cave air and hence CO_2-degassing precisely keeps pace with calcite precipitation; also, residence times are too short for either equilibration or evaporation effects to be noticeable (although equilibration is treated quantitatively in Dreybrodt & Scholz, 2011). Both Romanov (2008a, b) and Muhlinghaus et al. (2007) combined models of stalagmite growth beneath water films with their lateral evolution in $\delta^{13}C$ composition. Although the fractionation was expressed similarly, these authors differed in their treatment of the water film, choosing either to model it as a continuously moving film (Romanov), which is more applicable to faster drips (drip interval <50 s, Dreybrodt & Scholz, 2011), or as a lateral series of boxes which exchanged waters each time a new drop fell (Muhlinghaus), which is more similar to the situation with slow drips. In the latter case, the proportion of new water is given by the mixing coefficient ϕ where 0 (all old water) $\leq \phi \leq$ 1 (all new water).

The carbon isotope evolution is modelled as a Rayleigh fractionation process (see also section 5.5.1) as follows:

$$R^{13}_{HCO_3^-}(t) = R^{13}_{HCO_3^-}(0)\left(\frac{[HCO_3^-](t)}{[HCO_3^-](0)}\right)^{\bar{\alpha}-1} \qquad (8.12)$$

where R^{13} is the ratio $^{13}C/^{12}C$ either initially (0) or at time (t), $[HCO_3^-]$ refers to the concentration of HCO_3^- and $\bar{\alpha}$ is the mean fractionation factor for the reaction. HCO_3^- substitutes for DIC as at typical pH values of 8–8.4 it composes more than 95% of the DIC. For the reaction stoichiometry:

$$2HCO_3^- \rightarrow CaCO_3 + H_2O + CO_2 \qquad (8.13)$$

$\bar{\alpha}$ is simply the arithmetic mean of $\alpha_{HCO_3-CaCO_3}$ and $\alpha_{HCO_3-CO_2}$. Romanov et al. (2008b) gave separate treatments for three cases. In case A, the equilibrium fractionation factors were used, whereas in case B the loss of CO_2 was treated as a one-way reaction (i.e. without isotope exchange) and in case C both the loss of CO_2 and the precipitation of $CaCO_3$ were both treated as one-way reactions. As is shown in Fig. 8.17, B is the most extreme case, leading to significant enrichment in ^{13}C in the evolving solution because of the large fractionation between the DIC and CO_2 gas. Large changes in $\delta^{13}C$ were found within lateral distances of 10–20 cm.

Both Romanov et al. (2008b) and Mühlinghaus et al. (2007) examined the effects of varying the given parameters. Figure 8.17a illustrates both the degree of enrichment in ^{13}C along the growth lamina and the morphological consequences of particular conditions. In this diagram a long drip interval (unit b) is associated with slow growth and significant change from the input $\delta^{13}C$ value of −10‰ up to −5‰. Conversely, a short drip interval forms wide thick growth with little ^{13}C-enrichment (unit a). High temperature (unit e) leads to higher $\delta^{13}C$ than low temperature (unit d) but the temperature-sensitivity is weak. A low mixing coefficient (unit g) gives the most extreme effects of growth on the lateral flanks (increasing equilibrium radius) and ultra-high $\delta^{13}C$. Such a low coefficient might apply where splashing is strong, associated with a large drop fall height. Figure 8.17b illustrates the model of Romanov et al. (2008b) calibrated for the boundary conditions applying to a drip which was allowed to generate calcite precipitates on glass plates at Harrison's Cave, Barbados (Mickler et al., 2004). The lateral increase in $\delta^{13}C$ is well-matched by the model.

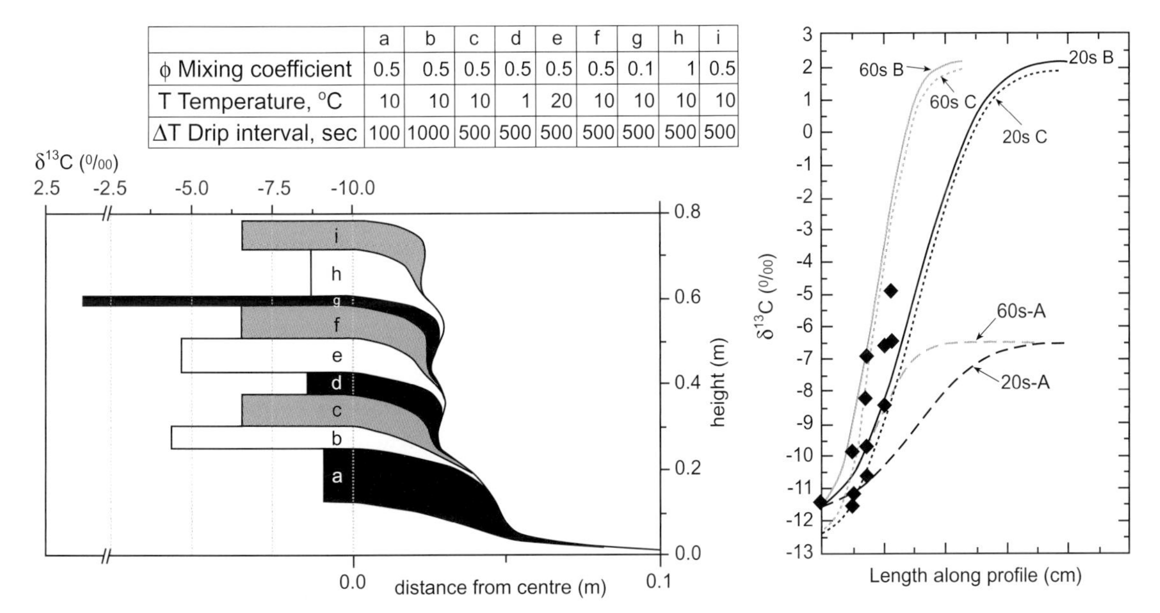

Fig. 8.17 (a) Summary of the consequences of changes in mixing coefficient, temperature and drip interval on speleothem morphology and $\delta^{13}C$ composition (after Mühlinghaus et al., 2007). The reference condition, with middle values of each parameter, is given by growth layers c, f and i. (b) Summary of model results from Romanov et al. (2008b) illustrating a good fit to field observations (diamonds) of lateral increase in $\delta^{13}C$ on glass plates beneath a drip in Harrison's Cave, Barbados (Mickler et al., 2004). Model lines shown are appropriate for field conditions: temperature (26 °C), initial calcium concentration (2.025 mmol L^{-1}), range of observed drip intervals (20–60 s), and assumed atmospheric level of P_{CO_2}. Fractionation is treated in terms of three models, A, B and C whose meaning is explained in the text.

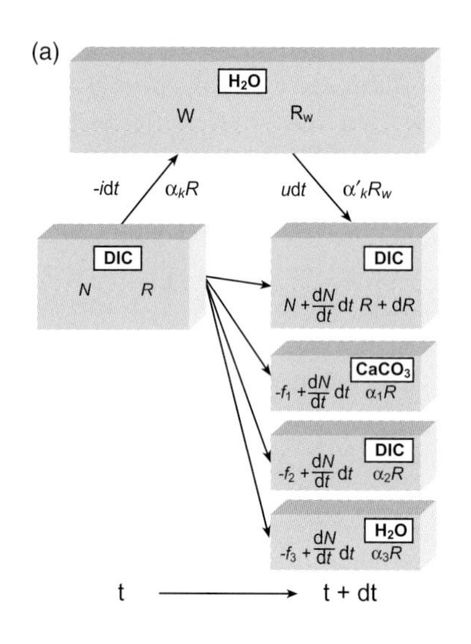

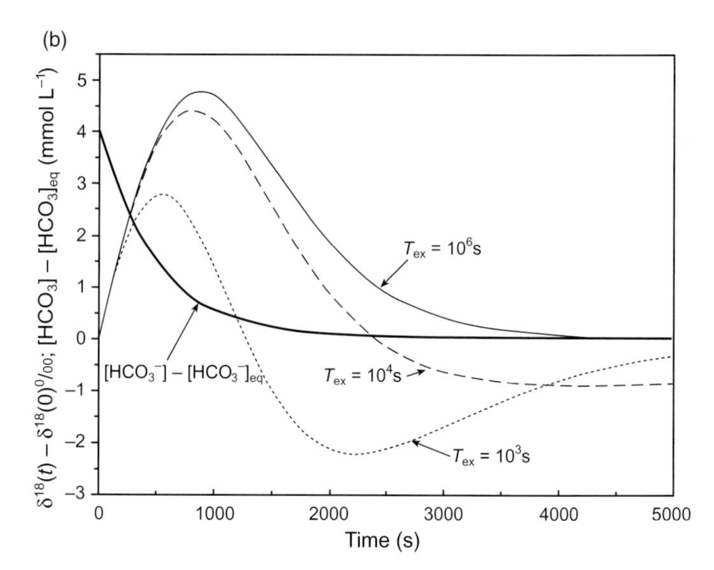

Fig. 8.18 (a) Scheme for mathematical treatment of kinetic evolution of $\delta^{18}O$ along a growth lamina (Scholz et al., 2009, after Mook & de Vries, 2000); N, number of molecules; R, ratio $^{18}O/^{16}O$; f_1, f_2 and f_3 refer to the fraction in each product. Oxygen in the initial DIC is partly lost to the products $CaCO_3$, CO_2, and H_2O, a process that competes kinetically with exchange with the existing H_2O reservoir. (b) Example of model-dependent evolution of oxygen isotopes as water flows over a stalagmite surface. The falling line illustrates the decrease in aqueous HCO_3^- concentration towards equilibrium with a time constant of around 500 s in this case. The curved lines showing the difference in $\delta^{18}O$ compositions of calcite compared with the composition at the speleothem apex, depending on the temperature-dependent oxygen exchange time (t_{ex})between HCO_3^- and water. Simplified from Dreybrodt and Scholz (2011).

The treatment of $\delta^{18}O$ variability (Fig. 8.18a) is more complex (Mook & de Vries, 2000; Dreybrodt, 2008; Scholz et al., 2009) for two reasons. One is that oxygen is partitioned between all three products in eqn. 8.13 (as discussed in Box 5.3), with the low $\delta^{18}O$ value of product water being responsible for the remaining DIC being enriched in ^{18}O. The second reason is that there will be exchange of oxygen atoms between the carbon species of the DIC and water molecules. Up to 2009, models used a characteristic time constant of exchange of 500–1000 s as given by Hendy (1971). However, Dreybrodt and Scholz (2011) considered the revised exchange times determined by Beck et al. (2005) which are much longer (66,000, 20,000 and 6200 s respectively at 5, 15 and 25 °C). Figure 8.18b illustrates that over shorter time periods of fluid evolution when most of the calcite deposition is expected to occur, the Rayleigh fractionation dominates and $^{18}O/^{16}O$ rises, whereas this ratio falls again over longer residence times when exchange has more impact.

Figure 8.19a combines results for $\delta^{18}O$ and $\delta^{13}C$ modelling by Mühlinghaus et al. (2009) to show that under a range of conditions, approximately equal rises in both would be expected laterally. This can be compared with Figure 8.19b, which illustrates experimental data of Wiedner et al. (2008). In these experiments, supersaturated solutions (mixtures of $CaCl_2$ and $NaHCO_3$) at atmospheric CO_2 were degassed by bubbling nitrogen and the water was then allowed to flow down a channel lined with glass fibre precipitating calcite as it did so. Although the thickness of the solution film was not characterized, there is reasonable agreement of modelled data (Fig. 8.19a), with field data (Mickler et al., 2006). However, Hendy tests on natural samples (see, for example, Spötl & Mangini, 2002; Asrat et al., 2007) show a somewhat larger range of $\delta^{13}C/\delta^{18}O$ slopes (0.6–2.9). Low gradients might

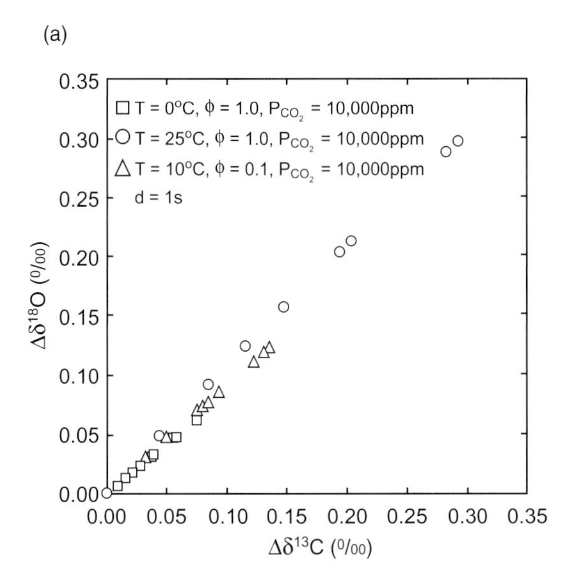

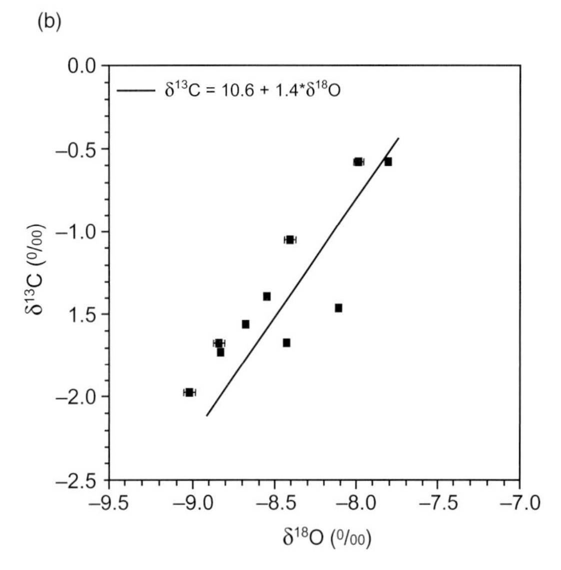

Fig. 8.19 (a) Model results for lateral changes in $\delta^{18}O$ and $\delta^{13}C$ away from the axial zone of a stalagmite during kinetic fractionation (Mühlinghaus et al., 2009). Note the greater change at 25 °C than at lower temperatures. The gradient is relatively independent of conditions. (b) Experimental results of increasing $\delta^{13}C$ and $\delta^{18}O$ in calcite laterally along a glass-fibre-floored channel (Wiedner et al., 2008). The gradient is comparable with, but shows slightly more change in $\delta^{13}C$ than, model results in (a).

conceivably reflect evaporative effects. Slopes of $\delta^{13}C/\delta^{18}O$ of 1.5 (at 10 °C) to 2.5 (at 23 °C) were obtained by Polag et al. (2010), using the same apparatus as Wiedner et al. (2008) and were used to argue for faster oxygen isotope exchange at higher temperature, although acknowledging that the experiments had not definitively verified the underlying theoretical treatment. Hence it is particularly interesting that a completely different theoretical analysis has been made by Guo (2009) who used transition state theory to predict C–O fractionations (including clumped isotopes) during degassing. He separately modelled two different degassing reactions:

$$\text{dehydration: } HCO_3^- + H^+ = \leftrightarrow (\text{fast}) H_2CO_3 \rightarrow CO_2 + H_2O \quad (8.14)$$

$$\text{decarboxylation: } HCO_3^- \rightarrow CO_2 + OH^- \quad (8.15)$$

The latter becomes more important at pH values greater than 8.67 (25 °C) to 8.8 (5–10 °C), but the former is completely dominant below pH 8. Guo (2009) calculated the following $\delta^{13}C/\delta^{18}O$ gradients.

1 Dehydration, CO_2-degassing only: 3.26.
2 Decarboxylation, CO_2-degassing only: 1.41.
3 Dehydration, $CaCO_3$-precipitation coupled to degassing: 2.34.
4 Decarboxylation, $CaCO_3$-precipitation coupled to degassing: 1.1.

This helps broaden the range of thinking on the issue of kinetic fractionations and suggests further useful observations and experiments that need to be made.

8.4.4 Clumped isotope geothermometry (Δ47 value)

A notable new development in recent years has been the study of the small and temperature-dependent preference for heavy nuclides to bond to each other, rather than to a lighter isotope: this field is referred to as clumped isotope geochemistry (Ghosh et al., 2006a; Eiler, 2007). For carbonates, a full analysis needs to consider the nuclides ^{18}O, ^{17}O, ^{16}O, ^{13}C and ^{12}C. A molecule with a particular combination of these nuclides is referred to as an *isotopologue*. In carbonate isotope analysis, $CaCO_3$ is decomposed into CaO and CO_2 by orthophosphoric acid and the CO_2 released has a dominant mass of 44 ($^{12}C^{16}O^{16}O$), but the masses 45 ($^{13}C^{16}O^{16}O$) and 46 ($^{12}C^{18}O^{16}O$) are measured in order to derive $\delta^{13}C$ and $\delta^{18}O$ signatures (correcting for the small contributions made by ^{17}O). Clumped isotope analysis focuses on mass 47, which is dominantly $^{13}C^{18}O^{16}O$. By making stringent high-precision (approximately ±0.005–0.02‰) analyses, the abundance of this isotopologue is found to be higher than the calculated random distribution; this is a thermodynamic effect reflecting the high stability of the ^{13}C–^{18}O bond. This abundance anomaly is referred to as Δ47, which is defined as:

$$\Delta 47 = [(R47/R47^{*} - 1) - (R46/R46^{*} - 1) - (R45/R45^{*} - 1)] \quad (8.16)$$

(multiplied by 1000 and expressed as per mil) where R is the ratio of the designated mass to mass 44 and the suffix '*' denotes the ratio expected with a random distribution of nuclides.

Δ47 can be calculated for a carbonate with specified $\delta^{18}O$ and $\delta^{13}C$ composition and is found to be a function of $1/T^2$ where T, the crystallization temperature, is in kelvins, with the value falling towards zero at high temperatures (Ghosh et al., 2006a; Schauble et al., 2006). It can also be calculated that the phosphoric acid reaction causes an offset in Δ47 of around 0.22‰ between the $CaCO_3$ and the evolved CO_2 which agrees with experimental observations (Guo et al., 2009). Early analytical results showed that natural and experimentally precipitated carbonates defined temperature closely, consistent with these calculations (Fig. 8.20a).

Two immediate applications were to constrain the Cenozoic rate of cooling via uplift of the Altiplano by analysis of soil carbonates (Ghosh et al., 2006b) and to support the relationship between calculated atmospheric P_{CO_2} and ocean temperature in Palaeozoic times by analysis of primary low-Mg calcite fossils (Came et al., 2007). These remarkable results represent palaeotemperature determination (with standard errors of typically 1–2 °C) from carbonate isotope composition without needing to know the water's isotope composition. Tripati et al.

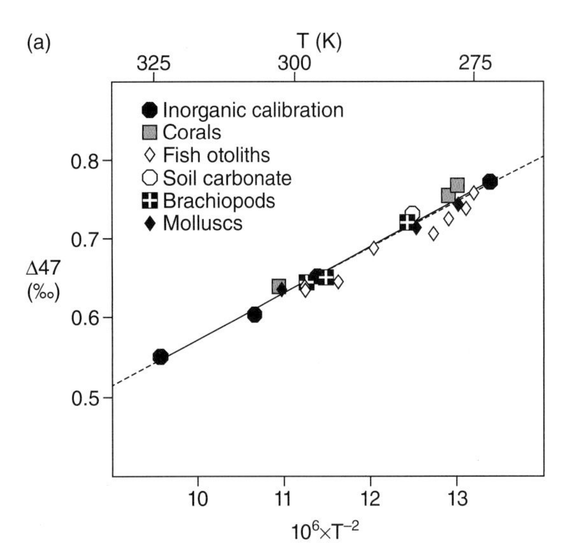

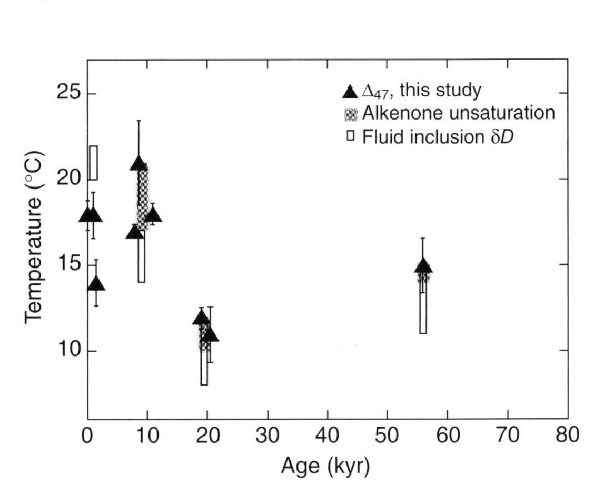

Fig. 8.20 (a) Relationship between measured Δ47 composition of natural and experimental carbonates and temperature (Eiler, 2007). (b) Inferred temperatures from Δ47 analysis of speleothems from Soreq Cave, Israel, compared with independent estimates using fluid inclusion δ^2H from speleothems at the same site, and alkenone compositions of contemporary marine sediments. Note that all the Δ47-derived temperatures have been corrected by −8 °C, this being the kinetic offset of a modern sample in the cave. From Affek et al. (2008).

(2010) have further established strong palaeotemperature functions from foraminifera and coccoliths which are slightly shallower than the inorganic calibration line shown on Fig. 8.20a. $\Delta 47$ in this case is not disturbed by vital effects that caused some deviations in $\delta^{13}C$ and $\delta^{18}O$ chemistry, attributed to the pH effect discussed earlier. There is also no difference between aragonite and calcite $\Delta 47$ values.

Sadly, the system is not so simple for speleothems because it turns out that $\Delta 47$ is particularly sensitive to the Hendy-type kinetic fractionations that arise during CO_2-degassing and $CaCO_3$ precipitation and hence $\Delta 47$ is lower and $\delta^{18}O$ higher in cave precipitates than predicted from subaqueous experiments (Daëron et al., 2011). Affek et al. (2008) found that a modern speleothem from Soreq Cave yielded formation temperatures 8 °C too high, attributing this to such kinetic effects. However, by assuming that this offset is constant for older samples from the same cave, they arrived at a set of results that are largely consistent with previous independent assessments of palaeotemperature (Fig. 8.20b). Although this is a good result for a first-of-its-kind study, a much more stringent test for the degree of kinetic modification would be needed for this method to become routine.

Guo's (2009) theoretical analysis indicates that the kinetic effect will vary, dependent on whether decarbonation or decarboxylation is involved and whether the fractionation is mostly related to initial CO_2-degassing or to simultaneous $CaCO_3$ precipitation and CO_2-degassing. For each 1‰ increase in $\delta^{18}O$, Guo (2009) calculated reductions in $\Delta 47$ of between 0.0175 and 0.029‰ for the four scenarios described at the end of the previous section. Guo et al. (2009 and unpublished) show, using data from Villars Cave, how geologically reasonable palaeotemperatures can be obtained from kinetically influenced speleothems, provided that fluid inclusion $\delta^{18}O$ is available. Given this information, the observed isotopic compositions can be plotted in $\Delta 47$–$\Delta 1000\ln\alpha^{18}O$ space, where the latter parameter represents the difference from expected equilibrium value for $\delta^{18}O$ of calcite. Depending on the degassing scenario as above, data can be projected along a line of appropriate slope to intersect the equilibrium line at the depositional temperature, albeit with relatively large errors of .±3–4 °C. A fuller account of the context of the Villars samples was given by Wainer et al. (2011), who explained that the samples displayed varying kinetic fractionation, obviating the approach of Affek et al. (2008). By using the method of Guo (2009), warming from late Marine Isotope Stage (MIS) 6 to MIS was constrained to be 13.9 ± 5.2 °C.

It is clear that $\Delta 47$ offers an independent window on the formation conditions of speleothems, but it is not a simple panacea for palaeotemperature work although further improvements in precision of fluid inclusion measurements would help (Wainer et al., 2011). It could be appropriate to target subaqueous deposits for analysis, because they do not depend on rapid degassing, although Daëron et al. (2011) cautioned that there can be a memory of previous kinetic effects.

8.5 Evolution of dripwater and speleothem chemistry along water flowlines

Here we summarize the range of processes that modify dripwater and speleothem compositions in the karst and cave environments, including issues discussed in Chapter 5. We base our discussion around Fig. 8.21 which offers some conceptual models with some indications of how solution and speleothem chemistries will vary along water flowlines. This complements the mathematical approach of Dreybrodt and Scholz (2011), which has characterized the relative timescale of several significant processes. To further progress quantitative modelling, it is necessary to take stock of the underlying assumptions and suggest new observations that could usefully be made.

Processes in the soil and epikarst zone determine the input water chemistry to the karst, but normally higher solute loads are expected in warmer climates, except where there is a strong soil moisture deficit. Other things being equal, higher solute loads lead to more supersaturated cave waters and are more likely to result in kinetic effects, and to strong changes along the water flowline. However,

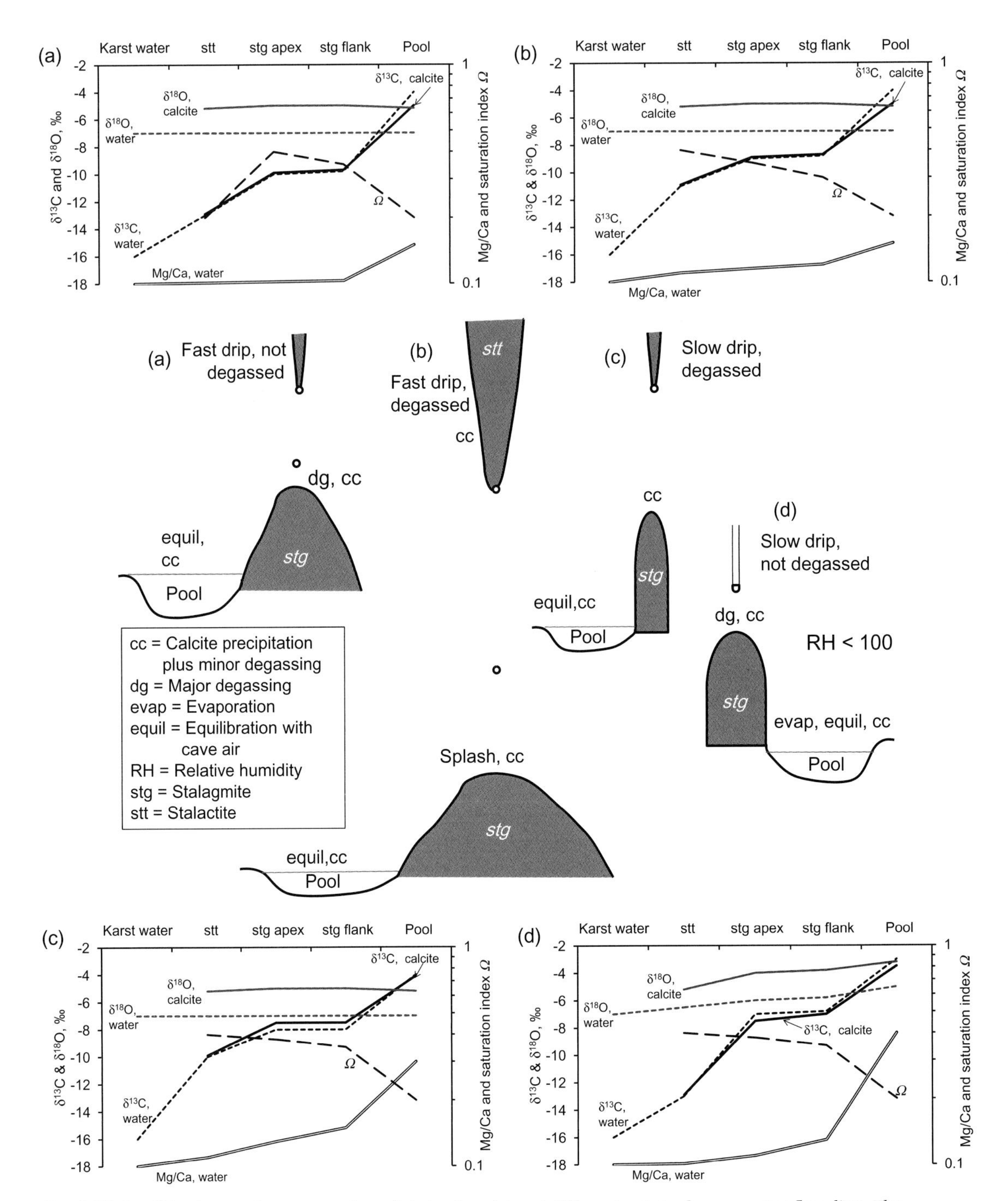

Fig. 8.21 Possible changes in water and precipitate chemistry at different points along a water flow line. The changing relative size of stalagmites and stalactites in relation to degassing was analysed by Baldini (2001). Calcite compositions are mean rather than instantaneous values. See text for discussion.

different temperature scenarios are not depicted in Fig. 8.21. Water is expected to be at equilibrium with calcite before entry to the cave, except for some fracture-fed water which might also not yet have equilibrated its $\delta^{18}O$ with DIC species (Dreybrodt & Scholz, 2011). In Fig. 8.21, indicative $\delta^{13}C$, $\delta^{18}O$ and Mg/Ca values are shown for these input karst waters; saturation index (Ω) is zero (but not shown because of the logarithmic scale).

The waters will tend to degas in the cave environment. In modelling the most common P_{CO_2} chosen is atmospheric, but the true variability was discussed in Chapter 4. Degassing will critically depend on the surface area to volume ratio of the fluid and may be episodic. In modelling, normally degassing is based on a thin film, where diffusion laws suggest it should be complete within a few seconds, although ceiling drops have a time constant of 1000 s (Dreybrodt, 2008), and within soda-straw stalactites, degassing may take up to thousands of seconds (Dreybrodt & Scholz, 2011) (see also Fig. 5.3). It may also be the case that water flows between air pockets of differing P_{CO_2} and that flow to stalactite tips preferentially follows narrow streams with thicker films. For this reason, Fig. 8.21 depicts water in a soda straw that is 'not degassed' (i.e. only partly degassed) contrasting with a larger stalactite where it is completely degassed, but allows for either possibility in the case of smaller solid stalactites. Water might be expected to be totally degassed on larger stalactites, unless preferential flow effects are important (Dreybrodt & Scholz (2011) calculated a 500 s travel time for water flowing down a 50 cm long stalactite, of 5 cm radius at the top). The extent of rise in the Ω value shown in Fig. 8.21a–d relates to the extent of degassing. The observations of Baldini (2001) from a limestone mine indicate that stalactite growth is optimized where drip rate is neither too fast (too little time for degassing) nor too slow (too little solute delivery).

Degassing is associated with loss of CO_2 with a low $\delta^{13}C$ value (but slightly higher $\delta^{18}O$ value compared with the DIC, Box 5.3). This loss is highly likely to be a kinetic (one-way) reaction (Dulinski & Rozanski, 1990; Dreybrodt & Scholz, 2011; Guo et al., 2011) which will tend to enhance the ^{13}C-enrichment in the resulting solution. Direct field evidence for kinetically enhanced $\delta^{13}C$ values was found by mass-balance calculations of data from Ernesto Cave by Frisia et al. (2011) and in the measurements by Spötl et al. (2005) and Baker et al. (2011), and is likely to be a major part of the explanation for the wide variety of modern $\delta^{13}C$ values found in modern precipitates in some caves (Baker et al., 1997a). Hence all four examples in Fig. 8.21 show significant changes in $\delta^{13}C$. Where extensive precipitation on the stalactite is occurring, these $\delta^{13}C$ rises are coupled with a rise in Mg/Ca and a further rise in $\delta^{13}C$ because of the associated loss of Ca and isotopically light carbon from solution. Figure 8.21c offers the closest to the ideal 'equilibrium' situation for stalagmite growth where the drip is slow enough that degassing is complete before dripping. Such stalagmites offer suitable material for a vegetation signal to be recorded in the $\delta^{13}C$ signature (provided that the degree of prior calcite precipitation (PCP) does not change over time).

If a degassed drop falls on to a stalagmite surface (Fig. 8.21b, c), there is less potential for further chemical change than where CO_2 is released explosively on impact (Fig. 8.21a, d). Hence, providing the drip rate is not too slow, Fig. 8.21d is a situation which optimizes the growth rate of the stalagmite, an example being the Crag Cave stalagmite 'Bilbo' of Baldini et al. (2008) and Wynn et al. (2008).

In addition to the immediate effects of CO_2-degassing on the $\delta^{13}C$ composition of the solution film and calcite in the axial zone of a stalagmite, Rayleigh fractionation can also have a role. This is apparent from the work of Henderson et al. (2008) and Hu et al. (2008) on the large stalagmite HS-4 from Heshang Cave in China, where evidence of PCP (increased Mg/Ca, Sr/Ca and $\delta^{13}C$) is seen even in the axial zone, which in this case is a zone 20 cm wide (Fig. 8.22a). This is because of the varying position in which drops fall, such that even in the axial zone, the mean distance that water has flowed is always greater than zero, giving an opportunity for Rayleigh fractionation. Such effects are, of course, accentuated towards the flanks of the structure giving rise to the classic Hendy relationship of covarying $\delta^{13}C$ and $\delta^{18}O$, but also increasing Mg/Ca and Sr/Ca to the extent that Ca is removed from

Fig. 8.22 Left, stalagmite HS-4 (2.5 m high) from Heshang Cave, China, with cross-sections of laminae shown on cut face to far left. Right, top plot shows a comparison of $\delta^{18}O$ series from Heshang and Dongge caves and lower plot shows their difference, $\Delta\delta^{18}O$, which is a linear function of annual rainfall in the range 1275–1425 mm in the region. Faint lines on lower plot show that shifting age models by 50 years either way makes little difference to the results. From Hu et al. (2008b).

the solution. During this process, equilibration of carbon and oxygen isotopes in solution between different species occurs over a time period of several tens of seconds to 100 s (Dreybrodt & Scholz, 2011).

The above discussion draws attention to a key variable in stalagmite formation which has not often been calculated in field situations: that is, the percentage of the solutes delivered by dripwater which are precipitated on the stalagmite surface. We have named this parameter the *solute removal index (SRI)*. On actively growing stalagmites in moist, low-temperature caves, values of the SRI range from 0.6% (discharge 0.7 L day^{-1}, Obir Cave (Fairchild et al., 2010)) to 3–18% (0.2–0.03 L day^{-1} discharge, Ernesto Cave (Miorandi et al., 2010)). The discharge relationship will also depend on the chemical drive for precipitation. A stalagmite forming from hyperalkaline dripwater at Poole's Cavern, UK, also with 0.7 L day^{-1} discharge, had an SRI of 20% (Hartland et al., 2010a): unpublished data on a slower drip indicate close to 100% precipitation. In principle the SRI could be calculated in any fossil situation where the Mg content of a stalactite tip was compared with the coeval stalagmite growth surface. Higher values would correspond to more PCP which can have a climatic significance (Fairchild et al., 2000; Johnson et al., 2006). This concept is analogous to that presented by Cohen and Gaetani (2010) for coral geochemistry, where they attribute changes to varying rates of precipitation from a specified fluid reservoir.

Figure 8.21 also illustrates the changes in chemistry that can be seen in pool waters and precipitates as compared with the nearby speleothems. These changes will be minimal where water flow rates are high, supersaturations are low and overflow or drainage of the pools occurs. Where residence time of the water is significant, however, then noticeable evolution of the water chemistry by carbonate precipitation is likely to occur (e.g. Gonzalez & Lohmann, 1988; Fairchild et al., 2000). Figure 8.21d depicts the case of a cave which has either limited dripwater, or strong air circulation, or both, leading to relative humidities sufficiently low as to allow significant evaporation. This will lead to changes in $\delta^{18}O$ plus further impact on $\delta^{13}C$, Mg/Ca and Sr/Ca because of calcite precipitation. More extreme changes in chemistry are likely in the splash zone or where there are helictites, and this can also be mapped out by the occurrence of aragonite in cave chambers with mixed calcite–aragonite mineralogy (Frisia et al., 2002).

A further process mentioned by Hendy (1971), but only recently incorporated into modelling (Scholz et al., 2009; Dreybrodt & Scholz, 2011), is equilibration between air and water. In the case of oxygen isotopes, over a period of weeks isotope compositions of experimental open-topped vessels converge predictably to the same value (Ingraham & Criss, 1993) indicative of the dynamics of water evaporation and condensation. Dreybrodt and Scholz (2011) calculated 3000 s as a representative exchange time between air and water relevant to carbon and oxygen isotopes. However, their statement that this invalidates palaeoclimate work on samples with longer exchange times is too pessimistic. Even in caves such as Carlsbad Caverns, where evaporative impacts are known, there may be limited evolution of pool chemistry at particular times (Ingraham et al., 1990) because of exchange with water vapour whose composition is in turn controlled by pools. Regarding carbon isotopes, there is still a need for more experimental work on carbon dioxide exchange, but there is clear field evidence for its impact on $\delta^{13}C$ in long residence time waters (Frisia et al., 2011). In a dynamically ventilated cave, the process of equilibration leads to a cave air control on $\delta^{13}C$ compositions, which varies seasonally inversely with CO_2 concentration (Spötl et al., 2005; Frisia et al., 2011). The implications for ^{14}C signatures are discussed in section 8.6.3.

As a result of the processes discussed above, characteristic variations and covariations are established.

1 Variation in $\delta^{13}C$ caused by degassing to approach chemical equilibrium with cave air. It can show kinetic enrichment and the amount of degassing is limited by cave air P_{CO_2} and hence the dynamics of cave ventilation.

2 Covariation in $\delta^{13}C$, Mg, Sr and Ba associated with Rayleigh fractionation and PCP along a flowline sufficiently slowly that $\delta^{18}O$ variations are absent. This type of covariation was not discussed by Hendy (1971), but is commonly present (e.g. McDermott et al., 1999, 2005; Johnson et al., 2006; Mattey et al., 2010; Fig. 5.24). It is kinetic in the sense that it is dependent on water flow rates and cave ventilation, but otherwise should show equilibrium fractionation and trace element partitioning.

3 Covariation in $\delta^{13}C$, $\delta^{18}O$, Mg and Sr related to kinetic fractionations, either related to degassing and $CaCO_3$ under a solution film, or related to evaporative processes (e.g. Père-Noël Cave, Belgium (Verheyden et al., 2008b)).

8.6 Process models of variability over time

In this concluding section, we summarize the manner of variation of speleothem chemistry on different timescales, drawing attention to the specific processes which give rise to these effects and the extent to which they are understood. In Chapter 1, we referred to five realms in which time-varying signals are generated or modified: (i) atmosphere, (ii) soil/epikarst, (iii) cave environment, (iv) $CaCO_3$ precipitation and (v) diagenetic alteration, which provide successively less transparent windows on past environments. The length-scale concept developed in Chapter 4 tells us that deeper into the karst environment signals are homogenized, particularly for hydrological effects, and to a lesser extent circulation changes driven by temperature variations. Hence the most responsive high-resolution sites are those accessible to fracture-fed water; other things being equal, this refers to shallower environments, but an exact depth scale cannot be generalized, as it is site-dependent. McDonald et al. (2004) have established the reality of this difference between more responsive and more stable sites at different depths by monitoring at the Australian Wombeyan caves. Fairchild and Treble (2009) have also commented that the host rock properties have a bearing, with more porous carbonates tending to producing more homogeneity of drip response. This issue is becoming important, because on the future research agenda is the ability to dip into long-term records and gain high-resolution (seasonally resolved) information.

8.6.1 Stadial- to glacial-length episodes

Profound changes to the Earth system occurred between cold-stages (glacials and the shorter stadials) and warm stages (interglacials and interstadials). In higher latitudes and at high altitude, the rate of growth as well as the geochemistry was typically affected. Conversely, in lower latitudes, changes related to atmospheric precipitation were the most important drivers; a prominent sub-set of the responses being the speleothem $\delta^{18}O$ records in monsoonally influenced regions (Wang et al., 2001; Fleitmann et al., 2007) as discussed in Chapter 12. A notable example of a multiproxy study is that of Cruz et al. (2007), who investigated the consequences of precessional forcing on climates over the last 120 kyr and their record from a slow-growing stalagmite from Botuverá Cave in southern Brazil (Fig. 8.23a). Present-day monitoring studies indicate a significant increase in PCP during the dry season and the strong covariation of Mg, Sr and $\delta^{13}C$ in the long-term record was thus reasonably interpreted as reflecting relative aridity, with the parallel $\delta^{18}O$ variation more directly reflecting rainfall amount. Closely similar styles of multiproxy variation over the Younger Dryas to Holocene interval were found in an Indonesian stalagmite by Griffiths et al. (2010a), indicating the way in which in-cave processes can change in parallel with large-scale external forcing and giving encouragement for the development of such studies in the tropics generally.

A soil-related control of chemical variability was demonstrated over the interval 70–280 ka from Buddha Cave in central China (Li et al., 2005). Variations in $^{87}Sr/^{86}Sr$, allowing for changes to other proxies, are attributed to increased intensity of leaching from silicate minerals in overlying wind-blown sediment during periods of more intense monsoonal activity (Fig. 8.23b, c). The authors paid careful attention to distinguishing between carbonate and silicate sources of Sr. This is a good illustration of the importance of characterizing the chemical composition and solubility of elements from the host rocks, soils and superficial sediments if one is to use any proxies other than $\delta^{18}O$ and growth rate.

The large changes in sea level made significant differences to boundary conditions for the speleothem factory in coastal regions, in addition to the direct effect on allowing or terminating speleothem growth as discussed in Chapters 7 and 12. A study by Goede et al. (1998) demonstrated that the exposure and aeolian transport of young carbonate sediments on the Tasman shelf during the last glacial period was reflected in the Sr isotope composition of contemporary speleothem deposition. This approach has been developed in the ongoing studies of early human activity along the southern coastline of South Africa that were mentioned in Chapter 2 (Marean et al., 2007; Bar-Matthews et al., 2010). One specific focus has been to use Sr isotopes to test physiographic models of landscape development across glacial-interglacial cycles. The question posed is whether $^{87}Sr/^{86}Sr$ can be used as a test of the reconstructed distance from the contemporary coastline. The basis is that when the coastline is close to the cave sites, and hence when fluxes from sea spray dominate the Sr budget, $^{87}Sr/^{86}Sr$ increasingly approaches the marine value of 0.7092. First results indicate that the premise is appropriate (Fisher et al., 2010).

A prominent gap in current literature at present is an attempt to understand changing seasonality from high-resolution record studies within records displaying glacial–interglacial variability. Such studies are certainly needed to help understand the changing distribution of rainfall through the year as monsoon conditions evolved, but they do require appropriate drip hydrology capable of resolving winter from summer rainfall.

8.6.2 Sub-millennial variation

Between the annual and the millennial timescales, the climate system displays a bewildering array of forcings and responses, as discussed in Chapters 3, 10 and 11. Because of this complexity, it can be difficult to confidently assign chemical changes to specific forcings, but speleothem geochemistry, constrained by excellent chronology, offers a unique window on certain events. Here we focus on the nature of the geochemical proxies, complementing the examples in Chapter 11.

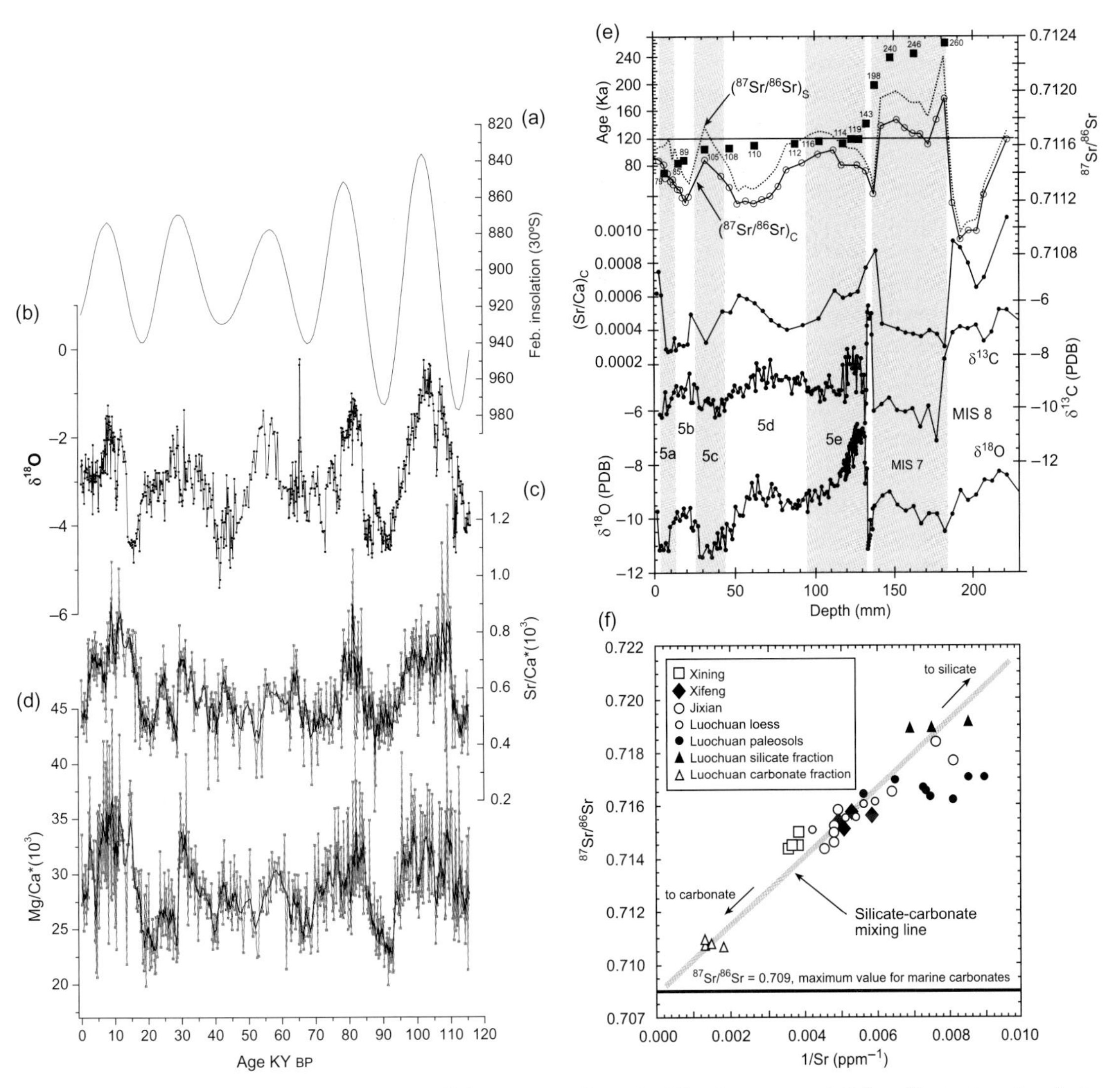

Fig. 8.23 (a)–(d) Solar insolation at 30 °S paralleling changes in $\delta^{18}O$, Sr and Mg from a stalagmite from Botuverá Cave, southern Brazil, over the past 120 kyr. Note the insolation axis is reversed. Trace element data are spot analysis obtained at 1 mm intervals by electron microprobe. The central portion of the stalagmite displays different crystal fabrics which may have led to second-order effects on trace elements. From Cruz et al. (2007). (e), (f) Li et al. (2005). (e) Speleothem record from Buddha Cave, central China, illustrating chemical variations in relation to cool (grey) and warm (white) marine isotope stages. $^{87}Sr/^{86}Sr$ isotopes are shown both as the measured value, $^{87}Sr/^{86}Sr_c$, and as the calculated value derived from the silicate fraction, $^{87}Sr/^{86}Sr_s$. (f) Inverse relationship of $^{87}Sr/^{86}Sr$ and Sr concentrations from loessic source materials in the Chinese Loess Plateau, explaining the relationship between $^{87}Sr/^{86}Sr$ and Sr/Ca in (e).

The study by Treble et al. (2007) built on the well-known record from Hulu Cave, eastern China (Wang et al., 2001) by focusing on the sharply defined 2‰ $\delta^{18}O$ shift which is correlated with the iceberg-rafting event Heinrich event 1 of the North Atlantic. Resolution was increased over conventional methods by the use of SIMS analysis and the full shift found to develop within 6 years (Fig. 8.24). Intriguingly, Mg follows the general trend of $\delta^{18}O$, both increasing together as expected from overall drier conditions, but Mg does not show the sudden step in $\delta^{18}O$, which could imply that the latter is a function of monsoon dynamics not reflected in rainfall amount. Comparison of different sites along the mean moisture flow direction is a more appropriate way of determining rainfall amount directly (Hu et al., 2008b). Figure 8.22b thus compares the Dongge and Heshang Cave records and the difference ($\Delta\delta^{18}O$) was calibrated by Hu et al. (2008b) from instrumental data in the past 50 years to be a linear function of annual rainfall in the region. $\Delta\delta^{18}O$ as proxy for reconstructed rainfall does not follow the insolation curve, but only shows a sub-millennial structure (Fig. 8.22b).

There are an increasing number of studies in which relatively short-lived events, in terms of Quaternary science, have been quite specifically characterized by speleothem work. Each of these features the role of PCP, as discussed in section 5.5. An example is the study of drought phases. Drysdale et al. (2006) used a combination of stable isotope, trace element and organic fluorescence data to characterize a drought phase around 4–4.2 ka that had previously been recognized in a series of sites from the Atlantic to Arabia and which is thought to coincide with a relatively high degree of ice-rafting in the North Atlantic. There is a similar broad geographic spread of evidence for an event around 1.1–1.2 ka which has been characterized in partially duplicated speleothem records from Clamouse Cave in southern France (McMillan et al., 2005). This is the last of several events in which a distinct episode of aragonite formation is triggered by a long-term drought, the progress of which can be traced by annual cycles in Mg and Sr, matching phenomena observed during cave monitoring (Fairchild et al., 2000). In this case, the relatively high porosity of the host rock may have provided favourable conditions for preservation.

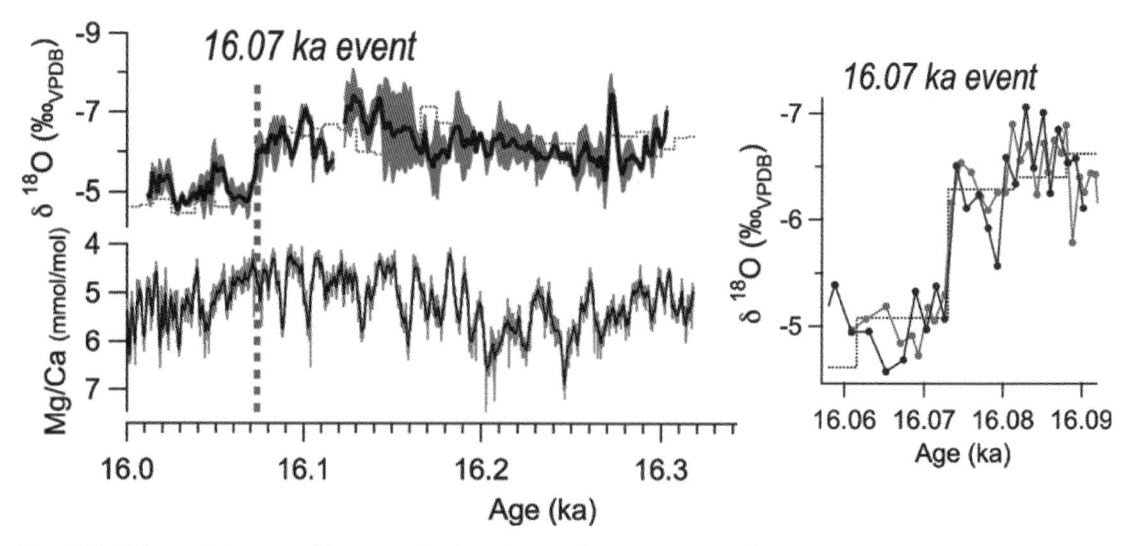

Fig. 8.24 High-resolution record from a speleothem from Hulu cave covering the time period at the onset of Heinrich event 1, with $\delta^{18}O$ analysis by SIMS (in black with uncertainty in grey and conventional analyses in dotted line)compared with Mg/Ca from laser ablation ICP–MS. Expanded view of the step at 16.07 ka shown to right. (From Treble et al., 2007.)

McMillan et al. (2005) used the extent of PCP to define an aridity index and Oster et al. (2010) did likewise for the late glacial to early Holocene conditions at a seasonally dry site in the Sierra Nevada, California, making use of a sophisticated modelling approach utilizing ^{14}C, $\delta^{13}C$ and Sr isotope data.

In contrast to the determination of arid events in caves already in seasonally dry climates, another recent development has been the exploration of the idea that high levels of chemical elements released by strong flushing of the soil zone might reflect periods of enhanced infiltration. Borsato et al. (2007) demonstrated that a cluster of elements, known to be transported as colloids, and liable to be leached from soil zones, defined annual peaks coincident with fluorescent laminae (Fig. 8.25). Elemental abundances were at a maximum in their 100 year record around the period of the First World War when the overlying forest was cut down, implying more effective infiltration was possible. Another example, presented by Fairchild and Hartland (2010) is an abundance of Pb and Zn, only during the initial phases of growth of a stalagmite in a limestone mine beneath ground that is known to have been cleared before forest regeneration over the past century, and for which isotopic forest regeneration signals have already been established (Baldini et al., 2005). Schimpf et al. (2011), in a study from the southern Andes, used Y as a tracer for detrital (including colloidal) transport and used it as a rainfall proxy, finding that it anti-correlated with Mg which in turn varied inversely with instrumental rainfall records over the past 90 years in two stalagmites. Other areas where future developments can be expected are recognition of the effects on speleothem chemistry of volcanic events (Frisia et al., 2005b, 2008; Siklosy et al., 2009), atmospheric pollution (Siklosy et al., 2011), and examination of the role of aerosol contributions to speleothem chemistry (Jeong et al., 2003; Fairchild et al., 2010).

8.6.3 Annual cycles

This section picks up on the material in Chapter 4 which deals with the dynamic annual variation of cave systems and illustrates how annual-scale studies can be used to gain detailed insights during snapshots of time within long-term records, as well as illuminating contemporary processes. The annual nature of the cycle can be confirmed by observation of modern growth rates, by comparing growth intervals with differences in U–Th dates, or by observation of the record of the late 20th century 'bomb spike' in ^{14}C created by atmospheric nuclear testing. Work on modelling the profile of bomb carbon records was pioneered by Genty et al. (1998, 2001a; see also Chapter 9) where it was found to be attenuated and lagged compared with the atmosphere, features which were interpreted in terms of models of turnover of soil carbon with some potential role for groundwater storage. An interesting result from studies at Lower St. Michael's Cave, Gibraltar (Mattey et al., 2008; Fig. 8.26) and Obir Cave (Smith et al., 2009) is the preservation of a less attenuated record than found at other sites, and one that shows little lag in its peak. This may well reflect the role of isotopic exchange between dripwater and a cave atmosphere which has a large component of external atmosphere with a contemporary ^{14}C signal.

Annual chemical signals in speleothems were first recognized by Roberts et al. (1998) and identified to be of general significance by Fairchild et al. (2001), Treble et al. (2003) and Desmarchelier et al. (2006), being the general manifestation of the phenomenon of annual lamination (Baker et al., 2008). Oxygen isotope variability has been specifically linked to changing dripwater composition through the year in drips with a fracture-fed component (Johnson et al., 2006). In the study of Mattey et al. (2008), annual $\delta^{13}C$ variations showed a clearer, simpler morphology than those of $\delta^{18}O$, and so were used to identify the chronology (Fig. 8.26). This then allowed the distinctive low $\delta^{18}O$ signature of winter to be objectively characterized and linked to climate as discussed in Chapter 10.

The variations in $\delta^{13}C$ and trace elements have been attributed to three types of variation (Fairchild et al., 2006a; Fairchild & Treble, 2009). Temperature-controlled patterns were discussed earlier in this chapter where they were noted to be preservable only near to cave entrances where annual temperature changes are large. The second and most important pattern is that of the changing elemental

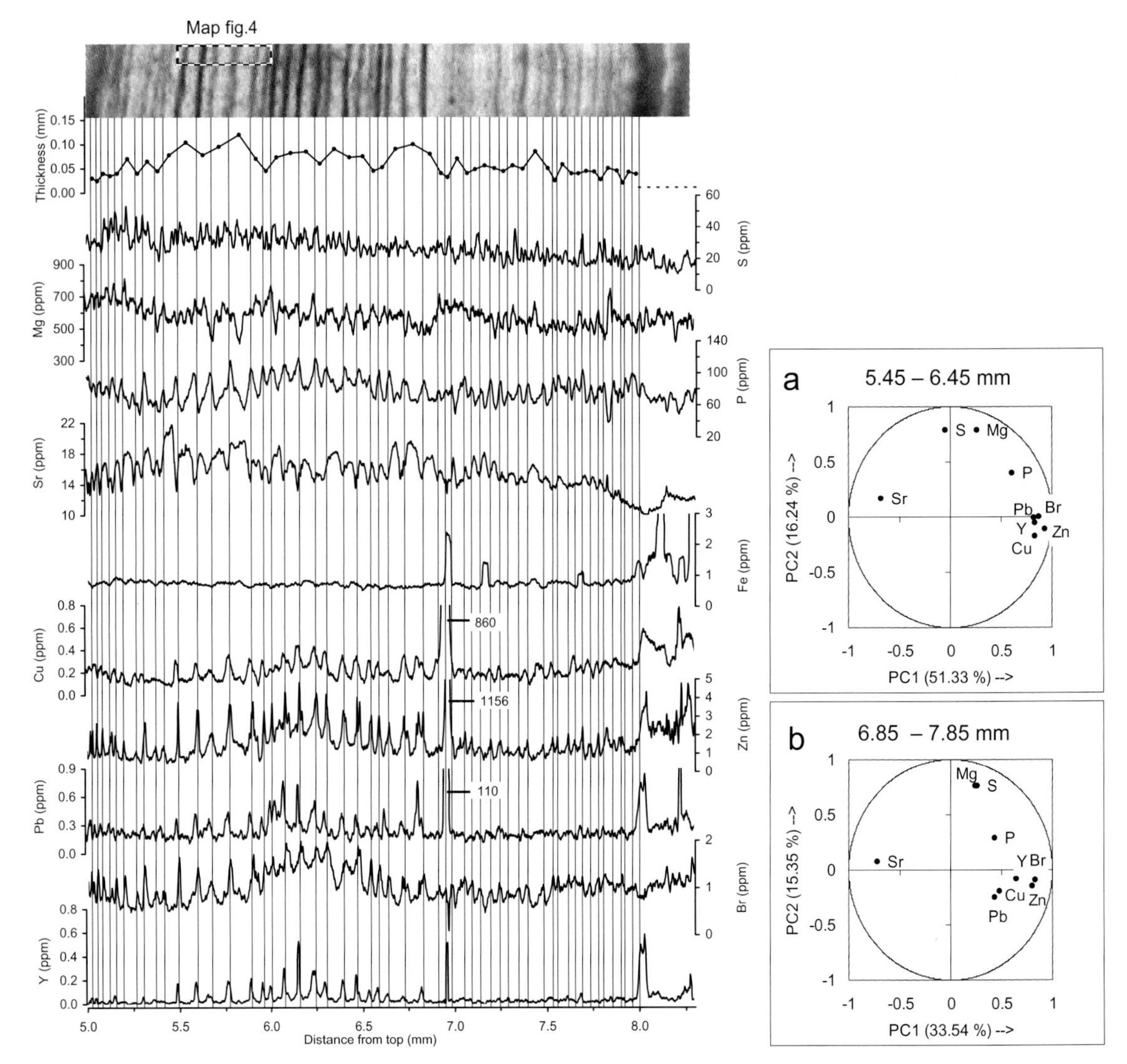

Fig. 8.25 Micro-X-ray fluorescence scans (1 μm resolution) of a stalagmite from Ernesto cave with annual peaks of several elements shown by vertical lines, and which coincide with visible laminae in the transmitted light image at the top. Rises in several elements (Cu, Pb, Zn, Br, Y, P) in the 6–6.5 mm interval coincides with prominent black laminae and a period of anthropogenic deforestation in the early 20th century. The diagrams (a) and (b) on the right are principal component analyses of different depth intervals demonstrating that the aforementioned elements cluster closely around the first principal component (PC1) opposite to Sr, whereas S and Mg define a second component (PC2). From Borsato et al. (2007).

composition of the dripwater through the year. There are two variants of this: one related to PCP and the other to flushing events.

As discussed earlier in this chapter and in section 5.5, PCP gives rise to chemical gradients that can vary seasonally. This will affect $\delta^{13}C$, Sr, Mg and other elements such as U and Ba which can substitute for Ca. Where these changes are large (>50% change in Tr/Ca ratio), they should be clearly preserved in the speleothem because such variations are greater than the uncertainties in the partition coefficients. In extreme cases, the change in fluid

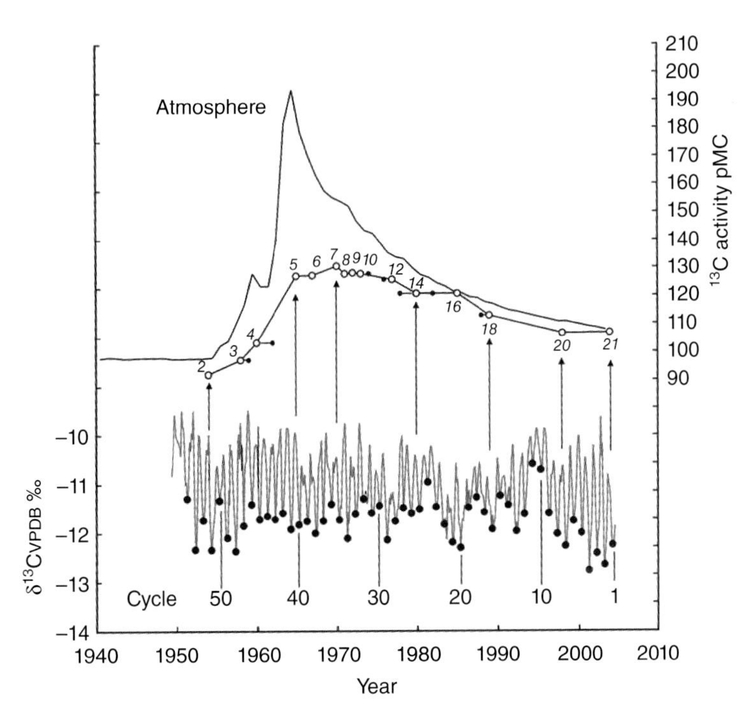

Fig. 8.26 Data from a speleothem in Lower St. Michael's Cave, Gibraltar (Mattey et al., 2008). Annual cycles in $\delta^{13}C$ compared with ^{14}C record from the stalagmite. The record is remarkable for the high levels of ^{14}C activity reached in relation to the bomb spike in the external atmosphere and for the lack of a lag in its peak. This not only verifies the annual chronology but also implies exchange of ^{14}C with the cave air.

composition can trigger a shift from calcite to aragonite mineralogy (Railsback et al., 1994). Caves marked by seasonal dryness show this phenomenon particularly well, although in principle seasonal changes in calcite growth rate related to degassing should also give rise to the same effect as has been clearly demonstrated by monitoring at sites in Texas by Banner et al. (2007) and in Katerloch Cave in Austria (Boch et al., 2011). In some cases these two phenomena are both present and coincide in the season of their maximum effect; this is particularly well-shown by the $\delta^{13}C$ data in Fig. 8.26 because the seasonal anomalies arise through a combination of degassing and calcite precipitation effects (Mattey et al., 2008, 2010). However, there are other sites where summer dryness and low P_{CO_2} in winter compete as drivers and this conflict needs more field data to understand the consequences.

Flushing or events are marked by *impulse laminae* (Fairchild et al., 2010) representing enrichments in colloid-transported events focused with distinct wet seasons, or during autumn infiltration via a component of fracture-fed flow. Where imaged at the highest resolution, they are seen to be composite (Fig. 8.2b–e) of successive events representing individual synoptic atmospheric patterns. Impulse laminae are typically also marked by inclusion of increased amounts of fluorescent organic matter. In some sites there are sinusoidal variations (rather than peaked variations) of elements that might be colloid-transported (Treble et al., 2003; Fairchild & Hartland 2010) and this could reflect aquifer storage and moderation of the infiltration events.

The third type of pattern is a crystallographic one, whereby there is a change in composition related to more complex factors such as a change in growth kinetics, incorporation of other trace species, or change in solution pH. Borsato et al. (2007) argued that the seasonal reductions in Sr in an Ernesto sample, clearly shown by the plots of Fig. 8.25, could reflect the incorporation of strongly

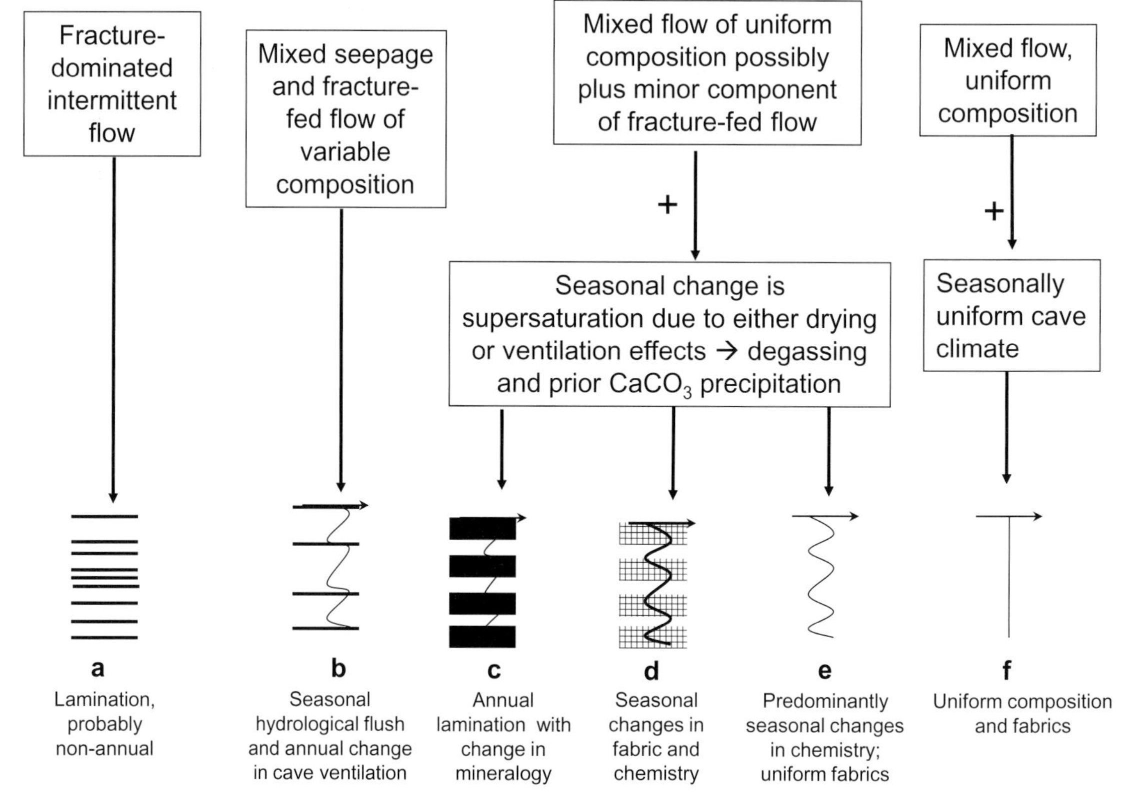

Fig. 8.27 Relationships between the flow-related and cave-climate factors and the high-resolution properties of speleothems (mainly stalagmites). Possible examples are (a) Ballynamintra, Ireland (Fairchild et al., 2001); (b) Tartair Cave (Fairchild et al., 2001), Ernesto and Obir caves (Fairchild et al., 2001; Huang et al., 2001; Frisia et al., 2005a; Borsato et al., 2007; Fairchild et al., 2010); (c) Drotsky's Cave, Botswana (Railsback et al., 1994); (d) Villars, France (Genty and Quinif, 1996; Genty et al., 1997; Baker et al., 2000); Clamouse (McDermott et al., 1999; Frisia et al., 2002; McMillan et al., 2005); Lower St. Michael's, Gibraltar (Mattey et al., 2008, 2010); Katerloch, Austria (Boch et al., 2011); inferred for tropical sites where lamination is present (see, for example, Rasbury & Aharon, 2006; Watanabe et al., 2010); (e) Crag stalagmite near 8.2 ka event (Baldini et al., 2002); (f) not recorded for certain: may only occur in some deep caves.

adsorbing species such as P, out-competing Sr for non-Ca sites. Indeed there is a general, although not universal, anti-correlation of P and Sr in speleothems. Also in Fig. 8.25, it can be seen that both S and Mg show a different pattern of variation. In the case of sulphate this is likely to be a response to seasonal variations in P_{CO_2} (and hence solution pH) because this affects the ratio of HCO_3^-/CO_3^{2-} and SO_4^{2-} substitutes for CO_3^{2-} (Frisia et al., 2005b). This pattern of sulphate variability was confirmed at Obir Cave (Fairchild et al., 2010). The Mg variation is more puzzling as it appears to increase at lower pH, which is the opposite of that found in relation to PCP or predicted by theory (Mg is more dehydrated at high pH), or found in marine foraminifera (Elderfield et al., 2006). However, this is a small effect which would not be noticeable at sites where PCP is important.

Finally, Fig. 8.27, which is refined from an earlier version presented in Fairchild et al. (2007), summarizes some of the key driving factors affecting the development of chemical variation and lamination within speleothems. The flushing events are represented in Fig. 8.27a (non-annual) and b (annual) whereas the seasonal variations in PCP are shown in Figs. 8.27c–e. The latter may also show some flushing characteristics given an appropriate climate and drip hydrology, but this has not often been shown (Baker et al., 1997b; Baldini et al., 2005). The control by hydrological factors is now well-established, but the role of the seasonal temperature cycle on P_{CO_2} needs more work. A way forward is to assess the use of sulphate variation as an index of the asymmetry of the seasonal temperature cycle which often drives the degree of cave ventilation.

CHAPTER 9
Dating of speleothems

9.1 Introduction

Speleothems have long been recognized as having relatively strong chronological control, especially over the time-period appropriate for uranium-series dating (approximately the past 500,000 years). This chapter considers the techniques available to date speleothems, and builds on material introduced in Chapter 2 (geology), Chapter 3 (atmosphere, soil and vegetation) and Chapter 5 (inorganic chemistry) as the sources of the elements used in the dating of speleothems.

In the same way that speleothem environmental and climate proxies are modified from the original source properties, in all cases the source signal used for speleothem dating is also altered by processes during the transfer to the speleothem. Table 9.1 presents the dating techniques that form the focus of this chapter, classified by source and transfer properties, and Fig. 9.1 and Table 9.2 the radioactive decay series and associated half-lives under consideration. We particularly focus on the relatively high-precision dating techniques that are likely to be of most widespread application. Section 9.2.1 therefore considers interval dating through annual lamina counting and section 9.2.2 considers the use of ^{14}C, especially in the 'post-bomb carbon' period for constraining modern growth. Section 9.2.3 reviews recent developments of U–Th dating, the most widely used radiometric dating technique, and section 9.2.4 considers the use of U–Pb dating of speleothems. Section 9.2 concludes with a brief consideration of other 'nonstandard' or less commonly applied dating techniques such as ^{210}Pb, ^{226}Ra–^{210}Pb, ^{231}Pa, amino-acid racemization, palaeomagnetism, and tephrochronology and other event markers. Techniques that we consider have been investigated in detail and been found to be inappropriate for the dating of speleothems owing to their low precision, such as thermoluminescence (TL) and electron spin residence (ESR) dating (Goslar & Hercmann, 1988; Grun, 1991), are not considered here. Many detailed reviews can be found elsewhere of the dating methods considered in this chapter, for example Richards and Dorale (2003) and Zhao et al. (2009) for uranium-series; Hua (2009) for radiocarbon; Tan et al. (2006) and Baker et al. (2008a) for interval counting; and Woodhead et al. (2006) and Rasbury and Cole (2009) for U–Pb dating. We refer the reader back to Chapter 8 for a consideration of appropriate sampling strategies for radiometric dating and section 4.4.2 for additional consideration of radon in modern cave environments.

Appropriate dating control of a speleothem proxy series also requires a consideration of the age–depth relationship. Except where continuous interval counting is possible, techniques are required to interpolate between dated sections in a speleothem. Historically, this has been undertaken using simple linear and nonlinear regression techniques, which may introduce significant additional uncertainty. Section 9.3 considers how speleothems accumulate over time, and the range of appropriate techniques

Speleothem Science: From Process to Past Environments, First Edition. Ian J. Fairchild, Andy Baker.

Table 9.1 The speleothem dating methods which are discussed in this chapter.

Dating technique	Source(s)	Transfer processes	Age range and precision	Optimum conditions for use
Interval counting (trace element and petro-graphic)	Bedrock, soil, cave atmosphere	Modulated by hydrology or cave climate; requires a rapid (faster than seasonal) discharge or cave climate variability. Can be imprinted owing to prior calcite precipitation.	Unlimited age range. Counting errors typically <3%.	Sites with seasonal variations in transport of organic matter and/or bedrock-derived trace elements.
(Fluorescence)	Soil	Modulated by hydrology; requires a rapid (less than seasonal) flow component.		Sites with seasonal 'flush' of organic matter.
^{14}C	Atmosphere, soil and bedrock	Atmospheric signal transferred via vegetation and soil systems. Combined atmosphere, soil and bedrock signals mixed by hydrology.	'Bomb ^{14}C' limited to post-1960 AD. Precision dependent on rate of C transfer. Use of 'pre-bomb' ^{14}C limited by uncertainty in percentage of dead carbon.	Modern samples with short residence time in soil and groundwater.
U–Th	Soil and bedrock	Signal transferred hydrologically. Detrital and organic matter bound Th can confound primary signal.	Past 600 kyr. Precision of up to 0.1%, typically ~1%, but much worse for detritally contaminated samples.	Samples with low detrital (soil or organic) component over the past ~500 kyr.
U–Pb	Bedrock and soil	Signal transferred hydrologically. Detrital and organic matter bound Pb can confound primary signal.	Ages reported to >400 Ma. Precision of 1–5%.	Samples with sufficient U–Pb variability. Low common Pb.

for generating appropriate age–depth relationships, focussing on cubic-spline and Markov Chain Monte Carlo (MCMC) approaches.

9.2 Dating techniques

9.2.1 Interval dating

Table 9.1 demonstrates that in many cases speleothems are likely to contain annual variations in trace elements, fluorescent organic matter or petrology. The processes generating this annual variability have been considered in sections 1.3, 5.4, 5.5, 6.3, 7.2 and 8.6.3. Reviews of annual laminae in speleothems include Tan et al. (2006) and Baker et al. (2008a); these focus on the more widely applied fluorescence and petrological laminae rather than annual trace element laminae. Confirmation of annual laminae in hand section (Broecker et al., 1960; Genty, 1992), and by fluorescence microscopy (Baker et al., 1993; Shopov et al., 1994), has led to their use in chronology building, with continuous lamina sequences of over 1000 yr in ideal samples (e.g. Proctor et al., 2002; Dasgupta et al., 2010). Automation of lamina counting has proved difficult, with manual lamina

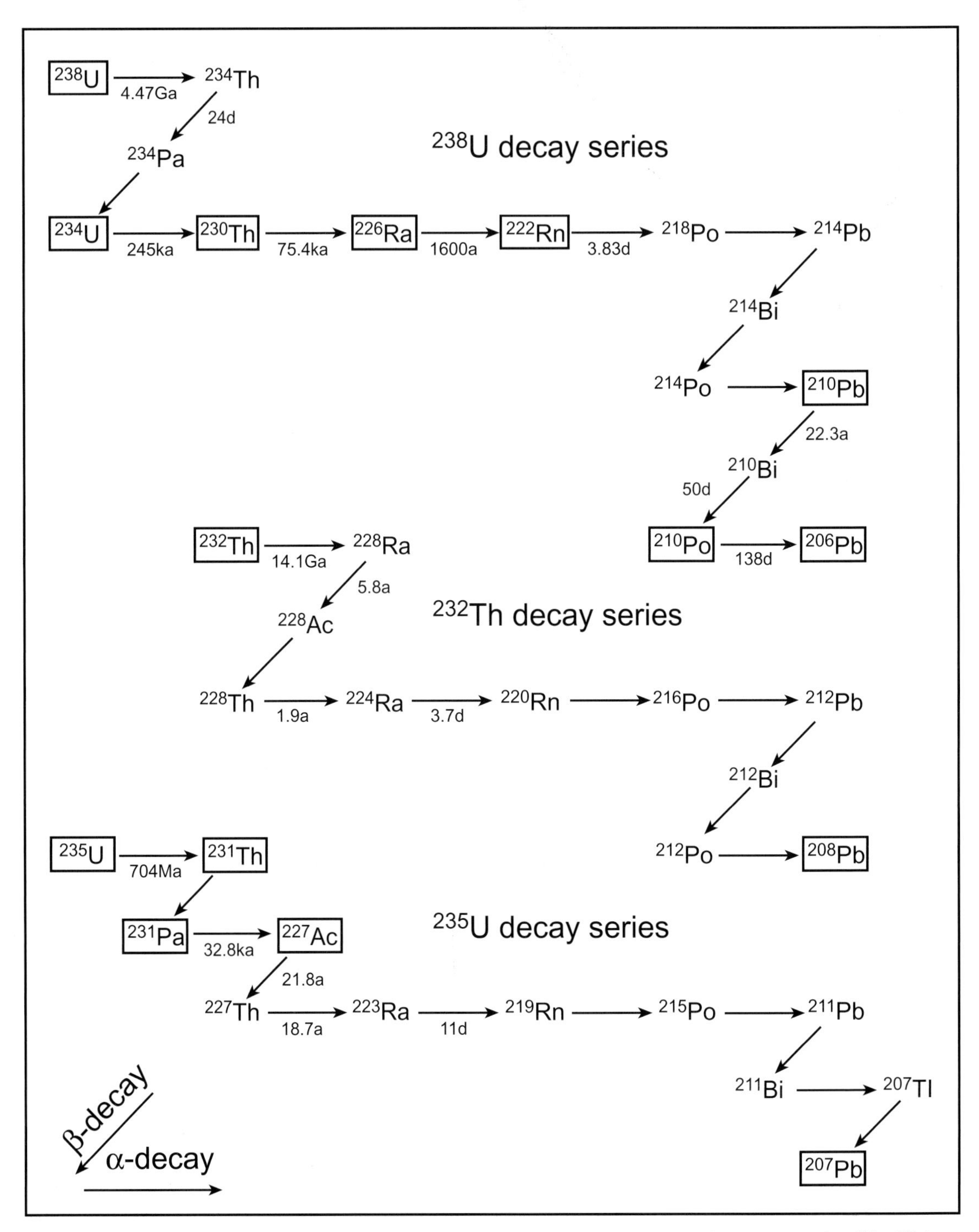

Fig. 9.1 The three principal decay series for uranium and thorium nuclides including those measured in ^{234}U–^{230}Th, ^{235}U–^{231}Pa, ^{210}Pb, ^{226}Ra, ^{238}U–^{206}Pb and ^{235}U–^{207}Pb dating. Half-lives are shown except where very short, and the more long-lived isotopes are enclosed in a box. The diagram is a composite of information from Dicken (2005) and Geyh and Schleicher (2000).

Table 9.2 Half-lives and decay constants of all isotopes discussed in this chapter.

Isotope	Half-life (yr)	Decay constant	$\pm 2\sigma$	Reference
^{238}U	4.4683×10^{9}	1.55125×10^{-10}	0.11%	Jaffey et al. (1971)
^{234}U	2.4525×10^{5}	2.8262×10^{-6}	0.20%	Cheng et al. (2000)
^{230}Th	7.569×10^{4}	9.158×10^{-6}	0.30%	Cheng et al. (2000)
^{235}U	7.0381×10^{8}	9.8485×10^{-10}	0.14%	Jaffey et al. (1971)
^{231}Pa	3.276×10^{4}	2.116×10^{-5}	0.7%	Robert et al. (1969)
^{232}Th	1.40×10^{10}	0.495×10^{-10}	0.5%	Holden (1990)
^{210}Pb	22.6	0.031083	0.4%	Holden (1990)
^{226}Ra	1599	0.000433	0.3%	Holden (1990)
^{14}C	5715	0.000121	0.5%	Holden (1990)

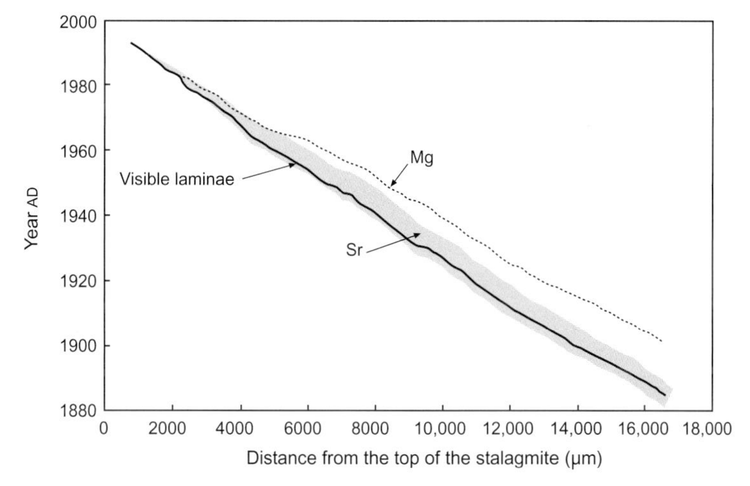

Fig. 9.2 Age–depth models for Obi84, Austria, comparing visible lamina counts (bold line) against trace element counts, simplified from Smith et al. (2009). The grey shade shows the upper and lower bounds of automated lamina counts (105–109 laminae counted) based on different threshold parameters used for Sr. Counting laminae based on P gives almost identical results, whereas use of Mg, which has a weaker annual signal, yields fewer peaks.

counts being the typical methodology for petrological and fluorescent laminae, limiting the use of this technique because of the many person–hours needed to construct chronologies. For trace element laminae, the problem to be overcome is the large amount of time series data that can be rapidly generated using ion microprobe, synchrotron and laser ablation approaches (Woodhead et al., 2009; Fairchild & Treble, 2009), and the need to confirm the annual periodicity of trace element variations. Meyer et al. (2006) and Smith et al. (2009) go some way to tackle these issues with the development of peak counting software, designed for use with trace element laminae but also applicable to visible and fluorescent laminae. Figure 9.2 compares visible lamina counts and automated trace element counts from an Austrian stalagmite (Smith et al., 2009). Although in reality any automated approach will benefit from validation against manually counted sequences where possible, it seems possible that rapid interval dating is within sight.

That interval-dating of speleothems can be achieved is now widely recognized, and it provides the most precise chronology possible for speleothem palaeoclimate and palaeoenvironmental reconstructions. Care should be taken to confirm that periodicity in a speleothem is likely to be annual from a process-based understanding of the surface climate and hydrology; for example, some regions experience two periods of groundwater recharge because of either two rainy seasons or snowmelt. For example, using a combination of lamina counting and U–Th dating, Linge et al. (2009a) give a demonstration of the latter from a speleothem in northern Norway. As well as providing annual-resolution chronology, interval dating can be used in combination with U–Th and U–Pb to better constrain these dating techniques, especially where isochron approaches are needed. This is discussed further in sections 9.2.3 and 9.2.4.

9.2.2 ^{14}C

The process of transferring carbon from the atmosphere, through the soil and bedrock to a speleothem is described in detail in Chapters 3, 5 and 8. Radiocarbon (^{14}C) in speleothems is originally sourced from the upper atmosphere where it is constantly produced by the interactions of cosmic rays with atmospheric nitrogen. The ^{14}C is dispersed widely through the atmosphere, and in the form of CO_2, is taken up by vegetation during photosynthesis and dissolved in rain and surface waters.

^{14}C has a half-life of 5715 years and its radioactivity was the property first used by speleothem researchers in an attempt to date speleothems (see, for example, Geyh, 1970). However, sources of carbon in a speleothem include bedrock, soil and in some cases cave atmosphere; bedrock derived carbon is 'dead carbon' in that all ^{14}C has long since decayed, and the soil carbon is derived from both plant root and soil microbial respired CO_2, as well as soil derived organic matter (see sections 3.3 and 5.4). Comparison with U–Th analyses demonstrated that the dead carbon percentage (dcp) in speleothems is in the range 5–40% (Genty & Massault, 1999; Genty et al., 1999, 2001a), with typical values 12–20% (Genty et al., 2001a) and with the observation that the dcp can vary over time. Genty et al. (2001a) report variations of dcp of the range 5–10% over time periods of 10^2–10^3 yr, therefore severely limiting the use of ^{14}C as a dating technique in speleothems where both the mean dcp is unknown (without an independent dating technique such as U–Th) as well as its variability over time. For samples which are otherwise poorly dated (e.g. samples with high detrital Th content that preclude precise U–Th dates, see next section), one might argue that ^{14}C dates with a prescribed error which includes any reasonable variation in dcp over time, might be acceptable. However, even in this case, the mean dcp is unlikely to be well constrained owing to the large uncertainty in any detritally corrected U–Th analyses. A variable dcp over time also has implications for the use of ^{14}C in speleothems, paired with U–Th analyses, to extend the radiocarbon calibration curve (e.g. Beck et al., 2001; Hoffmann et al., 2010) or make inferences about variations in atmospheric ^{14}C composition, as this approach has to assume a constant dcp or a constant error which incorporates the possible range of dcp (as modelled by Fohlmeister et al., 2011a).

A significant development has been the determination of ^{14}C in speleothems that post-date the atmospheric nuclear bomb testing era for the 1960s. These tests created a spike in atmospheric ^{14}C which is only just returning to pre-bomb levels (Fig. 9.3). This atmospheric ^{14}C signal is transformed through storage in both the vegetation and soil, and subsequent mixing of groundwater of different ages, leading to a damped and lagged increase in ^{14}C in speleothems. Genty and Massault (1999) and Genty et al. (1999) demonstrated the use of multiple ^{14}C determinations in modern stalagmites to not only confirm modern deposition through the presence of elevated ^{14}C (often greater than 100% modern carbon), but to determine the rate of transfer of carbon from the surface to the stalagmite. More recently it has also been identified that in some dynamically ventilated caves the atmospheric profile is transferred without any lag implying re-equilibration of drip waters with the cave air (Mattey et al., 2008; Smith et al., 2009). Demonstration of active or recent stalagmite deposition by ^{14}C analysis can be of great use in regions where drip waters are at, or close to, supersaturation and it is otherwise difficult to be sure that drip waters are actively precipitating calcite.

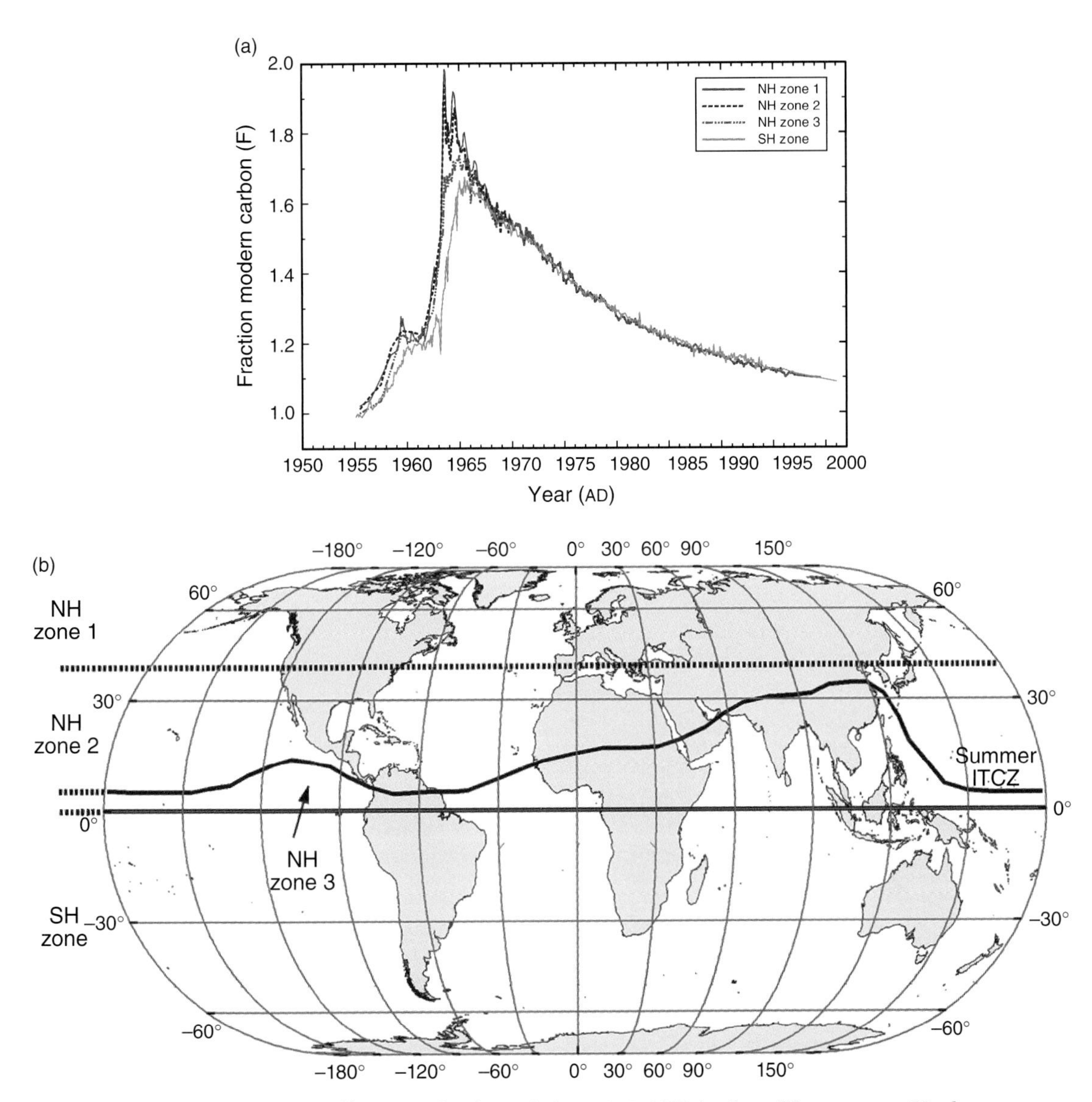

Fig. 9.3 (a) Regional tropospheric ^{14}C curves for the period AD 1955–2001 for four different zones (Northern Hemispheric zones 1–3 and Southern Hemispheric zone). (b) The four zones into which the tropospheric ^{14}C data have been grouped. From Hua (2009).

9.2.3 U–Th

U–Th dating has long been the basis of the chronology that is the foundation of speleothem palaeoclimate and environmental studies (see Chapter 1). Uranium is transported via the groundwater to a speleothem, where it is incorporated in a closed system and radioactively decays to thorium and other daughter products. The key assumption is that there is no transport of thorium in the groundwater owing to its low solubility. Figure 9.4 uses data from Chinese stalagmites to illustrate how over time the composition of a sample will progressively

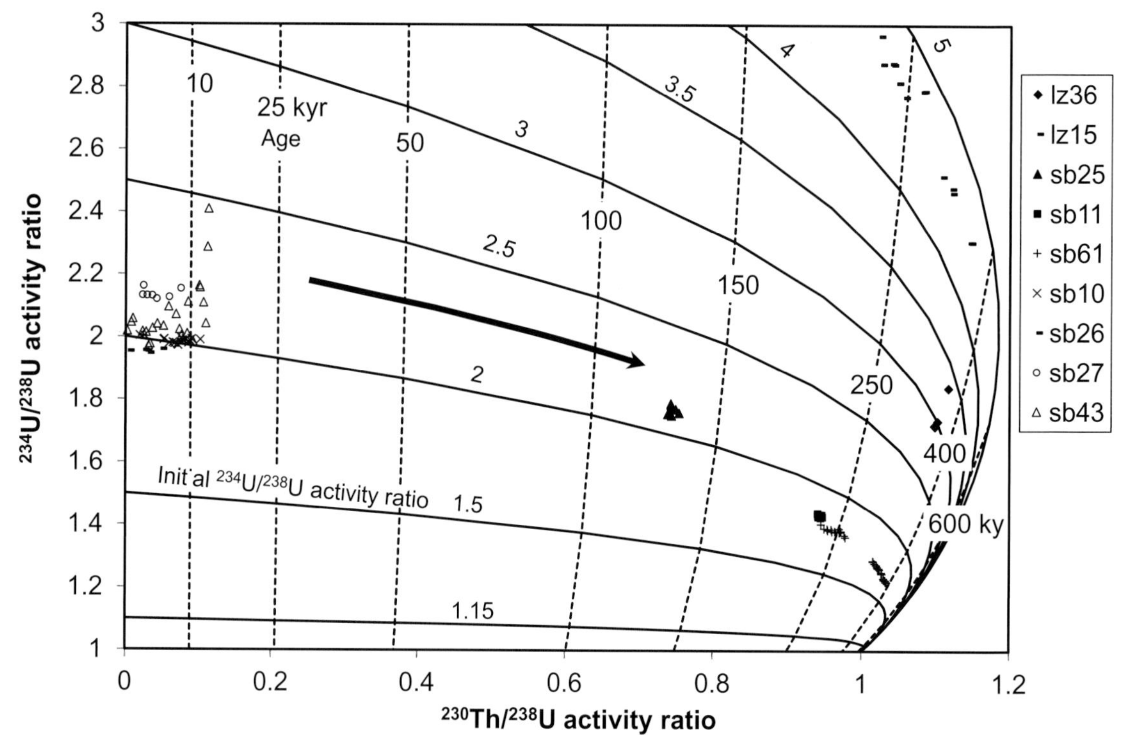

Fig. 9.4 Template to illustrate the principles of U–Th disequilibrium calculated from the decay equations, e.g. in Edwards et al. (1987). Speleothem calcite starts with a composition on the y-axis and its subsequent evolution follows the arrow shown, parallel to lines defining the ratio of radioactivity of ^{234}U to ^{238}U at the time of formation (this ratio falls over time because of faster decay of ^{234}U). Illustrative high-quality data from Sanbao (sb samples) and Linzhu (lz samples) caves, China (Cheng et al., 2009b; Dong et al., 2010) is included. The precision of measurement is worse for older samples but is still well within the size of the symbol used for plotting. The ultimate end-point is secular equilibrium, when the ratio of radioactivities of the isotopes in the decay chain $^{238}U \rightarrow {}^{234}U \rightarrow {}^{230}Th$ are one. Although samples from different caves tend to show characteristically different values for $^{234}U/^{238}U$ ratios (compare lz and sb data), the initial value of this ratio can clearly change over time (e.g. the oldest sb61 samples have lower ratios) and may be useful as a palaeoenvironmental variable.

evolve as the trapped uranium decays and to show how the precision of the method very rapidly deteriorates beyond 400 ka.

Earlier work from the late 1960s to the 1980s used alpha-spectrometry, with large sample sizes and slow counting times limiting the number of analyses that could be performed. By the 1990s, thermal ionization mass spectrometry (TIMS) revolutionized both sample size and counting time, and improved analytical precision allowed reliable dating at both young and old extremes of the U–Th dating range (Edwards et al., 1987). More recently, TIMS is being superseded by multi-collector inductively coupled plasma mass spectrometry (MC–ICPMS) (Hellstrom, 2003; Hoffmann, 2008), further decreasing the sample size needed (to typically 1 mg or less) as well as allowing simplification of the preparation chemistry. Both advances have also led to the ability to date speleothems with low ^{238}U concentrations (for example, late Holocene samples with $^{238}U \sim 10\,\mu g\,g^{-1}$; Mangini et al., 2005). The current state-of-art includes laser ablation MC–

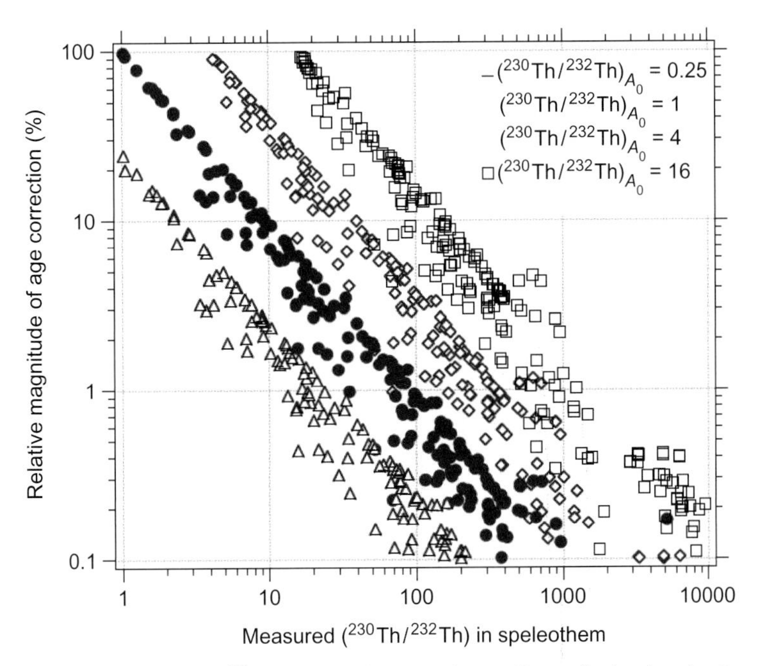

Fig. 9.5 The effect of correction for initial ^{230}Th illustrated using a large, diverse body of speleothem U–Th analyses previously undertaken at the University of Melbourne. The data are corrected assuming a range of initial [^{230}Th]/[^{232}Th], plotted as the percentage by which this correction changed the calculated age vs. measured [^{230}Th]/[^{232}Th]. From Hellstrom (2006).

ICPMS: Hoffman et al. (2009) demonstrate that a precision of 3–20% is possible on ^{230}Th/^{238}U by LA–MC–ICPMS, which is sufficient for rangefinder and screening ages. Multiple analyses on coeval samples increase the precision to low single figure; with up to 50 analyses per day possible. Also, Cheng et al. (2009) report U–Th ages with an order of magnitude reduction in analytical uncertainty.

The increasing precision in U–Th determinations has particularly focused attention on the uncertainty introduced into U–Th ages by the presence of Th transported by ground water. Typically called 'detrital' Th, it is transported with organic matter, colloidal material and fine sediments and, without correction for this Th input, can cause a substantial overestimate of the U–Th age. The major source of uncertainty is the range of values of detrital radiogenic to stable Th (^{230}Th/^{232}Th), which varies with host rock and soil characteristics, with a range of activity ratios quoted in the literature from 0.2 (Drysdale et al., 2006) to approximately 18 (Beck et al., 2001). Figure 9.5 shows the effect of variations in [^{230}Th]/[^{232}Th] on U–Th ages (Hellstrom, 2006), and demonstrates the need for independent determinations of the detrital component, or to focus on 'clean' samples where the ratio of [^{230}Th]/[^{232}Th] in the speleothem is greater than 300. A major unknown is also the variability in detrital [^{230}Th]/[^{232}Th] over time; typically the ratio is presumed to be constant and prescribed an arbitrary range of values (for example ~0.9 ± 0.45), where this uncertainty underestimates the actual possible range (Hellstrom, 2006). Combining U–Th determinations with interval counting would allow an assessment of the variability of ^{230}Th/^{232}Th over time and, for 'clean' samples, a reduction of age uncertainty through the use of the additional stratigraphic constraint on the U–Th ages provided by the interval count (Asrat et al., 2007; Dominguez-Villars et al., 2009b).

9.2.4 U–Pb

The most rapid advances in speleothem dating have arguably been in the field of U–Pb geochronology. Following the pioneering work of Richards et al. (1998), it has taken a decade for suitable methodologies to be developed. Woodhead et al. (2006) review the potential problems in U–Pb dating: the very low levels of Pb in speleothems, and the difficulty in obtaining a range of parent/daughter isotope ratios for isochron reconstruction without encountering the problem of variable initial Pb ('common Pb'). The former problem has been overcome by the development of MC-ICPMS approaches. However, our understanding of the range of samples that are suitable for U–Pb dating could still be improved. Woodhead et al. (2006) demonstrate one solution to the need to obtain a range of parent/daughter isotope ratios, using a range of isotopes which are well measured using MC–ICPMS. Using concordia diagrams, a concordia line is one along which concordant dates will lie, determined from independent radiometric methods. Figure 9.6 shows a Tera–Wasserburg isochron construction (Tera & Wasserburg, 1972), where $^{207}Pb/^{206}Pb$ is plotted against $^{238}U/^{206}Pb$. ^{207}Pb, ^{206}Pb and ^{238}U are all well measured using MC–ICPMS: for coeval samples the isotope ratios fall on a mixing line that has end-members of common Pb and pure radiogenic Pb, the latter falling on the Tera–Wasserburg concordia (the near horizontal line on Fig. 9.6). The U–Pb age is derived from the intersection of the mixing line and the Tera–Wasserburg concordia, and has more recently been applied by Polyak et al. (2008) to date water table speleothems, and Meyer et al. (2009) to date early Quaternary Alpine speleothems. Uncertainty in the U–Pb age using this approach in part derives from the fact that the Tera–Wasserburg plot requires knowledge of the initial $^{234}U/^{238}U$ activity ratio: the likely range of this parameter can be obtained from U–Th analyses of young samples or from $^{238}U/^{204}Pb$ isochrons for samples where ^{204}Pb can be precisely determined. Further applications of U–Pb dating of speleothems can be found in Chapter 12.

The major outstanding question related to U–Pb dating of speleothems is an understanding of the abundance of speleothem samples with variability

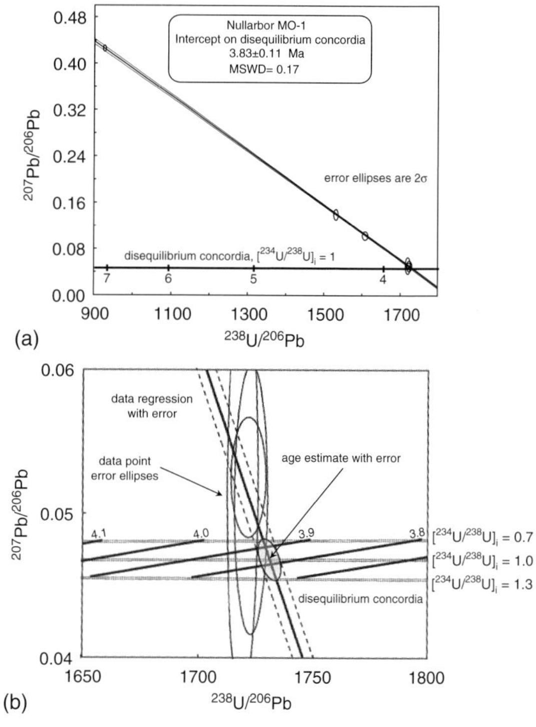

Fig. 9.6 (a) U–Pb data for Nullarbor sample LBCM01 ('MO-1') plotted using the Tera–Wasserburg construction. The quasi-horizontal line in this plot represents a disequilibrium concordia plotted for $[^{230}Th/^{238}U]_i = 0$, $[^{234}U/^{238}U]_i = 1$. Tick marks with numbers on this curve are ages in Ma. A linear regression with its associated 2σ uncertainty envelope is also shown passing through the individual blank-corrected U–Pb analyses. The calculated age is derived simply from the intersection of the two. (b) Detail of the area of intersection showing a family of possible disequilibrium concordia representing sample evolution under different $[^{234}U/^{238}U]_i$ conditions and chosen, in this case, to show the best estimate of the true value ±30%. Parallel lines in bold running across these concordia represent disequilibrium isochrons with ages marked in Ma. From Woodhead et al. (2006).

in Pb isotopes that render them suitable for U–Pb analysis, and the source of the U/Pb that is being dated. Figure 9.7 shows the range of estimated total common Pb plotted against ^{238}U, where total common lead is estimated from measured speleothem Th assuming a crustal Pb/Th ratio of 1.6 (Woodhead et al., 2006). A key requirement

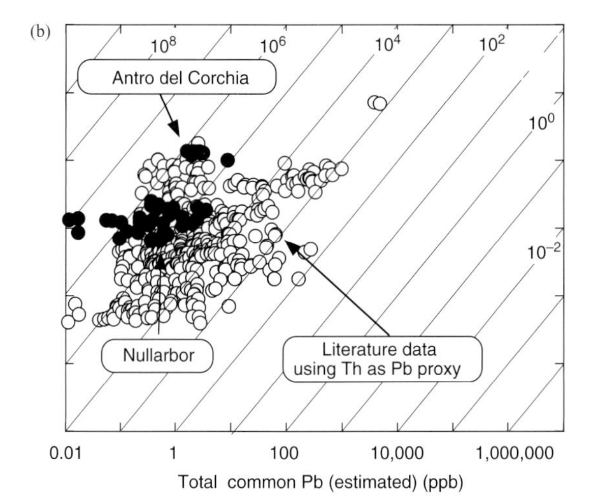

Fig. 9.7 U versus Pb concentration diagram from Woodhead et al. (2006). Diagonal lines represent the $^{238}U/^{204}Pb$ ratio. Open symbols represent literature data of U–Th-dated speleothems, where thorium has been used as a proxy for common Pb, assuming a crustal Pb/Th ratio of 1.6. Filled symbols are U–Pb-dated samples from Woodhead et al. (2006) from Nullarbor, Australia, and Corchia, Italy: all are characterized by very low common Pb contents.

appears to be speleothem material that is screened for low detrital Th (and therefore low common Pb), with successfully dated samples reported by Woodhead et al. (2006) having measured $^{230}Th/^{232}Th$ ratios of greater than 5000), which will therefore also have less common Pb. The source of this common lead is likely to be the same as that of detrital Th, namely Pb transported by organic matter, in fine colloids and as detrital material (see Chapter 8).

9.2.5 Other techniques

Interval counting, the detection of the atmospheric bomb 'spike' of ^{14}C in modern samples, U–Th and U–Pb provide a suite of dating techniques appropriate for most speleothem samples. The major limitation remains the dating of detritally contaminated samples, where high levels of Th or common Pb preclude precise chronological control, and in many cases such speleothems are probably best avoided for palaeoclimate and palaeoenvironmental reconstructions. However, some situations arise where such speleothems are the only specimens available and other techniques might need to be considered. Additionally, several alternative techniques are available which can be considered to be poorly developed or useful alternatives.

For modern speleothem deposition, Baskaran and Iliffe (1993), Tanakara et al. (1998) and Paulsen et al. (2003) successfully report total and excess ^{210}Pb profiles for an actively growing samples and use this to confirm modern growth. For suitable samples, ^{210}Pb can be considered an alternative methodology to the use of ^{14}C in modern stalagmites and one which is better at providing a radiometric age as it is less affected by storage in the overlying soil and vegetation. Condomines and Rih (2006) further report the first ^{226}Ra–^{210}Pb ages, although this technique is likely to be limited to ^{226}Ra rich waters such as found in hydrothermal environments.

For Late Quaternary age samples, Edwards et al. (1997) report the use of ^{231}Pa dating of carbonates. The technique can provide an independent cross-check of U–Th dates on speleothem samples but has not been widely adopted owing to the short half-life of the ^{233}Pa spike (27 days). Palaeomagnetism may also have use; variations in the magnetic susceptibility of detrital (and possibly authigenic) magnetic grains found in speleothems have been observed and could provide additional chronological control (Latham et al., 1986; Perkins, 1996). Over longer time periods, evidence of magnetic reversal has been used to provide a basic stratigraphic framework over 10^5–10^6-year timescales (for example, Baker et al., 1997c) and more recently palaeomagnetic analyses of cave sediments associated with U–Pb-dated speleothems has demonstrated great use in improving the temporal constraints on pre-Quaternary cave sediment sequences (Dirks et al., 2010; Pickering and Kramers, 2010). Under-used over all time periods is the use of event stratigraphy in speleothems: geochemical and physical horizons in speleothems that can be used to cross correlate between samples and provide a precise chronological control. Physical tephra horizons provide one extreme of such events, but geochemical markers might also be of equal use (Frisia et al., 2008).

9.3 Age–distance models

The reconstruction of an age–depth model of, in the case of speleothem research, an age–distance model, is of crucial importance before progressing to the generation of a proxy climate or environment time series. As stated by Telford et al. (2004), 'all age–depth models are wrong: but how badly?', and this is a useful starting point for speleothem researchers. The age–distance relationship may be very well understood from annually laminated stalagmites, where chronological uncertainties arise only from counting errors. Theoretical and actual stalagmite growth rate trends with time were presented in section 7.2.1. However, for most speleothem samples, a chronology will be made up of discontinuous sections of interval dating, radiometric dates such as U–Th and U–Pb with associated uncertainties in both age and distance axes, and event markers such as 'bomb' ^{14}C or other geochemical markers.

Annually laminated stalagmites provide a useful insight into the age–distance relationship of at least these types of stalagmite. Smith (2007) considered the statistical properties of seven series: petrological laminae from Oman (Burns et al., 2002), New Mexico (Polyak et al., 2001) and Ethiopia (Asrat et al., 2007), fluorescence lamina from Norway (Linge et al., 2009a) and NW Scotland (Proctor et al., 2000), and both ultraviolet and petrological laminae in stalagmites from Italy (Frisia et al., 2003) and China (Tan et al., 2003). Figure 9.8 shows age–depth relationships for the six stalagmites that have deposited over the past 2000 years. Time-series statistical analysis of the annual growth rate series demonstrates that they are statistically non-stationary over time, with periods of both high and low variance as often observed in other proxy series such as tree rings. All age–depth relationships exhibit statistically significant autocorrelation (see section 1.2.1), with the accumulation at time t_{+1} being correlated with that at time t: in many cases the autocorrelation persists well beyond time t_{+10}. Three of the series (China, Scotland and Norway) are statistically nonlinear, owing to sudden step changes in growth rate over time. Overall, stalagmite growth of these speleothems can be described

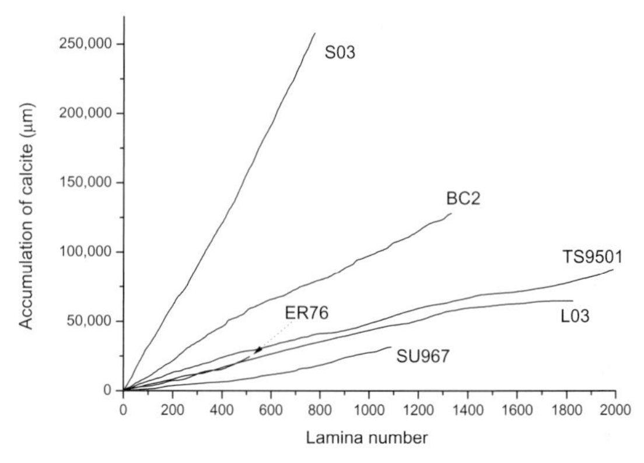

Fig. 9.8 Lamina thickness versus time for the six annually laminated stalagmites S03 (Oman), BC2 (New Mexico), TS9501 (China), L03 (Norway), SU967 (Scotland) and ER76 (Italy). All samples are plotted from either their start of growth, or for the past 2000 years before they were sampled for the longer series.

as being autocorrelated to long lag times, with sudden nonlinear jumps in growth which occur at irregular intervals. Such behaviour can be explained solely by changes in hydrological routing (section 4.2) as well as the effects of threshold responses to changing precipitation minus evaporation (P – E) and groundwater recharge. Despite this nonlinearity over short time periods, over longer periods such as the past 2000 years presented in Fig. 9.8, the age–depth relationships start to appear more linear, owing to the autocorrelation in accumulation rate from year to year. Age–depth relationships over this centennial-scale time range are visually similar to those over the millennial-scale as constrained by high precision ICP–MS dates (see Chapter 1).

However, persistent growth behaviour combined with nonlinear response means that speleothem age–depth curves are not suited to simple linear interpolation. This has been recognized for some years: spline functions have been recommended as appropriate for speleothems (Richards & Dorale, 2003) and Telford et al. (2004) recommend the cubic-spline function for lake sediments. Cubic spline functions are an appropriate approach where there are a large number of radiometric age determinations. However, a cubic spline approach does not take into consideration the known growth characteristics of speleothems (persistence, nonlinearity).

In reality, speleothem age–depth models are likely to be derived from a variety of sources, such as a combination of radiometric and interval analyses along with event horizons. Bayesian approaches which can combines all these sources of information would be powerful: such an approach has long been used by the radiocarbon community (for a review see Bronk Ramsey, 2008) with online and offline tools available (OxCal; http://c14.arch.ox.ac.uk/oxcal.html). In Oxcal, Markov chain Monte Carlo (MCMC) approaches are used, where age–depth relationships are built up from periods of defined deposition rate (e.g. annual laminated) as well as event horizons and radiometric (^{14}C) determinations. Another useful review of Bayesian approaches has been provided by Parnell et al. (2008), who provide examples of the BChron software package that is implemented in R and freely available (Haslett & Parnell, 2008). MCMC approaches have started to be used by the speleothem palaeoclimate community. For example, Drysdale et al. (2004) used an MCMC approach to fit an age–depth relationship to a Quaternary stalagmite from Italy, and Spötl et al. (2008) compared an MCMC approach (in this case Gibbs sampling which compares joint probabilities) with a mixed-effect regression model (Heegaard et al., 2005). Scholz and Hoffmann (2011) provided MCMC code for speleothem researchers using the *R* programming language.

Future speleothem age–depth modelling needs to develop the MCMC approaches above, integrating Bayesian statistics with realistic controls on the MCMC approach which account for the temporal autocorrelation and nonlinear jumps observed in speleothem lamina series. Such approaches are likely to be computationally expensive (for example, the BChron MCMC would typically run overnight) and specific to the speleothem research community and would have to be developed in-house.

9.4 Conclusions

Successful speleothem geochronology relies on an understanding of the wider speleothem incubator. Precise U–Th and U–Pb dating require minimal amounts of detrital Th and common Pb respectively, and are therefore likely to be more successful in samples which have low amounts of drip water organic matter, detrital material and fine colloids. Invariance in detrital $^{230}Th/^{232}Th$ activity ratio over time also relies on relatively stable soil and bedrock conditions. Interval dating via the presence of annual variations in fluorescent organic matter, petrological laminae or trace elements, is possible for speleothems with appropriate variability in hydrology, drip water chemistry or cave atmosphere, as discussed in previous chapters.

Age–depth models have developed in recent years with the improvement in computation power and the ability to routinely perform Bayesian statistics such as MCMC. These approaches allow the quantification of uncertainty in both *x*- and *y*-axes and a realistic attribution of uncertainty interpolated within radiometric age determinations. The combination of the suite of dating techniques and appropriate age–depth models allows palaeoenvironmental and palaeoclimate proxies to be placed on a precise timescale with a prescribed uncertainty. It is these palaeoclimate and environment reconstructions which are considered in detail in the following chapters.

IV
Palaeoenvironments

CHAPTER 10

The instrumental era: calibration and validation of proxy-environment relationships

The instrumental era is an essential period in understanding climate and environmental proxies in an Earth systems science context. Over this time period of approximately the past 100 years, speleothem chronology is optimally constrained owing to a combination of annual lamina counting, bomb pulse ^{14}C and U–Th analyses (see Chapter 9) and climate data are widely recorded (Fig. 10.1). Therefore the instrumental era is the best target against which to test speleothem proxies of climate and environment. Key avenues of investigation are (1) to determine whether there is any relationship between surface climate and a speleothem proxy; if so whether it is linear, and the extent to which is it is damped and smoothed in the proxy; (2) if not, to use the known modern variability of climate

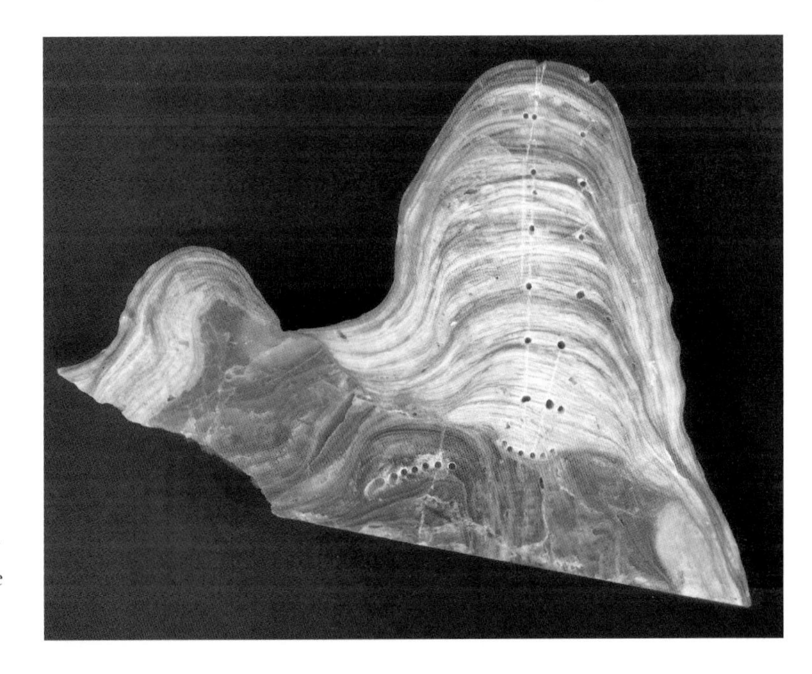

Fig. 10.1 La Faurie stalagmite Fau-stm6, Dordogne, France. Approximately 65 mm of modern annually laminated stalagmite deposition can be observed, with an average annual growth rate of 0.7 mm yr^{-1}. This stalagmite is one of six analysed in Genty et al. (2001b).

Speleothem Science: From Process to Past Environments, First Edition. Ian J. Fairchild, Andy Baker.

to ascribe a sensitivity of the proxy to past climate changes; and (3) through regression and other transfer function approaches, quantify uncertainties in the proxy climate reconstructions.

This chapter begins by reviewing available instrumental, interpolated and derived climate series that are of use to speleothem palaeoenvironmental research (section 10.1). It will cover methodologies for comparing speleothem proxies and instrumental data, reviewing methodologies used for other proxy types and appropriate methodologies for speleothems (section 10.2). The two subsequent sections focus on case studies that demonstrate how calibration against instrumental climate and environmental data series can help improve our understanding of the climate-proxy relationship, as well as improve process understanding. Section 10.3 focuses on the growth rate proxy: this proxy should have the most straightforward chronological control to undertake calibration. Section 10.4 focuses on chemical proxies, where sample resolution will be other than annual. The chapter concludes with an overview of the state-of-art and future directions.

10.1 Available instrumental and derived series

To compare stalagmite proxy data of climate and environmental change with modern data, one requires climate and environmental datasets that are, themselves, reliable. In an ideal world, this would mean continuous data that has a climate parameter with no measurement error and which is associated with a known time period. In reality, of course, one might expect a dataset that is discontinuous (probably with increasing data gaps as one goes back in time), which is associated with a well defined time period, and which has a measurement error (but one which is known or can be estimated). Table 10.1 lists the many different sources of modern climate and environmental data that might be of use to a researcher on speleothems or other archives, and Fig. 10.2 gives some examples. Datasets are grouped by whether they comprise direct observations, or whether they are interpolated products of these direct observations or whether they derive from reanalysis products which assimilate observational and satellite data or, finally, whether they are derived indices. Many of the series can be easily accessed from both the original data providers as well as through web-based tools.

The Climate Explorer toolbox (http://www.knmi.nl/publications/fulltexts/the_climate_explorer.pdf), developed by van Oldenborgh at the Netherlands Meteorological Institute, allows anyone to correlate station data, climate indices, observations, and reanalysis fields, and has been used extensively in this chapter. Climate Explorer is an internet-based tool for exploring modern climate data. Developed over the past decade, it contains many relevant climate time series. Measured temperature and precipitation data are available for stations in countries where this data has been made publicly accessible; this dataset therefore does not have full global coverage. However, interpolated and reanalysis series are available providing global coverage of a wide range of climate parameters. Examples of Climate Explorer output are shown in Figs. 10.3–10.5. A useful utility in Climate Explorer is the ability to upload your own time series: once this has been done then it can be correlated against any of the climate data within Climate Explorer. Examples of this functionality are provided in Fig. 10.8. If used, the reader is reminded to cite the use of Climate Explorer.

10.1.1 Directly measured data

Directly measured data comprise instrumental temperature, precipitation and precipitation $\delta^{18}O$ datasets. Instrumental temperature and rainfall series can be generated from observations made with thermometers and rain gauges that have used appropriate standard methodologies (for example, NOAA/NESDIS, 2002). Such series are often discontinuous, or undergo site changes, or may be affected by changes in site condition over time such as urbanization (which could increase temperature and shield rain gauges from precipitation). Because temperature is spatially homogenous over long distances, it is often possible to use a relatively distant, high quality, temperature series, corrected to the local site temperature. Precipitation is much more spatially variable, and a site just a few kilometres away from a study site might not accurately reflect precipitation at the site, especially in mountainous

Table 10.1 Summary of available modern climate series.

Parameter	Data source	Frequency	Time period
Directly measured data			
Temperature	Instrumental observations	Daily, often max and min	Longest series over 100 years, often discontinuous
Precipitation	Instrumental observations	Daily	Longest series over 100 years, often discontinuous
Precipitation $\delta^{18}O$	Water samples	Typically monthly	~1960s AD to present
Interpolated data products			
Temperature and precipitation	CRU[b], GISS[b], NCDC[b]	Monthly interpolation of instrumental series	~1850 AD to present
Sea surface temperature	Hadley Centre, NCDC	Monthly interpolation of instrumental series	~1880 AD to present
Sea level pressure	Various, interpolation of instrumental series	Monthly	1800 AD to present
Reanalysis data			
Temperature, precipitation, atmospheric pressure, wind speed and direction at various atmospheric levels	NCEP[b] and ERA[b]	Monthly	1957–2002 AD (ERA-40), 1958 AD—present (NCEP), 1989 AD—present (ERA-interim), 1878–2008 AD (NOAA Twentieth Century Reanalysis)
Indices[a]			
Water excess/PDSI[b]	Calculated from observations or as a reanalysis product	As frequent as observations: typically daily to monthly	As good as the instrumental series or length of reanalysis period
SOI[b]/ENSO[b]	Calculated from surface pressure or SST anomalies	Daily to annual	1850 AD to present
NAO[b]/AO[b]	Calculated from surface pressure	Daily to annual	~1820 AD to present

[a] For indices, time period refers to those based on instrumental series. Longer indices may be available which are also based on documentary or other proxy sources.
[b] NCEP, National Center for Environmental Prediction; NCAR, National Center for Atmospheric Research; ERA, European Reanalysis; CRU, Climate Research Unit; GISS, Godard Institute for Space Studies; NCDC, National Climate Data Center; NOAA, National Oceanic and Atmospheric Administration; PDSI, Palmer Drought Severity Index; SOI, Southern Oscillation Index; ENSO, El Niño/Southern Oscillation; NAO, North Atlantic Oscillation; AO, Arctic Oscillation.

terrain or where convective storms dominate the rainfall totals.

An extensive literature exists debating the relative merits of techniques which can test temperature and precipitation series for inhomogeneities, using methodologies such as non-parametric, fuzzy and Bayesian change point detection (Lanzante, 1996; Caussinus & Mestre, 2004; Dose & Menzel, 2004; Ruggieri et al., 2009). A range of these tools have been applied to various extents to interpolated data products such as gridded temperature and rainfall series; this is discussed more in the following section. Ultimately, even with these statistical tools, one cannot be certain that a series is of high quality without observing the metadata submitted at the time the data was recorded or the

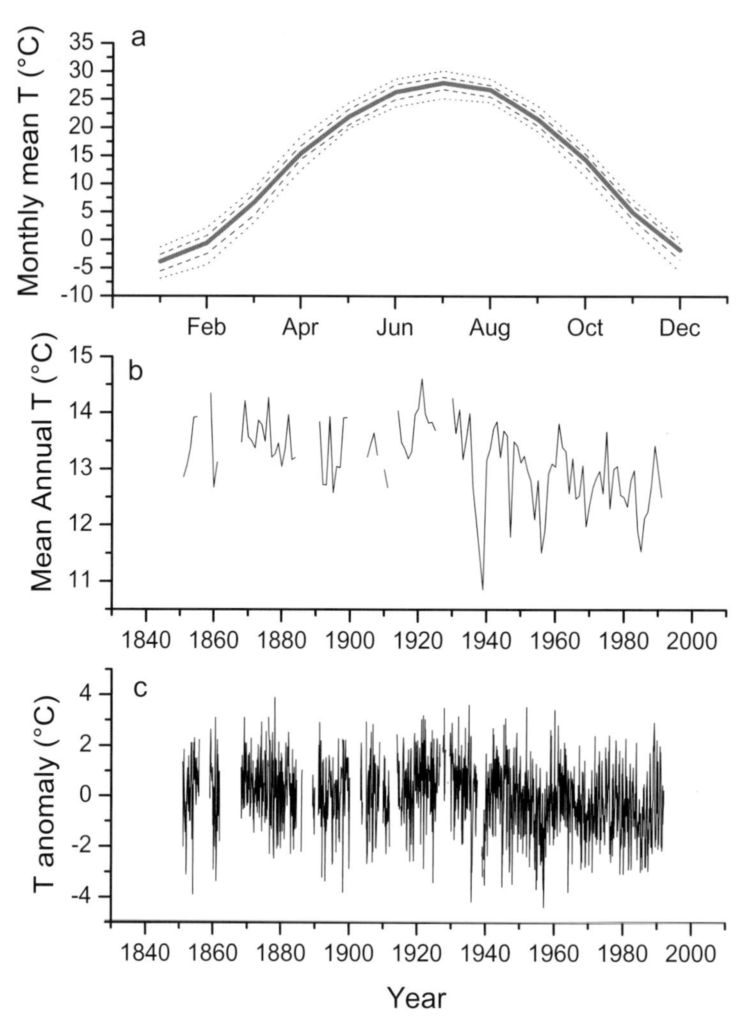

Fig. 10.2 Example of measured data series. (a) Beijing mean monthly temperature (representative of the climate at Shihua Cave and stalagmite records published by Tan et al. (2003, 2009); 17% and 2.5% percentiles of temperature also shown. (b) Annual mean temperature, 1841–1991 AD. (c) Temperature anomalies (monthly) with respect to the 1971–2000 AD mean. All plots use data directly exported from Climate Explorer.

original data returns. For example, in our own work in northwest Scotland, precipitation series were obtained for the 19th and 20th centuries from observations at Stornoway. In the 19th century, despite site changes, statistical analysis of the monthly mean precipitation total suggested a homogenous precipitation series could be developed for Stornoway, with no observed abrupt changes in rainfall amounts or variability which would suggest an inhomogeneous series. However, inspection of the original hand-written archived data demonstrated that incomplete returns were made for many years, with no daily rainfall records but just a monthly total, suggesting the data might not be reliable. Another interesting recent example is from the Swiss Alps, where a large body of proxy data (in this case temperatures determined from tree rings) has suggested that early instrumental temperature data has a greater offset than previously recognized owing to warm biases of early thermometer shelters (rather than Stevenson screens) (Frank et al., 2007). In a few studies, the uncertainty associated with instrumental data series, and its change through time, has been assessed. For example, the Central England Temperature (CET) series, compiled from a set of long instrumental temperature records and extending back to 1659 AD, has uncertainties decreasing from ±1 °C up to 1670 AD, then ±0.5 °C between 1670 and 1722 AD, and finally ±0.1 °C from 1722 AD (Parker et al., 1992).

A directly measured dataset of particular relevance to speleothem palaeoclimatology is a global dataset of isotopes in precipitation. The Global Network for Isotopes in Precipitation (GNIP), supported through the International Atomic Energy Agency (IAEA), has coordinated the collection and data archiving of 1-month integrated samples of precipitation from various global locations. Stations participating in the network collect monthly composite total rainfall, for tritium, deuterium and ^{18}O analysis (IAEA, 2009a). More recently, such data has been augmented in some areas by event rainfall data as well as river data (Global Network for Isotopes in Rivers (GNIR), which in regions of low evaporation will fall on the local meteoric water line, LMWL), all freely downloadable from a web interface. Both event and monthly mean precipitation δ^2H and $\delta^{18}O$ provide useful baseline datasets to determine the regional meteoric water line (RMWL) at a research site, often usefully augmented by a local rainfall isotope monthly mean series, collected using standard IAEA protocols (IAEA, 2009b) to obtain a LMWL. Raw data can be accessed via the IAEA website using an on-line interface called Water Isotope System for Data Analysis, Visualisation and Electronic Retrieval (WISER) (http://nds121.iaea.org/wiser/). Speleothem researchers collecting event or monthly precipitation samples to this standard are encouraged to submit their data to the IAEA database to enable their use by other researchers.

10.1.2 Interpolated data products

Several research groups have taken the measured data series and combined them using interpolation techniques to provide spatial surface climate datasets at a coarser resolution. Widely used series include those produced by the Climate Research Unit, University of East Anglia, UK; the National Aeronautics and Space Administration (NASA) Goddard Institute for Space Studies (GISS), USA; the National Oceanic and Atmospheric Administration's (NOAA) National Climatic Data Centre (NCDC); and the UK Hadley Centre. For each, significant effort was involved in acquiring measured series from meteorological organizations around the world, undertaking homogeneity tests and appropriate interpolation routines. For details of methodologies used, see for example Brohan et al. (2006), Hansen et al. (2001) and Smith et al. (2008). In all cases, the benefits of providing a globally gridded dataset has been at the expense of increased uncertainty, especially in regions where interpolation is less likely to produce acceptable results (e.g. regions of variable altitude or at the land–ocean interface). Originally designed as datasets against which general circulation model (GCM) output could be compared, interpolated products can be of use to speleothem researchers, especially when (1) original measured data series are not accessible or (2) as a reliable sea-surface temperature record, when teleconnections between speleothem proxies and ocean circulation are postulated. Examples of interpolated data products are presented in Fig. 10.3.

10.1.3 Reanalysis data

For recent decades, two significant research efforts have undertaken integration of both human-measured and satellite-measured data, to produce reanalysis data products. These provide gridded climate data at various atmospheric levels including surface pressure; the benefit over interpolated data products is therefore the availability of data at a range of atmospheric heights, but this comes with the penalty of shorter data series. The two widely used products, European Centre for Medium-Range Weather Forecasts' (ECMWF) ERA-40 and NOAA's NCEP/NCAR Reanalysis Project, cover approximately the past 50 years of climate history; for details see Uppala et al. (2005) and Kalnay et al. (1996). Recently, the Twentieth Century Reanalysis has become available, covering the period AD 1878–2008 (Compo et al., 2011). The

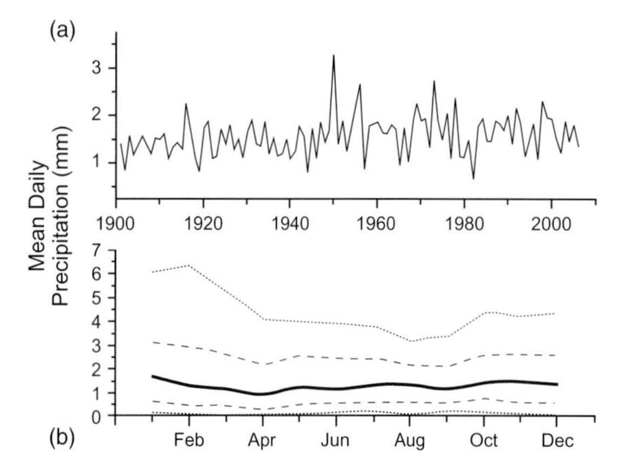

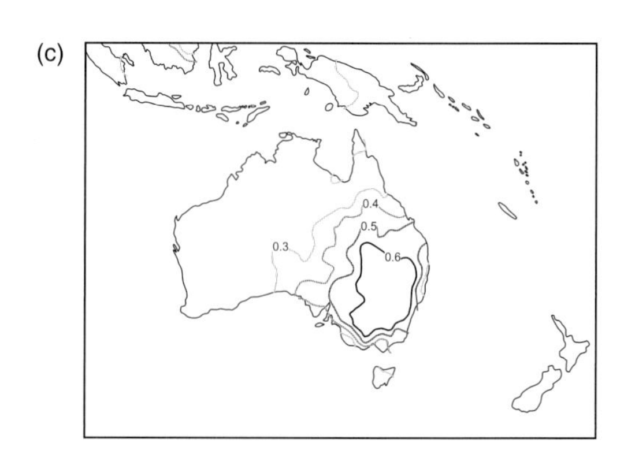

Fig. 10.3 Example of an interpolated climate series. Data are extracted from the CRU analysis TS3, for a 1 degree region (148–149 °E and 32–33 °S), which corresponds to the karst region of New South Wales, Australia. (a) Mean precipitation in mm/day. (b) Seasonal variability in precipitation. (c) Correlation between precipitation and gridded precipitation data in the same defined region in the CRU TS3 dataset, suggesting that speleothem records from the region would be representative of a large part of interior southeast Australia. Both graphs use data directly exported from Climate Explorer; map is directly exported from Climate Explorer.

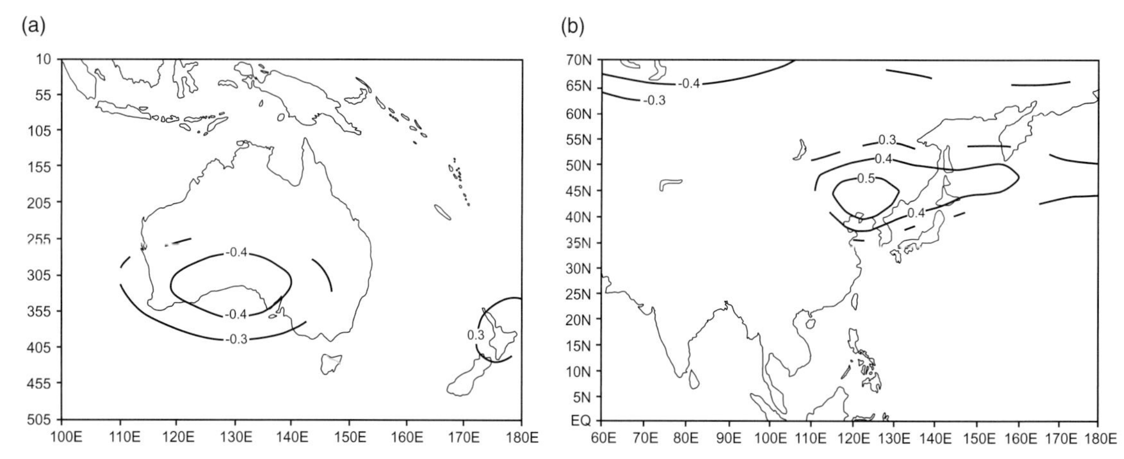

Fig. 10.4 Examples of the use of reanalysis data; all examples correlate measured or interpolated data with 500 mb pressure height for the period 1948–2006 AD. (a) Correlation with annual New South Wales (Australia) precipitation. (b) Correlation with Beijing mean monthly temperature. Contours are correlation coefficients and only those with greater than 90% significance are shown. Maps plotted using Climate Explorer.

advantage to the palaeoclimatologist is primarily the availability of a wider range of relevant meteorological parameters including zonal and meridional wind strength (which might correlate with rainfall source and therefore $\delta^{18}O$), or atmospheric pressure at the mid-atmosphere (500 mb), which may be more representative of the dynamical climatology. Examples of reanalysis product outputs are shown in Fig. 10.4.

10.1.4 Climate indices

Climate indices are series that have been derived from measured data; well-known indices that attempt to represent climate states such as the El Niño/

Southern Oscillation and North Atlantic Oscillation have been previously introduced in Chapter 3. Other indices of particular relevance to speleothem palaeoclimatologists and introduced in that chapter, include water excess or hydrologically effective precipitation, and various drought severity indices such as the Palmer Drought Severity Index (PDSI). All are available as long time series against which speleothem or other archives can be compared, and can be obtained either directly from the measured or interpolated series, or from reanalysis data.

10.2 Methodologies

10.2.1 Overview of methodologies used in other fields

Presuming one is in the possession of both a measured or derived instrumental series, as well as a stalagmite proxy series for which each proxy measurement can be ascribed an age, then one can consider a wide range of techniques which facilitate the comparison of proxy series to climate series. In this field, speleothem palaeoclimatology is many years behind comparable disciplines such as dendroclimatology and the use of documentary climate sources. Therefore in this section we review the established methodologies used in other fields, before moving on in section 10.2.2 to tackle the emerging methodologies appropriate to speleothem series. An overview is provided in Fig. 10.5.

Linear-regression-based techniques

The most apparently straightforward approach that can be used is a linear regression of the climate proxy against a target climate series or series of climate series. These series may be measured data, interpolated series, reanalysis data or indices. In all cases, the target of the regression should be a climate parameter for which there are physical, process-based reasons for a potential correlation. Care has to be taken that any long-term trends in either data series do not bias the regression analysis; regressions on de-trended data remove this possibility. Proxies are typically regressed against individual months from January to December, plus seasonal means and annual means. Errors associated with the regression can be used to ascribe uncertainties to the proxy climate reconstructions, and if the instrumental and proxy series are long enough, the series should be split into separate calibration and verification periods. For further reading, see Esper et al. (2005), Burger (2007), Christiansen et al. (2009) and Ammann et al. (2010). Where spatial (gridded) data are available, a further test is where one can plot the strength of any correlation over space, which should show a decreasing correlation between proxy and climate with distance away from the location of the proxy. In all cases, statistically significant correlations should show both temporal and spatial patterns that are logical e.g. if there is a strong correlation between July temperature and proxy, there should also be correlations between June and August temperature and the proxy, and for all three months, the strength of correlation should decrease with distance from the proxy record. Correlation must also be plotted as scatter plots to confirm that the basic assumptions of linear regression are being maintained e.g. that the series are linear and without outliers. One should note that the Pearson's product correlation coefficient is typically quoted in the subject area, despite the proxy series having temporal autocorrelation (a non-parametric correlation such as Kendall's tau or Spearman's correlation is more appropriate). The alternative approach often adopted, is for the statistical significance of a correlation coefficient to be adjusted to take account of this autocorrelation, by taking a lower number of degrees of freedom.

Building on the simple linear regression model approach, the dendroclimatology community have used autoregressive models to pre-whiten tree ring time series and better capture high frequency climate variability; specifically ARIMA (autoregressive integrated moving average) modelling is used to remove the serial correlation from the dataset (Cook, 1992). To capture low-frequency climate variability, a long-term regional growth trend is used as the de-trending procedure (the regional curve standardization approach, Briffa et al., 1996). For discontinuous proxy series, the regression estimation approach (RegEm) can be applied (Schneider, 2001; Rutherford et al., 2003). Most recently, the total least squares

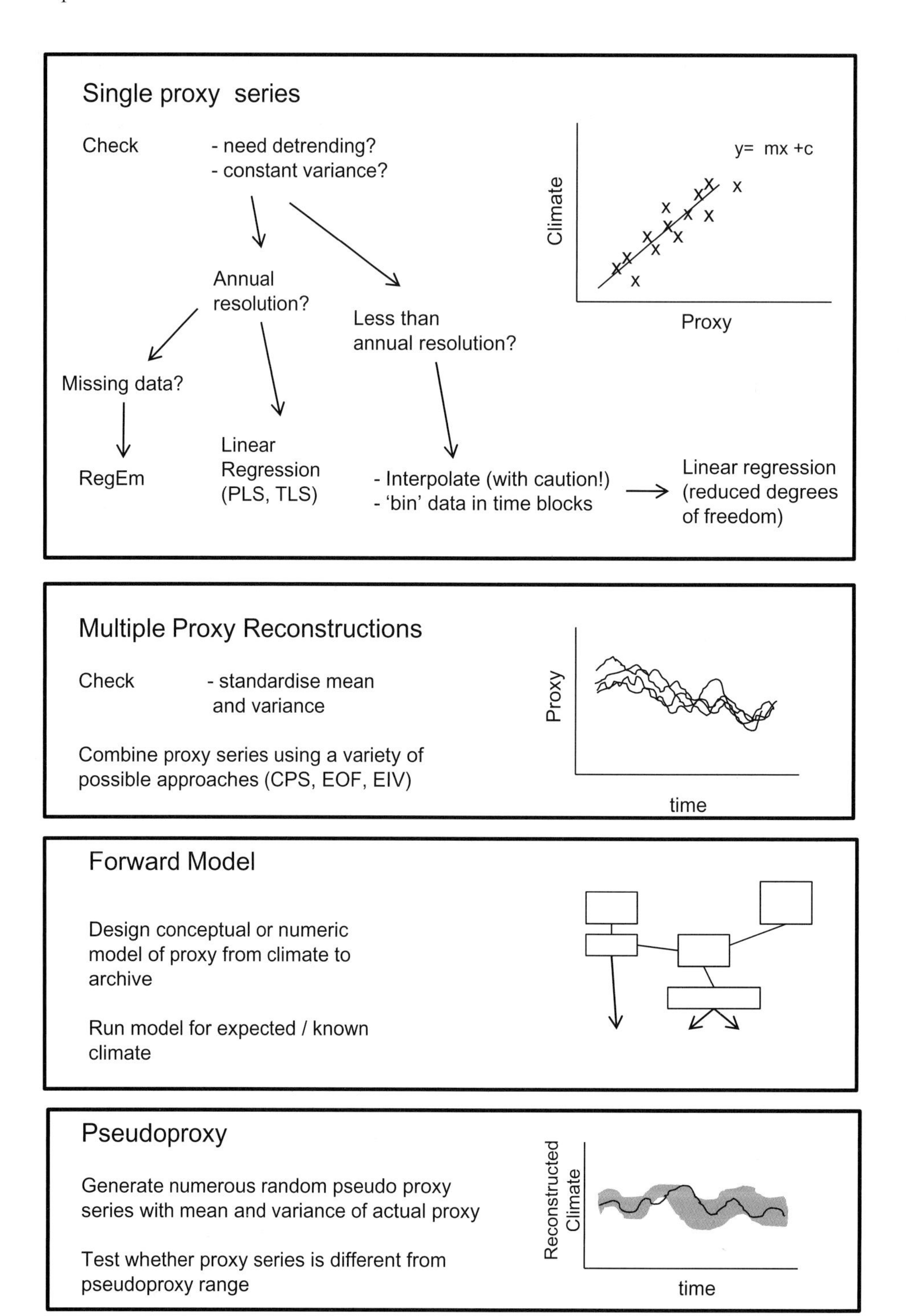

Fig. 10.5 Methodologies. PLS, partial least squares; TLS, total least squares; CPS, composite plus scale; EOF, empirical orthogonal function; EIV, error-in-variables; RegEm, regression estimation.

regression approaches have been adopted (Mann et al., 2008); these orthogonal regression methods better include errors in both *x*- and *y*-axes.

Compositing records

Once one has more than one proxy series with a common correlation with a climate parameter, then one can consider whether compositing the proxy series will generate an improved climate record with improved signal-to-noise ratio. The basic theory is that, with a common forcing signal, any non-climatic noise will be reduced and the climate signal strengthened. To do this, the different proxy series have to be standardized to a common mean and variance. Techniques such as empirical orthogonal functions (EOFs) can be used to determine the dominant mode of variability by obtaining principal components. Another approach is that of 'composite plus scale' (CPS; Lee et al., 2008), where each record is standardized to a mean of zero and a unit variance relative to a common period, and then composited by averaging the standardized series without weighting. A third approach that attempted to integrate both low- and high-frequency climate proxies in a calibration, used a wavelet-based approach (Moberg et al., 2005), but the methodological details are difficult to implement. The compositing of proxy series, and the distinctive 'hockey stick' Northern Hemisphere temperature records of the past approximately 1000 years, have been the focus of considerable scientific and media attention for over a decade. Expert analysis can be found at www.realclimate.org, with an excellent resource on all aspects of climate change at http://www.realclimate.org/index.php/archives/2007/05/start-here/. Speleothem proxy series first become composited within the Moberg et al. (2005) 1000 year climate reconstruction, and have subsequently featured in the composite series of Mann et al. (2008). For further details see section 11.2.

Forward modelling

Forward modelling is a process in which the climate series is taken as an input to a process model, as opposed to a target of an empirical regression equation. Uncertainties in the quantification of errors in regression-based approaches, for example owing to temporal autocorrelation in the proxy series, and the presence of both *x* and *y* errors on both proxy and climate, has led to interest in the development of process-based models. Driven by input climate data, forward models take an integrated earth systems approach to predict proxy series. As such, they also rely on good quality (long, continuous) measured or derived series, often including $\delta^{18}O$ of precipitation. Ideally, one could envisage forward models being driven by GCM output, producing pseudoproxy series (see next section) against which actual proxy series could be compared. Forward models published to date, outside of speleothem palaeoclimatology, include lake $\delta^{18}O$ models (Jones et al., 2005; Jones and Imbers, 2010).

Pseudoproxies

Pseudoproxies are statistical or modelled proxy series, and are a recent attempt to quantify the uncertainty associated with proxy climate reconstructions (Mann et al., 2005; Christiansen et al., 2009). For example, Moberg et al. (2008) used the statistical time series properties of different proxies used in the Moberg et al. (2005) reconstruction to generate a pseudoproxy reconstruction of the past 1000 year climate, where the pseudoproxy series have the same variance as the original proxy records. Pseudoproxy series can also be generated by the forward modelling approach as detailed previously, with model input parameters with a mean and variance, and autocorrelation of that to be expected over the model period. At the time of writing, this approach is postulated as a suitable test of palaeoclimate proxy series but rarely implemented (for a review and example, see Christiansen et al., 2009).

10.2.2 Appropriate methodologies for speleothem calibration

To calibrate speleothem proxies against modern climate data, three essential conditions need to be met. Firstly, the speleothem calcite (or aragonite) has to have been suitably sampled to prevent aliasing or other statistical issues arising (see Fairchild et al. (2006a) for detailed considerations of appropriate sampling methodologies). Secondly, it has to have been sampled at high temporal resolution so that there are enough data points for calibration, and so that the number of degrees of freedom within any statistical analyses are maximized. Thirdly, precise chronological control is essential, such as that provided by annual lamina counting. If these conditions are met, then one can consider using various appropriate methodologies. These are considered in detail in this section.

Linear-regression-based approaches

Section 10.2.1 introduced linear regression approaches as applied to the calibration of non-speleothem archives, in particular tree ring proxies. Proxies such as tree ring width and density, tree cellulose $\delta^{18}O$ and $\delta^{13}C$, proxies for peat water table, lake varve thickness, and ice core accumulation rate, typically correlate with the climate of a particular season of that year (in some biological proxies, also the previous year). This makes linear regression approaches relatively trivial, as one can regress the proxy against the target climate parameter for either the same year (t) or the previous year (t_{-1}). Speleothem-based regression approaches, however, are unlikely to be so straightforward, considering the variable amount of time it takes rainwater to reach a speleothem feed water. Unless the proxy is forced by the cave climate (in caves where climate varies seasonally, e.g. near entrance locations where cave air CO_2 or ventilation may drive growth rate or isotopic composition), it is probable that the mixing and storage of water in the karst aquifer means that the proxy will represent an integrated signal of the proceeding n years. Historically, a fixed value for n, such as 10 years, has been used to enable comparison with other decadal averaged proxies: however, given our knowledge of karst hydrology, such a fixed approach may increase the uncertainty associated with the regression.

One approach that can be used, to account for the mixing of ground water within the karst aquifer, is to take a transfer function approach (Baker et al., 2007; Baker and Bradley, 2010; Jex et al., 2010). This function takes the form

$$W_{t=n} = M\, I_n + (1-M)\sum I_{n-1 \text{ to } n-x} \qquad (10.1)$$

where $W_{t=n}$ is water at time n, where n is a preceding year, M is the proportion of fracture flow, I is the monthly instrumental climate parameter (e.g. total precipitation (P), mean temperature (T), total evapotranspiration) and x is the variable duration of storage flow component.

This simple mixing model presumes a simplified karst hydrology with just two flow components: a fracture flow component M that transfers the surface water to the cave drip in the same year, and a matrix flow component ($1 - M$) that has a slower transfer rate and mixes water of the preceding x years. Mixing models can be run for monthly, seasonal and annual climate parameters, to reflect the fact that water recharge to the aquifer can be highly seasonal, and result in a large number of transformed time series W, with variable x, M and I. Baker et al. (2007) and Jex et al. (2010) found the best correlations were observed when parameter M was set to <0.3 and $x < 10$ years. Several factors can guide the expected range of these parameters. The proportion of preferential (i.e. fracture-fed) to matrix flow may determine the stalagmite shape, with candlestick stalagmites expected to have M close to or equal to 0, and those with increasing widths a greater proportion of fracture-fed flow. The presence of hydrologically generated annual growth laminae, fluorescent organic matter or soil derived trace elements, would imply $M > 0.0$.

One of the disadvantages with any smoothed time series, whether a simple running mean or a more complex transfer function, is that this results in reduced degrees of freedom (df) within the regression model. In practical terms, this means that a much higher correlation coefficient is needed for a correlation to be statistically significant. For example, for most regions of the world, the length of instrumental climate series can be expected to be between 50 and 150 years. With $x = 1$ and an instrumental series of 100 years, then df = 99, and a regression yielding a correlation, $r = 0.50$, would be statistically significant at 99% confidence. With $x = 5$ and the same instrumental series, then df could be as low as 19, for which a statistically 99% confidence level significant correlation coefficient r would be 0.58. For many regions of the world, where the length of instrumental climate series is less than 100 years and with $5 < x < 20$, the df become unacceptably low and even when correlation coefficients are very high they would not achieve statistical significance.

Linear regression approaches, which consider a transfer function approach to account for the karst hydrology, may therefore suffer from yielding statistically insignificant correlations due to the reduced degrees of freedom. Additionally, other calibration approaches are needed that tackle the problems of equifinality and nonlinearity. For stalagmites, the extent to which a proxy time series may be reached by more than one climate input is not known, but the known complexities of karst hydrology mean that that the equifinality issue is one that cannot be ignored. Karst drip-water hydrology is known to be nonlinear, with under and overflow type behaviour that has been both conceptualized (Tooth & Fairchild, 2003) and observed (Genty & Deflandre, 1998; Baker & Brunsdon, 2003). In these circumstances, linear regression approaches will not provide an appropriate solution; however, the approach still has merit as it can improve the understanding of the climate–proxy transfer process. For example, through the analysis of several regressions between transformed instrumental series and a stalagmite proxy, one should expect to see similar correlations between regressions for adjacent calendar months, and between similar transformation functions x and M.

Compositing

Compositing of multiple speleothem proxies can use established techniques used within palaeoclimate research, the principal limitation at the time of writing being the paucity of high-resolution speleothem proxy archives. EOF and CPS approaches are particularly appropriate and applicable directly to proxy series when dating errors can be presumed to be negligible, for example for annually resolved series such as annual lamina thickness. When the proxy chronology is less secure, the 'binning' of standardized series into bins of prescribed time period is an approach that can be used to, in part, overcome the dating uncertainty. For example, Kaufman et al. (2009) binned various (non-speleothem) data into 200 10-year intervals over the past 2000 years. An example of a composited speleothem record is presented in section 10.3.1, and multiproxy composite records which include speleothem proxies in section 11.2.

Forward modelling

Forward modelling approaches to understanding the speleothem proxy—climate relationship have been developed in recent years (Baker & Bradley, 2010; Bradley et al., 2010; Baker et al., 2010; Wackerbarth et al., 2010; Dasgupta et al., 2010). Input data would typically include daily to monthly total precipitation and mean temperature, and used to calculate water excess using the Thornthwaite method. More complex models could use alternative methods to determine water excess, but Thornthwaite enables the significance of the ratio between precipitation and evapotranspiration to be explored. Working forward in time, recharge to the karst can be presumed to occur in months when the water balance was positive (soil moisture deficit is presumed negligible). This recharge water can be tagged with an appropriate proxy as the appropriate place in its flow path: for example $\delta^{18}O$ of the input precipitation or trace element composition derived from the bedrock, and then appropriately stored or transported through the karst aquifer, with fractionation processes implemented at relevant stages of the flow path and speleothem deposition. The ground water component of the forward model is typically a hydrological representation of a karst system, which could be a single linear reservoir (following Gilman & Newson, 1980); more complex reservoir approaches with overflow and underflow routing, or a series of linked reservoirs (see section 4.3). To date, researchers have not attempted to implement fully distributed hydrogeological models, but these are designed to determine the total flow field of an aquifer, not to deal with minor flow at a specific point in an extremely non-homogeneous aquifer. In all these lumped-parameter approaches, the modelled hydrology is just an approximation of the actual karst hydrogeology, and simple mixing models were used to modify the proxies under investigation. The models can incorporate in addition fractionation processes that might affect the proxy, either within the cave aquifer (e.g. prior calcite precipitation, microbial degradation) or between within the cave void. Ultimately, a complete Earth systems model can be envisaged which forward models both stalagmite morphology and geochemistry, taking into consideration all vegetation, soil, hydrogeological and cave environmental processes, although at the time of writing this is some way from being achieved.

Pseudoproxies

To date, the only speleothem pseudoproxy that has been generated is a statistical pseudoproxy of Shihua Cave annual laminae from Tan et al. (2003), as used in the past 1000 year reconstruction of Northern Hemisphere temperatures by Moberg et al. (2005). Pseudoproxy series in Moberg et al. (2008) were generated with an identical variance spectra to the original proxy, and then compared with the original Moberg et al. (2005) reconstruction, as well as GCM simulations and reconstructions based on using the CPS technique. This general approach presumes that the statistical properties of the pseudoproxy series actually reflect that of the true proxy. Moberg et al. (2008), for example, did not consider possible nonlinearities in the Shihua Cave stalagmite series, despite the fact that these are known to occur in stalagmite proxy records. Statistical speleothem pseudoproxies at present are unlikely to under-represent the variance to the actual proxy and require further research, especially the incorporation of nonlinear statistics.

10.3 Case studies of calibrated speleothem proxies

Having introduced appropriate methodologies, here we introduce selected case studies as examples of speleothem proxy records which have been calibrated against modern climate conditions. These will be considered proxy-by-proxy.

10.3.1 Annual lamina thickness

Stalagmite annual growth rate series were the first targets for instrumental calibration owing to the obvious benefits of having an immediately available annual chronology. A summary of calibrated records can be found in Table 10.2, which highlights the current lack of calibrated annual lamina thickness records derived from trace elements, despite the recognition of widespread annual cyclicity in this proxy. Additionally, a few records exist which cover the modern period but are uncalibrated (e.g. Fleitmann et al., 2004; Nott et al., 2007). Calibration against instrumental data has to date focused on linear-regression based approaches, and has yielded either a precipitation- or temperature-dominated signal, or one which is a mixed precipitation and temperature proxy. This is not surprising given the factors that determine speleothem growth rate introduced in section 7.3, which for well-ventilated caves, would be dominated by warm and wet conditions favourable for soil CO_2 production, and water

Table 10.2 Modern calibrated stalagmite annual growth lamina records.

Source	Lamina type	Location	Climate correlation	Correlant
Genty & Quinif (1996)	Visible laminae	Belgium	Water excess	r = 0.14–0.84 depending on stalagmite (annual), $n < 30$
Brook et al. (1999)	Aragonite-calcite pairs	Madagascar	SOI	r = 0.72 (annual), n = 44
Proctor et al. (2000)	Fluorescent laminae	Scotland	NAO	r = −0.70 (decadal), n = 96
Frisia et al. (2003)	Visible/fluorescent laminae	Italian Alps	Winter T	r = 0.28–0.56 depending on stalagmite, $n < 230$
Tan et al. (2003)	Fluorescent laminae	China	Summer T	r = 0.68 (annual), n = 55
Yadava et al. (2004)	Visible laminae	India	Annual P	r = 0.30 (decadal), n = 111
Baker et al. (2007)	Visible laminae	Ethiopia	Summer P	r = 0.50 (decadal)
Cai et al. (2010)	Visible laminae	Thailand	n/a	$r < -0.30$ (annual and 5-yr average), n = 91

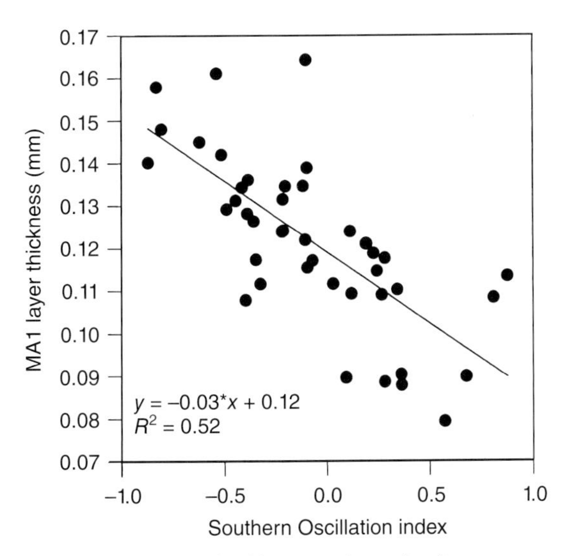

Fig. 10.6 Example of calibration of a speleothem proxy against climate; annual growth rate series MA1 versus the SOI (Brook et al., 1999).

supply to maintain speleothem drip rates throughout the year. For example, stalagmite MA1 from Madagascar (Brook et al., 1999) comes from a climate region with a distinct dry season and soil moisture deficit and where, depending on hydrological connectivity, dripwaters are likely to be supply-limited. Figure 10.6 shows the published correlation between MA1 and SOI. In contrast, stalagmite SU-96-7 from northwest Scotland (Proctor et al., 2000; Trouet et al., 2009) was formed under a peatland and much wetter conditions where a soil moisture deficit is only formed for short periods in the year; here soil CO_2 is determined by peat water table and predominantly rainfall and it is this factor which is dominating the growth rate–climate correlation (Table 10.2). In this region, precipitation is correlated with the NAO, and growth rate also exhibits strong correlations with the NAO (Proctor et al., 2000) and SST (Proctor et al., 2002). Stalagmite TS9501 from Shuihua Cave was shown by linear regression to calibrate against summer temperature (Tan et al., 2003). At this site close to Beijing, there is a seasonality of both temperature and rainfall in monsoon climate, and it was argued that summer warmth drives soil CO_2 production that dominates the growth rate signal. Frisia et al. (2003) reported a correlation for two out of three stalagmites between annual growth rate and winter temperature at the Ernesto Cave. At this seasonally snow-covered alpine location, it was argued that an annual recharge of snowmelt ensures that moisture is less likely to be limiting than temperature, which determines the length of the snow-free period and would be the primary driver of soil CO_2 productivity. Finally, Cai et al. (2010) failed to find any correlations between growth rate and climate parameters in a stalagmite from north-western Thailand which does show correlations with $\delta^{18}O$ (see next section).

The TS9501, SU-96-7 and ER76, 77 and 78 records have been publicly archived at http://www.ncdc.noaa.gov/paleo/paleo.html which allows one to reanalyse the linear regressions with rainfall and/or temperature. Figure 10.7 shows examples of such reanalyses which have simply sought to confirm the strength of the linear regressions through the analysis of spatial correlation maps. If a correlation is meaningful and realistic, it should show a spatial pattern associated with the proxy location. For example, Fig. 10.7a plots the correlation between SU-96-7 annual lamina thickness and winter SLP using the HadSLP dataset, and confirms the winter NAO-like pattern of sea level pressure in Proctor et al. (2000) and Trouet et al. (2009). In Fig. 10.7b, gridded surface temperature data are used to plot the spatial correlation of an Ernesto record (using ER76 as an example) and maximum temperature. The spatial correlations demonstrate that the winter temperature reconstruction from Ernesto annual lamina thickness should be applicable to the wider region of the European Alps.

Moving on from linear regressions and spatial correlations, Baker et al. (2007) used a transfer function approach to attempt to correlate stalagmite annual growth rate and surface climate for two Ethiopian stalagmites. In this region with two rainy seasons and a near-constant temperature, growth rate was compared against annual, seasonal, and monthly rainfall data. For both samples, correlations were weak and statistically insignificant for all months and seasons. Applying the climate transfer function, good correlations were observed for only one of the two samples, and this

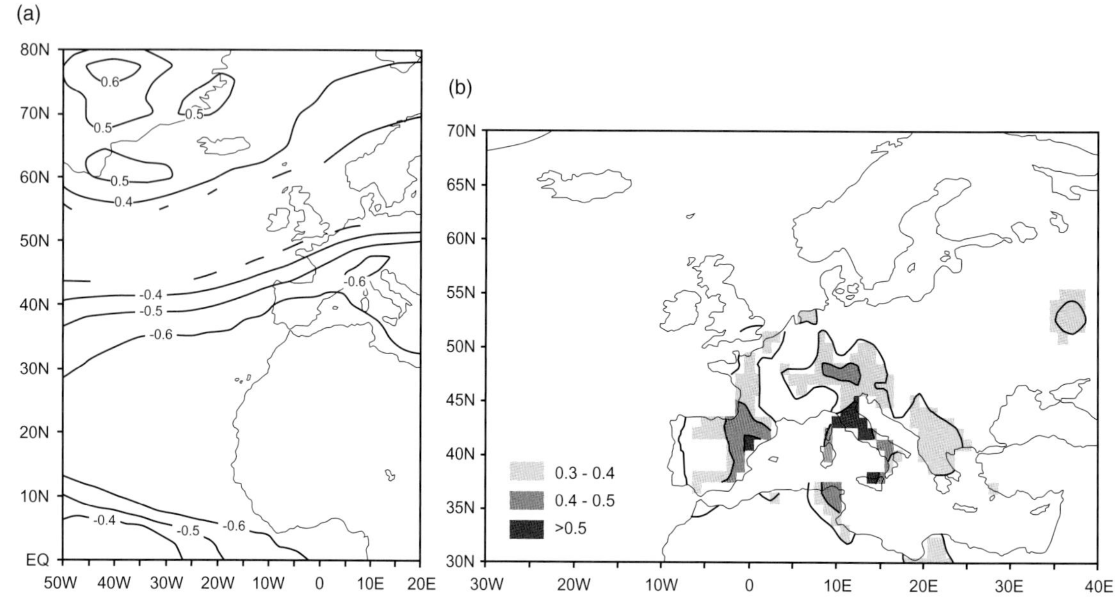

Fig. 10.7 Spatial correlations of stalagmite proxy series with modern climate data. (a) Correlation between annual growth rate in stalagmite SU-96-7, northwest Scotland and winter sea level pressure, demonstrating the classic NAO pressure pattern. (b) Correlation between annual growth rate of stalagmite ER76 and winter (NDJF) maximum temperature (CRU TS3 gridded temperature series), showing a correlation throughout Italy and the Austrian Alps. Maps redrawn from those exported from Climate Explorer.

was with summer rainfall, although the correlation was statistically similar to a simple decadal average ($r = 0.50$). For this one sample, growth rate correlated positively with high rainfall totals in May, June and July; but negatively with low rainfall totals in March and April. This pattern cannot be explained solely by a relationship with rainfall amount, because that should lead to a positive relationship between growth rate and both spring and summer rains. Many factors could be invoked, such as in years of lower spring rainfall, limited vegetation growth and soil CO_2 production could lead to spring rainfall derived dripwaters with relatively low dissolved calcium and magnesium concentrations. Irrespective of the mechanism, this case study highlights many of the benefits of comparison with instrumental data; the weak or non-existent correlations provide limits on the sensitivity of the proxy in these samples to temperature and precipitation (variations in growth rate outside those observed in the modern period might be expected to be forced by conditions not observed in the modern period). The differences between the two samples confirm that each stalagmite is likely to have a different correlation with surface climate depending on hydrology. Finally, in this region, caves are frequently poorly ventilated and therefore annual growth rate may also be limited by variations in cave air CO_2, which could also explain the poor correlations with surface climate.

A final example with respect to annual growth rates is the application of compositing of stalagmite growth rate records. With the increasing number of records being generated, this approach will gain increasing utility as a method to extract a common mode of variability within speleothem series. Although not exclusively limited to annual laminae series, it is these records that have first been composited by Smith et al. (2006) and then updated by Baker et al. (2008a). Smith et al. (2006) used all available annual lamina series from the Northern Hemisphere above 30 °N, the series from northwest Scotland, Beijing and Ernesto. Baker et al. (2008a) updated this with a second stalagmite series from northwest Scotland. In both cases, previous processing of the annual layer thickness data involved transforming the three time series so that the data were distributed with a mean of 0 and a variance of 1. The time series were then transformed using EOF-based techniques to extract the common mode of variability contained within the three records. In Smith et al. (2006), first EOF displayed remarkably high, positive loadings with all three series and explains 55% of the total variance. The effect of this transformation was an improved signal-to-noise ratio, and increased confidence that although the datasets are spatially sparse, the established links with Northern Hemispheric climate suggest this to be the common mode of variability contained within the three records. The reconstructed temperature series was derived using linear regression; that is the transformed dataset was calibrated to the instrumental temperature record (Jones et al., 1999; land areas north of 20 °N) over the period 1870–1960 AD, and validated using the remaining overlapping instrumental record, before extending back in time to encompass the past 500 years. The result of the Baker et al. (2008a) composite record is shown in Fig. 10.8. The stalagmite composite agrees within errors with other multiproxy reconstructions of Northern Hemispheric climate, but given the increasing awareness of the problems when trying to composite proxy records (Moberg et al., 2008; Christiansen et al., 2009), the further addition of annual laminated series are required to independently confirm that the common signal is robust. This is especially true, given the knowledge that in many cases the local calibrations appear to be dominated by correlations related to rainfall or atmospheric circulation. The increase in the number of stalagmite series from three to four in Baker et al. (2008a) increased the skill of the hindcast but 'flattened' the temperature reconstruction—addition of further series which continue to flatten the composite record would suggest that a common Northern Hemispheric temperature signal is not present in stalagmites. A vast amount of future research awaits!

10.3.2 $\delta^{18}O$

Instrumental calibration of $\delta^{18}O$ requires analysis of $\delta^{18}O$ at a resolution at least approaching annual and a well constrained chronology, typically using annual laminae. $\delta^{18}O$ series, unless drilled exactly to integrate one year of speleothem growth, will require appropriate treatment such as interpolation or smoothing (see Fairchild et al., 2006a). Several recent studies have used linear regression, transfer

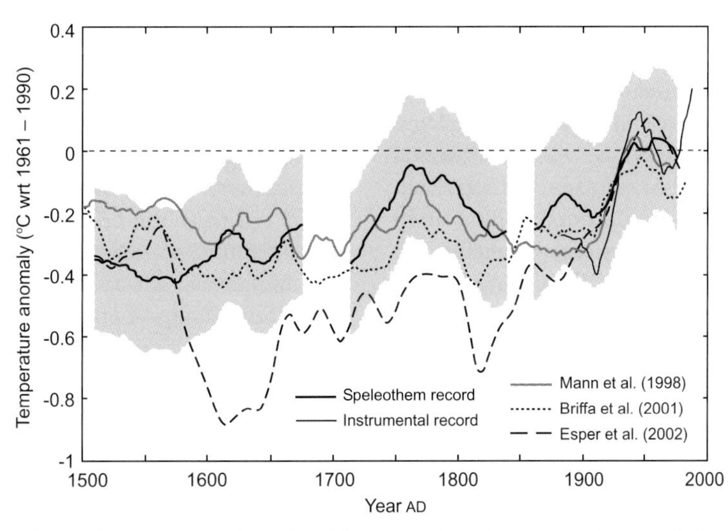

Fig. 10.8 Composite of annual lamina records updated from Smith et al. (2006) using an additional record in Baker et al. (2008a).

Table 10.3 Modern calibrated stalagmite $\delta^{18}O$ records.

Source	Location	Methodology	Climate correlant	Calibration length	Correlation
Yadava et al. (2004)	India	Linear regression	Annual rainfall	123 yr	−0.62 (decadal)
Baker et al. (2007)	Ethiopia	Linear Regression	Summer rainfall	94–103 yr	−0.22 to −0.32 (decadal)
Mattey et al. (2008)	Southwest Europe	Linear regression	Winter rainfall	42 yr	+0.68 (annual)
Cai et al. (2010)	Thailand	Linear regression	Summer monsoon	43 yr	−0.50 (5 yr average)
Jex et al. (2010)	Turkey	Linear regression Transfer function	Winter precipitation	43 yr	−0.71 (5 yr average) −0.72 (5 yr model)
Jex et al. (2011)	Turkey	Linear regression	Autumn–winter precipitation	66 yr	−0.52 (6 yr average)

function and forward modelling approaches to compare stalagmite $\delta^{18}O$ with modern climate data; these are summarized in Table 10.3. Many have again used linear regression or transfer function approaches, and typically stalagmite $\delta^{18}O$ has been shown to correlate with the $\delta^{18}O$ of the dominant recharge season. Yadava et al. (2004) demonstrated a strong negative correlation between decadal averaged annual total rainfall and $\delta^{18}O$ for a stalagmite from the Western Ghats, India, interpreted as being caused by an 'amount effect' in this monsoon climate region. Cai et al. (2010) demonstrated a negative correlation between 5-year averaged stalagmite $\delta^{18}O$ and the ratio of August–October (late monsoon season) to May–July (early monsoon season) rainfall. Jex et al. (2010) used an annually laminated stalagmite to demonstrate that $\delta^{18}O$ at a high altitude northeast Turkey site was correlated with winter precipitation (Fig. 10.9a), a logical finding given the predominance of snowmelt recharge at the location. Mattey et al. (2008) used a combination of annual laminae and annual $\delta^{13}C$ cycles to identify the winter $\delta^{18}O$ of precipitation preserved in a stalagmite from Gibraltar (Fig. 10.9b). In Ethiopia, where two rain seasons have different mean $\delta^{18}O$ and where stalagmite deposition is typically not in equilibrium, Baker et al. (2007, 2010) used both transfer function and forward modelling approaches to determine that a combination of the dominant recharge season and disequilibrium fractionation determined modern stalagmite $\delta^{18}O$. Figure 10.9c shows the comparison of forward modelled $\delta^{18}O$ and stalagmite $\delta^{18}O$.

Two modern calibration studies differ from those described above in that precipitation source or type were determined to be the driver of stalagmite $\delta^{18}O$. In Queensland, Australia, a simple conceptual model based on the fact that tropical cyclones comprise a light $\delta^{18}O$ end-member allowed a semi-quantitative reconstruction of cyclone frequency in an annually laminated stalagmite (Nott et al., 2007). Frappier et al. (2002) undertook sub-annual sampling of actively growing stalagmites from Belize and demonstrated a correlation between the amplitude of negative $\delta^{18}O$ excursions and tropical cyclone intensity.

The relatively few calibrated $\delta^{18}O$ stalagmite records all show a consistent pattern of calibrating against seasonal precipitation, typically the month(s) of ground water recharge, or the circulation pattern that generates a distinctive $\delta^{18}O$ source. An exception is Baldini et al. (2005), working at a site where vegetation re-growth was occurring after a long period of mine working. At this site (Brown's Folly Mine, Bath, UK), modern trends in $\delta^{18}O$ were hypothesized to be owing to increasing vegetation cover from thin unvegetated soils to dense secondary woodland, which affected the seasonality and amount of hydrologically effective precipitation.

In areas where disequilibrium deposition occurs, several studies have shown that primary climate signal is overprinted by kinetic fractionation or evaporation, and a positive offset in $\delta^{18}O$ results. The fractionation may vary over time in a manner which means that calibration is still possible. Modern calibration therefore suggests that quanti-

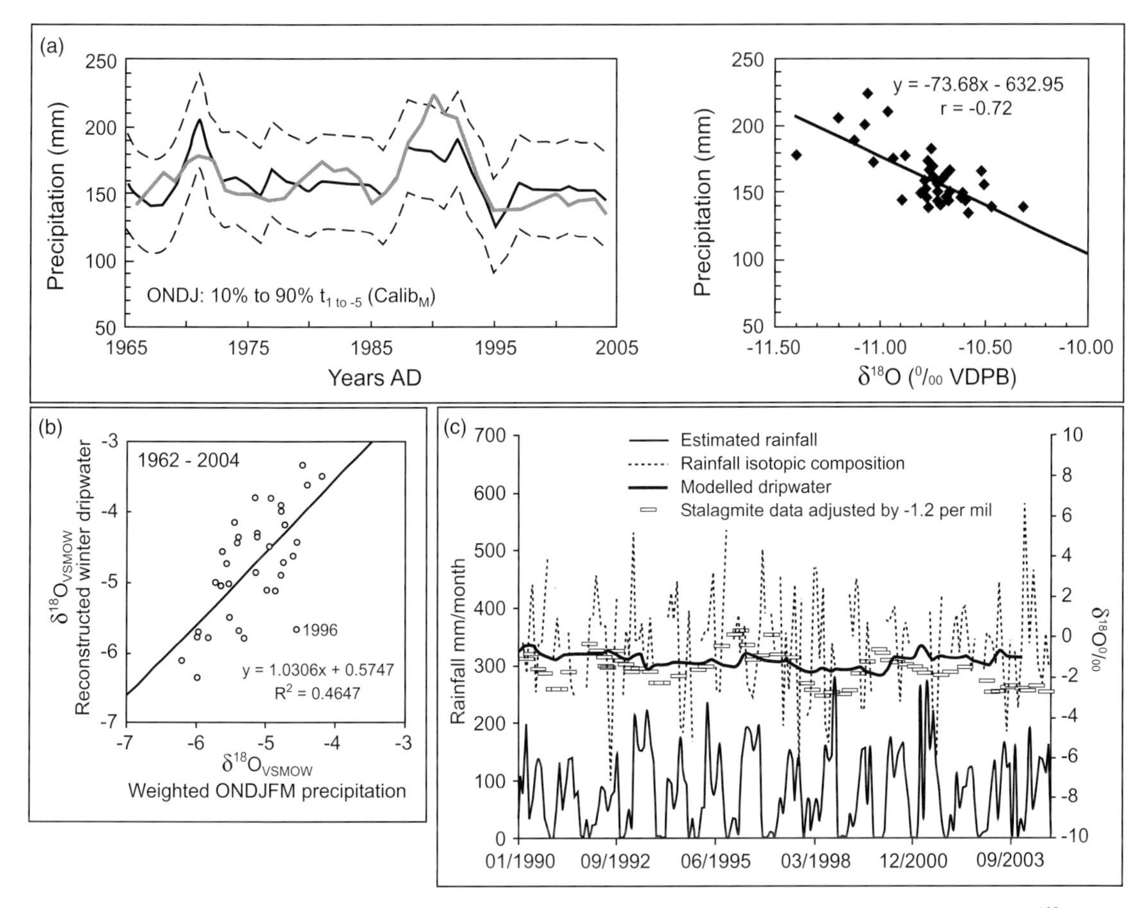

Fig. 10.9 Examples of $\delta^{18}O$ versus modern climate comparisons. (a) Regression between winter rainfall and $\delta^{18}O$ from a northeast Turkey speleothem, and comparison between reconstructed and actual winter precipitation from 1965–2005 AD. (b) Winter dripwater $\delta^{18}O$ (reconstructed from stalagmite $\delta^{18}O$) and actual winter precipitation $\delta^{18}O$, Gibraltar. (c) Forward-modelled dripwater $\delta^{18}O$ for an Ethiopian stalagmite.

fiable palaeoclimate reconstructions can be obtained from speleothems not at isotopic equilibrium. Notable also is the lack of any correlation with temperature (at least over this timescale of relatively low variation in mean annual temperature), the original target of stalagmite $\delta^{18}O$ when pioneering research was undertaken in the late 1960s.

10.3.3 Other proxies

Although a range of speleothem proxies besides $\delta^{18}O$ and annual growth rate can now be sampled at approximately annual resolution, only a few studies have yet to be reported in the literature. Possibly the most obvious target is $\delta^{13}C$, because data will be available paired with $\delta^{18}O$. Given the wide range of factors that affect $\delta^{13}C$ (see section 5.3), instrumental calibration might provide useful insights into the dominant processes. Although Frappier et al. (2002) demonstrate a semi-quantitative correlation between a modern Belize stalagmite $\delta^{13}C$ and ENSO, and Mattey et al. (2008) use the seasonal cyclicity in $\delta^{13}C$ to constrain their $\delta^{18}O$ reconstruction of winter precipitation $\delta^{18}O$, only one study (Jex et al., 2010) has demonstrated a direct linear correlation with climate, in this case a correlation with winter precipitation. This example is considered further in Chapter 11, along with the compositing of stalagmite $\delta^{13}C$ records in multiproxy reconstructions such as Mann et al. (2008). In many cases, a lack of correlation may be owing to the stronger autocorrelation in $\delta^{13}C$ series compared with $\delta^{18}O$ when the $\delta^{13}C$ signal is additionally smoothed in the soil (see section 3.3). This autocorrelation limits the number of degrees of freedom in the time series for regression analysis. A forward modelling approach might prove to be fruitful, Fohlmeister et al. (2011b) is a first step towards a carbon model for speleothem dripwaters.

Although there are several trace element records from modern sites, attempts to relate sub-annual to inter-annual changes to climate parameters are often confounded by other factors. For example, Fairchild and Treble (2009) reported that Mg in the same northwest Scotland sample studied by Proctor et al. (2000) did not clearly display the theoretically expected sub-annual temperature-dependent variations in a cave chamber with a 4.5 °C annual range (Fuller et al., 2008), perhaps because of inherent noise in the proxy; Smith et al. (2009) also reported that the trace element records from perennially wet Alpine sites did not show climate correlations over the instrumental period. Contents of colloidally transported elements, such as Y, Pb and Zn in a sample from the Ernesto cave, are high in layers deposited around the period of the First World War, but this is attributed to deforestation rather than a climatic anomaly (Borsato et al., 2007). Seasonally dry caves with prior calcite precipitation have been demonstrated from modern monitoring studies as discussed in Chapter 5 and have been shown to show coherent behaviour in Mg–Sr proxies in pre-instrumental period stalagmite growth (McMillan et al., 2005; Cruz et al., 2007), but there are few present-day data from modern caves of this type. An example is a record from the Moondyne Cave of southwest Australia, where distinct trace element changes were found (higher Mg, lower P and U) in the drier 1965–1990 AD period, as opposed to the wetter 1910–1965 AD period (Treble et al., 2003), and the interpretation is underpinned by a process understanding linked to the observation that there is a strong antipathetic relationship of Mg to seasonal rainfall.

One example that highlights the future potential of comparison with modern data is a comparison of multiple proxies. In an Ethiopian stalagmite, $\delta^{18}O$ and $\delta^{13}C$ were sampled at approximately annual resolution, and lipid biomarkers at approximately decadal resolution, in a modern, annually laminated stalagmite (Blyth et al., 2007), and from a site where the local vegetation history was known. In this sample, lipid biomarkers were shown to exhibit a rapid response to surface vegetation change, which could only be explained if the lipid signal imprinted in the stalagmites was dominated by an event-water, fissure-fed component which transported the biomarkers rapidly from the soil to the stalagmite (Fig. 10.10). A further implication of this finding is that lipids transported in the storage and slow flow compo-

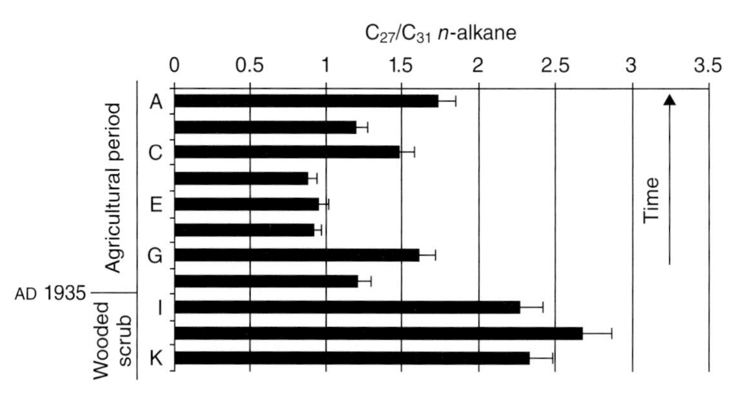

Fig. 10.10 Comparison of changes in speleothem organic biomarkers, in this example the ratio of C27–C31 alkanes, and surface vegetation change. From Blyth et al. (2007).

nents of the karst aquifer are biologically processed or chemically adsorbed before reaching the speleothem sample. Therefore, speleothems fed by long residence time waters alone might be expected to have lipid biomarker signatures that reflect microbial processing of the original soil lipids.

10.4 Questions raised and future directions

The period of instrumental climate data provides the best test of climate signal versus environmental noise in a stalagmite proxy, especially when combined with modern day monitoring of cave climate and hydrology. Even semi-quantitative calibrations, such as against extreme recharge events (Frappier et al., 2007; Nott et al., 2007; Dasgupta et al., 2010) or vegetation change (Baldini et al., 2005) provide insight into either likely dominant climate forcing functions or confounding environmental processes. Quantitative calibrations using linear regression approaches have also yielded useful quantified climate calibrations (for example, Proctor et al., 2000, Trouet et al., 2009); however, fundamental statistical issues with linear regression approaches relate to both the limitations raised when either autocorrelation or nonlinearities occur within the stalagmite proxy time series. Where transfer function approaches have been implemented, to date they suggest that stalagmite $\delta^{18}O$ and growth rate proxies reflect a mixing of waters of up to 10 years and with a proportion of event water of up to 30% (Baker & Bradley, 2010). All statistical calibrations of speleothem proxies have yet to adequately quantify the uncertainty associated with a proxy climate reconstruction (for example, how best to integrate both temporal and regression uncertainties; how to quantify uncertainties introduced by autocorrelation, etc.), issues that remain under active research in the wider palaeoclimate community (Moberg et al., 2008; Christensen et al., 2009).

Forward modelling approaches appear to provide future potential; this approach is most limited by the availability of suitable appropriate model input time-series (Bradley et al., 2010). Forward modelling allows the repeated simulation of a stalagmite proxy under varied karst hydrological regimes and can therefore attempt to capture the uncertainty in a speleothem proxy owing to hydrological variability. A future prospect is an integrated model of speleothem proxy series, including modelling of equilibrium and disequilibrium deposition (e.g. Romanov et al., 2008a, b; Scholz et al., 2009) and growth morphology models (e.g., Kaufman, 2003) with models of stable isotope fractionation (Mühlinghaus et al., 2009), hydrological models (e.g. Baker & Bradley, 2010) and soil-vegetation models.

Instrumentally calibrated stalagmite proxies series of annual lamina width have yielded quantified climate reconstructions of climate of the past millennium and longer, which have been archived for wider use by the palaeoclimate community in the World Data Center for Paleoclimatology. Their widespread use in climate reconstructions of the past 500–2000 years (for example, Moberg et al., 2005; Pauling et al., 2006; Mann et al., 2008; Goosse et al., 2010) is considered in section 11.2.

CHAPTER 11

The Holocene epoch: testing the climate and environmental proxies

The Holocene epoch can be viewed as both an additional test-bed to the modern period (introduced in Chapter 10) and a time where speleothem proxies can be tested in a period of relatively stable climate. Speleothem proxy calibrations, developed against instrumental climate data can be extrapolated, with care, back into the last millennium and earlier. Periods of Holocene environmental or climate change, whose timing or climate significance are relatively well understood from other proxies, can be used to assess the speleothem proxy response to these events. Arguably, it is also a time period where speleothems provide demonstrably high quality palaeoclimate and palaeoenvironmental information over other proxy archives. Carefully selected speleothems, and appropriate proxies, are able to yield new palaeoclimatic and palaeoenvironmental information throughout the Holocene, especially where annually resolved stalagmites are used to reconstruct precipitation-related reconstructions which are otherwise difficult to obtain. However, the anthropogenic impact on speleothem records always needs to be explicitly considered (Fig. 11.1).

In this chapter, each of these frameworks within which Holocene speleothem records might be investigated is considered. Section 11.1 provides a brief overview of the Holocene, focusing on changes in climate and environment which are relevant to speleothem study. Here, we refer back to the ideas of the speleothem factory and speleothem incubator introduced in Chapter 1. In section 11.2, we consider the use of speleothems to reconstruct climate and environmental histories over the past millennium, firstly focusing on stalagmites that have been calibrated against instrumental and cave

Fig. 11.1 Stalagmites in a low-level chamber in La Garma Cave, northern Spain. Human activity in this chamber occurred both in Palaeolithic times, when a lower entrance was open, and in Mediæval times, when a much more difficult descent was required. Speleothems were knocked over during historic visits, and the centre-left speleothem shows a narrower re-growth of white calcite, probably over the past few hundred years.

Speleothem Science: From Process to Past Environments, First Edition. Ian J. Fairchild, Andy Baker.

monitoring datasets, and the resultant climate reconstructions (section 11.2.1). We then consider how the data have been used by the wider palaeoclimate community in multi-proxy climate reconstructions, and proxy-model comparisons and assimilations (section 11.2.2). Section 11.3 considers the speleothem evidence of climate and environmental change over the Holocene as a whole. We start with the Early Holocene, specifically considering speleothem records that cover the period of the '8.2 ka event', arguably the largest transient Earth-system perturbation to have occurred in the Holocene, and one which tests the capability of speleothems to resolve responses on such timescales. The behaviour of speleothem proxies in relation to Holocene insolation variations is considered in section 11.3.2 for different regions of modern Earth climate. Climate variability on the decadal to centennial-scale deduced from annually resolved speleothem records is considered in section 11.3.3. We show the extent to which the use of speleothems has already shed light on our understanding of decadal-scale ocean-atmosphere processes over time and the possibility of detecting decadal- to centennial-scale forcing of climate. Finally, we focus on speleothem evidence of soil and vegetation change over the Holocene (section 11.3.4). The chapter concludes with the consideration of the questions raised by Holocene speleothem records (section 11.4).

11.1 A brief overview of the Holocene

A full review of Holocene climate and environmental change is beyond the scope of this text, but for detailed descriptions the reader is referred to Roberts (1998), Birks et al. (2003) and Battarbee and Binney (2008). Wanner et al. (2008) critically reviewed the state-of-knowledge of Mid- to Late Holocene climate change, a time period when the transition to an interglacial condition is complete. Figure 11.2 presents a spatial synthesis of global climate change for the period 6 ka compared with the pre-industrial Modern period (approximately 1700 AD), showing the variations in spatial structure driven by changes in orbital configuration

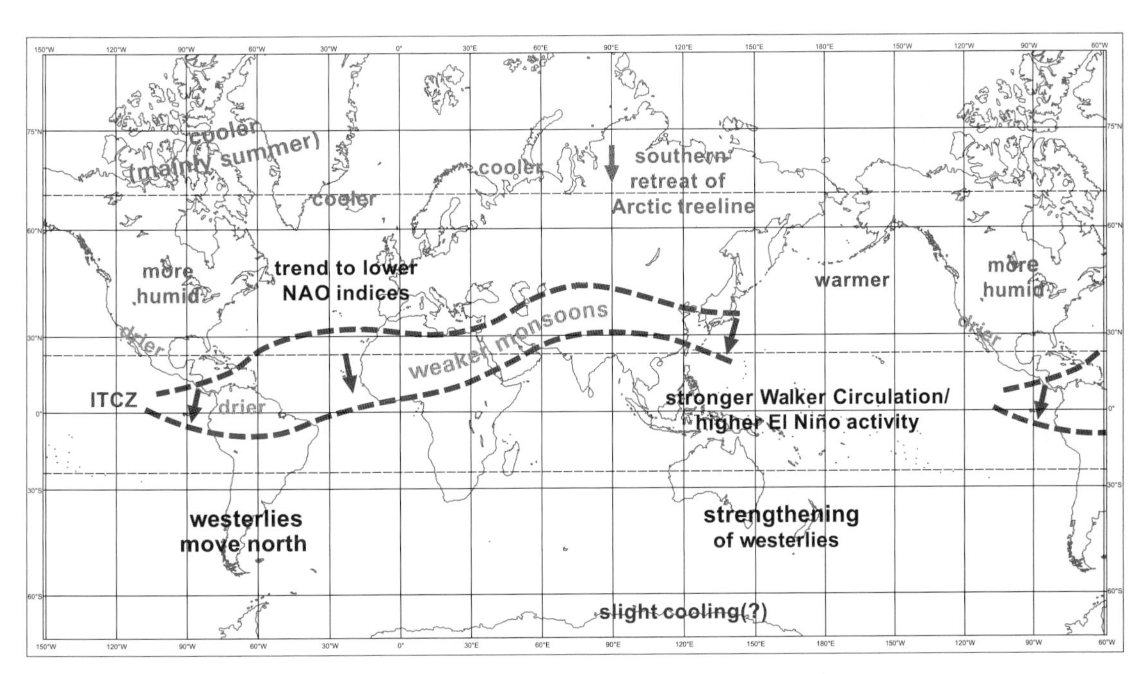

Fig. 11.2 Spatial synthesis: global climate change for the pre-industrial period (AD approximately 1700) compared with the Mid-Holocene (approximately 6 ka) (Wanner et al., 2008).

compared with the modern state (see sections 3.1.1 and 11.3.2). When considering multi-century-scale climate variability, however, Wanner et al. (2008) find the situation less clear, concluding that there is 'scant evidence for consistent periodicities and it seems that much of the higher-frequency variability observed is due to interval variability or complex feedback processes that would not be expected to show strict spectral coherence'. Clearly, precisely dated proxy records which record decadal to centennial-scale climate variability, and which can be related to the potential forcing mechanisms of orbital parameters, solar irradiance, explosive volcanic eruptions and greenhouse gases, are needed. Quasi-periodic, 1–2 kyr 'Bond cycles', originally identified from variations in abundance of ice-rafted debris in North Atlantic cores (Bond et al., 1997, 2001), are considered to be a possible exception to the lack of multi-centennial climate periodicity, but only for the Northern Hemisphere, and for selected time periods of the Holocene. At no time over the period 6 ka to present do Wanner et al. (2008) find consistent proxy evidence for rapid transitions in climate state. At decadal timescales, proxy and general circulation model (GCM) evidence suggests that orbital forcing over the Mid- to Late Holocene has led to an increase in El Niño/Southern Oscillation (ENSO) amplitude and a possible shift in the North Atlantic Oscillation (NAO) to more negative values.

Focusing on the environmental and climatic changes that occurred over the Holocene and which affected the stalagmite incubator, Fig. 11.3 presents a conceptual overview of the interrelation of these factors.

11.1.1 The Early Holocene

In a rapidly deglaciating Earth, the Earth system is moving towards a new dynamic equilibrium with respect to the retreating ice sheets. Changing global

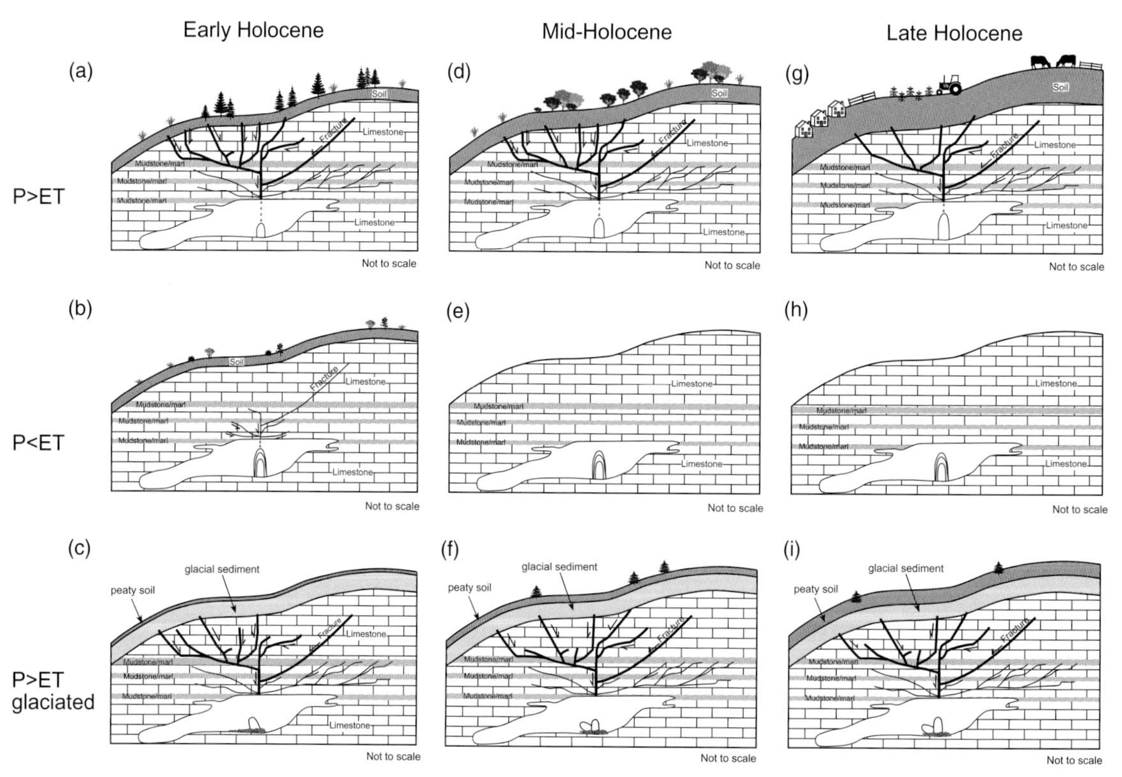

Fig. 11.3 Changes that may have occurred during the Holocene in (a, d, g) a lowland moist temperate site, in (b, e, h) an arid site, and (c, f, i) and upland temperate glaciated site.

(eustatic) and local (isostatic) sea levels generate new base-levels for fluvial systems and the (re) activation and (re)abandonment of cave systems. Relatively unstable soils may exist due to the presence of either a freshly deglaciated land surface (and associated pioneer ecosystems) or, in regions outside the glacial limit, changes in the climax vegetation as climate zones shift. Occasional meltwater pulses from ice-dammed lakes occur, some of which have wide expression (such as the 8.2 ka event, see section 11.3.3). New stalagmite incubators (Fig. 11.3) may become available, whereas others may be destroyed. Surface soil, vegetation and cave sediments are all likely to be in relative flux, and relatively unstable: therefore the 'factory' is in a state of often rapid and unpredictable (non-linear) flux. However, where aridity was experienced in the preceding glacial, speleothem growth is likely to restart. In caves which were hydrologically active in the glacial period, stalagmite growth may be on freshly deposited sediment and be relatively unstable. In areas that become progressively more arid, the time period marks the start of the 'death' of stalagmites as groundwater supply dries up. Overall, human impacts on the incubator are minimal and local.

11.1.2 The Mid-Holocene

As the Holocene progresses, it is marked by a slow decrease in Northern Hemisphere summer and autumn insolation, and an increasing stabilization of the ocean-atmosphere climate system. The season and latitude of maximum insolation varies through the Holocene, caused by changes in precession and obliquity. Essentially, from about 7 ka until present, the extra-tropics of the Northern Hemisphere have cooled while the tropics warmed (Lorenz et al., 2006). These insolation changes lead to southward movement in the Intertropical Convergence Zone (ITCZ) and the regional monsoon systems (Wanner et al., 2008). At the same time, there are indications that these changes in orbital forcing changed the behaviour of the internal climate system variability, with a possible decrease in ENSO activity and a shift in NAO index to more negative values (Wanner et al., 2008). By the Mid-Holocene, human impact may be seen to be increasingly important, with the first agricultural revolution starting and the first substantive deforestation in some regions. With respect to the cave environment, the Mid-Holocene can be viewed as a period of relatively stable internal cave environment, although with an unstable surface environment if human impacts are felt (Fig. 11.3). This relatively stable incubator preserves stalagmite growth, although in regions where the last major groundwater recharge was during the previous glacial period, long-term decreases in ground water may be visible in either baseline speleothem proxy data or may lead to the cessation of growth. In contrast, in monsoonal regions and regions marginal to the ITCZ, changes in insolation balance may lead to changes in recharge and stalagmite growth. Asrat et al. (2007) documented an example of the slow 'death' of a Mid-Holocene stalagmite from a loss of groundwater supply.

11.1.3 Late Holocene

Towards the end of the Holocene, human impact on firstly surface environments, then within cave systems, increases. Vegetation change and soil loss or improvement occur with the further intensification of agriculture and increased urbanization of the surface landscape. Summer insolation of the Northern Hemisphere continues to decrease, although the cooling effect is counteracted by increases in methane and carbon dioxide (Ruddiman, 2007) which, in the post-industrial period at least, represent a major anthropogenic disturbance. Regional or globally significant climate anomalies include the Mediaeval Climate Anomaly and the subsequent Little Ice Age. Speleothems in cooler regions may grow more slowly with the gradual cooling. Human impacts on stalagmite growth start to dominate over climate for the first time (Fig. 11.3).

11.2 The past millennium

The past 1000 years of Earth history has been the focus of global efforts to reconstruct climate variability using both proxy archives and GCMs (Jones et al., 2008; Wanner et al., 2008). During this time period, annually laminated proxies such as tree-ring, varved lake, ice-core, coral and speleothem records are available, which potentially give annually resolved (or better) proxy climate information. GCM simulations of the past 1000 years are also practical, and the time period includes the GCM spin-up and calibration periods for future model climate scenarios. Finally, the past 1000 years is a period for which we have the best information on changes in climate forcing through proxy records of solar output and volcanic activity. Therefore the past millennium can be used to test key hypotheses such as the following.

1 Is the late 20th century warmth unprecedented in the past millennium (and is it being recorded in speleothem proxy archives)?

2 To what extent are the Medieval Climate Anomaly and Little Ice Age global climate phenomena? What are the mechanisms (and are they recorded in speleothem proxy archives)?

3 What is the decadal climate variability in the pre-instrumental period? What can speleothems tell us about the mechanisms and forcings of this variability?

Chapter 10 demonstrated that instrumentally calibrated speleothem records can provide useful correlations with modern climate parameters such as rainfall amount or source or seasonal temperature. Here, we consider the extension of these calibrations to reconstruct the climates of the past millennium, remembering the caveats applied to this approach (e.g. Lee et al., 2008; Riedwyl et al., 2009; Amman et al., 2010). After consideration of the extension of these records (section 11.2.1) we then review their use in multi-proxy reconstructions (section 11.2.2).

11.2.1 Instrumentally calibrated speleothem climate reconstructions

The first instrumentally calibrated speleothem record that was extrapolated to provide a climate reconstruction over the past millennium, was from an annually laminated stalagmite from Madagascar (Brook et al., 1999), where lamina thickness was shown to correlate with the strength of the Southern Oscillation (SO), with thicker layers being deposited in wetter, low SO years (section 10.3.1). The full approximately 450-year reconstruction was shown to have good visual correlations in the low-frequency domain, with other proxy archives which also reflect ENSO state, such as a Galapagos coral $\delta^{18}O$ archive (Dunbar et al., 1994) and the Quelccaya ice cap accumulation rate (Fig. 11.4; Thompson et al., 1984). Further statistical comparison of the proxy series was not undertaken. In northwest Scotland, Proctor et al. (2000) demonstrated a correlation (based on linear regression) between annual lamina thickness, and total annual precipitation and mean annual temperature, with the correlation with precipitation being the stronger. Proctor et al. (2000) published a 1000-year reconstruction of total annual precipitation, based on the de-trended annual lamina thickness series, with later research updating the chronology and refining the calibration (Proctor et al., 2002). Plate 11.1a presents both raw and standardized and smoothed annual lamina thickness series for stalagmite SU-96-7, together with an additional stalagmite SU032 which grew for the past approximately 300 years (Fuller, 2006). Raw data show that each stalagmite has differences in mean and variance due to local (non-climatic) controls on annual lamina thickness, such as hydrological connectivity, flow switching and variations in overlying peat thickness. SU032, selected for analysis for its fast growth rate, although still instrumentally correlating with local climate, is relatively climate-insensitive. Smoothed and standardized data better reveal the underlying low-frequency climate signal contained in the records (Plate 11.1b). Owing to the sensitivity of precipitation in the region to the North Atlantic Oscillation, the stalagmite growth rate proxy can also be interpreted in a climate dynamical sense. The observed 60–80 year periodicity in annual lamina thickness is similar to that observed in ocean circulation models (Delworth & Mann 2000). The success of these stalagmite proxy climate reconstructions highlights the importance of demonstrating an instrumental calibration that agrees both with a process-understanding, as well as the generation of such proxy series in climatologically meaningful regions. Further use of the annual lamina thickness series of Proctor et al. (2000) in multiproxy reconstructions is detailed in the next section.

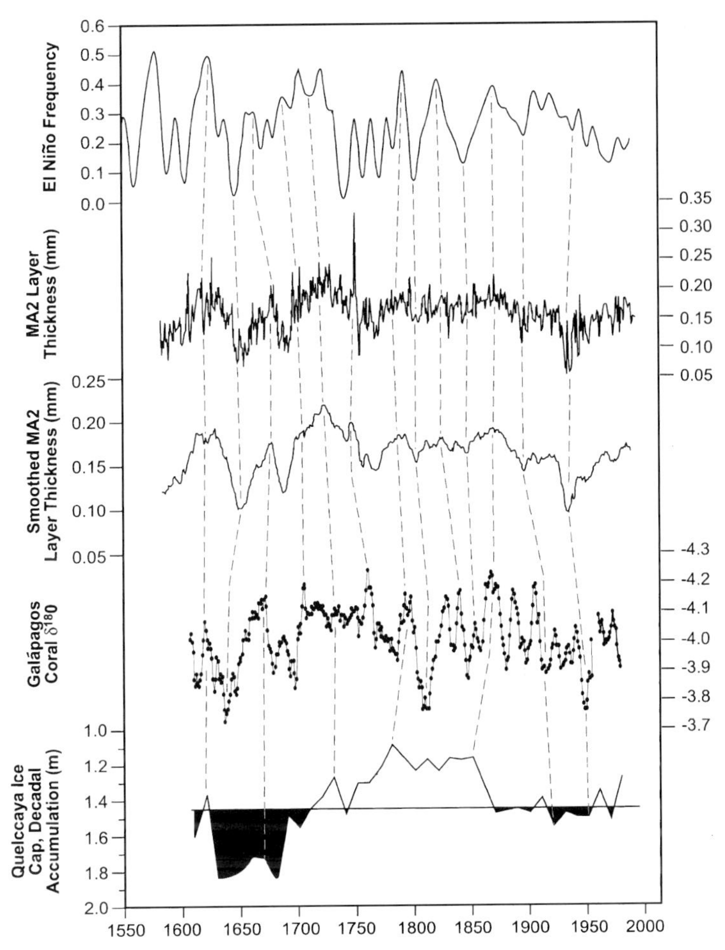

Fig. 11.4 Comparison of Madagascar stalagmite MA2 layer thickness with ice core and coral proxy records of ENSO state. (Brook et al., 1999).

Tan et al. (2003) presented an annual lamina thickness record for the past approximately 2650 years for a stalagmite from Shihua Cave, Beijing. Instrumental calibration demonstrates a dominant correlation with summer temperatures (and a minor negative correlation with summer precipitation).Tan et al. (2003) further tested the robustness of this correlation through the comparison of the reconstructed temperature from the stalagmite with other, hemispheric-scale, multi-proxy climate reconstructions of the past 1000 years, as well as Chinese historic records. As with the northwest Scotland reconstructions, the Shihua Cave record has been more recently used in multi-proxy climate reconstructions of the past 1000 years (see next section).

Frisia et al. (2003) presented correlations between annual lamina thickness and winter temperatures for three stalagmites from Ernesto Cave in the Italian Alps; Plate 11.1 shows the raw (Plate 11.1c) and decadal (plate 11.1d) smoothed, standardized, annual lamina series for the three stalagmite series, as archived at the World Data Centre for Paleoclimatology. The stalagmites, which had been growing discontinuously over the Holocene, all had a common modern growth period, and results are presented where instrumental calibration demonstrated some correlation ($r \sim 0.3$–0.4) with winter temperature. The notable observation in Frisia et al. (2003) is that all three stalagmites exhibited a strong slowing in growth rate between AD 1650–1713 and AD 1798–1840, periods that correspond to the well-known Maunder and Dalton solar minima, and confirming the importance of solar forcing as the first-order control on winter temperatures over the Late Holocene. However, divergence in growth rate between records, especially over the instrumental period, suggests that each stalagmite lamina thickness record is determined by more than temperature alone, as might be expected from a correlation of approximately 0.3–0.4 with temperature (suggesting approximately 85% of variance remains unexplained). Frisia et al. (2003) go on to undertake spectral and wavelet analysis of their growth rate series, and report several statistically significant peaks between 3.1 and 45 years in period, but emphasize particularly the presence of an approximately 11 year cyclicity in growth rate, interpreted as forced by solar activity modulating the seasonal duration of soil CO_2 production. However, although a peak between 10 and 12 years is present in each of the segments analysed, it is not usually the dominant one, and given the lack of correlation between solar output and instrumental winter temperature in this, or any other region over the modern period, this conclusion needs a higher burden of proof to be accepted.

A final instrumentally calibrated record is that of $\delta^{18}O$ and $\delta^{13}C$ in an annually laminated stalagmite from northern Turkey (Jex et al., 2010), the $\delta^{18}O$ record having been extended to reconstruct winter precipitation for the region for the past approximately 500 years (Jex et al., 2011). This 500 year reconstruction of precipitation, based on the regression of annually resolved stalagmite $\delta^{18}O$ against local instrumental precipitation (Plate 11.1e), demonstrates that there is little long-term variation in winter precipitation in the region over the past 500 years, with lower reconstructed precipitation for the period AD 1540–1560 featuring as being anomalous over the whole of the past 500 years. Plate 11.1f shows that modern day winter (October–January) precipitation is linked to the larger-scale regional circulation pattern of the North Sea Caspian Pattern (NCP). Increased October–January precipitation is thus suggested to be linked to reduced pressure over West Russia/Caspian Sea and increased pressure over northwest Europe during late autumn to winter, suggesting that northerly circulation leads to an increase in the isotopically light precipitation observed in northeast Turkey.

Finally, it is appropriate here to note an alternative calibration approach used by Lauritzen and Lundberg (1999) and Mangini et al. (2005) which does not involve direct regression with instrumental series. In both cases the lack of tight chronological control (lack of annual laminae), as well as very slow growth rates, precludes the use of instrumental calibration. Instead, the authors develop a transfer function between stalagmite $\delta^{18}O$ and a climate target, in both cases temperature, using a four to five point regression over a longer time period than the instrumental. This approach relies on using time periods of 'known temperature' both within and beyond the instrumental period. Figure 11.5 shows both transfer functions. For example, in the Alpine transfer function, the five data points are determined as (1) the modern cave temperature and the $\delta^{18}O$ of the most modern stalagmite growth; (2) stalagmite $\delta^{18}O$ in 1690s AD, which had the coldest temperatures in the Luterbacher et al. (2002) temperature reconstruction for the region; (3) the heaviest stalagmite $\delta^{18}O$, which owing to the *a priori* assumption of a correlation with temperature, is prescribed to be the coldest temperature at which stalagmite growth is possible (at or around 0 °C); (4, 5) stalagmite $\delta^{18}O$ from intermediate dates in the 19th and 20th centuries. Problems in using this approach are both practical and conceptual. Dating uncertainties generate large error bands in the *x*- (temperature) axis: these are shown, but not incorporated into the calibration uncertainty. The most positive $\delta^{18}O$ calibration point is based on a temperature of −0.2 °C, presuming a temperature correlation with $\delta^{18}O$, and that temperature is around freezing, neither of which is actually known. Conceptually, the approach forces a calibration to temperature, rather than any other parameter, and presumes a stationary relationship between $\delta^{18}O$ and temperature over time. With $n = 5$, as apparent in Fig. 11.5, the uncertainty associated with the gradient of the calibration is

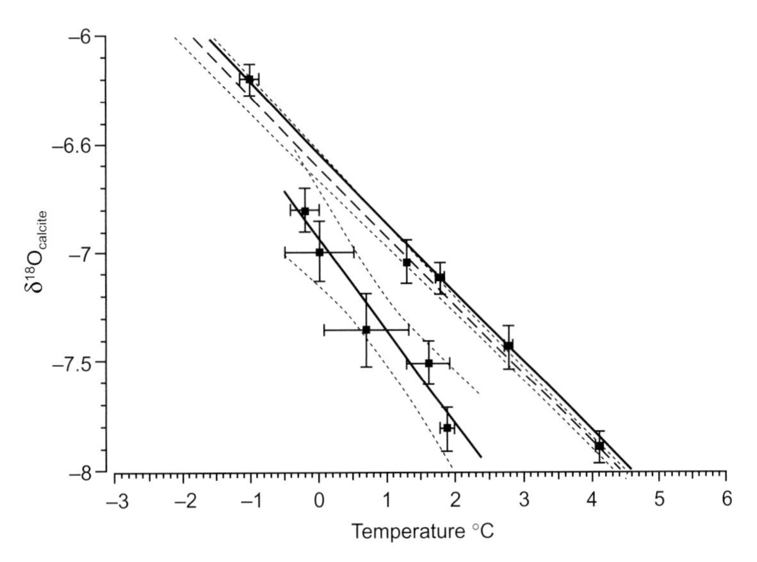

Fig. 11.5 Transfer functions developed for stalagmites from northern Norway (upper points, Lauritzen & Lundberg, 1999) and Austrian Alps (lower points, Mangini et al., 2005).

substantial (the confidence limits shown do not take into account the shown reconstruction uncertainty on either axes), and hence any reported temperature sensitivity has low reliability. Similar problems exist for the same methodology as used by Lauritzen and Lundberg (1999), compounded by additional dating uncertainty, as the sample was not actively growing when sampled, and the observation that more recent research into Norwegian stalagmite $\delta^{18}O$ (Linge et al., 2009b) over the Holocene suggests that between-sample variability in $\delta^{18}O$ is much greater than that observed by Lauritzen and Lundberg (1999).

11.2.2 Multi-proxy reconstructions and model-proxy comparisons

Subsequent to the publication and archiving of the stalagmite climate reconstructions of Lauritzen and Lundberg (1999), Tan et al. (2003), Proctor et al. (2000, 2002), and Mangini et al. (2005), stalagmite proxy climate reconstructions have been produced using multiproxy and model-proxy reconstructions. Some multiproxy studies have composited just one or two proxies from the same climate region or with the same climate forcing, to improve the climate signal-to-proxy noise in the reconstruction. For example, Tan et al. (2009) compared the Shihua Cave summer temperature reconstruction with a Qilian tree ring sequence and a GCM (ECHO-G) simulated millennial temperature record for China. The resulting combined summer temperature reconstruction (Fig. 11.6) shows a strong correlation with other multi-proxy temperature reconstructions (e.g. $r = 0.53$, 11-year smoothed with Esper et al., 2002). The combined temperature record is also correlated in the low frequency with a GCM simulation ($r = 0.61$, 31-year mean). Trouet et al. (2009) undertook a similar composition of two NAO-sensitive proxies: the northwest Scotland lamina thickness record and a Moroccan tree-ring composite record. The two sites, at opposite ends of the NAO dipole, would be expected to have inversely correlated wetness records, and the composited series exhibited strongest correlation in the low frequency (approximately 30-year average). The resultant long NAO reconstruction (NAO_{ms}, Fig. 11.7) was compared with GCM output (ECHAM4 and NCAR CCM) using a proxy-model analogue method. In this approach, data from GCM simulations are re-ordered to maximize the temporal agreement between the proxy records and matching data drawn from the model output. The re-ordered model data are then used to characterize more fully the climate patterns implied by

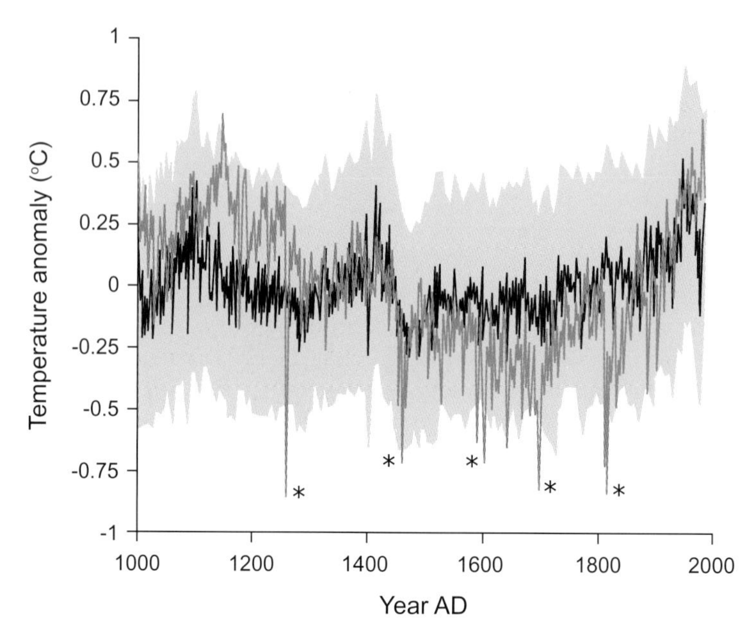

Fig. 11.6 Combined stalagmite / tree ring temperature reconstruction (black line) and uncertainty (grey infill) and GCM simulation (grey line). Asterisks highlight the periods of discrepancy between the model and palaeo-reconstructions. From Tan et al. (2009).

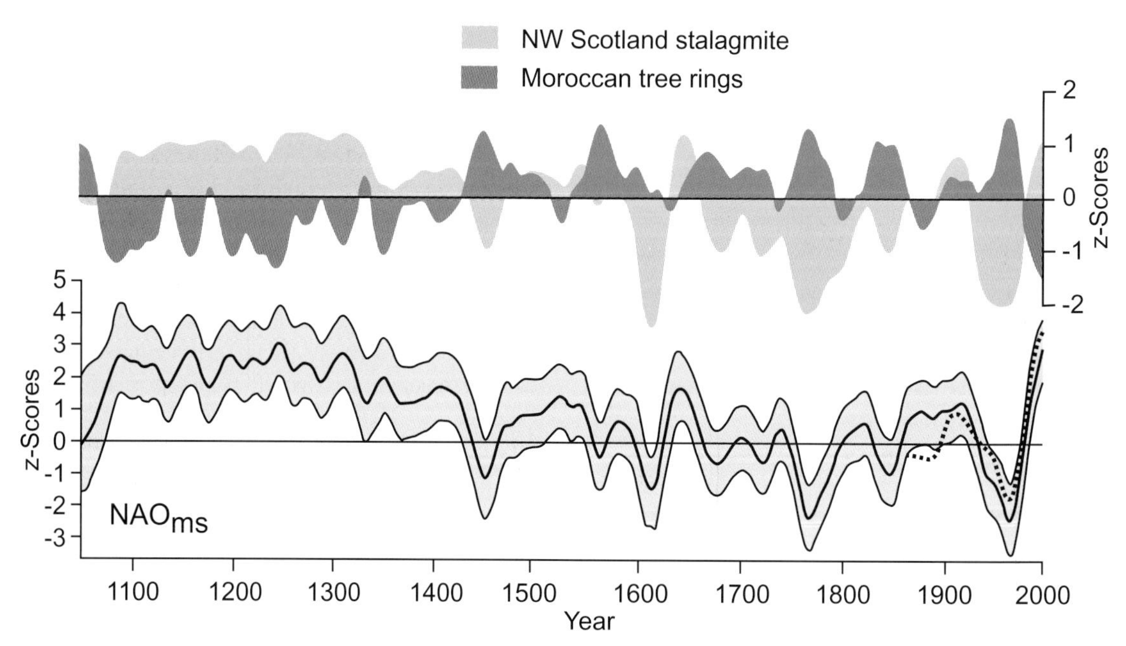

Fig. 11.7 Proxy-derived approximately 900-year NAO reconstruction. (top) reconstructed winter precipitation from northwest Scotland stalagmite annual lamina thickness and February–June Palmer Drought Severity Index (PDSI) from Moroccan tree rings. Records were normalized over the common period (1049–1995 AD) and smoothed using a 30-year cubic spline. (base) Winter NAO reconstruction (NAO_{ms}—black line) is the difference of the stalagmite and tree ring records. Shaded area is the estimated uncertainty. Grey line is the 30-year smoothed instrumental NAO series. From Trouet et al. (2009).

the sparse proxy series, in this case the NAO reconstruction.

The following are examples of multi-proxy reconstructions that include significant numbers of proxies, including speleothems: Moberg et al. (2005), Pauling et al. (2006) and Mann et al. (2008). The Moberg et al. (2005) Northern Hemisphere temperature reconstruction composites recordings including the annual lamina thickness record from Shihua Cave (Tan et al., 2003) and the Lauritzen and Lundberg (1999) North Norway $\delta^{18}O$ record (with an updated chronology). Plate 11.2 presents this reconstruction along with other multiproxy Northern Hemisphere temperature reconstructions; notable is the higher amplitude of temperature variability than other reconstructions which is a function of the wavelet-based function used to maintain the low-frequency signal. The more recent composite of Mann et al. (2008) is also shown in Plate 11.2; in this temperature reconstruction for the past 2000 years for both Northern and Southern Hemispheres which used over 450 screened data sources, several speleothem records were included. Screening was by regression of the proxy against local temperature over the instrumental period, and proxy series where $p < 0.1$ (using either annual or decadal mean temperature, depending on proxy resolution) were retained. Speleothem proxies all passed this screening. Archives that contained both high and low-frequency information included the northwest Scotland lamina thickness record (Proctor et al., 2002), lamina thickness, $\delta^{18}O$ and $\delta^{13}C$ from Oman stalagmite S3 (Fleitmann et al., 2004), $\delta^{13}C$ and $\delta^{18}O$ from South Africa (Lee-Thorp et al., 2001), and the Shihua Cave lamina thickness record (Tan et al., 2003). Speleothems which contained low-frequency information only included $\delta^{13}C$ and $\delta^{18}O$ from stalagmite D1 from Socotra, Oman (Burns et al., 2003), $\delta^{13}C$ and $\delta^{18}O$ from a stalagmite from Costa Rica (Burns, unpublished data), $\delta^{18}O$ from a Dongge Cave stalagmite (Wang et al., 2005) and $\delta^{13}C$ and $\delta^{18}O$ from another Shihua Cave stalagmite (Li et al., 1998). It is interesting to note that with this empirical approach to climate reconstruction, several $\delta^{13}C$ and $\delta^{18}O$ proxy series passed screening against instrumental temperature without this being the interpretation in the original paper, or a process-based understanding being articulated. Other records were retained as a temperature proxy despite stronger correlations with precipitation (e.g. northwest Scotland record of Proctor et al.(2002)). The latter record was used in a reconstruction of European precipitation for the past 500 years (Pauling et al., 2006) which featured instrumental, precipitation sensitive tree-ring and coral data.

The most recent use of proxy data is in proxy-model assimilations; Widmann et al. (2010) reviewed the range of possible data assimilation approaches. Proxy-model assimilations are one type, when GCM simulations use proxy climate information to 'direct' or inform the simulation throughout the GCM run. Typically, a simplified GCM will be used owing to the greater computation demands of this approach, which allows model runs to diverge at every proxy climate data point in the run. Skilful simulations disregard proxy data that are outside the bounds of the simulation and agree with the temperature-sensitive proxy climate data. Goosse et al. (2006, 2008) originally undertook simulations without speleothem proxy data but more recently have included speleothem series in a 600-year assimilated-GCM temperature reconstruction (Goosse et al., 2010). Interestingly, the Assynt stalagmite series of Proctor et al. (2000) was included, but growth rate is more dependent on precipitation, and the stalagmite exhibited a low correlation with the GCM temperature assimilations.

In this section we have focused on just the speleothem records of the past millennium that have either undergone some form of instrument or transfer function calibration, or which have been used in composite climate reconstructions. A vast number of other published studies present speleothem proxy climate data that cover the past millennium, but which have either a proxy sampling resolution, or dating uncertainty, that prevents them from providing useful climate information over this time period. Further research efforts in the generation of annual or near-annually resolved proxy records for well dated speleothems is likely to be of considerable use to the global palaeoclimate community. It is notable that it is the published records that have been publicly archived that are predominantly those which have been used in the community effort to reconstruct climates of the past 1000 years. However, it should be noted that in some cases the process-based interpretation in the original publication has been ignored; in others the application of local regression approaches has identified correlations between proxy and local climate. Archiving of substantive metadata would benefit the community in guiding non-experts in the use of the proxy series (see Appendix 1).

11.3 Holocene environmental changes: speleothem responses

11.3.1 The period of remnant ice sheets in the Early Holocene

The Early Holocene contains arguably the most significant Earth system event of the past 10,000 years in the form of the '8.2 ka event', owing to rapid melt water discharges from Arctic ice-sheet proglacial lakes. However, this event also occurred within the framework of ongoing multi-centennial climate variability and millennial-scale changes in orbital forcing, and is further combined with the Earth system response to the new interglacial state in the form of vegetation and surface and subsurface hydrological change. Within speleothem science, the time period is such that very precise U–Th analyses are possible to constrain the timing of proxy archives to typically less than 30 years. Combined with high-resolution proxy analyses and replicate samples, speleothem archives can provide critical insights into climate and environmental changes over this time period. However, it is only recently that enough speleothem records of suitable quality have become available to enable speleothem archives to drive forward understanding of environmental change in the wider research community. Here, we specifically consider two examples —the use of replicate stalagmite $\delta^{18}O_c$ records in a transect across the European Alps to better understand the Mediterranean Early Holocene sapropel S1, and a comparison of high resolution stalagmite records of the '8.2 ka event'.

The last Mediterranean sapropel

Sapropels, organic carbon-rich sediments found in marine cores, have been regularly deposited in the Mediterranean over the Quaternary (Meyers & Arnaboldi, 2008), with the last such depositional unit, sapropel S1, deposited in the Early Holocene. Conventional interpretation of sapropel formation is that they require an increased organic carbon input, for example via increased river discharge carrying organic matter, as well as suitable conditions for the preservation of organic matter such as anoxia. The formation of S1 in the Mediterranean over the period 8.2–7.4 ka is therefore interpreted as being caused by increased rainfall within catchments feeding the Mediterranean (including the Nile from East Africa). This increased discharge would lead to surface water dilution, a decrease in surface salinity, increased stratification, leading to anoxia and the deposition of sapropels.

Spötl et al. (2010) review evidence for changes in rainfall source in this time period using a transect of stalagmite $\delta^{18}O_c$ records. The transect crosses the Alps from south to north, with the mountain range forming a natural barrier between moisture sourced from the south and the Mediterranean region and that from the north and west from the Atlantic. Figure 11.8 shows both speleothem and lake records over the Early Holocene. At the southern extreme, Corchia Cave, Italy, which is just 40 km from the Mediterranean, shows a negative $\delta^{18}O_c$ excursion at the time of S1 (8.2–7.3 ka) and is interpreted as recording a precipitation amount signal from the Mediterranean source region (Zanchetta et al., 2007). Further north, Ernesto Cave and Katerloch (Boch et al., 2009) exhibit $\delta^{18}O_c$ excursions to heavier isotopes, indicative of an increased proportion of water sourced from the Mediterranean. Hölloch, at the furthest north on the transect, is outside the influence of Mediterranean moisture, and does not exhibit a change in $\delta^{18}O_c$ at this time, but does instead show a clear '8.2 ka event' which is indicative of its relationship to the North Atlantic source region. The Spötl et al. (2010) study demonstrates the potential of replicated speleothem records sampled at suitably high resolution and dating precision, as well as some potential problems with speleothem $\delta^{18}O_c$ archives. For example, within the transect, the stalagmite record from Spannagel (Vollweiler et al., 2006) shows regular multi-centennial variability and with nothing particularly distinctive about the $\delta^{18}O_c$ record at 8.2–7.3 ka.

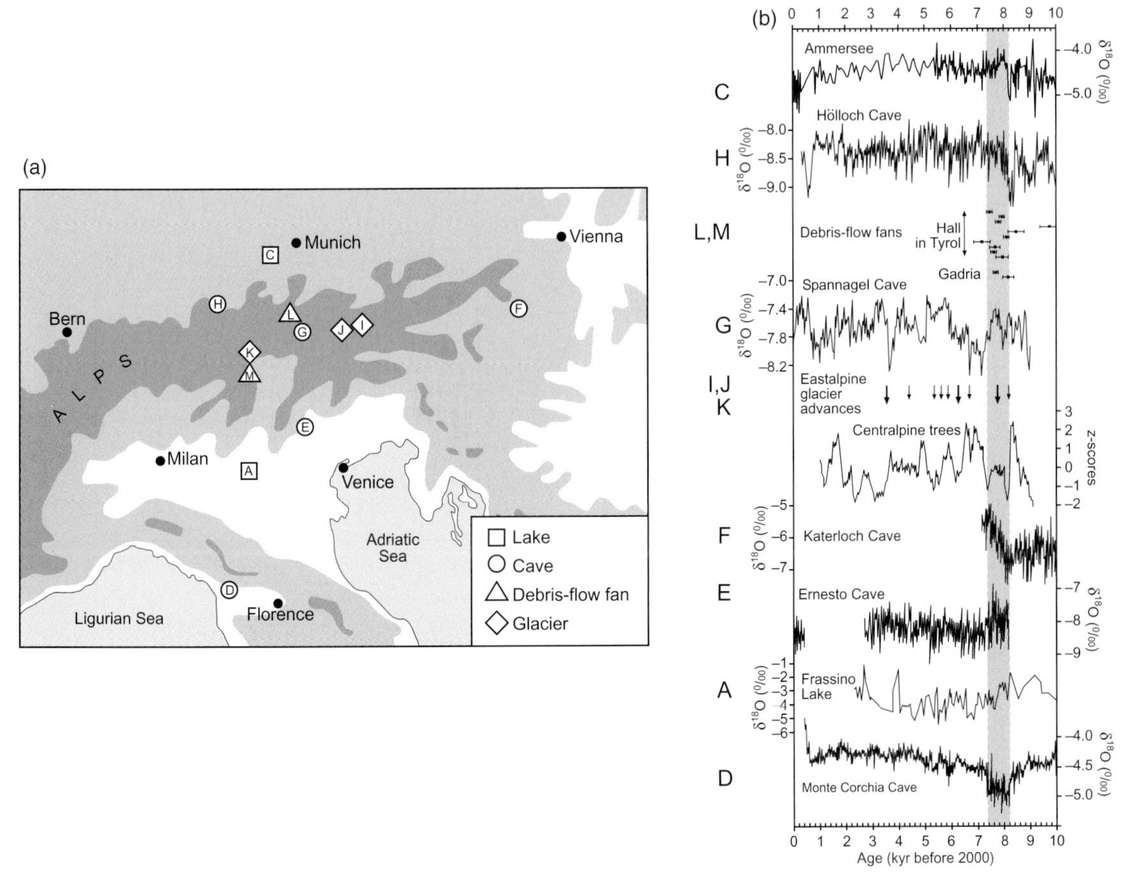

Fig. 11.8 Speleothem and lake archives during Mediterranean S1 deposition (after Spötl et al., 2010).

The '8.2 ka event'

The 8.2 ka event is clearly observed in the $\delta^{18}O$ records in Greenland ice cores, and is considered to have been caused by the rapid draining of glacial lakes related to the Laurentide ice sheet and an associated large meltwater pulse into the North Atlantic Ocean (Rohling & Palike, 2005; Alley & Agustdottir, 2005). The effect of this event, both in terms of sea level change and regional and global climate, is a focus of ongoing research using both proxy records and ocean–atmosphere GCMs. Despite being the most significant $\delta^{18}O$ event in the Holocene in the Greenland ice core records, the short-lived nature of the event (<200 years) requires proxy archives to have both high sampling resolution and annual chronological precision to be able to register its occurrence. The meltwater input into the North Atlantic would have led to a freshening, with potential effects on ocean circulation. LeGrande and Schmidt (2008) modelled the event and demonstrated that a weakening of the Atlantic Meridional Ocean Circulation was possible, and that North Atlantic surface waters would contain isotopically depleted $\delta^{18}O$ from the melt water for at least several decades. Therefore, proxy archives that might be expected to record an 8.2 ka event would be those sensitive to oceanic $\delta^{18}O$ variability in the North Atlantic region; evidence from other parts of the globe would confirm a wider climate impact.

In addition to ice cores (see, for example, Alley et al., 1997) and varved lakes (see, for example, Prasad et al., 2009), speleothems can provide a suitably high-resolution archive of the '8.2 ka

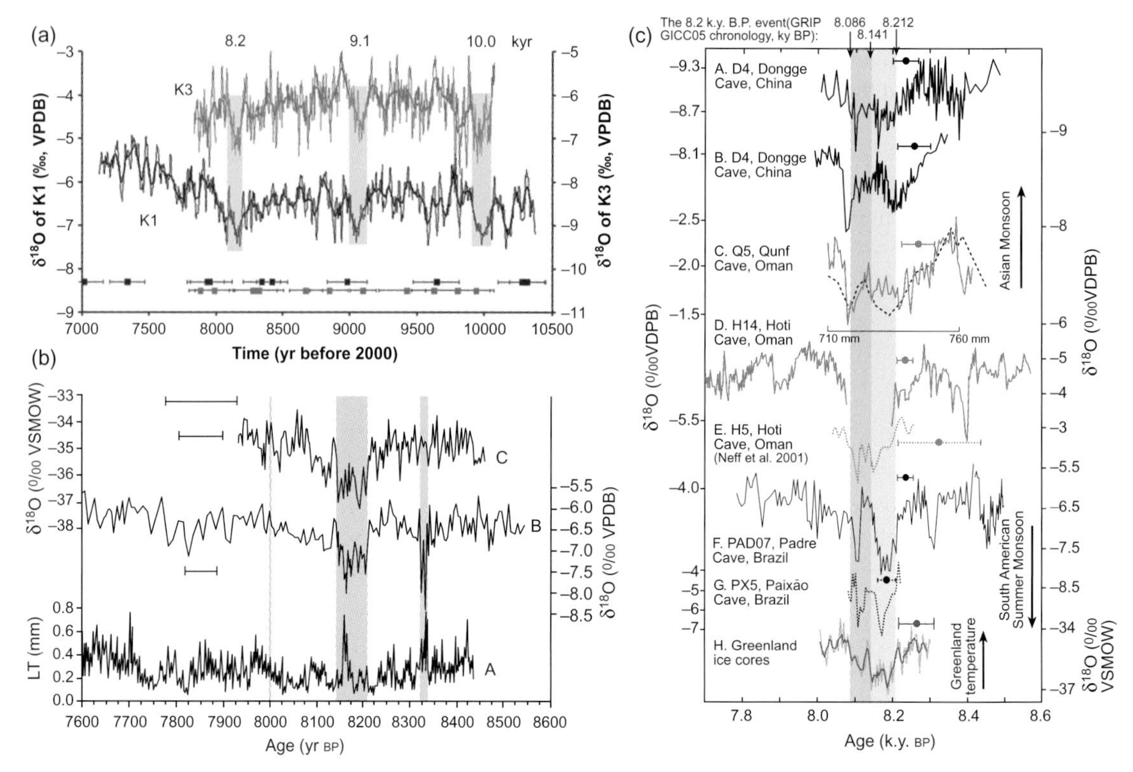

Fig. 11.9 (a) Katerloch stalagmite $\delta^{18}O$ for samples K1 and K3 (Boch et al., 2009). (b) Kaite Cave stalagmite LV5 $\delta^{18}O$ (B) and annual lamina thickness (A) (Domínguez-Villar et al., 2009b), along with the ice core composite of Thomas et al. (2007) (C). (c) Stalagmites D4 and DA (China), Q5, H4 and H5 (Oman), PAD07 and PX5 (Brazil) along with Thomas et al. (2007) ice core record (Cheng et al., 2009a). The $\delta^{18}O$ scales are reversed for Asian Monsoon records (China, Oman) as compared with South American Summer Monsoon (SASM) records from Brazil. Three arrows depict anti-phased changes between AM and SASM and changes in Greenland temperature. Error bars indicate typical dating errors (2σ) for each record around 8.2 ka event.

event' if sampled at suitable resolution (approximately annual) and dating precision, and if their hydrological connectivity to the surface is such that they are able to record short duration climate events. For example, stalagmite CC3 from Crag Cave, western Ireland, fails to show the event in its high-resolution $\delta^{18}O$ archive, despite a location close to the North Atlantic (Fairchild et al., 2006a). However, hydrologically this might be expected as the sample is characterized by a modern-day constant drip (Tooth & Fairchild, 2003), suggesting a significant groundwater storage component. In contrast, Domínguez-Villar et al. (2009b) saw a clear 8.2 ka event in an annually laminated and fast growing stalagmite from northern Spain (Fig. 11.9b). With near annual $\delta^{18}O$ resolution, stalagmite LV5 exhibits two $\delta^{18}O$ excursions, one of 1.0‰ at 8350–8340 years BP and 20 years' duration, and another of 0.7‰ starting at 8221–8211 ± 34 yr BP of 70 year duration. Annual growth rate, in contrast, does not record the 8.2 ka event, suggesting that any climatic effect in terms of a decrease in temperature, or change in the amount or seasonality of rainfall, were too small to be recorded in this proxy. Domínguez-Villar et al. (2009b) therefore suggest that at least some of the $\delta^{18}O$ signal is a direct representation of an isotopically depleted North Atlantic surface water $\delta^{18}O$.

Moving away from the North Atlantic, Boch et al. (2009) demonstrated a $\delta^{18}O$ minimum at 8175 years BP from two stalagmites at Katerloch, with the main ^{18}O-depletion occurring between 8196 and 8100 years BP and the whole $\delta^{18}O$ event occurring within the period 8246–8012 years BP (Fig. 11.9a). With near-annual resolution sampling for fast growing stalagmites, the maximum amplitude of $\delta^{18}O$ is 1.1‰. The timing of the event therefore matches other proxy archives of the 8.2 ka event, but in this Central European location the interpretation of the $\delta^{18}O$ excursion is likely to be a mixed signal of temperature, and changes in either rainfall source or amount. Katerloch stalagmites also demonstrate sensitivity to the proportion of Atlantic and Mediterranean moisture sources over the preceding Mediterranean sapropel S1, and show other short-lived $\delta^{18}O$ anomalies at 9.1 and 10.0 ka, which might also relate to North Atlantic melt water pulses.

Further afield, Cheng et al. (2009a) presented data documenting the 8.2ka event in stalagmites from Brazil, Oman and China, all regions which are influenced by summer monsoon rainfall (Fig. 11.9c). At these locations, preservation of the event in stalagmite proxy archives confirms a global climatic impact. In all cases, stalagmites are analysed for $\delta^{18}O$ at high resolution (2–5 years), with chronology provided by U–Th (and lamina counting in one instance). Cheng et al. (2009a) demonstrate that stalagmite $\delta^{18}O$ records an isotopic excursion starting at 8210 ± 20 years BP, with the first event over after 70 years, a subsequent rebound followed by a 29-year short event. The whole isotope excursion is over by 8080 ± 20 years BP. In these samples, the $\delta^{18}O$ excursion is in the direction of a weakening Asian monsoon ($\delta^{18}O$ rise of approximately 1‰) and strengthening South American Summer Monsoon ($\delta^{18}O$ fall of 2.5‰). The anti-phase $\delta^{18}O$ signal between the two monsoon regions is indicative of the impact of the North Atlantic melt water pulse, leading to a weakened Atlantic Meridional Circulation and a change in the mean latitudinal position of ITCZ affecting the monsoons.

In summary, concerted analytical effort over the past few years has demonstrated that stalagmite $\delta^{18}O$ preserves evidence of the 8.2 event when samples are analysed at near-annual resolution and are constrained by a combination of U–Th and annual lamina counting. Samples from Europe, Asia and South America demonstrate the existence of a $\delta^{18}O$ excursion starting around 8,200 years BP and lasting around 70 years and within a longer event of <200 years duration, agreeing with the ice-core composite of Thomas et al. (2007) of an event of 71 years within a total period of 161 years. One very high resolution and Atlantic marginal sample also demonstrates a short lived precursor event at 8350 years BP that is not observed in the ice core records (Domínguez-Villar et al., 2009b). Future research might focus on similar, high-resolution, sampling campaigns in other climate regions, or similar sampling strategies focused on other potential short-lived climate events in the Holocene including possible events at 9.3 and 10.0 ka as observed in Katerloch and in Greenland ice core (Rasmussen et al., 2007).

11.3.2 Orbital forcing over the Mid- to Late Holocene

Changes in the Earth's orbit relative to the sun is the most significant climate forcing in the Holocene, and Fig. 11.10 shows the changes in solar insolation reaching the top of the atmosphere over the past 6 ka as a function of latitude. The main features are (1) a steady decrease in June insolation in the high Northern Hemisphere latitudes with a total change of approximately 30 W m^{-2}, but with relatively little change over time in the Southern Hemisphere (Fig. 11.10a); (2) much less variability in December insolation (Fig. 11.10b). This results in a change in seasonality in insolation receipt, shown in Fig. 11.10c, with a decrease in the amount of seasonality in the Northern Hemisphere over the past 6 ka, and an increase in the seasonality of insolation in the Southern Hemisphere. Overall insolation change is shown in Fig. 11.10d, demonstrating a decrease in total insolation of approximately 100 W m^{-2} at highest Northern Hemisphere latitudes, whereas in the Southern Hemisphere insolation reaches a maximum increase of approximately 20 W m^{-2} which occurred at approximately 3 ka. The principal effects of changes in insolation over the Mid- to Late Holocene on the Earth's climate would therefore be a southward shift in the ITCZ (proxy evidence for which is demonstrated in

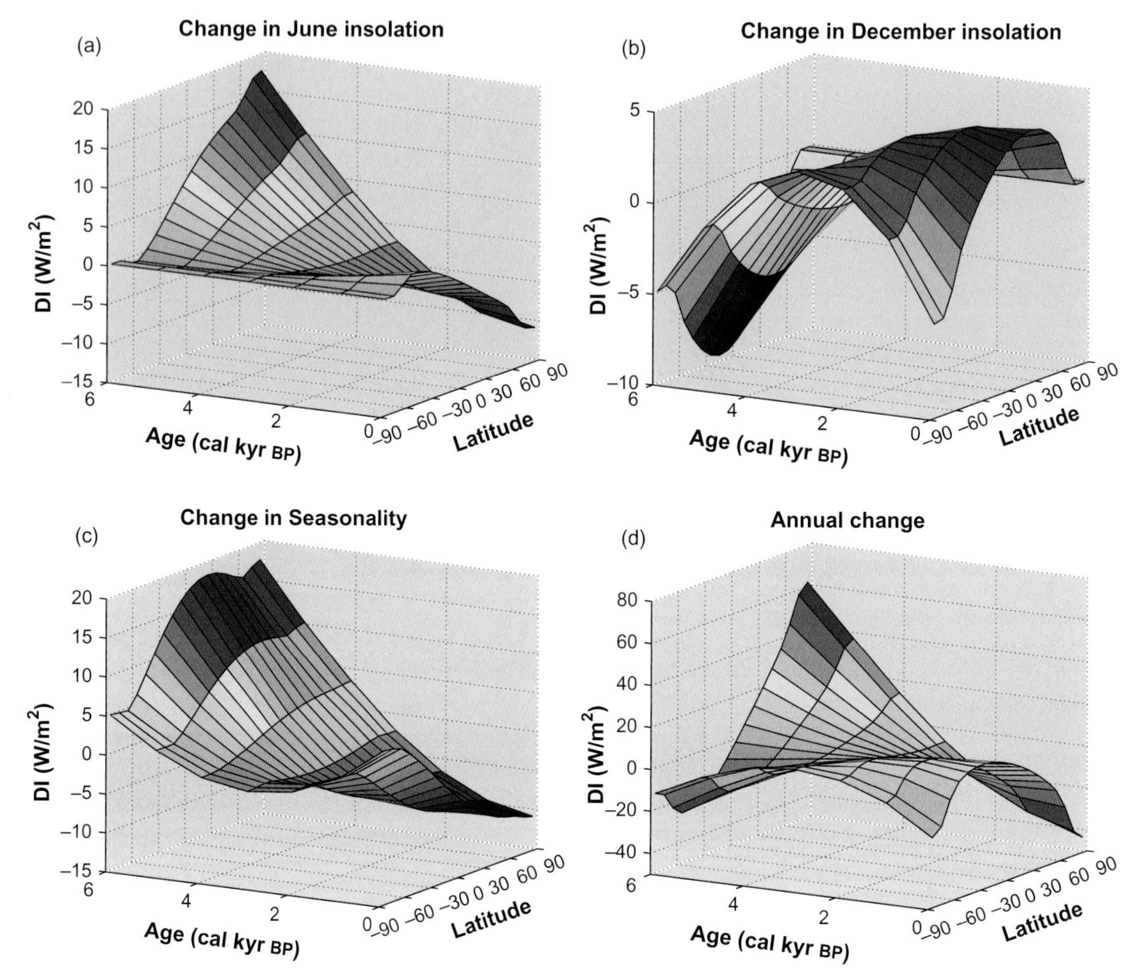

Fig. 11.10 Calculated deviations of the insolation from the long-term mean values ($W m^{-2}$) as a function of latitude for the past 6000 years. (a) June (Northern Hemisphere summer); (b) December (Northern Hemisphere summer); (c) seasonality (difference between June and December); (d) annual mean. From Wanner et al. (2008).

the Cariaco Basin sediments (Haug et al. (2001)) and a cooling in the Northern Hemisphere, mostly experienced in the Northern Hemisphere summer.

The most widely reported and spatially distributed stalagmite proxy for the past 6 ka is that of $\delta^{18}O_c$ (the oxygen isotope composition of calcium carbonate), which is a mixed proxy of temperature and precipitation source, trajectory and amount as detailed in sections 3.2, 8.6 and 10.3.2. Stalagmite series typically have isotope samples analysed every 5–30 years, often as point samples rather than time-averaged, a resolution only suitable for resolving low-frequency climate variations. For the time period of the past 6000 years, hemispheric mean temperature variability is low and therefore variations in stalagmite $\delta^{18}O_c$ can in many cases be interpreted as a local or regional atmospheric moisture response to solar insolation variations.

Partin et al. (2008) presented a compilation for a subset of data focused on low latitude $\delta^{18}O_c$ series (Fig. 11.11; additional $\delta^{18}O$ records have subsequently been published as described below). In China, data are now archived from Dongge Cave (two stalagmites; Dykoski et al., 2005; Wang et al., 2005), Heshang Cave (Hu et al., 2008b) and Sanbao Cave (five stalagmites; Dong et al., 2010). All show

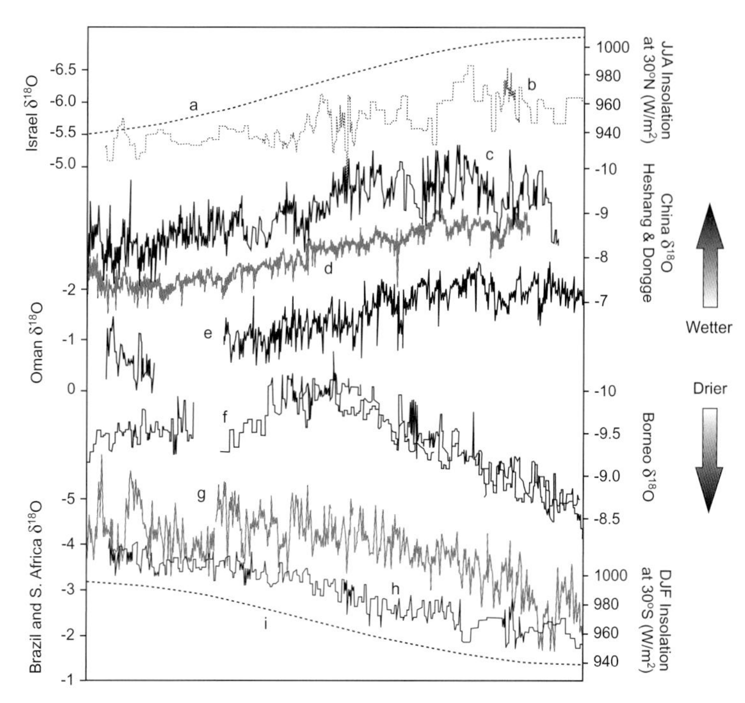

Fig. 11.11 Summer insolation curves and speleothem $\delta^{18}O$ records for the past 10kyr arranged from North to South. (a) Summer insolation at 30°N, (b) Soreq Cave, Israel; (c) Heshang Cave, China; (d) Dongge Cave, China; (e) Qunf Cave, Oman; (f) Snail Shell and Bukit Assam Caves, northern Borneo; (g) Cold Air Cave, South Africa; (h) Botuverá Cave, Brazil; (i) summer insolation at 30°S. From Partin et al. (2008).

similar trends, as shown in Fig. 11.11, of a decrease in $\delta^{18}O_c$ of approximately 1‰ from about 6ka to present, which is the same magnitude as that modelled for $\delta^{18}O_p$ (LeGrande & Schmidt, 2009). GCM simulations suggest that the change in $\delta^{18}O_p$ at these millennial timescales reflects the southward movement of the ITCZ and enhanced water vapour transport from the tropics, rather than a signal of local precipitation amount or intensity, with enhanced ocean–land water-vapour transport due to the enhanced Northern Hemisphere seasonality strengthening the monsoon effect. Site-by-site differences in mean $\delta^{18}O_c$ for the Chinese series reflect the different latitudes and significant differences in elevations of the sample sites. Figure 11.11 also illustrates that a stalagmite from Oman shows a very similar trend to the Chinese stalagmites. In this case, the speleothem from Qunf (Fleitmann et al., 2007) at 17°N is similarly affected by ocean–land water-vapour transport (the Indian Ocean monsoon) as well as the movement of the ITCZ.

In contrast to the Oman and Chinese records, in Indonesia the stalagmite $\delta^{18}O_c$ records from different parts of the region yield subtly contrasting responses to orbital forcing. Figure 11.11 presents data from Partin et al. (2008), where to the north of the equator in northern Borneo, stalagmite $\delta^{18}O_c$ exhibits an excursion to lighter isotopes at approximately 5ka. In a new record from Liang Lua in Flores, $\delta^{18}O_c$ falls back in time until approximately 7ka, which is interpreted as evidence of an intensification of monsoon, but one that started in the Early Holocene owing to sea level rise (providing a moisture source) rather than being driven by orbital forcing alone (Griffiths et al., 2009). At this site close to the equator, $\delta^{18}O_c$ remained constant from 7ka until present owing to the consistent influence of the ITCZ. Griffiths et al. (2010a) speculated that the differences between these two locations might be related to a northward shift in the ITCZ at that time. In Turkey, a stalagmite record from the north of the country also preserves a local

$\delta^{18}O$ signal. A trend to heavier isotopic composition in the Early Holocene and then relative stable $\delta^{18}O_c$ (Fleitmann et al., 2009) is observed to give a strong visual correlation with Black Sea $\delta^{18}O_c$ as recorded in marine core GeoB 7608-1 (Bahr et al., 2008). The Black Sea would appear to dominate the moisture source at this sample site, with the stalagmite $\delta^{18}O_c$ reflecting changes in Black Sea water level and source.

At mid-latitudes in both Northern and Southern Hemispheres, in regions where monsoon rainfall does not dominate, speleothem $\delta^{18}O$ demonstrates an even more complex response. In the USA, stalagmite samples from Mystery Cave (one stalagmite), Spring Valley Cavern (two samples, black) and Coldwater Cave (three samples, red) have divergent $\delta^{18}O_c$ (Fig. 11.12). Although the samples cover a wide region of the central north of the USA, the spatial and altitudinal spread of samples is similar to that of the Chinese stalagmites. Also notable is the observation that the $\delta^{13}C_c$ records in the samples are consistent between samples, and with surface vegetation records, owing to a C_3–C_4 vegetation change dominating this proxy archive (see section 11.3.4). The more positive $\delta^{18}O_c$ in some samples is explained by the authors as being caused by evaporative enrichment of $\delta^{18}O$ in the soil (see section 3.3) which was more likely to occur with changing land cover to a prairie vegetation, where soil water $\delta^{18}O$ would be more susceptible to evapotranspiration. Denniston et al. (1999) further demonstrated that at Coldwater Cave there is a relationship between surface topography (north versus south facing) and $\delta^{18}O_c$ enrichment.

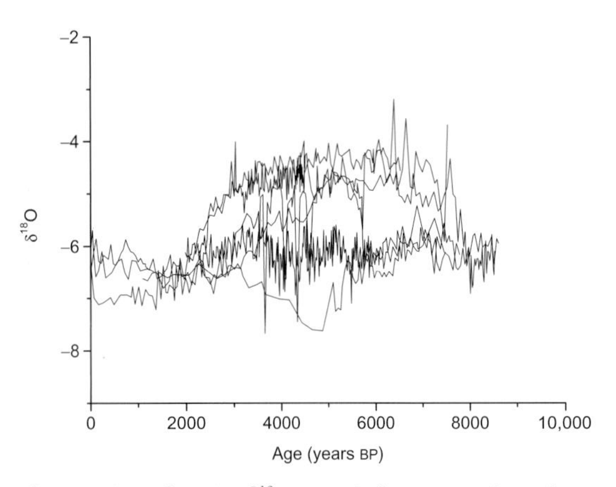

Fig. 11.12 Stalagmite $\delta^{18}O$ records from central north USA. From Denniston et al. (1999, 2000) and Dorale et al. (1992, using data archived at the World Data Centre for Paleoclimatology.

For Europe, McDermott et al. (2011) compiled $\delta^{18}O$ from 51 well-dated Holocene calcite stalagmites from the region. To analyse the low-frequency component of the different samples, which were of varying temporal resolution in $\delta^{18}O$ and timing of deposition, the authors analysed the isotope data at 1 ka time slices through the Holocene by taking averages of 50 year duration. McDermott et al. (2011) identified four effects on the low-frequency component of the stalagmite $\delta^{18}O$ records: (i) an east–west zonal pattern; (ii) a latitudinal effect north of 50 °N; (iii) an altitude effect above 400 m; and (iv) a Mediterranean moisture-source effect. McDermott et al. (2011) showed that for the European region south of 50 °N, and excluding sites proximal to the Mediterranean Sea or greater than 400 m altitude, stalagmite $\delta^{18}O$ over the Holocene can be explained by the longitudinal gradient. Plate 11.3 presents the predicted $\delta^{18}O$ for each 1 ka time slice (based on the regression of all stalagmite $\delta^{18}O$ versus longitude for each time slice, R^2 values for which were always >0.85) along with selected actual stalagmite $\delta^{18}O$ series. Plate 11.3 shows that the rate of change of $\delta^{18}O$ over time changes systematically from the Early to the Late Holocene, exhibiting a much steeper zonal gradient in the Early Holocene.

In New Zealand, Williams et al. (2004, 2005, 2010) have built a composite stalagmite $\delta^{18}O$ record from multiple samples covering the past approximately 30 ka. Focusing on the Holocene, Fig. 11.13 presents a comparison of data from both the central west North Island (seven speleothems with 34 dates and 934 stable isotopes over the period 0.5–14.6 ka), and northwest South Island (eight speleothems with 56 dates and 641 stable isotope data points over the period 0.5–30 ka BP). The two regions, either side of latitude of 40 °S, have internally consistent stalagmite $\delta^{18}O$ series but differing Holocene $\delta^{18}O$ trends between regions, especially in the Mid- to Late Holocene. Supported by alkenone sea-surface temperature data (Barrows

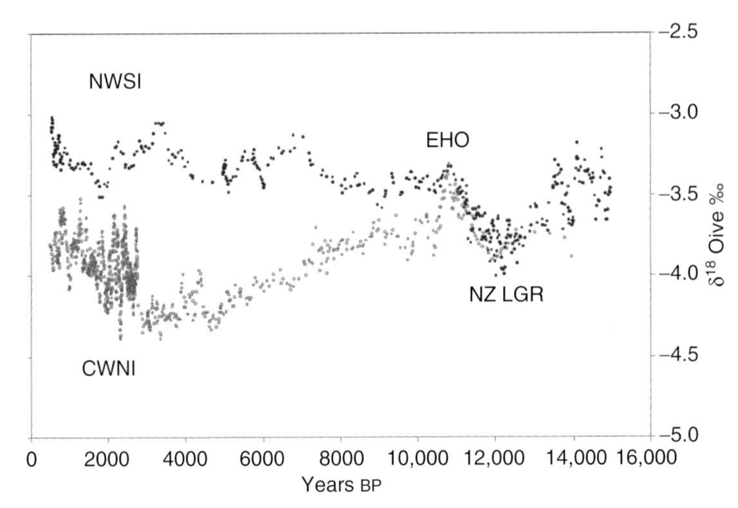

Fig. 11.13 Comparison of composite stalagmite records from New Zealand from northwest South Island (NWSI) and central west North Island (CWNI). Also labelled are the New Zealand Late Glacial Reversal (NZLGR) and Early Holocene Optimum (EHO). From Williams et al. (2010).

et al., 2007), Williams et al. (2010) interpreted this difference as representing the decline in influence of the warm West Auckland current in the central west North Island composite series.

Finally, in higher latitudes, Linge et al. (2009b) recently compiled stalagmite $\delta^{18}O_c$ records for a range of caves from Norway (Fig. 11.14) at latitude approximately 65 °N. Insolation receipt at this latitude is one of a continuous decline from approximately 6 ka to present, predominantly in summer. However, stalagmite response is again mixed, samples exhibiting both increasing and decreasing trends in $\delta^{18}O_c$ over this time period around differing mean $\delta^{18}O_c$. Sample-specific and local factors again appear to dominate stalagmite $\delta^{18}O_c$ rather than a regional response to incoming solar radiation over this time period.

An evaluation of stalagmite low-frequency $\delta^{18}O$ records for the Holocene demonstrates that, despite variations in incoming solar radiation being the primary climate forcing mechanism over this time period at a global scale, regional or local climate and environmental factors can dominate the stalagmite $\delta^{18}O$ archive. In most monsoon regions, consistent low-frequency trends in stalagmite $\delta^{18}O_c$ are observed which can be related to the strength of the monsoon over the Holocene, and demonstrate that potential for stalagmite $\delta^{18}O_c$ to provide records of water vapour transport in these regions over a range of frequencies. Reproducible, but regionally contrasting, series from Europe and New Zealand highlight the importance of regionally important climate processes in these regions. However, in other regions such as the central west USA and north Norway, stalagmite $\delta^{18}O_c$ records in the low-frequency domain are less replicable between samples. Local factors affecting stalagmite $\delta^{18}O_c$ may also dominate in moisture-limited regions where evapotranspired water is a significant component to the recharged groundwater, or where there is a local water source that provides a significant contribution to $\delta^{18}O$ of atmospheric precipitation. The variability of stalagmite $\delta^{18}O$ is therefore indicative of the 'mixed-proxy' nature of this record and is exemplified by the different modern calibrations of $\delta^{18}O_c$ against climate variables in section 10.3.2.

11.3.3 Evidence for multi-decadal and multi-centennial climate variability

In section 11.2 our current understanding of multi-decadal and multi-centennial climate variability in the Holocene was described. In précis, over multi-decadal timescales, external forcing due to

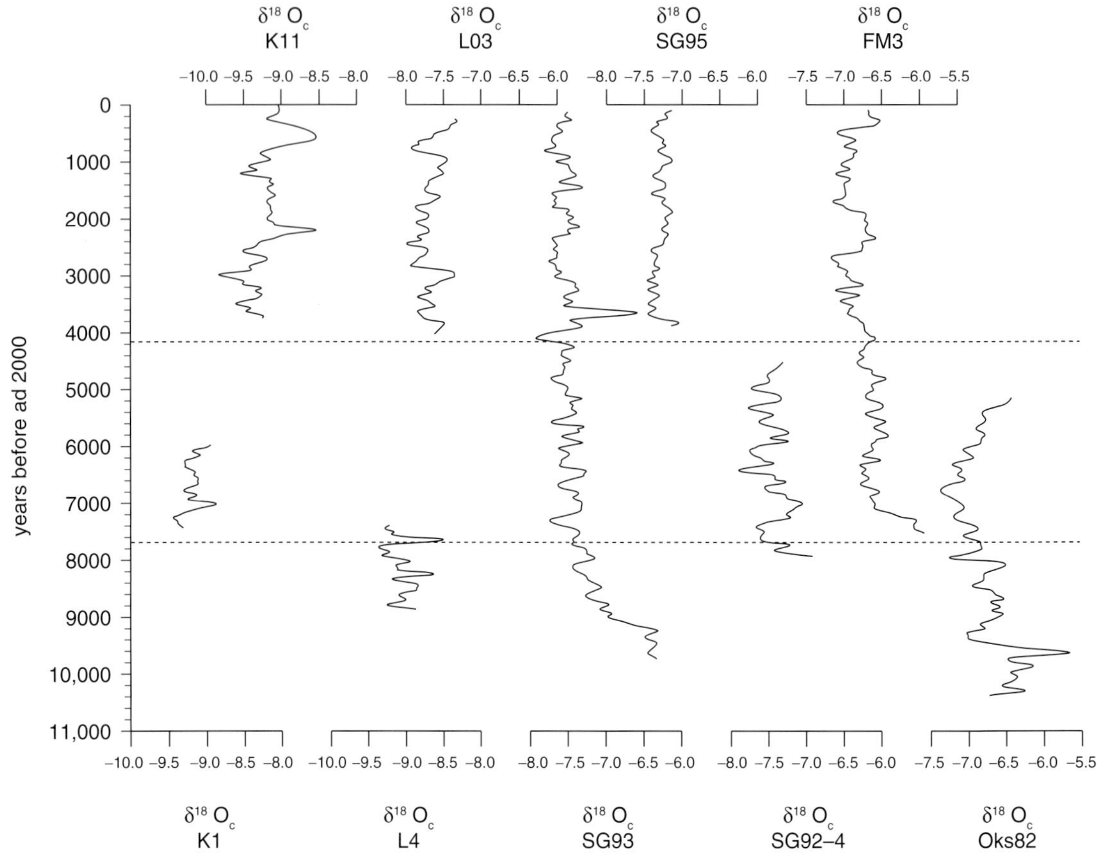

Fig. 11.14 Holocene stable oxygen isotope records from stalagmites between 64 and 67 °N. FM3 and Oks82 (Okshola, Fauske) from Linge et al. (2009b). SG93 (Lauritzen & Lundberg, 1999), SG95 (Linge et al., 2001), SG92-4 (Berstad, 1998) are all from Søylegrotta (Rana). L03 (Linge et al., 2009) is from Larshullet (Rana). K11 (Sundqvist et al., 2009) and K1 (Sundqvist et al., 2007) are from Korallgrottan, whereas L4 (Sundqvist et al., 2007) is from Labyrintgrottan. A Gaussian low-pass filter (100 year running mean) has been used to smooth the records. From Linge et al. (2009b).

solar variability is postulated, but demonstrable links via proxy archives to climate parameters are few, owing to the necessity of precise and high-resolution (approximately annual) chronologies combined with a strong proxy-climate relationship. Similarly, internal variability in the ocean–atmosphere–vegetation Earth system will also occur over multi-decadal time periods, and in the same way, proxy archives require high precision and a strong proxy–climate correlation. Feedbacks and interactions between the internal and external forcings are likely to be complex and not cyclical. Over multi-centennial timescales, several proxy archives suggest the presence of an approximately 1500 year quasi-periodic structure (the 'Bond cycles') over the Holocene. Archives with lower sampling resolution (e.g. approximately decadal), but precisely dated, should be capable of preserving climate variability over multi-centennial timescales if they contain a strong climate–proxy correlation. Box 11.1 details some of the time-series techniques that are commonly used to analyse the variability of proxy archives.

Speleothems have the necessary temporal resolution when sampled at suitably closely spaced intervals, to record multi-decadal climate variability.

Box 11.1 Times-series analysis of speleothems

Speleothem proxy time series are typically unevenly spaced in time, with the data points containing some form of internal structure. The goal of time-series analysis is to extract information about this structure. The deterministic component can include *trends*, *autocorrelation* or *periodic* behaviour (section 1.3 and Fig. 1.9) and can provide important information about the evolution of a variable over time and the possible physical mechanisms controlling its behaviour. In speleothem analysis, the time-series variable is typically a proxy such as $\delta^{18}O_c$, and the possible physical controlling mechanism is a climatic forcing. Speleothem proxy time series typically contain background trends, superimposed on which are data that have structure and exhibit persistence (e.g. consecutive values are often correlated), often also exhibiting nonlinear step changes between these periods of auto-correlated signal.

The extent to which a time series are time-invariant, or *stationary* (section 1.3), can dictate the presence of *trends* (section 1.3) within a speleothem time series. If a trend can be attributed to non-climatic forcings (e.g. Fig. 1.9c), it can be modelled and removed from a time series to enhance the climatic signal to noise ratio. This is a requirement of some statistical techniques such as spectral analysis (see below). This 'pre-processing' of a time series is used frequently in palaeoclimate research. Curve-fitting is a simple method of removing trends from a series. Commonly used functions include polynomials or exponential curves, which are fitted to the data using a least squares fitting procedure and then removed from the series. A more flexible, nonparametric alternative is to use a smoothing spline to de-trend a series. Cubic spline functions are readily available in packages such as R (the *rcs* and *ls* functions) and Matlab® (the *spline* function), and their use in constructing age-depth models was introduced in Chapter 9. *Singular spectrum analysis* (SSA) has also seen application to speleothem studies (Wang et al., 2005). The technique uses a lagged auto-covariance matrix where the leading principal component(s) can be used to describe the long-term trend within a series and therefore provides a means of de-trending a series. As well as easy implementation in R and Matlab, a freeware toolkit SSA-MTM is also available at http://www.atmos.ucla.edu/tcd/ssa/.

The assumption of independence between data points required for many statistical tests is frequently violated in speleothem time series where consecutive values are often strongly correlated. An example of such *autocorrelation* (section 1.3) can be seen in the dependence of stalagmite annual layer thickness in a particular year with that of the previous years (see Chapter 9). The autocorrelation of a series can be determined by statistics such as the Durbin–Watson statistic, and autocorrelation functions and autoregression models are readily available in Matlab® (the *autocorr* function) and R (the *ar* function). For unevenly spaced data such as that typically obtained from speleothem proxies, the REDFIT software is a useful tool which enables the fitting of a first-order autoregression function (code can be downloaded from http://www.geo.uni-bremen.de/geomod/staff/mschulz/). Autocorrelation is of particular relevance when considering regression analysis of speleothem time series (for example, when correlating a two autocorrelated speleothem proxies such as $\delta^{18}O$ versus $\delta^{13}C$), because autocorrelation can lead to the underestimation of the standard error of the regression and an over-estimate of the significance of the correlation. The calculation of the adjusted degrees of freedom of the time series, based on the autocorrelation function, allows a more realistic estimate of significance. The quantification of autocorrelation within a single proxy time series can also help elucidate the climate transfer process from surface to speleothem.

Linearity (section 1.3) is an underlying assumption of using modern analogues to understand the climate–speleothem relationship. However, nonlinear ties are often present in the karst system, especially introduced through flow switching and mixing during ground water transport. Tests for nonlinearity in speleothem proxy series can be performed using the White Neural Network Test, which is an appropriate test for time series that are autocorrelated. This test determines whether any nonlinear structure remains in the residuals of an autoregressive process fitted to the time series. This is easily implemented using the *wnntest* function within the R statistical software, and to date has only been applied to cave dripwater hydrology time series (Baker and Brunsdon, 2003).

Spectral analysis (e.g Fig. 12.7) is a widely used technique which is based upon the assumption that any process can be described as a linear combination of sine waves of varying frequencies (Weedon, 2003). The technique attempts to fit a series of sine curves with differing periods, phases and amplitudes to a set of data and is used to transform a time series from the time domain to the frequency domain. The traditional method of spectral analysis requires data that are evenly spaced in time. However, geochemical proxy data from speleothems are often non-uniformly sampled in the temporal domain, making the traditional method of spectral analysis inappropriate. The SPECTRUM program (Schulz and Stattegger, 1997; software can be downloaded from http://www.geo.uni-bremen.de/geomod/staff/mschulz/) has been specifically developed to compute both univariate and bivariate spectral analyses for unevenly sampled proxy time series. The output of the spectral analyses are depicted by the power spectrum, which is a plot of the squared amplitude, or power (the amount of energy per unit time), against the frequency (the number of cycles per unit time). Such a plot enables the identification of regular periodic or quasi-periodic com-

(Continued)

ponents within a time series, manifest as spectral peaks which correspond to concentrations of energy within the frequency domain. Bivariate analysis allows the comparison of the cross-spectrum of two series, allowing the analysis of the coherency and phase association between time series as a function of frequency.

Great care has to be taken when undertaking spectral analysis of speleothem proxy data. Data have to be detrended, and spectral analysis is particularly sensitive to the presence of outliers. The influence of the latter can be easily tested by their removal and reanalysis of the power spectrum. Artefacts can be introduced into a power spectrum by the presence of weakly nonlinear processes. This will produce peaks at multiples of the frequency of the actual peak, *f*, such that peaks are seen to occur at frequencies of 2*f*, 3*f*, 4*f*, . . . , particularly if the signal is of a non-sinusoidal nature. These spurious higher-frequency peaks are known as harmonics. Additional caution is needed to avoid the effects of *aliasing* (sections 1.3 and 8.1), which occurs when the sampling frequency is close to that being observed in the power spectrum. A rule-of-thumb is that no spectral frequencies can be determined below a value which is twice the sampling interval. Finally, the significance of spectral peaks is assessed by calculating confidence levels. Although the Fisher test is commonly used, this is applicable when a single spectral peak occurs above a white noise background. Other tests exist which might be more appropriate for a speleothem time-series where multiple spectral peaks are expected above a red-noise background (Weedon, 2003; Fairchild et al., 2006a).

While spectral analysis is useful for determining the frequency of signals, it does not provide any indication about the location of a particular frequency within a time series. In contrast, the *wavelet* transform provides an indication of local frequency behaviour and the temporal distribution of these frequencies. This property means wavelet analysis is of particular value to series which display transient or non-stationary behaviour. By taking advantage of a flexible window, wavelet analysis enables the examination of both large-scale features using a low-resolution window and smaller-scale events using a high-resolution window. Tests for statistical significance can be applied, and as with spectral analysis, care has to be taken to ensure that trends and outliers in the data do not affect the analysis. Software to undertake wavelet analysis in Matlab and Fortran is available at http://paos.colorado.edu/research/wavelets/ based on the work of Torrence and Compo (1998). Conventional wavelet analysis assumes evenly spaced data and therefore has rarely been applied to speleothem data where its main application would be the analysis of annual lamina width time-series (Smith et al., 2009). However, Thiebaut and Roques (2005) presented a solution for unevenly spaced data that uses a linear interpolation approach and which should provide wider use in the speleothem community. Cosford et al. (2008) provided an example of wavelet and cross-wavelet analysis of Chinese speleothem records.

Fairchild et al. (2006a) detailed the major sampling considerations necessary to prevent aliasing and other artefacts appearing in the proxy time series. The most important practical constraint is the cost of sampling a speleothem at suitably high temporal resolution for a long time period. For example, to resolve multi-decadal climate variability in the 50–100 year range, one would ideally want to observe at least seven times the time period of the variability and sample at approximately annual to biennial resolution, requiring between 200 and 700 analyses per speleothem. Fortunately, developments in analytical techniques including laser ablation inductively coupled plasma mass spectrometry (LA–ICPMS) for trace elements and rapid throughput mass spectrometers for $\delta^{18}O$, make this analytical load practical. Coupled with this is the requirement for high-precision dating which, given the typical uncertainties associated with radiometric ages on Holocene speleothems, is best achieved by a combination of annual lamina counting constrained by precise radiometric dating techniques, rather than by radiometric dates alone.

Several high-profile speleothem records have highlighted the potential for these archives to record multi-decadal or multi-centennial climate variability, as well as the potential problems faced when sampling frequency or dating precision is less than the optimal requirements as described above. Pioneering high $\delta^{18}O$ sampling resolution on Holocene stalagmites from Oman was undertaken around the turn of the millennium. For example, Neff et al. (2001) made over 800 $\delta^{18}O$ analyses over an approximately 3000-year growth period in the Early to Mid-Holocene, and Fleitmann et al. (2003a) produced a similarly high-resolution but discontinuous record from replicate stalagmites for theMid- to Late Holocene. In these Oman stalagmites, they observed a strong similarity between the atmospheric $\Delta^{14}C$ series (interpreted as an archive of solar variability) and oxygen isotopes (Fig. 11.15). Spectral analysis demonstrated multiple peaks which were statistically significant in both stalagmites, but mostly weakly so. Ignoring spectral frequencies that were close to the average sampling frequency (e.g. 4.1 years in Neff et al., 2001) or

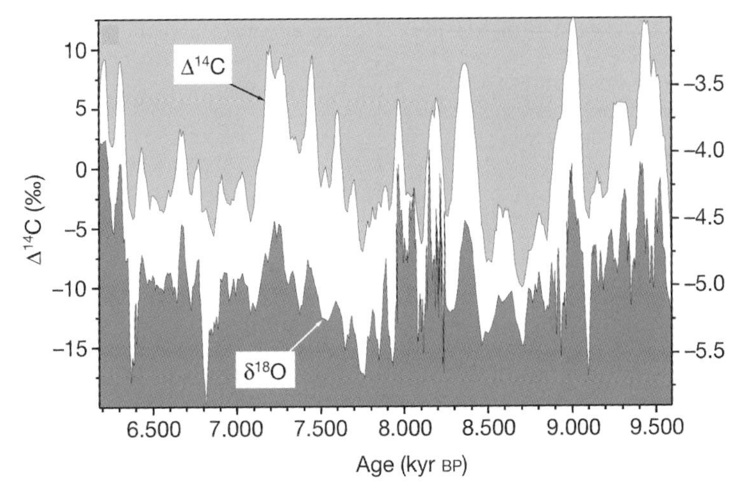

Fig. 11.15 Comparison of stalagmite $\delta^{18}O$ and atmospheric $\Delta^{14}C$. From Neff et al. (2001). The stalagmite $\delta^{18}O$ series is not tuned to the $\Delta^{14}C$ series.

which would have few return periods over the length of the $\delta^{18}O$ record, then statistically reliable peaks were observed at approximately 226 and approximately 35–28 years (Neff et al., 2001) and approximately 220, 140 and 107 years (Fleitmann et al., 2003a). Both Neff et al. (2001) and Fleitmann et al. (2003a) undertook further analyses by 'tuning' the chronology by peak matching the $\delta^{18}O_c$ to the $\Delta^{14}C$ target within the limits of the age errors of the dates. However, this presents a circular argument when trying to assess climate–proxy relationships and unsurprisingly, spectral peaks in the tuned $\delta^{18}O$ records more closely matched those in the $\Delta^{14}C$ record. Such a tuning approach between $\delta^{18}O_c$ and $\Delta^{14}C_{atm}$ was also undertaken by Niggemann et al. (2003) and Wang et al. (2005) and is not recommended: as well as the circularity argument it prevents the analysis of the lags between external climate forcing and internal system response (Jackson et al., 2008), and makes a likely incorrect assumption about a near instantaneous (e.g. approximately 1 year) stalagmite proxy response to surface climate forcing. Alternative tests of the relationship between $\delta^{18}O_c$ and Δ^{14}Catm series for stalagmites with relatively poor temporal control are considered below: ultimately, however, the climate–proxy relationship needs to be investigated from annually laminated deposits where errors in the age model can be best minimized. Nevertheless, Neff et al. (2001) and Fleitmann et al. (2003a) clearly showed a stalagmite $\delta^{18}O_c$ response over multi-decadal timescales, in a region sensitive to monsoon strength and where stalagmite $\delta^{18}O_c$ preserves a mixed record of moisture availability. Other archives which come tantalisingly close to the chronological and sampling precision required to precisely resolve multi-decadal climate variability include a 1270-year isotope record (Paulsen et al., 2003) and full Holocene isotope record (Wang et al., 2005), both from China and sampled at approximately 3- to 5-year resolution; the latter record is considered further below.

Before considering Holocene proxy climate archives with an annual chronological control, two investigations which attempt to resolve the issue of 'tuned' proxy archives are worth further discussion. Jackson et al. (2008) reanalysed the Niggemann et al. (2003) German $\delta^{18}O_c$ stalagmite record, focusing on the period 2.7–2.2 ka where there is a high amplitude fluctuation in the $^{14}C_{atm}$ record. They additionally measured ^{14}C in the stalagmite (Fig. 11.16), with the hypothesis that a ^{14}C excursion in the stalagmite at the same time could be used to independently correlate between the $^{14}C_{atm}$ and $\delta^{18}O_c$ records. However, lack of knowledge of the variability of the dead carbon proportion, and the age and rate of transfer of soil carbon to the stalagmite, together with observed higher amplitude

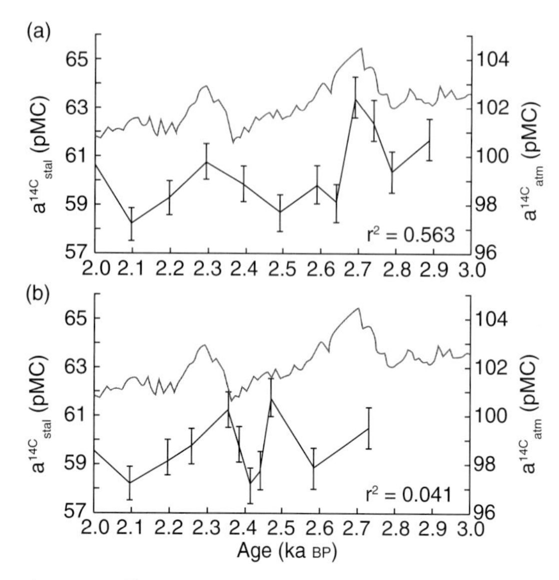

Fig. 11.16 ^{14}C for the stalagmite (points with error bars) and $^{14}C_{atm}$ for STAL-AH-1, where the a^{14}C is the initial activity of ^{14}C in percent modern carbon (pmc) in the stalagmite and contemporary atmosphere. Part (a) Assumes a linear growth model. Part (b) uses a point-to-point interpolation of U–Th analyses. The linear model shows a visible correlation, but with no lags between the atmospheric ^{14}C and stalagmite ^{14}C archive; the ^{14}C stalagmite record has greater amplitude than the ^{14}C atm. The point-to-point model has lower amplitude of a^{14}C and no clear correlation. From Jackson et al. (2008).

variability in stalagmite ^{14}C compared with $^{14}C_{atm}$, leads one to believe that this approach was not successful for this stalagmite. Jackson et al. (2008) also used a Monte Carlo approach to generate randomized time series of $\delta^{18}O_c$ to assess whether a relationship between $\delta^{18}O_c$ and $^{14}C_{atm}$ could have occurred by chance. Haam and Huybers (2010) also used Monte Carlo methods in their test for the presence of covariance between time series where the temporal control is uncertain, such as the relationship between $^{14}C_{atm}$ and a stalagmite proxy series. Their Maximum Covariance between Time uncertain series Test (MCTEST) was applied to the correlation between $\delta^{18}O_c$ and $^{14}C_{atm}$ in the Dongge Cave stalagmite record of Wang et al. (2005). Wang et al. (2005) had observed a correlation of 0.3 between stalagmite $\delta^{18}O_c$ and $^{14}C_{atm}$ for the Holocene when the $\delta^{18}O_c$ series was tuned within dating uncertainty. Haam and Huybers (2010) assumed that the $^{14}C_{atm}$ series has a fixed chronology and that the U–Th analyses on the stalagmite were time control points for the $\delta^{18}O_c$ series; the distributions of the maximum covariance for each dataset was evaluated using Monte Carlo simulations with the same distribution and autocovariance of the actual series, and it was demonstrated at the 95% level that there was no covariance between the two series, presuming no lags between $^{14}C_{atm}$ and stalagmite $\delta^{18}O_c$.

Surprisingly few Holocene speleothem records have both high-resolution proxy data as well as annual chronological control. Exceptions are a 2650-year record from Shihua Cave, China (Tan et al. (2003), previously discussed in section 11.2.1); a discontinuous 4000-year record of climate history from southwest USA (Polyak & Asmerom, 2001); the northwest Scotland stalagmite record of Proctor et al. (2000) covering the past approximately 1 ka (discussed briefly in section 11.2.1); and discontinuous stalagmite records covering the past 8 ka from Ethiopia (Asrat et al., 2007; Baker et al., 2010). Both the Shihua and northwest Scotland records have been discussed in section 11.2.1 in the context of instrumentally calibrated records of the past Millennium climate variability; Shihua Cave stalagmite TS9501 correlating with Beijing and Northern Hemisphere summer and mean annual temperature, and the northwest Scotland stalagmite SU-96-7 with winter precipitation and the NAO. Spectral frequencies identified in each of these stalagmite series are listed in Table 11.1. Tan et al. (2003) demonstrated that the 2650-year Shihua Cave record yields periodicities of approximately 206 and 325 years in the low-frequency domain, with a good visual correlation with the timing of the Chinese dynasties over the past 2.6 ka, but did not analyse the growth rate record at less than 100-year time periods. Reanalysis of the data archived http://www.ncdc.noaa.gov/paleo/paleo.html also yields statistically significant spectral power at the 33–45 year and more than 68-year periodicities, although most of the power exists at longer than 68-year periods. Northwest Scotland stalagmites yield consistent spectral frequencies in the range 50–70 years, identical to that observed in the North Atlantic region sea surface temperatures by Schlesinger and Ramankutty (1994). Such multidecadal variability has been observed in forced and unforced ocean–atmosphere models (Delworth et al., 1993; Delworth & Mann, 2000) and interpreted as being due to variations in ocean–atmosphere circulation (NAO and thermohaline circulation interactions). Polyak and Asmerom (2001) analysed annual lamina thickness records in five stalagmites from New Mexico. Although none were continuously forming over the modern period, precluding instrumental calibration, it appears likely that water supply is the major control on stalagmite growth in this region. Polyak and Asmerom (2001) also demonstrated a link between stalagmite growth and cultural periods in southwest USA. Further analysis of the lamina width data archived in the World Data Centre for Paleoclimatology demonstrates spectral frequencies of more than 73 years, 52–40 years and 22–28 years. Finally, Baker et al. (2010) presented a discontinuous annual lamina thickness and an approximately annual resolution $\delta^{18}O_c$ record from an Ethiopian stalagmite (Figure 11.17). Sampled from a region where $\delta^{18}O_c$ reflects the relative precipitation amount in the spring and summer rain seasons, both this stalagmite and a previous sample from the same region (Asrat et al., 2007) demonstrate

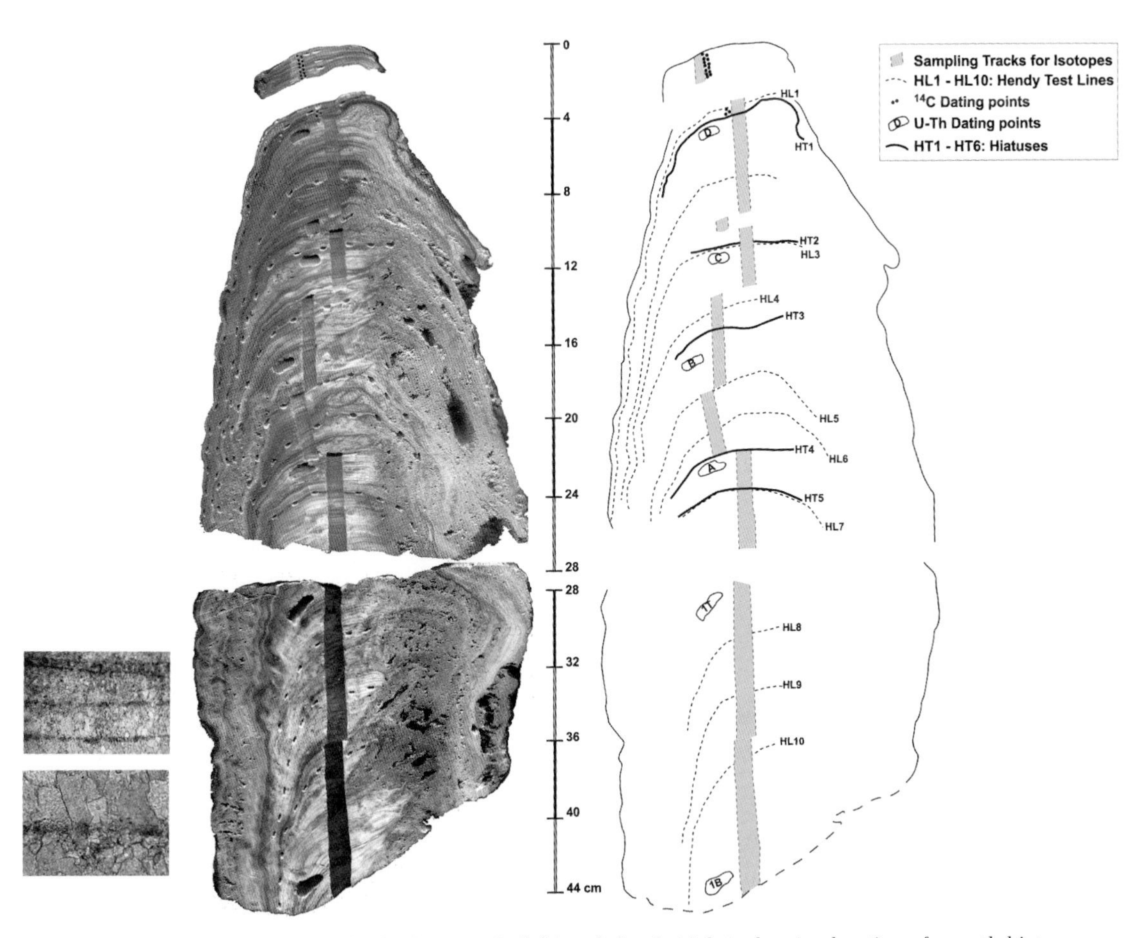

Fig. 11.17 Bero-1 hand-section in both photograph (left) and sketch (right), showing location of growth hiatuses, and sampling for U–Th, ^{14}C and $\delta^{13}C$ and $\delta^{18}O$ analyses. Bottom, thin section examples of annual laminae (bottom left, height of image 3.5 mm) and high-resolution image of one lamina showing seasonal change in fabric coinciding with deposition of organic matter (bottom right, height of image 0.85 mm). From Baker et al. (2010).

spectral frequencies of between 10 and 21 years in both $\delta^{18}O_c$ and growth rate. Decadal-scale rainfall in this region is influenced by both ocean and atmospheric process and their interactions, including the strength and position of the ITCZ, the East African Monsoon (Segele et al., 2009) and sea surface temperatures in both the Indian Ocean (Ummenhofer et al., 2009) and Atlantic (Chang et al., 1997).

Annually laminated stalagmites, providing both an annual growth rate proxy as well as excellent chronological precision for high-resolution isotope records, demonstrate the potential for speleothems to provide crucial insight into decadal-scale climate variability. The few studies which have the necessary resolution over the Holocene, demonstrate ubiquitous decadal-scale variability (Table 11.1) in regions as diverse as East Africa, southwest USA, Europe and China, with replicated samples in several locations, and from climate regions affected by the Asian Monsoon (Beijing, Oman), ENSO (New Mexico) and NAO (Scotland). This demonstrable ability to record low-frequency climate variability is of significance with the need to better quantify natural climate variability in the context of anthropogenic global warming. Despite the considerable analytical cost, further analytical effort in this area is likely to yield results of great significance.

Finally, multi-centennial climate variability over the Holocene has to be considered, and specifically the example of the possible presence of 1470 ± 500 year 'Bond cycles' over the Holocene. Observed as periods of enhanced ice rafted debris in North Atlantic ocean cores (Bond et al., 1997) over the Late Quaternary, their interpretation and cause over the Holocene remains controversial. Speleothem records from glacial marginal regions might be expected to preserve this periodicity if there was an associated temperature change or impact on atmospheric circulation, and records from wider locations would also be expected to capture hemispheric temperature or atmospheric circulation changes should the postulated events have a global impact. Speleothem records with a relatively low sampling frequency (e.g. decadal sampling) and chronological precision provided by U–Th analyses alone should contain evidence of Bond cycles if the proxy indicator is sensitive to the relevant climate change. However, stalagmite proxy archives are dominated by $\delta^{18}O_c$, often a proxy for variations in precipitation source or amount at a seasonal timescale, and to date demonstrable evidence of periodicity at this multi-centennial timescale in $\delta^{18}O_c$ is rare. Mangini et al. (2007) argued for a correlation between a combined stalagmite record from Spannagel Cave, central Alps (that of Vollweiler et al. (2006) with two extra samples to make a five-sample composite), and the de-trended and tuned haematite-stained grains HSG record of Bond et al. (2001). They additionally argued for a correlation between the bandpass filtered stalagmite composite record and the $^{14}C_{atm}$ record, with a phase lag at times. Figure 11.18 demonstrates that multi-centennial-scale $\delta^{18}O$ is present in this stalagmite $\delta^{18}O$ composite, but the uncertainty in the interpretation of the $\delta^{18}O_c$—climate relationship for each stalagmite in the composite, and the need to tune and de-trend to obtain a correlation with the haematite stained grains record of Bond et al. (2001), suggests that the climate forcing

Table 11.1 Annually resolved, annual chronology, Holocene stalagmite records.

Sample and age (ka)	Period (yr)	Proxy	Source
Mechara, Ethiopia, 7.8–4.2	8–9, 12–21, 24–34, 40–46	$\delta^{18}O$	Baker et al. (2010)
Ditto	10–15, 15–20, 30–43	growth rate	Baker et al. (2010)
Mechara, Ethiopia, 5.1–4.7	18–21	$\delta^{18}O$	Asrat et al. (2007)
Shuihua, China, 2.7– 0	33–45, >68	growth rate	Reanalysed after Tan et al. (2003)
New Mexico, USA	22–28, 40–52, >73	growth rate	Reanalysed after Polyak and Asmerom (2001)
Scotland, UK, 1.1–0	50–70	growth rate	Proctor et al. (2000)

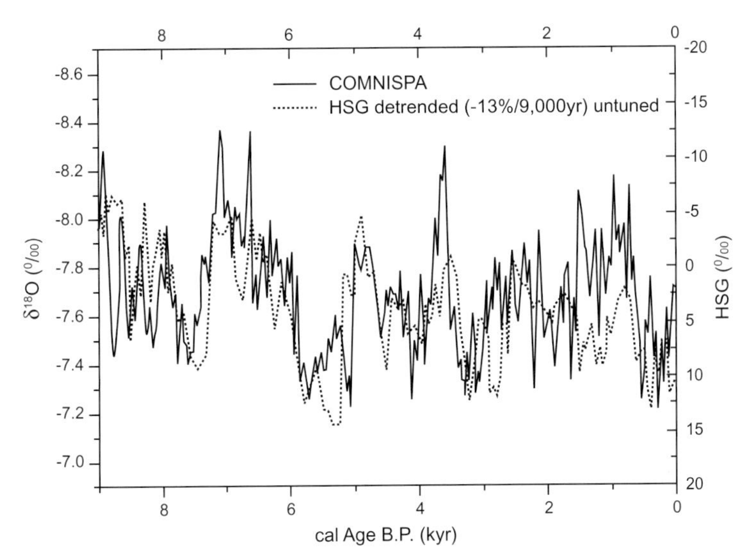

Fig. 11.18 The combined oxygen isotope signal from Spannagel, COMNISPA, together with the haematite stained grain (HSG) curve from core V29-191 (note reversed scale; Bond et al., 2001). Note that the HSG curve has been de-trended.

behind this millennial-scale variability in stalagmite $\delta^{18}O$ are still open to question. As presented in section 11.3.1, even the '8.2 ka event', nominally Bond event seven in the Holocene, requires high-resolution sampling and chronology. If Bond cycles did indeed occur over the Holocene, then one would presume that the events following the '8.2 ka event' would have a smaller isotopic signal and would also require high-resolution proxy analysis over the whole of the Holocene to be clearly observed in speleothem $\delta^{18}O_c$.

11.3.4 Speleothem evidence of Holocene soil and vegetation change

In contrast to the efforts to obtain high-resolution climate reconstructions from Holocene speleothems discussed previously in this chapter, speleothem records of environmental change such as soil development or vegetation change are less well developed. In part this is due to the abundance of other proxy archives such as lakes and wetlands which typically contain ubiquitous, direct evidence of surface environmental conditions. In contrast, speleothem evidence is limited to a range of proxies that are often only available at relatively low resolution owing to their low concentration (e.g. pollen, macromolecules), or which can be obtained at high resolution but whose interpretation is often ambiguous (e.g. $\delta^{13}C$, fluorescence intensity). The Holocene, therefore, is better considered as a test-bed against which speleothem proxies of soil and vegetation change can be compared with other well-established archives. For example, with respect to speleothem pollen analyses, the pioneering work of Bastin, which demonstrated that meaningful pollen records can be obtained from speleothems, was built upon his inter-comparison of Holocene speleothem and lake and peat records from Belgium (Bastin, 1978; Bastin et al., 1982; Bastin & Gewelt, 1986; Bastin, 1990) (see section 6.2.1).

In the north–central USA, a detailed picture of vegetation change has been developed using a mixture of stalagmite $\delta^{13}C$ and surface pollen records. Stalagmite $\delta^{13}C$ can be an ambiguous record of surface conditions (see section 5.4); however, the Holocene in this region experienced a change in ecotones with differing C_3 and C_4 photosynthetic pathways. Dorale et al. (1992) first postulated that changes in stalagmite $\delta^{13}C$ in the region reflected movement in the forest-prairie ecotone. Dorale et al. (1992) presented replicate records

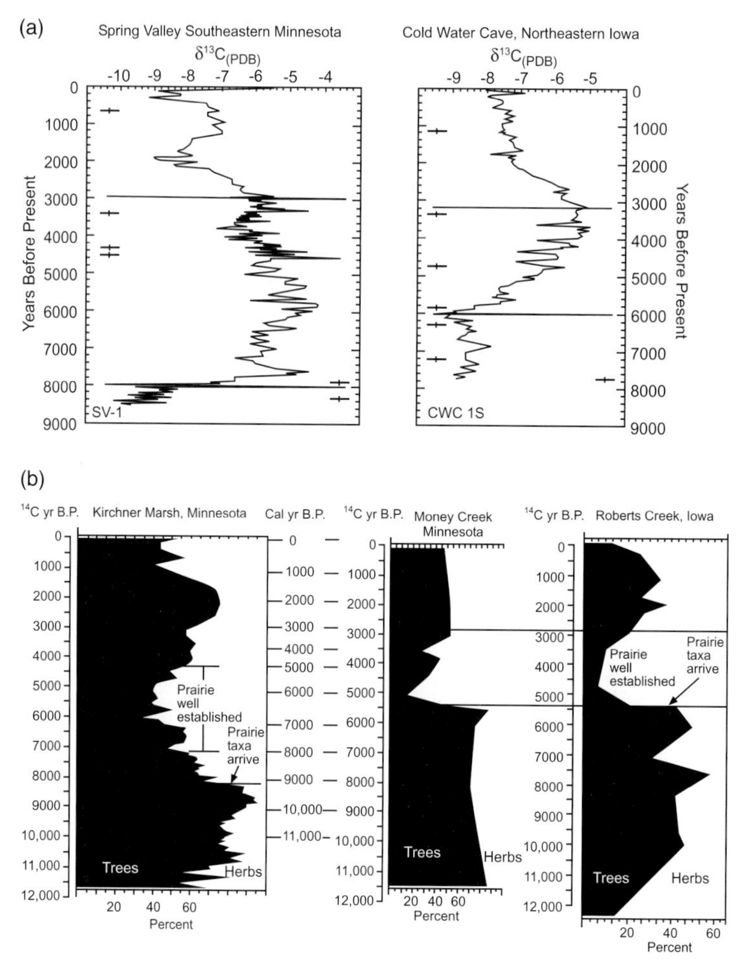

Fig. 11.19 Carbon and oxygen isotopes from stalagmites from Cold Water Cave and Spring Valley Cave (top) compared with the tree–herb percentages in marsh and creek deposits in the same ecotones. From Baker et al. (2002).

from Cold Water Cave, Iowa, which was supplemented by Denniston et al. (1999) with additional replicate material from Minnesota and comparison to surface wetland pollen records (R.G. Baker et al., 1996, 2002). Figure 11.19 summarizes the comparison of stalagmite $\delta^{13}C$ and surface pollen records from Minnesota and Iowa. In Iowa, a $\delta^{13}C$ transition occurs after approximately 5.9 ka and again after approximately 3.3 ka; this matches pollen records of the establishment of C_4 grass dominated prairie at this time. The same correlation is observed in Minnesota, with an earlier timing for the establishment of prairie, which again matches the surface pollen record. The use of replication and a multi-proxy approach, as well as the strength of the $\delta^{13}C$ signal during a transition between C_3 and C_4 photosynthetic pathway dominated ecotones, clearly overcomes most of the potential problems of ambiguity in the stalagmite $\delta^{13}C$ proxy.

An alternative multi-proxy approach to help understand stalagmite $\delta^{13}C$ records is the use of two proxies of surface environment in a single stalagmite. Baker et al. (2000) analysed organic matter fluorescence during the Holocene in the CC3 stalagmite from Crag Cave, Ireland, which had previously been analysed for $\delta^{18}O$ and $\delta^{13}C$ (McDermott

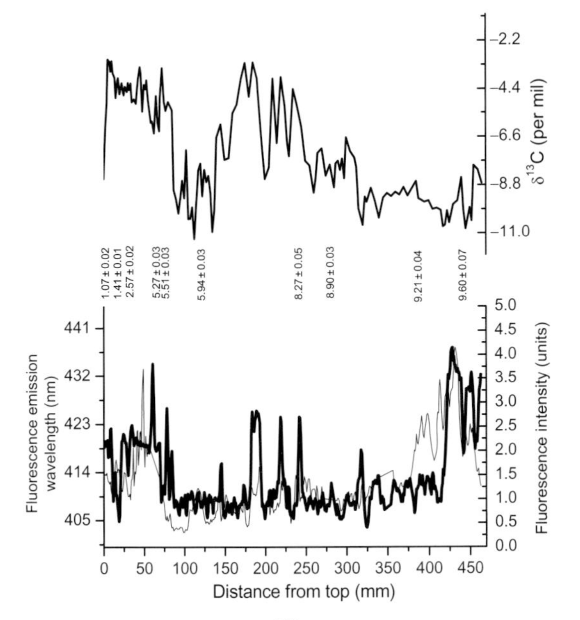

Fig. 11.20 Comparison of $\delta^{13}C$ and organic matter fluorescence intensity and emission wavelength for stalagmite CC-3, Ireland. Adapted from Baker et al. (2000).

et al., 1999). Figure 11.20 shows significant changes in fluorescence emission wavelength and intensity at 9.4 ± 0.05 ka and 4.0 ± 0.5 ka; a decrease in fluorescence wavelength suggests that the organic acids trapped within the stalagmite have a lower molecular weight or less aromatic structure. Compared with $\delta^{13}C$, no change is observed at 9.6 ka, suggesting that the climate and vegetation change at this time period did not affect vegetation photosynthetic pathways (e.g. C_3 vegetation was present). In contrast, a correlation between a $\delta^{13}C$ increase and fluorescence wavelength and intensity at 4.0 ka is observed. This change, probably from forest cover to grassland and agricultural use, was interpreted as a change in photosynthetic pathways as the region was settled for the first time. By implication, this multiproxy comparison indicates that $\delta^{13}C$ data post-4ka cannot be related to natural climate phenomena, but that earlier $\delta^{13}C$ data should reflect natural soil and vegetation processes.

A huge amount of speleothem $\delta^{13}C$ data has been collected by researchers interested in the $\delta^{18}O$ proxy, but is not always analysed and indeed is often not presented or archived. Despite the potential ambiguities in this proxy, Holocene multi-proxy approaches demonstrate its use at recording vegetation change over the Holocene where changes in overlying vegetation also reflects changing C_3 versus C_4 photosynthetic pathways. As a soil-derived signal, speleothem $\delta^{13}C$ records include any lags introduced through soil and vegetation carbon storage, but the soil and vegetation signals could be disentangled with appropriate use of replicate samples with different hydrological connectivity, and multiple soil and vegetation proxies. Such an approach would have great value in improving our understanding the soil and vegetation responses to climate and land use change.

11.4 Questions raised and future directions

This chapter, although considering the Holocene as a whole, has focused on the use of speleothem proxies to record climate change or variability over selected significant time periods. Over the past millennium, the longevity of speleothem growth assures that they have the potential to provide crucial information on decadal and longer timescale climate variability. Data useful to the wider palaeoclimate community require precise chronological control with the use of annually laminated stalagmites and approximately annual-resolution sampling (as a minimum). Even with this control, the fact that stalagmite proxies are typically a mixed signal of water availability and temperature; that they also exhibit a seasonal bias, typically towards the season of groundwater recharge; and that relatively low magnitude climate variability over the past 1000 years compared with potential variability introduced in the groundwater and cave environment, means that care is needed in proxy climate reconstructions. Instrumental calibration and cave monitoring (Chapters 8 and 10), as well as the choice of samples from climate sweet-spots (Chapter 3) obviously help in this regard, and many speleothem climate series are now becoming available and are being used in multiproxy climate reconstructions.

The strategy of high-resolution sampling combined with high precision dating and annual lamina counting has also yielded demonstrable rewards in the analysis of the spatial extent of the '8.2ka event'. However, the analytical effort to sample the whole Holocene at annual or better resolution has so far limited the reconstruction of whole Holocene records of climate variability. Snapshots of annually resolved proxy records over the Holocene, reveal a widespread multi-decadal variability from contrasting sites around the world. Further research in this area will yield further understanding of multi-decadal climate variability, with speleothems analysed in this detail unrivalled among archives capable of preserving low-frequency climate variability. In contrast, multi-centennial climate variability should be more easily extracted from long speleothem records; however, most records have been analysed with point samples at lower sampling resolution. Over the entire Holocene, at a global scale, stalagmite $\delta^{18}O$ records reflect the dominant insolation forcing, but centennial-scale variability, if present, is in most cases obscured by inadequate sampling strategies, a lack of replication, or is less important than decadal-scale variability.

Future research efforts, building high-resolution speleothem climate reconstructions, require an increase in analytical capability if research is to progress at a faster pace than previously. Currently, reconstructions typically require micromilled or hand-milled powders, which are subsequently presented to the appropriate analytical instrument. Annual chronology building remains an essentially manual process, with long lamina counts taking several weeks or months of research time. Improvements in *in situ* analytical techniques would greatly speed up analytical capabilities in isotope analysis as long as the resolution and precision can approach that of conventional approaches. Similarly, rapid annual lamina chronology building, for example using semi-automated approaches and increased use of annual variations in trace elements using *in situ* instrumentation, would be of great use to the research community. Future research targets, both climatic and environmental, might include high-resolution analyses focused on periods of known and variable climate forcing factors (volcanic eruptions postulated to have caused hemispheric cooling; periods of variable $^{14}C_{atm}$); events recorded in other archives but whose geographic impact and forcing is unknown (e.g. the '9.3 ka event', the 'Medieval Climate Anomaly' or the 'Bond cycles'); or periods of civilization change or collapse (for example, at 4.2 ka), where the combination of climate- and soil-derived proxies in a single well-dated archive might yield unique information about the relationship between climate change and soil carbon fluxes over the Holocene, for different soil types and stages in development. This important subject has barely been addressed by the speleothem research community.

CHAPTER 12

The Pleistocene and beyond

This final chapter considers speleothem archives that predate the Holocene, as well as providing a forward look to future research possibilities. Section 12.1 considers the Pleistocene and speleothem records of ice-age climate fluctuations. In particular, long, continuous and replicated stalagmite isotope series have recently been published from several regions of the world, which provide precisely dated records of the Late Quaternary climate. Speleothem archives during this period benefit from the greater magnitude of climate variability compared with the Holocene, such that the climate signal constrained within proxy archives is more likely to be greater than any environmental or hydrological uncertainty introduced in the climate transfer process. The considerable advantage of these speleothem archives is the precise chronology afforded by U–Th analysis, making them superior in chronological control to almost all other climate archives over this period, although the complexity of climate transfer process and the mixed nature of many stalagmite parameters make quantification of proxies in terms of a climate parameter difficult. The speleothem $\delta^{18}O$ proxy is the most widely used in Pleistocene samples, and in section 12.1 we review speleothem isotope records of Late Quaternary climate variability, with a specific focus on long replicated $\delta^{18}O$ records from China, the use of speleothems in identifying the global impact of interstadial events originally recognized from Greenland during the Pleistocene, as well as the nature and timing of ice-age deglaciations and sea level change.

Section 12.2 delves further back in time, where the recent developments in U–Pb dating have allowed the analysis of speleothems over the past 300 Myr of Earth history where they provide tantalizingly high-resolution snapshots of environmental and climate history (Fig. 12.1). Section 12.3 concludes our work, summarizing the questions raised and providing some suggestions for the future of speleothem palaeoenvironmental analysis.

12.1 Pleistocene proxy records (ice-age climate fluctuations defined and drawn)

12.1.1 Subaqueous speleothem records: Devils Hole, USA

In 1992, Winograd and co-workers published a 500,000 year $\delta^{18}O$ record from a sub-aqueous speleothem, which significantly raised the profile of speleothem palaeoclimate records (Ludwig et al., 1992; Winograd et al., 1992). Devils Hole (Nevada) is a fault-controlled fissure, collapsed near the surface, at least 130 m long, which intercepts the regional ground water at approximately 15 m below land surface (Riggs et al., 1994) and at the down-system end of the regional aquifer. The ground water temperature has remained constant over recent decades (between 33 and 34 °C), reflecting the large size of the aquifer, which is thought to be recharged by the Spring Mountains (80 km northeast) and the Nevada high ranges (400 km north–northeast). Aquifer transmissivity based on borehole data imply that the groundwater residence time is less than 1 kyr (Thomas et al., 1996) and ^{14}C analyses of groundwater dissolved

Speleothem Science: From Process to Past Environments, First Edition. Ian J. Fairchild, Andy Baker.

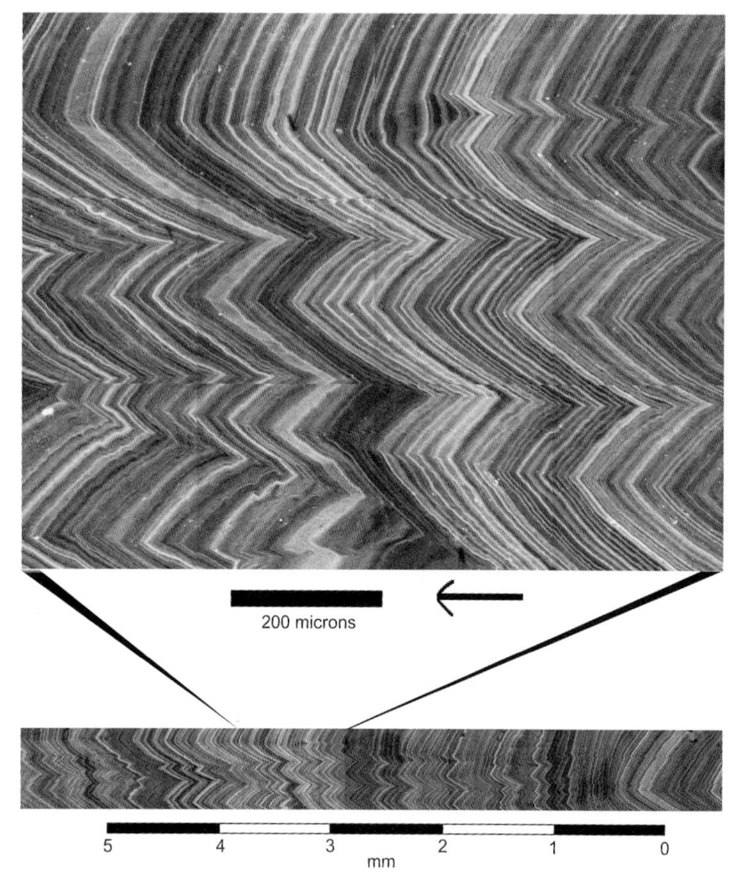

Fig. 12.1 Photomosaic of ultraviolet fluorescent annual bands from the Buffalo Cave Flowstone, South Africa, imaged using confocal microscopy. The flowstone grew from 2.0 to 1.5 Ma and is banded throughout; the layers shown here grew at approximately 1.74 Ma (Hopley et al., 2007b). Automated measurement and counting of the annual bands was performed using the peak counting software of Smith et al. (2009). The annual nature of the ultraviolet bands has been determined by comparison with the orbitally tuned oxygen isotope chronology. Average annual band thickness is 7 micrometres. Photograph courtesy of Phil Hopley.

organic carbon confirm an age of less than 7 ka (Thomas et al., 1996; Morse, 2002). Winograd et al. (2006) assessed the groundwater residence time to be less than 2 ka. Most of the Devils Hole is lined with approximately 30 cm thick mammillary calcite which has been deposited subaqueously. Thermal ionization mass spectrometry (TIMS) U–Th (Ludwig et al., 1992; Winograd et al., 2006) and ^{231}Pa (Edwards et al., 1997) dating have confirmed the chronology of the $\delta^{18}O$ record. Devils Hole speleothem DH-11 was first demonstrated to have been deposited from 566 ka to approximately 60 ka (Winograd et al., 1992), and re-sampling of mammillary calcite at greater depths has allowed an extension of the record to approximately 4.5 ka (samples DHC2-3 and DHC2-8; Winograd et al., 2006). The modern-day geochemistry of the groundwater at the site suggests that, with further sampling, further extension of this record to the present day is possible (Coplen, 2007). Figure 12.2 presents the younger more precisely dated part of the 566–4.5 ka $\delta^{18}O$ record, which is a composite of three mammillary speleothem $\delta^{18}O$ records.

The Winograd et al. (1992) record generated significant debate within the palaeoclimate community because the $\delta^{18}O$ clearly showed the imprint of a glacial–interglacial signal, but with a time of the glacial terminations earlier than that expected by the research community of the time. Coplen et al. (1994) later presented the $\delta^{13}C$ record for the period 566–60 ka, demonstrating that this inversely correlated with $\delta^{18}O$ ($r = -0.75$) with a lead of up to 7 ka: the $\delta^{13}C$ record was interpreted as a record of changes in vegetation extent or density. The early timing of glacial–interglacial transitions was particularly clear from the most precisely dated

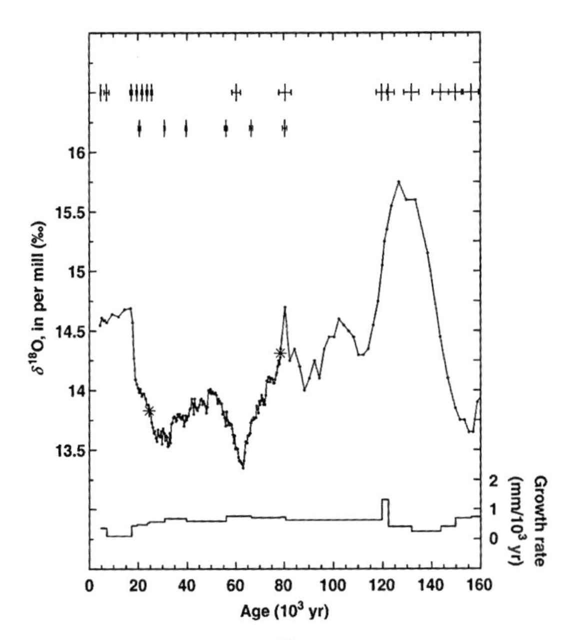

Fig. 12.2 The Devils Hole $\delta^{18}O$ series from 160 to 4.5 ka. $\delta^{18}O$ values are expressed relative to VSMOW. Asterisks mark the tie points between the three subaqueous speleothems. Location and uncertainty associated with U–Th analyses are shown above the $\delta^{18}O$ record and the mean deposition rates below. From Winograd et al. (2006).

section of the calcite: at the time of Termination II, $\delta^{18}O$ increased at 140 ± 3 ka (see Fig. 12.2) and $\delta^{13}C$ reached a minima at 133 ± 3 ka, both earlier than the anticipated timing of 128 ± 3 ka as observed in the marine orbitally tuned SPECMAP record (Martinson et al., 1987). Figure 12.2 furthermore shows that in the extended record (Winograd et al., 2006), the termination of the last glaciation is recorded as a 0.6‰ shift in $\delta^{18}O$ between 19 and 17 ka. Comparison of Devils Hole and marine records by Winograd et al. (2006) demonstrated that the Devils Hole $\delta^{18}O$ record resembles sea surface temperature records from marine cores off the California coast; at Termination II these also respond 10–15 ka before the insolation maximum (see, for example, Herbert et al., 2001). Devils Hole mammillary speleothem $\delta^{18}O$ therefore most likely integrates a regional groundwater response to rainfall in the groundwater source region, which is in turn forced by sea surface temperature warming in the California Current. $\delta^{13}C$ mostly likely reflects the vegetation response to the same forcing, but is better interpreted as a record of variations in soil microbial productivity and increase in organic matter sourced CO_2 within the catchment (section 5.4). The Devils Hole record highlights the importance of considering speleothem records in the context of climate hot-spots, before a speleothem proxy archive is interpreted as a global climate proxy (section 3.1.5). Devils Hole provides a unique proxy of sea surface temperatures related to the warming of the California Current, and a more precisely dated long record than can be obtained from marine archives. The teleconnection between the California Current and the global dynamics of ocean–atmosphere change over the last glacial–interglacial cycle has yet to be determined. Further discussion of the nature and timing of glacial terminations can be found in section 12.1.5.

The Devils Hole record highlights both the benefits and problems of working with subaqueous calcite deposits. Deposition rates are significantly lower than subaerial deposits, with the Devils Hole samples deposited at less than 1 μm per year. Milled at 250-μm resolution, $\delta^{18}O$ sample resolution is between 250 and 2500 years per sample, prohibiting the observation of millennial-scale climate variability. The slow growth rates also increase the time integration of TIMS U–Th samples. However, these problems are in part balanced by the length of palaeoclimate record contained in one sample, and the integrated 'catchment' nature of the record means that local and drip-specific factors affecting subaerial speleothems do not occur. Time will tell whether improvements in technology (such as inductively coupled plasma mass spectroscopy and laser ablation U–Th dating, Chapter 9; and improved resolution of micro-milling and the use of ion probe for $\delta^{18}O$, Chapter 5), make subaqueous speleothems a more favoured target for palaeoclimate research in the future.

12.1.2 Composite speleothem records: Soreq Cave, Israel

Soreq Cave, Israel, has been the focus of speleothem palaeoclimate studies by researchers at the Geological Survey of Israel and their collaborators.

Research at the site has continued over a 15 year period, and included initial monitoring of the modern-day climate and cave environment (Bar-Matthews et al., 1996; Ayalon et al., 1998), followed by speleothem palaeoclimate reconstructions for the past 185 kyr, and the comparison of the speleothem archives with other regional proxies. The latter initially focused on the use of $\delta^{18}O_c$ as a palaeoclimate proxy (Bar-Matthews et al., 1997,1999; Ayalon et al., 2002), with subsequent investigations additionally testing the trace element (Ayalon et al., 1999), fluid inclusion (Matthews et al., 2000; McGarry et al., 2004), clumped isotope (Affek et al., 2008) and ion microprobe $\delta^{18}O$ techniques (Kolodny et al., 2003; Orland et al., 2009). The research particularly highlights the insights that can be gained from long-term monitoring modern-day environmental parameters and in undertaking multi-proxy comparisons, both between speleothems proxies and between speleothems and other climate archives.

Soreq Cave was opened by mining activities in 1968 and is situated 40 km inland from the Mediterranean Sea. The climate is semi-arid, with between 500 and 600 mm of rainfall a year, with 95% of rainfall occurring in the winter wet season between November and April (Orland et al., 2009). Monitoring of rainfall $\delta^{18}O$ demonstrates an inverse relationship with total annual rainfall ($r = -0.70$), suggesting that speleothem $\delta^{18}O$ should be a proxy for winter rainfall amount if fractionation effects are negligible. Orland et al. (2009) demonstrate that this is

$$\delta^{18}Op \text{ (‰, Vienna Standard Mean Ocean Water (VSMOW))} = -0.0036 \text{ (annual precipitation, in millimetres)} - 3.9 \quad (12.1)$$

This indicates that a 1‰ change in $\delta^{18}O_p$ corresponds to approximately 280 mm less rain per year. However, in a semi-arid environment, evaporation of water in the soil or groundwater might occur, leading to isotope fractionation between rainfall and the cave. Modern-day monitoring of dripwaters in the cave suggest that dripwaters fall along the MMWL, indicating that evaporation effects are likely to be relatively insignificant. A second possible influence on speleothem $\delta^{18}O_c$ is fractionation caused by rapid degassing (see section 8.4.2). Affek et al. (2008) concluded that rapid degassing is important at the site based on analysis of Δ_{47} in Soreq Cave speleothems and a misfit between predicted and actual temperatures. This offset of approximately 1‰ between observed and calculated equilibrium $\delta^{18}O_c$ had not been previously observed using 'Hendy' tests or in the $\delta^{13}C$ data presented in Bar-Matthews et al. (1996), and highlights the advantages of analysing multiple proxies on speleothem samples. With this additional research, it allows the re-interpretation of the 185 ka $\delta^{18}O_c$ speleothem palaeoclimate record as one that is not necessarily a record of equilibrium deposition (*sensu* Bar-Matthews et al., 1996).

Figure 12.3 presents the most recent $\delta^{18}O$ and $\delta^{13}C$ records from Soreq Cave (Ayalon et al., 2002), which extend the record to 185 ka as a composite of 25 speleothems. The record was incrementally published over the preceding 5 years: Bar Matthews et al. (1997) presented the past 25 kyr from seven alpha-spectrometrically dated speleothems; Bar-Matthews et al. (1999) the past 60 kyr from 20 speleothems that had been TIMS U–Th dated; and Bar-Matthews et al. (2000) the past 145 kyr from 23 samples, all but one of which were TIMS U–Th dated. The final Ayalon et al. (2002) record shown in Fig. 12.3 comprised both stalagmites and stalactites, with samples drilled at 0.5–1.0 mm intervals, giving a decadal-scale sampling resolution. Where samples overlapped in time the authors state that they observed a good agreement in $\delta^{18}O$ and they spliced the records (although the precise methodology is not given). Bar-Matthews et al. (2003) further extend the regional stalagmite isotope record by combining isotope data from Soreq and Peqiin caves.

Figure 12.3 presents the speleothem composite record with marker excursions corresponding to organic-rich sediment (sapropel) deposition in Mediterranean Sea sediment cores. The Holocene sapropel record was considered in section 11.3.1, where sapropel S1 was preserved in Corchia Cave speleothem with a negative $\delta^{18}O$ isotope excursion, interpreted as a precipitation amount. The Soreq Cave record also exhibits negative $\delta^{18}O$ excursions

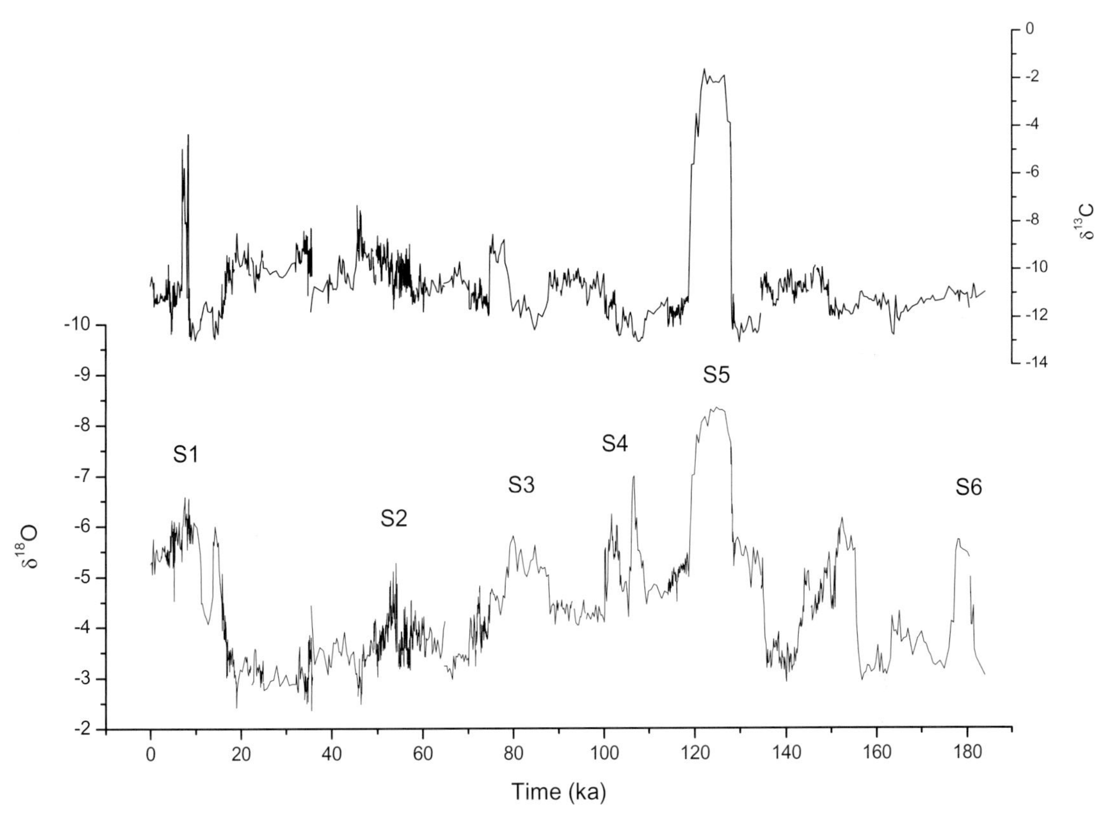

Fig. 12.3 The Soreq Cave $\delta^{18}O$ and $\delta^{13}C$ composite record, as archived at the World Data Centre for Paleoclimatology (ftp://ftp.ncdc.noaa.gov/pub/data/paleo/speleothem/israel/soreq_peqiin_2003.txt). The archived composite is made up of 21 speleothem samples constrained by 95 TIMS U–Th analyses and over 2000 stable isotope analyses. S1–S6 refer to sapropel periods (see text).

at the same time as sapropels S2–S6, with typically nonlinear responses at the start and end of the periods of sapropel formation. At Soreq Cave, presuming that fast degassing disequilibrium effects remain relatively constant over time, one can use the modern monitoring analogue to suggest that the Soreq Cave record also contains an amount effect, although changes in precipitation source and changes in the $\delta^{18}O$ of the Mediterranean Sea are also potential factors. The nonlinear jumps in $\delta^{18}O$ might reflect nonlinear responses of karst hydrology to climate variability at the site, such as an increased proportion of fissure-flow routed groundwater during these wetter periods. The $\delta^{13}C$ composite record is more ambiguous, with both positive and negative isotope excursions during sapropel deposition. The complexity of factors affecting $\delta^{13}C$ composition (see Chapters 3, 5 and 8), complicated further by the mixture of speleothem types (stalactites and stalagmites) at a site where rapid degassing is known to cause kinetic fraction, makes the interpretation of the $\delta^{13}C$ composite record difficult.

The publication of the Soreq Cave composite $\delta^{18}O$ record at the turn of the millennium further highlighted the potential of speleothems as a proxy climate archive. Ultimately, the composite $\delta^{18}O$ series has been recognized as a semi-quantitative and regional record of rainfall source or amount, reflecting conditions in the eastern Mediterranean region. Comparison with marine sediment cores in the Mediterranean Sea has confirmed the palaeoclimate signal contained within the $\delta^{18}O$ composite record, and it has helped constrain the timing of sapropel events which were otherwise beyond the range of dating of marine cores (Bar-Matthews et al., 2003; Almogi-Levin et al., 2009). The main weakness of the record is its composite nature, with a lack of clarity over the nature of the splicing of the speleothem records, the extent of inter-sample $\delta^{18}O$ variability, or a quantification of any additional $\delta^{18}O$ variability introduced by the use of both stalactite and stalagmite samples. However, alongside the Devils Hole record, research at Soreq Cave demonstrated the power of precisely dated $\delta^{18}O$ speleothem palaeoclimate reconstructions.

12.1.3 Palaeoclimate hotspots: the Asian monsoon and the nature of glacial terminations

Soreq and Devils Hole demonstrate how speleothem proxies record palaeoclimate change. In particular they show that over glacial–interglacial timescales, the magnitude of change in $\delta^{18}O_p$ is sufficient 'signal' to leave an imprint in speleothem $\delta^{18}O_c$. Both the Soreq Cave and Devils Hole records have demonstrated to be palaeoclimate records which are influenced by both global climate and regional ocean–atmosphere dynamics. They showed the potential for precisely dated $\delta^{18}O_c$ speleothem records of climate change over the Late Quaternary, but highlight the need for speleothem palaeoclimate records to be constructed at sites where regional climate variability is representative of global climate change. One such region has proven to be that of the Asian Monsoon. As discussed in section 11.3.2, LeGrande and Schmidt (2009) demonstrated, using Holocene general circulation model simulations, that the change in $\delta^{18}O_p$ in the Asian Monsoon region reflects the southward movement of the Intertropical Convergence Zone, and enhanced water vapour transport from the tropics, rather than a local signal of precipitation amount or intensity, with enhanced ocean–land water-vapour transport due to the enhanced Northern Hemisphere seasonality strengthening the monsoon effect. Over longer periods, one would now expect changes in insolation-driven changes in Northern Hemisphere seasonality would similarly affect $\delta^{18}O_p$ and become imprinted in speleothem $\delta^{18}O_c$.

Late Quaternary $\delta^{18}O_c$ stalagmite records have been progressively developed over the past decade (Wang et al., 2001; Dykoski et al., 2006; Cheng et al., 2009b; Liu et al., 2010). Plate 12.1 shows the compilation of stalagmite $\delta^{18}O$ presented in Cheng et al. (2009b), along with orbital forcing parameters and proxies from Antarctic ice and marine core ODP980. Multiple stalagmite samples from different caves within the Asian Monsoon region show a demonstrable replication of $\delta^{18}O$, with amplitude of approximately 4‰ over glacial–interglacial cycles. Over the past 350 kyr, Asian Monsoon stalagmite $\delta^{18}O$ shows a distinct correlation with Northern Hemisphere summer insolation. Superimposed on this are 'gouges' where $\delta^{18}O$ decreases and the Asian Monsoon weakens; these correlate with the timing of stadial events recorded in the Greenland ice cores and demonstrate a link between north Atlantic climate and the Asian Monsoon. This is considered in a global context in the next section.

Cheng et al. (2009b), Liu et al. (2010) and Broecker et al. (2010) particularly focused on the use of the stalagmite $\delta^{18}O$ records to better understand the nature of ice-age terminations. Figure 12.4 presents the $\delta^{18}O$ records for the last glacial termination (T1), the termination around 250,000 years BP (TIII) and the marine oxygen isotope stage 4/3 boundary, which all show a remarkable similarity in structure. Using terminology applied to T1, each have an initial Bølling-like warming, followed by an Older Dryas-like cooling, an Allerød-like warming and Younger Dryas-like cooling. This supports the idea of a prevalent self-similarity inherent to the climate system (Wolff et al., 2009) where the Younger Dryas is typical of ice sheet instability and its effect on ocean–atmosphere interactions during many Late Quaternary deglaciations (Broecker et al., 2010). High-resolution $\delta^{18}O$ speleothem records in the Asian Monsoon region have provided the essential evidence.

12.1.4 The timing of Greenland interstadials

The last glacial period is marked by rapid variations in climate classified as Greenland interstadials (GIS). Within the Greenland ice cores, these are represented as sawtooth-shaped variations in $\delta^{18}O$ of 3–4‰ amplitude and are also known as Dansgaard–Oeschger cycles (see Plate 12.2). Within

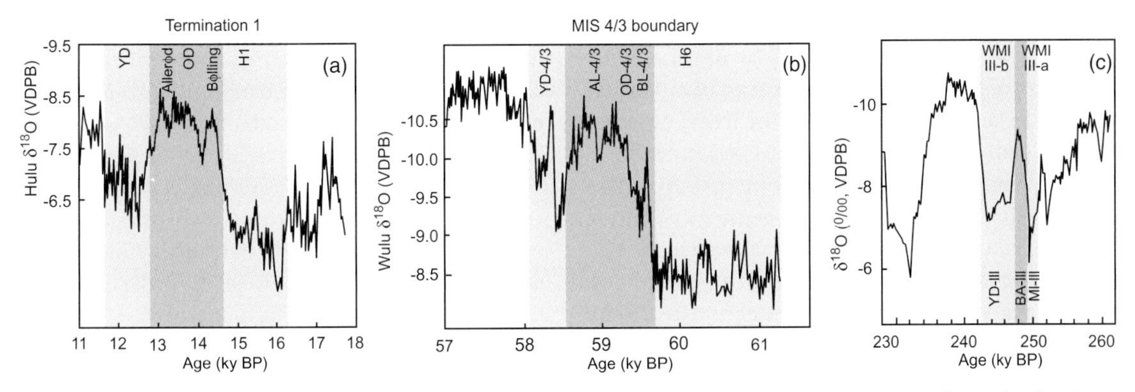

Fig. 12.4 Comparison of the structure of deglaciation at Termination III, the MIS 3/4 transition and Termination 1 determined from stalagmite $\delta^{18}O$ records. Redrawn from Liu et al. (2010) and Cheng et al. (2009b).

the adjacent North Atlantic, they have been observed as variations in the intensity of deposition of ice-rafted debris with a series of GIS interrupted by cold periods, culminating in extreme ice-rafting periods known as Heinrich events (Plate 12.2). GIS re-occur approximately every 1000–7000 years over the last glacial period, and the near-global extent of GIS is increasingly well documented (Voelker, 2002). However, their absolute timing has, until recently, remained uncertain. For example, within ice cores their chronology is based on layer counting and ice sheet accumulation models, and within other proxy archives chronological control is limited by the precision of associated radiocarbon analyses. Speleothems have the necessary chronological control and temporal resolution to contain the timing of GIS, as long as they leave an imprint in one of the high-resolution proxies such as growth rate, $\delta^{18}O$ or $\delta^{13}C$.

Plate 12.2 compiles $\delta^{13}C$ and $\delta^{18}O$ stalagmite records for the past 50 kyr that show a proxy response to GIS. Wang et al. (2001) first showed the potential of Chinese stalagmite $\delta^{18}O$ to record GIS over this period. They were followed by Burns et al. (2003), Wang et al. (2006) and Spötl et al. (2006) who all observed GIS in stalagmite $\delta^{18}O$ from regions as diverse as Socotra, Brazil, China, and the European Alps, confirming the global imprint of the GIS. In contrast, Genty et al. (2003) demonstrated that in southwest France the $\delta^{13}C$ proxy responded more strongly to GIS than $\delta^{18}O$. More recently, Fleitmann et al. (2009) present $\delta^{13}C$ and $\delta^{18}O$ GIS records from Turkey, and Wainer et al. (2009) and Genty et al. (2010), following Wainer et al. (2009), an updated composite $\delta^{13}C$ record from southwest France. Liu et al. (2010) have recently reported a relatively high-resolution (approximately decadal sampling frequency) $\delta^{18}O$ record from Wulu Cave, southwest China, demonstrating excellent agreement between the duration and structure of GIS in the North Greenland Ice Core Project (NGRIP) and replicate Chinese stalagmite $\delta^{18}O$ over the period 60–50 ka.

With this substantial stalagmite isotope archive of GIS, Fleitmann et al. (2009) could demonstrate agreement in the timing of GIS over the past 50 kyr between different speleothem archives e.g. between Hulu Cave, China, and Sofular Cave, Turkey. The speleothem GIS records can now be used to constrain revisions to the Greenland ice core chronologies, and they increase the precision of spectral analysis of the timing of GIS. The latter is particularly interesting as it allowed Fleitmann et al. (2009) to demonstrate the absence of any periodicities close to 1500 years (section 11.3.3). The Turkish stalagmite record is also useful in the investigation of isotope responses to climate change, as both $\delta^{13}C$ and $\delta^{18}O$ proxies demonstrate a rapid response to GIS. Fleitmann et al. (2009) report a mean transition time into a Greenland interstadial occurs in 62 ± 14 years recorded by $\delta^{18}O$ in NGRIP, within 121 ± 99 years recorded in stalagmite So-1 $\delta^{18}O$ and 252 ± 87 by $\delta^{13}C$; the slower response time of $\delta^{13}C$ likely reflecting soil organic matter turnover rates.

Research to date therefore demonstrates that in monsoonal climate regions such as Brazil, China and Socotra, GIS cause the weakening of monsoon circulation and that this has an impact on $\delta^{18}O_p$ in these regions such that it is preserved in most, if not all, climate-sensitive stalagmite $\delta^{18}O$ records. Replicability of $\delta^{18}O$ between stalagmites within a cave, and in the timing of GIS between different regions, demonstrates the potential for stalagmites to constrain the timing of GIS further for the period of ice core records, as well as providing new archives of GIS before the period of preserved Greenland ice. In contrast to $\delta^{18}O$, in regions where vegetation cover or productivity is climatically sensitive, such as in the Mediterranean region and southwest France, $\delta^{13}C$ can provide a sensitive proxy for GIS. High-resolution (approximately annual) analysis of $\delta^{18}O$ would further elucidate the relationship between centennial- and millennial-scale climate forcing (section 11.3.3).

12.1.5 Sea-level records from flooded caves

The timing and relative height of sea levels in the past can be constrained by two radiometrically dated archives—coral reefs and submerged speleothems—which provide excellent constraints on marine and ice core proxies of ice volume and relative sea level (Rohling et al., 2009; Siddall et al., 2009). Sea level variations are a mixture of *glacio-isostatic* (response of Earth to changes in surface loading of ice and geoid changes) and *eustatic* (global ice volume) factors. Precise timing of eustatic sea level rise resulting from change in ice volume, independent of orbitally tuned chronologies, provides a direct and independent measure of ice sheet build-up and decay, allowing the investigation of the relationship between insolation change and ice-age cycles. U–Th dating of corals has been successful over the past approximately 130 kyr, but careful sample selection is typically necessary for older samples (Stirling and Andersen, 2009). Speleothems currently provide some constraints over the past 300 kyr (Li et al., 1989; Lundberg & Ford, 1994); U–Pb dating has the potential to extend this further.

The classic methodology when using speleothems as archives of past sea levels is from dating the time of stalagmite growth in caves which have been, or are, currently submerged (see Fig. 7.27). Presuming that speleothem deposition is not constrained by climate (e.g. a lack of recharge), then the periods of growth cessation indicate the time when sea level rise flooded the cave (see Fig. 7.29). Spalding and Matthews (1972) showed the potential of this approach, dating a stalagmite from the Bahamas which grew in the last glacial and which is currently submerged. Harmon et al. (1978) built on this methodology, with alpha-spectrometric dating of stalagmites from submerged caves in Bermuda and Andros Island. More recently, researchers have analysed marine overgrowths on speleothems. These may be organic (e.g. worm casts, Antonioli et al., 2004) or inorganic overgrowths (e.g. 'phreatic overgrowths' of Vesica et al., 2000). Tuccimei et al. (2010) and Dorale et al. (2010) dated modern and Late Quaternary overgrowths in Mallorcan caves.

Despite the apparent simplicity of the approach and the importance of research questions that can be tackled, relatively few detailed studies have been undertaken owing to the technical difficulty in collecting speleothem samples from flooded caves and the relatively limited number of karst regions which are adjacent to the modern or palaeo-ocean. In addition, in tectonically unstable regions, the rate of tectonic uplift needs to be constrained if the timing of speleothem growth at an absolute depth below mean water level is to be converted into a past sea level. Despite this, the analysis of multiple samples from different relative elevations from one region, or long-lived samples growing through one or more high sea stands, has provided crucial insights into relative sea levels over the Late Quaternary. Li et al. (1989) first demonstrated the utility of using long-lived speleothems whose growth has been repeatedly interrupted by sea level rise. Flowstone DWBAH from the Grand Bahamas Island records high sea stands at approximately 233, 215, 125 and 100 ka, with sea levels higher than −10 to −15 m, the depth of the sample below modern mean sea level. Lundberg and Ford (1994) improved the dating of this flowstone and Dutton et al. (2009) recalculated the TIMS U–Th ages based on a revised detrital thorium isotope ratio. Dutton et al. (2009) have investigated relative sea level within marine isotope stage 7 at Argentarola Cave, Italy, using several long-lived speleothems from different relative depths. The use of long-lived specimens, by implication, suggests that their growth is relatively insensitive to surface climate variability, and that growth hiatuses are related to sea level rise. A contrasting approach is that of Richards et al. (1994), who dated multiple speleothems from Grand Bahama and South Andros islands covering a range of sea level elevations from 0 to −60 m and the period 80–10 ka. Further development of this archive, including the analysis of the isotope record contained within speleothem proxies during the period of growth, should yield unique data on former sea-levels.

12.2 Insights into pre-Quaternary palaeoenvironments

12.2.1 High-resolution snapshots of pre-Quaternary palaeoenvironments

The recently developed ability to routinely analyse ancient speleothem samples by the U–Pb technique has led to a series of attempts to interpret the inorganic and organic proxies preserved in the samples. In many cases, the main research aim was the use of speleothem material to obtain a U–Pb age to date an existing stratigraphic section, or to constrain landscape evolution or uplift rates. For example, Polyak et al. (2008) U–Pb dated subaqueous 'mammillary' speleothems in an attempt to constrain the age and evolution of the Grand Canyon (see also comments by Pederson et al. (2008) and Pearthree et al. (2008)) and Woodhead et al. (2006) report U–Pb dates on speleothem material from the Italian Alps which can be used to constrain the regional uplift history. The analysis of proxy information contained within the speleothems is often a secondary objective, but one which provides uniquely dated insights into palaeoenvironments in pre-Quaternary periods. Important considerations over this timescale are a demonstration that the proxy records within the speleothem samples have remained unaltered, and whether a modern analogue approach is valid to allow the interpretation of annual variability in high-resolution proxy time series extracted from the samples.

Woodhead et al. (2010) reported what is to date the oldest U–Pb age on a stalagmite, of 289 ± 0.68 Ma from a site in Oklahoma, where Ordovician limestone contains vertebrate rich palaeokarst cave fills and speleothems. Woodhead et al. (2010) observed 5- to 30-μm-thick laminae which are matched by trace element variations (Fig. 12.5) which, by comparison with Modern and Late Quaternary analogues, would suggest that these features are

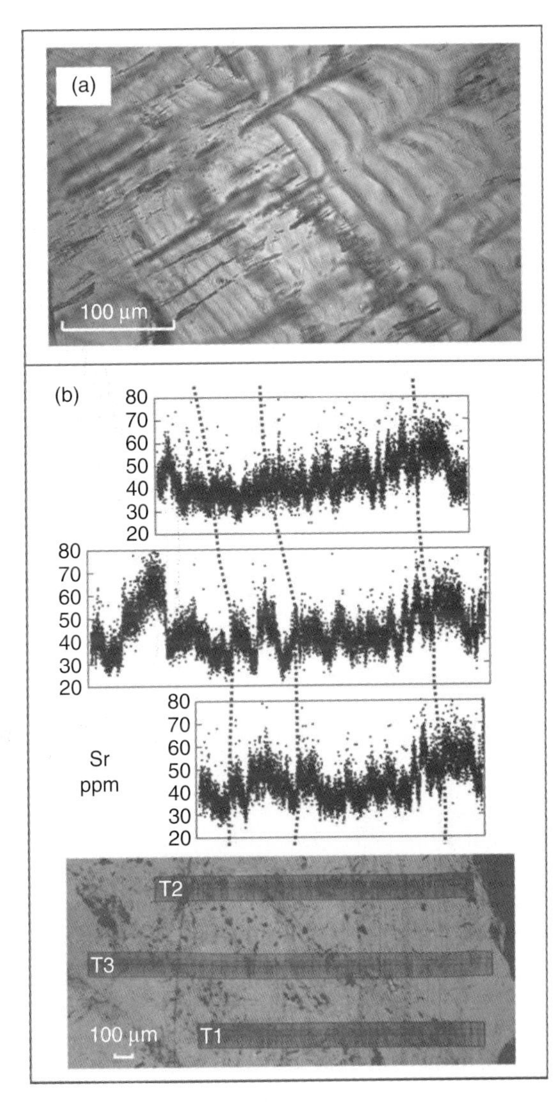

Fig. 12.5 Possible annual banding in a approximately 290 Ma old speleothem. (a) High-resolution image showing banding on scales of 5–30 μm. (b) Three laser traverses (T1–T3) all show consistent banding-scale Sr patterns, suggesting that this micrometre-scale variation is primary feature. From Woodhead et al. (2010).

annual. They also observe some secondary mobility of trace elements along fractures within the stalagmite, but in most cases the association between laminae and trace elements gives the authors confidence in interpreting $\delta^{18}O$, $\delta^{13}C$ and Ba, P, Mg and Sr variations in terms of palaeoclimate, with Ba and P correlate with $\delta^{18}O$. By analogy with the same observed correlation in Soreq Cave over the past 60kyr, these variations are interpreted as representing relatively moister and drier conditions. Meyer et al. (2009) used a modern analogue approach to investigate the palaeoenvironmental records contained in younger flowstone deposits. Sampled from approximately 2400m in the European Alps, the flowstones were dated to 2.019 + 0.037/−0.069 and 1.730 +0.032/−0.068 Ma (ALL1). A pollen sample extracted from WM1 contained Late Pliocene and Early Pleistocene pollen, matching the U–Pb date. Regular ultraviolet laminae were observed in the samples (Fig. 12.6), which were also brown in colour and contained plant macrofossils, and argued to be analogous to a seasonal climate (to generate ultraviolet laminae), and soil and vegetation in the catchment. Lamina measured at high resolution for a selected section of the flowstone demonstrated a growth rate similar to modern, as well as suggesting the presence of decadal and centennial-scale variability.

Both Woodhead et al. (2010) and Meyer et al. (2009) demonstrated the future possibilities of palaeoenvironmental interpretation of ancient speleothems. Although the annual nature of laminae in these samples cannot be demonstrated using radiometric means, comparison with trace element variations allows this to be demonstrated by analogy with contemporary cave processes. The problem then becomes one of resource; with the necessity to count annual lamina sequences to provide the chronology, as well as suitably high-resolution isotope or trace element analyses, to obtain climate records. Both Woodhead et al. (2010) and Meyer et al. (2009; see Fig. 12.6) analysed subsections of material to demonstrate the potential of high-resolution analysis of ancient samples). If sampled

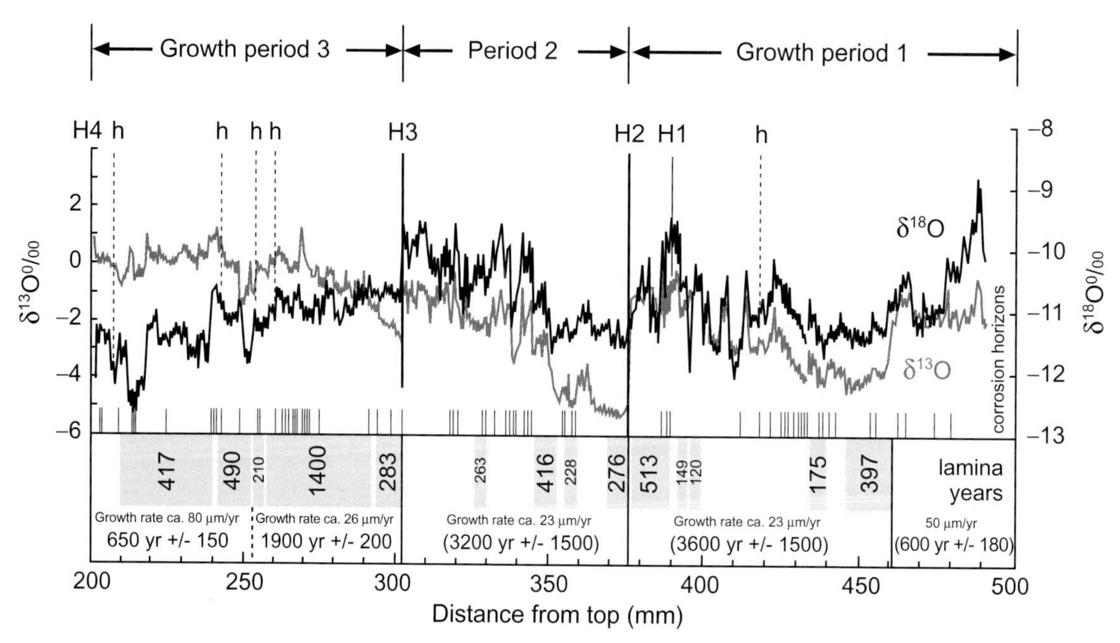

Fig. 12.6 Stable isotope and lamina count data from a 2 Ma flowstone WM-1 (redrawn from Meyer et al., 2009). Petrographic analysis demonstrates three growth periods, as well as several minor hiatuses (h) and corrosion horizons within the section of sample illustrated. Lamina counts on fluorescent laminae provide an indication of growth rate and duration and constrain the temporal variability in stable isotopes. For images of the fluorescent laminae, see Plate 7.5c.

at annual resolution, there is the possibility to undertake time series analysis to investigate climate variability in pre-Quaternary periods from decadal-scale variability up to the greater than 10^3 year timescale of Dansgaard–Oeschger cycles. These resource limitations are no different from those affecting researchers interested in Holocene or Late Quaternary climate change.

12.2.2 Dating archives of human evolution

The development of appropriate methodologies for the U–Pb dating of speleothems has led to a rush of analyses on samples associated with previously hard to date hominid remains of *Australopithicus*, a hominid genus that evolved in East Africa around 4 Ma, and which became extinct around 1.5 Ma. Various forms of australopithecids existed and it is widely considered that one evolved into the modern-day *Homo* genus. To date, researchers have focused on caves in the Sterkfontein region of South Africa and within the 'Cradle of Humankind World Heritage Site'. As in section 12.2.1, the primary aim of these studies is to constrain the timing of speleothem formation and therefore the associated sediment deposits. Pickering and Kramers (2010) reported new U–Pb dates on flowstones at Sterkfontein Caves, South Africa, a site containing the largest collection of *Australopithecus africanus* in the world. Here, the last appearance of *A. africanus* at 2.01 ± 0.05 Ma in Southern Africa was dated by U–Pb of associated flowstone. At Malapa Cave, in the same region, Dirks et al. (2010) reported U–Pb dates on a magnetically reversed flowstone of 2.026 ± 0.021 Ma and 2.024 ± 0.062 Ma (from different labs on a replicate sample), which underlies a magnetically normal sediment sequence containing *Australopithecus sediba*, therefore constrained to be 1.95–1.78 Ma. De Ruiter et al. (2009) reported U–Pb ages on speleothems from Cooper's Cave, again in the Sterkfontein region, with reported ages of 1.526 ± 0.088 Ma and approximately 1.4 Ma constraining deposits of *Australopithecus robustus*. De Ruiter et al. (2009) and Dirks et al. (2010) presented conventional models of cave formation, collapse and sediment entrapment, and Pickering and Kramers (2010) a discussion of sediment remobilization and inverse stratigraphy in ancient caves, all uniquely constrained by U–Pb analyses.

Although research to date within the Cradle of Humankind World Heritage Site has mostly focused on the use of speleothems as datable material for U–Pb (and palaeomagnetic) analyses, rather than as an archive of proxy climate information, an exception is the work of Hopley et al. (2007a, b), who considered the stable isotope composition of calcite and acid insoluble organic matter in two flowstones from the region. They undertook a relatively low-resolution (5 mm sample interval) $\delta^{18}O$ and $\delta^{13}C$ investigation on samples, whose ages are presently constrained by palaeomagnetic analysis on detrital grains within the flowstones and in the adjacent sediments. Spectral analysis of the $\delta^{18}O$ time series preserved in the flowstone from Buffalo Cave (approximate age 2.0–1.5 Ma) demonstrated significant peaks at 16 and 29 cm, the ratio of 1.8 between these was proposed by the authors to demonstrate an orbital forcing signature in $\delta^{18}O$ dominated by precessional cycles (22.8 and 19.8 kyr cycles relative to the 37.5 kyr cycles). The precessional cycles were sufficiently distinctive to allow orbital tuning, showing a remarkable fit to the relative intensities of precession in the interval 1.5–2 Ma calculated by Laskar et al. (2004). Current work uses lamina-counting to examine the cycles from within (Fig. 12.1). Hopley et al. (2007b) compared $\delta^{13}C$ signatures in this flowstone and one of approximate age 4–5 Ma from the nearby Collapsed Cone Cave. Unusually, Hopley et al. (2007b) analysed the $\delta^{13}C$ of acid insoluble organic matter within the two flowstones, which demonstrated the dominance of C_3 plants in the Collapsed Cone flowstone, whereas the Buffalo Cave flowstone had a mixture of C_3 and C_4 plants. The $\delta^{13}C$ of extracted OM was relatively sparsely sampled, but was used to calibrate the calcite $\delta^{13}C$ time series to a percentage C_4 grass and, (presuming constant dead carbon percentage, no change in degassing history, etc.), was inferred as evidence of a transition from C_3 to C_4 between approximately 4 and 2 Ma.

Hopley et al. (2007a, b) highlighted both the current problems and future potential for palaeoenvironmental and palaeoclimate reconstruction at

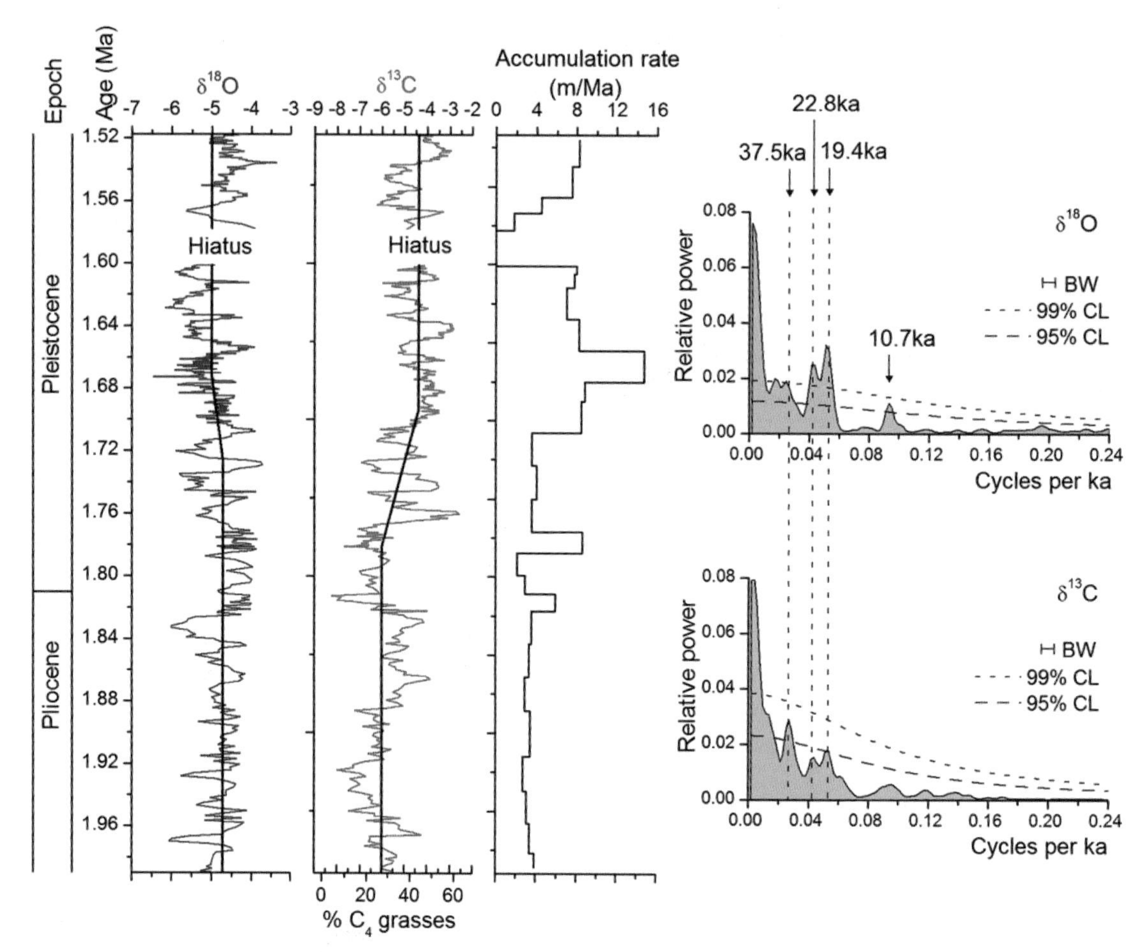

Fig. 12.7 Buffalo Cave flowstone $\delta^{18}O$ and $\delta^{13}C$ time series and associated spectra. The approximately 40 ka obliquity cycle is the dominant periodicity in the $\delta^{13}C$ record whereas the approximately 20 ka precessional cycles dominate the $\delta^{18}O$ record. The spectral peak for the $\delta^{18}O$ record labelled 10.7 ka represents the second harmonic of the precession cycles caused by non-sinusoidal oscillations. The percentage C_4 grasses values are determined using organic matter $\delta^{13}C$ values to correct for the host-rock carbon contribution to carbonate $\delta^{13}C$. The Buffalo Cave flowstone was formed in a mixed C_3 and C_4 plant environment in which the proportion of the two plant types was highly variable. The accumulation rate in metres per million years was determined from the orbital tuning. An inferred approximate doubling of accumulation rate occurs at 1.71 Ma, concurrent with an increase in the mean percentage of C_4 grasses. Thick lines on the $\delta^{18}O$ and $\delta^{13}C$ plots indicate the changing average values. BW, bandwidth; CL, confidence level. From Hopley et al. (2007a).

ancient sites where analyses are by necessity limited to flowstone samples with often fragmentary or discontinuous records. Difficulties include (1) the current focus on flowstones where differences in flow path could explain isotopic composition in $\delta^{13}C$ differences (section 8.5); (2) the problems of relatively low-resolution sampling, preventing the investigation of multi-decadal and multi-centennial-scale climate variability (see sections 11.3.3 and 12.2.1); (3) the difficulty in replicating either within or between sequences owing to their fragmented nature; (4) the need for U–Pb analyses on fragmented depositional units which cover many millions of years (see section 9.2.4).

12.3 Questions raised and looking to the future

In this book we have attempted to synthesize the current state-of-art in speleothem palaeoenvironmental research, with a focus on a process understanding of the relationship between surface climate and environments and the speleothem archive. It is the first research text in this field. Our final figure (Fig. 12.8) attempts to provide a historical overview, from which we can address the questions raised and attempt to look forward into speleothem research in the coming decades.

Figure 12.8 draws from the wider literature on the philosophy of science, in particular whether science advances through falsification of hypotheses and the cumulative accretion of knowledge (Popper, 1963), or by scientists working within paradigms (Kuhn, 1970), or research programmes (Lakatos, 1978). We consider that the fundamental advantage of working with speleothems is that they can be precisely radiometrically dated. Figure 12.8a demonstrates how it has been possible to investigate environmental and climate change over an increasing proportion of the geological record as the research discipline has developed over the past approximately 50 years. Initially, it was the ability to U–Th date speleothems that led to their use over the Late Quaternary, and Fig. 12.8b shows how U–Th dating has developed over time. Using a Kuhnian concept of paradigms (a set of practices that define a discipline at any one time) and 'paradigm shifts' (revolutionary advances in science that cause a new paradigm to come into existence), we suggest in Fig. 12.8b that the first paradigm shift in speleothem geochronology occurred around the start of the 1970s when the protocols for routine analysis using U–Th radiometric dating were first established. However, subsequent developments in U–Th analytical methods have led to a change in scientific practices, which in Kuhnian terms can therefore also be considered to have caused paradigm shifts, especially because they have facilitated the impact of speleothem research on the wider scientific community. More recently, other geochronological techniques (U–Pb, lamina counting) have also become significant. Because these complement the U–Th methods, they are more similar to the developments in Fig. 12.8c, where we consider the whole discipline of speleothem palaeoenvironmental analysis and the timing of major advances in our knowledge. Here, we argue that despite paradigm shifts in the field of geochronology, new ideas in the discipline as a whole have been developed incrementally without discarding the old ones completely. Where ruling ideas have been supplanted (e.g. the idea that $\delta^{18}O$ is a temperature proxy, that $\delta^{13}C$ reflects the C_4/C_3 plant ratio), they have not been completely falsified but instead been understood to still be possible, but as special cases. This accretion of understanding has had the effect of a gradual realization of the enormous number of potential proxies that exist. This lack of falsification does not reflect the inapplicability of logical rationalism in historical science as Popper (1963) believed, because the time-asymmetry of causation provides a basis for hypothesis testing analogous to experimental science (Cleland, 2001). Instead it reflects the site-specific nature of the way in which all but the most powerful forcings are transmitted to the speleothem, a complexity that becomes a strength in terms of the resources available for study.

A historical overview of the discipline also allows us to consider the development of the speleothem palaeoenvironmental research community. Here, we believe the major contribution has been the 'Climate Change: the Karst Record' conference series that was established in 1996 and which has enabled a research community to develop that can establish its own research programmes (*sensu* Lakatos, 1978). In particular, the first meetings determined the need to establish cave monitoring programmes to understand better the environmental signal constrained in speleothem proxies; later meetings championed the need for data archiving. The body of research knowledge and data accumulated over recent decades informs us that the combination of monitoring, developing long records and interdisciplinary approaches (i.e. interfacing with the different fields that are covered by this book) are vital. Although developments within the subject area continue to be very rapid, we suggest that this book marks a stage of maturity of the subject.

Given this historical overview, what can we speculate about the future?

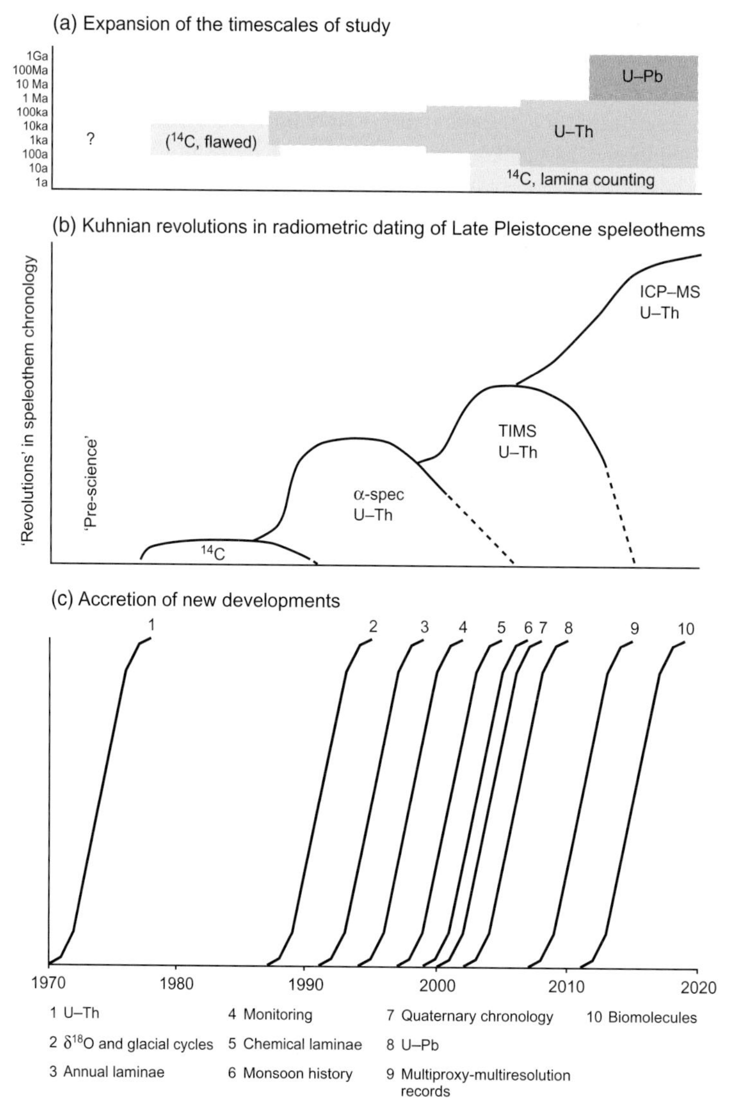

Fig. 12.8 (a) Expansion of the timescales of speleothem palaeoenvironmental research from approximately 1940 to present. Prior to the radiocarbon dating revolution of the 1950s, speleothems could not be dated and the discipline could be consider to be in a 'pre-science' stage. Radiocarbon analyses of speleothems was demonstrated to be flawed owing to the dead-carbon problem, although the use of the Modern atmospheric bomb ^{14}C signal is of use in constraining chronologies of the 20th century. (b) 'Revolutions' in U–Th dating of speleothems. After an initial revolution in the early 1970s, when U–Th dating made working on speleothem archives feasible, there have been paradigm shifts in technology which have allowed new kinds of scientific problems to be tackled. (c) The accretion of understanding in speleothem palaeoenvironmental research from the 1970s: demonstration of glacial cycles in $\delta^{18}O$ at Devils Hole; fluorescent and visible annual laminae; cave monitoring is championed through the 'Climate Change: the Karst Record' conference series ; the first demonstration of annual chemical laminae; demonstration of Late Quaternary monsoon histories in Chinese and Arabian stalagmite $\delta^{18}O$ records ; the use of $\delta^{13}C$, $\delta^{18}O$ and U–Th to constrain Quaternary ice-core chronologies; the development of routine U–Pb dating methods; development of multiproxy-multiresolution records; biomolecular palaeoenvironmental records are demonstrated.

1 With respect to U–Th dating, the technique now appears to have reached its limits in terms of the age range that can be covered, with the uncertainty of the ^{230}Th half-life the constraining factor preventing the determination of ages greater than 500 ka. Even without further developments in mass spectrometry, the diffusion of recent improvements in methodology (Cheng et al., 2009b) will set a standard for chronology that will facilitate many applications of speleothem work in Quaternary science.

2 Proxy records will be constructed with increasing temporal resolution. There is a need for high-resolution (annually resolved) archives of the Pleistocene, and earlier, to understand multi-decadal and multi-centennial climate variability. Historically, 'high resolution' has been used to describe speleothem records of varied temporal resolutions, but we suggest this should be reserved for records which are annual resolution or better.

3 Late Quaternary records such as those from Devils Hole, Soreq and Sofular have all demonstrated that exciting local-to-regional climate records can be obtained from speleothems (for example, Badertscher et al., 2011). But they also highlight the need to work with material from climate 'hotspots' or 'sweet-spots' if regional- to global-scale climate or environmental records are to be obtained.

4 The increased emphasis on cave monitoring programmes in the past decade has demonstrated that an understanding of the processes that potentially transform a proxy from the surface to the cave is invaluable, and essential when working on young samples. Future research will increasingly focus on the use of multi-proxy-multi-resolution records (where multiple proxies are used at a variety of resolutions from sub-annual to decadal or greater) in order to obtain the best possible palaeoenvironmental information. This use of multiple proxies at multiple resolutions will continue to be underpinned by varied modern process studies.

5 There is now a recognition, both within and outside the research community, that stalagmites are annually resolved archives which have the chronological precision and accuracy (using a combination of annual lamina counting and U–Th) to better understand climate variability over decadal-centennial timescales. In particular, speleothem palaeoenvironmental research can make huge contributions to our understanding of the last millennium or so of Earth's climate history, as there is a lack of continuous records which have the potential to preserve low frequency (decadal to centennial) climate variability.

6 We propose that significant advances in the use of speleothem biomarkers are possible. The fundamental ability to quantify several biomarkers is now established: future research will improve our understanding of the biomarker signal transfer from surface to cave, develop compound specific isotope analyses, and increase the biomarker toolbox to include proxies such as tetraether lipids and rDNA.

7 Speleothems have traditionally been viewed as being deposited in carbon and oxygen isotopic equilibrium. Increasingly, evidence from isotopologues and cave monitoring suggest that equilibrium deposition was more of a 'wish' than a reality. We suggest that future research will embrace disequilibrium deposition, understood to contain a quantifiable palaeoenvironmental signal, which can be determined using a cave monitoring and multi-proxy approach, including increased applications of fluid inclusion research.

8 We suggest a significant expansion of studies of pre-Quaternary material with the availability of U–Pb dating. Annually resolved records can be obtained using annual laminae, validated using a multiproxy approach. This allows the investigation of decadal to centennial-scale climate variability during periods of different palaeogeographies, opening up new research questions. For example, speleothems could complement marine records searching for identifying ocean–atmosphere interactions within pre-Quaternary palaeogeographies: were there NAO or precursor ENSO-like phenomena? Were there GIS-like events in previous ice ages? Gaining a deep understanding of the past widens our perspective and allows the present-day to be seen as a special case of the general behaviour of the Earth system. Such a broad perspective on what has been, and what might come to pass, is invaluable as our planet wends its way to its increasingly anthropogenically disturbed future.

APPENDIX 1
Archiving speleothems and speleothem data

Even a rudimentary understanding of speleothem-forming systems leads one to understand that they are vulnerable and that many speleothem samples, particularly stalagmites, are irreplaceable. Hence, at the present time, there is an urgent need for the community of scholars who study speleothems to adopt appropriate protocols for archiving speleothem samples and data, so that previous work is captured and future researchers can build on this work. This concern is part of the wider issue of minimizing destructive activity in caves and hence balancing scientific research with conservation (Frappier, 2008). Although archaeological investigations show that human disturbance and removal of speleothems has occurred since antiquity (e.g. Moyes, 2007), there has been a surge of specific targeted removal of material of palaeoclimatic interest from accessible sites since the 1960s. Currently, many of the original investigators are now retiring from their studies. In much of the world there are no specific legal restrictions on this activity, and the only protection for the deposits and the cave aesthetic by such sampling are the scruples of the scientists involved. On the other hand, in show caves and legally protected sites, a high degree of oversight of sampling is usually present, which in some cases restricts sampling to drilling of core material (methods for which are summarized by Spötl and Mattey (2012)) and/or removal from sites invisible to normal visitors, often coupled with requirements to restore the original visual appearance. Even in cases where the cave aesthetic damage is limited or temporary, the collected speleothems are scientifically a non-renewable resource. Most geologists have regarded speleothems simply as a type of rock sample and have been slow to grasp the necessity for conservation through archiving. If instead, one regards them as an archaeological resource, then the necessity for thorough documentation of the context and properties of the material is more obvious.

In the UK, good practice in archaeological archiving is summarized in Brown (2007) and has at its core the creation of a stable, ordered, accessible archive that is both documentary and material in nature. The recommendations go beyond what has been thought appropriate for speleothem studies, partly because of the complex teams that work at archaeological sites, but the salient features are still applicable. An overriding concern is that at the beginning of the investigation, the ultimate repository of the remaining speleothem and other samples, and the data derived from them, should be identified. Because it would be meaningless to deposit samples without information on their provenance and significance, attention is focused on the need to maintain documentation continually to accompany the samples. A tricky issue is that archaeological practice expects that written permission is given by the landowner (i.e. the owner at the time of the fieldwork) for the investigation and removal of material. Title is then normally formally

Speleothem Science: From Process to Past Environments, First Edition. Ian J. Fairchild, Andy Baker.

transferred to the receiving museum who require this to be clear in order to loan material to researchers in the future. In speleothem studies, two differences from archaeological studies are (i) it is unusual for formal written permission to have been obtained for sampling, and (ii) future researchers are much more likely than archaeologists to need to re-sample specimens destructively. For this reason, it is pragmatic to suggest that the final repository of collections should be, as at present, in university or research institute collections where specialist expertise exists, rather than necessarily in museums.

Recommendations on good practice are as follows.

1 For a given research collection, an electronic catalogue of samples should be maintained. This should summarize as a minimum the following: sample number/code, geographic coordinates of cave entrance, general location in cave system, nature of material and sampling (e.g. previously broken stalagmite, cored flowstone, etc.), date of collection and collectors, approximate age range of sample, cross-reference to archives of field data, laboratory studies and publications. In many countries, there is a formal (cadastral) numbering system for locations. If this catalogue is advertised on the World Wide Web, the chances of optimal collaborations and site conservation are maximized, with the expectation that the original collectors will have priority to complete their programme of study and maintain discretion over access.

2 A documentary archive should be created and maintained, including field notes and organized data on field relationships (e.g. cave plan showing exact surveyed position, stratigraphic/spatial relationships, drip hydrology and hydrogeochemistry, etc.), and data and meta-data to accompany the speleothem samples. The optimum specification for hard-copy archives to be useful in the long term is rigorous (Brown, 2007): the archive should be indexed, free from metal fastenings and self-adhesive labels, no folded pages, stored horizontally in acid-free cardboard boxes at a fixed temperature in the range 13–19 °C at a fixed humidity in the range 45–60%. Electronic data files, digital images including scans of speleothem samples at different stages of processing, and scanned copies of all documents should be stored and backed up at a different location. Given the frailty of all electronic media, there needs to be a copy on a permanent server that is regularly backed-up. A recent development is that some national funding agencies now make arrangements for data archiving; this is an extension of good practice from other related fields such as marine science where the necessity for data archiving has long been recognized.

3 The material archive should be stored in appropriate conditions for geological materials (for example as set out in Brunton et al. (1985) and Museums & Galleries Commission (1993)). Archaeological collections can include (i) bulk finds of common material, (ii) sensitive (high-value finds) and (iii) by-products of scientific sub-sampling. The equivalents in speleothem science might be (i) reworked, small or poorly preserved material that has no immediate value, (ii) archived portions (e.g. halves for speleothems) of samples (iii) labelled remains of sampled portions which, together with scans of specimens, allow the exact position of subsamples to be determined. Robust bulk finds can be stored together in boxes, but valuable material ought to be packed individually supported by inert plastic foam (e.g. Plastazote®) to prevent movement. Labelled multiple sub-samples are better unwrapped, but secured, for example in cut-outs in foam.

4 Responsibility for the archiving rests with the lead principal investigator, with aspects delegated to co-investigators, contract project scientists or technical staff where appropriate. If the final repository is a museum, assistance with some aspects of curation and archiving may be available.

5 Good practice in scientific research is that numerical data should be available to the community once presented in a publication. Some funding agencies go further in terms of placing data in an archive once organized and verified as analytically correct. Irrespective of the state of national data centres, in the case of palaeoclimate data there is an obvious repository for data from across the world, run by the United States National Climatic Data Center at http://www.ncdc.noaa.gov/paleo/paleo.html, from which data were used to produce several diagrams in this book. The speleothem section data are listed under investigator and geographic region. A key issue is the provision of adequate metadata to accompany the proxy time series data, including cave plans. It should be borne in mind that the data could well be accessed by users who are not particularly familiar with speleothem records, so a clear indication of uncertainties involved in their use should be provided. If the data are lodged at the time of publication, the latter will contain information by which the user can judge issues related to the quality of the age model or factors affecting the proxy variable.

An example of a country in which action is being taken is Switzerland. Firstly a meta-database, then a physical archive for cave samples of all types is being constructed, under the auspices of the speleological commission of the Swiss Academy of Sciences (M. Luetscher, personal communication, 2010).

References

Adamczyk, K., Prémont-Schwarz, M., Pines, D. et al. (2009) Real-time observation of carbonic acid formation in aqueous solution. *Science*, **326**, 1690–1694.

Adkins, J., Boyle, E., Curry, W. & Lutringer, A. (2003) Stable isotopes in deep-sea corals and a new mechanism for 'vital effects'. *Geochimica Cosmochimica Acta*, **67**, 1129–1143.

Aeschbach-Hertig, W., Peeters, F., Beyerle, U. & Kipfer, R. (2000) Palaeotemperature reconstruction from noble gases in ground water taking into account equilibration with entrapped air. *Nature*, **405**, 1040–1044.

Affek, H.P., Bar-Matthews, M., Ayalon, A. et al. (2008) Glacial/interglacial temperature variations in Soreq cave speleothems as recorded by 'clumped isotope' thermometry. *Geochimica Cosmochimica Acta*, **72**, 5351–5360.

Aggarwal, P.K., Fröhlich, K., Kulkarni, K.M. & Gourcy, L.L. (2004) Stable isotope evidence for moisture sources in the Asian summer monsoon under present and past climate regimes. *Geophysical Research Letters*, **31**, doi:10.1029/2004GL019911.

Alexandratos, V.G., Elzinga, E.J. & Reeder, R.J. (2007) Arsenate uptake by calcite: macroscopic and spectroscopic characterization of adsorption and incorporation mechanisms. *Geochimica Cosmochimica Acta*, **71**, 4172–4187.

Allen, J.R.L. (1985) *Principles of Physical Sedimentology*. London: George Allen & Unwin.

Allen, J.R.M., Brandt, U., Brauer, A. et al. (1999) Rapid environmental changes in southern Europe during the last glacial period. *Nature*, **400**, 740–743.

Alley, R.B. & Cuffey, K.M. (2001) Oxygen- and hydrogen-isotopic ratios of water in precipitation: beyond paleothermometry. *Reviews in Mineralogy and Geochemistry*, **43**, 527–553.

Alley, R.B. & Agustdottir, A.M. (2005) The 8k event: cause and consequences of a major Holocene abrupt climate change. *Quaternary Science Reviews*, **24**, 1123–1149.

Alley, R.B., Mayewski, P.A., Sowers, T. et al. (1997) Holocene climate instability: a prominent, widespread event 8200 yr ago. *Geology*, **25**, 483–486.

Allison, V.C. (1923) The growth of stalagmites and stalactites. *Journal of Geology*, **31**, 106–125.

Almogi-Levin, A., Bar-Matthews, M., Shriki, D., et al. (2009) Climatic variability during the last ~90 ka of the southern and northern Levantine Basin as evident from marine records and speleothems. *Quaternary Science Reviews*, **28**, 2882–2896.

Alvarez-Puebla, R.A., Valenzuela-Calahorro, C. & Garrido, J.J. (2006) Theoretical study on fulvic acid structure, conformation and aggregation. A molecular modelling approach. *Science of the Total Environment*, **358**, 243–254.

Amar, T. & de Freitas, C.R. (2005) Microclimate of a mid-latitude single entrance tourist cave. In: *Cave Management in Australia XVI. Proceedings of the Sixteenth Australian Conference on Cave and Karst Management*, 10–17 April 2005, Westport, New Zealand.

Ammann, C.M., Genton, M.G. & Li, B. (2010) Technical Note: correcting for signal attenuation from noise: sharpening the focus on past climate. *Climates of the Past*, **6**, 273–279.

Anderson, M.P. (2005) Heat as a ground water tracer. *Ground Water*, **43**, 951–968.

Anderson, S.P. (2007) Biogeochemistry of glacial landscape systems. *Annual Review of Earth and Planetary Sciences*, **35**, 375–399.

Ando, A., Kawahata, H. & Kakegawa, T. (2006) Sr/Ca ratios as indicators of varying modes of pelagic carbonate diagenesis in the ooze, chalk and limestone realms. *Sedimentary Geology*, **191**, 37–53.

Andreo, B., Carrasco, F. & Sanz de Galdeano, C. (1997) Types of carbonate aquifers according to the fracturation and the karstification in a southern Spanish area. *Environmental Geology*, **30**, 163–173.

Speleothem Science: From Process to Past Environments, First Edition. Ian J. Fairchild, Andy Baker.

Andrews, J.E. (2006) Palaeoclimatic records from stable isotopes in riverine tufas: synthesis and review. *Earth-Science Reviews*, **75**, 85–104.

Andrieux, C. (1969) Étude du climat de la grotte de Sainte-Catherine en Ariège selon le cycle 1967. *Annales de Spéléologie*, **24**, 19–74.

Angelova, D., Belfoul, M.A., Bouzid, S. et al. (2003) Paleoseismic phenomena in karst terrains in Bulgaria and Morocco. *Acta Carsologica*, **31**, 101–120.

Angert, A., Lee, J.-E. & Yakir, D. (2008) Seasonal variations in the isotopic composition of near-surface water vapour in the eastern Mediterranean. *Tellus*, **60B**, 674–684.

Anthony, D.M. & Granger, D.E. (2006) Five million years of Appalachian landscape evolution preserved in cave sediments. In: Harmon, R.S. & Wicks, C. (eds.) *Perspectives on Karst Geomorphology, Hydrology and Geochemistry—A Tribute Volume to Derek C. Ford and William B. White. Geological Society of America Special Paper* 404, 39–50.

Antonioli, F., Bard, E., Potter, E.-M. et al. (2004) 215-ka history of sea-level oscillations from marine and continental layers in Argentarola Cave speleothems (Italy). *Global and Planetary Change*, **43**, 57–78.

Appelo, C.A.J. & Postma, D. (2005) *Geochemistry, Groundwater and Pollution (2nd edition)*. A.A. Balkema, Leiden.

Aquilano, D., Rubbo, M., Catti, M. & Pavese, A. (1997) Theoretical equilibrium and growth morphology of $CaCO_3$ polymorphs. I. Aragonite. *Journal of Crystal Growth*, **182**, 168–184.

Aquilano, D., Coasta, E., Genovese, A. et al. (2003) Hollow rhombohedral calcite crystals encompassing CO_2 microcavities nucleated in solution. *Journal of Crystal Growth*, **247**, 516–522.

Araguas-Araguas, L., Froehlich, K. & Rozanski, K. (2000) Deuterium and oxygen-18 isotope composition of precipitation and atmospheric moisture. *Hydrological Processes*, **14**, 1341–1355.

Arnold, J., Clauser, C., Pechnig, R. et al. (2006) Porosity and permeability from mobile NMR core-scanning. *Petrophysics*, **47**, 306–314.

Arvidson, R.S., Ertan, I.E., Amonette, J.E. & Luttge, A. (2003) Variation in calcite dissolution rates: a fundamental problem? *Geochimica et Cosmochimica Acta*, **67**, 1623–1634.

Arvdison, R.S., Mackenzie, F.T. & Guidry, M.W. (2006) MAGic: a Phanerozoic model for the geochemical cycling of major rock-forming components. *American Journal of Science*, **306**, 135–190.

Asrat, A., Baker, A., Umer, M.M. et al. (2007) A high-resolution multi-proxy stalagmite record from Mechara, Southeastern Ethiopia: palaeohydrological implications for speleothem palaeoclimate reconstruction. *Journal of Quaternary Science*, **22**, 53–63.

Asrat, A., Baker, A., Leng, M.J. et al. (2008) Environmental monitoring in the Mechara caves, southeastern Ethiopia: implications for speleothem palaeoclimate studies. *International Journal of Speleology*, **37**, 207–220.

Astilleros, J.M., Pina, C.M., Fernández-Díaz, L. et al. (2006) Nanoscale phenomena during the growth of solid solutions on calcite $\{10\bar{1}4\}$ surfaces. *Chemical Geology*, **225**, 322–335.

Astilleros, J.M., Fernandez-Diaz, L. & Putnis, A. (2010) The role of magnesium in the growth of calcite: an AFM study. *Chemical Geology*, **271**, 52–58.

Atkinson, T.C. (1977a) Carbon dioxide in the atmosphere of the unsaturated zone: an important control of groundwater hardness in limestones. *Journal of Hydrology*, **35**, 111–123.

Atkinson, T.C. (1977b) Diffuse flow and conduit flow in limestone terrain in Mendip Hills, Somerset. *Journal of Hydrology*, **35**, 93–110.

Atkinson, T.C. (1983) Growth mechanisms of speleothems in Castleguard Cave, Columbian Icefields, Alberta, Canada. *Arctic and Alpine Research*, **15**, 523–536.

Atkinson, T.C., Smart, P.L. & Wigley, T.M.L. (1983) Climate and natural radon levels in Castleguard Cave, Columbia Icefields, Alberta, Canada. *Arctic and Alpine Research*, **15**, 487–502.

Atteia, O., Perret, D., Adatte, T. et al. (1998) Evolution of size distributions of natural particles during aggregation: modelling versus field results. *Colloids and Surfaces A: Physicochemical and Engineering Aspects*, **139**, 171–188.

Audra, P., Bini, A., Gabrovšek, F. et al. (2006) Cave genesis in the Alps between the Miocene and today: a review. *Zeitschrift für Geomorphologie N.F.*, **50**, 153–176.

Australian Bureau of Meteorology (2010) The Greenhouse Effect and Climate Change. http://www.bom.gov.au/info/GreenhouseEffectAndClimateChange.pdf. Accessed 6/2010.

Ayalon, A., Bar-Matthews, M. & Sass, E (1998) Rainfall-recharge relationships within a karstic terrain in the Eastern Mediterranean semi-arid region, Israel: $\delta^{18}O$ and δD characteristics. *Journal of Hydrology*, **207**, 18–31.

Ayalon, A., Bar-Matthews, M. & Kaufman, A. (1999) Petrography, strontium, barium and uranium concentrations, and strontium and uranium isotope ratios in speleothems as palaeoclimatic proxies: Soreq Cave, Israel. *The Holocene*, **9**, 715–722.

Ayalon, A., Bar-Matthews, M. & Kaufman, A. (2002) Climatic conditions during marine oxygen isotope stage 6 in the eastern Mediterranean region from the isotopic composition of speleothems of Soreq Cave, Israel. *Geology*, **30**, 303–306.

Ayele, A. & Kulhánek, O. (2000) Reassessment of source parameters for the major earthquakes in the East African rift from historical seismograms and bulletins. *Annali di Geofisica*, **43**, 81–94.

Ayora, C., Taberner, C., Saaltink, M.W. & Carrera, J. (1998) The genesis of dedolomites: a discussion based on reactive transport modeling. *Journal of Hydrology*, **209**, 346–365.

Babić, L., Lacković, D. & Horvatinčić, N. (1996) Meteoric phreatic speleothems and the development of cave stratigraphy: an example from Tounj Cave, Dinarides, Croatia. *Quaternary Science Reviews*, **15**, 1013–1022.

Baceta, J.J., Wright, V.P., Beavinton-Penney, S.J. & Pujalte, V. (2007) Palaeohydrogeological control of palaeokarst macro-porosity genesis during a major sea-level low-stand: Danian of the Urbasa-Andia plateau, Navarra, North Spain. *Sedimentary Geology*, **199**, 141–169.

Back, W. & Hanshaw, B.B. (1970) Comparison of chemical hydrogeology of the carbonate peninsulas of Florida and Yucatan. *Journal of Hydrology*, **10**, 330–368.

Badertscher, S., Fleitmann, D., Cheng, H., Edwards, R.L. et al. (2011) Pleistocene water intrusions from the Mediterranean and Caspian seas into the Black Sea. *Nature Geoscience*, **4**, 236–239.

Badino, G. (1995) Fisica del Clima Sotterraneo. *Memorie dell'Istituo Italiano di Speleologia*, **7**, serie II, Bologna.

Badino, G. (2005) Underground drainage systems and geothermal flux. *Acta Carsologica*, **34**, 277–316.

Badino, G. (2009) The legend of carbon dioxide heaviness. *Journal of Cave and Karst Studies*, **71**, 100–107.

Bahr, A., Lamy, F., Arz, H.W. et al. (2008) Abrupt changes of temperature and water chemistry in the late Pleistocene and early Holocene Black Sea. *Geochemistry Geophysics Geosystems*, Q01004 doi:10.1029/2007GC001683.

Baker, A. (2001) Fluorescence excitation-emission matrix characterisation of some sewage impacted rivers. *Environmental Science and Technology*, **35**, 948–953.

Baker, A. & Smart, P.L. (1995) Recent flowstone growth rates: field measurements in comparison to theoretical predictions. *Chemical Geology*, **122**, 121–128.

Baker, A. & Barnes, W.L. (1998) Comparison of the Luminescence Properties of Waters Depositing Flowstone and Stalagmites at Lower Cave, Bristol. *Hydrological Processes*, **9**, 1447–1459.

Baker, A. & Genty, D. (1999) Fluorescence wavelength and intensity variations of cave waters. *Journal of Hydrology*, **217**, 19–34.

Baker, A. & Brunsdon, C. (2003) Non-linearities in drip water hydrology: an example from Stump Cross Caverns, Yorkshire. *Journal of Hydrology*, **277**, 151–163.

Baker, A. & Spencer, R.G.M. (2004) Characterization of dissolved organic matter from source to sea using fluorescence and absorbance spectroscopy. *Science of the Total Environment*, **333**, 217–232.

Baker, A. & Bradley, C. (2010) Modern stalagmite $\delta^{18}O$: instrumental calibration and forward modelling. *Global and Planetary Change*, **71**, 201–206.

Baker, A., Smart, P.L., Edwards, R.L. & Richards, D.A. (1993) Annual banding in a cave stalagmite. *Nature*, **364**, 518–520.

Baker, A., Barnes, W.L. & Smart, P.L. (1996) Speleothem luminescence intensity and spectral characteristics: signal calibration and a record of palaeovegetation change. *Chemical Geology*, **130**, 65–76.

Baker, A., Ito, W., Smart, P.L. & McEwan, R.F. (1997a) Elevated and variable values of ^{13}C in speleothems in a British cave system. *Chemical Geology*, **136**, 263–270.

Baker, A., Barnes, W.L. & Smart, P.L. (1997b) Stalagmite drip discharge and organic matter fluxes in Lower Cave, Bristol. *Hydrological Processes*, **11**, 1541–1555.

Baker, A., Caseldine, C.J., Hatton, J. et al. (1997c) A Cromerian complex stalagmite from the Mendip Hills, England. *Journal of Quaternary Science*, **12**, 533–537.

Baker, A., Genty, D., Dreybrodt, W. et al. (1998a) Testing theoretically predicted stalagmite growth rate with Recent annually laminated samples: implications for past stalagmite deposition. *Geochimica et Cosmochimica Acta*, **62**, 393–404.

Baker, A., Genty, D. & Smart, P.L. (1998b) High-resolution records of soil humification and palaeoclimate change from speleothem luminescence excitation-emission wavelength variations. *Geology*, **26**, 903–906.

Baker, A., Proctor, C.J., & Barnes, W.L. (1999) Variations in stalagmite luminescence laminae structure at Poole's Cavern, England, AD1910 to AD 1996: calibration of a palaeoprecipitation proxy. *The Holocene*, **9**, 683–688.

Baker, A., Genty, D. & Fairchild, I.J. (2000a) Hydrological characterisation of stalagmite dripwaters at Grotte de Villars, Dordogne, by the analysis of inorganic species and luminescent organic matter. *Hydrology and Earth System Sciences*, **4**, 439–449.

Baker, A., Bolton, L., Brunsdon, C., Charlton, M. & McDermott, F. (2000b) Visualisation of luminescence excitation-emission timeseries: palaeoclimate implications from a 10,000 year stalagmite record from Ireland. *Geophysical Research Letters*, **27**, 2145–2148.

Baker, A., Asrat, A., Fairchild, I.J. et al. (2007) Analysis of the climate signal contained within $\delta^{18}O$ and growth rate parameters in two Ethiopian stalagmites. *Geochimica et Cosmochimica Acta*, **71**, 2975–2988.

Baker, A., Smith, C.L., Jex, C. et al. (2008a) Annually laminated speleothems: a review. *International Journal of Speleology*, **37**, 193–206.

Baker, A., Tipping, E., Thacker, S.A. & Gondar, D. (2008b) Relating dissolved organic matter fluorescence and functional properties. *Chemosphere*, **73**, 1765–1772.

Baker, A., Asrat, A., Fairchild, I.J. et al. (2010) Decadal-scale rainfall variability in Ethiopia recorded in an annually laminated, Holocene-age, stalagmite. *The Holocene*, **20**, 827–836.

Baker, A., Wilson, R., Fairchild, I.J. et al. (2011) High resolution $\delta^{18}O$ and $\delta^{13}C$ records climate from an annually laminated Scottish stalagmite and relationship with last millennium climate. *Global and Planetary Change*, doi: 10.1016/j.gloplacha.2010.12.007.

Baker, R.G., Betts, E.A., Schwert, D.P. et al. (1996) Holocene paleoenvironments of northeast Iowa. *Ecological Monographs*, **66**, 203–234.

Baker, R.G., Bettis, E.A, Denniston, R.F. et al. (2002) Holocene paleoenvironments in southeastern Minnesota —chasing the prairie-forest ecotone. *Palaeogeography, Palaeoclimatology, Paleoecology*, **177**, 103–122.

Baldini, J. (2001) Morphologic and dimensional linkage between recently deposited speleothems and drip water from Browns Folly Mine, Wiltshire, England. *Journal of Cave and Karst Studies*, **63**, 83–90.

Baldini, J.U.L. (2010) Cave atmosphere controls on stalagmite growth rate and palaeoclimate records. In: Pedley, H.M. & Rogerson, M. (eds.) *Tufas and Speleothems: Unravelling the Microbial and Physical Controls. Geological Society Special Publication* **336**, 283–294.

Baldini, J.U.L., McDermott, F. & Fairchild, I.J. (2002) Structure of the 8200–year cold event revealed by a speleothem trace element record. *Science*, **296**, 2203–2206.

Baldini, J.U.L., McDermott, F., Baker, A. et al. (2005) Biomass effects on stalagmite growth and isotope ratios: a 20th century analogue from Wiltshire, England. *Earth and Planetary Science Letters*, **240**, 486–494.

Baldini, J.U.L., Baldini, L.M., McDermott, F. & Clipson, N. (2006a) Carbon dioxide sources, sinks, and spatial variability in shallow temperate zone caves: evidence from Ballynamintra Cave, Ireland. *Journal of Cave and Karst Studies*, **68**, 4–11.

Baldini, J.U.L., McDermott, F. & Fairchild, I.J. (2006b) Spatial variability in cave drip water hydrochemistry: implications for stalagmite paleoclimate records. *Chemical Geology*, **235**, 390–404.

Baldini, J.U.L., McDermott, F., Hoffmann, D.K., Richards, D.A. & Clipson, N. (2008) Very high-frequency and seasonal cave atmosphere PCO_2 variability: implications for stalagmite growth and oxygen isotope-based paleoclimate records. *Earth and Planetary Science Letters*, **272**, 118–129.

Baldini, L.M., McDermott, F., Foley, A.M. & Baldini J.U.L. (2008) Spatial variability in the European winter precipitation $\delta^{18}O$-NAO relationship: implications for reconstructing NAO-mode climate variability in the Holocene. *Geophysical Research Letters* **35**, doi:10.1029/2007GL032027.

Baldini, L.M., McDermott, F., Baldini, J.U.L. et al. (2010) The role of atmospheric circulation, air mass trajectory, and moisture source region in determining Irish precipitation $\delta^{18}O$ values. *Climate Dynamics*, **35**, 977–993.

Ball, T.K., Cameron, D.G., Colman, T.B. & Roberts, P.D. (1991) Behaviour of radon in the geological environment—a review. *Quarterly Journal of Engineering Geology*, **21**, 161–182.

Banks, E.D., Taylor, N.M., Gulley, J. et al. (2010) Bacterial calcium carbonate precipitation in cave environments: a function of calcium homeostasis. *Geomicrobiological Journal*, **27**, 444–454.

Banner, J.L. (2004) Radiogenic isotopes: systematics and applications to earth surface processes and chemical stratigraphy. *Earth-Science Reviews*, **65**, 141–194.

Banner, J.L., Musgrove, M.L., Asmerom, Y. et al. (1996) High-resolution temporal record of Holocene groundwater chemistry: tracing links between climate and hydrology. *Geology*, **24**, 1049–1053.

Banner, J.L., Guilfoyle, A., James, E.W. et al. (2007) Seasonal variations in modern speleothem calcite growth in central Texas, U.S.A. *Journal of Sedimentary Research*, **77**, 615–622.

Bar-Matthews, M., Matthews, A. & Ayalon, A. (1991) Environmental controls of speleothem mineralogy in a karstic dolomitic terrain (Soreq Cave, Israel). *Journal of Geology*, **99**, 187–207.

Bar-Matthews, M., Ayalon, A., Matthews, A., Sass, E. & Halicz, L. (1996) Carbon and oxygen isotopic study of the active water-carbonate system in a karstic Mediterranean cave: implications for palaeoclimate research in semiarid regions. *Geochimica et Cosmochimica Acta*, **60**, 337–347.

Bar-Matthews, M., Ayalon, A. & Kaufman, A. (1997) Late Quaternary paleoclimate in the Eastern Mediterranean region from stable isotope analysis of speleothems at Soreq Cave, Israel. *Quaternary Research*, **47**, 155–168.

Bar-Matthews, M., Ayalon, A. & Kaufman, A. (1998) Middle to late Holocene (6500 Yr. period) paleoclimate in the Eastern Mediterranean regions from stable isotope composition of speleothems from Soreq Cave, Israel. In Isaar, A.S., Brown, N. (Eds.), *Water, Climate and Society in Time of Climate Change*. Kluwer, Dordrecht, Netherlands, pp. 203–214.

Bar-Matthews, M., Ayalon, A., Kaufman, A. & Wasserburg, G.J. (1999) The Eastern Mediterranean paleoclimate as a reflection of regional events: Soreq Cave, Israel. *Earth and Planetary Science Letters*, **166**, 85–95.

Bar-Matthews, M., Ayalon, A. & Kaufman, A. (2000) Timing and hydrological conditions of Sapropel events in the Eastern Mediterranean, as evident from speleothems, Soreq cave, Israel. *Chemical Geology*, **169**, 145–156.

Bar-Matthews, M., Ayalon, A, Gilmour, M., et al. (2003) Sea-land oxygen isotopic relationships from planktonic foraminifera and speleothems in the Eastern Mediterranean region and their implications for paleorainfall during interglacial intervals. *Geochimica et Cosmochimica Acta*, **67**, 3181–3199.

Bar-Matthews, M., Marean, C.W., Jacobs, Z. et al. (2010) A high resolution and continuous isotopic speleothem record of palaeoclimate and paleoenvironment from 90 to 53 ka from Pinnacle Point on the south coast of South Africa. *Quaternary Science Reviews*, **29**, 2131–2145.

Barrows, T.T., Lehman, S.J., Fifield, L.K. & De Deckker, P. (2007) Absence of cooling in New Zealand and the adjacent ocean during the Younger Dryas chronozone. *Science*, **318**, 86–88.

Barry, R.G. & Chorley, R.J. (2009) *Atmosphere, Weather and Climate*. Routledge, London.

Barton, H.A. (2006) An introduction to cave microbiology: a review for the non-specialist. *Journal of Cave and Karst Studies*, **68**, 43–54.

Barton, H.A., Spear, J.R. & Pace, N.R. (2001) Microbial life in the underworld: biogenicity in secondary mineral formations. *Geomicrobiology Journal*, **18**, 359–368.

Barton, H.A. & Northrup, D.E. (2007) Geomicrobiology in cave environments: past, current and future perspectives. *Journal of Cave and Karst Studies*, **69**, 163–178.

Baskaran, M. & Iliffe, T.M. (1993) Age determination of recent cave deposits using excess ^{210}Pb—A new technique, *Geophysical Research Letters*, **20**, 603–606.

Bastin, B. (1978) L'analyse pollinique des stalagmites: une nouvelle possibilité d'approach des fluctuations climatiques du Quaternaire. *Annales Société Géologique de Belgique*, **101**, 13–19.

Bastin, B. (1990) L'analyses pollinique des concretions stalagmitiques: méthodologies et resultants en provenance des grottes Belges. *Karstologica Mémoires*, **2**, 3–10.

Bastin, B. & Gewelt, M. (1986) Analyses pollinique et datation14C de concretions stalagmitiques Holocènes: apports complementaires des deux methods. *Geographie Physique et Quaternaire*, **40**, 185–196.

Bastin, B., Dupuis, C. & Quinif, Y. (1982) Étude microstratigraphique et palynologique d'une croûte stalagmitique de la Grotte de la Vilaine Source (Arbe Belgique): méthodologie et resultants. *Revue Belge de Geographie*, **106**, 109–120.

Bastin, B., Cordy, J.-M., Gewelt, M. & Otte, M. (1986) Fluctuations climatique enregistrées depuis 125,000 ans dans les couches de remplissage de la Grotte Scandina (Provence de Namur, Belgique). *Bulletin de l'Association Française pour l'etude du Quaternaire*, 1/2, 168–177.

Bastin, B., Quinif, Y., Dupuis, C. & Gascoyne, M. (1988) La sequence sedimentaire de la Grotte de Bohon (Belgique). *Annales de la Société Géologique de Belgique*, **111**, 51–60.

Battarbee, R. & Binney, H. (eds.) (2008) *Natural Climate Variability and Global Warming: A Holocene Perspective*. Blackwell, Oxford.

Beck, J.W., Richards, D.A., Edwards, R.L et al. (2001) Extremely large variations of atmospheric C-14 concentration during the last glacial period. *Science*, **292**, 2453–2458.

Beck, W., Grossman, E. & Morse, J. (2005) Experimental studies of oxygen isotope fractionation in the carbonic acid system at 15 °C, 25 °C and 40 °C. *Geochimica et Cosmochimica Acta*, **69**, 3493–3503.

Becker, A., Ferry, M., Monecke, K. et al. (2005) Multiarchive paleoseismic record of late Pleistocene and Holocene strong earthquakes in Switzerland. *Tectonophysics*, **400**, 153–177.

Becker, A., Davenport, C., Eichenberger, U. et al. (2006) Speleoseismology: a critical perspective. *Journal of Seismology*, **10**, 371–388.

Beltrami, H. (2001) Surface heat flux histories from inversion of geothermal data: energy balance at the Earth's surface. *Journal of Geophysical Research*, **106**(B10), 21979–21993.

Beltrami, H. & Kellman, L. (2003) An examination of short- and long-term air-ground temperature coupling. *Global and Planetary Change*, **38**, 291–303.

Benavente, J., Vadillo, I., Carrasco, F. et al. (2010) Air carbon dioxide contents in the vadose zone of a Mediterranean karst. *Vadose Zone Journal*, **9**, 126–136.

Benedetti, M.F., Van Riemsdijk, W.H., Koopal, L.K. et al. (1996) Metal ion binding by natural organic matter: from the model to the field. *Geochimica et Cosmochimica Acta*, **60**, 2503–2513.

Bennett, R.J. & Chorley, R.J. (1978) *Environmental Systems: Philosophy, Analysis and Control*. London: Methuen.

Bernal, J.P., Lachniet, M., McCulloch, M. et al. (2011) A speleothem record of Holocene climate variability from southwestern Mexico. *Quaternary Research*, **75**, 104–113.

Berner, R.A. (1980) *Early Diagenesis. A Theoretical Approach*. Princeton University Press, New Jersey.

Berner, R.A. (2004) *The Phanerozoic Carbon Cycle: CO_2 and O_2*. Oxford: Oxford University Press.

Berner, R.A., Lasaga, A.C. & Garrels, R.M. (1983) The carbonate-silicate geochemical cycle and its effects on atmospheric carbon-dioxide over the past 100 million years. *American Journal of Science*, **283**, 641–683.

Berstad, I. (1998) *Uranseriedatering og stabilisotopanalyse av speleothemer fra Søylegrotta, Mo i Rana*. Cand. scient. thesis, University of Bergen, Bergen, 122 pp.

Bertalanffy, L. von (1950) An outline of general system theory. *British Journal of Philosophical Science*, **1**, 134–165.

Bertalanffy, L. von (1968) *General System Theory: Foundations, Development, Applications*. New York: George Braziller.

Bertaux, J., Sondag, F., Santos, R. et al. (2002) Paleoclimatic record of speleothems in a tropical region: study of laminated sequences from a Holocene stalagmite in Central-West Brazil. *Quaternary International*, **89**, 3–16.

Bhattacharya, S.K., Froehlich, K., Aggarwal, P.K. & Kulkarni, K.M. (2003) Isotopic variation in Indian Monsoon precipitation: records from Bombay and New Delhi. *Geophysical Research Letters*, **30**, doi:10.1029/2003GL018453.

Bickle, M.J., Chapman, H.J., Bunbury, J. et al. (2005) The relative contributions of silicate and carbonate rocks to riverine Sr fluxes in the headwaters of the Ganges. *Geochimica et Cosmochimica Acta*, **69**, 2221–2240.

Bini, A., Quinif, Y., Sules, O. & Uggeri, A. (1992) Les mouvements tectoniques récents dans les grottes du Mont Campo dei Fiori (Lombardie, Italie). *Karstologia*, **19**, 23–30.

Binladen, J., Gilbert, M.T.P. & Willerslev, E. (2007) 800,000 year old mammoth DNA, modern elephant DNA or PCR artefact? *Biology Letters*, **3**, 55–56.

Bird, M.I., Boobyer, E.M., Bryant, C. et al. (2007) A long record of environmental change from bat guano deposits in Makangit Cave, Palawan, Philippines. *Earth and Environmental Science Transactions of the Royal Society of Edinburgh*, **98**, 59–69.

Birks, J., Battarbee R., Mackay A. & Oldfield F. (2003) *Global Change in the Holocene*. Hodder Arnold, London.

Black, J.R., Epstein, E., Rains, W.D. et al. (2008) Magnesium-isotope fractionation during plant growth *Environmental Science and Technology*, **42**, 7831–7836.

Blisniuk, P.M. & Stern, L.A. (2005) Stable isotope paleoaltimetry: a critical review. *American Journal of Science*, **305**, 1033–1074.

Blyth, A.J. & Frisia, S. (2008) Molecular evidence for bacterial mediation of calcite formation in cold high-altitude caves. *Geomicrobiology Journal*, **25**, 101–111.

Blyth, A.J. & Watson J.S. (2009) Thermochemolysis of organic matter preserved in stalagmites: a preliminary study. *Organic Geochemistry*, **40**, 1029–1031.

Blyth, A.J., Asrat, A., Baker, A. et al. (2007) A new approach to detecting vegetation and land-use change using high-resolution lipid biomarker records in stalagmites. *Quaternary Research*, **68**, 314–324.

Blyth, A.J., Baker, A., Collins, M.J. et al. (2008) Molecular organic matter in speleothems and its potential as an environmental proxy. *Quaternary Science Reviews*, **27**, 905–921.

Blyth, A.J., Watson, J.S., Woodhead, J. & Hellstrom, J. (2010) Organic compounds preserved in a 2.9 million year old stalagmite from the Nullarbor Plain, Australia. *Chemical Geology*, **279**, 101–105.

Blyth, A.J., Baker, A., Thomas, L.E. & Van Calsteren, P. (2011) A 2000–yr lipid biomarker record preserved in a stalagmite from north-west Scotland. *Journal of Quaternary Science*, **26**, 326–334.

Boch, R., Spötl, C. & Kramers, J. (2009) Coincident high-resolution isotope records of early Holocene climate change from two stalagmites of Katerloch Cave, Austria. *Quaternary Science Reviews*, **28**, 2527–2538.

Boch, R., Spötl, C. & Frisia, S. (2011) Origin and palaeoenvironmental significance of lamination in stalagmites from Katerloch Cave, Austria. *Sedimentology*, **58**, 508–531.

Bögli, A. (1964) Mischungskorrosion—ein Beiträg zur Verkarstungsproblem. *Erdkunde*, **18**, 83–92.

Bögli, A. (1980) *Karst Hydrology and Physical Speleology*. Springer-Verlag, Berlin.

Bond, G., Showers, W., Cheseby, M. et al. (1997) A pervasive millennial-scale cycle in North Atlantic and glacial climates. *Science*, **278**, 1257–1266.

Bond, G., Kromer, B., Beer, J. et al. (2001) Persistent solar influence on north Atlantic climate during the Holocene. *Science*, **294**, 2130–2136.

Borsato, A. (1997) Dripwater monitoring at Grotta di Ernesto (NE-Italy): a contribution to the understanding of karst hydrology and the kinetics of carbonate dissolution. *6th Conference on limestone hydrology and fissured media, La Chaux-de-Fonds, Switzerland 1997, Proceedings Volume 2*, 57–60.

Borsato, A. & Miorandi, R. (2003) Concentrazione di CO_2 e O_2 in cavità del Trentino. *Atti del XIX Congresso Nazionale di Speleologia, Bologna, 27–31 Agosto 2003*, pp. 169–176.

Borsato, A., Frisia, S., Jones, B. & van der Borg, K. (2000) Calcite moonmilk: crystal morphology and environment of formation in caves in the Italian Alps. *Journal of Sedimentary Research*, **70**, 1179–1190.

Borsato, A., Frisia, S., Fairchild, I.J. et al. (2007) Trace element distribution in annual stalagmite laminae mapped by micrometer-resolution X-ray fluorescence: implications for incorporation of environmentally significant species. *Geochimica et Cosmochimica Acta*, **71**, 1494–1512.

Bosák, P., Pruner, P. & Kadlec, J. (2003) Magnetostratigraphy of cave sediments: application and limits. *Studia Geophysica et Geodaetica*, **47**, 301–330.

Bosch, R.F. & White, W.B. (2004) Lithofacies and transport of sediments in karstic aquifers. In: Sasowksy, I.D. & Mylroie, J. (eds.) *Studies of Cave Sediments. Physical and Chemical Records of Paleoclimate*. New York: Kluwer, pp. 1–22.

Bosence, D. (2005) A genetic classification of carbonate platforms based on their basinal and tectonic settings in the Cenozoic. *Sedimentary Geology*, **175**, 49–72.

Böttcher, M.E. (1993) *The experimental investigation of ore-deposit relevant metal enrichment reactions from aqueous solutions with special regards to the formation of rhodochrosite ($MnCO_3$) [in German]*. PhD thesis (unpublished), University of Göttingen, Germany.

Böttcher, M.E. & Dietzel, M. (2010) Metal-ion partitioning during low-temperature precipitation and dissolution of anhydrous carbonates and sulphates. In: Stoll, H. & Prieto, M., editors. *Ion partitioning in ambient temperature aqueous systems: from fundamentals to applications in climate proxies and environmental geochemistry*. EMU Notes in Mineralogy, **10**, 139–187.

Bottrell, S.H. & Atkinson T.C. (1992) Tracer study of flow and storage in the unsaturated zone of a karstic limestone aquifer. In: Hötzl H. & Werner A. (eds.), *Tracer Hydrology*. Balkema, Rotterdam, pp. 207–211.

Bottrell, S.H., Smart, P.L., Whitaker, F. & Raiswell, R. (1991) Geochemistry and isotope systematics of sulfur in the mixing zone of Bahmian Blue Holes. *Applied Geochemistry*, **6**, 97–103.

Bourges, F., Mangin, A., & d'Hulst, D. (2001) Carbon dioxide in karst cavity atmosphere dynamics: the example of the Aven d'Orgnac (Ardeche). *Comptes Rendus de L'Academie Des Sciences Serie II Fascicule A—Sciences de la Terre et des Planetes*, **333**, 685–692.

Bourges, F., Genthon, P., Mangin, A. & d'Hulst, D. (2006a) Microclimates of L'Aven d'Orgnac and other French limestone caves (Chauvet, Esparros, Marsoulas). *International Journal of Climatology*, **26**, 1651–1670.

Bourges, F., Mangin, A., D'Hulst, D., & Genthon P. (2006b) La conservation de l'art parietal préhistorique des grottes, les raisons d'un miracle. *Bulletin de la Société Préistorique Ariège-Pyrénées, tome LXI*, 43–50.

Boutton, T.W., Archer, S.R., Midwood, A.J., Zitzer, S.F. & Bol, R. (1998) $\delta^{13}C$ values of soil organic carbon and their use in documenting vegetation change in a subtropical savannah ecosystem. *Geoderma*, **82**, 5–41.

Bowen, G.J. (2008) Spatial analysis of the intra-annual variation of precipitation isotope ratios and its climatological corollaries. *Journal of Geophysical Research*, **113**, D05113, doi:10.1029/2007JD009295.

Bowen, G.J. & Wilkinson, B. (2002) Spatial distribution of $\delta^{18}O$ in meteoric precipitation. *Geology*, **30**, 315–318.

Bradl, H.B. (2004) Adsorption of heavy metal ions on soils and soils constituents. *Journal of Colloid and Interface Science*, **277**, 1–18.

Bradley, C., Baker, A., Jex, C. & Leng, M.J. (2010) Hydrological uncertainties in the modelling of cave drip-water $\delta^{18}O$ and the implications for stalagmite palaeoclimate reconstructions. *Quaternary Science Reviews* **29**, 2201–2214.

Bradley, R.S. (1999) *Quaternary Palaeoclimatology*, 2nd edition. Academic Press, 613pp.

Brady, P.V. & Walther, J.V. (1990) Kinetics of quartz dissolution at low-temperatures. *Chemical Geology* **82**, 253–264.

Brandt, C.J. (1990) Simulation of the size distribution and erosivity of raindrops and throughfall drops. *Earth Surface Processes and Landforms*, **15**, 687–698.

Brantley, S.L. & White, A.F. (2009) Approaches to modeling weathered regolith. *Reviews in Mineralogy and Geochemistry*, **70**, 435–484.

Brantley, S.L., Chesley, J.T. & Stillings, L.L. (1998) Isotopic ratios and release rates of strontium measured from weathered feldspars. *Geochimica et Cosmochimica Acta*, **62**, 1493–1500.

Brasier, A.T. (2011) Searching for travertines, calcretes and speleothems in deep time: processes, appearances, predictions and the impact of plants. *Earth-Science Reviews*, **104**, 213–239.

Bridge, J.S. (2003) *Rivers and Floodplains: Forms, Processes and the Sedimentary Record*. Oxford: Blackwell Science.

Brassell, S.C., Eglinton, G., Marlowe, I.T. et al. (1986) Molecular stratigraphy: a new tool for climatic assessment. *Nature*, **320**, 129–133.

Bray, E.E. & Evans, E.D. (1961) Distribution of n-paraffins as a clue to recognition of source beds. *Geochimica et Cosmochimica Acta*, **22**, 2–15.

Braybrook, A.L., Heywood, B.R., Jackson, R.A. & Pitt, K. (2002) Parallel computational and experimental studies of the morphological modification of calcium carbonate by cobalt. *Journal of Crystal Growth*, **243**, 336–344.

Brenninkmeijer, C.A.M., Van Geel, B. & Mook, W.G. (1982) Variations in the D/H and O-18/O-16 ratios in cellulose extracted from a peat bog core. *Earth and Planetary Science Letters*, **61**, 283–290.

Briffa K.R., Jones, P.D., Schweingruber, F.H. et al. (1996) Tree-ring variables as proxy-climate indicators: problems with low-frequency signals. In: Jones, P.D., Bradley, R.S. & Jouzel, J. (eds) *Climatic variations and forcing mechanisms of the last 2000 years*. Springer-Verlag, Berlin, pp. 9–41.

Broecker, W.S., Olson, E.A., & Orr, P.C. (1960) Radiocarbon measurement and annual rings in cave formations. *Nature*, **185**, 93–94.

Broecker, W.S., Denton, G.H., Edwards, R.L. et al. (2010) Putting the Younger Dryas cold event into context. *Quaternary Science Reviews*, **29**, 1078–1081.

Brohan, P., Kennedy, J.J., Harris, I. et al. (2006) Uncertainty estimates in regional and global observed temperature changes: a new dataset from 1850. *Journal of Geophysical Research*, **111**, D12106, doi:10.1029/2005JD006548.

Bronk Ramsey, C. (2008) Deposition models for chronological records. *Quaternary Science Reviews*, **27**, 42–60.

Brook, G.A. & Nickmann, R.J. (1996) Evidence of Late Quaternary environments in northwestern Georgia from sediments preserved in Red Spider cave. *Physical Geography*, **16**, 69–78.

Brook, G.A., Folkoff, M.E. & Box, E.O. (1983) A world model of soil carbon dioxide. *Earth Surface Processes and Landforms*, **8**, 79–88.

Brook G.A., Burney, D.A. & Cowart, J.B. (1990) Desert paleoenvironmental data from cave speleothems with examples from the Chihuahuan, Somali-Chalbi, and Kalahari Deserts. *Palaeogeography, Palaeoclimatology, Palaeoecology* **76**, 311–329.

Brook, G.A., Rafter, M.A., Railsback, L.B. et al. (1999) A high-resolution proxy record of rainfall and ENSO since AD 1550 from layering in stalagmites from Anjohibe Cave, Madagascar. *The Holocene*, **9**, 695–705.

Brooks, J.R., Barnard, H.R., Coulombe, R. & McDonnell, J. (2010) Ecohydrologic separation of water between trees and streams in a Mediterranean climate. *Nature Geoscience*, **3**, 100–104.

Broughton, P.L. (1983a) Lattice deformation and curvature in stalactitic carbonate. *International Journal of Speleology*, **13**, 19–30.

Broughton, P.L. (1983b) Environmental implications of competitive growth fabrics in stalactitic carbonate. *International Journal of Speleology* **13**, 31–41.

Brown, D., Worden, J. & Noone, D. (2008) Comparison of atmospheric hydrology over convective continental regions using water vapour isotope measurements from space. *Journal of Geophysical Research* **113**, D15124, doi:10.1029/2007JD009676.

Brown, D.H. (2007) *Archaeological Archives. A guide to best practice in creation, compilation, transfer and curation*. Archaeological Archives Forum. Available at: http://www.britarch.ac.uk/archives/.

Brunsden, D. & Thornes, J.B. (1979) Landscape Sensitivity and Change. *Transactions of the Institute of British Geographers*, **4**, 463–484.

Brunton, C.H.C., Besterman, T.P. & Cooper, J.A. (1985) *Guidelines for the curation of Geological Materials. Geological Curators Group*, Geological Society of London, Miscellaneous Papers 17. Available online at: http://www.geocurator.org/pubs/pubs.htm.

Buckley, H.E. (1951) *Crystal Growth*. Wiley, New York.

Budd, D. (1988) Aragonite-to-calcite transformation during fresh-water diagenesis of carbonates: insights from pore-water chemistry. *Geological Society of America Bulletin*, **100**, 1260–1270.

Buddemeier, R.W. & Oberdorfer, J.A. (1997) Hydrogeology of Enewatak Atoll. In: Vacher, H.L. & Quinn, T. (eds.) *Geology and Hydrogeology of Carbonate Islands*. Amsterdam: Elsevier, pp. 667–692.

Buecher, R.H. (1999) Microclimate study of Karchner Caverns. *Journal of Cave and Karst Studies*, **61**, 108–120.

Buffle J., Wilkinson K.J., Stoll S. et al. (1998) A generalized description of aquatic colloidal interactions: the three-colloidal component approach. *Environmental Science and Technology*, **32**, 2887–2899.

Buhl D., Immenhauser A., Smeulders G. et al. (2007) Time series δ^{26}Mg analysis in speleothem calcite: kinetic versus equilibrium fractionation, comparison with other proxies and implications for palaeoclimate research. *Chemical Geology*, **244**, 715–729.

Buhmann, D. & Dreybrodt, W. (1985) The kinetics of calcite dissolution and precipitation in geologically relevant situations of karst areas: I. Open system. *Chemical Geology*, **48**, 189–211.

Bui-Thi-Mai & Girard, M. (1988) Apports actuels et anciens de pollens dans le grotte de Foissac (Aveyron, France). *Institut français de Pondichéry, Travaux de la Section Scientifique et Technique*, **25**, 43–53.

Bull, I.D., van Bergen, P.F., Nott, C.J. et al. (2000) Organic geochemical studies of soils from the Rothamsted classical experiments—V. The fate of lipids in different long-term experiments. *Organic Geochemistry*, **31**, 389–408.

Bull, P.A. (1981) Some fine-grained sedimentation phenomena in caves. *Earth Surface Processes and Landforms*, **6**, 11–22.

Bullen, T., White, A., Blum, A. et al. (1997) Chemical weathering of a soil chronosequence on granitoid alluvium. 2. Mineralogic and isotopic constraints on the behaviour of strontium. *Geochimica et Cosmochimica Acta*, **61**, 291–306.

Burger, G. (2007) On the verification of climate reconstructions. *Climate of the Past*, **3**, 397–409.

Burney, D.A. & Burney, L. (1993) Modern pollen deposition in cave sites: experimental results from New York State. *New Phytologist*, **124**, 523–535.

Burney, D.A., Brook, G.A. & Cowart, J.B. (1994) A Holocene pollen record for the Kalahari Desert of Botswana from a U-series dated speleothem. *The Holocene*, **4**, 225–232.

Burns, S.J., Fleitmann, D., Mudelsee, M. et al. (2002) A 780–yr annually resolved record of Indian Ocean monsoon precipitation from a speleothem from south Oman. *Journal of Geophysical Research*, **107**, D20, 4434 doi:10.1029/2001JD001281.

Burns, S.J., Fleitmann, D., Matter, A., et al. (2003) Indian Ocean climate and an absolute chronology over Dansgaard/Oeschger events 9 to 13. *Science*, **301**, 1365–1367.

Burton, W.K., Cabrera, N. & Frank, F.C. (1951) The growth of crystals and the equilibrium structure of their surfaces. *Philosophical Transactions of the Royal Society of London*, A243, 299–358.

Busenberg, E. & Plummer, L.N. (1985) Kinetic and thermodynamic factors controlling the distribution of SO_4^{2-} and Na^+ in calcites and selected aragonites. *Geochimica et Cosmochimica Acta*, **49**, 713–725.

Cabrol, P. & Coudray, J. (1982) Climatic fluctuations influence the genesis and diagenesis of carbonate speleothems in southwestern France. *National Speleological Society Bulletin*, **44**, 112–117.

Cacchio, P., Contento, R., Ercole, C., Cappuchio, G., Martinez, M.P. & Lepidi, A. (2004) Involvement of microorganisms in the formation of carbonate speleothems in the Cervo Cave (L'Aquila-Italy). *Geomicrobiology Journal*, **21**, 497–509.

Cai, B.G., Pumijumnong, N., Tan, M., et al. (2010) Effects of intraseasonal variation of summer monsoon rainfall on stable isotope and growth rate of a stalagmite from northwestern Thailand. *Journal of Geophysical Research-Atmospheres*, **115**, D21104, doi:10.1029/2009JD013378.

Calaforra, J.M., Forti, P. & Fernandez-Cortes, A. (2008) Speleothems in gypsum caves and their paleoclimatological significance. *Environmental Geology*, **53**, 1099–1105.

Came, R.E., Eiler, J.M., Veizer, J. et al. (2007) Coupling of surface temperatures and atmospheric CO_2 concentrations during the Palaeozoic era. *Nature*, **449**, 198–202.

Campbell, N.A. (1996) *Biology*, 4th Edition. Benjamin/Cummings, Menlo Park, California.

Carew, J.L. & Mylroie, J.E. (1997) Geology of the Bahamas. In: Vacher, H.L & Quinn, T. (eds.) *Geology and Hydrology of Carbonate Islands*. Amsterdam: Elsevier, pp. 91–139.

Carrión, J.S. & Scott, L. (1999) The challenge of pollen analysis in paleoenvironmental studies of hominid beds: the record from Sterkfontein Caves. *Journal of Human Evolution*, **36**, 410–408.

Carslaw, H.S. & Jaeger, C. (1959) *Conduction of heat in solids* (2nd edition). Oxford University Press, New York.

Caseldine, C.J., McGarry, S.F., Baker, A., Hawkesworth, C. & Smart, P.L. (2008) Late Quaternary speleothem pollen in the British Isles. *Journal of Quaternary Science*, **23**, 193–200.

Castanier, S., Métayer-Levrel, G. & Perthuisot, J.-P. (1999) Ca-carbonates precipitation and limestone genesis—the microbiogeologist point of view. *Sedimentary Geology*, **126**, 9–23.

Catuneau, O., Abreu, V., Bhattacharya, J.P. et al. (2009) Towards the standardization of sequence stratigraphy. *Earth-Science Reviews*, **92**, 1–33.

Caussinus, H. & Mestre, O. (2004) Detection and correction of artificial shifts in climate series. *Journal of the Royal Statistical Society Series C-Applied Statistics*, **53**, 405–425.

Celle-Jeanton, H., Travi, Y. & Blavoux, B. (2001) Isotopic typology of the precipitation in the Western Mediterranean region at three different time scales. *Geophysical Research Letters*, **28**, 1215–1218.

Cenki-Tok, B., Chabaux, F., Lemarchand, D. et al. (2009) The impact of water-rock interaction and vegetation on calcium isotope fractionation in soil- and stream waters of a small, forested catchment (the Strengbach case). *Geochimica et Cosmochimica Acta*, **73**, 2215–2228.

Cerling, T.E. (1984) The stable isotopic composition of modern soil carbonate and its relationship to climate. *Earth and Planetary Science Letters*, **71**, 229–240.

Cerling, T.E., Wang, Y. & Quade, J. (1993) Expansion of C4 ecosystems as an indicator of global ecological change in the late Miocene. *Nature*, **361**, 344–345.

Chacko, T. & Deines, P. (2008) Theoretical calculation of oxygen isotope fractionation factors in carbonate systems. *Geochimica Cosmochimica Acta*, **72**, 3642–3660.

Chameides, W.L. & Perdue, E.M. (1997) *Biogeochemical Cycles*. Oxford: Oxford University Press.

Chan, P.Y. & Goldenfeld, N. (2007) Steady states and linear stability analysis of precipitation patterns formation at geothermal hot springs. *Physical Review E* **76**, 046104, doi: 10.1103/PhysRevE.76.046104.

Chang, P., Ji, L. & Li, H. (1997) A decadal climate variation in the tropical Atlantic Ocean from thermodynamic air-sea interactions. *Nature*, **385**, 516–518.

Chapman, J.B., Ingraham, N.L. & Hess, J.W. (1992) Isotopic investigation of infiltration and unsaturated zone processes at Carlsbad cavern, New Mexico. *Journal of Hydrology*, **133**, 343–363.

Chapman, P.J., Shand, C.A., Edwards, A.C. & Smith, S. (1997) Effect of storage and sieving on the phosphorus composition of soil solution. *Soil Science Society of America Journal*, **61**, 315–321.

Charman, D.J., Caseldine, C.J., Baker, A. et al. (2001) Paleohydrological records from peat profiles and speleothems in Sutherland, NW Scotland. *Quaternary Research*, **55**, 223–234.

Cheng, H., Edwards, R.L., Hoff, J. et al. (2000) The half-lives of uranium-234 and thorium-230. *Chemical Geology*, **169**, 17–33.

Cheng H., Fleitmann, D., Edwards, R.L. et al. (2009a) Timing and structure of the 8.2 ka event inferred from $\delta^{18}O$ records of stalagmites from China, Oman and Brazil. *Geology*, **37**, 1007–1010.

Cheng, H., Edwards, R.L., Broecker, W.S. et al. (2009b) Ice age terminations. *Science*, **326**, 248–252.

Chester, R. (1990) *Marine Geochemistry*. Unwin Hyman, Boston.

Choppy, J. (1986) Dynamique de l'air. (2nd edition) *'Syntheses Karstiques', Série 11: Processus Climatiques dans les vides karstiques*. Spéleo Club de Paris.

Choquette, P.W. & Pray, L.C. (1970) Geologic nomenclature and classification of porosity in sedimentary carbonates. *American Association of Petroleum Geologists Bulletin*, **54**, 207–250.

Chorley, R.J. & Kennedy, B.A. (1971) *Physical Geography: a systems approach*. London: Prentice-Hall.

Chou K., Garrels R.M. & Wollast, R. (1989) Comparative study of the kinetics and mechanisms of dissolution of carbonate minerals. *Chemical Geology*, **78**, 269–282.

Christiansen, B., Schmith, T. & Thejll, P. (2009) A surrogate ensemble study of climate reconstruction methods: stochasticity and robustness. *Journal of Climate*, **22**, 951–976.

Ciais, P. & Jouzel, J. (1994) Deuterium and oxygen-18 in precipitation—isotopic model, including mixed cloud processes. *Journal of Geophysical Research*, **99**, 16793–16803.

Cigna, A.A. (1967) An analytical study of air circulation in caves. *International Journal of Speleology*, **3**, 41–54.

Cigna, A.A. (2002) Modern trend in cave monitoring. *Acta Carsologica*, **31**, 35–54.

Cigna, A.A. (2005) Radon in caves. *International Journal of Speleology*, **34**, 1–18.

Clark, I.D. & Fritz, P. (1997) *Environmental Isotopes in Hydrogeology*. Lewis, Boca Raton, New York.

Clark, I.D., Fontes, J.-C. & Fritz, P. (1992) Stable isotope disequilibria in travertine from high pH waters: laboratory investigations and field observations from Oman. *Geochimica et Cosmochimica Acta*, **56**, 2041–2050.

Claussen, M. (1997) Modeling bio-geophysical feedback in the African and Indian monsoon region. *Climate Dynamics*, **13**, 247–257.

Cleland, C.E. (2001) Historical science, experimental science, and the scientific method. *Geology*, **29**, 987–990.

Cobb, K.M., Adkins, J.F., Partin, J.W. & Clark, B. (2007) Regional-scale climate influences on temporal variations of rainwater and cave dripwater oxygen isotopes in northern Borneo. *Earth and Planetary Science Letters*, **263**, 207–220.

Coble, P.G. (1996) Characterization of marine and terrestrial DOM in seawater using excitation-emission matrix spectroscopy. *Marine Chemistry*, **51**, 325–346.

Cockburn, H.A.P. & Summerfield, M.A. (2004) Geomorphological applications of cosmogenic isotope analysis. *Progress in Physical Geography*, **28**, 1–42.

Cohen, A.L. & Gaetani, G.A. (2010) Ion partitioning and the geochemistry of coral skeletons: solving the mystery of the vital effect. In: Stoll, H. & Prieto, M., editors. *Ion partitioning in ambient temperature aqueous systems: from fundamentals to applications in climate proxies and environmental geochemistry*. EMU Notes in Mineralogy, **10**, 399–415.

Coles, G.M., Gilbertson, D.D, Hunt, C.O. & Jenkinson, R.D.S. (1989) Taphonomy and the palynology of cave deposits. *Cave Science*, **16**, 83–89.

Collister, C. & Mattey, D. (2008) Controls on water drop volume at speleothem drip sites: an experimental study. *Journal of Hydrology*, **358**, 259–267.

Compo, G.P., Whitaker, J.S., Sardeshmukh, P.D., et al. (2011) The Twentieth Century Reanalysis Project. *Quarterly Journal of the Royal Meteorological Society*, **137**, 1–28.

Condomines, M. & Rihs, S. (2006) First ^{226}Ra-^{210}Pb dating of a young speleothem. *Earth and Planetary Science Letters*, **250**, 4–10.

Conn H.W. (1966) Barometric wind in Wind and Jewel caves, South Dakota. *Bulletin of the National Speleological Society*, **28**, 55–69.

Cook, E.R. (1992) Using tree rings to study past El Nino/Southern Oscillation influences on climate. *In 'El Nino: Historical and Paleoclimatic Aspects of the Southern Oscillation.'* (H.F. Diaz, and V. Markgraf, Eds.), pp. 204–214. Cambridge University Press.

Cooke, M.L., Simo, J.A., Underwood, C.A. & Rijken, P. (2006) Mechanical stratigraphic controls on fracture patterns within carbonates and implications for groundwater flow. *Sedimentary Geology*, **184**, 225–239.

Coplen, T.B. (2007) Calibration of the calcite-water oxygen-isotope geothermometer at Devil's Hole, Nevada, a natural laboratory. *Geochimica et Cosmochimica Acta*, **71**, 3948–3957.

Coplen, T. (2008) Explanatory glossary of terms used in expression of relative isotope ratios and gas ratios. (IUPAC Recommendations 2008). *International Union of Pure and Applied Chemistry Inorganic Chemistry Division Commission on Isotopic Abundances and Atomic Weights, 1–27.*

Coplen, T.B., Winograd, I.J., Landwehr, J.M. & Riggs A.C. (1994) 500,000 year stable carbon isotopic record from Devils Hole, Nevada. *Science*, **263**, 361–365.

Corbella, M., Ayora, C. & Cardellach, E. (2004) Hydrothermal mixing, carbonate dissolution and sulphide precipitation in Mississippi Valley-type deposits. *Mineralium Deposita*, **39**, 344–357.

Cosford, J., Qing H., Eglington, B. et al. (2008) East Asian monsoon variability since the Mid-Holocene recorded in a high-resolution, absolute-dated aragonite speleothem from eastern China. *Earth and Planetary Science Letters*, **275**, 296–307.

Covington, M.D., Saar, M., Wicks, C.M. & Gabrovšek, F. (2011) Dimensionless metrics that characterize the relationships between signals observed at springs and karst aquifer geometry. *Proceedings H2Karst, 9th Conference on Limestone Hydrogeology, Besancon (France)*, 1–4 September 2011, 107–110.

Covington, M.D., Luhmann, A.J., Wicks, C.M. & Saar, M.O. (2012) Process length scales and longitudinal damping in karst conduits. *Journal of Geophysical Research - Earth Surface* (in press).

Cox, G., James, J.M., Osborne, R.A.L. & Leggett, K.E.A. (1989) Stromatolitic crayfish-like stalagmites. *Proceedings of the University of Bristol Spelaeological Society*, **18**, 339–358.

Craig, H. (1961) Isotopic variations in meteoric waters. *Science*, **133**, 1702–1703.

Craig, H. & Gordon, L.I. (1965) Deuterium and oxygen 18 variations in the ocean and the marine atmosphere, in Tongiorgi E. (ed.) *Stable Isotopes in Oceanographic Studies and Palaeotemperatures*, Consiglio Nazionale della RicercheLaboratoria di Geologia Nucleare, Pisa, Italy, pp. 9–130.

Crouzeix, C., Le Mouël, J-L., Perrier, F. et al. (2003) Long-term thermal evolution and effect of low power heating in an underground quarry. *Comptes Rendus Geoscience*, **335**, 345–354.

Crowther, J. (1989) Groundwater chemistry and cation budgets of tropical karst outcrops, peninsular Malaysia, I. calcium and magnesium. *Journal of Hydrology*, **107**, 169–192.

Cruz, F.R., Burns, S.J., Jercinovic, M. et al. (2007) Evidence of rainfall variations in Southern Brazil from trace element ratios (Mg/Ca and Sr/Ca) in a Late Pleistocene stalagmite. *Geochimica et Cosmochimica Acta*, **71**, 2250–2263.

Cuezva, A., Sanchez-Moral, Saiz-Jimenez, C. & Cañaveras, J.C. (2009) Microbial communities and associated mineral fabrics in Altamira Cave, Spain. *International Journal of Speleology*, **38**, 83–92.

Cuezva, S., Fernandez-Cortes, A., Benavente, D. et al. (2011) Short-term CO_2(g) exchange between a shallow karstic cavity and the external atmosphere during summer: role of the surface soil layer. *Atmospheric Environment*, **45**, 1418–1427.

Cui, J., An, S., Wang, Z. et al. (2009) Using deuterium excess to determine the sources of high-altitude precipitation: implications in hydrological relations between sub-alpine forests and alpine meadows. *Journal of Hydrology*, **373**, 24–33.

Cui, J.-W., Huang, J-H., Meyers, P.A. et al. (2010) Variation in solvent-extractable lipids and n-alkane compound-specific carbon isotopic compositions with depth in a Southern China karst area soil. *Journal of Earth Sciences*, **21**, 382–391.

Cullingford, C.H.D. (1953) *British Caving. An Introduction to Speleology*. Routledge, London.

Culver, DC & White, W.B. (2005) *Encyclopedia of Caves*. Elsevier, Amsterdam.

Cunningham, K.I. & LaRock, E.J. (1991) Recognition of Microclimate Zones Through Radon Mapping, Lechuguilla Cave, Carlsbad Caverns National Park, New Mexico. *Health Physics*, **61**, 493–500.

Curl, R.L. (1972) Minimum Diameter Stalactites. *Bulletin of the National Speleological Society*, **34**, 129–136.

Curl, R.L. (1973) Minimum Diameter Stalagmites. *Bulletin of the National Speleological Society*, **35**, 1–9.

Curl, R.L. (1974) Deducing flow velocity in cave conduits from scallops. *Bulletin of the National Speleological Association*, **36**, 1–5.

Curti, E. (1999) Coprecipitation of radionuclides with calcite: estimation of partition coefficients based on a review of laboratory investigations and geochemical data. *Applied Geochemistry*, **14**, 433–445.

Daëron, M., Guo, W., Eiler, J. et al. (2011) $^{13}C^{18}O$ clumping in speleothems: observations from natural caves and precipitation experiments. *Geochimica et Cosmochimica Acta*, **75**, 3303–3317.

Daley, T.J., Barber, K.E., Street-Perrott, F.A. et al. (2010) Holocene climate variability revealed by oxygen isotope analysis of Sphagnum cellulose from Walton Moss, northern England. *Quaternary Science Reviews*, **29**, 1590–1601.

Dandurand, J.L., Gout, R., Hoefs, J. et al. (1982) Kinetically controlled variations of major components and carbon and oxygen isotopes in a calcite-precipitating spring. *Chemical Geology*, **36**, 299–315.

Danielli, H.M.C. & Edington, M.A. (1983) Bacterial calcification in limestone caves. *Geomicrobiology Journal*, **3**, 1–16.

Dansgaard, W. (1964) Stable isotopes in precipitation. *Tellus*, **16**, 436–468.

Dasgupta, S., Saar, M.O., Edwards, R.L. et al. (2010) Three thousand years of extreme rainfall events recorded in stalagmites from Spring Valley Caverns, Minnesota. *Earth and Planetary Science Letters*, **300**, 46–54.

Davey, M.E. & O'Toole, G.A. (2000) Microbial biofilms: from ecology to molecular genetics. *Microbiology and Molecular Biology Reviews*, **64**, 847–867.

Davidson, E.A., Janssens, I.A. & Luo, Y.-Q. (2006) On the variability of respiration in terrestrial ecosystems: moving beyond Q_{10}. *Global Change Biology*, **12**, 154–164.

Davidson, G.R. (1995) The stable isotopic composition and measurement of carbon in soil CO_2. *Geochimica et Cosmochimica Acta*, **59**, 2485–2489.

Davies, W.E. (1951) Mechanics of cavern breakdown. *Bulletin of the National Speleological Society*, **13**, 36–42.

Davies, W.M. (1930) Origin of limestone caverns. *Bulletin of the Geological Society of America*, **41**, 475–628.

Davin, L.B. & Lewis, N.G. (2005) Lignin primary structures and dirigent sites. *Current Opinion in Biotechnology*, **16**, 407–415.

Davis, K.J., Dove, P.M. & De Yoreo, J.J. (2000) The role of Mg^{2+} as an impurity in calcite growth. *Science*, **290**, 1134–1137.

Davis, K.J., Dovae, P.M., Wasylenki, L.E. & De Yoreo, J.J. (2004) Morphological consequences of differential Mg^{2+} incorporation at structurally distinct steps on calcite. *American Mineralogist*, **89**, 714–720.

Davis, K.J., Nealson, K.H. & Lüttge, A. (2007) Calcite and dolomite dissolution rates in the context of microbe-mineral surface interactions. *Geobiology*, **5**, 191–205.

Day, C.C. & Henderson, G.M. (2011) Oxygen isotopes in calcite grown under cave-analogue conditions. *Geochimica et Cosmochimica Acta*, **75**, 3956–3972.

de Choudens-Sánchez, V. & González, L.A. (2009) Calcite and aragonite precipitation under controlled instantaneous supersaturation: elucidating the role of $CaCO_3$ saturation state and Mg/Ca ratio on calcium carbonate polymorphism. *Journal of Sedimentary Research*, **79**, 363–376.

de Freitas, C.R. & Littlejohn, R.N. (1987) Cave climate: assessment of heat and moisture exchange. *Journal of Climatology*, **7**, 553–569.

de Freitas, C.R. & Schmekal, A. (2003) Condensation as a microclimate process: measurement, numerical simulation and prediction in the Glowworm Cave, New Zealand. *International Journal of Climatology*, **23**, 557–575.

de Freitas, C.R.; Littlejohn, R.N.; Clarkson, T.S. & Kristament, I.S. (1982) Cave climate: assessment of airflow and ventilation. *Journal of Climatology*, **2**, 383–397.

de Ruiter, D.J., Pickering, R., Steininger, C.M. et al. (2009) New Australopithecus robustus fossils and associated U-Pb dates from Cooper's Cave (Gauteng, South Africa). *Journal of Human Evolution*, **56**, 497–513.

de Vicente, I., Ortega-Retuerta, E., Romera, O., Morales-Baquero, R. & Reche, I. (2009) Contribution of transparent exopolymer particles to carbon sinking flux in an oligotrophic reservoir. *Biogeochemistry*, **96**, 13–23.

de Vicente, I., Ortega-Retuerta, E., Mazuecos, I.P., Pace, M.L., Cole, J.J. & Reche, I. (2010) Variation in transparent exopolymer particles in relation to biological and chemical factors in two contrasting lake districts. *Aquatic Sciences*, **72**, 443–453.

de Yoreo, J.J., Zepeda-Ruiz, L.A., Friddle, R.W. et al. (2009) Rethinking classical crystal growth models through molecular scale insights: consequences of kink-limited kinetics. *Crystal Growth and Design*, **9**, 5135–5144.

Degryse, F., Smolders, E & Parker, D.R. (2009) Partitioning of metals (Cd, Co, Cu, Ni, Pb, Zn) in soils: concepts, methodologies, prediction and applications—a review. *European Journal of Soil Science*, **60**, 590–612.

Deines, P., Langmuir, D. & Harmon, R.S. (1974) Stable carbon isotope ratios and the existence of a gas phase in the evolution of carbonate ground waters. *Geochimica et Cosmochimica Acta*, **38**, 1147–1164.

Delaby, S. (2001) Palaeoseismic investigations in Belgian caves. *Netherlands Journal of Geosciences*, **80**, 323–332.

Delworth, T., Manabe, S. & Stouffer, R.J. (1993) Interdecadal variations of the thermohaline circulation in a coupled ocean-atmosphere model. *Journal of Climate*, **6**, 1993–2011.

Delworth, T.L. & Mann, M.E. (2000) Observed and simulated multidecadal variability in the Northern Hemisphere. *Climate Dynamics*, **16**, 661–676.

Demény, A., Kele, S. & Siklósy, Z. (2010) Empirical equations for the temperature dependence of calcite-water oxygen isotope fractionation from 10 to 70 °C. *Rapid Communications in Mass Spectrometry*, **24**, 3521–3526.

Dennis, P.F., Rowe, P.J. & Atkinson, T.C. (2001) The recovery and isotopic measurement of water from fluid inclusions in speleothems. *Geochimica et Cosmochimica Acta*, **65**, 871–884.

Denniston, R.F., González, L.A., Asmerom, Y. et al. (1999) Evidence for increased cool season moisture during the middle Holocene. *Geology*, **27**, 815–818.

Denniston, R.F., González, L.A., Asmerom, Y. et al. (2000) Speleothem carbon isotopic records of Holocene environments in the Ozark Highlands, USA. *Quaternary International*, **67**, 61–67.

Desmarchelier, J.A., Hellstrom, J.C. & McCulloch, M.T. (2006) Rapid trace element analysis of speleothems by ELA-ICP-MS. *Chemical Geology*, **31**, 102–117.

Dicken, A.P. (2005) *Radiogenic Isotope Geology*. Cambridge University Press, 2nd Edition.

Dickson, J.A.D. (1983) Graphical modelling of crystal aggregates and its relevance to cement diagnosis. *Philosophical Transactions of the Royal Society of London A*, **309**, 465–502.

Dickson, J.A.D. (1990) Carbonate mineralogy and chemistry. In: Tucker, M.E. & Wright, V.P. *Carbonate Sedimentology*. Oxford, Blackwell, pp. 284–313.

Dickson, J.A.D. (1993) Crystal-growth diagrams as an aid to interpreting the fabrics of calcite aggregates. *Journal of Sedimentary Petrology*, **63**, 1–17.

Dickson, J.A.D. (1997) Synchronous intracrystalline $\delta^{13}C$ and $\delta^{18}O$ differences in natural calcite crystals. *Mineralogical Magazine*, **61**, 243–248.

Dietzel, M., Tang, J.W., Lies, A. & Kohler S.J. (2009) Oxygen isotope fractionation during inorganic calcite precipitation—Effects of temperature, precipitation rate and pH. *Chemical Geology*, **268**, 107–115.

Dirks, P.H.G.M., Kibii, J.M., Kuhn, B.F. et al. (2010) Geological setting and age of Australopithicus sediba from Southern Africa. *Science*, **328**, 205–208.

Doctor, D.H., Kendall, C., Sebestyen, S.D. et al. (2008) Carbon isotope fractionation of dissolved inorganic carbon (DIC) due to outgassing of carbon dioxide from a headwater stream. *Hydrological Processes*, **22**, 2410–2423.

Doerr, S.H., Davies, R.R., Lewis, A., Lewis, I. & Pilkington, G. (2006) Known Nullarbor caves—just scratching the surface? Quantifying unexplored cave volume using microgravity and draught measurements. (abstract) *Cave and Karst Science*, **32**, 44.

Domínguez-Villar, D., Fairchild, I.J., Baker, A. et al. (2009a) Decrease in cave temperature instrumentally recorded during the last 30 years in Eagle Cave, Spain. In: *7th International Conference on Geomophology*, 6–11 July, Melbourne, Australia.

Domínguez-Villar, D., Fairchild, I.J., Baker, A. et al. (2009b) Oxygen isotope precipitation anomaly in the North Atlantic region during the 8.2 ka event. *Geology*, **37**, 1095–1098.

Domínguez-Villar, D., Fairchild, I.J., Carrasco, R.M. et al. (2010) The effect of visitors in a tourist cave and the resulting constrains on natural thermal conditions for palaeoclimate studies (Eagle Cave, central Spain). *Acta Carsologica*, **39**, 491–502.

Dong J.G., Wang, Y.J., Cheng H. et al. (2010) A high-resolution stalagmite record of the Holocene East Asian monsoon from Mt Shennongjia, central China. *Holocene*, **20**, 257–264.

Dorale, J.A. & Liu, Z. (2009) Limitations of Hendy test criteria in judging the palaeoclimatic suitability of speleothems and the need for replication. *Journal of Cave and Karst Studies*, **71**, 73–80.

Dorale, J.A., Gonzalez, L.A., Reagan, M.K. et al. (1992) A high-Resolution Record of Holocene Climate Change in Speleothem Calcite from Cold Water Cave, Northeast Iowa, *Science*, **258**, 1626–1630.

Dorale, J.A., Edwards, R., Ito, E. & Gonzalez, L.A. (1998) Climate and vegetation history of the midcontinent from 75 to 25 ka: a speleothem record from Crevice Cave, Missouri, USA. *Science*, **282**, 1871–1874.

Dorale, J.A., Onac, B.P., Fornos, J.J. et al. (2010) Sea-level highstand 81,000 years ago in Mallorca. *Science*, **327**, 860–863.

Dose, V. & Menzel, A. (2004) Bayesian analysis of climate change impacts in phenology. *Global Change Biology*, **10**, 259–272.

Dove, P.M. & Hochella, M.F. (1993) Calcite precipitation mechanisms and inhibition by orthophosphate—insitu observations by scanning force microscopy. *Geochimica et Cosmochimica Acta*, **57**, 705–714.

Drake, J.J. (1983) The effects of geomorphology and seasonality on the chemistry of carbonate groundwater. *Journal of Hydrology*, **61**, 223–236.

Draxler, R.R. & Rolph, G.D. (2003) *HYSPLIT (Hybrid Single-Particle Lagrangian Integrated Trajectory) Model*. NOAA Air Resources Laboratory, Silver Spring, MD. http://ready.arl.noaa.gov/HYSPLIT.php

Drever, J.I. (1982) *The Geochemistry of Natural Waters*. Prentice-Hall, New Jersey.

Dreybrodt, W. (1980) Deposition of calcite from thin films of calcareous solutions and the growth of speleothems. *Chemical Geology*, **29**, 89–105.

Dreybrodt, W. (1981) The kinetics of calcite precipitation from thin films of calcareous solutions and the growth of speleothems: revisited. *Chemical Geology*, **32**, 237–245.

Dreybrodt, W. (1988) *Processes in Karst Systems*. Springer-Verlag, Berlin.

Dreybrodt, W. (1996) Principles of early development of karst conduits under natural and man-made conditions revealed by mathematical analysis of numerical models. *Water Resources Research*, **32**, 2923–2935.

Dreybrodt, W. (1999) Chemical kinetics, speleothem growth and climate. *Boreas*, **28**, 347–356.

Dreybrodt, W. (2008) Evolution of the isotopic composition of carbon and oxygen in a calcite precipitating H_2O-CO_2-$CaCO_3$ solution and the related isotopic composition of calcite in stalagmites. *Geochimica et Cosmochimica Acta*, **72**, 4712–4724.

Dreybrodt, W. & Lamprecht, G. (1981) Computer-simulation des wachstums von stalagmiten. *Die Höhle*, **31**, 11–21.

Dreybrodt, W. & Franke, H.W. (1987) Wachstumsgeschwindigkeiten und durchmesser von kerzenstalagmiten. *Die Höhle*, **38**, 1–6.

Dreybrodt, W. & Romanov, D. (2008) Regular stalagmites: the theory behind their shape. *Acta Carsologica*, **37**, 175–184.

Dreybrodt, W. & Gabrovšek, F. (2009) Small-scale terraces and isolated rimstone pools on stalagmites in caves exhibit striking similarity to large-scale terrace landscapes at hot springs. *Acta Carsologica*, **38**, 19–26.

Dreybrodt, W. & Scholz, D. (2011) Climatic dependence of stable carbon and oxygen isotope signals recorded in speleothems: from soil water to speleothem calcite. *Geochimica et Cosmochimica Acta*, **75**, 734–752.

Dreybrodt, W., Lauckner, J., Liu, Z. et al. (1996) The kinetics of the reaction $CO_2 + H_2O \rightarrow H^+ + HCO_3^-$ as one of the rate-limiting steps for the dissolution of calcite in the system H_2O-CO_2-$CaCO_3$. *Geochimica et Cosmochimica Acta*, **60**, 3375–3381.

Dreybrodt, W., Eisenlohr, L., Madry, B. & Ringer, S. (1997) Precipitation kinetics of calcite in the system $CaCO_3$–H_2O–CO_2: the conversion to CO_2 by the slow process $H^+ + HCO_3^- \rightarrow CO_2 + H_2O$ as a rate limiting step. *Geochimica et Cosmochimica Acta*, **61**, 3897–3904.

Dreybrodt, W., Gabrovšek, F. & Perne, M. (2005) Condensation corrosion: a theoretical approach. *Acta Carsologica*, **34**, 317–348.

Drysdale, R.N., Zanchetta, G., Hellstrom, J.C. et al. (2004) Palaeoclimatic implications of the growth history and stable isotope ($\delta^{18}O$ and $\delta^{13}C$) geochemistry of a Middle to Late Pleistocene stalagmite from central-western Italy. *Earth and Planetary Science Letters*, **227**, 215–229.

Drysdale, R., Zanchetta, G. & Hellstrom, J. (2006) Late Holocene drought responsible for the collapse of old world civilisations is recorded in an Italian flowstone. *Geology*, **34**, 101–104.

Dubey, S.K., Tripathi, A.K. & Upadhyay, S.N. (2006) Exploration of soil bacterial communities for their potential as bioresource. *Bioresource Technology*, **97**, 2217–2224.

Dublyansky, Y.V. & Spötl, C. (2009) Hydrogen and oxygen isotopes of water from inclusions in minerals: design of a new crushing system and on-line continuous-flow isotope ratio mass spectrometric analysis. *Rapid Communications in Mass Spectrometry*, **23**, 2605–2613.

Dublyansky, Y., Spötl, C., Luetscher, M. & Boch, R. (2010) Temperature dependence of oxygen isotopes in speleothems: experimental vs. empirical approach (abstract). *3rd Daphne workshop, 30th June-2nd July 2010, Innsbruck, Austria*, 63–64.

Dueñas, C., Frenández, M.C., Cañete, S. et al. (1999) ^{222}Rn concentrations, natural flow rate and the radiation exposure levels in the Nerja Cave. *Atmospheric Environment*, **33**, 501–510.

Dulinski, M. & Rozanski, K. (1990) Formation of 13C/12C isotope ratios in speleothems: a semi-dynamic model. *Radiocarbon*, **32**, 7–16.

Dunbar, R.B., Wellington, G.M., Colgan, M.W. & Glynn, P.W. (1994) Eastern Pacific sea surface temperature since 1600 AD: the $\delta^{18}O$ record of climate variability in Galápagos corals. *Paleoceanography*, **9**, 291–315.

Dunham, R.J. (1962) Classification of carbonate rocks according to depositional texture. In: Ham, W.E. (editor), *Classification of Carbonate Rocks. Memoir of the American Association of Petroleum Geologists*, **1**, 108–121.

Duplessy, J.C., Labeyrie, J., Lalou, C. & Nguyen, H.V. (1970) Continental climatic variations between 130,000 and 90,000 years BP. *Nature*, **226**, 631–633.

Dutton, A., Bard, E., Antonioli, F. et al. (2009) Phasing and amplitude of sea-level and climate change during the penultimate interglacial. *Nature Geoscience*, **2**, 355–359.

Dykoski, C.A., Edwards, R.L., Cheng, H. et al. (2005) A high-resolution, absolute-dated Holocene and deglacial Asian monsoon records from Dongge cave, China. *Earth and Planetary Science Letters*, **233**, 71–86.

Ebinger, C.J. & Cassey, M. (2001) Continental breakup in magmatic provinces: an Ethiopian example. *Geology*, **29**, 527–530.

Edwards, E.J., Osborne, C.P., Strömberg, C.A.E., Smith, S.A. and the C4 Grasses Consortium (2010) The origins of C4 grasslands: integrating evolutionary and ecosystem science. *Science*, **328**, 587–591.

Edwards, R.L., Chen, J.H. & Wasserburg, G.J. (1987) 238U- 234U- 230Th- 232Th systematics and the precise measurements of time over the past 500,000 years. *Earth and Planetary Science Letters*, **81**, 175–192.

Edwards, R.L., Cheng, H., Murrell, M.T. & Goldstein, S.J. (1997) Protactinium-231 dating of carbonates by thermal ionisation mass spectrometry: implications for Quaternary climate change. *Science*, **276**, 782–786.

Eglinton, G. & Hamilton, R.J. (1967) Leaf epicuticular waxes. *Science*, **156**, 1322–1335.

Ehlers, E. & Krafft, T. (eds.) (2001) *Understanding the Earth System*. Berlin: Springer.

Ehrenberg, S.N. & Nadeau, P.H. (2005) Sandstone vs. carbonate petroleum reservoirs: A global perspective on porosity-depth and porosity-permeability relationships. *American Association of Petroleum Geologists Bulletin*, **89**, 435–445.

Eiler, J.M. (2007) 'Clumped-isotope' geochemistry—The study of naturally-occurring, multiply-substituted isotopologues. *Earth and Planetary Science Letters*, **262**, 309–327.

Einseidl, F. & Mayer B. (2005) Sources and processes affecting sulphate in a karstic groundwater system of the Franconian Alb, southern Germany. *Environmental Science and Technology* **39**, 7118–7125.

Ek, C. & Gewelt M. (1985) Carbon dioxide in cave atmospheres. New results in Belgium and comparison with some other countries. *Earth Surface Processes and Landforms*, **10**, 173–187.

Elderfield, H., Yu, J., Anand, P. et al. (2006) Calibrations for benthic foraminiferal Mg/Ca paleothermometry and the carbonate ion hypothesis. *Earth and Planetary Science Letters*, **250**, 633–649.

Ellwood, B.B., Petruso, K.M. & Harrold, F.B. (1997) High-resolution paleoclimatic trends for the Holocene identified using magnetic susceptibility data from archaeological excavations in caves. *Journal of Archaeological Science*, **24**, 569–573.

Ellwood, B.B., Harrold, F.B., Benoist, S.L. et al. (2001) Paleoclimate and intersite correlations from Late Pleistocene/Holocene cave sites: results from southern Europe. *Geoarchaeology*, **16**, 433–463.

Elzinga, E.J., Reeder, R.J., Withers, S.H. et al. (2002) EXAFS study of rare-earth element coordination in calcite. *Geochimica et Cosmochimica Acta*, **66**, 2875–2885.

Embry, A.F. & Klovan, J.E. (1971) A Late Devonian reef tract on northeastern Banks Island, Northwest Territories. *Bulletin of the Canadian Petroleum Geologists*, **19**, 730–781.

Emery, D. & Myers, K. (1996) *Sequence Stratigraphy*. Blackwell: Oxford.

Epstein S., Buchsbaum R., Lowenstam H. & Urey H.C. (1953) Revised carbonate-water isotopic temperature scale. *Bulletin of the Geological Society of America*, **64**, 1315–1326.

Esper, J., Cook, E.R. & Schweingruber, F.H. (2002) Low-frequency signals in long tree-ring chronologies for reconstructing past temperature variability. *Science*, **295**, 2250–2253.

Esper, J., Frank, DC, Wilson, R.J.S. et al. (2005) Effect of scaling and regression on reconstructed temperature amplitude for the past millennium. *Geophysical Research Letters*, **32**, L07711. doi: 10.1029/2004GL021236.

Esteban, M. & Klappa, C.F. (1983) Subaerial exposure environment. In: Scholle, P.A., Bebout, D.G. & Moore, C.H. (eds.) *Carbonate Depositional Environments*. Tulsa: American Association of Petroleum Geologists, Memoir **33**, 1–54.

Fabricius, I.L. (2003) How burial diagenesis of chalk sediments controls sonic velocity and porosity. *American Association of Petroleum Geologists Bulletin*, **87**, 1755–1778.

Faimon J., Štelcl J. & Sas D. (2006) Anthropogenic CO_2-flux into cave atmosphere and its environmental impact: a case study in the Císarská Cave (Moravian Karst, Czech Republic). *Science of the Total Environment*, **369**, 231–245.

Fairchild I.J. & Killawee J.A. (1995) Selective leaching in glacierized terrains and implications for retention of primary chemical signals in carbonate rocks. In: Kharaka, Y.K. & Chudaev, O.V. (eds.) *Water-Rock Interaction. Proceedings of the 8th International Symposium on Water-Rock Interaction—WRI-8, Vladivostok, Russia, 15–19 August 1995*. A.A. Balkema, Rotterdam, pp. 79–82.

Fairchild, I.J. & Kennedy, M.J. (2007) Neoproterozoic glaciation in the Earth System. *Journal of the Geological Society, London*, **164**, 895–921.

Fairchild, I.J. & McMillan, E.A. (2007) Speleothems as indicators of wet and dry periods. *International Journal of Speleology*, **36**, 79–84.

Fairchild I.J. & Treble P.C. (2009) Trace elements in speleothems as recorders of environmental change. *Quaternary Science Reviews*, **28**, 449–468.

Fairchild, I.J. & Hartland, A. (2010) Trace element variations in stalagmites: controls by climate and by karst system processes. In: Stoll, H. & Prieto, M., editors. *Ion partitioning in ambient temperature aqueous systems: from fundamentals to applications in climate proxies and environmental geochemistry*. EMU Notes in Mineralogy, **10**, 259–287.

Fairchild, I.J., Quest, M., Tucker, M.E. & Hendry, G.L. (1988) Chemical analysis of sedimentary rocks. In: Tucker, M.E. (Ed.) *Techniques in Sedimentology*, Blackwell, Oxford, pp. 274–354.

Fairchild, I.J., Bradby, L., Sharp, M. & Tison, J.-L. (1994) Hydrochemistry of carbonate terrains in Alpine glacial settings. *Earth Surface Processes and Landforms*, **19**, 33–54.

Fairchild, I.J., Killawee, J.A. Hubbard, B. & Dreybrodt, W. (1999a) Interactions of calcareous suspended sediment with glacial meltwater: a field test of dissolution behaviour. *Chemical Geology*, **155**, 243–263.

Fairchild, I.J., Killawee, J.A., Sharp, M.J. et al. (1999b) Solute generation and transfer from a chemically reactive Alpine glacial-proglacial system. *Earth Surface Processes and Landforms*, **24**, 1189–1211.

Fairchild, I.J., Borsato, A., Tooth, A.F. et al. (2000) Controls on trace element (Sr-Mg) compositions of carbonate cave waters: implications for speleothem climatic records. *Chemical Geology*, **166**, 255–269.

Fairchild, I.J., Baker, A., Borsato, A. et al. (2001) Annual to sub-annual resolution of multiple trace element trends in speleothems. *Journal of the Geological Society*, **158**, 831–841.

Fairchild, I.J., Smith, C.L., Baker, A. et al. (2006a) Modification and preservation of environmental signals in speleothems. *Earth-Science Reviews*, **75**, 105–153.

Fairchild, I.J., Tuckwell, G.W., Baker, A. & Tooth, A.F. (2006b) Modelling of drip water hydrology and hydrogeochemistry in a weakly karstified aquifer (Bath, UK): implications for climate change studies. *Journal of Hydrology*, **321**, 213–231.

Fairchild, I.J., Baker, A., Fuller, L. et al. (2006c) Speleophysiology: a key to understanding high-resolution information in speleothems. In: Onac, B.P., Tămas, T., Constantin, S. & Perşoiu, A. (eds.) *Archives of Climate Change in Karst*. Proceedings of the symposium Climate Change: The Karst Record (IV), Băile Herculane, Romania 26–29 May 2006. Leesburg, Virginia: Karst Waters Institute (Special Publication 10), pp. 4–5.

Fairchild, I.J., Frisia, S., Borsato, A. & Tooth, A.F. (2007) Speleothems. In: *Geochemical Sediments and Landscapes* (ed. Nash, D.J. & McLaren, S.J.), Blackwell, Oxford, pp. 200–245.

Fairchild, I.J., Loader, N.J., Wynn, P.M., et al. (2009) Sulfur fixation in wood mapped by synchrotron X-ray studies: implications for environmental archives. *Environmental Science and Technology*, **43**, 1310–1315.

Fairchild, I.J., Spötl, C., Frisia, S., et al. (2010) Petrology and geochemistry of annually laminated stalagmites from an Alpine cave (Obir, Austria): seasonal cave physiology. In: Pedley, H.M. & Rogerson, M. (eds) Tufas and Speleothems: unravelling the Microbial and Physical Controls. *Geological Society, London, Special Publication*, **336**, 295–321.

Fang, C. & Moncrieff, J.B. (1999) A model for soil CO_2 production and transport 1: model development. Agric. *Forest Meteorology*, **95**, 225–236.

Fang, C. & Moncrieff J.B. (2001) The dependence of soil CO_2 efflux on temperature. *Soil Biology and Biochemistry*, 155–165.

Fantle, M.S. & DePaolo, D.J. (2007) Ca isotope in carbonate sediment and pore fluid from ODP Site 807A: the Ca^{2+} (aq)-calcite equilibrium fractionation factor and calcite recrystallization rates in Pleistocene sediments. *Geochimica et Cosmochimica Acta*, **71**, 2524–2546.

Farrand, W.R. (2001) Sediments and stratigraphy in rock-shelters and caves: a personal perspective on principles and pragmatics. *Geoarchaeology*, **16**, 537–557.

Farrant, A.R., Smart, P.L., Whitaker, F.F. & Tarling, D.H. (1995) Long-term Quaternary uplift rates inferred from limestone caves in Sarawak, Malaysia. *Geology*, **23**, 357–360.

Faulkner, T. (2006) Tectonic inception in Caledonide marbles. *Acta Carsologica*, **35**, 7–21.

Faulkner, T. (2007) The one-eighth relationship that constrains deglacial seismicity and cave development in Caledonide marbles. *Acta Carsologica*, **36**, 195–202.

Faulkner, T. (2009) The endokarstic erosion of marble in cold climates: Corbel revisited. *Progress in Physical Geography*, **33**, 805–814.

Feng, X., Faiia, A.M. & Posmentier, E.S. (2009) Seasonality of isotopes in precipitation: a global perspective. *Journal of Geophysical Research* **114**, D08116, doi:10.1029/2008 JD011279.

Fernandez, P.L., Gutierrez, I., Quindós, L.S. et al. (1986) Natural ventilation of the Paintings Room in the Altamira cave. *Nature*, **321**, 586–588.

Fernandez-Cortes, A., Calaforra, J.M. & Sanchez-Martos, F. (2006) Spatiotemporal analysis of air conditions as a tool for the environmental management of a show cave (Cueva del Agua, Spain). *Atmospheric Environment*, **40**, 7378–7394.

Fernandez-Cortes, A., Sánchez-Moral, S., Cuezva, S. et al. (2009) Annual and transient signatures of gas exchange and transport in the Castañar de Ibor cave (Spain). *International Journal of Speleology*, **38**, 153–162.

Fernandez-Cortes A., Sánchez-Moral S., Cuezva S. et al. (2010) Characterization of trace gases fluctuations on a cave using entropy of curves. *International Journal of Climatology*, **31**, 127–143.

Fernandez-Cortes, A., Cuezva, S., Sanchez-Moral, S. et al. (2011) Detection of human-induced environmental disturbances in a show-cave. *Environmental Science and Pollution Research* (in press).

Filipponi, M, Jeannin, P-Y. & Tacher, L. (2009) Evidence of inception horizons in karst conduit networks. *Geomorphology*, **106**, 86–99.

Filipponi, M., Jeannin, P.Y. & Tacher, L. (2010) Understanding cave genesis along favourable bedding planes. The role of the primary rock permeability. *Zeitshcrift für Geomorphologie*, **54** (supplement 2), 91–114.

Finch, A.A., Shaw, P.A., Weedon, G.P. & Holmgren, K. (2001) Trace element variation in speleothem aragonite: potential for paleoenvironmental reconstruction. *Earth and Planetary Science Letters*, **186**, 255–267.

Finch, A.A., Shaw, P.A., Holmgren K., & Lee-Thorp J. (2003) Corroborated rainfall records from aragonitic stalagmites. *Earth and Planetary Science Letters*, **215**, 265–273.

Fischer, M.J. & Treble, P.C. (2008) Calibrating climate-$\delta^{18}O$ regression models for the interpretation of high-resolution speleothem $\delta^{18}O$ time series. *Journal of Geophysical Research*, **113**, D17103, doi:10.1029/2007 JD009694.

Fisher, E.C., Bar-Matthews, M., Jerardino, A. & Marean, C.W. (2010) Middle and Late Pleistocene paleoscape modelling along the southern coast of South Africa. *Quaternary Science Reviews*, **29**, 1382–1398.

Fleitmann, D., Burns, S.J., Mudelsee, M. et al. (2003a) Holocene forcing of the Indian monsoon recorded in a stalagmite from Southern Oman. *Science*, **300**, 1737–1739.

Fleitmann, D., Burns, S.J., Neff, U. et al. (2003b) Changing moisture sources over the last 330,000 years in Northern Oman from fluid-inclusion evidence in speleothems. *Quaternary Research*, **60**, 223–232.

Fleitmann, D., Burns, S.J., Neff, U. et al. (2004) Paleoclimate interpretation of high-resolution oxygen isotope profiles derived from annually laminated speleothems from Southern Oman. *Quaternary Science Reviews*, **23**, 935–945.

Fleitmann, D., Burns, S.J., Mangini, A. et al. (2007) Holocene ITCZ and Indian monsoon dynamics recorded in stalagmites from Oman and Yemen (Socotra). *Quaternary Science Reviews*, **26**, 170–188.

Fleitmann, D, Cheng, H, Badertscher, S. et al. (2009) Timing and climatic impact of Greenland interstadials recorded in stalagmites from northern Turkey. *Geophysical Research Letters*, **36**, L19707, doi:10.1029/2009GL040050.

Florea, L.J., Vacher, H.L., Donahue, B. & Naar, D. (2007) Quaternary cave levels in peninsular Florida. *Quaternary Science Reviews*, **26**, 1344–1361.

Florence, R.G. (1996) *Ecology and Silviculture of Eucalypt Forests*. Canberra: CSIRO.

Flores, O., Gritti, E.S. & Jolly, D. (2009) Climate and CO_2 modulate the C3/C4 balance and $\delta^{13}C$ signal in simulated vegetation. *Climates of the Past*, **5**, 431–440.

Fohlmeister, J., Kromer, B. & Mangini, A. (2011a) The influence of soil organic matter age spectrum on the reconstruction of atmospheric ^{14}C levels via stalagmites. *Radiocarbon*, **53**, 99–115.

Fohlmeister, J., Scholtz, D., Kromer, B. & Mangini, A. (2011b) Modelling carbon isotopes of carbonates in cave drip water. *Geochimica et Cosmochimica Acta*, **75**, 5219–5228.

Folk, R.L. (1962) Spectral subdivision of limestone types. In: Ham, W.E. (editor), *Classification of Carbonate Rocks. Memoir of the American Association of Petroleum Geologists*, **1**, 62–84.

Folk, R.L. & Assereto R. (1976) Comparative fabrics of length-slow and length-fast calcite and calcitzed aragonite in a Holocene speleothem, Carlsbad Caverns, New Mexico. *Journal of Sedimentary Petrology*, **46**, 486–496.

Ford, D C. (1965) The origin of limestone caverns: a model from the central Mendip Hills, England. *Bulletin of the National Speleological Society of America*, **27**, 109–132.

Ford, DC (1980) Threshold and limit effects in karst geomorphology. In: Coates, D.R. and Vitek, J.D. (eds.) *Thresholds in Geomorphology*. London: George Allen & Unwin, pp. 345–362.

Ford, DC & Ewers, R.O. (1978) The development of limestone cave systems in the dimensions of length and breadth. *Canadian Journal of Earth Sciences*, **15**, 1783–1798.

Ford, DC & Williams, P.W. (1989) *Karst Geomorphology and Hydrogeology*. London: Unwin Hyman.

Ford, D. & Williams, P. (2007) *Karst Hydrogeology and Geomorphology*. Chichester: John Wiley.

Forti, P. (2001) Seismotectonic and paleoseismic studies from speleothems: the state of the art. *Geologica Belgica*, **4**, 175–185.

Frank, D., Büntgen, U., Böhm, R. et al. (2007) Warmer instrumental measurements versus colder reconstructed temperatures: shooting at a moving target. *Quaternary Science Reviews*, **26**, 3298–3310.

Franke, F.W. (1965) The theory behind stalagmite shapes. *Studies in Speleology*, **1**, 89–95.

Frappier, A.B. (2008) A stepwise screening system to select storm-sensitive stalagmites: taking a targeted approach to speleothem sampling methodology. *Quaternary International*, **187**, 25–39.

Frappier, A., Sahagian, D., González, L.A. & Carpenter, S.J. (2002) El Niño events recorded by stalagmite carbon isotopes. *Science*, **298**, 565.

Frappier, A.B., Sahagian, D., Carpenter, S.J. & Gonzalez, L.A. (2007) Stalagmite stable isotope record of recent tropical cyclone events. *Geology*, **35**, 111–114.

Fricke, H.C. & O'Neil, J.R. (1999) The correlation between $^{18}O/^{16}O$ ratios of meteoric water and surface temperature: its use in investigating terrestrial climate change over geologic time. *Earth and Planetary Science Letters*, **170**, 181–196.

Friedman, I. & O'Neil, J.R. (1977) Compilation of stable isotope fractionation factors of geochemical interest. *US Geological Survey Professional Paper* 440–KK, 49p.

Frisia, S. (1996) Petrographic evidences of diagenesis in speleothems: some examples. *Speleochronos*, **7**, 21–30.

Frisia, S. & Borsato, A. (2010) Karst. In: Alonso-Zarza, A.M. & Tanner, L.H. (eds.) *Carbonates in Continental Settings*. Elsevier, Amsterdam, pp. 269–318.

Frisia, S., Bini, A. & Quinif, Y. (1993) Morphologic, crystallographic and isotopic study of an ancient flowstone (Grotta di Cunturines, Dolomites)—implications for Paleoenvironmental reconstructions. *Speleochronos*, **5**, 3–18.

Frisia S., Borsato, A., Fairchild, I.J. & McDermott, F. (2000) Calcite fabrics, growth mechanisms and environments of formation in speleothems from the Italian Alps and southwest Ireland. *Journal of Sedimentary Research*, **70**, 1183–1196.

Frisia, S., Borsato, A., Fairchild, I.J. et al. (2002) Aragonite-calcite relationships in speleothems (Grotte de Clamouse, France): environment, Fabrics and carbonate geochemistry. *Journal of Sedimentary Research*, **72**, 687–699.

Frisia, S, Borsato, A, Preto, N. & McDermott, F. (2003) Late Holocene annual growth in three Alpine stalagmites records the influence of solar activity and the North Atlantic Oscillation on winter climate. *Earth and Planetary Science Letters*, **216**, 411–424.

Frisia, S., Borsato, A., Spötl, C. et al. (2005a) Climate variability in the SE Alps of Italy over the past 17000 years reconstructed from a stalagmite record. *Boreas*, **34**, 445–455.

Frisia, S., Borsato, A., Fairchild, I.J. & Susini, J. (2005b) Variations in atmospheric sulphate recorded in stalagmites by synchrotron micro-XRF and XANES analyses. *Earth and Planetary Science Letters*, **235**, 729–740.

Frisia, S., Borsato, A. & Susini, J. (2008) Synchrotron radiation applications to past volcanism archived in speleothems: an overview. *Journal of Volcanology and Geothermal Research*, **177**, 96–100.

Frisia, S., Fairchild, I.J., Fohlmeister, J. et al. (2011) Carbon mass-balance modelling and carbon isotope exchange processes in dynamic caves. *Geochimica et Cosmochimica Acta*, **75**, 380–400.

Frumkin, A. & Stein, M. (2004) The Sahara-East Mediterranean dust and climate connection revealed by strontium and uranium isotopes in a Jerusalem speleothem. *Earth and Planetary Science Letters*, **217**, 451–464.

Frumkin, A., Karkanas, P., Bar-Matthews, M., et al. (2009) Gravitational deformations and fillings of aging caves: the example of Qesem karst system, Israel. *Geomorphology*, **106**, 154–164.

Fuller, L. (2006) *High resolution multiproxy geochemical Holocene climate records from 1000–year old Scottish stalagmites*. PhD Thesis, University of Birmingham.

Fuller, L., Baker, A., Fairchild, I.J., Spötl, C., Marca-Bell, A., Rowe, P. and Dennis, P.F. (2008) Isotope hydrology of dripwaters in a Scottish cave and implications for stalagmite palaeoclimate research. *Hydrology and Earth System Sciences*, **12**, 1065–1074.

Gabitov, R.I. & Watson, E.B. (2006) Partitioning of strontium between calcite and fluid. *Geochemistry, Geophysics, Geosystems*, **7**, Q11004, doi:10.1029/2005GC001216.

Galy, A., Bar-Matthews, M., Halicz, L. & O'Nions, R.K. (2002) Mg isotopic composition of carbonate: insight from speleothem formation. *Earth and Planetary Science Letters*, **201**, 105–115.

Gamble, D.W. Dogwiler, J.T. & Mylroie, J. (2000) Field assessment of the microclimatology of tropical flank margin caves. *Climate Research*, **16**, 37–50.

Gams, I. (1981) Contribution to morphometrics of stalagmite. *Proceedings of the 8th International Congress of Speleology*, 276–278.

Ganor, J., Reznik, I.J. & Rosenberg, Y.O. (2009) Organics in water-rock interactions. *Reviews in Mineralogy and Geochemistry*, **70**, 259–369.

Garofalo, P.S., Fricker, M.B., Günther, D. et al. (2010) Climatic control on the growth of gigantic gypsum crystals within hypogenic caves (Naica mine, Mexico)? *Earth and Planetary Science Letters*, **289**, 560–569.

Garrels, R.M. & Christ, C.L. (1965) *Solutions, Minerals, and Equilibria*. Harper & Row, New York.

Garrels, R.M. & McKenzie, F.T. (1971) *Evolution of Sedimentary Rocks*. W.W. Norton, New York.

Garzione, C.N., Quade, J., DeCelles, P.G. & English, N.B. (2000) Predicting paleoelevation of Tibet and the Himalaya from $\delta^{18}O$ vs. Altitude gradients in meteoric water across the Nepal Himalaya. *Earth and Planetary Science Letters*, **183**, 215–229.

Gascoyne, M. (1983) Trace element partition coefficients in the calcite-water system and their palaeoclimatic significance in cave studies. *Journal of Hydrology*, **61**, 213–222.

Gat, J.R. (1996) Oxygen and hydrogen isotopes in the hydrological cycle. *Annual Review of Earth and Planetary Sciences*, **24**, 225–262.

Gat, J.R. (2000) Atmospheric water balance—the isotopic perspective. *Hydrological Processes* **14**, 1357–1369.

Genty, D. (1992) Les spéléothèmes du tunnel de Godarville (Belgique)—un exemple exceptionnel de concrétionnement moderne—intérêt pour l'étude de la cinétique de la précipitation de la calcite et de sa relation avec les variations d'environnement. *Speleochronos*, **4**, 3–29.

Genty, D. (1993) Mise en évidence d'alternances saisonnières dans la structure interne des stalagmites. Intérêt pour la reconstitution des paléoenvironnements continentaux. *Comptes Rendus Academie Sciences Paris*, **317**, Série II, 1229–1236.

Genty, D. (2008) Palaeoclimate research in Villars Cave (Dordogne, SW-France). *International Journal of Speleology*, **37**, 173–191.

Genty, D. & Quinif, Y. (1996) Annually laminated sequences in the internal structure of some Belgian stalagmites-importance for paleoclimatology. *Journal of Sedimentary Research*, **66**, 275–288.

Genty, D. & Deflandre, G. (1998) Drip flow variations under a stalactite of the Pere Noel cave (Belgium). Evidence of seasonal variations and air pressure constraints. *Journal of Hydrology*, **211**, 208–232.

Genty, D. & Massault, M. (1999) Carbon transfer dynamics from bomb ^{14}C and $d^{13}C$ times series of a laminated stalagmite from SW France: modelling and comparison with other stalagmites. *Geochimica Cosmochimica Acta*, **63**, 1537–1548.

Genty, D., Baker, A. & Barnes, W.L. (1997) Comparison of annual luminescent and visible laminae in stalagmites. *Comptes Rendus Academie Sciences Paris*, Serie II, **325**, 193–200.

Genty, D., Vokal, B., Obelic, B. & Massault, M. (1998) Bomb ^{14}C time history recorded in two modern stalagmites—importance for soil organic matter dynamics and bomb ^{14}C distribution over continents. *Earth and Planetary Science Letters*, **160**, 795–809.

Genty, D., Massault, M., Gilmour, M. et al. (1999) Calculation of past dead carbon proportion and variability by the comparison of AMS ^{14}C and TIMS U/Th ages on two Holocene stalagmites. *Radiocarbon*, **41**, 251–270.

Genty, D., Baker, A., Massault, M. et al. (2001a) Dead carbon in stalagmites: carbonate bedrock paleodissolution vs. aging of soil organic matter. Implications for ^{13}C variations in speleothems. *Geochimica et Cosmochimica Acta*, **65**, 3443–3457.

Genty, D., Baker, A. & Vokal, B. (2001b) Inter and intra annual growth rates of European stalagmites. *Chemical Geology*, **176**, 193–214.

Genty, D., Diot, M.-F. & O'Yl, W. (2001c) Sources of pollen in stalactite drip water in two caves in southwest France. *Cave and Karst Science*, **28**, 59–66.

Genty, D., Blamart, D, Ouahdi, R. et al. (2003) Precise dating of Dansgaard-Oeschger climate oscillations in western Europe from stalagmite data. *Nature*, **421**, 833–837.

Genty, D., Combourieu-Nebout, N., Peyron, O. et al. (2010) Isotopic characterisation of rapid climate events during OIS3 and OIS4 in Villars cave stalagmites (SW-France) and correlation with Atlantic and Mediterranean pollen records. *Quaternary Science Reviews*, **29**, 2799–2820.

Gewelt, M. & Ek, C. (1983) L'Evolution saissonièrre de la teneur en CO_2 de l'air de deux grottes Belges: Ste-Ann et Brialmont, Tilff. In Patterson, K.N & Sweeting M.M. (eds.), *New Directions in Karst*. Norwich, Geo Books.

Geyh, M.A. (1970) Zeitliche abrenzung von klimaänderungen mit ^{14}C-daten von kalksinter und organischen substanzen. *Beihefte zum geologischen Jahrbuch*, **98**, 15–22.

Geyh, M.A. & Schleicher, H. (2000) *Absolute Age Determination*. Springer-Verlag.

Ghinassi, M., Colonese, A.C., di Giuseppe, Z. et al. (2009) The Late Pleistocene clastic deposits in the Romito Cave, southern Italy: a proxy record of environmental changes and human presence. *Journal of Quaternary Science*, **24**, 383–398.

Ghosh, P., Adkins, J., Affek, H. et al. (2006a) ^{13}C-^{18}O bonds in carbonate minerals: a new kind of paleothermometer. *Geochimica et Cosmochimica Acta*, **70**, 1439–1456.

Ghosh, P., Garzione, C.N. & Eiler, J.M. (2006b) Rapid uplift of the Altiplano revealed through ^{13}C-^{18}O bonds in paleosol carbonates. *Nature*, **311**, 511–515.

Gilbert, P.U.P.A., Abrecht, M. & Frazer, B.H. (2005) The organic-mineral interface in biominerals. *Reviews in Mineralogy and Geochemistry*, **59**, 157–185.

Gill, B.C., Lyons, T.W. & Frank, T.D. (2008) Behavior of carbonate-associated sulphate during meteoric diagenesis and implications for the sulphur isotope paleoproxy. *Geochimica et Cosmochimica Acta*, **72**, 4699–4711.

Gilli, E. (1986) Néotectonique dans les massifs karstiques. Un exemple dans les Préalpes de Nice: la Grotte des Deux Gordes. *Karstologia*, **8**, 51–52.

Gilli, E. (1992) Au coeur d'un chevauchement dans les gouffres du Calernaum et des Baoudillouns. *Karstologia*, **19**, 39–48.

Gilli, E. (1999) Research on the February 18 (1996) earthquake in the caves of Saint-Paul-de-Fenouillet area (eastern Pyrenees, France). *Geodinamica Acta*, **12**, 143–158.

Gilli, E. & Serface, R. (1999) Evidence of palaeoseismicity in the caves of Arizona and New Mexico (USA). *Comptes Rendus Academie Sciences*, **329**, 31–37.

González, L.A., Carpenter, S.J. & Lohmann, K.C. (1992) Inorganic calcite morphology: roles of fluid chemistry and fluid flow. *Journal of Sedimentary Research*, **62**, 382–399.

Gillieson, D. (1986) Cave sedimentation in the New Guinea Highlands. *Earth Surface Processes and Landforms*, **11**, 533–543.

Gillieson, D. (1996) *Caves: process, development, and management*. Blackwell, London.

Gilman, K. & Newson, M.D. (1980) Soil pipes and pipeflow—a hydrological study in upland Wales. *British Geomorphological Research Group, Research Monograph Series 1*, Geo Books, Norwich, 110pp.

Giorgi, F. (2006) Climate change hot-spots. *Geophysical Research Letters*, **33**, L08707, doi:10.1029/2006GL025734.

Goede, A., McCulloch, M., McDermott, F. & Hawkesworth, C. (1998) Aeolian contribution to strontium and strontium isotope variations in a Tasmanian speleothem. *Chemical Geology*, **149**, 37–50.

Goldberg, P. & Sherwood, S.C. (2006) Deciphering human prehistory through the geoarcheological study of cave sediments. *Evolutionary Anthropology*, **15**, 20–36.

Gonzalez, L.A. & Lohmann, K.C. (1988) Controls on mineralogy and composition of spelean carbonates: Carlsbad Cavern, New Mexico, in James N.P. and Choquette P.W. eds., *Paleokarst*. New York, Springer-Verlag, pp. 81–101.

Goosse H., Renssen, H., Timmermann, A. et al. (2006) Using paleoclimate proxy-data to select an optimal realisation in an ensemble of simulations of the climate of the past millennium. *Climate Dynamics*, **27**, 165–184.

Goosse, H., Mann, M.E. & Renssen, H. (2008) What we can learn from combining paleoclimate proxy data and climate model simulations of past centuries? In: Battarbee R. & Binney H. (eds.) *Natural Climate Variability and Global Warming: a Holocene Perspective*, 163–188, Blackwell, Oxford.

Goosse, H., Crespin, E., de Montety, A. et al. (2010) Reconstructing surface temperature changes over the past 600 years using climate model simulations with data assimilation. *Journal of Geophysical Research*, **115**, D09108, doi:10.1029/2009JD012737.

Goslar, T. & Hercmann, H. (1988) TL and ESR dating of speleothems and radioactive disequilibrium in the uranium series. *Quaternary Science Reviews*, **7**, 423–427.

Gospardovič, R. (1976) Razvoj jam med Pivško kotlino in Planinskim poljem v kvartarju (The Quaternary caves development between the Pivka basin and Polje of Planina). *Acta Carsologica*, **VII**, 9–135 & 20 Plates (In Slovenian with English summary pp. 121–135).

Goudie, A. (2006) *The human impact on the natural environment: past, present, and future*. Blackwell, Oxford.

Gouin, P. (1979) *Earthquake History of Ethiopia and the Horn of Africa*. International Development Research Center, Ottawa, Canada.

Gradstein, F., Ogg, J., Smith A. et al. (2004) *A Geologic Time Scale 2004*. Cambridge: Cambridge University Press.

Graham Wall, B.R. (2006) Influence of depositional setting and sedimentary fabric on mechanical layer evolution in carbonate aquifers. *Sedimentary Geology*, **184**, 203–224.

Griffiths, M.L., Drysdale, R.N., Gagan, M.K. et al. (2009) Increasing Australian-Indonesian monsoon rainfall linked to early Holocene sea-level rise. *Nature Geoscience*, **2**, 636–639.

Griffiths, M.L., Drysdale, R.N., Gagan, M.K. et al. (2010a) Evidence for Holocene changes in Australian-Indonesian monsoon rainfall from stalagmite trace element and stable isotope ratios. *Earth and Planetary Science Letters*, **295**, 30–36.

Griffiths, M.L., Drysdale, R.N., Vonhof, H.B. et al. (2010b) Younger Dryas-Holocene temperature and rainfall history of southern Indonesia from $\delta^{18}O$ in speleothem calcite and fluid inclusions. *Earth and Planetary Science Letters*, **295**, 30–36.

Grigor'ev, D.P. (1965) *Ontogeny of Minerals* (translated from the Russian by Brenner, Y.) Israel Program for Scientific Translations, Jerusalem.

Grun, R. (1991) Potential and problems of ESR dating. *Nuclear Tracks and Radiation Measurements*, **18**, 143–153.

Guan, H., Simmons, C.T. & Love, A.J. (2009) Orographic controls on rain water isotope distribution in the Mount Lofty Ranges of South Australia. *Journal of Hydrology*, **374**, 255–264.

Gunn, J. (1974) A model of the karst percolation system of waterfall swallet, Derbyshire. *Transactions of the British Cave Research Association*, **3**, 159–164.

Gunn, J. (1981) Limestone solution rates and processes in the Waitomo district, New Zealand. *Earth Surface Processes and Landforms*, **6**, 427–445.

Gunn, J. (1983) Point-recharge of limestone aquifers—a model from New Zealand karst. *Journal of Hydrology*, **61**, 19–29.

Gunn, J. (ed) (2004) *Encyclopedia of Caves and Karst Science*. Fitzroy Dean, New York.

Gunn, J., Fletcher, S. & Prime, D. (1991) Research on radon in British limestone caves and mines 1970–1990. *Cave Science*, **18**, 63–67.

Guo, W. (2009) Carbonate clumped isotope thermometry: application to carbonaceous chondrites and effects of kinetic isotope fractionation, PhD thesis. California Institute of Technology.

Guo, W., Mosenfelder, J.L., Goddard, W.A. & Eiler, J.M. (2009) Isotopic fractionations associated with phosphoric acid digestion of carbonate minerals: insights from first-principles theoretical modelling and clumped isotope measurements. *Geochimica et Cosmochimica Acta*, **73**, 7203–7225.

Gupta P., Noone D., Galewsky J., Sweeney C. & Vaughn B.H. (2009) Demonstration of high-precision continuous measurements of water vapour isotopologues in laboratory and remote field deployments using wavelength-scanned cavity ring-down spectroscopy (WS-CRDS) technology. *Rapid Communications in Mass Spectrometry*, **23**, 2534–2542.

Haam, E. & Huybers, P. (2010) A test for the presence of covariance between time-uncertain series of data with application to the Dongge Cave speleothem and atmospheric radiocarbon records. *Paleoceanography*, **25**, PA2209.

Haeuselmann, P., Granger, D.E., Jeannin, P.-Y. & Lauritzen, S.-E. (2007) Abrupt glacial valley incision at 0.8 Ma dated from cave deposits in Switzerland. *Geology*, **35**, 143–146.

Hajna, N.Z. (2003) *Incomplete solution: weathering of cave walls and the production, transport and deposition of carbonate fines*. Ljubljana: ZRC Publishing.

Hajna, N.Z., Mihevc, A., Pruner, P. & Bosák, P. (2008a) *Palaeomagnetism and Magnetostratigraphy of Karst Sediments in Slovenia*. Ljubljana: ZRC Publishing.

Hajna, N.Z., Pruner, P., Mihevc, A., Schnabl, P. & Bosák, P. (2008b) Cave sediments from the Postojnska-Planinska cave system (Slovenia): evidence of multi-phase evolution in epiphreatic zone. *Acta Carsologica*, **37**, 63–86.

Hakl, J., Hunyadi, I., Csige, I. et al. (1997) Radon transport phenomena studied in karst caves—international experiences on radon levels and exposures. *Radiation Measurements*, **28**, 675–684.

Halicz, L., Segal, I., Fruchter, N. et al. (2008) Strontium stable isotopes fractionate in the soil environments? *Earth and Planetary Science Letters*, **272**, 406–411.

Hamato, H., Landsberger, S., Harbottle, G. & Panno, S. (1995) Studies of radioactivity and heavy metals in phosphate fertilizer. *Journal of Radioanalytical and Nuclear Chemistry*, **194**, 331–336.

Hansen, J.E., Ruedy, R., Sato, M. et al. (2001) A closer look at United States and global surface temperature change. *Journal of Geophysical Research*, **106**, 23947–23963.

Harmon, R.S., Schwarcz, H.P. & Ford, DC (1978) Late Pleistocene sea level history of Bermuda. *Quaternary Research*, **9**, 205–218.

Hartland, A. (2011) *Speciation and hydrogeochemistry of trace metal-NOM complexes in speleothem-forming groundwaters*. PhD Thesis, University of Birmingham, UK.

Hartland, A., Fairchild, I.J., Lead, J.R. (2010a) The dripwaters and speleothems of Poole's Cavern: a review of recent and ongoing research. *Cave and Karst Science*, **36**, 37–46.

Hartland, A., Fairchild, I.J., Lead, J.R. & Baker, A. (2010b) Fluorescent properties of organic carbon in cave dripwaters: effects of filtration, temperature and pH. *Science of the Total Environment*, **408**, 5940–5950.

Hartland, A., Fairchild, I.J. Lead, J.R. et al. (2011) Size, speciation and lability of NOM-metal complexes in hyperalkaline cave dripwater. *Geochimica et Cosmochimica Acta*, **75**, 7533–7551.

Hartman, H.L. (1961) *Mine ventilation and air conditioning*. New York: Ronald Press.

Hartman, P. (1987) Modern PBC theory, in Sunagawa, I. (ed.), *Morphology of Crystals*: Terra Scientific Publishing Company, Tokyo, 269–319.

Haslett, J. & Parnell, A. (2008) A simple monotone process with application to radiocarbon dated depth chronologies. *Journal of the Royal Statistical Society C*, **57**, 1–20.

Hatté, C., Rousseau, D.-D. & Guiot, J. (2009) Climate reconstruction from pollen and $\delta^{13}C$ records using inverse vegetation modeling—Implication for past and future climates. *Climate of the Past*, **5**, 147–156.

Haug, G.H., Hughen, K.A., Sigman D.M. et al. (2001) Intertropical convergence zone through the Holocene. *Science*, **293**, 1304–1308.

Hays, P.D. & Grossman, E.L. (1991) Oxygen isotopes in meteoric calcite cements as indicators of continental paleoclimate. *Geology*, **19**, 441–444.

He, H. & Smith, R.B. (1999) Stable isotope composition of water vapour in the atmospheric boundary layer above the forests of New England. *Journal of Geophysical Research—Atmospheres*, **104**, 11657–11673.

He, Y., Pang, H., Theakstone, W.H. et al. (2006) Isotopic variations in precipitation at Bangkok and their climatological significance. *Hydrological Processes*, **20**, 2873–2884.

Heathwaite, A.L. (1997) Sources and pathways of phosphorus loss from agriculture. In: (Tunney, H., Careton, O.T., Brookes, P.C. & Johnstone, A.E., editors) *Phosphorus loss from soil to water*. CAB International, Wallingford, UK, pp. 205–223.

Hedges, J.I. & Mann, DC (1979) The characterization of plant tissues by their cupric oxide oxidation products. *Geochimica et Cosmochimica Acta*, **43**, 1803–1807.

Hedges, J.I., Blanchette, R.A., Weliky, K. & Devol, A.H. (1988) Effects of fungal degradation on the CuO oxidation products of lignin: a controlled laboratory study. *Geochimica et Cosmochimica. Acta*, **52**, 2717–2726.

Heegaard, E., Birks, H.J.B. & Telford, R.J. (2005) Relationships between calibrated ages and depth in stratigraphic sequences: an estimation procedure by mixed-effect regression. *The Holocene*, **15**, 612–618.

Hellstrom, J. (2003) Rapid and accurate U/Th dating using parallel ion-counting multi-collector ICP-MS. *Journal of Analytical Atomic Spectrometry*, **18**, 1346–1351.

Hellstom, J. (2006) U-Th dating of speleothem with high initial ^{230}Th using stratigraphic constraint. *Quaternary Geochronology*, **1**, 289–295.

Hellstrom, J.C., McCulloch, M.T. & Stone J. (1998) A 31,000 year high-resolution record of southern hemisphere maritime climate changes, from the stable isotope geochemistry of New Zealand speleothems. *Quaternary Research*, **50**, 167–178.

Henderson, G.M. (2002) New oceanic proxies for paleoclimate. *Earth and Planetary Science Letters*, **203**, 1–13.

Henderson, G.M. (2006) Caving in to new chronologies. *Science*, **313**, 620–622.

Henderson, G.M., Hu, C.Y. & Johnson, K.R. (2008) Controls on trace elements in stalagmites derived from in situ growth in a Chinese cave. [abstract]. *Geochimica et Cosmochimica Acta*, **72**, A366.

Henderson-Sellers, A. & Robinson, P.J. (1999) *Contemporary Climatology*. Prentice Hall, New York.

Hendy, C.H. (1971) The isotopic geochemistry of speleothems—I. The calculation of the effects of different modes of formation on the isotopic composition of speleothems and their applicability as palaeoclimatic indicators. *Geochimica et Cosmochimica Acta*, **35**, 801–824.

Hendy, C.H. & Wilson, A.T. (1968) Palaeoclimate data from speleothems. *Nature*, **219**, 48–51.

Herbert, T.D., Schuffert, J.D., Andreasen, D. et al. (2001) Collapse of the California Current during glacial maxima linked to climate change on land. *Science*, **293**, 71–76.

Hill, C.A. & Forti, P. (1997) *Cave Minerals of the World*. National Speleological Society, Huntsville, Alabama.

Hoffmann, D.L. (2008) ^{230}Th isotope measurements of femtogram quantities for U-series dating using multi ion counting (MIC) MC-ICPMS. *International Journal of Mass Spectrometry*, **275**, 75–79.

Hoffmann, G., Werner, M. & Heimann, M. (1998) Water isotope module of the ECHAM atmospheric general circulation model: a study of timescales from days to several years, *Journal of Geophysical Research*, **103**, 16871–16896.

Hoffmann, D.L., Prytulak, J., Richards, D.A. et al. (2007) Procedures for accurate U and Th isotope measurements by high precision MC-ICPMS. *International Journal of Mass Spectrometry*, **264**, 97–109.

Hoffmann, D.L., Spötl, C. & Mangini, A. (2009) Micromill and in situ laser ablation sampling techniques for high spatial resolution MC-ICPMS U-Th dating of carbonates. *Chemical Geology*, **259**, 253–261.

Hoffmann, D.L., Beck, J.W., Richards, D.A. et al. (2010) Towards radiocarbon calibration beyond 28 ka using speleothems from the Bahamas. *Earth and Planetary Science Letters*, **289**, 1–10.

Holden, N.E. (1990) Total half-lives for selected nuclides, *Pure and Applied Chemistry*, **62**, 941–958.

Holdsworth, G. (2008) A composite isotopic thermometer for snow. *Journal of Geophysical Research*, **113**, D08102, doi:10.1209/2007JD008634.

Holland, H.D. (2005) Sea level, sediments and the composition of seawater. *American Journal of Science*, **305**, 220–239.

Holland, H.D., Kirsipu, T.V., Huebner, J.S. & Oxburgh, U.M. (1964) On some aspects of the chemical evolution of cave waters. *Journal of Geology*, **72**, 36–67.

Hopley, P.J., Marshall, J.D., Weedon, G.P. et al. (2007a) Orbital forcing and the spread of C4 grasses in the late Neogene: stable isotope evidence from South African speleothems. *Journal of Human Evolution*, **53**, 620–634.

Hopley, P.J., Weedon, G.P., Marshall, J.D. et al. (2007b) High- and low-latitude orbital forcing of early hominin habitats in South Africa. *Earth and Planetary Science Letters*, **256**, 419–432.

Hopley, P.J., Marshall, J.D. & Latham, A.G. (2009) Speleothem preservation and diagenesis in South African hominin sites: implications for paleoenvironments and geochronology. *Geoarchaeology*, **24**, 519–547.

Hopmans, E.C., Weijers, J.W.H., Schefuss, E., et al. (2004) A novel proxy for terrestrial organic matter in sediments based on branched and isoprenoid tetraether lipids. *Earth and Planetary Science Letters*, **224**, 107–116.

Hoyos, M., Soler, V., Cañavera, J.C. et al. (1998) Microclimatic characterization of a karstic cave: human impact on microenvironmental parameters of a prehistoric rock art cave (Candamo Cave, northern Spain). *Environmental Geology*, **33**, 231–242.

Hu, C., Huang, J., Fang, N. et al. (2005) Adsorbed silica in stalagmite carbonate and its relationship to past rainfall. *Geochimica et Cosmochimica Acta*, **69**, 2285–2292.

Hu, C., Henderson, G.M., Huang, J.H. et al. (2008a) Quantification of Holocene Asian monsoon rainfall from spatially separated cave records. *Earth and Planetary Science Letters*, **266**, 221–232.

Hu, C., Henderson, G.M., Huang, J. et al. (2008b) Report of a three-year monitoring programme at Heshang Cave, Central China. *International Journal of Speleology*, **37**, 143–151.

Hua, Q. (2009) Radiocarbon: a chronological tool for the recent past. *Quaternary Geochronology*, **4**, 378–390.

Huang, X.-Y., Cui, J.-W., Pu, Y., Huang, J.-H. & Blyth, A.J. (2008) Identifying 'free' and 'bound' lipid fractions in stalagmite samples: an example from Heshang Cave, Southern China. *Applied Geochemistry*, **23**, 2589–2595.

Huang, Y. & Fairchild, I.J. (2001) Partitioning of Sr^{2+} and Mg^{2+} into calcite in karst-analogue experimental solutions. *Geochimica et Cosmochimica Acta*, **65**, 47–62.

Huang, Y., Fairchild, I.J., Borsato, A. et al. (2001) Seasonal variations in Sr, Mg and P in modern speleothems (Grotta di Ernesto, Italy). *Chemical Geology*, **175**, 429–448.

Hudson, N.J., Baker, A. & Reynolds, D. (2007) Fluorescence analysis of dissolved organic matter in natural, waste and polluted waters—a review. *River Research and Applications*, **23**, 631–649.

Huggett, R.J. (1985) *Earth Surface Systems*. Berlin: Springer-Verlag.

Hughes, S.G., Taylor, E.L., Wentzell, P.D. et al. (1994) Models of conductance measurements in quality assurance of water analysis. *Analytical Chemistry*, **66**, 830–835.

IAEA (2009a) GNIP programme. Global Network of Isotopes in Precipitation. http://www-naweb.iaea.org/napc/ih/GNIP/IHS_GNIP.html (accessed 20/10/09).

IAEA (2009b) IAEA-WMO Programme on Isotopic Composition of Precipitation: Global Network of

Isotopes in Precipitation (GNIP) Technical Procedure for Sampling. http://www-naweb.iaea.org/napc/ih/GNIP/userupdate/sampling.pdf.

Ihlenfeld, C., Norman, M.D., Gagan, M.K. & Drysdale, R.N. (2003) Climatic significance of seasonal trace element and stable isotope variations in a modern freshwater tufa. *Geochimica et Cosmochimica Acta*, **67**, 2341–2357.

Immenhauser, A., Buhl, D., Richter, D. et al. (2010) Magnesium-isotope fractionation during low-Mg calcite precipitation in a limestone cave—Field study and experiments. *Geochimica et Cosmochimica Acta*, **74**, 4346–4364.

Ingraham N.L. & Taylor B.E. (1991) Light isotope systematics of large-scale hydrologic regimes in California and Nevada. *Water Resources Research*, **27**, 77–90.

Ingraham, N.L. & Criss, R.E. (1993) Effects of surface area and volume on the rate of isotopic exchange between water and water vapour. *Journal of Geophysical Research*, **98**, 20547–20553.

Ingraham, N.L., Chapman, J.B. & Hess, J.W. (1990) Stable isotopes in cave pool systems: Carlsbad Cavern, New Mexico, U.S.A. *Chemical Geology*, **86**, 65–74.

Inskeep, W.P. & Bloom, P.R. (1986) Kinetics of calcite precipitation in the presence of water-soluble ligands. *Soil Science Society of America Journal*, **50**, 1167–1172.

IPCC (Intergovernmental Panel on Climate Change) (2007) *Climate Change 2007: the Physical Science Basis*. Cambridge: Cambridge University Press.

Jackson, A.S., McDermott, F. & Mangini, A. (2008) Late Holocene climate oscillations and solar fluctuations from speleothem STAL-AH-1, Sauerland, Germany: A numerical perspective. *Geophysical Research Letters*, **35**, L06702, doi:10.1029/2007GL032689.

Jacob, J., Disnar, J-R., Boussafir, M. et al. (2007) Contrasted distributions of triterpene derivatives in the sediments of Lake Caçó reflect palaeoenvironmental changes during the last 20,000 yrs in NE Brazil. *Organic Geochemistry*, **38**, 180–197.

Jaffey, A.H., Flynn, K.F., Glendenin, L.E. et al. (1971) Precision measurement of half-lives and specific activities of ^{235}U and ^{238}U, *Physical Review C*, **4**, 1889–1906.

James J.M. (1977) Carbon dioxide in the cave atmosphere. *Transactions of the British Cave Research Association*, **4**, 417–429.

James, J.M., Pavey, A.J. & Rogers, A.F. (1975) Foul air and the resulting hazards to cavers. *Transactions of the British Cave Research Association*, **2**, 79–88.

James, N.P. & Choquette, P.W. (eds.) (1988) *Paleokarst*. New York: Springer-Verlag.

James, N.P. & Choquette, P.W. (1990a) Limestones—the sea-floor diagenetic environment. In: McIlreath, I.A. & Morrow, D.W. (eds.) *Diagenesis*. St. John's, Newfoundland: Geological Association of Canada, pp. 13–34.

James, N.P. & Choquette, P.W. (1990b) Limestones—the meteoric diagenetic environment. In: McIlreath, I.A. & Morrow, D.W. (eds.) *Diagenesis*. St. John's, Newfoundland: Geological Association of Canada, pp. 35–73.

Jennings, J.N. (1968) Syngenetic karst in Australia: contributions to the study of karst. Canberra: Australian National University. *Department of Geography Publication Number* G/5, 41–110.

Jennings, J.N. (1985) *Karst Geomorphology*. Oxford: Basil Blackwell.

Jenson, J.W., Keel, T.M., Mylroie, J.R. et al. (2006) Karst of the Mariana Islands: the interaction of tectonics, glacioeustasy, and freshwater/seawater mixing in island carbonates. In: Harmon, R.S. & Wicks, C. (eds.) *Perspectives on Karst Geomorphology, Hydrology and Geochemistry—A Tribute Volume to Derek C. Ford and William B. White. Geological Society of America Special Paper* 404, 129–138.

Jeong, G.Y., Kim, S.J. & Chang, S.J. (2003) Black carbon pollution of speleothems by fine urban aerosols in tourist caves. *American Mineralogist*, **88**, 1872–1878.

Jernigan, J.W. & Swift, R. (2001) A mathematical model of air temperature in Mammoth Cave, Kentucky. *Journal of Cave and Karst Studies*, **61**, 3–8.

Jex, C., Claridge, E., Baker, A. & Smith, C. (2008) Hyperspectral imaging of speleothems. *Quaternary International*, **187**, 5–14.

Jex, C., Baker, A., Fairchild, I.J. et al. (2010) Calibration of speleothem δ^{18}O with instrumental climate records from Turkey. *Global and Planetary Change*, **71**, 207–217.

Jex, C.N., Baker, A., Eden, J.M., et al. (2011) A 500 yr speleothem-derived reconstruction of late autumn-winter precipitation, North East Turkey. *Quaternary Research*, **75**, 399–405.

Jiménez-López, C., Caballero, E., Huertas, F.J. & Romanek, C.S. (2001) Chemical, mineralogical and isotope behaviour, and phase transformations during the precipitation of calcium carbonate minerals from intermediate ionic solutions at 25 °C. *Geochimica et Cosmochimica Acta*, **65**, 3219–3231.

Jimenez-Lopez, C., Rodriguez-Navarro, A., Dominguez-Vera, J.M. & Garcia-Ruiz, J.M. (2003) Influence of lysozyme on the precipitation of calcium carbonate: a kinetic and morphologic study. *Geochimica Cosmochimica Acta*, **67**, 1667–1676.

Jiménez-Sánchez, M., Stoll, H., Vadillo, I. et al. (2008) Groundwater contamination in caves: four case studies in Spain. *International Journal of Speleology*, **37**, 53–66.

Jin, Z., Wang, S., Zhang, F. & Shi, Y. (2010) Weathering, Sr fluxes, and controls on water chemistry in the Lake

Qinghai catchment, NE Tibetan Plateau. *Earth Surface Processes and Landforms*, **35**, 1057–1070.

Jo, K.N., Woo, K.S., Hong, G.H. et al. (2010) Rainfall and hydrological controls on speleothem geochemistry during climatic events (droughts and typhoons): an example from Seopdong Cave, Republic of Korea. *Earth and Planetary Science Letters*, **295**, 441–450.

Johnsen, S.J., Dansgaard, W. & White, J.W.C. (1989) The origin of Arctic precipitation under present and glacial conditions. *Tellus*, **41B**, 452–468.

Johnson, C.M., Beard, B.L. & Albarede, F. (2004) *Geochemistry of non-traditional stable isotopes. Review in Mineralogy and Geochemistry*, Volume **55**. Mineralogical Society of America, Washington DC.

Johnson, K.R. & Ingram, B.L. (2004) Spatial and temporal variability in the stable isotope systematics of modern precipitation in China: implications for paleoclimate reconstructions. *Earth and Planetary Science Letters*, **220**, 365–377.

Johnson, K.R., Hu, C., Belshaw, N.S. & Henderson, G.M. (2006) Seasonal trace-element and stable isotope variations in a Chinese speleothem: the potential for high-resolution paleomonsoon reconstruction. *Earth and Planetary Science Letters*, **244**, 394–407.

Jones, B. & Kahle, C.F. (1993) Morphology, relationship, and origin of fiber and dendrite calcite crystals. *Journal of Sedimentary Research*, **63**, 1018–1031.

Jones, B. (1987) The alteration of sparry calcite crystals in a vadose setting, Grand Cayman Island. *Canadian Journal of Earth Sciences*, **24**, 2292–2304.

Jones, B. (1989) Calcite rafts, peloids, and micrite in cave deposits from Cayman Brac, British West Indies. *Canadian Journal of Earth Sciences*, **26**, 654–664.

Jones, B. (1992) Caymanite, a cavity-filling deposit in the Oligocene-Miocene Bluff Formation of the Cayman Islands. *Canadian Journal of Earth Sciences*, **29**, 720–736.

Jones, B. (2009) Phosphatic precipitates associated with actinomycetes in speleothems from Grand Cayman, British West Indies. *Sedimentary Geology*, **219**, 302–317.

Jones, B. (2010) Microbes and calcite speleothems. In: Pedley, H.M.& Rogerson, M. (eds) *Tufas and Speleothems: Unravelling the Microbial and Physical Controls*. Geological Society, London, Special Publication, **336**, 7–30.

Jones, M.D. & Imbers, J. (2010) Modelling Mediterranean lake isotope variability. *Global and Planetary Change*, **171**, 193–200.

Jones, M.D., Leng, M.J., Roberts, C.N. et al. (2005) A coupled calibration and modelling approach to the understanding of dry-land lake oxygen isotope records. *Journal of Paleolimnology*, **34**, 391–411.

Jones, P.D, New, M., Parker, D.E., Martin, S. & Rigor, I.G. (1999) Surface air temperature and its changes over the past 150 years. *Reviews of Geophysics*, **37**, 173–199.

Jones, P.D., Briffa, K.R., Osborn, T.J. et al. (2008) High-resolution paleoclimatology of the last millennium: a review of the current status and future prospects. *The Holocene*, **19**, 3–49.

Joussaume, S., Sadourny, R. & Jouzel, J. (1984) A general-circulation model of water isotope cycles in the atmosphere. *Nature*, **311**, 24–29.

Jouzel, J. & Merlivat, L. (1984) Deuterium and oxygen 18 is precipitation: modeling of the isotopic effects. *Journal of Geophysical Research*, **89**, 11749–11757.

Kagan, E.J., Agnon, A., Bar-Matthews, M. & Ayalon, A. (2005) Dating large infrequent earthquakes by damaged cave deposits. *Geology*, **33**, 261–264.

Kajikawa, Y., Wang, B. & Yang, J. (2009) A multi-time scale Australian monsoon index, *International Journal of Climatology*, **30**, 1114–1120.

Kalnay, E., Kanamitsu, M. & Kistler, R. (1996) The NCEP/NCAR 40–year reanalysis project. *Bulletin of the American Meteorological Society*, **77**, 437–471.

Kaplan, J.O., Bigelow, N.H., Prentice, I.C. et al. (2003) Climate change and arctic ecosystems: 2. Modeling, paleodata-model comparisons, and future projections. *Journal of Geophysical Research* **108**, 8171. doi: 10.1029/2002JD002559.

Kappes, R., Rifai, H., Selg, M. & Wonik, T. (2007) Carbonate porosities and pore types determined using NMR spectroscopy on the basis of examples from the Wilsengen karst borehole (Swabian Alb). *Zeitschrift der Deutschen Gesellschaft für Geowissenschaften*, **158**, 1011–1023.

Karkanas, P., Bar-Yosef, O., Goldberg, P., & Weiner, S. (2000) Diagenesis in prehistoric caves: the use of minerals that form in situ to assess the completeness of the archaeological record. *Journal of Archaeological Science*, **27**, 915–929.

Karkanas, P., Schepartz, L.A., Miller-Antonio, S., Wang, W. & Huang, W. (2008) Late Middle Pleistocene climate in southwestern China: inferences from the stratigraphic record of Panxian Dadong Cave, Guizhou. *Quaternary Science Reviews*, **27**, 1555–1570.

Karmann, I., Cruz, F.W., Viana, O. & Burns, S.J. (2007) Climate influence on trace element geochemistry of waters from Santana–Pérolas cave system, Brazil. *Chemical Geology*, **244**, 232–247.

Kaufman, A., Wasserburg, G.J., Porcelli, D. et al. (1998) U-Th isotope systematics from the Soreq cave, Israel and climatic correlations. *Earth and Planetary Science Letters*, **156**, 141–155.

Kaufman, A., Bar-Matthews, M., Ayalon, A. & Carmi, I. (2003) The vadose flow above Soreq Cave, Israel: a tritium study of the cave waters. *Journal of Hydrology*, **273**, 155–163.

Kaufman, D.S., Schneider, D.P., McKay, N.P. et al. (2009) Recent warming reverses long-term Arctic cooling. *Science*, **325**, 1236–1239.

Kaufmann, G. (2003) Stalagmite growth and paleoclimate: the numerical perspective. *Earth and Planetary Science Letters*, **214**, 251–266.

Kaufmann, G. (2009) Modelling karst geomorphology on different time scales. *Geomorphology*, **106**, 62–77.

Kaufmann, G. & Dreybrodt, W. (2004) Stalagmite growth and paleoclimate: an inverse approach. *Earth and Planetary Science Letters*, **224**, 529–545.

Kaufmann, G. & Romanov, D. (2008) Cave development in the Swabian Alb, south-west Germany: a numerical perspective. *Journal of Hydrology*, **349**, 302–317.

Kaufmann, G., Romanov, D. & Hiller, T. (2010) Modeling three-dimensional karst aquifer evolution using different matrix flow contributions. *Journal of Hydrology*, **388**, 241–250.

Kelleher, B.P. & Simpson, A.J. (2006) Humic substances in soils: are they really chemically distinct? *Environmental Science and Technology*, **40**, 4605–4611.

Kendall, A.C. (1985) Radiaxial fibrous calcite: a reappraisal. In: Schneidermann, N. & Harris, P.M. (eds.) (1985) *Carbonate Cements*. Tulsa: Society of Economic Paleontologists and Mineralogists, Special Publication **36**, 59–77.

Kendall, A.C. & Broughton, P.L. (1978) Origin of fabrics in speleothems composed of columnar calcite crystals. *Journal of Sedimentary Petrology*, **48**, 519–538.

Kendall, T.A. & Martin, S.T. (2005) Mobile ions on carbonate surfaces. *Geochimica et Cosmochimica Acta*, **69**, 3257–3263.

Kerans, C. & Donaldson, J.A. (1988) Proterozoic paleokarst profile, Dismal Lakes Group, N.W.T., Canada. In: James, N.P. & Choquette, P.W. (eds.) *Paleokarst*. Springer, New York, pp. 167–182.

Kiessling, W., Aberhan, M. & Villier, L. (2008) Phanerozoic trends in skeletal mineralogy driven by mass extinctions. *Nature Geoscience*, **1**, 527–530.

Kigoshi, K. (1971) Alpha-recoil thorium-234: dissolution into water and the uranium-234/uranium-238 disequilibrium in nature. *Science*, **173**, 47–48.

Killawee, J.A., Fairchild, I.J., Tison, J.-L. et al. (1998) Segregation of solutes and gases in experimental freezing of dilute solutions: implications for natural glacial systems *Geochimica et Cosmochimica Acta*, **62**, 3637–3655.

Kim, S.T. & O'Neil, J.R. (1997) Equilibrium and nonequilibrium oxygen isotope effects in synthetic carbonates. *Geochimica et Cosmochimica Acta*, **61**, 3461–3475.

Kim, S.-T., O'Neil, J.R., Hillaire-Marcel, C. & Mucci, A. (2007) Oxygen isotope fractionation between synthetic aragonite and water: influence of temperature and Mg^{2+} concentration. *Geochimica et Cosmochimica Acta*, **71**, 4704–4715.

Kim, Y.-S. & Sanderson, D.J. (2010) Inferred fluid flow through fault samage zones based on the observation of stalactites in carbonate caves. *Journal of Structural Geology*, **32**, 1305–1316.

Kinniburgh, D.G., van Riemsdijk, W.H., Koopal, L.K. et al. (1999) Ion binding to natural organic matter: competition, heterogeneity, stoichiometry and thermodynamic consistency. *Colloids and Surfaces A-Physicochemical and Engineering Aspects*, **151**, 147–166.

Kinsman, D.J.J. (1976) Evaporites: relative humidity control of primary mineral facies. *Journal of Sedimentary Petrology*, **46**, 273–279.

Kita, N.T., Ushikubo, T., Fu, B., Valley, J.W. (2009) High precision SIMS oxygen isotope analysis and the effect of sample topography. *Chemical Geology*, **264**, 43–57.

Klimchouk, A. (2009) Morphogenesis of hypogenic caves. *Geomorphology*, **106**, 100–117.

Klimchouk, A. & Ford, D. (2000a) 3.1. Types of karst and evolution of hydrogeologic setting. In: Klimchouk, A.B., Ford, DC, Palmer, A.N. & Dreybrodt, W. (eds.) *Speleogenesis. Evolution of Karst Aquifers*. Huntsville: National Speleological Society, pp. 47–53.

Klimchouk, A. & Ford, D. (2000b) 3.2. Lithologic and structural controls of dissolutional cave development. In: Klimchouk, A.B., Ford, DC, Palmer, A.N. & Dreybrodt, W. (eds.) *Speleogenesis. Evolution of Karst Aquifers*. Huntsville: National Speleological Society, pp. 54–64.

Klimchouk, A.B., Ford, D.C, Palmer, A.N. & Dreybrodt, W. (2000) (eds) *Speleogenesis: Evolution of Karst Aquifers*. Huntsville: National Speleological Society.

Kluge, T., Marx, T., Scholz, D. et al. (2008) A new tool for palaeoclimate reconstruction: Noble gas temperatures from fluid inclusions in speleothems. *Earth and Planetary Science Letters*, **269**, 407–414.

Kluge, T., Riechelmann, D.F.C., Wieser, M., et al. (2010) Dating cave drip water by tritium. *Journal of Hydrology*, **394**, 396–406.

Kober, B., Schwalb, A., Schettler, G. & Wessels, M. (2007) Constraints on paleowater dissolved loads and on catchment weathering over the past 16 ka from $^{87}Sr/^{86}Sr$ ratios and Ca/Mg/Sr chemistry of freshwater ostracodes tests in sediments of Lake Constance, Central Europe. *Chemical Geology*, **240**, 361–376.

Kohn, M.J. & Welker, J.M. (2005) On the temperature correlation of $\delta^{18}O$ in modern precipitation. *Earth and Planetary Science Letters*, **231**, 87–96.

Kolodny, Y., Bar-Matthews, M., Ayalon, A. et al. (2003) A high spatial resolution delta O-18 profile of a speleothem using an ion-microprobe. *Chemical Geology*, **197**, 21–28.

Koshikawa, T., Kido, Y. & Tada, R. (2003) High-resolution rapid elemental analysis using an XRF microscanner. *Journal of Sedimentary Research*, **73**, 824–829.

Koster, R.D., de Valpine, P. & Jouzel, J. (1993) Continental water recycling and $H_2^{18}O$ concentrations. *Geophysical Research Letters*, **20**, 2215–2218.

Kottek, M., Grieser, J., Beck, C., Rudolf, B. & Rubel, F. (2006) World Map of the Köppen-Geiger climate classification updated. *Meteorologie Zeitschrift.*, **15**, 259–263.

Kowalczk, A.J. & Froelich, P.N. (2010) Cave air ventilation and CO_2 outgassing by radium-222 modeling: how fast do caves breathe? *Earth and Planetary Science Letters*, **289**, 209–219.

Kowalski, A.S. & Sánchez-Cañete, E.P. (2010) A new definition of the virtual temperature, valid for the atmosphere and the CO_2- rich air of the vadose zone. *Journal of Applied Meteorology and Climatology*, **49**, 1692–1695.

Kowalski, A.S., Serrano-Ortiz, P., Janssens, I.A. et al. (2008) Can flux tower research neglect geochemical CO_2 exchange? *Agricultural and Forest Meteorology*, **148**, 1045–1054.

Kranjc, A. & Opara, B. (2002) Temperature monitoring in Škocjanskih Jamah. *Acta Carsologica*, **31**, 85–96.

Krawczyk, W.E. & Ford, DC (2006) Correlating specific conductivity with total hardness in limestone and dolomite karst waters. *Earth Surface Processes and Landforms*, **31**, 221–234.

Krüger, Y., Stoller, P., Rička, J. & Frenz, M. (2007) Femtosecond lasers in fluid inclusion analysis: overcoming metastable phase states. *European Journal of Mineralogy* **19**, 693–706.

Krüger, Y., Marti, D., Hidalgo, R. et al. (2011) Liquid-vapour homogenisation of fluid inclusions in stalagmites: evaluation of a new thermometer for paleoclimate research. *Chemical Geology*, **289**, 39–47.

Kuhlbrodt, T., Griesel, A., Montoya, M. et al. (2007) On the driving processes of the Atlantic meridional overturning circulation. *Reviews of Geophysics* **45**, RG2001 doi:10.1029/2004RG000166.

Kuhn, T.S. (1970) *The Structure of Scientific Revolutions*. 3rd Ed. Chicago and London: University of Chicago Press.

Kulik, D.A. (2010) Geochemical thermodynamic modelling of ion partitioning. In: Stoll, H. and Prieto, M., editors. *Ion partitioning in ambient temperature aqueous systems: from fundamentals to applications in climate proxies and environmental geochemistry*. EMU Notes in Mineralogy, **10**, 65–138.

Kump, L.R., Kasting, J.F. & Crane, R.G. (2009) *The Earth System*, 3rd edition. New Jersey: Prentice-Hall.

Kurita, N., Ichiyanagi, K., Matsumoto, M.D. & Ohata, T. (2009) The relationship between the isotopic content of precipitation and the precipitation amount in tropical regions. *Journal of Geochemical Exploration*, **102**, 113–122.

Labat, D., Ababou, R. & Mangin A. (2000a) Rainfall-runoff relations for karstic springs. Part I: convolution and spectral analysis. *Journal of Hydrology*, **238**, 123–148.

Labat, D., Ababou, R. & Mangin, A. (2000b) Rainfall-runoff relations for karstic springs. Part II: continuous wavelet and discrete orthogonal multi-resolution analyses. *Journal of Hydrology*, **238**, 149–178.

Labat, D., Mangin, A. & Ababou, R. (2002) Rainfall-runoff relations for karstic springs: multifractal analysis. *Journal of Hydrology*, **256**, 176–195.

Labourdette, R., Lascu, I., Mylroie, J. & Roth, M. (2007) Process-like modelling of flank-margin caves: from genesis to burial evolution. *Journal of Sedimentary Research*, **77**, 965–979.

Lacave, C. & Koller, M.G. (2004) What can be concluded about seismic history from broken and unbroken speleothems. *Journal of Earthquake Engineering*, **8**, 431–455.

Lacelle, D. (2007) Environmental setting (micro)morphologies and stable C-O isotope compositions of cold climate carbonate precipitates—a review and evaluation of their potential as palaeoclimatic proxies. *Quaternary Science Reviews*, **26**, 1670–1689.

Lachniet, M.S. (2009) Climatic and environmental controls on speleothem oxygen-isotope values. *Quaternary Science Reviews*, **28**, 412–432.

Lachniet, M.S. & Patterson, W.P. (2006) Use of correlation and stepwise regression to evaluate physical controls on the stable isotope values of Panamanian rain and surface waters. *Journal of Hydrology*, **324**, 115–140.

Lachniet, M.S. & Patterson, W.P. (2009) Oxygen isotope values of precipitation and surface water in northern Central America (Belize and Guatemala) are dominated by temperature and amount effects. *Earth and Planetary Science Letters*, **284**, 435–446.

Lakatos, I. (1978) *The Methodology of Scientific Research Programmes: Philosophical Papers Volume 1*. Cambridge: Cambridge University Press.

Lakshtanov, L.Z. & Stipp, S.L.S. (2007) Experimental study of nickel(II) interaction with calcite: adsorption and coprecipitation. *Geochimica et Cosmochimica Acta*, **71**, 3686–3697.

Lambert, W.J. & Aharon, P. (2011) Controls on dissolved inorganic carbon and $\delta^{13}C$ in cave waters from DeSoto Caverns: implications for speleothem $\delta^{13}C$ assessments. *Geochimica et Cosmochimica Acta*, **75**, 753–768.

Langmuir, D. (1997) *Aqueous Environmental Chemistry*. Prentice-Hall, New Jersey.

Lanzante, J.R. (1996) Resistant, robust and non-parametric techniques for the analysis of climate data: theory and examples, including applications to historical radiosonde station data. *International Journal of Climatology*, **16**, 1197–1226.

Lardge, J.S., Duffy, D.M. & Gillan, M.J. (2009) Investigation of the interaction of water with the calcite (10.4) surface using ab initio simulation. *Journal of Physical Chemistry C*, **113**, 7207–7212.

Lario, J., Sánchez-Moral, S., Ceuzva, S. et al. (2006) High ^{222}Rn levels in a show cave (Castañar de Ibor, Spain): proposal and application of management measures to minimize the effects on guides and visitors. *Atmospheric Environment*, **40**, 7935–7400.

Lasaga, A.C. & Luttge, A. (2001) Variation of crystal dissolution rate based on a dissolution stepwave model. *Science*, **291**, 2400–2404.

Laskar, J., Robutel, P., Joutel, F. et al. (2004) A long term numerical solution for the insolation quantities of the Earth. *Astronomy and Astrophysics*, **428**, 261–285.

Latham, A.G., Schwarcz, H.P. & Ford, DC (1986) The paleomagnetism and U-Th dating of Mexican stalagmite DAS2. *Earth and Planetary Interiors*, **79**, 195–207.

Lauritzen, S.-E. & Lundberg, J. (1999) Calibration of the speleothem delta function: an absolute temperature record for the Holocene in northern Norway, *The Holocene*, **9**, 659–669.

Lauritzen, S.-E. & Lundberg, J. (2000) 6.1 Solutional and erosional morphology. In: Klimchouk, A.B., Ford, D.C, Palmer, A.N. & Dreybrodt, W. (eds), *Speleogenesis: Evolution of Karst Aquifers*. Huntsville: National Speleological Society, pp. 408–426.

Lauritzen, S.E., Løvlie, R. Moe, D. & Østbye, E. (1990) Paleoclimate deduced from a multidisciplinary study of a half-million year old stalagmite from Rana, Northern Norway. *Quaternary Research*, **34**, 306–316.

Lauritzen, S.E., Haugen, J.E., Løvlie, R. & Giljenielsen, H. (1994) Geochronological potential of Isoleucine epimerization in calcite speleothems. *Quaternary Research*, **41**, 52–58.

Lawrence, J.R. & Gedzelman, S.D. (1996) Low stable isotope ratios of tropical cyclone rains. *Geophysical Research Letters* **23**, 527–530.

Lawrence, J.R. & White, J.W.C. (1991) The elusive climate signal in the isotopic composition of precipitation. In: Taylor, H.P., O'Neil, J.R., Kaplan, I.R. (eds), *Stable Isotope Geochemistry: A Tribute to Samuel Epstein*. Geochemical Society Special Publication 3, St. Louis, Missouri, pp. 169–185.

Le Treut, H., Somerville, R., Cubasch, U., et al. (2007) Historical Overview of Climate Change. In: *Climate Change 2007: The Physical Science Basis. Contribution of Working Group I to the Fourth Assessment Report of the Intergovernmental Panel on Climate Change*. Solomon, S., Qin, D., Manning, M. et al. (eds.). Cambridge University Press.

Lead, J.R. & Wilkinson, K.J. (2006) Aquatic colloids and nanoparticles: current knowledge and future trends. *Environmental Chemistry*, **3**, 159–171.

Lee, J.-E. & Fung, I. (2008) 'Amount effect' of water isotopes and quantitative analysis of post-condensation processes. *Hydrological Processes*, **22**, 1–8.

Lee, T.C.K., Zwiers, F.W. & Tsao, M. et al. (2008) Evaluation of proxy-based millennial reconstruction methods. *Climate Dynamics*, **31**, 263–281.

Lee, Y.J., Elzinga, E.J. & Reeder, R.J. (2005) Cu(II) adsorption at the calcite-water interface in the presence of natural organic matter: kinetic studies and molecular-scale characterization. *Geochimica et Cosmochimica Acta*, **69**, 49–61.

Leeder, M.R. (1999) *Sedimentology and Sedimentary Basins: from Turbulence to Tectonics*. Wiley, N.Y.

Lee-Thorp, J.A., Holmgren, K., Lauritzen, S.E. et al. (2001) Rapid climate shifts in the southern African interior throughout the mid to late Holocene. *Geophysical Research Letters*, **28**, 4507–4510.

LeGrande A.N. & Schmidt G.A. (2006) Global gridded data set of the oxygen isotopic composition in seawater. *Geophysical Research Letters* **33**, L12604, doi:10.1029/2006GL026011.

LeGrande, A.N. & Schmidt, G.A. (2008) Ensemble, water isotope-enabled, coupled general circulation modeling insights into the 8.2 ka event, *Paleoceanography*, **23**, PA3207. doi: 10.1029/2008PA001610.

LeGrande, A.N. & Schmidt, G.A. (2009) Sources of Holocene variability of oxygen isotopes in paleoclimate archives. *Climate of the Past*, **5**, 441–455.

Lehmann, J., Solomon, D., Kinyangi, J. et al. (2008) Spatial complexity of soil organic matter forms at nanometre scales. *Nature Geoscience* **1**, 238–242.

Legatzki, A., Ortiz, M., Neilson, J.W. et al. (2011) Bacterial and archaeal community structure of two adjacent calcite speleothems in Kartchner Caverns, Arizon, USA. *Geomicrobiology Journal*, **28**, 99–117.

Lemarchand, D., Wasserburg, G.J. & Papnastassiou, D.A. (2004) Rate-controlled calcium isotope fractionation in synthetic calcite. *Geochimica et Cosmochimica Acta*, **68**, 4665–4678.

Lerman, A. (1994) Surficial weathering fluxes and their geochemical controls. In: *National Research Council Material Fluxes on the Surface of the Earth*. Washington, DC: National Academy Press.

Lewin, J. & Woodward, J.C. (2009) Karst geomorphology and environmental change. In: Woodward, J.C. (ed.) *The Physical Geography of the Mediterranean*. Oxford: Oxford University Press, 287–317.

Lewis, T.J. & Wang, K. (1998) Geothermal evidence for deforestation induced warming: implications for climatic impact of land development. *Geophysical Research Letters*, **25**, 535–538.

Li, H.C., Gu, D.L. & Stott, L.D. (1998) Applications of interannual-resolution stable isotope records of speleothem: climate changes in Beijing and Tianjin, China during the past 500 years—the delta O-18 record. *Science in China, Series D*, **41**, 362–368.

Li, H.-C., Ku, T.-H., You, C.-F. et al. (2005) $^{87}Sr/^{86}Sr$ and Sr/Ca in speleothems for paleoclimate reconstruction in Central China between 70 and 280 kyr ago. *Geochimica et Cosmochimica Acta*, **69**, 3933–3947.

Li, T.-Y., Shen, C.-C., Li, H.-C. et al. (2011) Oxygen and carbon isotopic systematics of aragonite speleothems and water in Furong Cave, Chongqing, China. *Geochimica et Cosmochimica Acta*, doi:10.1016/j.gca.2011.04.003.

Li, W.-X., Lundberg, J., Dickin, A.P. et al. (1989) High precision mass-spectrometric uranium-series dating of cave deposits and implication for palaeoclimate studies. *Nature*, **339**, 334–336.

Li, X., Wang, C, Huang, J. et al. (2011) Seasonal variation of fatty acids from drip water in Heshang Cave, central China. *Applied Geochemistry*, **26**, 341–347.

Lin, Y.-P. & Singer, P.C. (2006) Inhibition of calcite precipitation by orthophosphate: speciation and thermodynamic considerations. *Geochimica et Cosmochimica Acta*, **70**, 2530–2539.

Liñán Baena, C., Andreo Navarro, B., Carrasco Cantos, F. & Vadillo Pérez, I. (2000) Consideractiones acerca de la influencia del CO_2 en las hydroquímica de las aguas de goteo de la Cueva de Nerja (Provincia de Málaga). *Geotemas*, **1**, 341–344.

Liñan, C., Vadillo, I. & Carrasco, F. (2008) Carbon dioxide concentration in air within the Nerja Cave (Malaga, Andalusia, Spain). *International Journal of Speleology*, **37**, 99–106.

Linden, P.F. (1999) The fluid mechanics of natural ventilation. *Annual Review of Fluid Mechanics*, **31**, 201–238.

Linge, H., Lauritzen, S.E., Lundberg, J. et al. (2001) Stable isotope stratigraphy of Holocene speleothems: examples from a cave system in Rana, northern Norway. *Palaeogeography, Palaeoclimatology, Palaeoecology*, **167**, 209–224.

Linge, H., Baker, A., Andersson, C. & Lauritzen, S.E. (2009a) Variable luminescent lamination and initial $^{230}Th/^{232}Th$ activity ratios in a late Holocene stalagmite from northern Norway. *Quaternary Geochronology*, **4**, 181–192.

Linge, H., Lauritzen, S.-E., Andersson, J.K. et al. (2009b) Stable isotope records for the last 10 000 years from Okshola cave (Fauske, northern Norway) and regional comparisons. *Climate of the Past*, **5**, 667–682.

Lismonde, B. (2004) Le flux géothermique avec circulation d'eau profonde dans les karsts: la surprise des transitoires. *Karstologia*, **44**, 51–55.

Liu, D., Wang, Y., Cheng, H. et al. (2008) A detailed comparison of Asian Monsoon intensity and Greenland temperature during the Allerød and Younger Dryas events. *Earth and Planetary Science Letters*, **272**, 691–697.

Liu, D, Wang, Y, Cheng, H. et al. (2010) Sub-millennial variability of Asian monsoon intensity during the early MIS 3 and its analogue to the ice age terminations. *Quaternary Science Reviews*, **29**, 1107–1115.

Liu, Q., Wang, H., Zhao, R. et al. (2010) Bacteria isolated from dripping water in the oligotrophic Heshang Cave in Central China. *Journal of Earth Science (China)*, **21**, 325–328.

Liu, Z., Tian, L., Chai, X. & Yao T. (2008) A model-based determination of spatial variation of precipitation $\delta^{18}O$ over China. *Chemical Geology*, **249**, 203–212.

Liu, Z., Dreybrodt, W. & Wang, H. (2010) A new direction in effective accounting for the atmospheric CO_2 budget: considering the combined action of carbonate dissolution, the global water cycle and photosynthetic uptake of DIC by aquatic organisms. *Earth-Science Reviews*, **99**, 162–172.

Lorens, R.B. (1981) Sr, Cd, Mn and Co distribution coefficients in calcite as a function of calcite precipitation rate. *Geochimica et Cosmochimica Acta*, **45**, 553–561.

Lorenz, S.J. Kim, J.H., Rimbu, N. et al. (2006) Orbitally driven insolation forcing on Holocene climate trends: evidence from alkenone data and climate modeling, *Paleoceanography*, **21**, PA1002. doi: 10.1029/2005PA001152.

Loucks, R.G. (1999) Paleocave carbonate reservoirs: origins, burial-depth modifications, spatial complexity, and reservoir implications. *American Association of Petroleum Geologists Bulletin*, **83**, 1795–1834.

Lovelock, J. (1988) *The Ages of Gaia*. London: W.W. Norton.

Lowe, D.J. (1992) *The origin of limestone caverns: an inception horizon hypothesis*. Unpublished PhD Thesis, Manchester Metropolitan University.

Lowe, D.J. (2000) 3.3. Role of stratigraphic elements in speleogenesis: the speleoinception concept. In Klimchouk, A.B., Ford, D.C, Palmer, A.N. & Dreybrodt, W. (eds), *Speleogenesis: Evolution of Karst Aquifers*. Huntsville: National Speleological Society, pp. 65–76.

Lowe, D.J. & Gunn, J. (1997) Carbonate speleogenesis: an inception horizon hypothesis. *Acta Carsologica*, **26**, 457–488.

Ludwig, K.R., Simmons, K.R., Szabo, B.J. et al. (1992) Mass-spectrometric 230Th-234U-238U dating of the Devils Hole calcite vein. *Science*, **258**, 284–287.

Luetscher, M. & Jeannin, P-Y. (2004) Temperature distribution in karst systems: the role of air and water fluxes. *Terra Nova*, **16**, 344–350.

Luetscher, M., Lismonde, B. & Jeannin, P-Y. (2008) Heat exchanges in the heterothermic zone of a karst system: Monlesi cave, Swiss Jura Mountains. *Journal of Geophysical Research*, **113**, F02025. doi: 10.1029/2007JF000892.

Luetscher, M., Hoffmann, D.L., Frisia, S. & Spötl, C. (2011) Holocene glacier history from alpine speleothems, Michbach cave, Switzerland. *Earth and Planetary Science Letters*, **302**, 95–106.

Lundberg, J. & Ford, DC (1994) Late Pleistocene sea level change in the Bahamas from mass spectrometric U-series dating of submerged speleothem. *Quaternary Science Reviews*, **12**, 1–14.

Lundberg, J. & McFarlane, D.A. (2007) Pleistocene depositional history in a periglacial terrane: a 500 k.y. record from Kents Cavern, Devon, United Kingdom. *Geosphere*, **3**, 199–219.

Lundberg, J., Lord, T.C. & Murphy, P.J. (2010) Thermal ionization mass spectrometer U-Th dates on Pleistocene speleothems from Victoria Cave, North Yorkshire, UK: implications for paleoenvironment and stratigraphy over multiple glacial cycles. *Geosphere*, **6**, 379–395.

Luo, W.-J. & Wang, S.-J. (2008) Transmission of oxygen isotope signals of precipitation-soil water-drip water and its implications in Liangfeng Cave of Guizhou, China. *Chinese Science Bulletin*, **53**, 3364–3370.

Luterbacher, J., Xoplaki, E., Rickli, R. et al. (2002) Reconstruction of Sea Level Pressure fields over the eastern North Atlantic and Europe back to 1500. *Climate Dynamics*, **18**, 545–561.

Ma, Q.L., Mottos-Ros, V., Lie, W.Q. et al. (2010) Multi-elemental mapping of a speleothem using laser-induced breakdown spectroscopy. *Spectrochimica Acta B*, **65**, 707–714.

Mackenzie, F.T., Arvidson, R.S. & Guidry, M. (2008) Chemostatic modes of the ocean-atmosphere-sediment system through Phanerozoic time. *Mineralogical Magazine*, **72**, 329–332.

MacNeil, A.J. & Jones, B. (2006) Sequence stratigraphy of a Late Devonian ramp-situated reef system in the Western Canada Sedimentary Basin: dynamic responses to sea-level change and regressive reef development. *Sedimentology*, **53**, 321–359.

Maher, B.A. (2008) Holocene variability of the East Asian summer monsoon from Chinese cave records: a re-assessment. *The Holocene*, **18**, 861–866.

Maliva, R.G. & Siever, R. (1988) Diagenetic replacement controlled by force of crystallization. *Geology*, **16**, 688–691.

Maltsev, V.A. (1998) Stalactites with 'internal' and 'external' feeding. *Proceedings of the University of Bristol Speleological Society*, **21**, 148–158.

Mangini, A., Spötl, C. & Verdes, P. (2005) Reconstruction of temperature in the Central Alps during the past (2000) yr from a $\delta^{18}O$ stalagmite record. *Earth and Planetary Science Letters*, **235**, 741–751.

Mangini, A., Verdes, P., Spötl, C. et al. (2007) Persistent influence of the North Atlantic hydrography on European winter temperature during the last 9000 years. *Geophysical Research Letters*, **34**, L02704. doi: 10.1029/2006GL028600.

Mann, M.E., Rutherford, S., Wahl, E. et al. (2005) Testing the fidelity of methods used in proxy-based reconstructions of past climate. *Journal of Climate*, **18**, 4097–4107.

Mann, M.E., Zhang, Z., Hughes, M.K. et al. (2008) Proxy-based reconstructions of hemispheric and global surface temperature variations over the past two millennia. *Proceedings of the National Academy of Sciences of the USA*, **105**, 13252–13257.

Marean, C.W., Bar-Matthews, M., Bernatchez, J. et al. (2007) Early human use of marine resources and pigment in South Africa during the Middle Pleistocene. *Nature*, **449**, 905–908.

Marseille, F., Disnar, J.R., Guillet, B. & Noack, Y. (1999) n-Alkanes and free fatty acids in humus and A1 horizons of soils under beech, spruce and grass in the Massif-Central (Mont-Lozère), France. *European Journal of Soil Science*, **50**, 433–441.

Martín-García, R., Alonso-Zarza, A.M. & Martín-Pérez, A. (2009) Loss of primary texture and geochemical signatures in speleothems due to diagenesis: evidences from Castañar Cave, Spain. *Sedimentary Geology*, **221**, 141–149.

Martinson, D.G., Pisias, N.G., Hays, J.D. et al. (1987) Age dating and the orbital theory of the ice ages:

development of a high-resolution 0 to 300000 year chronostratigraphy. *Quaternary Research*, **27**, 1–29.

Mason, H.E., Frisia, S., Tang, Y. et al. (2007) Phosphorus speciation in calcite speleothems determined from solid-state NMR spectroscopy. *Earth and Planetary Science Letters*, **254**, 313–322.

Masson-Delmotte, V., Jouzel, J., Landais, A. et al. (2005) GRIP deuterium excess reveals rapid and orbital-scale changes in Greenland moisture origin. *Science*, **306**, 118–121.

Mattey, D.M., Lowry, D., Duffet, J. et al. (2008) A 53 year seasonally resolved oxygen and carbon isotope record from a modern Gibraltar speleothem: reconstructed drip water and relationship to local precipitation. *Earth and Planetary Science Letters*, **269**, 80–95.

Mattey, D.P., Fairchild, I.J., Atkinson, T.C. et al. (2010) Seasonal microclimate controls on calcite fabrics, stable isotopes and trace elements in modern speleothems from St. Michaels Cave, Gibraltar. In: Pedley, H.M. & Rogerson, M. (eds.) *Tufas and Speleothems. Geological Society of London Special Publication*, **336**, 323–344.

Matthews, A., Ayalon, A. & Bar-Matthews, M. (2000) D/H ratios of fluid inclusions of Soreq cave (Israel) speleothems as a guide to the Eastern Mediterranean Meteoric Line relationships in the last 120 ky. *Chemical Geology*, **166**, 183–191.

Mavrocordatos, D., Mondi-Couture, C., Atteia, O. et al. (2000) Formation of a distinct class of Fe-Ca(-Corg)-rich particles in a complex peat-karst system. *Journal of Hydrology*, **237**, 234–247.

Mazarrón, F.R. & Cañas, I. (2009) Seasonal analysis of the thermal behaviour of traditional underground wine cellars in Spain. *Renewable Energy*, **34**, 2484–2492.

Mazzullo, S.J. & Harris, P.M. (1992) Mesogenetic dissolution: its role in porosity development in carbonate reservoirs. *American Association of Petroleum Geologists*, **76**, 607–620.

McArthur, J.M., Howarth, R.J. & Bailey, T.R. (2001) Strontium isotope stratigraphy: LOWESS version 3: best fit to the marine Sr-isotope curve for 0–509 Ma and accompanying look-up table for deriving numerical age. *Journal of Geology*, **109**, 155–170.

McCarthy, J.F. & Shevenell, L. (1998) Processes controlling colloid composition in a fractured and karstic aquifer in eastern Tennessee, USA. *Journal of Hydrology*, **206**, 191–218.

McClain, M.E., Swart, P.K. & Vacher, H.L. (1992) The hydrogeochemistry of early meteoric diagenesis in a Holocene deposit of biogenic carbonates. *Journal of Sedimentary Petrology*, **62**, 1008–1022.

McCrea, J.M. (1950) On the isotopic chemistry of carbonates and a paleotemperature scale. *Journal of Chemical Physics*, **18**, 849–853.

McDermott, F. (2004) Palaeo-climate reconstruction from stable isotope variations in speleothems: a review. *Quaternary Science Reviews*, **23**, 901–918.

McDermott, F., Frisia, S., Huang, Y. et al. (1999) Holocene climate variability in Europe: evidence from $\delta^{18}O$, textural and extension-rate variations in three speleothems. *Quaternary Science Reviews*, **18**, 1021–1038.

McDermott, F., Mattey D.P. & Hawkesworth C. (2001) Centennial-scale Holocene climate variability revealed by a high-resolution speleothem delta O-18 record from SW Ireland. *Science*, **294**, 1328–1331.

McDermott, F., Schwarcz, H.P. & Rowe, P.J. (2005) 6. Isotopes in speleothems. In: Leng, M.J. (Ed.) *Isotopes in Palaeoenvironmental Research*. Springer, Dordrecht, The Netherlands, pp. 185–225.

McDermott, F., Atkinson, T.C., Fairchild, I.J., Baldini, L.M. & Mattey, D.P. (2011) A first evaluation of the spatial gradients in $\delta^{18}O$ recorded by European Holocene speleothems. *Global and Planetary Change*, **75**, 275–287.

McDonald, J., Drysdale, R. & Hill, D. (2004) The 2002–2003 El Niño recorded in Australian cave drip waters: implications for reconstructing rainfall histories using stalagmites. *Geophysical Research Letters*, **31**, L22202, doi:10.1029/2004GL020859.

McDonald, J., Drysdale, R., Hill, D. et al. (2007) The hydrochemical response of cave drip waters to sub-annual and inter-annual climate variability, Wombeyan Caves, SE Australia. *Chemical Geology*, **244**, 605–623.

McGarry, S.F. (2000) *Multiproxy Quaternary Palaeoenvironmental Records from Speleothem Pollen and Organic Acid Fluorescence*. Unpublished PhD, University of Exeter.

McGarry, S. & Baker, A. (2000) Organic acid fluorescence: applications to speleothem palaeoenvironmental reconstruction. *Quaternary Science Reviews*, **19**, 1087–1101.

McGarry, S.F. & Caseldine, C.J. (2004) Speleopalynology: a neglected tool in British Quaternary studies. *Quaternary Science Reviews*, **23**, 2389–2404.

McGarry S.F., Bar-Matthews M., Matthews A. et al. (2004) Constraints on hydrological and palaeotemperature variations in the Eastern Mediterranean region in the last 140 ka given by the δD values of speleothem fluid inclusions. *Quaternary Science Reviews*, **23**, 919–934.

McGillen, M. & Fairchild, I.J. (2005) An experimental study of the controls on incongruent dissolution of $CaCO_3$ under analogue glacial conditions. *Journal of Glaciology*, **51**, 383–390.

McIntyre, W.L. (1963) Trace element partition coefficients—a review of theory and application to geology. *Geochimica et Cosmochimica Acta*, **27**, 1209–1264.

McKnight, D.M., Boyer, E.W., Westerhoff, P.K. et al. (2001) Spectrofluorometric characterization of dissolved organic matter for indication of precursor organic material and aromaticity. *Limnology and Oceanography*, **46**, 38–48.

McMillan, E.A., Fairchild, I.J., Frisia, S. et al. (2005) Annual trace element cycles in calcite-aragonite speleothems: evidence of drought in the western Mediterranean 1200–1100 yr BP. *Journal of Quaternary Science*, **20**, 423–433.

Meakin, P. & Jamtveit, B. (2010) Geological pattern formation by growth and dissolution in aqueous systems. *Proceedings of the Royal Society A*, **466**, 659–694.

Medina-Elizade, M., Burns, S.J., Lea, D.W. et al. (2010) High resolution stalagmite climate record from the Yucatán Peninsula spanning the Maya terminal classic period. *Earth and Planetary Science Letters*, **298**, 255–262.

Melim, L.A. (1996) Limitations on lowstand meteoric diagenesis in the Pliocene-Pleistocene of Florida and Great Bahama Bank: implications for eustatic sea-level models. *Geology*, **24**, 893–896.

Melim, L.A. & Masaferro, J.L. (1997) Geology of the Bahamas: subsurface geology of the Bahama Banks. In: Vacher, H.L & Quinn, T. (eds.) *Geology and Hydrology of Carbonate Islands*. Amsterdam: Elsevier, pp. 161–182.

Menendez, B., David, C. & Nistal, A.M. (2001) Confocal scanning laser microscopy applied to the study of pore and crack networks in rocks. *Computers & Geosciences*, **27**, 1101–1109.

Merlivat L. & Jouzel, J. (1979) Global climatic interpretation of the deuterium-oxygen 18 relationship for precipitation. *Journal of Geophysical Research*, **84**, 5029–5033.

Meyer, H.J. (1984) The influence of impurities on the growth rate of calcite. *Journal of Crystal Growth*, **66**, 639–649.

Meyer, M.C., Faber, R. & Spötl, C. (2006) The WinGeol Lamination Tool: new software for rapid, semi-automated analysis of laminated climate archives. *The Holocene*, **16**, 753–761.

Meyer, M.C., Cliff, R.A., Spötl, C. et al. (2009) Speleothems from the earliest Quaternary: snapshots of paleoclimate and landscape evolution at the northern rim of the Alps. *Quaternary Science Reviews*, **26**, 1374–1391.

Meyers, P. & Arnaboldi, M. (2008) Paleooceanographic implications of nitrogen and organic carbon excursions in mid-Pleistocene sapropels from the Tyrrhenian and Levantine basins, Med Sea. *Palaeogeography, Palaeoecology, Palaeoclimatology*, **266**, 112–118.

Miall, A.D. (1996) *The Geology of Fluvial Deposits*. Berlin: Springer-Verlag.

Michaelis, J., Usdowski, E. & Menschel, G. (1985) Partitioning of ^{13}C and ^{12}C on the degassing of CO_2 and the precipitation of calcite—Rayleigh-type fractionation and a kinetic model. *American Journal of Science*, **285**, 318–327.

Mickler, P.J., Banner, J.L., Stern, L., et al. (2004) Stable isotope variations in modern tropical speleothems: evaluating equilibrium vs. kinetic effects. *Geochimica et Cosmochimica Acta*, **68**, 4381–4393.

Mickler, P.J., Stern, L. & Banner, J.L. (2006) Large kinetic isotope effects in modern speleothems. *Geological Society of America Bulletin*, **118**, 65–81.

Milanolo, S. & Gabrovšek, F. (2009) Analysis of carbon dioxide variations in the atmosphere of Srednja Bikambarska Cave, Bosnia and Herzegovina. *Boundary-Layer Meteorology*, **131**, 479–493.

Milne, C.J., Kinniburgh, D.G. & Tipping, E. (2001) Generic NICA-Donnan model parameters for proton binding by humic substances. *Environmental Science and Technology*, **35**, 2049–2059.

Milne, C.J., Kinniburgh D.G., van Riemsdijk W.H. & Tipping E. (2003) Generic NICA-Donnan model parameters for metal-ion binding by humic substances. *Environmental Science and Technology*, **37**, 958–971.

Miorandi, R., Borsato, A., Frisia, S. et al. (2010) Epikarst hydrology and implications for stalagmite capture of climate changes at Grotta di Ernesto (N.E. Italy): results from long-term monitoring. *Hyrological Processes*, **24**, 3101–3114.

Moberg, A., Soneckin, D.M., Holmgren, K., Datsenko, N.M., & Karlen, W. (2005) Highly variable Northern Hemisphere temperatures reconstructed from low- and high-resolution proxy data. *Nature* **433**, 613–617.

Moberg, A., Mohammad, R. & Mauritsen, T. (2008) Analysis of the Moberg et al. (2005) hemispheric temperature reconstruction. *Climate Dynamics*, **31**, 957–971.

Monteith, D.T., Stoddard, J.L., Evans, C.D. et al. (2007) Dissolved organic carbon trends resulting from changes in atmospheric deposition chemistry. *Nature*, **450**, 537–541.

Mook, W.G. (2001) *Environmental Isotopes in the Hydrological Cycle. Volume I (Theory Methods Review); Volume II (Atmospheric Water); Volume III (Surface Water)*. UNESCO, Paris/IAEA Vienna. Full text available on-line at: http://www.iaea.org/programmes/ripc/ih/volumes/volumes.html.

Mook, W. & de Vries, J.J. (2000) *Environmental Isotopes in the Hydrological Cycle. Principles and Applications. Volume 1:*

Introduction—Theory, Methods, Review. International Atomic Energy Agency, Vienna.

Mook, W.G., Bommerson, J.C. & Staverman, W.H. (1974) Carbon isotope fractionation between dissolved bicarbonate and gaseous carbon dioxide. *Earth and Planetary Science Letters*, **22**, 169–176.

Moore, C. (2001) *Carbonate Reservoirs. Porosity Evolution and Diagenesis in a Sequence Stratigraphic Framework*. Amsterdam: Elsevier.

Moore, G.W. (1952) Speleothem—a new cave term. *National Speleological Society News*, **10**(6), 2.

Moore, G.W. (1956) Aragonite speleothems as indicators of paleotemperature. *American Journal of Science*, **254**, 746–753.

Moore, G.W. (1962) The growth of stalactites. *National Speleological Society Bulletin*, **24**, 95–106.

Moore, G.W. & Nicholas, B.G. (1964) *Speleology. The Study of Caves*. DC Heath, Boston.

Moore, G.W. & Sullivan, G.N. (1978) *Speleology The study of Caves*. Zephyrus Press, Teaneck (NY).

Morse, B.S. (2002) Radiocarbon dating of groundwater using paleoclimate constraints and dissolved organic carbon in the southern Great Basin, Nevada and California. Unpublished Thesis, University of Nevada at Reno, 63 pp.

Morse, J.W. & Casey, W.H. (1988) Ostwald processes and mineral paragenesis in sediments. *American Journal of Science*, **288**, 537–560.

Morse, J.W. & Mackenzie, F.T. (1990) *Sedimentary Carbonate Minerals*. Elsevier, Amsterdam.

Morse, J.W. & Arvidson, R.S. (2002) The dissolution kinetics of major sedimentary carbonate minerals. *Earth-Science Reviews*, **58**, 51–84.

Morse, J.W., Arvidson, R.S. & Lüttge, A. (2007) Calcium carbonate formation and dissolution. *Chemical Reviews* **107**, 342–381.

Motyka, J. (1998) A conceptual model of hydraulic networks in carbonate rocks, illustrated by examples from Poland. *Hydrogeology Journal*, **6**, 469–482.

Moyes, H. (2007) The Late Classic drought cult: ritual activity as a response to environmental stress among the ancient Maya. In: *Cult in Context: Reconsidering Ritual in Archaeology*, edited by D. Barrowclough & C. Malone, Oxbow Books, Oxford, pp. 217–228.

Mucci, A. & Morse, J.W. (1983) The incorporation of Mg^{2+} and Sr^{2+} into calcite overgrowths—influences of growth rate and solution composition. *Geochimica et Cosmochimica Acta*, **47**, 217–233.

Mühlinghaus, C., Scholz, D. & Mangini, A. (2007) Modelling stalagmite growth and $\delta^{13}C$ as a function of drip interval and temperature. *Geochimica et Cosmochimica Acta*, **71**, 2780–2790.

Mühlinghaus, C., Scholz, D. & Mangini, A. (2009) Modelling fractionation of stable isotopes in stalagmites. *Geochimica et Cosmochimica Acta*, **73**, 7275–7289.

Müller, W., Shelley, M., Miller, P. et al. (2009) Initial performance metrics of a new custom-designed ArF excimer LA-ICPMS system coupled to a two-volume laser-ablation cell. *Journal of Analytical Atomic Spectrometry*, **24**, 209–214.

Muramatsu, H., Tashiro, Y., Haegawa, N. et al. (2002) Seasonal variations of ^{222}Rn concentrations in the air of a tunnel located in Nagano city. *Journal of Environmental Radioactivity*, **60**, 263–274.

Murphy, D.M., Solomon, S., Portmann, R.W., Forster, P.M. & Wong, T. (2009) An observationally based energy balance for the Earth since 1950. *Journal of Geophysical Research*, **114**, D17107, doi:10.1029/2009JD012105.

Murphy, P.J. & Lundberg, J. (2009) Uranium series dates from the windy pits of the North York Moors, United Kingdom: implications for late Quaternary ice cover and timing of speleogenesis. *Earth Surface Processes and Landforms*, **34**, 305–313.

Murray, J.W. (1954) The deposition of calcite and aragonite in caves. *Journal of Geology*, **62**, 481–492.

Museums & Galleries Commission (1993) *Standards in the Museum 3. Care of Geological Collections 1993*, London.

Musgrove, M. & Banner, J.L. (2004) Controls on the spatial and temporal variability of vadose dripwater geochemistry: Edwards Aquifer, central Texas. *Geochimica et Cosmochimica Acta*, **68**, 1007–1020.

Mutchler, C.K. and Larson, C.L. (1971) Splash amounts from waterdrop impact on a smooth surface. *Water Resources Research*, **7**, 195–200.

Mylroie, J.E. & Carew, J.L. (2000) 5.1 Speleogenesis in coastal oceanic settings. In: Klimchouk, A.B., Ford, DC, Palmer, A.N. & Dreybrodt, W. (eds.) *Speleogenesis. Evolution of Karst Aquifers*. Huntsville: National Speleological Society, pp. 226–233.

Myneni, S.C.B. (2002) Soft X-ray Spectroscopy and spectromicroscopy studies of organic molecules in the Environment. In: Fenter, P.A. Rivers, M.L., Sturchio, N.C. & Sutton, S.R. (editors) *Application of Synchrotron Radiation in Low –Temperature Geochemistry and Environmental Science*. Reviews in Mineralogy & Geochemistry 49, Mineralogical Society of America, Washington, pp. 485–579.

Navarro, C., Carrión, J.S., Navarro, J., Munuea, M. & Prieto, A.R. (2000) An experimental approach to the palynology of cave deposits. *Journal of Quaternary Science*, **15**, 603–619.

Neff, U., Burns, S.J., Mangini, A. et al. (2001) Strong coherence between solar variability and the monsoon in Oman between 9 and 6 kyr ago. *Nature*, **411**, 290–293.

Neuser, R.D. & Richter, D.K. (2007) Non-marine radiaxial fibrous calcites—examples of speleothems proved by electron backscatter diffraction. *Sedimentary Geology*, **194**, 149–154.

Niggemann, S., Mangini, A, Richter D.K. & Würth G. (2003) A paleoclimate record of the last 17,600 years in stalagmites from the B7 cave, Sauerland, Germany. *Quaternary Science Reviews*, **22**, 555–567.

Nitoiu, D. & Beltrami, H. (2005) Subsurface thermal effect of land use changes. *Journal of Geophysical Research*, **110**, F01005. doi: 10.1029/2004JF000151.

NOAA/NESDIS (2002) US Climate Reference Network Site Information Handbook. NOAA/NEDIS CRN Series X030. http://www1.ncdc.noaa.gov/pub/data/uscrn/documentation/program/X030FullDocumentD0.pdf.

Nordstrom, D.K., Plummer, L.N., Langmuir, D. et al. (1990) Revised chemical equilibrium data for major water-mineral reactions and their limitations. In: Melchior DC & Bassett R.L. (eds.) *Chemical Modeling of Aqueous Systems II. American Chemical Society Series* **416**, Washington, DC, pp. 398–413.

Nott, J., Haig, J., Neil, H. et al. (2007) Greater frequency of landfalling tropical cyclones at centennial compared with seasonal and decadal scales. *Earth and Planetary Science Letters*, **255**, 367–372.

Nur, A. (2008) *Apocalypse. Earthquakes, Archaeology and the Wrath of God*. New Jersey: Princeton University Press.

O'Brien, B.J. (1956) 'After-glow' of cave calcite. *Bulletin of the National Speleological Society*, **18**, 50–51.

O'Neil, J.R., Clayton, R.N. & Mayeda, T.K. (1969) Oxygen isotope fractionation in divalent metal carbonates. *Journal of Chemical Physics*, **51**, 5547–5558.

Onac, B.P. (1997) Crystallography of speleothems. In: Hill, C. & Forti, P. (Eds.) *Cave Minerals of the World*, 2nd edition, National Speleological Society, Huntsville, Alabama, 230–236.

Onac, B.P., Constantin, S., Lundberg, J. & Lauritzen, S.-E. (2002) Isotopic climate record in a Holocene stalagmite from Ursilor Cave (Romania). *Journal of Quaternary Science*, **17**, 319–327.

Orland, I.J., Bar-Matthews, M., Kita, N.T. et al. (2009) Climate deterioration in the Eastern Mediterranean as revealed by ion microprobe analysis of a speleothem that grew from 2.2 to 0.9 ka in Soreq Cave, Israel. *Quaternary Research*, **71**, 27–35.

Ortega, R., Maire, R., Deves, G. & Quinif, Y. (2005) High-resolution mapping of uranium and other trace elements in recrystallized aragonite-calcite speleothems from caves in the Pyrenees (France): implication for U-series dating. *Earth and Planetary Science Letters*, **237**, 911–923.

Ortega-Retuerta, E., Passow, U., Duarte, C.M. & Reche, I. (2009) Effects of ultraviolet radiation on (not so) transparent exopolymer particles. *Biogeosciences*, **6**, 3071–3080.

Ortega-Retuerta, E., Duarte, C.M. & Reche, I. (2010) Significance of bacterial activity for the distribution and dynamics of transparent exopolymer particles in the Mediterranean Sea. *Microbial Ecology*, **59**, 808–818.

Osborne, R.A.L. (2000) 3.7 Paleokarst and its significance for speleogenesis. In: Klimchouk, A.B., Ford, DC, Palmer, A.N. & Dreybrodt, W. (eds.) *Speleogenesis. Evolution of Karst Aquifers*. Huntsville: National Speleological Society, pp. 113–123.

Osborne, R.A.L. (2004) The trouble with cupolas. *Acta Carsologica*, **33**, 9–36.

Osborne, R.A.L. (2008) Cave turbidites. *Acta Carsologica*, **37**, 41–50.

Oster, J.L., Montanez, I.P., Guilderson, T.P. et al. (2010) Modeling speleothem $\delta^{13}C$ variability in a central Sierra Nevada cave using ^{14}C and $^{87}Sr/^{86}Sr$. *Geochimica et Cosmochimica Acta*, **74**, 5228–5242.

Palmer, A.N. (1991) Origin and morphology of limestone caves. *Geological Society of America Bulletin*, **103**, 1–21.

Palmer, A.N. (2007) *Cave Geology*. Cave Books, Dayton, Ohio.

Pancost, R.D., Bass, M., van Geel, B. & Sinninghe-Damsté, J.S. (2002) Biomarkers as proxies for plant inputs to peats: an example from a subboreal ombrotrophic bog. *Organic Geochemistry*, **33**, 675–690.

Panno, S.V., Lundstrom, C.C., Hackley, K.C. et al. (2009) Major earthquakes recorded by speleothems in Mid-western U.S. Caves. *Bulletin of the Seismological Society of America*, **99**, 2147–2154.

Pape, J.R., Banner, J.L., Mack, L.E. et al. (2010) Controls on oxygen isotope variability in precipitation and cave drip waters, central Texas, USA. *Journal of Hydrology*, **385**, 203–215.

Paquette, J. & Reeder, R.J. (1990) New type of compositional zoning in calcite: insights into crystal growth mechanisms. *Geology*, **18**, 1244–1247.

Paquette, J. & Reeder, R.J. (1995) Relationship between surface structure, growth mechanism, and trace element incorporation in calcite. *Geochimica et Cosmochimica Acta*, **59**, 735–749.

Parker, D.E., Legg, T.P. & Folland, C.K. (1992) A New Daily Central England Temperature Series, 1772–1991. *International Journal of Climatology*, **12**, 317–342.

Parks, C.D. (1991) A review of the possible mechanisms of cambering and valley bulging. In: *Quaternary Engineering Geology*, Foster, A., Culshaw, M.G., Cripps, J.C., Little, J.A., Moon, C.F. (eds). *Engineering Geology*

Special Publication 7. Bath: Geological Society, pp. 373–380.

Parnell, A.C., Haslett, J., Allen, J.R.M. et al. (2008) A flexible approach to assessing synchroneity of past events using Bayesian reconstructions of sedimentation history. *Quaternary Science Reviews*, **27**, 1872–1885.

Partin J.W., Cobb K.M. & Banner J.L. (2008) Climate variability recorded in tropical and sub-tropical speleothems. *PAGES News*, **16**, 9–10.

Pausata, F.S.R., Battisti, D.S., Nisancioglu, K.H. & Bitz, C.M. (2011) Chinese stalagmite $\delta^{18}O$ controlled by changes in the Indian monsoon during a simulated Heinrich event. *Nature Geoscience*, **4**, 474–480.

Paterson, R.J., Whitaker, F.F., Smart, P.L., Jones, G.D. & Oldham, D. (2008) Controls on early diagenetic overprinting in icehouse carbonates: insights from modelling hydrological zone residence times using CARB3D+. *Journal of Sedimentary Research*, **78**, 258–281.

Pauling, A., Luterbacher, J., Casty, C. & Wanner, H. (2006) Five hundred years of gridded high-resolution precipitation reconstructions over Europe and the connection to large-scale circulation. *Climate Dynamics*, **26**, 387–405.

Paulsen, D.E., Li, H.C. & Ku, T.L. (2003) Climate variability in central China over the last 1270 years revealed by high-resolution stalagmite records. *Quaternary Science Reviews*, **22**, 691–701.

Pearthree, P.A., Spencer, J.E., Faulds, J.E & House, P.K. (2008) Comment on 'Age and Evolution of the Grand Canyon Revealed by U-Pb Dating of Water Table–Type Speleothems'. *Science*, **321**, 1634.

Pederson, J., Young, R., Lucchitta, I. et al. (2008) Comment on 'Age and Evolution of the Grand Canyon Revealed by U-Pb Dating of Water Table–Type Speleothems'. *Science*, **321**, 1634.

Pédrot, M., Dia, A., Davranche, M. et al. (2008) Insights into colloid-mediated trace element release at the soil-water interface. *Journal of Colloid and Interface Science*, **325**, 187–197.

Perkins, A.M. (1996) Observations under electron microscopy of magnetic minerals extracted from speleothems. *Earth and Planetary Science Letters*, **139**, 281–289.

Perrette, Y. & Jaillet, S. (2010) Spatial distribution of soda straws growth rates of the Coufin Cave (Vercors, France). *International Journal of Speleology*, **39**, 61–70.

Perrette, Y., Delannoy, J.J., Desmet, M. et al. (2005) Speleothem organic matter content imaging. The use of a Fluorescence Index to characterise the maximum emission wavelength. *Chemical Geology*, **214**, 193–208.

Perrette Y., Poulenard J., Saber A-l., et al. (2008) Polycyclic Aromatic Hydrocarbons in stalagmites: occurrence and use for analyzing past environments. *Chemical Geology*, **251**, 67–76.

Perrier, F., Morat, P. & Le Mouël, J-L. (2001) Pressure induced temperature variations in an underground quarry. *Earth and Planetary Science Letters*, **191**, 145–156.

Perrier, F., Richon, P., Crouzeix, C. et al. (2004) Radon-222 signatures of natural ventilation regimes in an underground quarry. *Journal of Environmental Radioactivity*, **71**, 17–32.

Perrier, F. Le Mouël, J-L. Poirier, J-P. & Shnirman, M.G. (2005) Long-term climate change and surface versus underground temperature measurements in Paris. *International Journal of Climatology*, **25**, 1619–1631.

Perrier, F., Richon, P., Gautam, U. et al. (2007) Seasonal variations of natural ventilation and radon-222 exhalation in a slightly rising dead-end tunnel. *Journal of Environmental Radioactivity*, **97**, 220–235.

Perrin, J. & Luetscher, M. (2008) Inference of the structure of karst conduits using quantitative tracer tests and geological information: example of the Swiss Jura. *Hydrogeology Journal*, **16**, 951–967.

Petit, J.R., White, J.W.C., Young N.W. et al. (1991) Deuterium excess in recent Antarctic snow. *Journal of Geophysical Research*, **96**, 5113–5122.

Petkovšek, Z. (1968) Climatic conditions in the shallow-holes at cave entrances. *Proceedings of the 4th International Congress of Speleology 1965*, **3**, 181–188, Ljubljana.

Pezdič, J., Šušteršič, F. & Misič, M. (1998) On the role of clay-carbonate reactions in speleo-inception: a contribution to the understanding of the earliest stage of karst channel formation. *Acta Carsologica*, **27**, 187–200.

Pfahl, S. & Wernli, H. (2009) Lagrangian simulations of stable isotopes in water vapour: an evaluation of nonequilibrium fractionation in the Craig-Gordon model. *Journal of Geophysical Research* **114**, D20108, doi:10.1029/2009JD012054.

Pflitsch, A. & Piasecki, J. (2003) Determination of an airflow system in Niedzwiedzia (Bear) Cave, Kletno, Poland. *Journal of Cave and Karst Studies*, **65**, 160–173.

Phillips, B.L., Lee, Y.J. & Reeder, R.J. (2005) Organic coprecipitates with calcite: NMR spectroscopic evidence. *Environmental Science & Technology*, **39**, 4533–4539.

Phillips, J.D. (2009) Changes, perturbations, and responses in geomorphic systems. *Progress in Physical Geography*, **33**, 17–30.

Piccini, L., Zanchetta, G., Drysdale, R.N. et al. (2008) the environmental features of the Monte Corchia cave system (Apuan Alps, central Italy) and their effects on speleothem growth. *International Journal of Speleology*, **37**, 153–172.

Pickering, R. & Kramers, J.D. (2010) Re-appraisal of the stratigraphy and determination of new U-Pb dates for the Sterkfontein hominin site, South Africa. *Journal of Human Evolution*, **59**, 70–86.

Pingitore, N.E. (1976) Vadose and phreatic diagenesis: processes, products and their recognition in corals. *Journal of Sedimentary Petrology*, **46**, 785–1006.

Pingitore, N.E. & Eastman, M.P. (1986) The coprecipitation of Sr^{2+} with calcite at 25 °C and 1 atmosphere. *Geochimica et Cosmochimica Acta*, **50**, 2195–2203.

Pingitore, N.E., Lyttle, F.W., Davies B.M. et al. (1992) Model of incorporation of Sr^{2+} in calcite: determination by X-ray absorption spectroscopy. *Geochimica et Cosmochimica Acta*, **56**, 1531–1538.

Pingitore, N., Meitzner, G. & Love, K. (1995) Identification of sulphate in natural carbonates by X-ray absorption spectroscopy. *Geochimica et Cosmochimica Acta*, **59**, 2477–2483.

Pionke, H.B., Gburek, W.J., Sharpley, A.N. & Zollweg, J.A. (1997) Hydrological and chemical controls on phosphorus loss from catchments. In: (Tunney, H., Careton, O.T., Brookes, P.C. & Johnstone, A.E., editors) *Phosphorus loss from soil to water*. CAB International, Wallingford, UK, pp. 225–242.

Pitty, A. (1966) An approach to the study of karst water. *University of Hull Occasional Papers in Geography*, 5.

Pitty, A.F. (1968) Calcium carbonate content of karst water in relation to flow-through time. *Nature*, **217**, 939–940.

Plan, L., Filipponi, M., Behm, M., Seebacher, R. & Jeutter, P. (2009) Constraints on alpine speleogenesis from cave morphology—A case study from the eastern Totes Gebirge (Northern Calcareous Alps, Austria). *Geomorphology*, **106**, 118–129.

Plummer, L.N., Parkhurst, D.L. & Kosiur, D.R. (1975) MIX2, a computer program for modeling chemical reactions in natural waters. *U.S. Geological Survey, Water Resources Investigations Report 61*.

Plummer, L.N., Wigley, T.M.L. & Parkhurst, D.L. (1978) The kinetics of calcite dissolution in CO_{2-} water systems at 5° to 60 °C and 0.0 to 1.0 atm CO_2. *American Journal of Science*, **278**, 179–216.

Plummer, L.N., Parkhurst, D.L. & Wigley, T.M.L. (1979) Critical review of the kinetics of calcite dissolution and precipitation. In: Jenne, E.A. (Editor), *Chemical modelling in aqueous systems*. American Chemical Society, Washington, 537–573.

Plummer, L.N., Busenberg, E. & Riggs, A.C. (2000) In-situ of calcite at Devils Hole, Nevada: comparison of field and laboratory rates to a 500,000 year record of near-equilibrium calcite growth. *Aquatic Geochemistry*, **6**, 257–274.

Poage, M.A. & Chamberlain, C.P. (2001) Empirical relationships between elevation and the stable isotope composition of precipitation and surface waters: considerations for studies of paleoelevation change. *American Journal of Science*, **301**, 1–15.

Polag, D., Scholz, D., Mühlinghaus, C. et al. (2010) Stable isotope fractionation in speleothems: laboratory experiments. *Chemical Geology*, **279**, 31–39.

Pollack, H.N. & Huang, S. (2000) Climate reconstruction from subsurface temperatures. *Annual Review of Earth and Planetary Science*, **28**, 339–365.

Pollack, H.N., Smerdon, J.E. & van Keken, P.E. (2005) Variable seasonal coupling between air and ground temperatures: a simple representation in terms of subsurface thermal diffusivity. *Geophysical Research Letters*, **32**, L15405. doi: 10.1029/2005GL023869.

Polyak, V.J. & Asmerom, Y. (2001) Late Holocene climate and cultural changes in the southwestern United States. *Science*, **294**, 148–151.

Polyak, V.J., Cokendolpher, J.C., Norton, R.A. & Asmerom, Y. (2001) Wetter and cooler late Holocene climate in the southwestern United States from mites preserved in stalagmites. *Geology*, **29**, 643–646.

Polyak, V., Hill, C. & Asmerom, Y. (2008) Age and evolution of the Grand Canyon revealed by U-Pb dating of water table-type speleothems. *Science*, **319**, 1377–1380.

Popper, K.R. (1963) *Conjectures and Refutations: The Growth of Scientific Knowledge*.

Porter, M.L. (2007) Subterranean biogeography: what we have learnt from molecular techniques. *Journal of Cave and Karst Studies*, **69**, 179–186.

Portillo, M.C., Porca, E., Cuezva, S. et al. (2009) Is the availability of different nutrients a critical factor for the impact of bacteria on subterraneous carbon budgets? *Naturwissenschaften*, **96**, 1035–1042.

Postpischl, D., Agostini, S., Forti, P. & Quinif, Y. (1991) Palaeoseismicity from karst sediments: the 'Grotta del Cervo' cave case study (central Italy). *Tectonophysics*, **193**, 33–44.

Pote, D.H., Daniel, T.C., Nichols, D.J. et al. (1999) Seasonal and soil-drying effects on runoff phosphorus relationships to soil phosphorus. *Journal of Environmental Quality*, **28**, 170–175.

Prasad, S., Witt, A., Kienel, U. et al. (2009) The 8.2 ka event: evidence for seasonal difference and the rate of climate change in western Europe. *Global and Planetary Change*, **67**, 218–226.

Pratt, B.R., James, N.P. & Cowan, C.A. (1992) 16. Peritidal Carbonates. In: Walker, R.G. & James, N.P. (eds.) *Facies Models. Response to Sea Level Change*. St. John's Newfoundland: Geological Association of Canada.

Prentice, I.C., Cramer, W., Harrison, S.P. et al. (1992) A global biome model based on plant physiology and dominance, soil properties and climate. *Journal of Biogeography* **19**, 117–134.

Prieto, M. (2010) Thermodynamics of ion partitioning in solid solution—aqueous solution systems. In: Stoll, H. and Prieto, M., editors. *Ion partitioning in ambient temperature aqueous systems: from fundamentals to applications in climate proxies and environmental geochemistry*. EMU Notes in Mineralogy, **10**, 1–42.

Proctor, C.J., Baker, A., Barnes, W.L. & Gilmour, M.A. (2000) A thousand year speleothem proxy record of North Atlantic climate from Scotland. *Climate Dynamics*, **16**, 815–820.

Proctor, C.J., Baker, A. & Barnes, W.L. (2002) A three thousand year record of north Atlantic climate. *Climate Dynamics*, **19**, 449–454.

Prokoph, A., Shields, G.A. & Veizer, J. (2009) Compilation and time-series analysis of a marine carbonate $\delta^{18}O$, $\delta^{13}C$, $^{87}Sr/^{86}Sr$ and $\delta^{34}S$ database through Earth history. *Earth-Science Reviews*, **87**, 113–133.

Przylibski, T.A. (1999) Radon concentration changes in the air of two cave in Poland. *Journal of Environmental Radioactivity*, **45**, 81–94.

Pulido-Bosch, A., Martín-Rosales, W. López-Chicano, M. et al. (1997) Human impact in tourist karstic cave (Aracena, Spain). *Environmental Geology*, **31**, 142–149.

Pulido-Bosch, A., Motyka, J., Pulido-Leboeuf, P. & Borczak, S. (2004) Matrix hydrodynamic properties of carbonate rocks from the Betic Cordillera (Spain). *Hydrological Processes*, **18**, 2893–2906.

Pumpanen, J., Ilvesniemi, H., & Hari, P. (2003) A Process-Based Model for Predicting Soil Carbon Dioxide Efflux and Concentration. *Soil Science America Journal*, **67**, 402–413.

Putnis, A. (2010) Effects of kinetics and mechanisms of crystal growth on ion-partitioning in solid solution-aqueous solution (SS-AS) systems. In: Stoll, H. and Prieto, M., editors. *Ion partitioning in ambient temperature aqueous systems: from fundamentals to applications in climate proxies and environmental geochemistry*. EMU Notes in Mineralogy, **10**, 43–64.

Quinn, T.M. & Saller, A.H. (1997) Geology of Anewatak atoll, Republic of the Marshall Islands. In: Vacher, H.L. & Quinn, T. (eds.) *Geology and Hydrogeology of Carbonate Islands*. Amsterdam: Elsevier, pp. 637–666.

Rahmstorf, S. (1995) Bifurcations of the Atlantic thermohaline circulation in response to changes in the hydrological cycle. *Nature*, **378**, 145–149.

Raich, J.W. & Potter, C.S. (1995) Global patterns of carbon dioxide emissions from soils. *Global Biogeochemical Cycles*, **9**, 23–36.

Raich, J.W. & Tufekcioglu, A. (2000) Vegetation and soil respiration: correlations and controls. *Biogeochemistry*, **4**, 71–90.

Railsback, L.B. (2003) An earth scientist's periodic table of the elements and their ions. *Geology*, **31**, 737–740 (and insert).

Railsback, L.B. (2008) Speciation of inorganic carbon in aqueous solution. http://www.gly.uga.edu/railsback/Fundamentals/815CO2(g)toCO3(aq)04P.pdf

Railsback, L.B., Brook, G.A., Chen, J. et al. (1994) Environmental controls on the petrology of a late Holocene speleothem from Botswana with annual layers of aragonite and calcite. *Journal of Sedimentary Research A*, **64**, 147–155.

Railsback, L.B., Dabous, A.A., Osmond, J.K. & Fleischer, C.J. (2002) Petrographic and geochemical screening of speleothems for U-series dating: an example from recrystallized speleothems from Wadi Sannur Cavern, Egypt. *Journal of Cave and Karst Studies*, **64**, 108–116.

Railsback, L.B., Liang, F., Vidal Romaní, J.R. et al. (2011) Petrographic and isotopic evidence for Holocene long-term climate change and shorter-term environmental shifts from a stalagmite from the Serra do Courel of northwestern Spain, and implications for climatic history across Europe and the Mediterranean. *Palaeogeography, Palaeoclimatology, Palaeoecology*, **305**, 172–184.

Raiswell, R., Brimblecombe, P., Dent, D.L. & Liss, P.S. (1980) *Environmental Chemistry: The Earth-Air-Water Factory*. London: Arnold.

Ramseyer, K., Miano, T.M., D'Orazio, V. et al. (1997) Nature and origin of organic matter in carbonates from speleothems, marine cements and coral skeletons. *Organic Geochemistry*, **26**, 361–378.

Rasbury, E.T. & Cole, J.M. (2009) Directly dating geologic events: U-Pb dating of carbonates. *Reviews of Geophysics*, **47**, RG3001. doi: 10.1029/2007RG000246.

Rasbury, M. & Aharon, P. (2006) ENSO-controlled rainfall variability records archived in tropical stalagmites from the mid-ocean island of Niue, South Pacific. *Geochemistry, Geophysics, Geosystems*, **7**, Q07010, doi:10.1029/2005GC001232.

Rasmussen, S.O., Andersen, K.K., Svensson, A.M. et al. (2006) A new Greenland ice core chronology for the last glacial termination. *Journal of Geophysical Research—Atmospheres*, **111**, D06102. doi:10.1029/2005JD006079.

Rasmussen, S.O., Vinther, B.M., Clausen, H.B. & Andersen, K.K. (2007) Early Holocene climate oscillations recorded in three Greenland ice cores. *Quaternary Science Reviews*, **26**, 1907–1914.

Rauch, H.W. & White, W.B. (1970) Lithologic controls on the development of solution porosity in carbonate aquifers. *Water Resources Research*, **6**, 1175–1192.

Rauch, H.W. & White, W.B. (1977) Dissolution kinetics of carbonate rocks. 1. Effects of lithology on dissolution rate. *Water Resources Research*, **13**, 381–394.

Rautaray, D., Ahmad, A. & Sastry, M. (2003) Biosynthesis of $CaCO_3$ crystals of complex morphology using a fungus and an actinomycete. *Journal of the American Chemical Society Communications*, **125**, 14656–14657.

Ready, C.D. & Retallack, G.J. (1995) Chemical composition as a guide to paleoclimate of paleosols. *Abstracts of the Annual Meeting of the Geological Society of America*, New Orleans **27**, pp. A237.

Reddy, M.M. & Nancollas, G.H. (1970) The crystallisation of calcium carbonate I. *Journal of Colloid and Interface Science*, **36**, 166–171.

Reddy, M.M. & Nancollas, G.H. (1971) The crystallisation of calcium carbonate II. *Journal of Colloid and Interface Science*, **37**, 824–829.

Reeder, R.J. (1993) Crystal chemistry of the rhombohedral carbonates. *Reviews in Mineralogy*, **11**, 1–47.

Reeder, R.J. (1996) Interaction of divalent cobalt, zinc, cadmium, and barium with calcite surface during layer growth. *Geochimica et Cosmochimica Acta*, **60**, 1543–1552.

Reeder, R.J. & Grams J.C. (1987) Sector zoning in calcite cement crystals: implications for trace element distributions in carbonates. *Geochimica et Cosmochimica Acta*, **51**, 187–194.

Reeder, R.J., Valley, J.W., Graham, C.M. & Eiler, J.M. (1997) Ion microprobe study of oxygen isotopic compositions of structurally non-equivalent growth surfaces on synthetic calcite. *Geochimica et Cosmochimica Acta*, **61**, 5057–5063.

Reeder, R.J., Lamble, G.M. & Northrup, P.A. (1999) XAFS study of the coordination and local relaxation around Co^{2+}, Zn^{2+}, Pb^{2+}, and Ba^{2+} trace elements in calcite. *American Mineralogist*, **84**, 1049–1060.

Reeder, R.J., Nugent, M., Lamble, G.M. et al. (2000) Uranyl incorporation into calcite and aragonite: XAFS and luminescence studies. *Environmental Science and Technology*, **34**, 638–644.

Reeder, R.J., Nugent, M., Tait C.D. et al. (2001) Coprecipitation of Uranium (VI) with calcite: XAFS, micro-XAS, and luminescence characterisation. *Geochimica et Cosmochimica Acta*, **65**, 3491–3503.

Refsnider, K.A. (2010) Dramatic increase in late Cenozoic alpine erosion rates recorded by cave sediments in the southern Rocky Mountains. *Earth and Planetary Science Letters*, **297**, 505–511.

Rehrl, C., Birk, S. & Klimchouk, A.B. (2008) Conduit evolution in deep-seated settings: conceptual and numerical models based on field observation. *Water Resources Research*, **44**, WII425, doi:10.1029/2008WR006905.

Reichstein, M., Tenhunen, J.D., Roupsard, O. et al. (2002) Severe drought effects on ecosystem CO_2 and H_2O fluxes at three Mediterranean evergreen sites: revision of current hypotheses? *Global Change Biology*, **8**, 999–1017.

Rekolainen, S., Ekholm, P., Ulén, B. & Jordan, C. (1997) Phosphorus losses from agriculture to surface waters in the Nordic countries. In Tunney, H., Careton, O.T., Brookes, P.C. & Johnstone, A.E. (1997). *Phosphorus loss from soil to water*. CAB International, Wallingford, UK, pp. 77–93.

Renault, P. (1967–1968) Contribution a l'étude des actions mécaniques et sédimentologiques dans la spéléogenèse. *Annales de Spéleologie*, 22–23, 1–337.

Retallack, G.J. (1994) The environmental factor approach to the interpretation of paleosols. In R. Amundson, J. Harden & M. Singer (editors), Factors in soil formation——a fiftieth anniversary perspective. *Special Publication of the Soil Science Society of America*, Madison, **33**, 31–64.

Retallack, G.J. (1998) Fossil soils and completeness of the rock and fossil record. In S.K. Donovan & C.R.C. Paul (editors), *The Adequacy of the Fossil Record*. John Wiley and Sons, Chichester, 131–162.

Retallack, G.J. (2005) Pedogenic carbonate proxies for amount and seasonality of precipitation in paleosols. *Geology*, **33**, 333–336.

Reynard, L.M., Day, C.C. & Henderson, G.M. (2011) Large fractionation of calcium isotopes during cave-analogue calcium carbonate growth. *Geochimica et Cosmochimica Acta*, **75**, 3726–3740.

Reynolds, D.M. & Ahmad, S.R. (1997) Rapid and direct determination of wastewater BOD values using a fluorescence technique. *Water Research*, **31**, 2012–2018.

Ribes, A.C., Lundberg, J., Waldron, D.J et al. (2000) Photoluminescence imaging of speleothem microbanding with a high-resolution confocal scanning laser microscope. *Quaternary International*, **68**, 253–259.

Richards, D.A. & Dorale, J. (2003) Uranium-series Chronology and Environmental Applications of Speleothems. *Reviews in Mineralogy and Geochemistry* **52**, 407–460.

Richards, D.A., Smart, P.L. & Edwards, R.L. (1994) Maximum sea levels for the last glacial period form U-series ages of submerged speleothems. *Nature*, **367**, 357–360.

Richards, D.A., Bottrell, S.H., Cliff, R.A. et al. (1998) U-Pb dating of a speleothem of Quaternary age. *Geochimica et Cosmochimica Acta*, **62**, 3683–3688.

Richards, K. & Clifford, N. (2008) Science, systems and geomorphologies: why LESS may be more. *Earth Surface Processes and Landforms*, **33**, 1323–1340.

Richon P., Perrier F., Pili E. & Sabroux J.C. (2009) Detectability and significance of 12h barometric tide in radon-222 signal, dripwater flow rate, air temperature and carbon dioxide concentration in an underground tunnel. *Geophysical Journal International*, **176**, 683–694.

Richter, D.K., Gotte, T., Niggemann, S. & Wurth, G. (2004) REE^{3+} and Mn^{2+}-activated cathodoluminescence in lateglacial and Holocene stalagmites of central Europe: evidence for climatic processes? *The Holocene*, **14**, 759–767.

Riedwyl, N., Küttel, M., Luterbacher, J. & Wanner, H. (2009) Comparison of climate field reconstruction techniques: application to Europe. *Climate Dynamics*, **32**, 381–395.

Rieley, G., Collier, R.J., Jones, D.M. & Eglinton, G. (1991) The biogeochemistry of Ellesmere Lake, U.K.–I: source correlation of leaf wax inputs to the sedimentary lipid record. *Organic Geochemistry*, **17**, 901–912.

Riggs, A.C., Carr, W.J., Kolesar, P.T & Hoffman, R.J. (1994) Tectonic speleogenesis of Devils Hole, Nevada, and implications for hydrogeology and the development of long, continuous paleoenvironmental records. *Quaternary Research*, **42**, 241–254.

Rimstidt, J.D., Balog, A. & Webb, J. (1998) Distribution of trace elements between carbonate minerals and aqueous solutions. *Geochimica et Cosmochimica Acta*, **62**, 1851–1863.

Risi C., Ony S. & Vimeux F. (2008) Influence of convective processes on the isotopic composition ($\delta^{18}O$ and δD) of precipitation and water vapour in the tropics: 2. Physical interpretation of the amount effect. *Journal of Geophysical Research* **113**, D19306, doi:10.1029/2008JD009943.

Robert, J., Miranda, C.F. & Muxart, R. (1969) Mesure de la periode du protactinium-231 par microcaloimetrie, *Radiochimica Acta*, **11**, 104–108.

Roberts, N. (1998) *The Holocene: An Environmental History*, 2nd edition. Oxford: Blackwell.

Roberts, M.S., Smart, P.L. & Baker, A. (1998) Annual trace element variations in a Holocene speleothem. *Earth and Planetary Science Letters*, **154**, 237–246.

Roberts, N., Jones, M.D., Benkaddour, A. et al. (2008) Stable isotope records of Late Quaternary climate and hydrology from Mediterranean lakes: the ISOMED synthesis. *Quaternary Science Reviews*, **27**, 2426–2441.

Rode, S., Oyabu, N., Kobayashi, K. et al. (2009) True atomic-resolution imaging of $\{10\bar{1}4\}$ calcite in aqueous solution by frequency modulation atomic force microscopy. *Langmuir*, **25**, 2850–2853.

Rodhe, H. (1992) 4. Modeling biogeochemical cycles. In: Butcher, S.S., Charlson, R.J., Orians, G.H. & Wolfe, G.V. (eds.) *Global Biogeochemical Cycles*, Academic Press, London, 55–71.

Rogerson, M. Pedley H.M. & Wadhawan, J.D., et al. (2008) New insights into biological influence on the geochemistry of freshwater carbonate deposits. *Geochimica et Cosmochimica Acta*, **72**, 4976–4987.

Rohling, E.J. & Palike, H. (2005) Centennial-scale climate cooling with a sudden cold event around 8200 years ago. *Nature*, **434**, 975–979.

Rohling, E.J., Grant, K., Bolshaw, A.P. et al. (2009) Antarctic temperature and global sea level closely coupled over the past five glacial cycles. *Nature Geoscience*, **2**, 500–504.

Rollion-Bard, C., Blamart, D., Cuif, J.-P. & Dauphin, Y. (2010) *In situ* measurements of oxygen isotopic composition in deep-sea coral, *Lophelia pertusa*: re-examination of the current geochemical models of biomineralization. *Geochimica et Cosmochimica Acta*, **74**, 1338–1349.

Romanov, D., Kaufmann, G. & Dreybrodt, W. (2008a) Modelling stalagmite growth by first principles of chemistry and physics of calcite precipitation. *Geochimica et Cosmochimica Acta*, **72**, 423–437.

Romanov, D., Kaufmann, G. & Dreybrodt, W. (2008b) $\delta^{13}C$ profiles along growth layers of stalagmites: comparing theoretical and experimental results. *Geochimica et Cosmochimica Acta*, **72**, 438–448.

Roques, H. (1969) A review of present-day problems in the physical chemistry of carbonates in solution. *Transactions of the Cave Research Group of Great Britain*, **11**, 139–163.

Rossum, J.R. (1975) Checking the accuracy of water analyses through the use of conductivity. *Journal of the American Water Works Association*, **67**, 204–205.

Rousseau, L., Pepe, C. & DeLumley, H. (1992) Revelation of fossil activity in the middle-Pleistocene flowstones through biogeochemical markers. *Comptes Rendus de L'Academie des Sciences*, **315**, 1819–1825.

Rousseau, L., Laafar, S., Pepe, C & DeLumley, H. (1995) Sterols as biogeochemical markers—results from ensemble-E of the stalagmitic floor, Grotte-du-Lazaret, Nice, France. *Quaternary Science Reviews*, **14**, 51–59.

Rowley, D.B. & Garzione, C.N. (2007) Stable isotope-based paleoaltimetry. *Annual Review of Earth and Planetary Sciences*, **35**, 463–508.

Rozanski K., Araguás-Araguás, L. & Gonfiantini, R. (1993) Isotopic patterns in modern global precipitation. In Swart, P.K. et al. (eds.) *Climate Change in Continental Isotopic Records*, Geophysical Monograph Series **78**, American Geophysical Union, Washington, DC, 1–36.

Rubel, F. & Kottek, M. (2010) Observed and projected climate shifts 1901–2100 depicted by world maps of the

Köppen-Geiger climate classification. *Meteorologie Zeitschrift*, **19**, 135–141.

Ruddiman, W.F. (2007) The early anthropogenic hypothesis: challenges and responses. *Reviews of Geophysics*, **45**, RG4001. doi: 10.1029/2006RG000207.

Ruddiman, W.F. (2008) *Earth's Climate. Past and Future*. (2nd edition). Freeman, New York.

Rühling, A. & Tyler, G. (2004) Changes in the atmospheric deposition of minor and rare elements between (1975) and (2000) in south Sweden, as measured by moss analysis. *Environmental Pollution*, **131**, 414–423.

Ruggieri, E., Herbert, T., Lawrence, K.T. et al. (2009) Change point method for detecting regime shifts in paleoclimatic time series: application to $\delta^{18}O$ time series of the Plio-Pleistocene. *Paleoceanography*, **24**, PA1204. doi: 10.1029/2007PA001568.

Ruiz-Agudo, E., Mees, F., Jacobs, P. & Rodriguez-Navarro, C. (2007) The role of saline solution properties on porous limestone salt weathering by magnesium and sodium sulfates. *Environmental Geology*, **52**, 305–317.

Ruiz-Agudo, E., Putnis, C.V., Jiménez-López, C. & Rodriguez-Navarro, C. (2009) An atomic force microscopy of calcite dissolution in saline solutions: the role of magnesium ions. *Geochimica et Cosmochimica Acta*, **73**, 3201–3217.

Rushdi, A.I., Clark, P.U., Mix, A.C. et al. (2011) Composition and sources of lipid compounds in speleothem calcite from southwestern Oregon and their paleoenvironmental implications. *Environmental Earth Science*, **62**, 1245–1261.

Rutherford, S., Mann, M.E., Delworth, T.L. & Stouffer, R. (2003) Climate field reconstruction under stationary and nonstationary forcing. *Journal of Climate*, **16**, 462–479.

Rzonca, B. (2008) Carbonate aquifers with hydraulically non-active matrix: a case study from Poland. *Journal of Hydrology*, **355**, 202–213.

Salve, R., Krakauer, N.Y., Kowalsky, M.B. & Finsterle, S. (2008) A quantitative assessment of microclimatic perturbations in a tunnel. *International Journal of Climatology*, **28**, 2081–2087.

Sanchez, P.A., Ahamed, S., Carré, F. et al. (2009) Digital Soil Map of the World. *Science*, **325**, 380–381.

Sánchez-Moral S., Soler V., Cañaveras J.C. et al. (1999) Inorganic deterioration affecting the Altamira Cave, N Spain: quantitative approach to wall-corrosion (solution-etching) processes induced by visitors. *Science of the Total Environment*, **243**, 67–84.

Sandberg, P. (1985) Aragonite cements and their occurrence in ancient limestone. In: Schneidermann, N. & Harris, P.M. (eds.) (1985) *Carbonate Cements*. Tulsa: Society of Economic Paleontologists and Mineralogists, Special Publication **36**, 33–57.

Sarbu, S.M. & Lascu, C. (1997) Condensation corrosion in Movile cave, Romania. *Journal of Cave and Karst Studies*, **59**, 99–102.

Sasowksy, I.D. & Mylroie, J. (eds.) (2004) *Studies of Cave Sediments. Physical and Chemical Records of Paleoclimate*. New York: Kluwer.

Sasowsky, I.D., White, W.B. & Schmidt, V.A. (1995) Determination of stream-incision rate in the Appalachian plateaus by using cave-sediment magnetostratigraphy. *Geology*, **23**, 415–418.

Saurer, M., Schweingruber, F., Vaganov, E.A. et al. (2002) Spatial and temporal oxygen isotope trends at the northern tree-line in Eurasia. *Geophysical Research Letters*, **29**. doi: 10.1029/2001GL013739.

Schauble, E.A., Ghosh, P. & Eiler J.M. (2006) Preferential formation of ^{13}C–^{18}O bonds in carbonate minerals, estimated using first-principles lattice dynamics. *Geochimica et Cosmochimica Acta*, **70**, 2510–2529.

Scheidegger, Y., Baur, H., Brennwald, M.S. et al. (2010) Accurate analysis of noble gas concentrations in small water samples and its application to fluid inclusions in stalagmites. *Chemical Geology*, **272**, 31–39.

Schettler, G., Romer, R.L., Qiang, M. et al. (2009) Size-dependent geochemical signatures of Holocene loess deposits from the Hexi Corridor (China). *Journal of Asian Earth Sciences* **35**, 103–136.

Schimpf, D., Kilian, R., Kronz, A. et al. (2011) The significance of chemical, isotopic, and detrital components in three coeval stalagmites from the superhumid southernmost Andes (53°S) as high-resolution palaeo-proxies. *Quaternary Science Reviews*, **30**, 443–459.

Schlager, W. (2005) *Carbonate Sedimentology and Sequence Stratigraphy*. SEPM: Tulsa, Oklahoma.

Schlesinger, M.E. & Ramankutty, N. (1994) An oscillation in the global climate system of period 65–70 years. *Nature*, **367**, 723–726.

Schmidt G.A., LeGrande A.N. & Hoffmann G. (2007) Water isotope expressions of intrinsic and forced variability in a coupled ocean-atmosphere model. *Journal of Geophysical Research*, **112**, D10103, doi:10.1029/2006JD007781.

Schneider, T. (2001) Analysis of incomplete climate data: estimation of mean values and covariance matrices and imputation of missing values. *Journal of Climate*, **14**, 853–871.

Schneidermann, N. & Harris, P.M. (eds.) (1985) *Carbonate Cements*. Tulsa: Society of Economic Paleontologists and Mineralogists, Special Publication 36.

Schneider, D.P. & Steig, E.J. (2008) Ice cores record significant 1940s Antarctic warmth related to tropical climate variability. *Proceedings of the National Academy of Sciences of the USA*, **105**, 12154–12158.

Scholle, P.A. (1978) *A Color Illustrated Guide to Carbonate Rock Constituents, Textures, Cements, and Porosities*. Tulsa: American Association of Petroleum Geologists, Memoir 27.

Scholle, P.A. & Schluger, P.R. (eds.) (1979) *Aspects of Diagenesis*. Tulsa: Society of Economic Paleontologists and Mineralogists, Special Publication, 26.

Scholle, P.A. & Halley, R.B. (1985) Burial diagenesis: out of sight, out of mind! In: Schneidermann, N. & Harris, P.M. (eds.) *Carbonate Cements*. Tulsa: Society of Economic Paleontologists and Mineralogists, Special Publication 36, 309–334.

Scholle, P.A., Bebout, D.G. & Moore, C.H. (eds.) (1983) *Carbonate Depositional Environments*. Tulsa: American Association of Petroleum Geologists, Memoir 33.

Scholz, D. & Hoffmann, D.L. (2011) StalAge—an algorithm designed for construction of speleothem age models. *Quaternary Geochronology*, **6**, 369–382.

Scholz, D., Mühlinghaus, C. & Mangini, A. (2009) Modelling $\delta^{13}C$ and $\delta^{18}O$ in the solution layer on stalagmite surfaces. *Geochimica et Cosmochimica Acta*, **73**, 2592–2602.

Schott J., Pokrovsky O.S. & Oelkers E.H. (2009) The link between mineral dissolution/precipitation kinetics and solution chemistry. *Reviews in Mineralogy and Geochemistry* **70**, 207–258.

Schouten, S., Huguet, C., Hopmans, E. et al. (2007) Analytical Methodology for TEX(86) Paleothermometry by High-Performance Liquid Chromatography/Atmospheric Pressure Chemical Ionization-Mass Spectrometry. *Analytical Chemistry*, **79**, 2940–2944.

Schulz, M. & Stattegger, K. (1997) SPECTRUM: spectral analysis of unevenly spaced paleoclimatic time series. *Computers and Geosciences*, **23**, 929–945.

Schuster, S.C. (2008) Next-generation sequencing transforms today's biology. *Nature Methods*, **5**, 16–18.

Schwarcz, H.P. & Rink, W.J. (2001) Dating methods for sediments of caves and rockshelters with examples from the Mediterranean region. *Geoarcharchaeology*, **16**, 355–371.

Schwarcz, H.P., Harmon, R.S., Thompson, P. & Ford, DC (1976) Stable isotope studies of fluid inclusions in speleothems and their palaeoclimatic significance. *Geochimica et Cosmochimica Acta*, **40**, 657–665.

Šebela, S. (2008) Broken speleothems as indicators of tectonic movements. *Acta Carsologica*, **37**, 51–62.

Segele, Z.T., Lamb, P.J & Leslie, L.M. (2009) Seasonal-to-interannual variability of Ethiopia / Horn of Africa Monsoon. Part I: associations of wavelet-filtered large-scale atmospheric circulation and global sea surface temperature. *Journal of Climate*, **22**, 3396–3421.

Self, C.A. & Hill, C.A. (2003) How speleothems grow: an introduction to the ontogeny of cave minerals. *Journal of Cave and Karst Studies*, **65**, 130–151.

Sellers, W.D. (1965) *Physical Climatology*, University of Chicago Press.

Senesi, N., Miano, T.M., Provenzano, M.R. & Brunetti, G. (1989) Spectroscopic and compositional comparative characterisation of IHSS reference and standard fulvic and humic acids of various origins. *Science of the Total Environment*, **81/82**, 143–156.

Senesi, N., Miano, T.M., Provenzano, M.R. & Brunetti, G. (1991) Characterisation, differentiation, and classification of humic substances by fluorescence spectroscopy. *Soil Science*, **152**, 259–271.

Sengupta, S. & Sarkar, A. (2006) Stable isotope evidence of dual (Arabian Sea and Bay of Bengal) vapour sources in monsoonal precipitation over north India. *Earth and Planetary Science Letters*, **250**, 511–521.

Shahack-Gross, R., Berna, F., Karkanas, P. & Weiner, S. (2004) Bat guano and preservation of archaeological remains in cave sites. *Journal of Archaeological Science*, **31**, 1259–1272.

Sharp, M., Parkes, J., Cragg, B. et al. (1999) Widespread bacterial populations at glacier beds and their relationship to rock weathering and carbon cycling. *Geology*, **27**, 107–110.

Sharp, Z. (2007) *Principles of Stable Isotope Geochemistry*. Pearson Prentice-Hall, NJ.

Sharpley A.N. & Rekolainen S. (1997) Phosphorus in agriculture and its environmental implications. In (eds. Tunney H., Careton O.T., Brookes P.C. & Johnstone A.E.) *Phosphorus loss from soil to water*. CAB International, Wallingford, UK, pp. 1–53.

Sheldon, N.D. & Tabor, N.J. (2009) Quantitative paleoenvironmental and paleoclimatic reconstruction using paleosols. *Earth-Science Reviews*, **95**, 1–52.

Sheldon, N.D., Retallack, G.J. & Tanaka, S. (2002) Geochemical climofunctions from North America soils and application to paleosols across the Eocene-Oligocene boundary in Oregon. *Journal of Geology*, **110**, 687–696.

Shinker, J.J. (2007) *Global Climate Animations. Digital Library for Earth System Education*. http://www.dlese.org/library/catalog_DLESE-000-000-001-774.htm.

Shiraishi, F., Reimer, A., Bissett, A. et al. (2008) Microbial effects on biofilm calcification, ambient water chemistry and stable isotope records in a highly supersaturated setting (Westerhöfer Bach, Germany). *Palaeogeography, Palaeoclimatology, Palaeoecology*, **262**, 91–106.

Shopov, Y.Y. (2006) Speleothem paleoluminescence—the past twenty years. In: Harmon, R.S. & Wicks, C. (eds.) *Perspectives on karst geomorphology, hydrology, and goechemsitry—A tribute volume to Derek C. Ford and William B. White. Geological Society of America Special Paper 404*, 319–330.

Shopov, Y.Y., Ford, DC & Schwarz, H.P. (1994) Luminescent microbanding in speleothems—high-resolution chronology and paleoclimate. *Geology*, **22**, 407–410.

Short, M., Baygents, J., Beck, J.W. et al. (2005) Stalactite growth as a free-boundary problem: a geometric law and its platonic ideal. *Physical Reviews Letters*, **94**, 018501, doi: 10.1103/PhysRevLett.94.018501.

Siddall, M., Chappell, J. & Potter, E.-K. (2009) Eustatic sea level during past interglacials. Chapter 7. *Reviews of Quaternary Science*, 75–92.

Siklosy, Z., Demény, A., Vennemann, T.W. et al. (2009) Bronze Age volcanic event recorded in stalagmites by combined isotope and trace element studies. *Rapid Communications in Mass Spectrometry*, **23**, 801–808.

Siklosy, Z., Kern, Z., Demeny, A. et al. (2011) Speleothems and pine trees as sensitive indicators of environmental pollution—A case study of the effect of uranium-ore mining in Hungary. *Applied Geochemistry*, **26**, 666–678.

Silvestru, E. (1999) Perennial ice in caves in temperate climate and its significance. *Theoretical and Applied Karstology*, **11–12**, 83–93.

Silvia, M.O. & Ignacio, C.G. (2005) Comparison of hygrothermal conditions in underground wine cellars from a Spanish area. *Building and Environment*, **40**, 1384–1394.

Simms, M.J. (1994) Emplacement and preservation of vertebrates in caves and fissures. *Zoological Journal of the Linnean Society*, **112**, 261–283.

Simms, M.J. (2004) Tortoises and hares: dissolution, erosion and isostasy in landscape evolution. *Earth Surface Processes and Landforms*, **29**, 477–494.

Simon, K.S., Pipan, T. & Culver DC (2007) A conceptual model of the flow and distribution of organic carbon in caves. *Journal of Cave and Karst Studies*, **69**, 279–284.

Simunek, J. & Suarez, D.L. (1993) Modeling of carbon dioxide transport and production in soil 1. Model development. *Water Resources Research*, **29**, 487–497.

Sinclair, D.J. (2011) Two mathematical models of Mg and Sr partitioning into solution during incongruent calcite dissolution: implications for dripwater and speleothem studies. *Chemical Geology*, **283**, 119–133.

Skinner, L. (2008) Facing future climate change: is the past relevant? *Philosophical Transactions of the Royal Society A*, **366**, 4627–4645.

Smart, P.L. & Friederich, H. (1987) Water movement and storage in the unsaturated zone of a maturely karstified carbonate aquifer, Mendip Hills, England. *Proceedings of the Conference on Environmental Problems in Karst Terranes and their Solutions*. National Water Well Association, Dublin, Ohio, 59–87.

Smart, P.L. & Whitaker, F.F. (1991) Karst processes, hydrology and porosity evolution. In: Wright, V.P., Esteban, M. & Smart, P.L. (eds.) *Palaeokarsts and Palaeokarstic Reservoirs, P.R.I.S. Occasional Publication Series*, **2**, 1–55.

Smart, P.L., Beddows, P.A., Coke, J. et al. (2006) Cave development on the Caribbean coast of the Yucatan Peninsula, Quintana Roo, Mexico. In: Harmon, R.S. & Wicks, C. (eds.) *Perspectives on Karst Geomorphology, Hydrology and Geochemistry—A Tribute Volume to Derek C. Ford and William B. White. Geological Society of America Special Paper* **404**, 105–128.

Smerdon, J.E., Pollack, H.N., Cermak, V. et al. (2006) Daily, seasonal and annual relationships between air and subsurface temperatures. *Journal of Geophysical Research*, **111**, D07101. doi: 10.1029/2004JD005578.

Smith, C.L. (2007) *The Statistical Analysis of Speleothem Paleoclimate Records*. Unpublished PhD Thesis, University of Birmingham, UK.

Smith, C.L., Baker, A., Fairchild, I.J. et al. (2006) Reconstructing hemispheric scale climates from multiple stalagmite records. *International Journal of Climatology*, **26**, 1417–1424.

Smith, C.L., Fairchild, I.J., Spötl, C. et al. (2009) Chronology-building using objective identification of annual signals in trace element profiles of stalagmites. *Quaternary Geochronology*, **4**, 11–21.

Smith, D.I., Atkinson, T.C. & Drew, D.P. (1976) 6. The hydrology of limestone terrains. In: Ford, T.D. & Cullingford, C.H.D. (eds.) *The Science of Speleology*. London: Academic Press, pp. 179–212.

Smith, G.K. (1997) Carbon dioxide, caves and you. http://www.wasg.iinet.net.au/ Accessed 28/10/09.

Smith, M.P., Soper, N.J., Higgins, A.K., Rasmussen, J.A. & Craig, L.E. (1999) Palaeokarst systems in the Neoproterozoic of eastern North Greenland in relation to extensional tectonics on the Laurentian margin. *Journal of the Geological Society*, **156**, 113–124.

Smith, T.M., Reynolds, R.W., Thomas C. et al. (2008) Improvements to NOAA's historical merged land-ocean surface temperature analysis (1880–2006). *Journal of Climate*, **21**, 2283–2296.

Smithson, P.A. (1991) Inter-relationships between cave and outside air temperatures. *Theoretical and Applied Climatology*, **44**, 65–73.

Sodemann, H., Masson-Delmotte, V., Schwierz, C., Vinther, B.M. & Wernli, H. (2008) Interannual variability of Greenland winter precipitation sources: 2. Effects of North Atlantic Oscillation variability of stable isotopes in precipitation. *Journal of Geophysical Research*, **113**, D12111, doi:10.1029/2007JD009416.

Solomon, D.K. & T.E. Cerling. (1987) The annual carbon dioxide cycle in a montane soil: observations, modeling, and implications for weathering. *Water Resources Research*, **23**, 2257–2265.

Solomon, S., Qin, D., Manning, M. et al. (2007) *Climate Change 2007: The Physical Science Basis*. Cambridge University Press.

Spalding, R.F. & Matthews, T.D. (1972) Submerged stalagmites from caves in the Bahamas: indicators of low sea level stand. *Quaternary Research*, **2**, 470–472.

Spero, H.J., Bijma, J., Lea, D.W. & Bemis, B.E. (1997) Effect of seawater carbonate concentration on planktonic foraminiferal carbon and oxygen isotopes. *Nature*, **390**, 497–500.

Sposito, G. (1989) *The Chemistry of Soils*. Oxford University Press, New York.

Spötl, C. (2005) A robust and fast method of sampling and analysis of delta C-13 of dissolved inorganic carbon in ground waters. *Isotopes in Environmental and Health Studies*, **41**, 217–221.

Spötl, C. & Mangini, A. (2002) Stalagmite from the Austrian Alps reveals Dansgaard-Oeschger events during isotope stage 3: implications for the absolute chronology of Greenland ice cores. *Earth and Planetary Science Letters*, **203**, 507–518.

Spötl, C. & Mattey, D. (2006) Stable isotope microsampling of speleothems: a comparison of drill, micromill and laser ablation techniques. *Chemical Geology*, **235**, 48–58.

Spötl, C. & Mattey, D. (2012) Scientific drilling of speleothems—a technical note. *International Journal of Speleology*, **41**, 29–34.

Spötl, C., Fairchild, I.J., & Tooth, A.F. (2005) Cave air control on dripwater geochemistry, Obir caves (Austria): implications for speleothem deposition in dynamically ventilated caves. *Geochimica et Cosmochimica Acta*, **69**, 2451–2468.

Spötl, C., Mangini, A. & Richards, D.A. (2006) Chronology and paleoenvironment of marine isotope stage 3 from two high-elevation speleothems, Austrian Alps, *Quaternary Science Reviews*, **25**, 1127–1136.

Spötl, C., Scholz, D. & Mangini, A. (2008) A terrestrial U/Th-dated stable isotope record of the Penultimate Interglacial. *Earth and Planetary Science Letters*, **276**, 283–292.

Spötl, C., Nicolussi, K., Patzelt, G. et al. (2010) Humid climate during the deposition of sapropel 1 in the Mediterranean Sea: assessing the influence on the Alps. *Global and Planetary Change*, **71**, 242–248.

Stanley, S.M. (2006) Influence of seawater chemistry on biomineralization throughout Phanerozoic time: paleontological and experimental evidence. *Palaeogeography, Palaeoclimatology, Palaeoecology*, **232**, 214–236.

Staudt, W.J., Reeder, R.J. & Schoonen, M.A.A. (1994) Surface structural controls on compositional sector zoning of SO_4^{2-} and SeO_4^{2-} in synthetic calcite crystals. *Geochimica et Cosmochimica Acta*, **58**, 2087–2098.

Stepanov, V.I. (1997) Notes on mineral growth from the archive of V.I. Stepanov (1924–1988). *Proceedings of the University of Bristol Spelaeological Society*, **21**, 25–42.

Stephenson, A.E., DeYoreo, J.J., Wu, L. et al. (2008) Peptides enhance magnesium signature in calcite: insights into origins of vital effects. *Science*, **322**, 724–727.

Stevens, M.B., González-Rouco, J.F. & Beltrami, H. (2008) North American climate of the last millennium: underground temperatures and model comparison. *Journal of Geophysical Research*, **113**, F01008. doi: 10.1029/2006JF000705.

Stewart, J.W.B. & Tiessen, H. (1987) Dynamics of soil organic phosphorus. *Biogeochemistry* **4**, 41–60.

Stipp, S.L.S., Gutmannsbauer, W. & Lehmann, T. (1996) The dynamic nature of calcite surfaces in air. *American Mineralogist*, **81**, 1–8.

Stirling, C.H. & Andersen, M.B. (2009) Uranium-series dating of fossil coral reefs: extending the sea-level record beyond the last glacial cycle. *Earth and Planetary Science Letters*, **284**, 269–283.

Stock, G.M., Granger, D.E., Sasowsky, I.D. et al. (2005) Comparison of U-Th, paleomagnetism, and cosmogenic burial methods for dating caves: implications for landscape evolution studies. *Earth and Planetary Science Letters*, **236**, 388–403.

Strnad, L., Ettler, V., Mihaljevic, M., et al. (2009) Determination of Trace Elements in Calcite Using Solution and Laser Ablation ICP-MS: calibration to NIST SRM Glass and USGS MACS Carbonate, and Application to Real Landfill Calcite. *Geostandards and Geoanalytical Research*, **33**, 347–355.

Strong, M., Sharp, Z.D. & Gutzler, D.S. (2007) Diagnosing transport using D/H ratios of water vapour. *Geophysical Research Letters* **34**, doi:10.1029/2006GL028307.

Stumm, W. & Morgan, J.J. (1996) *Aquatic Chemistry*. J. Wiley, New York.

Sturm, C., Vimeux, F. & Krineer, G. (2007) Intraseasonal variability in South America recorded in stable water isotopes. *Journal of Geophysical Research* **112**, doi:10.1029/2006JD008298.

Sunagawa, I. (2005) *Crystals. Growth, Morphology and Perfection*. Cambridge University Press.

Sundqvist, H.S., Baker, A. & Holmgren, K. (2005) Luminescence variations in fast-growing stalagmites from Uppsala, Sweden. *Geografiska Annaler*, **87A**, 539–548.

Sundqvist, H.S., Holmgren, K. & Lauritzen, S.E. (2007) Stable isotope variations in stalagmites from the northwestern Sweden document climate and environmental changes during the early Holocene. *The Holocene*, **17**, 259–267.

Svensson, A., Andersen, K.K., Bigler, M. et al. (2008) A 60000 year Greenland stratigraphic ice core chronology. *Climate of the Pats*, **4**, 47–57.

Sverdrup, H.U. & Warfvinge, P. (1995) Estimating field weathering rates using laboratory kinetics. *Reviews in Mineralogy*, **31**, 485–541.

Sverjensky, D.A. (1984) Prediction of Gibbs free energies of calcite-type carbonates and the equilibrium distribution of trace elements between carbonates and aqueous solutions. *Geochimica et Cosmochimica Acta*, **48**, 1127–1134.

Sweeting, M.M. (1972) *Karst Landforms*. London: Macmillan.

Sylvester, P. (ed.) (2008) *Laser Ablation ICP-MS in the Earth Sciences: Current Practices and Outstanding Issues*. Mineralogical Association of Canada, Short Course Series, Volume **40**, Vancouver, B.C.

Szabo, B.J., Kolesar, P.T., Riggs, A.C. et al. (1994) Paleoclimatic inferences form a 120,000–yr calcite record of water-table fluctuation in Browns Room of Devils Hole, Nevada. *Quaternary Research*, **41**, 59–69.

Szeidovitz, G., Surányi, G., Gribovszki, K. et al. (2008) Estimation of an upper limit on prehistoric peak ground acceleration using the parameters of intact speleothems in Hungarian caves. *Journal of Seismology*, **12**, 21–33.

Tan, M., Qin, X.G., Shen, L.M., et al. (1999) Bioptical microcycles of laminated speleothems from China and their chronological significance. *Chinese Science Bulletin*, **44**, 1604–1607.

Tan, M., Liu, T., Hou, J. et al. (2003) Cyclic rapid warming on centennial scale revealed by a 2650–year stalagmite record of warm season temperature. *Geophysical Research Letters* **30**, 1617. doi:10.1029/2003GL017352.

Tan, M., Baker, A., Genty, D. et al. (2006) Applications of stalagmite laminae to paleoclimate reconstructions: comparison with dendrochronology/climatology. *Quaternary Science Reviews*, **25**, 2103–2117.

Tan, M., Shao, X.-M., Liu, J. & Cai, B. (2009) Comparative analysis between a proxy-based climate reconstruction and GCM-based simulation of temperatures over the last millennium in China. *Journal of Quaternary Science*, **24**, 547–551.

Tanahara, A., Taira, H. & Takemura, M. (1997) Radon distribution and the ventilation of a limestone cave on Okinawa. *Geochemical Journal*, **31**, 49–56.

Tanaka, K., Takahashi, Y. & Shimizu, H. (2007) Determination of rare earth element in carbonate using laser-ablation inductively-coupled plasma mass spectrometry: an examination of the influence of the matrix on laser-ablation inductively-coupled plasma mass spectrometry analysis. *Analytica Chimica Acta*, **583**, 303–309.

Tanakara, A., Taira, H., Yamakawa, K. & Tsuha, A. (1998) Application of excess ^{210}Pb dating method to stalactites. *Geochemical Journal*, **32**, 183–187.

Tang, J., Köhler, S.J. & Dietzel, M. (2008a) Sr^{2+}/Ca^{2+} and $^{44}Ca/^{40}Ca$ fractionation during inorganic calcite formation: I. Sr incorporation. *Geochimica et Cosmochimica Acta*, **72**, 3718–3732.

Tang, J., Dietzel, M., Böhm, F. et al. (2008b) Sr^{2+}/Ca^{2+} and $^{44}Ca/^{40}Ca$ fractionation during inorganic calcite formation: II. Ca isotopes. *Geochimica et Cosmochimica Acta*, **72**, 3733–3745.

Tang, K. & Feng, X. (2001) The effect of soil hydrology on oxygen and hydrogen isotopic compositions of plants' source water. *Earth and Planetary Science Letters*, **185**, 355– 367.

Telford, R.J., Heegaard, E. & Birks, H.J.B. (2004) All age-depth models are wrong: but how badly? *Quaternary Science Reviews*, **23**, 1–5.

Teng, H.H. (2004) Controls by saturation state on etch pit formation during calcite dissolution. *Geochimica et Cosmochimica Acta*, **68**, 253–262.

Teng, H.H., Dove, P.M. & De Yoreo, J.J. (1999) Reversed calcite morphologies induced by microscopic growth kinetics: insight into biomineralization. *Geochimica et Cosmochimica Acta*, **63**, 2507–2512.

Teng, H.H., Dove, P.M. & De Yoreo, J.J. (2000) Kinetics of calcite growth: surface processes and relationships to macroscopic rate laws. *Geochimica et Cosmochimica Acta*, **64**, 2255–2266.

Tera, F. & Wasserburg, G.J. (1972) U-Th-Pb systematics in three Apollo 14 basalts and the problem of initial Pb in lunar rocks. *Earth and Planetary Science Letters*, **14**, 281–304.

Tesoriero, A.J. & Pankow, J.F. (1996) Solid solution partitioning of Sr^{2+}, Ba^{2+} and Cd^{2+} into calcite. *Geochimica et Cosmochimica Acta*, **60**, 1053–1063.

Tharp, T.M. (1995) Design against collapse of limestone caverns. In: Beck, B.F. (ed.) *Karst Geohazards*. Rotterdam: A.A. Balkema, pp. 817–824.

Thiebaut, C. & Roques, S. (2005) Time-scale and time-frequency analysis of irregularly sampled astronomical time series. *Journal of Applied Signal Processing*, **15**, 2486–2499.

Thomas, E.R., Wolff, E.W., Mulvaney, R. et al. (2007) The 8.2ka event from Greenland ice cores. *Quaternary Science Reviews*, **26**, 70–81.

Thomas, J.M., Welch, A.H., & Dettinger, M.D. (1996) Geochemistry and isotope hydrology of representative aquifers in the Great Basin region of Nevada, Utah, and adjacent states. *U.S. Geological Survey Professional Paper 1409–C.*

Thompson, L.G., Mosley-Thompson, E. & Arnao, B.M. (1984) El Niño-Southern Oscillation events recorded in the stratigraphy of the tropical Quelccaya ice cap, Peru. *Science*, **234**, 361–64.

Thornthwaite, C.W. (1948) An approach toward a rational classification of climate. *Geographical Review*, **38**, 55–94.

Thorstenson, D. & Parkhurst, D. (2004) Calculation of individual equilibrium constants for geochemical reactions. *Geochimica Cosmochimica Acta*, **68**, 2449–2465.

Thrailkill, J. (1971) Carbonate deposition in Carlsbad Caverns. *Journal of Geology*, **79**, 683–695.

Tiljander, M.I.A., Saarnisto, M., Ojala, A.K.E. & Saarinen, T. (2003) A 3000–year palaeoenvironmental record from annually laminated sediment of Lake Korttajarvi, central Finland. *Boreas*, **32**, 566–577.

Tipping, E. (1994) WHAM—A chemical equilibrium model and computer code for waters, sediments, and soils incorporating a discrete site electrostatic model of ion-binding by humic substances. *Computers and Geosciences* **20**, 973–1023.

Tipping, E. (1998) Humic ion-binding model VI: an improved description of the interactions of protons and metal ions with humic substances. *Aquatic Geochemistry*, **4**, 3–48.

Tipping, E. (2002) *Cation Binding by Humic Substances*. Cambridge University Press.

Tipping, E. & Hurley, M.A. (1992). A unifying model of cation binding by humic substances. *Geochimica Cosmochimica Acta*, **56**, 3627–3641.

Tooth, A.F. (2000) *Controls on the geochemistry of speleothem-forming karstic drip waters*. PhD thesis, Keele University.

Tooth, A.F. & Fairchild, I.J. (2003) Soil and karst aquifer hydrological controls on the geochemical evolution of speleothem-forming drip waters, Crag Cave, southwest Ireland. *Journal of Hydrology*, **273**, 51–68.

Torrence, C. & Compo, G.P. (1998) A practical guide to wavelet analysis. *Bulletin of the American Meteorological Society*, **79**, 61–78.

Tourney, J. & Ngwenya, B.T. (2009) Bacterial extracellular polymeric substances (EPS) mediate $CaCO_3$ morphology and polymorphism. *Chemical Geology*, **262**, 138–146.

Treble, P., Shelley, J.M.G., & Chappell, J. (2003) Comparison of high resolution subannual records of trace elements in a modern (1911–1992) speleothem with instrumental climate data from southwest Australia. *Earth and Planetary Science Letters*, **216**, 141–153.

Treble, P.C., Chappell, J. & Shelley, J.M.G. (2005a) Complex speleothem growth processes revealed by trace element mapping and scanning electron microscopy of annual layers. *Geochimica et Cosmochimica Acta*, **69**, 4855–4863.

Treble, P.C., Chappell, J., Gagan, M.K., et al. (2005b) In situ measurement of seasonal $\delta^{18}O$ variations and analysis of isotopic trends in a modern speleothem from southwest Australia. *Earth and Planetary Science Letters*, **233**, 17–32.

Treble, P., Budd, W.F., Hope, P.K. & Rustomji, P.K. (2005c) Synoptic-scale climate patterns associated with rainfall delta O-18 in southern Australia. *Journal of Hydrology* **302**, 270–282.

Treble, P.C., Schmitt, A.K., Edwards, R.L. et al. (2007) High resolution Secondary Ionisation Mass Spectrometry (SIMS) $\delta^{18}O$ analyses of Hulu Cave speleothem at the time of Heinrich Event 1. *Chemical Geology*, **238**, 197–212.

Treble P., Fairchild I.J. & Fischer M.J. (2008) Understanding climate proxies in southwest-Australian speleothems. *PAGES News*, **16**, 17–19.

Tripati, A.K., Eagle, R.A., Thiagarajan, N. et al. (2010) ^{13}C-^{18}O isotope signatures and 'clumped isotope' thermometry in foraminifera and coccoliths. *Geochimica et Cosmochimica Acta*, **74**, 5697–5717.

Troester, J.W. & White, W.B. (1984) Seasonal fluctuations in the carbon dioxide partial pressure in a cave atmosphere. *Water Resources Research*, **20**, 153–156.

Trombe, F. (1952) *Traité de Spéleologie*. Payot, Paris.

Trouet, V., Esper, J., Graham, N.E. et al. (2009) Persistent positive North Atlantic Oscillation mode dominated the Medieval Climate Anomaly. *Science*, **324**, 78–80.

Tuccimei, P., Soligo, M., Ginés, J. et al. (2010) Constraining Holocene sea levels using U-Th age of phreatic overgrowths on speleothems from coastal caves in Mallorca (Western Mediterranean). *Earth Surface Processes and Landforms*, **35**, 782–790.

Tucker, M.E. (1991) *Sedimentary Petrology* 2nd edition. Blackwell: Oxford.

Tucker, M.E. (1993) Carbonate diagenesis and sequence stratigraphy. In: Wright, V.P. (ed.) *Sedimentology Review*, **1**, 51–72.

Tucker, M.E. & Wright, V.P. (1990) *Carbonate Sedimentology*. Oxford: Blackwell.

Tucker, M.E., Gallagher, J. & Leng, M.J. (2009) Are beds in shelf carbonates millennial-scale cycles? An example from the mid-Carboniferous of northern England. *Sedimentary Geology*, **214**, 19–34.

Turin, H.J. & Plummer, M.A. (2000) Lechuguilla cave pool chemistry, 1986–1999. *Journal of Cave and Karst Studies*, **62**, 135–143.

Tyler, G. (2004) Ionic charge, radius, and potential control root/soil concentration ratios of fifty cationic elements in the organic horizon of a beech (*Fagus sylvaticus*) forest podzol. *Science of the Total Environment*, **329**, 231–239.

Tyler, G. (2005) Changes in the concentrations of major, minor and rare-earth elements during leaf senescence and decomposition in a *Fagus sylvatica* forest. *Forest Ecology and Management*, **206**, 167–177.

Tyler, G. & Olsson, T. (2001) Concentrations of 60 elements in the soil solution as related to the soil acidity. *European Journal of Soil Science*, **52**, 151–165.

Uemura, R., Matsui, Y., Yoshimura, K., Motoyama, H. & Yoshida, N. (2008) Evidence of deuterium excess in water vapor as an indicator of ocean surface conditions. *Journal of Geophysical Research* **113**, D19114, doi:10.1029/2008JD010209.

Ummenhofer, C.C., Sen Gupta, A., England, M.H. et al. (2009) Contributions of Indian Ocean sea surface temperatures to enhanced East African rainfall. *Journal of Climate*, **22**, 993–1013.

Uppala, S.M., Kållberg, P.W., Simmons, A.J. et al. (2005) The ERA-40 re-analysis. *Quarterly Journal of the Royal Meteorological Society*, **131**, 2961–3012.

Urey, H.C. (1947) The thermodynamic properties of isotopic substances. *Journal of the Chemical Society*, 562–581.

Usdowski, E., Hoefs, J. & Menschel, G. (1979) Relationship between ^{13}C and ^{18}O fractionation and changes in major element composition in a recent calcite-depositing spring—a model of chemical variations with inorganic $CaCO_3$ precipitation. *Earth and Planetary Science Letters*, **42**, 267–276.

Vacca, E. & Delfino, V.P. (2004) Three-dimensional topographic survey of the human remains in Lamalunga Cave (Altamura, Bari, Southern Italy). *Collegium Antropologicum* **28**, 1, 113–119.

Vacher, H.L. (1988) Dupuit-Ghyben-Herzberg analysis of strip-island lenses. *Geological Society of America Bulletin*, **100**, 580–591.

Vacher, H.L. & Mylroie, J.E. (2002) Eogenetic karst from the perspective of an equivalent porous medium. *Carbonates and Evaporites*, **17**, 182–196.

Vacher, H.L., Bengtsson, T.O. & Plummer, L.N. (1990) Hydrology of meteoric diagenesis: residence time of meteoric ground water in island fresh-water lenses with application to aragonite-calcite stabilization rate in Bermuda. *Geological Society of America Bulletin*, **102**, 223–232.

Vahrenkamp, V.C., Swart, P.K. & Ruiz, J. (1991) Episodic dolomitization of late Cenozoic carbonates in the Bahamas: evidence from strontium isotopes. *Journal of Sedimentary Petrology*, **61**, 1002–1014.

Vail, P.R., Mitchum, R.M. & Thompson, S. (1977) Seismic stratigraphy and global changes of sea-level, part 4: global cycles of relative changes of sea-level. *Memoir of the American Association of Petroleum Geologists*, **26**, 83–97.

Valen, V., Laurtizen, S.-E. & Løvlie, R. (1997) Sedimentation in a high-latitude karst cave: Sirijordgrotta, Nordland, Norway. *Geologisk Tidsskrift*, **77**, 233–250.

van Bergen, P.F., Nott, C.J., Bull, I.D. et al. (1998) Organic geochemical studies of soils from the Rothamsted Classical Experiments—IV. Preliminary results from a study of the effect of soil pH on organic matter decay. *Organic Geochemistry*, **29**, 1779–1795.

van Beynen, P., Ford, D & Schwarcz, H. (2000) Seasonal variability in organic substances in surface and cave waters at Marengo Cave, Indiana. *Hydrological Processes*, **14**, 1177–1197.

van Beynen, P. & Febbroriello, P. (2006) Seasonal isotopic variability of precipitation and cave drip water at Indian Oven Cave, New York. *Hydrological Processes*, **20**, 1793–1803.

van Breukelen, M.R., Vonhof, H.B., Hellstrom, J.C. et al. (2008) Fossil dripwater in stalagmites reveals Holocene temperature and rainfall variation in Amazonia. *Earth and Planetary Science Letters*, **275**, 54–60.

van Campo, M. & Leroi-Gourham, A. (1956) Note préliminaire à l'étude des pollens fossils de diférents niveaux des grottes D'Arcy-sur-Cure. *Bulletin Muséum Société Préhistorie Francais*, **28**, 326–330.

van Geel, B. (1986) *Application of fungal and algal remains and other microfossils in palynological analyses*. In Berglund, B.E. (ed) Handbook of Holocene Palaeoecology and Palaeohydrology. Wiley, Chichester. pp. 497–505.

van Geel, B. (2001) *Non-pollen palynomorphs*. In: J.P. Smol, H.J.B. Birks and W.M. Last (eds.) Tracking environmental change using lake sediments.; Volume **3**: Terrestrial, algal and silicaceous indicators. Kluwer, Dordrecht, pp. 99–119.

van Geel, B., Pals J.P., van Reenen G.B.A. & van Huissteden J. (1995) The indicator value of fossil fungal remains, illustrated by a palaeoecological record of a Late Eemian/Early Weichselian deposit in the Netherlands. In Herngreen, G.F.W. & L. van der Valk (eds) *Neogene and Quaternary Geology of North-West Europe. Mededelingen Rijks Geologische Dienst* **52**, 297–315.

Van Gundy, J.J. & White, W.B. (2009) Sediment flushing in Mystic Cave, West Virginia, USA, in response to the (1985) Potomac Valley flood. *International Journal of Speleology*, **38**, 103–109.

Vandycke, S. & Quinif, Y. (2001) Recent active faults in Belgian Ardenne revealed in Rochefort karstic network (Namur Province, Belgium). *Netherlands Journal of Geosciences*, **80**, 297–304.

Vaughan, M.J., Maier, R.M. & Pryor, B.M. (2011) Fungal communities on speleothem surfaces in Kartchner Caverns, Arizona, USA. *International Journal of Speleology*, **40**, 65–77.

Vesica, P.L., Tuccimei, P., Turi, B., et al. (2000) Late Pleistocene paleoclimates and sea-level change in the Mediterranean as inferred from stable isotope and U-series studies of overgrowths on speleothems, Mallorca, Spain. *Quaternary Science Reviews*, **19**, 865–879.

Verdugo, P., Orellana, M.V., Chin, W-C., Petersen, T.W., van den Eng, G., Benner, R. & Hedges, J.I. (2008) Marine biopolymer self-assembly: implications for carbon cycling in the ocean. *Faraday Discussions*, **139**, 393–398.

Verheyden, S., Keppens, E., Fairchild, I.J. et al. (2000) Mg, Sr and Sr isotope geochemistry of a Belgian Holocene speleothem: implications for paleoclimate reconstructions. *Chemical Geology*, **169**, 131–144.

Verheyden, S., Genty, D., Cattani, O. & van Breukelen, M.R. (2008a) Water release patterns of heated speleothem calcite and hydrogen isotope compositions of fluid inclusions. *Chemical Geology*, **247**, 266–281.

Verheyden, S., Genty, D., Deflandre, G. et al. (2008b) Monitoring climatological, hydrological and geochemical parameters in the Père-Noël cave (Belgium): implication for the interpretation of speleothem isotopic and geochemical time series. *International Journal of Speleology*, **37**, 221–234.

Vesica, P.L., Tuccimei, P., Turi, B. et al. (2000) Late Pleistocene paleoclimates and sea-level change in the Mediterranean as inferred from stable isotope and U-series studies of overgrowths on speleothems, Mallorca, Spain. *Quaternary Science Reviews*, **19**, 865–879.

Vieira, G.T., Mora, C. & Ramos, M. (2003) Ground temperature regimes and geomorphological implication in a Mediterranean mountain (Serra da Estrela, Portugal). *Geomorphology*, **52**, 57–72.

Villegas-Jiménez, A., Mucci, A. & Paquette, J. (2009a) Proton/calcium ion exchange behavior of calcite *Physical Chemistry Chemical Physics*, **11**, 8895–8912.

Villegas-Jiménez, A., Mucci, A., Pokrovsky, O.S. & Schott J. (2009b) Defining reactive sites on hydrated mineral surfaces: rhombohedral carbonate minerals. *Geochimica et Cosmochimica Acta*, **73**, 4326–4345.

Voelker, A.H.L. (2002) Global distribution of centennial-scale records for Marine Isotope Stage (MIS) 3: a database. *Quaternary Science Reviews*, **21**, 1185–1212.

Vogel, J.C., Grootes, P.M. & Mook, W.G. (1970) Isotopic fractionation between gaseous and dissolved carbon dioxide. *Zeitschrift für Physik*, **230**, 225–238.

Vokal, B. (1999) *The carbon transfer in karst areas- an application to the study of environmental changes and paleoclimatic reconstruction*. PhD. Thesis, Polytechnic of Nova Gorica. 90pp. Nova Gorica (Slovenia).

Vollweiler, N., Scholz, D., Muhlinghaus, C. et al. (2006) A precisely dated climate record for the last 9 kyr from three high alpine stalagmites, Spannagel cave, Austria. *Geophysical Research Letters*, **33**, L20703, doi:10.1029/2006GL027662.

Vonhof, H.B., van Breukelen, M.R., Postma, O. et al. (2006) A continuous-flow crushing device for on-line δ^2H analysis of fluid inclusion water in speleothems. *Rapid Communications in Mass Spectrometry*, **20**, 2553–2558.

Vuille, M. & Werner, M. (2005) Stable isotopes in precipitation recording South American summer monsoon and ENSO variability: observations and model results. *Climate Dynamics*, **25**, 410–413.

Vuille, M., Werner, M., Bradley, R.S., et al. (2005) Stable isotopes in East African precipitation record Indian Ocean zonal mode. *Geophysical Research Letters*, **32**, doi:10.1029/2005GL023876.

Wackerbarth, A.K., Scholz, D., Fohlmeister, J. & Mangini, A. (2010) Modelling the δ^{18}O of cave drip water and speleothem calcite. *Earth and Planetary Science Letters*, **299**, 287–297.

Wainer, K., Genty, D., Blamart, D. et al. (2009) A new stage 3 millennial climatic variability record from a SW France speleothem. *Paleogeography, Paleoclimatology, Paleoecology*, **271**, 130–139.

Wainer, K., Genty, D., Blamart, D. et al. (2011) Speleothem record of the last 180 ka in Villars cave (SW France): investigation of a large δ^{18}O shift between MIS6 and MIS5. *Quaternary Science Reviews*, **30**, 130–146.

Walker, J.D. & Geissman, J.W. (compilers) (2009) Geologic Time Scale: Geological Society of America. doi: 10.1130/2009.CTSO04R2C.

Wang, B. & Fan, Z. (1999) Choice of South Asian summer monsoon indices. *Bulletin of the American Meteorological Society*, **80**, 629–638.

Wang, B., Wu R. & Lau K.-M. (2001) Interannual variability of Asian summer monsoon: contrast between the

Indian and western North Pacific-East Asian monsoons. *Journal of Climate*, **14**, 4073–4090.

Wang, H., Zhao, X. Liu, Q., et al. (2008) Microbial communities from the DNA record at HS4 stalagmite in Heshang Cave, Qingjiang, Hubei, P R China. *Proceedings of the 5th Conference: Climate Change—the Karst Record*. Chongqing, China.

Wang, H., Zeng, C., Liu, Q. et al. (2010) Calcium carbonate precipitation induced by a bacterium strain isolated from an oligotrophic cave in Central China. *Frontiers in Earth Science, China*, **4**, 148–151.

Wang, Y. & Xu, H. (2001) Prediction of trace metal partitioning between minerals and aqueous solutions: a linear free energy correlation approach. *Geochimica et Cosmochimica Acta*, **65**, 1529–1543.

Wang, Y.J., Cheng, H., Edwards, R.L. et al. (2001) A high resolution absolute dated late Pleistocene monsoon record from Hulu Cave, China. *Science*, **294**, 2345–2348.

Wang, Y.J., Cheng, H., Edwards, R.L. et al. (2005) The Holocene Asian monsoon: links to solar changes and North Atlantic climate. *Science*, **308**, 854–857.

Wang, Y.J., Cheng, H., Edwards, R.L. et al. (2008) Millennial- and orbital-scale changes in the East Asian monsoon over the past 224,000 years. *Nature*, **451**, 1090–1093.

Wang, X.F., Auler, A.S., Edwards, R.L. et al. (2006) Interhemispheric anti-phasing of rainfall during the last glacial period. *Quaternary Science Reviews*, **25**, 3391–3403.

Wanner, H., Beer, J., Butikofer, J. et al. (2008) Mid- to Late Holocene climate change: an overview. *Quaternary Science Reviews*, **27**, 1791–1828.

Warren, J.K. (1989) *Evaporite Sedimentology*. New Jersey: Prentice Hall.

Warwick, G.T. (1950) Calcite bubbles—a new cave formation? *National Speleological Society Bulletin*, **12**, 38–42.

Wasylenki, L.E., Dove, P.M., Wilson, D.S. & De Yoreo, J.J. (2005a) Nanoscale effects of strontium on calcite growth: an in situ AFM study in the absence of vital effects. *Geochimica et Cosmochimica Acta*, **69**, 3017–3027.

Wasylenki, L.E., Dove, P.M. & De Yoreo, J.J. (2005b) Effects of temperature and transport conditions on calcite growth in the presence of Mg^{2+}: implications for paleothermometry. *Geochimica et Cosmochimica Acta*, **69**, 4227–4236.

Watanabe, Y., Matsuoka, H., Sakai, S. et al. (2010) Comparison of stable isotope time series of stalagmite and meteorological data from West Java, Indonesia. *Palaeogeography, Palaeoclimatology, Palaeoecology*, **293**, 90–97.

Waters, J.R. (2003) *Energy Conservation in Buildings: a guide to part L of the Building Regulations*. Wiley-Blackwell, London.

Watson, E.B. (2004) A conceptual model for near-surface kinetic controls on the trace-element and stable isotope composition of abiogenic crystals. *Geochimica et Cosmochimica Acta*, **68**, 1473–1478.

Webb, R.H. (1996) Confocal optical microscopy. *Reports on Progress in Physics*, **59**, 427–471.

Webster, P.J. & Yang, S. (1992) Monsoon and ENSO: selectively interactive systems. *Quarterly Journal of the Royal Meteorological Society*, **118**, 877–926.

Weedon, G.P. (2003) *Time-Series Analysis and Cyclostratigraphy*. Cambridge University Press, Cambridge.

Weijers, J.W.H., Schouten, S., van den Donker, J.C. et al. (2007) Environmental controls on bacterial tetraether membrane lipid distribution in soils. *Geochimica et Cosmochimica Acta*, **71**, 703–713.

Weiner S. & Dove P.M. (2003) An overview of biomineralization processes and the problem of the vital effect. In: Dove P.M., Weiner S. & De Yoreo J.J. (eds.) *Biomineralization. Reviews in Mineralogy and Geochemistry*, **54**, 1–29.

Wells, M.L. (1998) Marine colloids: a neglected dimension. *Nature*, **391**, 530–531.

Weltje, G.J. & Tjallingii, R. (2008) Calibration of XRF core scanners for quantitative geochemical logging of sediment cores: theory and application. *Earth and Planetary Science Letters*, **274**, 423–438.

Wen, X.-F., Sun, X.M., Zhang, S.C. et al. (2008) Continuous measurement of water vapour D/H and O-18/O-16 isotope ratios in the atmosphere. *Journal of Hydrology*, **349**, 489–500.

Westaway, K.E., Sutikna, T., Morwood, M.J. & Zhao, J.-X. (2010) Establishing rates of karst landscape evolution in the Tropics: a context for the formation of archaeological sites in western Flores, Indonesia. *Journal of Quaternary Science*, **25**, 1018–1037.

Westphal, H., Surholt, I., Kiesl, C., Thern, H.F. & Kruspe, T. (2005) NMR measurements in carbonate rocks: problems and an approach to a solution. *Pure and Applied Geophysics*, **162**, 549–570.

Weyhenmeyer, C.E., Burns, S.J., Waber, H.N. et al. (2002) Isotope study of moisture sources, recharge areas, and groundwater flow paths within the eastern Batinah coastal plain, Sultanate of Oman. *Water Resources Research*, **38**, 1184, doi:10.1029/2000WR000149.

Weyl, P.K. (1958) Solution kinetics of calcite. *Journal of Geology*, **66**, 163–176.

Whitaker, F.F. & Smart, P.L. (1997) Hydrogeology of the Bahamian Archipelago. In: Vacher, H.L & Quinn, T.

(eds.) *Geology and Hydrology of Carbonate Islands*. Amsterdam: Elsevier, pp. 183–216.

Whitaker, F.F. & Smart, P.L. (2007a) Geochemistry of meteoric diagenesis in carbonate islands of the northern Bahamas: 1. Evidence from field studies. *Hydrological Processes*, **21**, 949–966.

Whitaker, F.F. & Smart, P.L. (2007b) Geochemistry of meteoric diagenesis in carbonate islands of the northern Bahamas: 2. Geochemical modelling and budgeting of diagenesis. *Hydrological Processes*, **21**, 949–966.

White A.F. & Brantley S. (2003) The effect of time on the weathering of silicate minerals: why do weathering rates differ in the laboratory and field? *Chemical Geology*, **202**, 479–506.

White, E.L. & White, W.B. (2000) 6.2 Breakdown morphology. In: Klimchouk, A.B., Ford, DC, Palmer, A.N. & Dreybrodt, W. (eds.) *Speleogenesis. Evolution of Karst Aquifers*. Huntsville: National Speleological Society, pp. 427–429.

White, I.D., Mottershead, D.N. & Harrison, S.J. (1992) *Environmental Systems. An introductory text*. 2nd edition. London: Chapman & Hall.

White, W.B. (1977) Role of solution kinetics in the development of karst aquifers. In Tolson, J.S. & Doyle, F.L. (eds.) *Karst Hydrogeology: International Association of Hydrogeologists, Memoir*, **12**, 176–187.

White, W.B. (1988) *Geomorphology and Hydrology of Karst Terrains*. Oxford: Oxford University Press.

White, W.B. (1999) Conceptual models for karstic aquifers. In: Palmer, A.N., Palmer, M.V. & Sasowsky, I.D. (eds.) *Karst Modeling, Special Publication 5*. Charles Town, Wet Virginia: Karst Waters Institute, 11–16.

White, W.B. (2006) Fifty years of karst hydrology and hydrogeology: 1953–2003. In: Harmon, R.S. & Wicks, C. (eds.) *Perspectives on Karst Geomorphology, Hydrology and Geochemistry—A Tribute Volume to Derek C. Ford and William B. White. Geological Society of America Special Paper*, **404**, 139–152.

White, W.B. & Brennan, E.S. (1989) Luminescence of speleothems due to fulvic acid and other activators. *Proceedings of the 10th International Congress of Speleology*, 212–214.

Widmann, M., Goosse, H., van der Schrier, G. et al. (2010) Using data assimilation to study extratropical Northern Hemisphere climate over the last millennium. *Climate of the Past*, **6**, 627–644.

Wiedner, E., Scholtz, D., Mangini, A. et al. (2008) Investigation of the stable isotope fractionation in speleothems with laboratory experiments. *Quaternary International*, **187**, 15–24.

Wiener, N. (1948) *Cybenetics or Control and Communication in the Animal and the Machine*. New York: John Wiley.

Wiesenberg, G.L.B., Schwarzbauer, J., Schmidt, M.W.I. & Schwark, L. (2004) Source and turnover of organic matter in agricultural soils derived from n-alkane/n-carboxylic acid compositions and C-isotope signatures. *Organic Geochemistry*, **35**, 1371–1393.

Wigley, T.M.L. (1967) Non-steady flow through a porous medium and cave breathing. *Journal of Geophysical Research*, **72**, 3199–3205.

Wigley, T.M.L. & Brown, M.C. (1971) Geophysical applications of heat and mass transfer in turbulent pipe flow. *Boundary Layer Meteorology*, **1**, 300–320.

Wigley, T.M.L. & Brown, M.C. (1976) The Physics of Caves. In: Ford, T.D. & Cullingford, C.H.D. (eds) *The Science of Speleology*, Academic Press, London, pp. 329–358.

Wigley, T.M.L. & Plummer, L.N. (1976) Mixing of carbonate waters. *Geochimica et Cosmochimica Acta*, **40**, 989–995.

Wilkening, M.H. & Watkins, D.E. (1976) Air exchange and ^{222}Rn concentrations in the Carlsbad Caverns. *Health Physics*, **31**, 139–145.

Willerslev, E., Hansen, A.J., Binladen, J.G. et al. (2003) Diverse plant and animal genetic records from Holocene and Pleistocene sediments. *Science*, **300**, 791–795.

Williams, P.W. (2008) The role of the epikarst in karst and cave hydrogeology: a review. *International Journal of Speleology*, **37**, 1–10.

Williams, P.W. (2009) *World Map of Carbonate Rock Outcrops version 3.0* http://www.sgges.auckland.ac.nz/research/karst.shtml. Accessed 13th June 2010.

Williams, P.W. & Fowler, A. (2002) Relationship between oxygen isotopes in rainfall, cave percolation waters and speleothem calcite at Waitomo, New Zealand. *Journal of Hydrology (NZ)*, **41**, 53–70.

Williams, P.W., King, D.N.T., Zhao, J.-X. & Collerson, K.D. (2004) Speleothem master chronologies: combined Holocene $\delta^{18}O$ and $\delta^{13}C$ records from the North Island of New Zealand and their palaeoenvironmental interpretation. *The Holocene*, **14**, 194–208.

Williams, P.W., King, D.N.T., Zhao, J.-X. & Collerson, K.D. (2005) Late Pleistocene to Holocene composite speleothem chronologies from South Island, New Zealand—did a global Younger Dryas really exist? *Earth and Planetary Science Letters*, **230**, 301–317.

Williams, P.W., Neil, H.L. & Zhao, J.-X. (2010) Age frequency distribution and revised stable isotope curves for New Zealand speleothems: palaeoclimatic implications. *International Journal of Speleology*, **39**, 99–102.

Wimpenny, J., Manz, W. & Szewzyk, U. (2000) Heterogeneity in biofilms. *FEMS Microbiology Reviews*, **24**, 661–671.

Winograd, I.J., Coplen, T.B., Landwehr, J.M. et al. (1992) Continuous 500,000–year climate record from vein calcite in Devils Hole, Nevada. *Science*, **258**, 255–260.

Winograd, I.J., Landwehr, J.M., Coplen, T.B. et al. (2006) Devils Hole, Nevada, $\delta^{18}O$ record extended to the mid-Holocene. *Quaternary Research*, **66**, 202–212.

Witkamp, M, & Frank, M.L. (1969) Evolution of CO_2 from litter, humus, and subsoil of a pine stand. *Pedobiologia*, **9**, 358–365.

Wolff, E.W., Fischer, H. & Rothlisberger, R. (2009) Glacial terminations as southern warmings without northern control. *Nature Geoscience*, **2**, 206–209.

Wong, C.I., Banner, J.L. & Musgrove, M.L. (2011) Seasonal dripwater Mg/Ca and Sr/Ca variations driven by cave ventilation: implications for and modeling of speleothem palaeoclimate records. *Geochimica et Cosmochimica Acta*, **75**, 3514–3529.

Woodward, J.C. & Goldberg, P. (2001) The sedimentary records in Mediterranean rockshelters and caves: archives of environmental change. *Geoarchaeology*, **16**, 327–354.

Woodhead, J., Hellstrom, J. & Maas, R. (2006) U-Pb geochronology of speleothems by MC-ICPMS. *Quaternary Geochronology*, **1**, 208–221.

Woodhead, J.D., Hellstrom, J., Hergt, J.M. et al. (2009) Isotopic and elemental imaging of geological materials by laser ablation inductively coupled plasma-mass spectrometry. *Geostandards and Geoanalytical Research*, **31**, 331–343.

Woodhead, J., Reisz, R., Fox, D., et al. (2010) Speleothem climate records from deep time? Exploring the potential with an example from the Permian. *Geology*, **38**, 455–459.

Worden, J., Noone, D., Bowman, K. et al. (2007) Importance of rain evaporation and continental convection in the tropical water cycle. *Nature*, **445**, 528–532.

Worthington, S.R.H. (2001) Depth of conduit flow in unconfined carbonate aquifers. *Geology*, **29**, 355–358.

Worthington, S.R.H., Ford, DC & Beddows, P.A. (2000) 8.1 Porosity and permeability enhancement in unconfined carbonate aquifers as a result of solution. In: Klimchouk, A.B., Ford, DC, Palmer, A.N. & Dreybrodt, W. (eds.) *Speleogenesis. Evolution of Karst Aquifers*. Huntsville: National Speleological Society, pp. 463–472.

Wynn, J.G. & Bird, M. (2008) Environmental controls on the stable carbon isotopic composition of soil organic matter: implications for modelling the distribution of C3 and C4 plants, Australia. *Tellus*, **60B**, 604–621.

Wynn, J.G., Harden, J.W. & Fries, T.L. (2006) Stable carbon isotope depth profiles and soil organic carbon dynamics in the lower Mississippi Basin. *Geoderma*, **131**, 89–109.

Wynn, P.M., Fairchild, I.J., Baker, A. et al. (2008) Isotopic archives of sulphur in speleothems. *Geochimica et Cosmochimica Acta*, **72**, 2465–2477.

Wynn, P.M., Fairchild, I.J., Frisia S. et al. (2010) High-resolution sulphur isotope analysis of speleothem carbonate by secondary ionisation mass spectrometry. *Chemical Geology*, **271**, 101–107.

Xia, K., Bleam, W. & Helmke, P.A. (1997a) Studies of the nature of Cu^{2+} and Pb^{2+} binding sites in soil humic substances using X-ray absorption spectroscopy. *Geochimica et Cosmochimica Acta*, **61**, 2211–2221.

Xia, K., Bleam, W. & Helmke, P.A. (1997b) Studies of the nature of binding sites of first row transition elements bound to aquatic and soil humic substances using X-ray absorption spectroscopy. *Geochimica et Cosmochimica Acta*, **61**, 2223–2235.

Xie, S., Yi, Y., Huang, J. et al. (2003) Lipid distribution in a subtropical southern China stalagmite as a record of soil ecosystem response to palaeoclimate change. *Quaternary Research*, **60**, 340–347.

Yadava, M.G., Ramesh, R. & Pant, G.B. (2004) Past monsoon rainfall variations in peninsula India recorded in a 331–year old speleothem. *The Holocene*, **14**, 517–524.

Yang, H., Ding, W., Zhang, C.L. et al. (2011) Occurrence of tetraether lipids in stalagmites: implications for sources and GDGT-based proxies. *Organic Geochemistry*, **42**, 108–115.

Yang, M.J., Stipp, S.L.S. & Harding, J. (2008) Biological Control on Calcite Crystallization by Polysaccharides. *Crystal Growth and Design*, **8**, 4066–4074.

Yeh, S.W., Kug, J.S., Dewitte, B. et al. (2009) El Niño in a changing climate. *Nature*, **461**, 511–514.

Yuan D., Cheng H., Edwards R.L. et al. (2004) Timing, duration, and transitions of the last interglacial Asian monsoon. *Science* **304**, 575–578.

Zachara, J.M., Cowan, C.E. & Resch, C.T. (1991) Sorption of divalent metals on calcite. *Geochimica et Cosmochimica Acta*, **55**, 1549–1562.

Zak K., Onac B.P. & Perşoiu A. (2008) Cryogenic carbonates in cave environments: a review. *Quaternary International* **187**, 84–96.

Zanchetta, G., Drysdale, R.N., Hellstrom, J.C. et al. (2007) Enhanced rainfall in the Western Mediterranean during deposition of sapropel S1: stalagmite evidence from

Corchia cave (C Italy) *Quaternary Science Reviews*, **26**, 279–286.

Zare, R.N., Kuramoto, D.S., Haase, C. et al. (2009) High-precision optical measurements of $^{13}C/^{12}C$ isotope ratios in organic compounds at natural abundance. *Proceedings of the National Academy of Science*, **106**, 10928–10932.

Zeebe, R.E. (1999) An explanation of the effect of seawater carbonate ion concentration on foraminiferal stable isotopes. *Geochimica et Cosmochimica Acta*, **63**, 2001–2007.

Zeebe, R.E. (2007) An expression for the overall oxygen isotope fractionation between the sum of dissolved inorganic carbon and water. *Geochemistry, Geophysics, Geosystems*, **8**, Q09002. doi:10.1029/2007GC001663.

Zhao, J.-X., Yu, K.-F. & Feng, Y.-X. (2009) High precision ^{238}U ^{234}U ^{230}Th disequilibrium dating of the recent past: a review. *Quaternary Geochronology*, **4**, 423–433.

Zhong, S. & Mucci, A. (1995) Partitioning of rare earth elements (REEs) between calcite and seawater solution at 25 °C and 1 atm, and high dissolved REE concentrations. *Geochimica et Cosmochimica Acta*, **55**, 1549–1562.

Zhornyak, L.V., Zanchetta, G., Drysdale, R.N. et al. (2011) Stratigraphic evidence for a 'pluvial phase' between ca 8200–7100 ka from Renella cave (Central Italy). *Quaternary Science Reviews*, **30**, 409–417.

Zhou, H., Greig, A., You, C.-F. et al. (2011) Arsenic in a speleothem from central China: stadial–interstadial variations and implications. *Environmental Science & Technology*, **45**, 1278–1283.

Zhou, J., Lundstrom, C.C., Fouke, B. et al. (2005) Geochemistry of speleothem records from southern Illinois: development of $(^{234}U)/(^{238}U)$ as a proxy for paleoprecipitation. *Chemical Geology*, **221**, 1–20.

Zimmerman, U., Ehhalt, D. & Munnich, K.O. (1967) Soil water movement and evapotranspiration: changes in the isotopic composition of the water. *Proceedings of the Isotope Hydrology Symposium, 14–18, International Atomic Energy Authority, Vienna*, 567–584.

Zimmermann, M., Meir, P., Bird, M.I., Malhi, Y. & Ccahuana, A.J.Q. (2009) Climate dependence of heterotrophic soil respiration from a soil-translocation experiment along a 3000 m tropical forest altitudinal gradient. *European Journal of Soil Science*, **60**, 895–906.

Index

plain type indicates page number in text
italics indicates figure number
bold indicates plate number

Speleothem Science: From Process to Past Environments, First Edition. Ian J. Fairchild, Andy Baker.